Bacteriology, Virology and
for Students of Medicine

First published in 1925 as A Handbook of Bacteriology for Students and Practitioners of Medicine, *this well known textbook was written by Professor Joseph W. Bigger, of the University of Dublin. It quickly achieved wide acceptance and was revised five times by the original author, the sixth edition appearing in 1949.*

Professor F. S. Stewart, who had succeeded Professor Bigger as Professor of Bacteriology and Preventive Medicine in Dublin, then took over the book and was solely responsible for the seventh, eighth and ninth editions, this last being published in 1968 under the title Bacteriology and Immunology for Students of Medicine.

For the tenth edition Professor Stewart has been joined by Professor T. S. L. Beswick, Professor of Virology in the University of Manchester, and to indicate the scope of the book which has expanded with the subject the title has again been changed to include virology.

The immense advances that have taken place in recent years in microbiology and immunology have necessitated an unusually full revision for this edition and the sections covering immunity and virology have been almost entirely rewritten. Nevertheless, the book remains primarily a textbook for students of medicine; it should also be of value to undergraduates taking Honours Science courses, and as a work of basic reference for medical graduates.

Bacteriology, Virology and Immunity for Students of Medicine

BACTERIOLOGY VIROLOGY AND IMMUNITY

For Students of Medicine

TENTH EDITION

F.S. STEWART MD, FRCPI, FCPath, MRIA

Member of the Royal Irish Academy
Fellow Emeritus of Trinity College, Dublin
Formerly Professor of Bacteriology and Preventive Medicine, University of Dublin

T.S.L. BESWICK MA, MSc, MD, FRCP, FRCPath

Professor of Virology, University of Manchester
Honorary Consultant Virologist, Manchester North and Central Districts, Salford Area Health Authority and North Manchester Regional Virus Laboratory

Baillière Tindall · London

A BAILLIÈRE TINDALL book published by
Cassell & Collier Macmillan Publishers Ltd
35 Red Lion Square, London WC1R 4SG
and at Sydney, Auckland, Toronto, Johannesburg
an affiliate of
Macmillan Publishing Co. Inc.
New York

First published 1925
Ninth edition 1968
Tenth edition 1977
ISBN 0 7020 0637 8

English Language Book Society edition (of Tenth edition) 1977

Printed in Great Britain by
The Whitefriars Press Ltd, London and Tonbridge

Contents

Colour Plates

between pages 24 and 25

Preface

In the nine years that have elapsed since the publication of the ninth edition of this book microbiology and immunology have continued to grow at an explosive rate and this has necessitated a very full revision of the text. As in previous editions, however, the emphasis is on the general principles of microbiology and immunity and on the aspects of these subjects that are of importance in relation to infectious disease. As a result there has been a further reduction in material of a purely technical nature, for which several excellent texts are currently available. Thus the systematic bacteriology has been reduced by about one-quarter and the chapters on diagnosis and on the microscopic examination and cultivation of bacteria have been combined in one. Nevertheless there has inevitably been some overall increase in the text owing to the incorporation of additional material, particularly in immunity, which has been virtually completely rewritten (by F.S.S.) and in virology (by T.S.L.B.), the areas in which growth has been most rapid. No part of the book, however, has escaped major revision. An important change has been its division into four sections, general microbiology, immunity, systematic bacteriology and virology. Although the book has been prepared with the requirements of the medical student particularly in view, it should also be of value to undergraduates taking honours science courses, to postgraduates working for higher degrees and as a work of reference for medical graduates.

The authors are indebted to many people for assistance in respect of certain sections of the text, particularly their colleagues and former colleagues in Dublin and Manchester. Professor McEntegart of the Department of Medical Microbiology in the University of Sheffield kindly offered the hospitality of his department to one of them (T.S.L.B.) during a two months leave of absence granted to him by the University of Manchester and devoted to a final revision of the text. The authors also wish to thank Dr W. Hodgkiss of the Torrey Research Station, Aberdeen, for a number of the electron micrographs and Dr R. Ollerenshaw and his colleagues, especially the late Mr J. Kilshaw, of the Department of Medical Illustration in Manchester, and Mrs F. S. Stewart for new line-drawings. Finally we would like to thank the editorial staff of Baillière Tindall for their patience and understanding during the long period in which the work was in preparation.

February 1977

F. S. STEWART
T. S. L. BESWICK

Part I

GENERAL MICROBIOLOGY

I.1

The General Properties of Bacteria

The term bacteria is applied in a general sense to a somewhat heterogeneous collection of minute unicellular organisms of relatively simple structure. As a group bacteria are closely related both to the simpler plants (the microscopic algae and fungi) and to the simpler animals (the protozoa) but may be regarded as even simpler and more primitive. On the whole their resemblance to the algae and fungi is greater than to the protozoa and they are therefore usually classified in the plant kingdom in which they are assigned to the class Schizomycetes or fission fungi.

In certain respects, however, bacteria obviously possess properties normally associated with the animal kingdom, many being actively motile and owing their motility to the possession of organelles (flagella) similar to those found in motile protozoa. In fact van Leeuwenhoek, the discoverer of bacteria, referred to them because of their frequent motility as little animals or 'animalcules'. In order to deal with this dilemma some microbiologists, following the proposal of Haeckel, a nineteenth-century biologist, assigned bacteria and unicellular micro-organisms in general to a kingdom distinct from both the animal and plant kingdoms—the protista. The protista have in turn been divided into higher and lower protists, differentiated by the presence or absence of a nuclear membrane. Protozoa and fungi which possess this membrane are classified as higher protists, while others, such as bacteria and blue–green algae, are classed as lower protists.

The number of types of bacteria is enormous. The majority are free-living organisms which play a vital role in the economy of nature—in the circulation of carbon, nitrogen and other elements. In fact it can be said that without the activities of the free-living bacteria, other forms of life would be impossible. They are also the agents in a diversity of natural processes of great social and economic importance—the souring of milk, the ripening of cheese, the curing of tobacco. Many of the free-living bacteria have been exploited by man because of the diversity of metabolic end-products they produce—acetic acid, citric acid, butanol, acetone and antibiotics, to name but a few. Here, however, we are concerned with the parasitic bacteria whose life is associated with the animal body and, more particularly, with the pathogenic bacteria which produce disease; these constitute only a small minority of the total bacterial population.

The classification of the various types of bacteria presents considerable difficulties and will be considered more fully in a later chapter (Chapter II.1). For the moment, however, we may note that on purely morphological grounds they may be subdivided into six main groups.

1. *Bacteria proper or true bacteria.* These are rigid unbranched free-living organisms. To the medical bacteriologist they are the most important of the Schizomycetes. They may be further divided according to their shape into cocci, bacilli, vibrios and spirilla.

Cocci are more or less spherical organisms. They are given descriptive names depending on the ways in which the individual organisms are arranged. Diplococci are arranged in pairs; streptococci in chains and staphylococci in clusters like bunches of grapes. Tetrads are cocci arranged in groups of four cells and sarcinae in cubical packets of eight cells.

Bacilli are rod-shaped bacteria whose length is at least twice their width. Some of the smaller bacilli may, however, assume an almost coccal

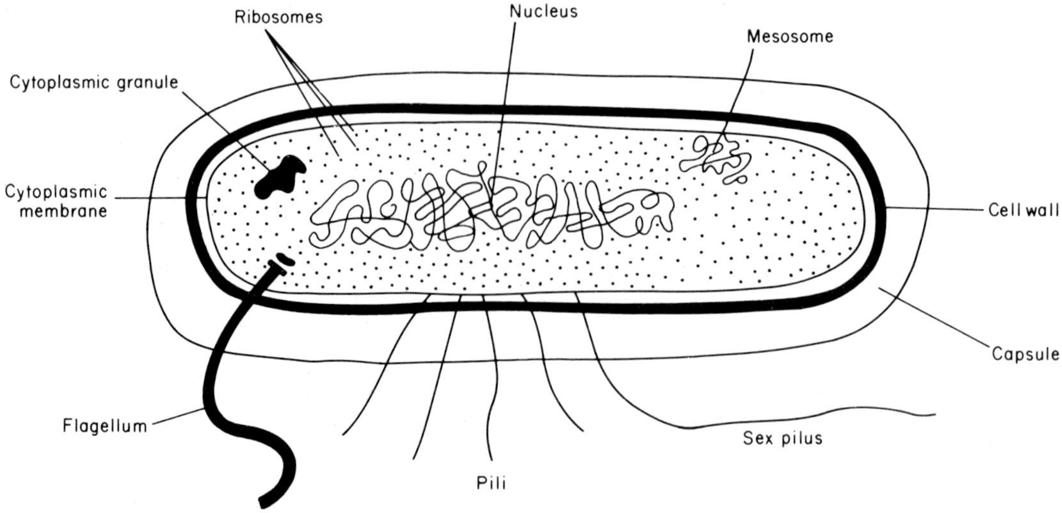

Fig. 1.1. The structure of a bacterial cell.

appearance. Vibrios are curved rods the curve of which forms less than one complete spiral. Longer rigid curved organisms usually with several spirals are known as spirilla.

2. *Actinomycetes.* These, like the bacteria proper, are rigid organisms but resemble the fungi in that they may show true branching and in tending to form filaments.

3. *Spirochaetes.* These are flexuous non-branched organisms of spiral shape.

4. *Mycoplasmas.* These are highly pleomorphic organisms of indefinite shape. In general they resemble bacteria which have been deprived of their rigid cell walls (p. 5).

5. *Rickettsiae and chlamydiae.* These are very small organisms which were at one time regarded as being intermediate between viruses and bacteria. The chlamydiae have frequently been described as large viruses, but are now recognized as bacteria, though differing from the true bacteria in being so highly parasitic that they can multiply only within other living cells. This obligatory parasitism is due to a deficiency in the full complement of metabolic enzymes necessary for the maintenance of a free-living existence. Nevertheless, though highly parasitic, these organisms have some independent metabolic capability, and can be regarded as living organisms, in a sense in which viruses (p. 337) manifestly cannot. It seems likely that both rickettsiae and chlamydiae have developed from true bacteria by a process of retrograde evolution.

6. *Miscellaneous higher bacteria.* These have been very inadequately studied and are of doubtful status. They are free-living organisms of unusual morphology, comprising stalked, sheathed, budding and slime forms. Though of great intrinsic interest, they are of no importance in medical microbiology.

STRUCTURE OF BACTERIA

Bacteria are the smallest free-living organisms; their size is measured in micrometres (μm) 1 μm being 10^{-6} m (1/25 000 in). Because of this a microscope affording a considerable degree of magnification is necessary for their demonstration. Most of the cocci have a diameter of about 1 μm in length and from 0·5 to 1 μm in width. Different species of bacteria tend to have characteristic average dimensions but individual strains may show appreciable deviation from these.

When examined in the living condition bacteria are seen to be transparent, colourless and homogeneous or finely granular. In unstained wet preparations the contrast between bacteria and background is poor. They are therefore usually examined in stained preparations. Staining not only increases contrast and therefore facilitates the visualization of bacteria but also, when stains specific for these are employed, permits the

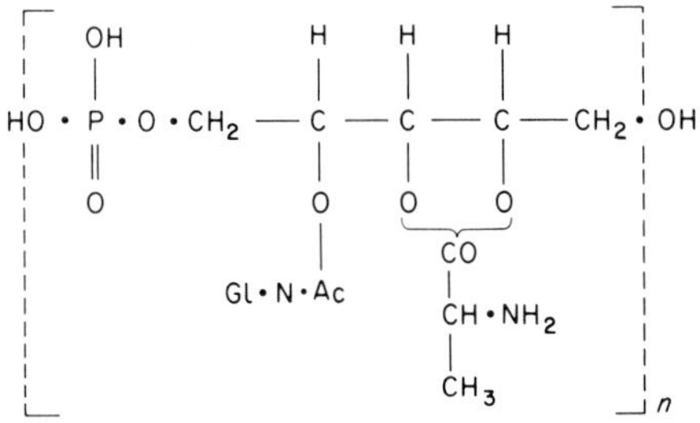

Fig. I.2. Ribitol techoic acid from a strain of *Staphylococcus aureus*. The molecule contains one alanine residue for every two ribitol residues. Gl.n.Ac=N-acetyl glycosamine. $n = 12–16$.

recognition of certain structural features. For general purposes the stain most frequently employed is Gram's stain (p. 26). On the basis of their reactions in this stain bacteria may be classified into two large groups—Gram-positive and Gram-negative. This classification is of great importance in relation to bacterial identification.

In spite of their small size bacteria have a well organized cellular structure. Like the cells of higher organisms the bacterial cell possesses a *nucleus*, a *cytoplasm* and a *cytoplasmic membrane*. The true bacteria and the actinomycetes resemble plant cells in possessing, in addition, a *cell wall* which is distinct from the cytoplasm and cytoplasmic membrane.

The cell wall is a complex rigid structure which gives bacteria their definite shape. It must possess considerable strength so as to enable it to withstand the osmotic pressure—in some cases this is as much as 20 atmospheres—of the intracellular constituents. Were it not in fact for the cell wall, bacteria would burst, since the delicate cytoplasmic membrane would by itself be unable to withstand this high internal pressure. The relationship of the cell wall to the underlying cytoplasmic membrane may be thought of as being like that of a bicycle tyre to its inner tube. The thickness of most bacterial cell walls under normal conditions of growth is in the range of from 10 to 20 nm, the cell walls of Gram-positive bacteria being in general somewhat thicker than those of Gram-negative bacteria.

The rigidity of the bacterial cell wall is due to the presence of a basal three-dimensional enveloping structure known as *peptidoglycan* or *murein*. The cell wall peptidoglycans so far examined have been found to be composed chemically of a backbone consisting of the amino sugars N-acetyl glucosamine and N-acetyl muramic acid, to the muramic acid residues of which are attached peptide side-chains comprising a limited number of amino acids, D- and L-alanine, D-glutamic acid and either L-lysine, diaminopimelic acid, diaminobutyric acid or ornithine. The latter two dibasic amino acids have only recently been recognized in cell walls. In some species either glycine or aspartic acid is also present. In the intact cell the three-dimensional structure of the peptidoglycan is completed by extensive cross-linking between neighbouring peptide chains. In *Staphylococcus aureus* this appears to occur through a glycine pentapeptide, linking lysine and terminal D-alanine residues.

In the case of certain bacteria, notably *Bacillus* species, *Micrococcus lysodeikticus* and *Str. faecalis*, the wall mucopeptide of the intact cell is susceptible to the action of lysozyme. As a result of this action the wall is broken down leaving a structure known as a *protoplast*, which is bounded only by the cytoplasmic membrane. Protoplasts are osmotically sensitive and undergo rapid lysis in isotonic or hypotonic solutions. Because of this they can only be maintained in intact form in hypertonic solutions. Protoplasts may also be obtained from certain bacteria by growing them in the presence of penicillin or other antibiotics which are capable of inhibiting cell wall synthesis. Protoplasts are much more readily prepared from Gram-positive than from Gram-negative bacteria. This is not surprising, as the peptidoglycan is a relatively minor component of the Gram-negative cell wall. In fact, protoplasts obtained from Gram-negative organisms appear to retain a good deal of the cell wall lipopolysaccharide and lipoprotein.

Peptidoglycan is the major constituent of the cell walls of Gram-positive bacteria constituting from 50% to as much as 90% of the wall. In Gram-negative species, however, mucopeptide constitutes only from 5 to 10% of the wall, but even under these circumstances it is responsible for the osmotic stability of the cell. This is because the osmotic pressure of the Gram-negative cell appears to be considerably less than that of the Gram-positive cell. In addition to mucopeptide, Gram-positive cell walls contain polysaccharides and teichoic acids. The latter are polymeric complexes of ribitol phosphate or glycerol phosphate with simple sugar or amino sugar

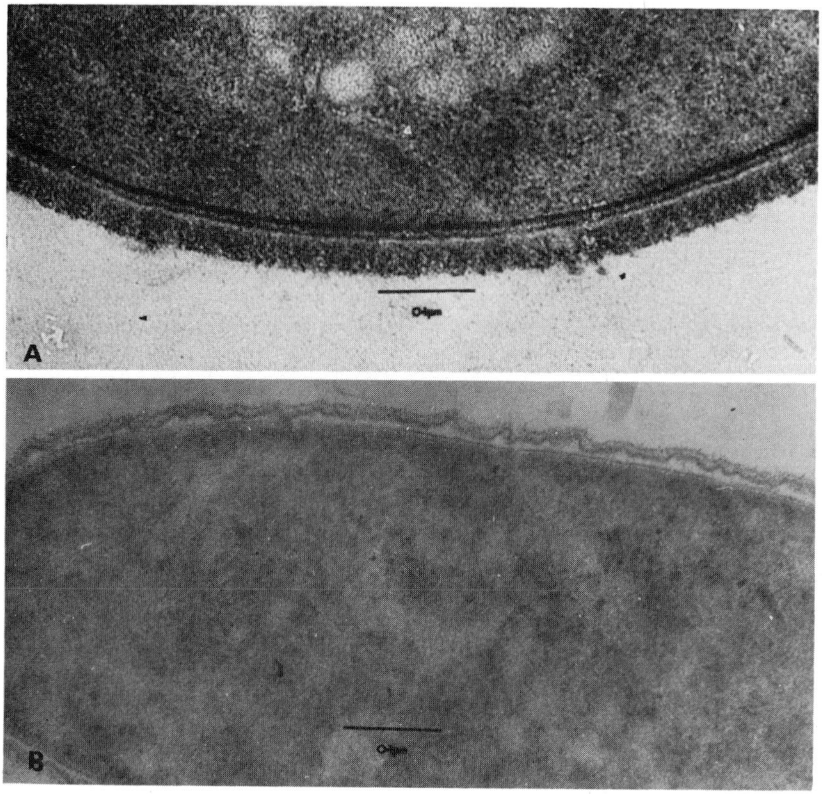

Fig. 1.3. Electron micrographs of the cell wall in Gram-positive (A) and Gram-negative (B) bacteria. *(W. Hodgkiss.)*

residues, together with alanine in the unnatural D-configuration. It now seems possible that some at least of these additional components are bound to the mucopeptide by covalent bonds. Their precise significance in cell wall structure is at present unknown. Gram-positive cell walls do not appear to contain protein as an internal component of the wall but in some cases protein may be present in a microcapsular or capsular layer, e.g. the M proteins of the streptococci (p. 213), and the glutamyl polypeptides of *Bacillus* species (p. 226).

Gram-negative cell walls are considerably more complex biochemically than Gram-positive cell walls, and unlike the latter contain a high concentration of lipid and protein. The precise anatomical arrangement of the various components in the Gram-negative cell wall is not fully understood. It would appear, however, that the cell wall mucopeptide occurs as an inner layer, outside which is the lipoprotein and lipopolysaccharide, each possibly occurring as further distinct layers.

From the marked difference in cell wall composition shown between the Gram-positive and Gram-negative bacteria it might be expected that the chemical composition of the cell wall could be useful as a taxonomic criterion. That this is the case was first shown by Cummins and Harris who found that the sugar composition of the cell walls of various Gram-positive bacteria was of particular value for differentiation at both generic and species level. Thus the cell walls of *Staph. aureus* contain hexosamine but no simple sugars. Rhamnose is found in the cell walls of many streptococci but is absent from the pneumococcus cell wall, and arabinose is apparently a universal component of the cell walls of the corynebacteria. Recent evidence suggests that the presence of diaminobutyric acid may be associated significantly with plant pathogenicity in the corynebacteria.

Beneath the cell wall, as an anatomically distinct

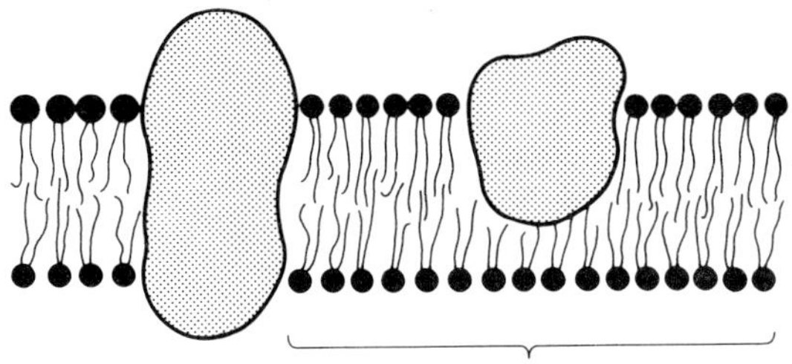

Fig. I.4. A cross-section of a cytoplasmic membrane.

structure, is the cytoplasmic membrane. The membrane is semi-permeable, allowing the passage of water but impermeable to large molecular weight compounds and also to many small molecular weight compounds, e.g. simple sugars and amino acids. Because of this it constitutes the osmotic barrier of the cell, controlling the passage of nutrients into the cytoplasm and of end products of metabolism out of it.

Compared with the cell wall the cytoplasmic membrane is very thin, from 6 to 10 nm in cross-section. The structure of the cytoplasmic membrane as proposed by Singer and Nicolson is now generally accepted. This envisages the membrane as having a bimolecular leaflet structure consisting of a sandwich of two layers of lipid, the hydrophobic side-chains of which are on the inside and the hydrophilic residues on the outside, in contact on the one hand with the external environment and on the other with the cytoplasm. Interspersed in the lipid bilayers are molecules of protein with their hydrophilic residues projecting on the outside and inside of the membrane. The molecules of which the membrane is composed have considerable translational mobility through the membrane and also rotational mobility in the transverse axis of the membrane. The entire structure is consequently described as being a fluid mosaic.

There is considerable evidence that the cytoplasmic membrane, together with its cytoplasmic extensions (see below), may be the functional equivalent of the mitochondria of higher cells. Thus in *Staph. aureus* and *Micrococcus lysodeikticus* it has been found to be the main location of Krebs cycle enzymes (p. 18), of the electron transport chain and of oxidative phosphorylation. It also contains *permeases* which are necessary for the active transport of many small molecular weight organic compounds across the membrane. In addition, recent evidence indicates that, at least in Gram-positive cells, the membrane participates in the synthesis of the cell wall.

Morphologically, the bacterial cytoplasm is much simpler than the cytoplasm of higher organisms. Electron micrographs of bacteria have failed to reveal structures analogous to mitochondria and the complex membranous system of higher organisms known as endoplasmic reticulum and the Golgi apparatus are notably absent. However, certain bacteria, particularly Gram-positive bacteria, contain invaginations of the cytoplasmic membrane which have sometimes a lamellar and sometimes a vesicular form. These structures, known as *mesosomes*, can sometimes be seen attached to the bacterial chromosome; they are often closely related to the septum of dividing cells. It is thought that they may have a regulatory role in cell division, ensuring that each daughter cell receives a nucleus. Some may be involved in the secretion of extracellular products.

The main structural components of the bacterial cytoplasm would appear to be the ribonucleoprotein granules, known as *ribosomes*. These measure from 10 to 20 nm in diameter, and are the sites of protein synthesis. It has been estimated that a single bacterium may contain upwards of 10 000 ribosomes. The ribosomes have been shown to occur in groups, known as *polysomes*, linked together like beads on a chain by messenger RNA (p. 19). The individual ribosomes have a sedimentation coefficient of 70 Svedberg units. Each 70S ribosome is a complex of two smaller units with sedimentation coefficient of 50S and 30S. In the presence of magnesium these form stable complexes to yield a 70S ribosome. In low magnesium concentrations, however, they dissociate and in the absence of magnesium further degradation occurs. Ribosomes appear to be the points of attack of certain antibiotics which act on bacteria by interfering with protein synthesis (Chapter I.5).

Although it was for a long time disputed, bacteria are now accepted to possess a structure which is the morphological and functional equivalent of the nucleus of higher organisms. Like the latter it carries the genetic blueprint of the cell coded in the nucleotide sequence of its desoxyribonucleic acid (DNA).

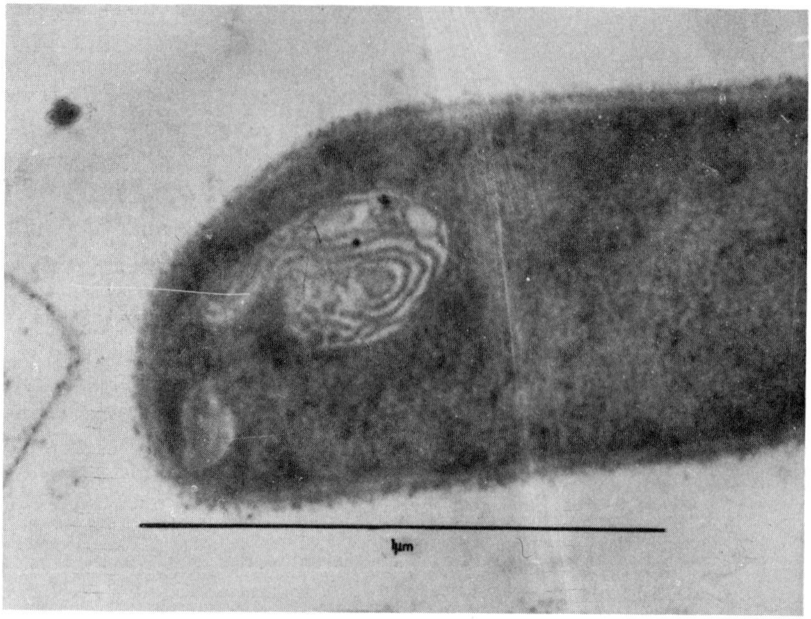

Fig. I.5. The mesosome of *Clostridium bifermentans*. (*W. Hodgkiss*.)

As seen in electron micrographs of very thin sections of bacteria, nuclei appear as central areas of lower electron density than the rest of the cell. These areas consist of interlacing bundles of very thin fibrils, about 2 to 6 nm in diameter, in a clear structureless matrix. There is convincing autoradiographic evidence that, in the case of *E. coli* and probably other organisms as well, the nuclear DNA occurs in the form of a single chromosomal thread of about 2000 μm in length. Except when mobilized for transfer, it exists in the cell as a circular closed loop. One can imagine the coiling necessary to fit this long thread into the nuclear area of a cell which is in itself only 2 to 3 μm in length. It is this folding which gives the appearance of bundles of fibrils seen in electron micrographs. This concept of a single chromosomal thread is in agreement with the genetic evidence indicating the determinants of transmissible characters on a single linkage group. The bacterial cell thus differs genetically from the cells of other organisms in being genetically haploid and in possessing only a single chromosome. There is, therefore, no necessity for the occurrence in bacteria of the complicated series of changes associated with mitosis and meiosis in higher organisms.

Table I.1. DNA and protein composition of an average bacterial cell

DNA	*Protein*
Single molecule 2000 μm long	2×10^6 molecules
Weight 5×10^6 daltons	Weight 5×10^{10} daltons
$1{\cdot}6 \times 10^7$ nucleotides	8×10^8 amino acids
2500 nucleotides per gene	400 amino acids per molecule
6400 genes	Possibly 1000 molecular species

Volume 10^{-12} ml; dry weight $1{\cdot}5 \times 10^{-11}$ daltons.

The bacterial nucleus differs from the nucleus of higher cells in two further respects:

1. It does not possess a nuclear membrane, and is therefore in direct contact with the cytoplasm of the cell. This may well facilitate rapid transfer of messenger RNA from the nucleus to the ribosomes. Cells lacking a nuclear membrane are sometimes referred to as *procaryotic* in contrast to *eucaryotic* cells which possess a nuclear membrane.
2. The bacterial DNA does not appear to be

associated with a basic protein, though the presence of some protein in the nucleus has not been excluded.

Many bacteria possess in addition to the above structure *flagella, fimbriae* or *pili, capsules, spores* and various types of *granules*. Since these structures are found only in certain species their demonstration is of importance in bacterial identification.

Flagella are the locomotor organs of bacteria. They are long, contractile hair-like filaments measuring from 12 to 19 nm in diameter, and usually several micrometres in length. In some bacterial species, e.g. the enterobacteriaceae, they are arranged along the sides of the organism. This is known as *peritrichous* flagellation. In others, such as *Pseudomonas,* flagella occur singly or in tufts at one or both ends—*polar* flagellation. Individually, flagella are too thin to be seen by ordinary light microscopy but they can be stained by silver impregnation methods. They are best demonstrated by electron microscopy when they appear, particularly those from peritrichously flagellated bacteria, as helical or wavelike structures. In electron micrographs the flagella can be seen to pass through the cell wall and to originate either from the cytoplasm or from the cytoplasmic membrane. Though they are not attached to the cell wall, flagella can apparently initiate movement only when the cell wall is present. Thus protoplasts prepared from *Bacillus* species by lysozyme treatment and possessing apparently normal flagella, are non-motile.

From studies with a variety of peritrichously flagellated species it would appear that the standard method of attachment is through a hook at the proximal end of the flagellum. The hook appears to be attached to a disc-like or spherical basal structure, the whole complex being closely associated with, and probably directly attached to, the cytoplasmic membrane. The basal structure is responsible for the motility of flagella in a manner which has not been fully explained. This appears, however, to be due to the first instance to the passage of helical waves of contraction down the flagellum, the effect of which is to propel the bacteria forward. Relative to their size, the speed with which the bacteria can be propelled is considerable; actively motile species are capable of travelling at up to 50 μm/sec. For a typical Gram-negative bacillus of approximately 5 μm in length, this would be equivalent to a man performing a 100 m dash in 5 sec.

Chemically, flagella are composed mainly, if not entirely, of protein. They consist of strands wound round each other in helical forms which in certain species, e.g. *Vibrio*, are surrounded by sheaths. The strands are made up of a globular protein called *flagellin*, disposed helically around a hollow core. Under acid conditions the subunits dissociate, but on reversal of these conditions they reaggregate to form intact flagella. This process is analogous to the self-assembly of viral capsids and suggests that under *in vivo* conditions flagella may spontaneously assemble from preformed units.

Fimbriae or *pili* are hair-like processes arising like flagella from granules in the cytoplasmic membrane and are only found in certain bacteria, particularly in Gram-negative bacilli. They differ from flagella in being shorter and thinner, straight, and play no part in the motility of the organism. Fimbriae cannot be demonstrated with the light microscope but can be seen in suitably prepared electron micrographs. They are best developed when the bacteria are grown in fluid media. They are probably structures by which bacteria can attach themselves to cells and to other particulate material. Certain bacteria possess specialized fimbriae or pili which are longer and thinner than the common type. These appear to be hollow and to constitute conjugation tubes through which DNA is transferred from one organism to another during conjugation; they are consequently referred to as sex pili.

Bacteria of the genera *Bacillus* and *Clostridium* produce highly resistant dormant forms known as *spores*. These appear first as round or oval bodies which form within the cytoplasm of the organism. They are highly refractile but stain with great difficulty. In simple Gram-stained preparations they appear as clear unstained areas. When stained, however, e.g. by Möller's method, they hold the stain firmly and are not readily decolorized. The spore may be situated in the centre of the bacillus equatorial, near one end-subterminal, or protruding from one end-terminal, the precise location being constant within a single species. In the *Bacillus* genus the diameter of the spore is normally less than that of the bacterium but in the genus *Clostridium* the spore is usually wider than the bacterium.

The first stage in the formation of the spore is a condensation of chromatin, apparently

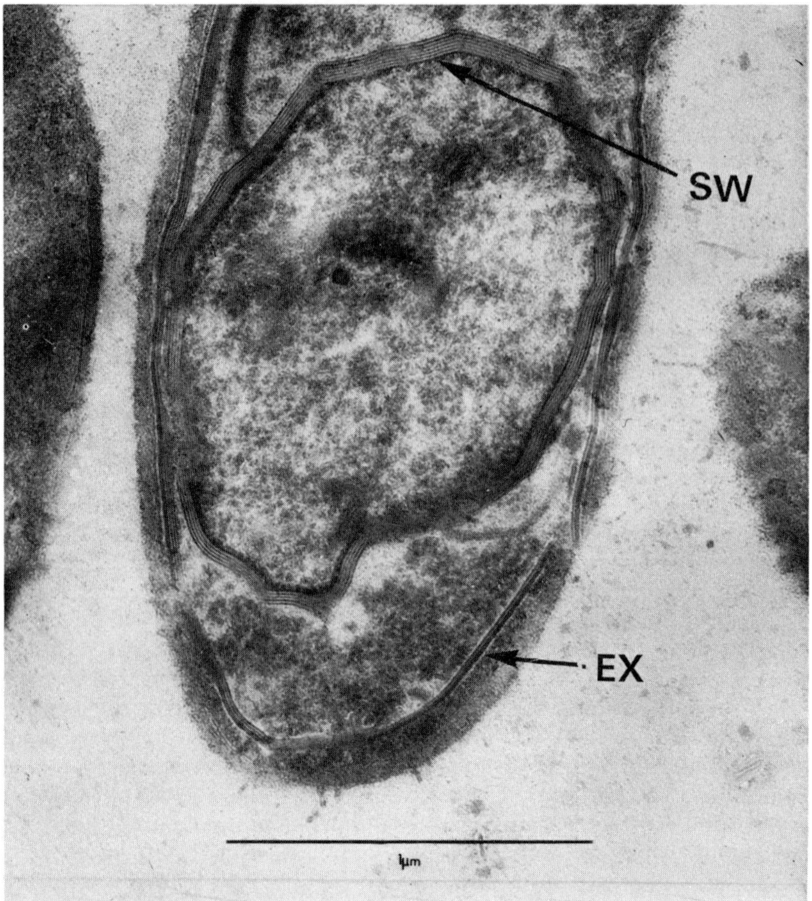

Fig. I.6. The developing spore and exosporium of *Clostridium pasteurianum*. EX, exosporium; SW, spore wall. *(W. Hodgkiss.)*

representing half the cellular DNA, in an area at one end of the bacterium known as the *spore field* which is destined to be the site of formation of the new spore. At the same time a transverse septum derived from the cytoplasmic membrane forms by a process of invagination, and divides the developing spore, now known as the *forespore*, from the rest of the bacillus, known as the *sporangium*. From this stage on, the two areas of the cell are functionally distinct, the sporangium continuing to produce components characteristic of the vegetative form, and the forespore producing components characteristic of the spore. The dividing septum eventually completely encircles the forespore as a double-layered membrane, the inner layer of which becomes the spore wall. The spore *cortex* is then laid down between the inner and outer layers and the process completed by transformation of the outer layer into the spore coats and *exosporium*.

Finally, when the spore is fully formed, the body of the organism degenerates and the spore is set free, the entire process culminating in the release of the mature spore taking from four to eight hours to complete. During sporulation major changes are observed in energy metabolism. Thus in *Bacillus* species there is an activation of the enzymes of the tricarboxylic and glyoxylic acid cycles which are present, though dormant, in the vegetative cells. The primary function of this activation appears to be the provision of Krebs cycle intermediates for the synthesis of the characteristic spore component, *dipicolinic acid*. In this process poly-β-hydroxy

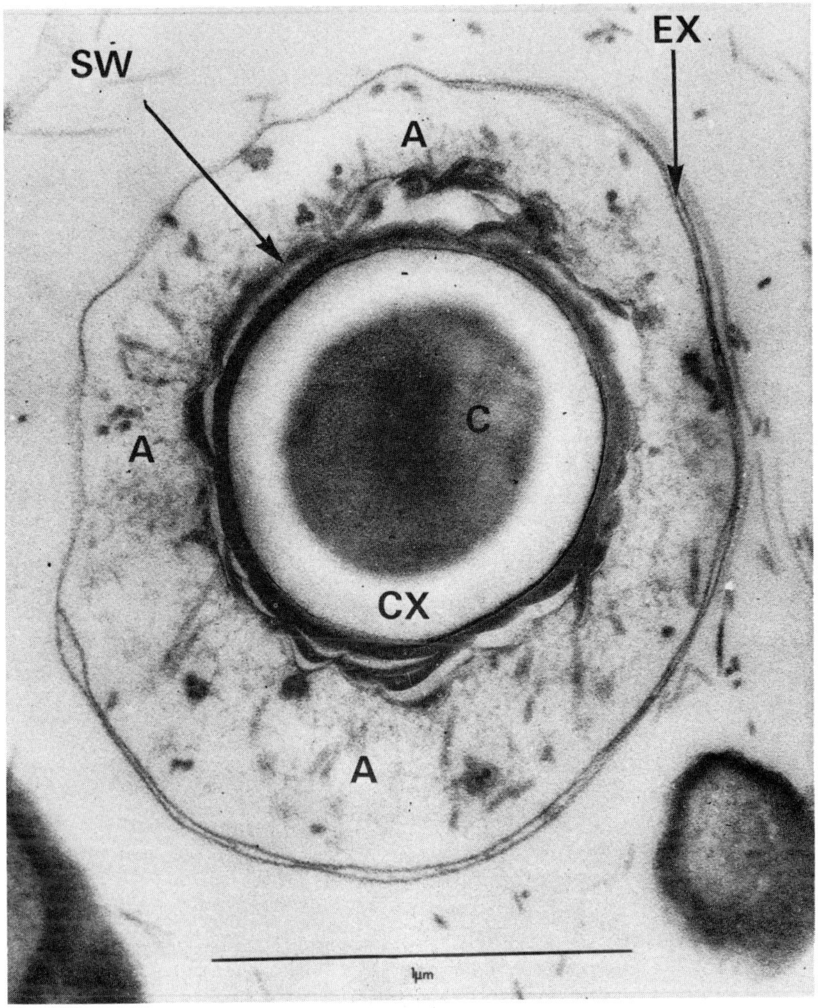

Fig. I.7. The spore and exosporium of *Clostridium botulinum* Type E. EX, exosporium; A, appendages; SW, spore wall; CX, cortex; C, core.

butyrate, which is synthesized during the early stages of sporulation and which disappears during the later stages, appears to play an important intermediary role.

Spores are dormant bacterial forms; they are non-metabolizing and non-reproducing. Spore DNA is not replicated, messenger RNA is not produced and the spore protein synthesizing system is defective as judged by in vitro incorporation experiments. Their biological importance derives from their high resistance to a variety of agencies to which the vegetative form is susceptible—heat, ultra-violet irradiation, mechanical disruption and most chemical disinfectants. They can, as a result, survive for much longer periods than the vegetative form in environments which would be completely inimical to the latter. Spores have, in fact, been recovered from dried plants which have been preserved for some 300 years. Of particular importance to medicine is their high resistance to heat; some will resist boiling for several hours. This necessitates the employment, as in the autoclave, of temperatures of over 100°C for the sterilization of sporulating species.

Spores show marked differences from vegetative

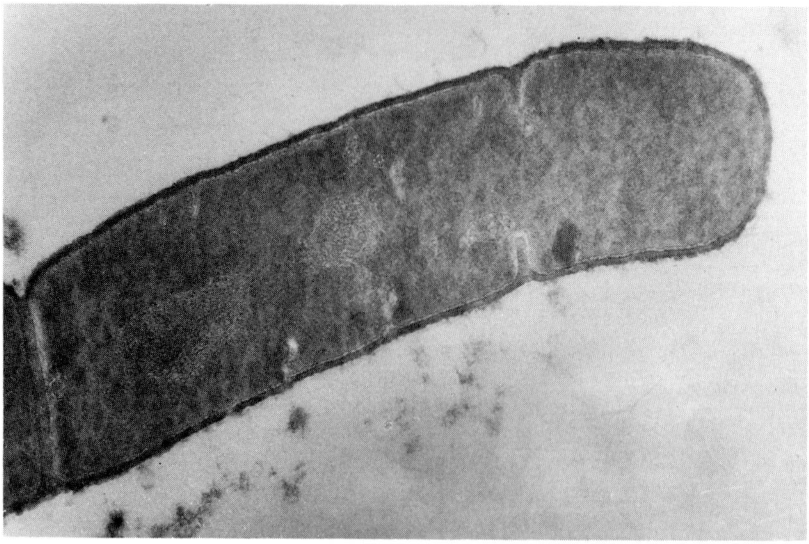

Fig. I.8. An electron micrograph of a thin section of a bacillus. Note the cell wall, transverse septum and nucleoides. (*From Glauert* et al. *(1961) Expt. cell Res., 22, 73.*)

organisms in morphology, chemical composition and antigenic structure. The structure of the spore shows up well in electron micrographs of very thin sections. In the centre is the cytoplasm of the spore which appears to be much more homogeneous than the cytoplasm of the vegetative form. The cytoplasm is bounded by a delicate membrane or spore wall; outside the spore wall is a thick *cortex* of low density and low affinity for dyes. The cortex appears to contain most of the mucopeptide of the spore. This is surrounded by the coats of the spore of which up to three may be present. Finally, outside the spore coats there is in many cases a loosely attached outer layer or *exosporium*. Recent electron micrographic evidence indicates a nap of numerous, very thin filaments on the exosporia of certain spores.

The most striking chemical differences between spores and vegetative cells are their lower free water content, and the presence of a large amount—on average about 10% of the mass of the spore—of an unusual compound, dipicolinic acid (DPA), occurring as its calcium salt. Spores have been shown to possess a number of spore-specific antigens and to produce a variety of enzymes differing both in antigenic and physical properties from the corresponding enzymes of the vegetative form. The spores are unusually rich in sulphur-containing amino acids, notably cysteine, which appears to be located mainly in the spore coat. The

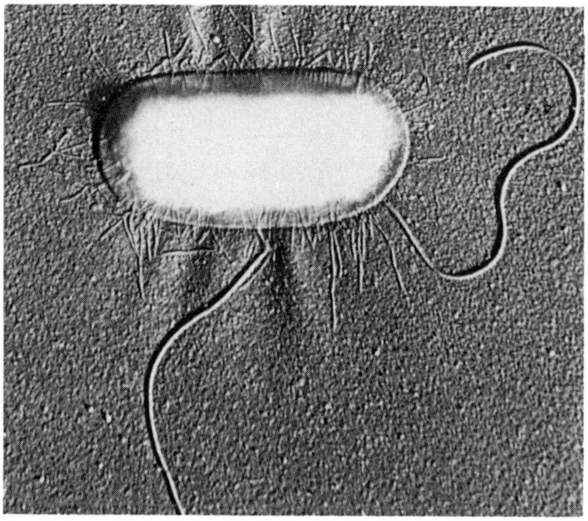

Fig. I.9. An electron micrograph of *Escherichia coli* showing fimbriae and two long wavy flagella. (*From Duguid (1955) J. Path. Bact., 70, 335.*)

disulphide linkages derived from these are thought to be responsible for the increased radiation resistance of the spore. Most workers believe that the heat resistance of the spore is in some way related to the presence of a calcium dipicolinic acid chelate.

In general, sporulation is initiated in conditions where, as a result of nutrient limitation, rapid growth is terminated; the precise mechanism by which this occurs is not known. It would appear, however, that sporing bacteria possess in effect two genomes, one for the vegetative cell and one for the spore. During vegetative growth transcription of the spore genome is inhibited, and as sporulation proceeds the transcription of the vegetative genome is inhibited. Evidence has been obtained by Halvorson for the presence during sporulation in *Bacillus* of a *sporulation factor* which he believes may have the effect of derepressing the transcription of the spore genome. In addition he has postulated the production by the spore genome of a substance which inhibits the transcription of the vegetative genome. This latter substance may be identical for the *Bacillus* genus with an antibiotic compound shown by Balassa to be produced during sporulation.

Under suitable conditions the spore regenerates the vegetative form. This process, known as *germination*, is initiated apparently by the stimulation of certain so-called trigger nutrients, usually simple sugars, amino acids and ribosides present in the environment. It can be also be induced in some cases by unnatural means, e.g. by mechanical abrasion or by exposure to certain surface active compounds. In general, the readiness with which generation may be induced under such conditions is greatly increased by prior mild heat treatment.

The *capsule* is an outer covering of jelly-like material surrounding the cell wall. It is not essential to the life of the cell since it may be lost spontaneously by mutation and, in some cases, removed enzymically without loss of viability. Occasionally, the outer limit of the capsule may be poorly defined; it is then usually referred to as a *slime layer*. The capsules of most of the pathogenic bacteria are viscous polysaccharide gums which give the colony of the organism a mucoid appearance. Capsules have a low affinity for aniline dyes, and are most reliably demonstrated by 'negative staining' procedures. Polysaccharide capsules are readily washed off the cell with water and water must therefore be avoided in capsule stains. Organisms of the *Bacillus* genus are unusual in that their capsules are composed largely of polypeptide. In pathogenic species the capsules inhibit the engulfment of the bacteria by the phagocytes, and play an important part in determining virulence; if the capsule is lost by either mutation or enzymic degradation the organism becomes avirulent.

Certain bacteria are known to possess surface antigens, which exist in a distinct layer external to the cell wall, though insufficiently thick to be seen by ordinary microscopic methods. This layer is sometimes referred to as a *microcapsule*. Important examples of microcapsular antigens are the M antigens of the streptococci and the Vi antigens of *Salm. typhi*. Of these, the M antigens of the streptococci have probably most claim to be described as microcapsular, since they can be removed enzymically without affecting the viability of the organism or the integrity of the cell wall. In the intact cell they are normally covered by a true capsule.

Intracellular *granules* may be observed in many species of bacteria. The most important of these are the *metachromatic* or *volutin* granules which are highly characteristic of the corynebacteria and which are of practical importance in the identification of these organisms. Metachromatic granules stain very intensely with basic dyes; because of this it was at one time thought that they consisted largely of ribonucleic acid. It is now known, however, that they are in fact granules of highly polymerized metaphosphate. Metachromatic granules accumulate in the bacterial cell only when the latter is grown on relatively rich media. Certain bacterial species show the presence of granules of glycogen like polysaccharide and of lipid. Their demonstration is not of any diagnostic importance.

Many bacteria when grown under unsuitable environmental conditions assume an unusual and frequently bizarre morphology; this may take the form of spherical, lemon-shaped, boat-shaped or irregular coarse spiral structures usually known as *involution forms*. Most involution forms appear to be non-viable under ordinary conditions of cultivation. They are probably forms in which the synthesis of the bacterial cell wall is defective and therefore analogous to sphaeroplasts and L forms (p. 284).

GROWTH OF BACTERIA

Bacteria multiply by simple division or *fission*. Fission occurs when the cell reaches a critical size, but the immediate stimulus of the division process is unknown. The first stage in fission consists of a division of the nucleus, followed by division of the cell. Nuclear division generally occurs in phase with cell division but when cells are growing rapidly it tends to outpace cell division. Consequently during rapid growth the cells usually show two or more nuclei, and one or more transverse septa prior to division. The rate at which fission occurs varies with different species. Many species, e.g. *E. coli* and many Gram-negative bacilli, multiply very rapidly, dividing, in their most rapid growth phase and under optimal conditions, once in every 20 to 30 minutes. Some organisms, however, e.g. tubercle bacilli, grow very slowly, dividing only once or twice in 24 hours.

Two main types of cell division have been described. During the first, characteristic of Gram-negative bacteria, the cell appears to divide by a constrictive or pinching process. In the second, there is an ingrowth of the cytoplasmic membrane to form a transverse septum, the new transverse cell wall being laid down between adjacent layers of membrane.

The growth of individual cells has recently been studied by the fluorescent antibody technique. This has shown two main types of growth mechanism:

1. The cell wall may grow by intercalation, i.e. by the insertion at several foci throughout the wall of segments of new material. There is some evidence that this involves a preliminary autolysis at the sites of fresh synthesis. This type of growth appears to be common in Gram-negative bacteria.
2. The new material is laid down largely at the point at which cell division is going to occur. This seems to occur particularly in Gram-positive organisms.

Bacteria may be cultivated in the laboratory by providing them with suitable nutrient material. This may be supplied either in fluid or in solid media. When inoculated into fluid media most species grow diffusely throughout the medium but some, particularly those which prefer a high oxygen tension, grow mainly on the surface of the medium while others grow mainly at the bottom. A fluid culture of a rapidly growing organism such as *E. coli* may contain after overnight incubation as many as 10^9 to 10^{10} organisms per ml. On solid media bacteria grow in the form of colonies, each colony consisting of the descendants of a single cell deposited on the surface of the medium. If a large number of organisms is seeded onto the medium the colonies will fuse together and produce a confluent sheet of growth.

When bacteria are inoculated into a fluid medium there is a delay of some hours before they begin to multiply. This period of delay is known as the *lag phase*. Although there is no obvious multiplication during the lag phase, bacteria are nevertheless highly active in it. From the beginning they increase in size and synthesize cellular material. This synthetic activity is clearly shown if the nitrogen concentration of a culture, which is a measure of the amount of bacterial protein present, is plotted against time.

The lag phase is followed by the *logarithmic phase*, during which the bacteria are multiplying at their maximum rate. The time required for one bacterial division during the logarithmic phase is known as the *generation time*. In the log phase the number of organisms present in each generation period is virtually twice that in the previous period. Consequently if the logarithm of the number of organisms is plotted against time the result is a straight line. The phase is of relatively short duration, lasting at most for some hours. In its termination three factors are of importance:

1. Exhaustion of nutrients. This factor is probably of much greater importance to bacteria growing naturally than under the usual conditions of a laboratory culture.
2. The accumulation of toxic metabolic end-products which inhibit bacterial growth.
3. The achievement of a maximum population density. It would appear that even under conditions of optimal nutrition, and in the absence of significant inhibition by metabolic end-products, there is a minimal amount of living space in which a bacterium will grow.

Following the log phase there is a *stationary phase*, during which the bacterial population remains roughly constant. In this phase a balance is apparently struck between bacterial multiplication and bacterial death. The stationary phase is followed by the *phase of decline*, during which the number of organisms progressively decreases, the culture eventually becoming sterile. Complete sterility may not be achieved, however, for a considerable time.

NUTRITIONAL REQUIREMENTS

During the growth of bacteria there occurs an active synthesis of the complex constituents of the cell. In order that this synthesis may take place, the elements which are ultimately incorporated in the cell protoplasm must be present in an assimilable form in the medium in which the organism is growing. Phosphorus, sulphur, sodium, potassium, iron and other elements required in small amounts are readily assimilated from simple salts present in the environment of the organism. Different types of bacteria, however, show considerable differences in the nature of the food material which they can accept as sources of the elements carbon and nitrogen. These differences correspond in general with differences in the synthetic power of the bacteria. Some organisms—*autotrophs*—are endowed with considerable synthetic power and can utilize very simple inorganic compounds, such as carbon dioxide and nitrates or ammonium salts, as sources of carbon and nitrogen respectively. From these they can build up their complex carbohydrates, proteins, fats and other essential constituents.

The bacteria in which we are mainly interested are classed as *heterotrophs*. These differ from the autotrophs in requiring organic sources of carbon such as are found in carbohydrates, proteins, amino acids and fatty acids, for growth. There is in fact virtually no type of organic compound which cannot be utilized by some species of bacteria. These compounds serve not only as sources of energy but also as sources of the carbon required for the synthesis of cell material. Among the heterotrophic bacteria, however, there are marked differences in nutritional requirements. Some, such as *E. coli*, are relatively non-exacting in their demands, being capable of growing on fairly simple media containing glucose or other carbohydrate as a source of energy and common inorganic salts as a source of nitrogen, phosphorous, sulphur and other mineral requirements. Many of the parasitic bacteria, however, are much more fastidious than *E. coli* and will not grow on a simple glucose salt medium such as described above, but require the addition of complex organic compounds, which they are unable to synthesize for themselves. Such compounds are known as *growth factors*. In general, the nutritional demands of bacteria—that is, the range of growth factors they require—can be correlated with their degree of parasitism. As a result of long habituation, the more highly parasitic bacteria have become adapated to the well-stocked nutritional environment of the animal body and, finding many of their complex organic requirements preformed, do not need the elaborate synthetic powers of the less exacting free-living bacteria. This development of nutritional needs and the loss of synthetic power may therefore be considered as a sort of retrograde evolution during which bacteria have lost more and more of their synthetic capacity and have in consequence become increasingly dependent on an adequate supply of preformed growth factors.

Bacterial growth factors can be conveniently classified in three large groups:

1. Amino acids.
2. Bacterial vitamins.
3. A miscellaneous group of compounds of unrelated chemical composition.

Amino acid requirement may vary from that of comparatively non-exacting organisms, such as *Salm. typhi*, which needs only a single amino acid—tryptophane—to that of highly exacting organisms such as *Leuconostoc mesenteroides*, which requires 17 amino acids, none of which it can synthesize. In general, the requirement for amino acids is most frequently for the aromatic type, since these are the most difficult to synthesize. Gram-positive bacteria are much more exacting in their demands for amino acids than are Gram-negative bacteria. On the other hand, since they must obtain their amino acid requirements preformed from the environment, Gram-positive bacteria have had to develop the capacity to maintain a high intracellular concentration of amino acid against a concentration gradient. Even where considerable amino acid requirement exists, bacteria are nevertheless able to meet some of their needs by synthesis from ammonium salts. Nitrate can be utilized as an alternative source of nitrogen by some of the heterotrophic bacteria. The organism must, however, first reduce it to ammonia through the action of the enzyme nitratase. In some cases amino acid requirement depends on the vitamin composition of the medium. If the medium is deficient in a vitamin required for amino acid synthesis, the vitamin being one that the organism cannot synthesize, the appropriate corresponding amino acid must be supplied preformed.

Bacterial vitamins play a catalytic role in bacterial metabolism, generally as constituents of coenzymes, which—in contrast to the amino acids—are required by bacteria in very low concentrations. The most important of the bacterial vitamins are those belonging to the B group of animal vitamins—biotin, nicotinic acid, pantothenic acid, pyridoxine, paraminobenzoic acid, riboflavine, thiamine and folic acid. Their role in bacterial metabolism is essentially similar to their role in tissue metabolism. Other important bacterial vitamins are lipoic acid, which is required in the oxidative decarboxylation of pyruvic acid, and haematin which is a component of bacterial cytochromes and of the enzymes peroxidase and catalase. The need for purines and pyrimidines is related to the growth factor composition of the medium. Oleic acid is required by the corynebacteria and *Erysipelothrix*, and the polyamines putrescine and spermidine by *H. influenzae* and *H. parainfluenzae*.

It is probable, however, that most bacteria require all or practically all these factors and that they can synthesize the majority of those they require from simpler substances. Those they cannot synthesize must be supplied preformed in the medium. The power of bacteria to synthesize growth factors is of great importance in connection with animal nutrition, since animals can utilize vitamins produced by bacteria in the intestine.

Bacteria in the body have an ample supply of their required growth materials. In the laboratory, however, it is essential, when attempting to grow a particular organism, to ensure that the medium satisfies all its nutritional requirements. Blood or other body fluids can usually be relied on to supply a considerable variety of food materials such as amino acids and growth factors lacking in the simpler laboratory media. When the needs of an organism are exactly known, it is possible to prepare a *synthetic medium* satisfying these and containing only known constituents. Such defined media have been particularly valuable in the production of pure bacterial toxins where it is desirable to reduce unwanted material to a minimum. They have little, if any, application in routine bacteriological work.

All bacteria require for the initiation of growth a low concentration of carbon dioxide; this is essential to heterotrophic organisms as a means of replenishing the dicarboxylic acids of the Krebs cycle. Sufficient carbon dioxide both to initiate and to maintain growth is usually obtained from endogenous metabolism. Large amounts of carbon dioxide are, however, indispensable to gonococci and meningococci, and to the bovine variety of *Br. abortus*.

Bacteria show considerable differences in their requirement for and tolerance of molecular oxygen. Some, the *obligatory aerobes*, grow only in its presence. Organisms of this type, e.g. gonococci, meningococci and *V. cholerae,* grow poorly in the depths of fluid media where the oxygen tension is insufficiently high for significant growth. Others, the *obligatory anaerobes*, grow only when all traces of oxygen are absent from the medium. The majority of pathogenic bacteria are *facultative anaerobes*, i.e. organisms which will grow in both aerobic and anaerobic conditions. Why oxygen inhibits the growth of some organisms is not fully understood. It has been suggested that it is due to the production of hydrogen peroxide, a substance highly toxic to bacteria, coupled with a lack of the detoxicating enzyme, *catalase*, of aerobic organisms. An alternative possibility is that the obligatory anaerobes possess enzymes which, for activity, must be in the reduced state. Most if not all, bacteria appear to require carbon dioxide in small amounts and some, such as *Br. abortus*, in relatively small amounts. As carbon dioxide is produced by the metabolism of bacteria, enough is usually present in a normal inoculum to initiate growth.

Each type of bacterium multiplies best within a restricted temperature range. For most of the pathogenic bacteria optimum growth temperature (i.e. the temperature at which the organism multiplies at a maximum rate but which is not necessarily the temperature of maximum yield) is 37°C, with upper and lower growth limits of 40 to 50°C and 15 to 20°C respectively. In general, the more highly parasitic a bacterium the closer does its temperature range of growth approximate to 37°C, the temperature to which it has become accustomed by association with the animal body. Organisms with an optimal temperature of about 37°C are known as *mesophiles*.

Some bacteria, notably certain free-living members of the *Pseudomonas* genus, have an optimum temperature of around 20°C, and are capable of significant growth in temperatures below 10°C, a temperature at which mesophiles will not grow. Organisms of this type, which are known as *psychrophiles*, are widely distributed in nature; they are not of pathogenic importance though they may be responsible for food decomposition. The precise reason for the low growth range of the psychrophiles is unknown. In some cases it has been suggested that this is due to the inactivity of certain permease mechanisms at low temperatures but in other cases no such correlation has been established.

At the other extreme are the *thermophiles*, many of which have optimum temperatures in the range of 50 to 60°C, some being capable of growing up to 75°C. It is probable that the unusual behaviour of the thermophilic bacteria is due to the high heat resistance of their enzyme proteins. Thermophilic bacteria, though not pathogenic, are of importance since they may be responsible for spoilage in food that has been processed by heat.

BACTERIAL METABOLISM

The mechanisms involved in the breakdown and utilization of foodstuffs by bacteria are basically similar to those occurring in higher organisms. The similarity is in fact so great that bacteria are frequently the material of choice for the investigation of general biochemical mechanisms. Bacteria nevertheless show certain important metabolic differences. First, they are metabolically much more active. In general the metabolic activity of an organism depends on the ratio of its surface area to its volume, a ratio which in the case of bacteria is extremely high. Thus the surface area/volume ratio for man is 0·024; for bacteria it is in the region of 5000. Bacteria therefore possess, relatively, a very large surface through which nutrients may be absorbed and through which metabolic end-products may be released into the environment and this in turn results in a high metabolic rate. Second, bacteria differ from other organisms in the types of substrate that they can utilize for the production of energy and for the synthesis of cell material; amongst different bacteria there are in addition important species differences in the types of substrates that can be metabolized—differences which are of considerable importance in diagnostic bacteriology. Thus the capacity to utilize and break down lactose is of importance in the differentiation of the enterobacteriaceae. Third, bacteria differ in the types of end-product which they produce. Here again differences of diagnostic importance are encountered amongst various bacterial species.

In order that a molecule can be metabolized it must first gain access to the cell. The cell wall and cytoplasmic membrane of bacteria appear to be normally completely impermeable to large molecular weight compounds such as proteins, polysaccharides and lipids. Such molecules must therefore be broken down in the extracellular environment to smaller units before they can be assimilated; this the cell achieves by the secretion of various hydrolytic enzymes—proteases, polysaccharidases and lipases. The component parts into which large molecules are thus broken down, namely the amino acids, fatty acids and simple sugars, constitute the fundamental building blocks of the cell.

Nutrient material which is absorbed into the cell is utilized by the organism for two purposes. First, for the production of energy which is required for the synthesis of cell material and in the case of motile organisms for motility. A considerable amount of energy is, of course, lost to the cells as heat through the large surface of the organism. Second, the nutrient materials are utilized for the synthesis of the various structural components of the cell and of the enzymes and coenzymes required in bacterial metabolism.

The energy required for bacterial growth is supplied mainly by oxidation. As in other organisms bacterial oxidation consists in the removal of hydrogen from an oxidizable substrate; essentially the process is one of removing electrons which is followed by hydrogen transfer. This is affected by the agency of various dehydrogenases which are specific for the substrate attacked. The hydrogen is then transferred to an appropriate hydrogen acceptor, the process therefore being one of a coupled oxidation and reduction. Most oxidations involve the transfer of hydrogen through a series of such acceptors. The energy released by oxidation is used in the first place to effect the synthesis of high energy bonds. These are the readily hydrolysable phosphate anhydride bonds of adenosine diphosphate (ADP) and adenosine triphosphate (ATP). The energy stored in such compounds is released by rupture of the phosphate bond, and in this way can be used to drive endergonic synthetic reactions. Energy is also transferred by the high energy C—S bonds of acetyl coenzyme A and acetyl lipoic acid.

Two main types of oxidation may be distinguished:

1. Oxidation occurring under aerobic conditions and involving the uptake of molecular oxygen. In this process, which is known as *respiration*, the hydrogen is transferred through a series of hydrogen acceptors and ultimately to molecular oxygen. Most pathogenic bacteria can effect this change and as in mammalian tissues it is dependent on the process of the respiratory pigment *cytochrome*. Respiration is a highly efficient energy-yielding mechanism since by it the substrate can be completely metabolized; in the case of glucose this results in the complete breakdown of the glucose molecule to yield carbon dioxide and water.

Certain bacteria, however, notably the streptococci and the clostridia, do not possess any cytochrome components. As a result such organisms cannot obtain energy by aerobic oxidation. In the presence of oxygen the streptococci can, however, effect some aerobic oxidation through the agency of flavoprotein. The energy released by this process is, however, apparently not utilizable for growth. Oxidation by flavoprotein *per se* results in the production of hydrogen peroxide, a substance which is highly toxic to many species of bacteria. Some streptococci appear to break down the hydrogen peroxide thus formed by the action of peroxidase. As a general rule, however, aerobic bacteria which produce hydrogen peroxide during aerobic growth break it down by the action of catalase.

2. The second main type of oxidation—that which is not associated with the uptake of molecular oxygen—is known as *fermentation*. In this case hydrogen is transferred to some other organic compound which is correspondingly reduced. Thus in the bacterial breakdown of glucose by the Embden–Meyerhof–Parnas route, the oxidation of 3-phosphoglyceraldehyde to 1,3-diphosphoglyceric acid is associated with a concomitant reduction of pyruvic to lactic acid. The breakdown of the substrate is not as complete in fermentation as it is in respiration; fermentation is therefore a much less efficient energy-yielding process.

The most important groups of substrates used for the production of energy by the pathogenic bacteria are the carbohydrates. The carbohydrate most usually available is glucose and this can be utilized by most of the heterotrophic bacteria. The main mechanism involved in the bacterial breakdown of glucose is by the Embden–Meyerhof–Parnas route identical with that occurring in yeast and muscle. Other mechanisms by which glucose may be broken down are the Entner–Doudoroff pathway and the pentose pathway or hexose monophosphate shunt. The latter is of importance in supplying the intermediate erythrose-4-phosphate required for the synthesis of the aromatic amino acids. All three of these pathways involve the initial phosphorylation of glucose to glucose-6-phosphate, and the production of pyruvic acid—a key intermediary in bacterial metabolism—as a principal end-product. The fate of pyruvate is varied. It can be completely oxidized via the Krebs citric acid cycle, coenzyme and cytochrome to carbon dioxide and water or it can be broken down by changes of fermentative type to give a variety of end-products—formic, acetic, propionic, lactic, butyric and succinic acids, ethyl, propyl and butyl alcohols, acetyl methyl carbinol and 2,3-butylene glycol, hydrogen and carbon dioxide. The precise end-products formed by the fermentative breakdown of pyruvate differ with different organisms and in certain cases these differences are of diagnostic importance.

It would be beyond the scope of this chapter to attempt to review the multiplicity of mechanisms involved in bacterial syntheses. It is sufficient to say that bacteria can build up complex molecules from simple precursors by a series of elaborate and sequential biosynthetic pathways. These are mediated by enzymes the synthesis of which is controlled directly by the bacterial genome, the energy for synthesis being dervied by breakdown of the high energy phosphate bonds generated from bacterial oxidation. The enzymes involved in metabolism and on which the life of the cell depends, are themselves synthesized from their constituent amino acids under the direct control of chromosomal DNA; this carries in the sequence of its nucleotide bases the genetic code indicating the sequence of incorporation of amino acids into protein. There is now conclusive evidence for a triplet code, each amino acid being denoted by a nucleotide sequence or word of three bases. Different amino acids are represented by different triplets produced by various combinations of the four DNA bases—adenine, guanine, cytosine and thymine. The triplet code corresponding to all the commonly occurring amino acids has now been determined.

In the translation of the genetic code into protein, three phases may be distinguished:

1. The base sequence of one of the strands of the chromosomal DNA is transcribed into a complementary sequence of bases in a high molecular weight RNA known as messenger RNA (mRNA). Its composition is determined as in the case of the DNA double helix by hydrogen-bonding properties. Thus guanine pairs with cytosine, RNA uracil pairs with DNA adenine and DNA thymine with RNA adenine.

Messenger RNA is synthesized on a template of chromosomal DNA through the activity of the enzyme RNA polymerase. On completion, the messenger RNA becomes detached from the template and migrates to the cytoplasm where it becomes associated with a group of ribosomes to form a polysome (p. 1). As a component of the polysome it serves as a template for the assembly of the amino acids into proteins.

2. Each amino acid activated by reaction with ATP combines with a low molecular weight RNA known as transfer RNA (tRNA) to form an amino acyl RNA complex. These reactions are catalyzed by enzymes—amino acyl RNA synthetases—specific for each amino acid. Each amino acid has a specific transfer RNA, characterized by a specific site for the corresponding amino acyl tRNA synthetase, and also by a specific triplet of nucleotide bases through which it pairs with messenger RNA.

3. The final stage is initiated by the alignment of transfer RNA molecules with their associated amino acids, along the messenger RNA in a sequence determined by the complementarity of their bases with those of messenger RNA. As a result of this alignment the amino acids are arranged in a sequence determined by the base sequence of chromosomal DNA. Two sites, A and P, may be distinguished in each 70S ribosome, the sites being shared between the 30S and 50S ribosomes. In the synthesis of the peptide chain the amino acyl tRNA binds first to the A site, while the elongating chain is bound to the P site. A peptide bond is then formed between the amino acyl tRNA and the peptidyl transfer RNA on the P site. The RNA attached to the latter is released and the peptidyl transfer RNA now located at the A site is transmitted to the P site and the messenger RNA moves with it. This process is repeated until the peptide chain is complete.

BACTERIAL GENETICS

As we have seen, bacteria possess a single circular chromosome. In the case of *E. coli* and *Salm. typhimurium* the location of many of the genes on the chromosome has been determined and detailed chromosome maps prepared. The chromosome nucleic acid, which has a molecular weight of approximately 5×10^9, contains about 6000 genes, each containing an average of 1200 nucleotide pairs and capable of coding at three base pairs per amino acid, for proteins with an average composition of 400 amino acids and corresponding to a molecular weight of about 100 000. Although our knowledge of the details of chromosome replication is incomplete it is clear that the first stage must be a separation of the two DNA strands, new complementary strands being laid down on each of the original strands acting as an initiator by a process of base pairing. The new DNA strand is synthesized from nucleotide triphosphates through the action of a DNA polymerase. There is, in addition, evidence that the DNA double helix is attached at some point to the cytoplasmic membrane, possibly to a mesosome. This attachment appears to constitute the point, known as the replicator, at which replication of the DNA strand commences.

Synthesis of DNA has been found to take place in short segments which are then joined together by a ligase. Three polymerases, I, II and III, have been described. Polymerase I appears to be a repair enzyme and polymerase III to be involved principally in DNA synthesis. It operates in a 5′ to 3′ direction along the polynucleotide chain. In order that synthesis should be initiated, unwinding of the double helix must occur. This appears to be achieved through the action of a swivel enzyme which breaks single strands so as to permit unwinding and then repairs the break. It is probable that the entire replication process in bacteria is achieved through a series of such breaks, but many of the details of this process are obscure.

This model of replication in which one strand of the original duplex is retained in the daughter strands is known as semiconservative. In bacteria, replication from the growing point is unidirectional; in bacteriophages it may occur in two directions (divergent replication). In related bacterial strains growth always occurs from the same point on the chromosome, as defined by its relation to genetic markers.

Extrachromosomal genetic determinants

Many species of bacteria possess extra-chromosomal genetic determinants. These are known as plasmids. They range in molecular weight from 1×10^6 daltons to 100×10^6 daltons. Small plasmids (those with a M.W. of approximately 25×10^6 daltons) can exist in up to ten copies per cell; large plasmids (over 25×10^6 daltons) usually occur in only a single copy per cell. Plasmids may be conveniently classified according to the functions they determine, as follows:

1. F or fertility factors, which were the first plasmids discovered. They were found by Hayes, who showed that they determined the capacity of a cell to act as a chromosome donor in conjugation (p. 21). They are large plasmids (M.W. approx. 55×10^6 daltons).
2. Bacteriocinogeny factors, which control the production of certain bacteriocines by bacteria (p. 461).
3. R or resistance factors, which carry determinants for resistance to various drugs. They are particularly numerous among the Gram-negative intestinal bacteria, but are found in a variety of other organisms as well.
4. Plasmids which determine a variety of bacterial characteristics, e.g. urease and hydrogen sulphide production by *Proteus*, lactose fermentation by a variety of Gram-negative bacteria and enterotoxin production by *E. coli*.
5. Cryptic or silent plasmids, for which a function has not yet been identified.

Most large plasmids are spontaneously transmissible from a donor to an acceptor cell; small plasmids are not spontaneously transmissible. Transmissibility is dependent on the presence of transfer factor genes, at least 15 of which have been identified in *E. coli*. Several of these determine the occurrence of sex pili through which the genetic material passes from the donor to the recipient in conjugation. Non-transmissible plasmids can be transferred from donor to acceptor by transducing bacteriophages (p. 461). Staphylococcal plasmids are uniformly of this type.

Plasmids appear to have specific attachment sites on the bacterial cell membrane. Consequently a single bacterium cannot harbour two plasmids competing for the same membrane site and such plasmids are described as incompatible. They may, however, undergo hybrid formation to create a single complex plasmid which may persist as such in the cell. In addition to the F factor, certain other plasmids can also integrate into the chromosome and, acting as a sex factor, can mediate its transfer into a recipient cell.

At normal cell division, large plasmids divide coordinately with the bacterial chromosome and distribute at cell division into the daughter cells. Small plasmids undergo what is described as relaxed multiplication, giving rise to multiple copies per cell. Large plasmids undergo stringent multiplication so that only one or two copies are found per cell. At conjugation, i.e. when the pilus of a plasmid-bearing cell makes contact with a recipient which does not contain the plasmid, replication is triggered. One strand of the duplex does not replicate, however, and it passes into the recipient cell where it circularizes and then undergoes replication. Under certain circumstances, e.g. on transfer from *E. coli* to *Proteus*, the transfer factor genes are not replicated and there is instead a multiple replication of the non-transfer factor genes carried by the plasmid. Such plasmids are not spontaneously transferable.

There is direct evidence that F and other plasmids are composed of DNA (F has a M.W. of 62×10^6 daltons). For a part of the time plasmids exist in the cell as closed supercoiled circles. They are as a result less readily broken by shearing forces. This facilitates their separation from chromosomal DNA by physical means.

BACTERIAL VARIATION

When maintained in the same environment bacteria, like other organisms, exhibit a high degree of stability in their properties. This stability is due to the fact that their properties are controlled by genes which are replicated at cell division—an identical set of genes being transmitted to each daughter cell. When, however, the environmental conditions are altered, bacteria exhibit a marked capacity to change their properties in such a way as to ensure their survival in the altered environment. This change in properties is known as *variation*. Variation may occur in two ways:

1. As a result of a genotypic change. Changes of this type are heritable, i.e. are transmitted to daughter cells.
2. As a result of a phenotypic change. This is a change in the behaviour of the cell which is not associated with a change in genotype. Phenotypic changes are therefore non-heritable.

Genotypic variation

There are two main varieties of genotypic variation: mutation and various forms of genetic transfer or exchange; in which genetic material is transferred from one bacterial cell to another.

Mutation

This, as in other organisms, is the occurrence of a change in the structure of a single gene leading to an alteration or loss in its function and as a result an alteration of the properties of the cell. The chemical basis for this change is an alteration in the base sequence of the chromosomal DNA. Such a change may occur spontaneously or as a result of the action of mutagenic agents, the most important of which are base analogues, alkylating agents and ultra-violet irradiation. Spontaneous mutations are rare events usually occurring with a frequency of about 1 in 10^7 to 10^{10} cell divisions. The change in base sequence may be due to the substitution of one base pair in the DNA double helix for another, the deletion of a base pair or the insertion of a new base pair.

Under natural conditions mutation is by far the most important source of genotypic variation in bacteria. Since bacteria are haploid, a mutated gene has considerable freedom of expression as no complementary gene exists—as in diploid organisms—to which it may be recessive. Moreover, bacteria are relatively simple in structure and a mutated gene may be able to express itself without necessity for alteration in the other properties of the organism. As a result of these factors, bacteria show much greater potential for variation than do higher organisms. They are consequently less dependent than diploid organisms on genetic exchanges for the degree of variation which will permit successful environmental adaptation.

Mutations may occur in relation to most, if not all, of the observable properties of bacteria—morphology, colonial appearance, biochemical activity, antigenic structure, drug sensitivity and pathogenicity. As a general rule, provided the environmental conditions remain unchanged, mutants do not replace the parent strain. The environment may, however, be such as to select certain mutants. In such circumstances the mutant will tend to replace the parent strain. Its success in doing so will depend on its relative superiority in the selective environment. Thus mutants to drug resistance, for example, rapidly replace their drug-sensitive parents in the presence of the drug. In this case the mutant has an absolute superiority over the parent in an environment in which the drug is present. As in other organisms, the environmental selection of chance mutations form the basis of bacterial evolution.

Genetic exchange

This denotes the change in the properties of a cell by transfer to it of genetic material from another cell. Four types of genetic exchange can be distinguished—*transformation, conjugation, phage-mediated transfer* and *plasmid transfer*. Transformation and conjugation are processes involving the transfer of chromosomal genes while plasmid transfer is the spontaneous transfer of extra-chromosomal genes. In phage-mediated transfer, bacteriophages act as carriers of genetic material from one cell to another; this is considered in more detail in Chapter IV.13.

Conjugation

This phenomenon was first described by Lederberg and Taum in 1946, as occurring among certain nutritionally exacting or auxotrophic mutants of a strain of *Escherichia coli* K12 and is the nearest equivalent to sexuality amongst bacteria. It differs from other types of exchange involving chromosomal genes in requiring physical contact between the conjugating pair and in that the participants exhibit sexual polarization. Lederberg and Tatum showed that when two auxotrophic K12 mutants with complementary nutritional deficiencies were mixed together, a small number of hybrids without nutritional defect, known as prototrophs, were produced. These would grow on an unsupplemented minimal medium on which the parental strains would not grow. It was subsequently shown by Hayes that in the parental strains participating in conjugation, one of the parents functioned as a male or donor depending on the sex or fertility factor (F) in the cytoplasm. Strains possessing this factor are known as F plus (F^+) and behave as males. Strains lacking this factor are known as F minus (F^-) and behave only as females. Several functions can be ascribed to the F factor:

1. It determines the production of a specific antigen on the surface of F^+ strains. This antigen diminishes the charge on the surface of the F^+ cells, so that on random collision they tend to adhere to F^- cells.

2. It controls the synthesis of specialized pili through which a chromosome fragment can be injected from the male into the female cell. These pili have been found to carry receptors for certain bacteriophages which are specific for F^+ cells.

3. It mobilizes the F^+ chromosome for transfer. In the normal F^+ *E. coli*, the chromosomal DNA exists as a closed, circular loop and the F agent occurs free in the cytoplasm. Before conjugation can occur, the F agent must migrate from its normal cytoplasmic location to the chromosome into which it becomes inserted. Cells with F factors integrated in this way are known as Hfr (high frequency) cells. In the absence of conjugation the chromosome replicates normally. If contact with an F^- cell is established an abnormal replication process is triggered. Only one of the original strands is replicated and this forms a new chromosome in the daughter donor cell. The other strand is transferred through the sex pilus into the recipient. In this process the proximal end of the chromosome, known as the origin, carries only a small moiety of the replicated F factor, the major portion of which is at the distal end of the now linear chromosome. The size of the chromosomal fragment transferred depends on random breakage of the chromosome during transfer. Consequently, the closer to the origin a particular gene is situated, the more frequently will the marker property controlled by that gene be transferred in conjugation. Frequency of transfer in conjugation can therefore be used as a basis for the preparation of a chromosome map. The position of a marker can also be determined by interrupted mating experiments in which mating cells are separated mechanically in a Waring blendor. The earlier the entry of a marker the closer it is to the origin. The transferred fragment is then incorporated into the homologous portion of the recipient's chromosome. The process appears to have some analogy with the crossing over mechanism in diploid cells, but the eliminated chromosomal fragment does not form a reciprocal recombinant with the complementary section of the donor chromosome. The bacterial male, therefore, is rather like the male of the praying mantis which dies in the act of copulation.

Complete transfer of the chromosome of *E. coli* takes approximately 80 minutes under optimal conditions. This permits the construction of a

Table I.2. Conjugation of auxotrophic mutants of *Escherichia coli* K12 to yield prototrophs

Mutant	can synthesize	cannot synthesize
58.161 F^+	threonine, leucine, B_1	methionine
W.677 F^-	methionine	threonine, leucine, B_1

Neither mutant will grow on 'minimal' ammonium salt-glucose medium, lacking threonine, leucine, methionine and B_1. When conjugation occurs however and W.677 receives a section of the 58.161 chromosome which carries genes for the synthesis of threonine, leucine and B_1 the resultant hybrid will grow on minimal medium.

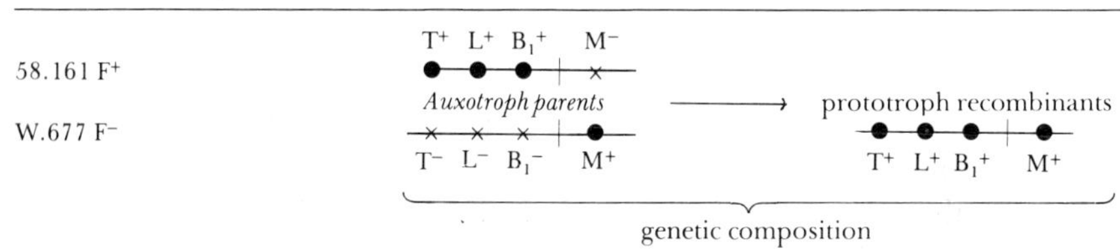

chromosome map in which the time taken for individual markers to be transferred can be plotted, this time being directly related to the distance of the marker from the origin.

Occasionally integrated F agents are ejected from the Hfr chromosome and are found to have incorporated some chromosomal genes through recombination with an area contiguous to the F locus. Such modified F factors are known as F prime factors or *F genotes*, and their transfer coupled with the transfer of the associated chromosomal marker is known as F-duction or sexduction.

Though first described in *E. coli*, intraspecific and sometimes interspecific conjugation has now been described in many bacterial species. For some of these at least chromosomal transfer does not involve the Hfr transformation.

Transformation

This phenomenon, discovered in the pneumococcus by Griffith in 1928, was the first demonstration that genetic material could be transferred from one bacterium to another. Griffith observed that mice which had been injected with live, non-capsulated variants of capsulated type 2 pneumococci and heat-killed pneumococci of capsular type 1, died of a septicaemia due to capsulated type 1 strains. It was apparent that the non-capsulated type 2 organisms had in some way been transformed into capsulated type 1 organisms. This transformation was subsequently demonstrated under in vitro conditions and the general principle established that non-capsulated pneumococci derived from one type could be transformed into capsulated pneumococci of another, by exposure to a transforming agent present in cell-free filtrates or extracts of the capsulated strain. The transforming agent was later identified by Avery, McLeod and McCarty as DNA.

Intraspecific transformation has been recognized in strains of *H. influenzae, Bacillus subtilis, Neisseria meningitidis* and various streptococci, as well as the pneumococcus. It has been studied most in the pneumococcus, a species in which a wide range of readily recognizable characteristics may be transferred. Transforming agents appear to consist of large fragments of DNA equivalent to about 1% of the total bacterial genome. Under optimal conditions each cell is capable of taking up about ten such molecules. Consequently, in the presence of an excess of transforming agent, about 10% of the cells of a transformable culture can be shown to be transformed in respect of a specific marker characteristic. There is clear evidence that transformation is due to the incorporation of the DNA fragment, or more probably a single strand of the DNA double helix, into the chromosome of the recipient where it exchanges for the corresponding segment of the latter and becomes thenceforth an integral part of the recipient's genome. In the pneumococcus the capacity to be transformed, known as *competence*, only appears transitorily in a culture, found only for about 40 minutes towards the end of the logarithmic phase of growth. This is presumably the only time during growth when the cells are permeable to the large DNA molecule. There is some reason to believe that this permeability may arise from a physical breakdown of the cell membrane. Another possibility is that competence may require the synthesis of appropriate receptor sites on the bacterial surface, and that this may only occur at a particular stage in growth.

Phenotypic variation

This defines a change in the behaviour of the cell produced in response to some environmental stimulus. Such changes do not involve the genome, and are temporary in character—reversal to the normal pattern occurring when the environmental stimulus producing them has been withdrawn. In contrast to mutational changes they are acquired by the bacterial population as a whole rather than by a few mutant cells. Changes of this type may be broadly described as adaptive. They may be divided into two main categories.

Environmental control of the rate of enzyme synthesis

Many bacterial enzymes are *inducible*, the synthesis of the enzyme only occurring to a significant extent in the presence of the enzyme substrate or a substrate analogue. Inducible enzymes are usually, though not invariably, enzymes involved in the breakdown of exogenous substrates. Non-inducible enzymes, whose rate of synthesis is independent of the presence of substrates, are

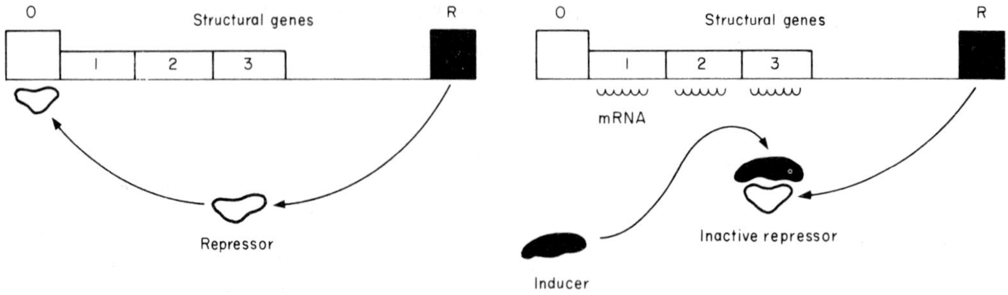

Fig. I.10. The Jacob-Monod model for enzyme induction. *Left,* The repressor combines with the operator and prevents transcription of structural genes into mRNA. *Right,* The inducer combines with the repressor and prevents its combination with the operator. Structural genes are transcribed. O = operator (site of initiation of transcription of structural genes under its control). R = regulator gene (codes for synthesis of repressor).

referred to as *constitutive*. Notable examples of inducible enzymes are: (*a*) those involved in the transport and breakdown of lactose by *E. coli*, namely *β-galactoside permease* and *thiogalactoside transacetylase*, all of which are inducible by lactose; and (*b*) the extracellular *penicillinases* produced by various *Bacillus* species and by *Staph. aureus* which are inducible by penicillin.

The model currently invoked to explain enzyme induction is the one proposed by Jacob and Monod. It postulates the participation in enzyme production of three distinct components:

1. A structural gene, or genes, which code for the synthesis of the enzyme protein and of any other protein which forms a functional unit with the enzyme. Thus in *E. coli* the synthesis of β-galactosidase is coordinated with the synthesis of β-galactoside permease and thiogalactoside transacetylase.

2. The transcription of the structural genes into messenger RNA is controlled by an operator gene, which when active, as it were, switches on the transcription mechanism.

3. A *regulator* gene which is responsible for the production of a cytoplasmic *repressor* whose function is to inhibit the action of the *operator*. When the operator is inhibited the messenger RNA providing the information for the synthesis of the enzyme is not transcribed, and the enzyme is not produced.

If, however, the substrate, i.e. lactose or certain substrate analogues such as thiogalactosides, are present they combine with the repressor and prevent it from inhibiting the activity of the operator. There is now considerable evidence that where, as in lactose fermentation, a number of enzymes are involved in the metabolism of a compound, the genes responsible for determining their synthesis are arranged sequentially on the chromosome and are under the control of a single operator. Such a coordinated genetic unit is referred to as an *operon*. Mutations may occur at various sites in such a system. The regulator gene may fail to produce a repressor or the operon may become insensitive to the repressor; under these circumstances the production of the enzyme becomes constitutive rather than inducible. On the other hand, the regulator gene may produce repressor incapable of being inhibited by inducer or the structural genes may become insensitive to the operon. In these cases no enzyme is produced and the presence of the corresponding structural genes can only be demonstrated in recombination experiments. The lactose repressor has recently been demonstrated directly; in general it would appear to possess the properties of an allosteric protein (see below), capable of combining specifically both with the operator and with the inducer. However, the production of repressor is not inhibited by exposure of the cell to agents which inhibit protein synthesis. This suggests the possibility that the repressor might not be a protein. Nevertheless, inhibition could fail to occur because only a few molecules of repressor per cell may be required for effective repression.

The process of enzyme induction obviously gives bacteria great flexibility in their capacity to respond to changes in the nutritional composition of an environment. Enzymes involved in biosynthetic pathways, e.g. in the synthesis of amino acids, are non-inducible, unlike the enzymes considered above. Their production may, however, be inhibited by the biosynthetic end product. This was first shown by Cohn and Monod in 1953. They found that the production of tryptophane synthetase and methionine synthetase was repressed if the organisms were supplied with tryptophane and methionine respectively. Where a number of enzymes are involved in a synthesis the end product appears to be capable of repressing the production of most, if not all, of the enzymes involved simultaneously. This mechanism is easily explained on the basis of a model similar to that devised by Jacob and Monod for inducible enzymes. The enzymes involved in the synthetic sequence are arranged sequentially in an operon under the control of an operator gene. In this case, however, the regulator gene is postulated to produce an inactive *apo-repressor* which is incapable of combining with the operator gene. It is transformed into a fully active repressor by combination with the end product of the biosynthetic pathway. This mechanism permits a great economy for the cell in the synthesis of enzyme protein since it allows the production of a particular biosynthetic enzyme to be switched off when the enzyme is no longer required.

The compound adenosine 3′,5′ mono-phosphate (cAMP) plays an important role in the control of metabolic processes of bacteria. Thus in the provision of energy, glucose is more efficiently used than lactose. Consequently, in the presence of glucose the production of the enzyme β-galactosidase, which breaks down lactose, is inhibited. In this phenomenon, which is an example of catabolite repression, cAMP plays a central role. Its production is repressed in the presence of glucose and the metabolic inhibition produced by glucose is reversed by the addition of cAMP. The intracellular level of cAMP almost certainly affects the synthesis of many other bacterial enzyme systems in a similar manner.

Feedback inhibition

A second major way in which bacteria may respond to environmental changes is the process known as *feedback inhibition*, in which the activity, but not the production, of a biosynthetic enzyme is damped down by the biosynthetic end-product. In this case the end-product appears to combine with the first enzyme of the metabolic pathway so as to inhibit its activity. The end-product responsible for feedback inhibition shows no structural resemblance to the substrate of the inhibited enzyme. Thus threonine, which is synthesized from aspartic acid, will inhibit the formation of an aspartokinase which is involved in the formation of aspartyl phosphate from aspartate. The inhibited enzyme must therefore have two distinct combining specificities: for its normal substrate and for the biosynthetic end-product. For proteins of this type Jacob and Monod have proposed the name of *allosteric* proteins. Feedback inhibition may be thought of primarily as a fine control of metabolism, in contrast to coordinated enzyme repression which may be regarded as a coarse control.

I.2

Bacteriological Techniques

THE USE OF THE MICROSCOPE

Bacteria may be examined microscopically either in unstained wet preparations or in stained preparations.

Wet preparations

Unstained preparations have a restricted application in bacteriology. They are mainly used for the demonstration of bacterial motility and for the demonstration of spirochaetes.

Motility is best demonstrated microscopically in hanging drop preparations in which a drop of culture or other bacteria-containing fluid is deposited on the centre of a coverslip which is then inverted over the concavity of a specially prepared glass slide or over a metal ring attached by petroleum jelly to the surface of an ordinary slide. Alternatively, the fluid may be placed on the surface of a slide and a coverslip dropped over it.

For the demonstration of spirochaetes, which are not visible under direct illumination, dark ground illumination is employed. This requires the use of a special condenser which reflects the light on the object obliquely so that only the rays refracted by the object enter the objective. With this technique the organisms are seen brightly outlined against a dark background.

Stained preparations

Preparation of films

A film of the preparation to be examined is first made on a glass slide by spreading with a wire loop. If the material to be examined is from a solid culture, it is first emulsified in a drop of water on the slide and the film is then spread. The technique requires practice and only with experience can films be prepared, which are not too thick or too thin.

Before staining, films must be dried and fixed. Drying is best carried out by holding the slide between fingers and thumb and moving gently to and fro above a Bunsen flame. The film is then fixed by passing it three times through the flame. It is of the greatest importance that during drying and fixation the film should not be overheated.

A variety of staining methods is available for the demonstration of special structures—spores, flagella, capsules, various types of intracellular granule, nuclei—and for the demonstration of particular organisms—spirochaetes, fungi, chlamydiae and rickettsiae. The most important of the stains employed are the Gram stain and the Ziehl–Neelsen stain and only these will be described here.

Gram's method

This is the staining method most frequently used in diagnostic bacteriology. The organisms are first stained with an aniline dye of the violet series—usually crystal violet—and are then treated with an iodine solution which acts as a mordant. Next they are washed in alcohol and finally counterstained with a dye of contrasting colour, e.g. fuchsin. On the basis of their reaction to the Gram stain bacteria can be divided into two great groups—Gram-positive and Gram-negative. Gram-positive organisms retain the violet stain after decolorization with alcohol: with Gram-negative organisms, on the other hand, the violet stain is removed by alcohol treatment and the organisms are seen to be stained with the counterstain employed. Gram-positive organisms are, therefore, purple in colour and when fuchsin is used as counterstain Gram-negative organisms are pink.

The determination of the Gram character of an organism is of great importance in identification. The pathogenic cocci, with the exception of the neisseriae, are Gram-positive. The pathogenic bacilli are Gram-negative, except the following genera: *Corynebacterium, Bacillus, Clostridium, Actinomyces, Listeria* and *Erysipelothrix* which are Gram-positive. Mycobacteria stain with difficulty by Gram's method but when stained are also Gram-positive.

In the past, numerous attempts have been made to attribute the Gram-positive character to particular chemical compounds, notably nucleoproteins, nucleic acids and polyglycerophosphate in the Gram-positive cell. In none of these cases, however, has a satisfactory correlation between staining properties and chemical composition been established. On the whole it seems more likely that the difference between Gram-positive and Gram-negative cells is due to a difference in the permeability of their cell membranes, in particular, that the membranes of Gram-positive organisms are less permeable to the dye iodine complex in alcoholic solution.

The distinction between Gram-positive and Gram-negative shows a high degree of correlation with certain biological properties.

1. As previously mentioned the cell walls of Gram-positive bacteria differ considerably in chemical composition from those of Gram-negative bacteria (p. 5).
2. Gram-positive organisms are much less susceptible to mechanical damage than are Gram-negative, but show on the whole a greater susceptibility to inimical chemical agencies.
3. Gram-positive organisms are on the whole more nutritionally exacting than Gram-negative.
4. The production of potent exotoxins (Chapter I.4) is almost exclusively a property of Gram-positive organisms.

There are many modifications of the original Gram's method, but the one here described, that of Jensen, is probably the simplest and most reliable.

1. Stain with crystal violet for half a minute.
2. Pour off excess stain and wash off residual stain with Lugol's iodine. Cover with fresh iodine and allow to act for one minute.
3. Pour the iodine solution off the slide and shake it so as to free it as completely as possible of the solution.
4. Flood with alcohol, rock and pour off. Repeat until alcohol removes no further stain from the film.
5. Wash with water.
6. The film is then counterstained. The object of this is to render visible Gram-negative bacteria and cells from which the alcohol has removed the violet colour. The counterstain must contrast in colour with the violet of Gram-positive bacteria and should not be too intense. Dilute carbol fuchsin is the counterstain most commonly used. For the demonstration of intracellular Gram-negative organisms, e.g. gonococci and meningococci, neutral red is better since it gives sharper differentiation between the bacteria and the cell nucleus.
7. Wash with water. Blot. Dry.

Ziehl–Neelsen method

Next to the Gram stain this is the method most frequently used in diagnostic bacteriology. It is the method routinely employed for the demonstration of tubercle bacilli—organisms which are not normally stainable by Gram's method.

1. The fixed film is covered with strong carbol fuchsin which should be filtered onto the slide through a filter paper. Heat is then applied below the slide either with a small flame of the Bunsen, a spirit lamp or spirit torch. The heating should be sufficient to cause steam to rise, but boiling the stain must be avoided. Staining should last for at least ten minutes, heat being applied occasionally. Care must be taken to prevent the stain drying on the slide, fresh stain being added if necessary.
2. Wash well in water.
3. Flood the slide with 20% sulphuric acid in water and leave on for one or two minutes. In the acid the pink colour changes to yellow or brown.
4. Wash in water. The pink colour may return. If so, steps 3 and 4 are repeated as many times as necessary, until the film has not more than a faint pink tinge.
5. Wash in 95% alcohol for one minute.
6. Wash in water.

7. Counterstain. Löffler's methylene blue applied for 30 seconds is commonly used.
8. Wash with water. Blot. Dry.

Steps 3 and 5 may be replaced with acid alcohol. After treatment for five minutes, the film is rinsed with water and, if the pink colour returns, the acid alcohol is replaced. It may be necessary to alternate acid alcohol and water several times before the film has been sufficiently decolorized.

Tubercle bacilli resist decolorization with acid and appear bright red, because of which they are referred to as acid-fast; other bacteria and cells take the colour of the counterstain used.

After films have been examined microscopically the objective should be wiped free of oil. If this is not done, tubercle bacilli may be transferred in the oil to the films subsequently to be examined. For the same reason it is important that when the oil is placed on the slide prior to examination, the glass rod used should not be allowed to touch the slide.

THE CULTIVATION OF BACTERIA

As we have seen, the nutritional requirements of the pathogenic bacteria may be very complex, these being capable of growing only when supplied with a variety of complex growth factors. Their needs can be met by preparing completely defined or *synthetic* media made from the particular compounds, e.g. amino acids, etc., required by the organisms. Such media are used considerably in research and in preparation of bacterial end-products as free as possible from unnecessary foreign material. The requirements of the pathogenic bacteria for complex substances are, however, more easily met from naturally occurring material such as blood, serum, yeast, meat extract and egg. Such substances are therefore used as the basis of the media most frequently employed in diagnostic bacteriology.

Both solid and fluid media are used. Solid media are essential for the isolation of bacteria in pure culture. A single organism deposited on the surface of a solid medium will multiply to form a *colony*. This colony, which is invisible to the naked eye, is composed of millions of daughter cells of the original organism. It is, however, a pure culture and can be picked off and seeded into or onto other media for further investigation. In a fluid medium, on the other hand, the bacteria grow diffusely; consequently, if it has been inoculated with material containing more than one species, these will grow as a mixed culture.

Fluid media, although of great importance in pure culture study of bacteria, are not by themselves suitable for the isolation of organisms from mixtures. They may, however, be used as *enrichment* media prior to plating. Thus, if material containing a small number of organisms is inoculated into a fluid medium and the latter, after incubation, is subcultured onto a solid medium, the chances of isolation will be much greater, since the organisms will have multiplied in the fluid medium.

Many media incorporate substances which render them *differential* or *selective*. A differential medium is one which contains some substance on which a particular organism or group of organisms has a characteristic action. The alteration in the medium resulting from this action enables one to recognize the presence, in the material cultured, of this organism or group of organisms. Thus by adding lactose and an indicator, broth is rendered a differential medium for lactose-fermenting bacteria. If these are present they ferment the lactose, producing acid which changes the colour of the indicator; if absent, the indicator does not change.

A selective medium is one which inhibits the growth of all bacteria except those of a particular type or group. A medium is usually made selective by adding to it some substance which tends to prevent the growth of unwanted types while having little or no effect on those which are desired. Selective media are of particular importance in the isolation of bacteria from some site in the body, such as the throat or intestinal tract, which has an extensive normal flora.

Most of the pathogenic bacteria will grow only in the range of pH 6 to pH 8. Optimum growth is usually restricted to the range pH 7 to pH 7·4, which corresponds to the pH of the body fluids in which these organisms are accustomed to live. The media used for their cultivation are therefore usually adjusted by suitable buffers to a pH within this range.

The media must be sterile, i.e. they must contain no living organisms. Sterilization of media is usually carried out in the *autoclave*. Certain media, however, deteriorate at the high temperature of the autoclave and for these the *steamer* is used. Some of the substances used in the preparation of media may be adversely affected by any form of heat sterilization. These ingredients can be sterilized by *filtration*.

A great variety of media are used for the isolation and cultivation of individual groups of bacteria. Some are referred to in individual sections, but only a few basic media will be covered here.

Types of media employed in the cultivation of the pathogenic bacteria

Nutrient broth

Nutrient broth is the basis of many of the media used in the cultivation of pathogenic bacteria. By itself it is a valuable subculture medium for bulk preparation of organisms for a variety of purposes, e.g. to provide antigens for serological examination, as inocula for sensitivity testing and in vaccine production. Three types of broth are used: *infusion* broth, *extract* broth and *digest* broth. Infusion broth and extract broth contain as their main ingredients peptone and meat extract. Fresh extracts of meat, usually of ox heart, are used in infusion broth, and commercial extracts, e.g. Lab Lemco, in extract broth.

Digest broth is a special type which is made from and enzymatic digest of meat. Peptone is not required since the digest contains a considerable amount of protein breakdown products. Digest broth is a much richer medium than nutrient broth and is of particular value in the cultivation of pneumococci and streptococci.

Nutrient agar

By the addition of agar to nutrient broth, a solid medium, nutrient agar, is obtained. Agar is an extract of certain seaweeds and consists mainly of polysaccharide. By itself it has virtually no nutrient properties.

Nutrient agar is a gel which can be melted by heating to 100°C and becomes solid at about 45°C. It is not liquefied by the action of any of the pathogenic bacteria. Nutrient agar is usually used in Petri dishes in quantities of 15 to 20 ml of agar. For subcultures and for the preparation of suspensions for slide agglutination tests, agar slopes or slants containing about 5 ml of agar, solidified in a sloped position, are commonly employed.

Blood agar

By itself nutrient agar is not a rich medium. Its nutrient properties are, however, greatly improved by the addition of blood and the resulting medium, which is known as blood agar, is the standard medium for the isolation of many pathogens. In addition to supporting their growth it will indicate whether they are haemolytic or not—an important diagnostic criterion. Citrated or defibrinated human and horse blood are most frequently employed, usually in a concentration of from 5 to 10%.

Culture of anaerobic bacteria

Anaerobic bacteria are usually isolated by inoculation onto an appropriate solid medium, e.g. blood agar, which is then incubated in a McIntosh and Fildes anaerobic jar. This is a jar from which the air may be exhausted and replaced with hydrogen or nitrogen. Anaerobic bacteria may also be grown under aerobic conditions in fluid media containing reducing substances. Two important media of this type are *Robertson's meat medium* and *Brewer's sodium thioglycollate medium*.

Sugar media

These media are used to determine the capacity of different strains of bacteria to ferment various sugars. They can only be used for organisms which are in unambiguously pure culture. The most commonly used sugar media are fluid media containing the particular substance under test, together with peptone as a source of nitrogen.

The medium must also contain a pH indicator for which neutral red, phenol red or, more commonly, Andrade's indicator is generally employed. The medium is distributed in test tubes on the bottom of which rest, in an inverted position, very small inner tubes known as Durham tubes. Fermentation of the carbohydrate may result in the formation of acid alone or of acid accompanied by the evolution of gas. The production of acid is shown by a change in the colour of the medium—this is to a reddish pink if Andrade's indicator is used; if gas is also produced it collects at the top of the Durham tube.

Peptone water sugars are not satisfactory for certain organisms, e.g. streptococci, pneumococci and neisseriae, with exacting nutritional requirements. For streptococci and pneumococci Hiss's serum water sugars are usually employed; these contain serum in a concentration of 25% sugar and an indicator. Fermentation of the sugar is shown by a change in the colour of the indicator and by coagulation of the serum. The fermentation reactions of the gonococci and meningococci, which are highly aerobic and therefore do not grow well in liquid media, are best determined on agar media containing the appropriate carbohydrate and 5% of rabbit serum.

Dehydrated media

Many media can now be obtained commercially in dehydrated form. These are particularly valuable for small laboratories which do not have the facilities, or in fact the need, for media preparation on a large scale.

Cultural technique

In most cases media are inoculated with a wire loop which is sterilized in the Bunsen flame before and after use. Inoculation of plates is usually by the parallel streak method. The first series of streaks is charged from the material under examination. The second and third series, each at right angles to the previous series, are made without recharging but after sterilization of the loop. In this way the inoculum becomes successively diluted and, provided it has not been too large initially, well isolated colonies are usually found in one quadrant of the plate. Large quantities of material are usually inoculated with a Pasteur pipette as in the isolation of tubercle bacilli.

For the great majority of the pathogenic bacteria the optimum temperature of cultivation is 37°C, the customary period of incubation being from 18 to 48 hours, except for particularly slow-growing organisms such as tubercle bacilli. Plates are incubated in the inverted position, i.e. with the lid downwards so that the water of condensation does not flow over the surface of the medium.

When bacteria from an originally mixed inoculum have been grown in plate culture, pure cultures of individual species may be obtained by picking off single colonies with a wire loop and inoculating them onto an appropriate subculture medium. However, where there has initially been a considerable mixture of organisms, as in the isolation of enteric pathogens from faeces, single colonies from the primary plate should be subcultured to a further plate; single colonies from the latter can usually be assumed to be free from contaminants.

DIAGNOSTIC BACTERIOLOGY

The primary object of diagnostic bacteriology is to establish that the causative organism of a disease is present in the body of the patient. This may be attempted in one or both of two ways: (1) *directly* by microscopy and cultural isolation or (2) *indirectly* by the demonstration that the patient has developed an antibody response to the organism.

Collection of material for bacteriological examination

The first step in the demonstration and/or isolation of a pathogenic organism is the collection of a satisfactory specimen from the patient. This should be taken with full aseptic precautions to prevent contamination by adventitious bacteria. It should be adequate in amount and be transmitted to the laboratory in an appropriate sterile container, clearly labelled and accompanied by a statement of the patient's main symptoms and of the possible clinical diagnosis. This statement is of considerable value to the bacteriologist in providing a guide to the type of laboratory examination required. Since delay may cause the death of delicate organisms the specimen should be transmitted to the laboratory as rapidly as possible. In the case of particularly delicate organisms, e.g. *Neisseriea, Haemophilus* and *Bordetella*, cultures should be inoculated directly from the patient.

Inspection

Before any specific bacteriological examination is undertaken the specimen should be carefully inspected. Direct visual examination may reveal, for example, the presence in pus of the granules of *Actinomyces israelii* or of pus and mucus in a dysenteric stool. The colour of pus may suggest infection with *Ps. aeruginosa* or its putrid smell the presence of certain anaerobic bacteria.

Microscopy

Unstained wet preparations which will reveal pus cells or red cells are important in the examination of urine and cerebrospinal fluid. For the demonstration of bacteria, however, wet preparations are of limited value, their most important use being in the demonstration of *Treponema pallidum* in primary syphilitic lesions by dark ground examination.

In the examination of some material (faeces, for example, for the presence of bacilli of the enteric or dysentery groups) the examination of the bacteria in stained films does not yield any information of value; but in the great majority of cases films should be prepared as a routine and stained by Gram's method. These will not only be of value in establishing the diagnosis but will frequently indicate the use of a particular cultural method and will often help materially in the selection of suitable chemotherapy. Although the Gram stain may indicate the general nature of the organism responsible for the infection it will only rarely permit a completely definite immediate diagnosis. It is most valuable in doing so in infections due to organisms with a characteristic morphology, e.g. in gonorrhoea in the male, Vincent's angina and meningococcal meningitis.

For the demonstration of tubercle bacilli, films are stained by the Ziehl–Neelsen method. This should be used routinely for pus, sputum and serous fluids. It must be remembered, however, that acid-alcohol-fast organisms are not necessarily tubercle bacilli. A particular danger lies in the use of the Ziehl–Neelsen method for demonstration of tubercle bacilli in urine which has not been taken by catheter. In these specimens *Myco. smegmatis* may be present and may be confused with the tubercle bacillus. In some cases special staining methods, such as that to demonstrate capsules, may be employed on films made from the material under examination.

In recent years the fluorescent antibody technique has been increasingly used for the direct demonstration of organisms in clinical material. It has been particularly valuable as a diagnostic procedure in conditions in which specific identification would not have been possible by the use of conventional microscopic techniques. In such cases the method has permitted a much more rapid diagnosis than cultural examination. Currently, the procedure is used, with appropriately absorbed sera—so as to eliminate non-specific reactions—for the identification of group A streptococci in throat swabs, of enteropathogenic strains of *E. coli* in faecal smears and of gonococci in urethral exudates. The method also appears to be of some value in the demonstration of *Bordetella pertussis* in nasopharyngeal swabs and in the examination of the cerebrospinal fluid from cases of acute pyogenic meningitis due to *H. influenzae, Neisseria meningitidis* and the pneumococcus. Other applications under investigation are the demonstration of typhoid bacilli and dysentery bacilli in faecal material and of diphtheria bacilli in throat swabs. There is little doubt that in the near future the range of usefulness of this technique will be considerably increased.

Cultural examination

As a routine medium for the cultivation of many of the organisms causing disease in man, blood agar is used. Where, however, the clinical history of the case, the source of the material or the preliminary examination points to a certain organism or group of organisms, media appropriate for these must be employed.

If the specimen is fluid it should be centrifuged and cultures and films prepared from the centrifuged deposit. In the case of urine it is now a common practice to prepare cultures from the uncentrifuged fluid in such a way as to provide an estimate of the number of organisms present. Such an estimate can be simply obtained by inoculating plates with a single standardized loopful of the urine.

Plates are incubated aerobically as a routine in all cases. When, however, the presence of anaerobic organisms is suspected anaerobic cultures are obligatory. They are desirable though not essential in the routine cultivation of material, e.g. pus and other exudates, from which aerobic bacteria are usually isolated but in which anaerobes, notably anaerobic streptococci, are occasionally present. In cases of suspected *Neisseria* or *Brucella* infection a plate should also be incubated in an atmosphere of 5 to 10% carbon dioxide.

Blood culture

Isolation of bacteria from the blood is an important method of diagnosis in a number of conditions, in particular the coccal septicaemias, subacute bacterial endocarditis, brucellosis and typhoid fever.

The blood is inoculated into bottles or tubes of suitable fluid media and after incubation these are subcultured onto appropriate solid media.

In the taking of blood for culture great care must be exercised to avoid contamination of the blood by organisms present in the air or on the patient's skin. The blood culture bottle is inoculated at the bedside and should be warmed to body temperature before use. Since bacteria may be present in the blood only in small numbers a large volume of blood—5 to 10 ml—should be used as inoculum.

Subcultures should be made every 48 hours for a week and at intervals up to at least three weeks from receipt. This long period of incubation is necessary because some organisms—viridans streptococci and brucellae—may grow very slowly. When the bottles are removed from the incubator they should not be allowed to cool before subculture.

Species identification

This may require the investigation of a considerable number of bacterial characteristics for which subcultures will be required. For this purpose a pure culture of the suspected pathogen is prepared by picking off a single colony from the isolation medium and inoculating it onto or into an appropriate subculture medium, e.g. agar slope, broth or blood agar plate. In the isolation of non-lactose fermenters from faeces it is desirable to subculture a single colony from the selective isolation medium onto a MacConkey plate, all further investigation being carried out with cultures derived from a single colony on the latter.

Microscopy

Gram's method is used as a routine and will show whether the organisms under examination are cocci or bacilli, whether they exhibit any characteristic arrangement and whether they are Gram-positive or Gram-negative. When necessary stains for capsules, spores and metachromatic granules are employed. Motility may be determined by the examination of wet preparations from fluid cultures or by the determination of a swarming or spreading growth on semisolid agar.

Colonial appearance

The colour, size, shape, outline and surface of the colony of the organism is of considerable value in its identification. If blood agar is employed it should be noted whether the colonies are surrounded by a clear zone of haemolysis—β haemolysis, in which the blood is not only lysed but digested—or, as occurs with the pneumococci and viridans streptococci, by a zone of greenish discoloration—α haemolysis.

Growth requirements

Growth requirements are of value in some cases. Thus in the identification of *H. influenzae* it may be necessary to determine the organism's requirements for haematin and cozymase. The demonstration of a requirement for carbon dioxide is of value in the recognition of *Br. abortus*. If an acid-fast organism grows freely on blood agar it can be identified as a saprophytic mycobacterium—not a tubercle bacillus. It frequently is possible to show that an organism is a saprophytic contaminant, because it grows better at a low temperature, e.g. 25°C, than at 37°C. The capacity to utilize citrate as sole source of carbon and to grow in the presence of potassium cyanide is important in the identification of the enterobacteriaceae.

Biochemical reactions

Many species have specific biochemical characteristics which are of value in identification. The biochemical characteristics most commonly examined are: (*a*) the presence of particular enzymes in bacterial cultures and (*b*) the presence in cultures of certain compounds as metabolic end-products. These are, of course, produced as a result of enzyme action but are for convenience classified separately.

The enzymes most frequently looked for specifically are gelatinase, urease, oxidase, coagulase, lecithinase and catalase.

For the detection of *gelatinase* a tube of gelatin is inoculated with a straight wire and incubated at 22°C. In the case of the anthrax bacillus a characteristic type of gelatin liquefaction occurs, having the appearance of an inverted fir tree.

Urease is detected by growing the organism in a urea-containing medium such as Christensen's. If the urea is decomposed, ammonia is liberated and the indicator changes to a purple-pink colour.

Catalase may be detected by adding a few drops of hydrogen peroxide (20 volumes) to a broth culture of the organism or to the growth on a solid medium. If the organism produces catalase there will be a vigorous evolution of bubbles of oxygen due to decomposition of the peroxide.

The demonstration of *oxidase, coagulase* and *lecithinase* is of importance in the recognition of the gonococcus, *Staphylococcus aureus* and *Clostridium perfringens* respectively; the tests used are considered in the appropriate sections.

The most important tests in which metabolic end-products are demonstrated are the sugar fermentation reactions. They are of great value in the identification of many species.

For the differentiation of the enteric bacteria a great variety of biochemical tests may be employed. The most important of these are the following:

1. The production of indole in 24-hour broth or peptone water culture prepared from a peptone rich in the amino acid tryptophane.
2. The production of acetyl methylcarbinol (acetoin) in buffered glucose broth cultures is known as the Voges Proskauer (VP) reaction.
3. The methyl red test. This depends on the final reaction of the medium after glucose fermentation. Buffered glucose broth is inoculated with the organism and incubated for four days. A few drops of methyl red solution are added and the resulting colour noted. If red (showing high and persisting acidity), the test is positive; if yellow (indicating low or transient acidity), it is negative.
4. Production of phenylalanine deaminase.

Antigenic structure

In many cases specific identification requires the use of serological methods. Serological identification is usually carried out by the agglutination technique. In the ordinary diagnostic laboratory agglutination tests are mainly required for the identification of *Salmonella* and *Shigella* species. Diagnostic antisera specific for the more important of these are available commercially. For bacterial identification agglutination tests are carried out in the first place by the slide method using living suspensions. Positive results are then confirmed by the tube technique. The most important application of the precipitin reaction is in the identification of the different varieties of haemolytic streptococci.

Pathogenicity tests

Many of the bacteria pathogenic for man are pathogenic also for laboratory animals. In such cases virulence tests in animals may be of considerable value in species identification. They have been used most for the identification of tubercle bacilli, anthrax bacilli, diphtheria bacilli, pasteurellae and clostridia. In the case of exotoxin-producing organisms, viz. diphtheria bacilli and the clostridia, the identity of the organisms can be completely established by showing that their lethal effect on laboratory animals is neutralized by the appropriate antitoxin. Virulence tests are now less frequently used than formerly, more reliance being placed on in vitro characteristics, but in particular instances they may still be necessary.

Differentiation of types

Many bacterial species can be subdivided into types and on occasion the bacteriologist may have to determine not only the species but the type within a species to which a particular organism belongs.

The criteria on which type differentiation is made vary with different species. In some, e.g. *Str. pyogenes* and the pneumococci, it is possible to distinguish a number of serologic types which differ from one another by the possession of type-specific antigens. *Staph. aureus* and some of the salmonellae, notably *Salm. typhi*, may be divided into types on the basis of their susceptibility to various bacterial viruses—bacteriophages—which are capable of lysing them (Chapter IV.13). The diphtheria bacilli are divided into three types—*gravis, intermedius* and *mitis*—on the basis of morphology and cultural properties. In a few species type differentiation may be based on biochemical properties; this method, however, is not as valuable as those indicated above. When a species can be subdivided into types by different criteria the types are usually quite independent of each other.

Typing by serologic methods and by bacteriophage has been found to be of considerable value in epidemiological investigations. By the use of these methods it may be possible to trace the source of an infection or to follow the spread of an organism in the community with a degree of certainty which could not be achieved by a mere species identification.

Serological diagnosis

In certain conditions, it may be difficult or impossible to demonstrate the organism responsible by either microscopic or cultural methods. In these circumstances, serological methods, if available, may be employed. In these, attempts are made to demonstrate the presence of antibodies to a suspected organism in the patient's serum, and from the presence of antibody, or from the presence of a large amount of antibody, we infer that the patient is infected with that organism. Ideally, a rising antibody titre should be demonstrated, because only in this way is there any certainty that the antibody demonstrable has developed in response to the current infection. For this purpose two specimens of serum must be examined, one taken at an early stage of the disease and the second a week or ten days later. In many cases, however, this is not feasible and the diagnosis must, if possible, be made on the results of a single serological examination. The interpretation of single tests will depend on the normal level of antibody in the population, the stage in the disease at which the specimen is taken and on whether the patient has been previously immunized by a vaccine which stimulates production of the antibody in question.

Antibodies may be demonstrated by a variety of serological techniques (Chapter III.5). The most important of these for diagnostic bacteriology are the complement fixation test, which is of particular value in the diagnosis of syphilis and of viral infections, and the agglutination reaction, which is the most generally useful technique for the demonstration of antibody responses in bacterial infections. Agglutination tests are most frequently carried out in the ordinary clinical laboratory for the diagnosis of enteric infection and in this context are known collectively as the Widal test. They are also of considerable value in the diagnosis of brucellosis (p. 266) of various rickettsial diseases and, using the MG streptococcus as test antigen, of primary atypical pneumonia (p. 283).

Various techniques may be used but the principle of each is the same. Serial dilutions are made of the patient's serum and these are mixed with standard killed suspensions of the organism against which antibody is being measured. The result is expressed as a titre which is the highest dilution of the organism giving standard agglutination.

I.3

Disinfection and Sterilization

The terms *disinfection* and *sterilization* are generally used—rather imprecisely—to indicate the treatment of material so as to destroy or otherwise eliminate any living organisms present. There is, however, a tendency to prefer the term sterilization for physical methods of treatment and the term disinfectant for chemical methods. Thus chemicals which will kill bacteria are known as *disinfectants*. Material that has been treated in such a way that it contains no living organisms is said to be *sterile*.

Disinfectant and sterilizing procedures are of great practical importance. Their main applications are as follows:

1. *For disinfection in the literal sense of rendering material non-infective.* The treatment of material so as to kill any pathogenic organisms present, e.g. in the sterilization of surgical instruments or the disinfection of clothing, is of the greatest value in our efforts to prevent the spread of infection.

2. *In the treatment of food so as to kill, or inhibit, the growth of the organisms responsible for food decomposition.* Some of the methods used for this purpose, e.g. drying, salting and pickling, are of great antiquity and date as traditional and empirical procedures from the beginnings of civilized society.

3. *In the bacteriological laboratory, for the sterilization of media and glassware which are to be used in the cultivation of bacteria.*

Natural disinfection

Although the deliberate application of disinfectant techniques is of great importance in medicine, appreciable disinfection occurs spontaneously under natural conditions—pathogenic organisms which have become adapted to a parasitic role in the animal body as a rule surviving poorly outside it. The more important factors operating under such conditions and leading to the death of bacteria are deprivation of nutriment and the disinfectant action of light and desiccation. In certain natural environments other special factors come into play; thus in water bacteria are ingested by algae and protozoa, while in the soil they are exposed to the action of antibiotic substances produced by various soil organisms.

PHYSICAL METHODS OF STERILIZATION

Mechanical trauma

Bacteria can be killed by simple mechanical trauma, e.g. by grinding, by shaking with small particles, by ultrasonic irradiation and by repeated freezing and thawing. Mechanical methods, however, have no practical application in disinfection—although they are of considerable value as research procedures for the rupture of bacteria, so as to release intracellular constituents.

Heat

The application of heat is the most important of all the methods of disinfection and provided the heat used is adequate it is the most certain and rapid; it is also easily controlled and unlike chemical disinfection leaves no potentially harmful residue.

As with all other types of disinfection the sterilization of a bacterial population by heat is a gradual process. Throughout most of the process the rate of disinfection is approximately a logarithmic one, i.e. if the logarithm of the number of survivors is plotted against time the resultant curve is a straight line.

The time required for sterilization is inversely related to the temperature of exposure—the higher the temperature the shorter the time required. Thus a culture which requires one hour for sterilization at 60°C will be sterilized in only a few minutes at 80°C; at 100°C sterilization will be virtually instantaneous.

Different types of bacteria show considerable differences in heat susceptibility. In general most vegetative bacteria are killed by exposure, under moist conditions, to temperatures of 60 to 65°C for half an hour. A few species have, however, an unusually high degree of heat susceptibility. Spirochaetes, for example, are killed in a very short time at 40°C and many of the streptococci, with the notable exception of *Strep. faecalis*, are killed rapidly in the range of 50 to 55°C. Sporing organisms, on the other hand, have a very high resistance to heat, most spores being capable of withstanding a temperature of 100°C for very long periods. This is why high temperature sterilization, i.e. by the use of temperatures over 100°C, is obligatory for hospital sterilization and for the sterilization of many types of preserved foods.

Dry heat

Dry heat is the preferred method of sterilization for glassware, e.g. glass syringes, and for materials such as oils, jellies and powders which are impervious to steam. It is unsuitable for material, e.g. fabrics, which may be damaged by heat. As compared with steam sterilization it has the disadvantages that a longer time and a higher temperature are required. The most widely used type of dry heat sterilizer is the hot air oven, which is usually electrically heated. In the simpler gravity convection ovens the air circulates by convection. This form of oven is unsatisfactory since it is difficult to ensure that there has been adequate air circulation. The mechanical convection type of oven in which the air is circulated by a fan is much more efficient. Suitable sterilizing times in the hot air oven are three hours at 140°C, one hour at 160°C and 20 to 30 minutes at 180°C.

Moist heat

As previously mentioned most vegetative bacteria can be fairly rapidly killed by temperatures in the range of 60 to 65°C. The most important applications of temperatures in this range are in the pasteurization of milk and in the preparation of bacterial vaccines.

Boiling is frequently used for the sterilization of glass syringes, surgical instruments and small pieces of apparatus. Since, however, many spores will withstand boiling for a considerable time, boiling must be regarded as inadequate for the sterilization of such materials.

Steam sterilization

Exposure to steam is the most widely used and the most effective technique of moist heat sterilization. Steam is a much more efficient sterilizing agent than hot air at the same temperature, for the following reasons. First, bacteria are intrinsically more susceptible to moist than to dry heat. Second, steam is a more rapid sterilizing agent than hot air because on condensation it gives up its latent heat of vaporization thereby rapidly heating the objects on which it condenses. Third, steam has a much greater power of penetrating porous material. This is due to the fact that on condensation a partial vacuum is created which has the effect of sucking in more steam from outside until the material becomes thoroughly permeated.

Steam may be employed in three ways.

Steam at 100°C. This is achieved in a steamer in which the steam is generated at atmospheric pressure by electricity or gas from water in the bottom of the chamber. The materials to be sterilized are placed on perforated shelves through which the steam passes to escape through an opening at the top. The method is mainly used for the sterilization of bacteriological media which would be damaged at the high temperature of the autoclave. The procedure of 'intermittent' or 'fractional' sterilization is usually employed; this ensures the destruction of spores which might not be killed by a single steaming. The procedure involves steaming for short periods on each of three successive days. It should be noted that the method can be applied only to nutrient media in which spores can germinate in the intervals between steaming.

Low temperature steam. This method has recently been advocated for the sterilization of materials such as blankets and polythene tubing, which would be damaged by steam at high temperatures. Saturated steam at temperatures of from 70 to 90°C may be generated in sterilizers operating at pressures of 30 to 65 kPa (10 to 20 in Hg). Under these conditions steam has been found to be a much more effective sporicidal agent than water at the same temperature. The efficiency of the process can be significantly increased by the injection of a small amount of formalin along with the steam.

Steam at temperatures greater than 100°C. This is achieved by the use of the autoclave. The use of such high temperatures is necessitated by the high thermal resistance of bacterial spores (p. 9).

Essentially an autoclave is a very strong boiler or cylinder with a door which can be hermetically sealed and constructed sufficiently strongly to withstand the high pressures required. The pressure is maintained at the appropriate level during sterilization by means of a safety valve set to blow off at a predetermined pressure. In construction, autoclaves vary enormously from the simpler laboratory types—in which the steam is generated in the autoclave itself, and which are essentially little more than glorified pressure cookers—to the highly sophisticated hospital sterilizers with virtually automatic control of the sterilizing cycle.

Cold

Bacteria may also be killed by exposure to cold. Two different circumstances must be distinguished.

Cold shock

This is a sudden reduction in temperature without actual freezing. Some species are highly susceptible to cold shock. Thus a 95% drop in the number of viable *Escherichia coli* has been reported to occur following sudden chilling from 45 to 15°C. The mechanism of this effect is unknown but it is possibly due to a difference in the rate of contraction on cooling of different intracellular constituents, with a resultant disorganization of cellular structure.

Freezing

It was at one time thought that the lethal effect of freezing was due to physical damage to the cell membrane and/or cell wall by ice crystals. It is now generally accepted that such a mechanism can play only a minor role, if any, in the death of frozen organisms. There are probably two major factors involved:

1. The formation of ice crystals outside the cell by the withdrawal of water from the cell interior increases the intracellular salt concentration. This would in turn be capable of causing considerable damage by protein denaturation and is probably a major cause of death when bacteria are frozen to temperatures of not lower than –30 to –35°C.
2. Formation of ice inside the cell. There is evidence that this can only occur if the bacteria are frozen to temperatures lower than –35°C and, even then, only under conditions of rapid cooling. The lethal effect of intracellular ice does not appear to be due to a direct action of the ice on the cytoplasm but only occurs during defreezing. It appears to be maximal when the bacteria are heated slowly and minimal when they are heated rapidly.

Ultra-violet radiation

Light has considerable disinfectant properties and plays a very important part in the spontaneous sterilization which occurs under natural conditions. Its disinfectant action is due mainly to the ultra-violet rays, most of which are screened out by glass and also to a considerable extent by smoky and foggy atmospheres.

The most active of the ultra-violet rays are those with wave-lengths in the region 240 to 280 nm. Radiation in this region is, however, present to only a small extent in the solar radiation which succeeds in penetrating the earth's atmosphere, the latter containing little radiation with wave-length below 290 nm.

Ultra-violet radiation can be produced artificially by special lamps. Those most commonly employed are of the low pressure mercury vapour type. With these, over 95% of the radiation emitted is of wave-length 253·7 nm. Germicidal UV lamps must be used with care since the radiation has a damaging effect on the eyes; they must consequently be sited in such a way that the eyes are not exposed directly to the radiation.

Most vegetative bacteria appear to exhibit much the same order of susceptibility to ultra-violet irradiation; Gram-positive species, however, appear to show a slighly greater resistance than Gram-negative. Spores are highly resistant and the susceptibility of the viruses is variable. As a practical method of disinfection, however, ultra-violet irradiation has the serious disadvantage that its penetrating power is very slight. It is therefore only of value for the sterilization of surfaces, of very thin fluid films and of the air.

There seems little doubt that the lethal effect of ultra-violet irradiation in the most actively bactericidal range is due to its strong absorption by, and resultant damage to, the nuclear DNA. The most important effect of its action on DNA is the production of *thymine dimers*. Dimer formation occurs in the first place between adjacent thymines of the same DNA strand, and later between thymines on opposite strands. The former leads to a copying error when the DNA is replicated, as a result of interference with hydrogen-bonding relationships. Dimer formation between thymines on opposite strands might be expected to result in an inhibition of the capacity of the nucleic acid to replicate.

Bacteria possess mechanisms which will repair the damage caused by ultra-violet radiation. The most studied of these mechanisms is photo-reactivation. This is a temperature-dependent mechanism involving enzymatic monomerization of thymine dimers in situ. Photoreactivating enzymes have been identified in a wide range of bacterial species as well as in mycoplasmas, certain fungi, protozoa and cell lines of chicken tissues. They have not so far been described in mammalian cell lines. The yeast photoreactivating enzymes are particularly suitable for experimental work since they are not contaminated, as are bacterial preparations, by nucleases which degrade transforming DNA. Photoreactivating enzymes bind to ultra-violet irradiated DNA in the dark, but on subsequent exposure to visible light, which shows peaks of activity at 360 to

Table I.3. Bactericidal activity of electromagnetic radiations

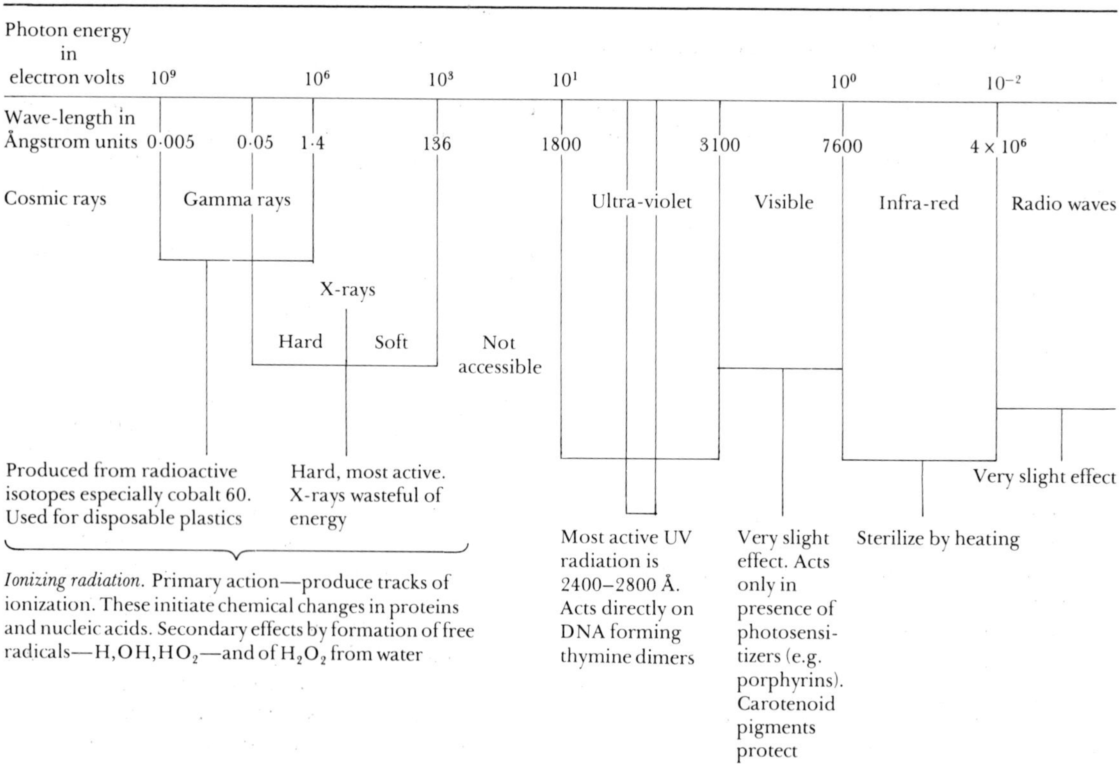

390 nm, the enzyme is released, leaving a repaired DNA to which it can no longer bind. Exposure to visible light possibly results in the activation of the enzyme so that it can effect the repair. When this is completed the enzyme is released. The repair is effected in situ in the DNA strand and there is no evidence of dimer excision. The specificity of the photoreactivating enzymes have been extensively investigated.

Bacteria can also repair ultra-violet damage to the DNA in the dark, the most important mechanism being excision repair. The currently accepted model is as follows:

1. Incision of the phosphodiester backbone to produce a single strand break on the 5′ side of the pyrimidine dimer. Incision requires a prior recognition of the radiation damage and is effected by an endonuclease specific for ultra-violet damaged DNA.

2. A new DNA strand is synthesized using the undamaged strand as template. Synthesis is effected by the Kornberg enzyme, DNA polymerase 1, which is a specific repair enzyme and is not involved in normal DNA replication. The enzyme also possesses exonuclease activity, first making an excision in the DNA backbone to release a dimer containing tri- or tetranucleotides. Excision and polymerization proceed past this point, however, releasing further single nucleotides.

3. Finally the 3′ end of the new strand is joined to the adjacent 5′ end of the original strand by the action of the polynucleotide ligase. The extent to which excision repair occurs is affected by environmental conditions. It is greater during the stationary than during the logarithmic phase of growth and is enhanced if the organisms are held in buffer solution after irradiation—this is known as liquid holding recovery—or if the amino acid content of the medium is restricted in the case of amino acid requiring organisms or if they are exposed to visible radiation of 334 nm before ultra-violet irradiation. All these procedures retard DNA replication but have no effect on the repair system, consequently allowing the latter time in which to operate. Excision repair is of considerable general importance since it will repair damage due to spontaneous errors in replication, ionizing radiations and radiomimetic chemicals.

A further repair mechanism is known as post-replication repair. In this, DNA replication proceeds normally but the new DNA fragments are lacking in large gaps corresponding to the dimer sites. These gaps are made good by cross-over events with intact segments produced by normal replication of undamaged strands.

Effect of visible light

Although most of the disinfectant action of sunlight is due to radiation in the near ultra-violet range, radiation in the visible range has some disinfectant properties. For this, however, the presence of a photodynamic sensitizing substance is required. The phenomenon of photodynamic sensitization can be demonstrated with dyes, e.g. eosin, which fluoresce on expsore to visible light. The phenomenon was first demonstrated at the beginning of the century with *Paramecium*. Oxygen is essential and the mechanism involved appears to be a photo-oxidation. Under natural conditions bacterial porphyrins and flavins are believed to be of importance as photodynamic sensitizers. Evidence has been presented by Stanier that bacteria may be protected from photo-oxidation by naturally occurring carotenoid pigments. Non-pigmented mutants of naturally pigmented species show a great increase in susceptibility to visible light.

Ionizing radiations

Ionizing radiations have considerable disinfectant action. They have a much higher energy content than ultra-violet radiation and have consequently a much greater capacity to induce lethal chemical changes. These changes are primarily the result of the production by the radiation of tracks of ionization in the cells. In the killing of bacteria the changes induced in the nucleic acids—particularly in DNA—are probably of greatest importance. Ionizing radiations also affect the cell through the production from water molecules both extracellularly and intracellularly of hydrogen peroxide and of oxidizing free radicals.

Of the various types of ionizing radiation only cathode rays and gamma rays would appear to have any reasonable prospect of practical application. Both these types of radiation have received considerable attention in recent years as possible disinfecting agents. Generators and accelerators are now available which can produce cathode rays on a commercial scale and radioactive isotopes—in particular, cobalt 60 which produces gamma radiation—are now available in substantial quantities as by-products of atomic energy installations.

Ionizing radiations are being used on an increasing scale for the sterilization of surgical materials, particularly of disposable items made of rubber or plastic. They are of particular importance in the sterilization of disposable syringes. Other materials which may be sterilized in this way are surgical catgut, bone and tissue grafts and adhesive dressings.

Desiccation

Although lethal to bacteria, drying is not applicable as a practical method of sterilization. Bacterial species differ considerably in their susceptibility to drying. Some, such as gonococci and the cholera vibrio, are rapidly killed. Tubercle bacilli and streptococci, on the other hand, are resistant and may persist in the dried state, e.g. in dust or in books, for long periods. The effect of drying depends on a number of factors—a low temperature, a protein medium and the absence of oxygen affording bacteria considerable protection. Bacteria that survive the drying process are in a state of suspended animation and in this condition may remain viable for years. Consequently, drying is widely used as a method of preserving bacterial cultures. The initial mortality is consderably diminished if the cultures are dried in vacuo from the frozen state. This process is known as *lyophilization* or *freeze drying*.

Filtration

Filtration has a number of bacteriological applications. It may be used for the preparation from cultures of cell-free bacterial products, e.g. toxins and enzymes, to free virus-containing fluids from bacteria and for the sterilization of media or media ingredients which would be damaged by heating. For these purposes filters with pores sufficiently small to hold back bacteria must be employed.

Filtration is usually carried out under negative pressure, the fluid being sucked through the filter into a receiving flask, which is connected to an exhaust pump. On exhaustion of the flask by the pump the fluid is sucked through the filter.

During filtration there may be appreciable adsorption of material from the fluid being filtered onto the surface of the filter. This adsorption occurs because filters are not merely mechanical sieves but carry electrical charges by virtue of which they attract material carrying an opposite charge. Adsorption may result in the clogging of the filter or, more serious still, in the retention of substances, e.g. enzymes, which it is desired to obtain in the filtrate.

For laboratory use Seitz filters made of asbestos and filters of diatomaceous earth (Berkfeld and Mandler), unglazed porcelain (Chamberland and Doulton), sintered glass and collodion are employed. Collodion or Gradocol filters are for many purposes the most satisfactory; they are virtually non-adsorptive and can be made in accurately graded pore sizes.

Filtration is of considerable importance in the purification of water supplies destined for human consumption. Two methods are employed.

Slow sand filtration. This method employs large filter beds made, from the bottom up, of clinker, gravel and sand. The essential filtering agent, however, is the vital or zoogloeal layer, consisting mainly of algae and protozoa, which forms on the surface of the sand when the water has been allowed to flow through the filter for some time. Until this layer forms the filter cannot be used. Eventually, after prolonged use, the surface of the filter becomes too thick and must be scraped off and allowed to reform.

Rapid sand filtration. In this method alum is added to the water. The alum precipitates, carrying with it many of the bacteria present. The water is then passed through a sand filter on the surface of which the alum accumulates and acts like the vital layer of a slow sand filter.

Of the two methods the slow sand filter is considerably more efficient but neither can be relied on to render water completely sterile. To reduce its bacterial content still further water after filtration is treated with chlorine. For the filtration of small household supplies unglazed porcelain filters are frequently employed. To maintain their efficiency these must be cleaned at regular intervals.

Another application of filtration of practical importance is for the removal of micro-organisms from the air. Air filtration is of value in industries utilizing fermentation processes, e.g. those concerned with the manufacture of antibiotics in which the control of extraneous contamination is essential. Air filtration is also being employed on an increasing scale in hospital operating theatres and dressing rooms as a means of controlling hospital cross-infection. Filters of slag wool, glass wool and spun glass fibres have been mainly used. Cotton wool is also an extremely efficient air filter but is only effective as long as it can be kept dry. It is not used in air filtration on a large scale but is of course of the greatest value to the bacteriologist as a method of preserving the sterility of culture media.

CHEMICAL DISINFECTION

Chemical agents exhibit two distinct types of antibacterial effect: a *bactericidal* or killing effect and a *bacteriostatic* or growth-inhibiting effect. Bactericidal activity may be demonstrated by mixing a suspension of bacteria with the disinfectant and then after a period of contact under defined conditions the mixture is subcultured to a nutrient medium. If the compound has a bactericidal action no growth of the organism will occur on subculture. To demonstrate bacteriostatic activity the disinfectant is incorporated in a suitable nutrient medium, e.g. broth which is then inoculated with the organism and incubated. A bacteriostatic effect is shown by the absence of growth after incubation.

Most of the chemical compounds which can kill bacteria exhibit a bacteriostatic effect in concentrations lower than those required to kill. The relationship between bactericidal and bacteriostatic concentrations is not, however, the same for all disinfectants. Some compounds, e.g. the mercurials and the quaternary ammonium compounds, have very marked bacteriostatic properties inhibiting bacterial growth in dilutions very much higher than those required to kill. Others, e.g. the halogens, which of the commonly used disinfectants are unique in this respect, appear to be exclusively or almost exclusively bactericidal.

On the whole we know relatively little of the precise ways in which disinfectants kill bacteria; it is probable that in most cases killing occurs through some form of enzyme inactivation either by protein denaturation, oxidation or by a combination of the antibacterial agent with specific groups of enzyme proteins. Some of the compounds used as disinfectants, e.g. the quaternary ammonium compounds, are capable of damaging the bacterial cell membrane and thus permitting leakage of essential intracellular compounds from the cell into its environment. This type of effect is believed to be of primary importance in the killing action of such compounds.

Kinetics of disinfection

When a bacterial population is exposed to a disinfectant there is a progressive reduction with time in the number of surviving organisms. Disinfection is therefore a gradual process. In this connection it is important to remember that when we study the killing of bacteria by a disinfectant the observable effect is the effect on a bacterial population not the effect on a single bacterial cell. It follows therefore that the larger the bacterial population exposed the longer is the time required for its sterilization.

It is generally accepted as a working hypothesis that the rate at which disinfection occurs is uniform in the sense that during each unit of time a constant proportion of the organisms present at the beginning of that time is killed. In this event if the logarithm of the number of survivors after exposure to a disinfectant is plotted against the time of exposure the resultant curve is a straight line. This relationship is expressed mathematically in the following formula:

$$K = 1/t \log B/b,$$

where $B =$ the initial number of organisms and $b =$ the number of after time t. There is, however, considerable doubt as to whether the logarithmic curve, though mathematically convenient, is in fact a correct general expression of the process of disinfection. It appears to be approximately correct when relatively high concentrations of disinfectant are used. With lower concentrations, however, many bacteriologists claim that the disinfection curve has a signmoid character, the rate being slow in the early stages then proceeding rapidly for most of the disinfection process and finally slowing down at the end.

As might be expected the rate of disinfection varies with the concentration of disinfectant employed. The effect of concentration on rate, however, is not constant and varies considerably with different disinfectants. With phenolic compounds, for example, a change in the concentration of the disinfectant has a marked effect on the disinfection rate. Thus with phenol, halving the concentration of disinfectant increases approximately 64-fold the time required for sterilization. With most other disinfectants changes in concentration have a very much less dramatic effect on sterilizing time than in the case of phenol. For example with the mercurials halving the concentration of disinfectant only doubles the time required for sterilization.

Factors affecting disinfectant action

The efficiency of a disinfectant is affected by a variety of factors:

1. Since disinfection is a chemical process its rate is considerably affected by temperature. In most cases the higher the temperature the more rapid the rate and consequently the shorter the time required for sterilization.

2. Different organisms show considerable differences in their susceptibility to disinfection. Spores in particular are highly resistant, possibly because the relatively impermeable spore coverings form an effective barrier to the passage of the disinfectant into the cell. Of the commonly used disinfectants only the halogens and formaldehyde possess any useful sporicidal activity. Tubercle bacilli also exhibit a high resistance to chemical disinfection, particularly to disinfectants in aqueous solution. This lack of susceptibility is usually attributed to the high lipid content of the tubercle bacillus. On the whole Gram-positive organisms appear to be more susceptible to the action of disinfectants than Gram-negative. Of the latter *Pseudomonas* appears to be one of the most resistant.

3. The activity of disinfectants may be markedly affected by substances present in the environment. Most disinfectants are to a greater or lesser extent inhibited by organic material with a high protein content, e.g. serum, blood and pus. This inhibition is most marked in the case of aniline dyes, the mercurials and the cationic detergents. Certain disinfectants, depending on their nature, are inhibited by specific compounds. The quaternary ammonium compounds, for example, are markedly inhibited by soaps and lipids and the mercurials by compounds containing sulphydryl (SH) groups. The pH of the environment may also have considerable effect. Some compounds, e.g. the cationic detergents and aniline dyes, are most active under alkaline conditions while others, notably the phenols and chlorine, are most active at an acid pH.

Acids and alkalis

The disinfectant activity of strong acids and alkalis depends in general on their acid or basic strength. Many weak organic acids have, however, a higher activity than would be expected from their dissociation constants. This activity is due to the toxicity of the organic anion.

The tubercle bacillus is appreciably more resistant than other organisms to disinfection by acid and alkali. Consequently material from which it is proposed to culture tubercle bacilli may be treated with acid or alkali to free it from other bacteria present.

Salts

All salts have some degree of toxicity for bacteria. The most toxic are those of the heavy metals, mercury and silver, and the least toxic the salts of sodium and potassium.

Of the heavy metals the most frequently used is mercury. Two types of compound are employed—the *inorganic salts* and the *organic mercurials* in which mercury is combined with an organic radical. Both types of compound have very marked bacteriostatic properties but cannot be regarded as good disinfectants since their bactericidal activity is relatively slight. Mercury compounds appear to act on bacteria by combining with sulphydryl (SH) groups of bacterial proteins and other essential intracellular compounds. With these they form a relatively loose combination from which they can be dissociated by the addition of SH-containing compounds. Since pus and exudates contain a variety of such compounds they antagonize the action of the mercurials. This constitutes a further serious limitation to their medical utility.

The anti-infective properties of silver, as reflected in the use of silver urns as domestic utensils, have been known from antiquity. Metallic silver is itself bacteriostatic in a very high dilution. This effect is known as the *oligodynamic* effect and is probably due to free silver ions released in low concentrations from the metal. Of the silver salts, silver nitrate was widely used before the discovery of antibiotics for prophylaxis of ophthalmia neonatorum by instillation into the infant's eyes.

Halogens

Chlorine and *iodine* are the only halogens which have any practical application as disinfectants. For certain purposes—chlorine as a water disinfectant and iodine as a skin disinfectant—they are unequalled. Their great value as disinfectants is due to the following:

1. They are bactericidal in a very high dilution. They are in fact unique amongst the disinfectants in that their activity is practically exclusively bactericidal.
2. Their action is very rapid.
3. They possess considerable activity against sporing organisms.

In addition to chlorine itself three types of chlorine compound—the hypochlorites, the inorganic chloramines and the organic chloramines—are available. The disinfectant action of all chlorine compounds is due to their capacity to liberate free chlorine. In solution the liberated chlorine forms hypochlorous acid which in its non-ionized form is believed to be the active disinfectant agent. Since hypochlorous acid is least dissociated at acid pH values chlorine disinfectants are in general most active under acid conditions.

For large-scale water disinfection chlorine is usually employed in the form of the compressed gas, but for the small-scale purification of domestic supplies it is usually employed in the form of chlorinated lime (calcium hypochlorite). When water is treated with chlorine it is first necessary to determine its 'chlorine demand'. This is due to a variety of substances which may be present in the water which are capable of combining with chlorine. Although some of the compounds thus formed, in particular the organic chloramines, may have appreciable disinfectant action, their action is much slower than that of free chlorine. Consequently it is usual to add sufficient chlorine to satisfy the chlorine demand of the water and at the same time to provide enough residual chlorine for active disinfection. A residual chlorine concentration of from 0·2 to 0·4 parts per million is quite adequate for this purpose. In the case of waters which contain a large amount of organic material a considerable amount of chlorine may have to be added initially to leave this amount of residual chlorine.

Hypochlorites have a wide application in the dairy industry for the disinfection of equipment, for which they are possibly the best compounds at present available. Calcium hypochlorite is frequently employed for the disinfection of faeces, urine and sanitary utensils and is highly effective for this purpose. It has however little, if any, activity against tubercle bacilli. Sodium hypochlorite in the form of an aerosol disseminated by spraying has been used extensively as an air disinfectant but its value for this purpose is doubtful.

Inorganic chloramines, which are compounds of chlorine with ammonia, of varying chlorine composition, have appreciable disinfectant action. In solution they are more stable than the hypochlorites releasing their available chlorine more slowly. Consequently they have a much more sustained action but, as might be expected, are active in lower dilution. Because of their persistent action they are of special value for the sterilization of water when there is a delay in its distribution.

Iodine is used almost exclusively as a skin disinfectant and is the only disinfectant commonly used for this purpose which is active against sporing organisms. It is usually employed as a tincture. This is an alcoholic solution containing 2·5% iodine and 2·5% potassium iodide in 90% alcohol. Lugol's solution (a watery solution containing 5% iodine and 10% potassium iodide) may also be employed for skin disinfection and has the advantage that it is less irritant to the tissues than the tincture. In the concentrations in which it is normally used iodine is not appreciably affected by organic matter. Like chlorine it is most active at an acid pH. The active agent in disinfection is the iodine molecule. An occasional disadvantage encountered in the use of iodine as a skin disinfectant is that it can produce skin rashes in individuals who are unduly susceptible to it. In the form of a 2% aqueous solution iodine is probably the best agent available for the sterilization of clinical thermometers.

Mixtures of iodine with various surface active

agents, which act as carriers for the iodine, are known as iodophors; they have been widely used in the United States especially for the sterilization of dairy equipment. Compounds of this type, particularly those formed with non-ionic detergents, are claimed to be more active than the alcoholic and aqueous solutions.

Phenols

Phenolic compounds are probably the most widely used of all disinfectants. The disinfectant activity of phenol itself was first discovered by Lister whose use of it to control hospital infections laid the foundation of modern surgical asepsis. Phenol is, however, no longer used as a general disinfectant. It is expensive and compared with its modern derivatives has a relatively low degree of activity. It is in addition not suitable for application to the skin or mucous membranes since it is readily absorbed with the production of severe toxic symptoms. Its main importance is as a standard with which other disinfectants may be compared and as a bacteriostatic agent in materials to be used for injection.

The antibacterial activity of phenol is greatly increased by various forms of substitution in the phenol nucleus—the compounds which are of most importance being the *alkyl-* and *chloro-* derivatives and the *diphenyls.* In addition the more active compounds are appreciably less toxic than phenol, being much less readily absorbed from the skin and mucous membranes. High activity is, however, associated with certain disadvantages. Whereas the simpler compounds—phenol itself and the cresols—are not significantly affected by the presence of organic matter and have much the same degree of activity against Gram-positive and Gram-negative organisms, the activity of the more active high molecular weight compounds is appreciably diminished by the presence of organic material and is considerably more marked against Gram-positives than against Gram-negatives.

The simplest of the alkyl phenols are the cresols, obtained industrially by the distillation of coal tar and the principal ingredients of the lower boiling point fractions. The cresols resemble phenol closely in their antibacterial effects but are appreciably more active. Lysol, one of the most commonly used general purpose disinfectants, consists of 50% cresol in soap. Like phenol it is too toxic for application to the skin, but is widely used in laboratories for the preliminary sterilization of infected glassware and in hospitals for the disinfection of excreta.

The xylenols or methyl cresols have provided two extremely important disinfectants—*p*-chloro-*sym-m*-xylenol (PCMX) and 2,4-dichloro-*sym-m*-xylenol (DCMX). Of these PCMX has been more used; it is the main component of a number of commercial preparations of which the best known and most widely used is probably Dettol. PCMX has considerable activity against both Gram-positive and Gram-negative organisms. Its capacity to retain its activity for a considerable time on the skin and other surfaces is of special value in medicine. It is of low toxicity and is not markedly inhibited by organic matter.

The most important of the diphenyl compounds are the chloro substituted diphenylmethanes—dichlorophane, tetrachlorophane and hexachlorophane. The bactericidal activity of these compounds is somewhat less than that of the chloroxylenols but they are bacteriostatic in very high dilutions. In fact because of their marked bacteriostatic potency they were at first considerably over-rated as bactericidal agents. Hexachlorophane is probably the best known. Unlike other phenols it retains its activity in the presence of large amounts of soap and germicidal soaps containing low concentrations of hexachlorophane have achieved a considerable popularity. Hexachlorophane soaps when frequently used cause a marked reduction of the bacterial flora of the skin and for this reason have been strongly recommended for habitual use by surgeons. Hexchlorophane is appreciably more active against Gram-positive than against Gram-negative organisms.

Chlorhexidine (1,6-di-4′-chlorphenyldiguanido-hexane) has come into extensive use in recent years as a skin disinfectant. It is actively bactericidal for a wide range of Gram-positive and Gram-negative species and because of its rapid action and low toxicity is of special value for clinical use. A 0·5% solution of the digluconate in 70% ethyl alcohol is reported to be as effective for skin disinfection as tincture of iodine and has the advantage that it lacks the irritative properties of the latter.

Soaps

Though soaps possess disinfectant properties their

activity is of a comparatively low order; it is consequently unlikely that much disinfectant action occurs during the short time taken in washing the hands. Nevertheless the process of cleaning with soap and water mechanically removes organisms from the skin and has in effect a disinfectant action.

Alcohols

Ethyl alcohol is non-specific in action, being active against both Gram-positive and Gram-negative organisms. It possesses in addition appreciable activity against tubercle bacilli. It has no significant sporicidal activity; in fact the recovery of anthrax spores after 20 years' exposure to ethyl alcohol has been recorded.

Ethyl alcohol almost certainly acts by protein denaturation and exhibits a bactericidal effect only in the presence of water. Appreciable disinfectant activity only occurs in concentrations of from 40 to 95%—the optimum range being from 50 to 70%.

In general the disinfectant activity of the alcohols increases with increase in chain length. Above ten carbon atoms, however, the solubility of the compound becomes too low for practical use. Apart from ethyl alcohol the only alcohols which have so far had any significant application have been isopropyl alcohol and the dihydric alcohols ethylene and propylene glycol. Isopropyl alcohol, which is appreciably more active and less volatile than ethyl alcohol, has been recommended for the sterilization of clinical thermometers.

Dyes

Two groups of dyes, the *aniline* dyes and the *acridines*, have been considerably used as skin and wound antiseptics. Both groups are bacteriostatic in high dilution but are of low bactericidal activity. Of the aniline dyes brilliant green, malachite green and crystal violet (an impure form of the latter is known as gentian violet) have been mainly used. All are amino derivatives of triphenylmethane. The aniline dyes are highly selective being much more active against Gram-positive than against Gram-negative organisms.

The acridines have, for clinical purposes, two important advantages over the aniline dyes: (*a*) though somewhat more active against Gram-positives than against Gram-negatives they are not as selective as the aniline dyes and (*b*) they are little if at all affected by the presence of organic matter. The following compounds all of which are amino substituted acridines are of clinical importance: proflavine, acriflavine, euflavine and aminacrine.

Quaternary ammonium compounds

This group of compounds has the general formula NR_4X where R stands for an alklyl or phenyl radical and X an inorganic radical. They may therefore be regarded as substituted ammonium salts. In solution they ionize to yield an organic cation and an inorganic anion. The organic cation of the quaternaries has lipophilic properties and confers on the compounds marked surface activity. The more important compounds now in use are cetrimide, benzalkonium chloride and domiphen bromide.

The quaternary ammonium compounds have very marked antibacterial properties and are bactericidal for a wide range of organisms. Although active against both Gram-positive and Gram-negative species they are appreciably more active against the Gram-positives. They are virtually without activity against tubercle bacilli and spores. A special feature of the group, and one in which they resemble the mercurials, is that they are bacteriostatic in very high dilution. This can be attributed to the fact that because of their high surface activity they are readily adsorbed to the bacterial surface even when present in very low concentration. Their antibacterial activity, like that of the aniline dyes, is much more marked at alkaline than at acid pH values. Although strongly haemolytic for red cells under in vitro conditions they appear to possess little in vivo toxicity. Unfortunately they are markedly inhibited by organic matter, particularly by proteins, phospholipids and fats. In this connection it has been suggested that the relatively low susceptibility of the Gram-negatives is due to their high lipid content. The quaternary ammonium compounds are also antagonized by soaps and other anionic compounds. The possibility of antagonism by soap should be borne in mind when any member of this group is being used for the disinfection of the skin.

The readiness with which quaternaries are adsorbed to surfaces is of practical importance in disinfection for the following reasons:

1. When applied to the skin they form an antibacterial surface film which persists for a considerable time. There is some doubt, however, as to their value as skin disinfectants since the residual detergent may not be actively bactericidal.

2. They are readily adsorbed to glassware and to fabrics. Consequently they are widely used for the sterilization of food utensils in restaurants and hotels, for the sterilization of milk containers in the. airy industry and for the disinfection of blankets in hospitals.

Formaldehyde

When dispersed in gaseous form formaldehyde has considerable bactericidal activity and is one of the very limited number of disinfectants active against spores. The sterilizing efficiency of formaldehyde is very low under dry conditions but increases with increase in moisture content of the air up to a level of around 50% relative humidity. Sterilization is not strictly gaseous but is as a result of condensation of the gas on exposed surfaces. Consequently formaldehyde has relatively poor penetrating power and this constitutes a considerable limitation to its practical use. In liquid form (Formalin) formaldehyde is used to sterilize materials such as wool and hides which may harbour anthrax spores. It is also employed for the disinfection of the shoes and footwear of persons suffering from athlete's foot since it is very active against the fungi responsible for this condition.

Glutaraldehyde

This relatively new disinfectant has a wide range of action including a marked lethal effect on spores and tubercle bacilli. It is usually employed as a 2% buffered solution (Cidex) which is claimed to be stable for two weeks at 20°C. It has been recommended for the sterilization of cystoscopes and other endoscopic instruments which cannot be subjected to heat sterilization.

Ethylene oxide

Ethylene oxide is a cyclic ether $(CH_2)_2O$ which boils at 10·8°C and which therefore is a gas at ordinary room temperatures. By itself it is highly inflammable but forms safe non-inflammable mixtures with carbon dioxide and fluorinated hydrocarbons. It is toxic to man by inhalation and on contact with the skin has a vesicant action. It cannot be used, therefore, for sterilizing the air of occupied rooms. Ethylene oxide is active against all types of bacteria including spores and tubercle bacilli but its action is slow. It can be used for sterilizing a wide range of materials but is of particular value for materials which would be damaged by heat, e.g. polythene tubing. It has been of special value in the sterilization of heart–lung machines.

I.4

Bacteria in Health and Disease

Most bacteria are free-living organisms deriving their nourishment from inert organic or inorganic materials. Although often of great importance in the economy of nature, these free-living bacteria are of little interest to the medical bacteriologist, who is concerned primarily with the limited numbers of species which are parasitic for man.

In respect of their capacity to produce disease the parasitic bacteria form a continuous series, ranging from those with little or no power of producing disease to those that when present in the body almost invariably produce disease. From the evolutionary point of view the non-pathogenic or occasionally pathogenic bacteria must be regarded as being more successful than the more highly pathogenic varieties. They are able to establish a harmless parasitic relationship with their hosts, sometimes to the advantage of the latter, which ensures their own chances of survival. Highly pathogenic bacteria can rarely if ever establish such a relationship and, by frequently killing the host, diminish their own chances of survival.

THE NORMAL BODY FLORA

At birth the skin and mucous membranes are, as might be expected, sterile. It is in fact possible by the adoption of stringent isolation procedures to maintain laboratory animals in a germ-free state. Such animals have been of considerable value for research, particularly in connection with the development of immunity mechanisms.

In the normal animal, however, the surfaces of the body which are in contact with the environment, namely the skin and mucous membranes, rapidly become colonized by organisms present in the environment. The organisms that can establish themselves in this way differ in the various parts of the body, which consequently show variations in their normal flora. Such differences are due to the operation of local and environmental factors which favour and tend to select certain species. We know relatively little of the precise nature of these environmental factors and of the way in which they tend to favour particular bacteria. We do know, however, that changes in environmental conditions may result in marked changes in local flora.

The organisms of this normal flora are known as *commensals* and obtain their nutriment from the secretions and waste products of the body. Occasionally species of relatively high pathogenicity may appear in the normal flora without causing disease. When this happens the individual in whose body the pathogen is found is known as a *carrier*. The carrier state is, however, unusual—the highly pathogenic organisms either initiating an infection or being rapidly eliminated from the body.

A knowledge of the types of bacteria found in different areas is of great importance in diagnostic bacteriology and these will be briefly considered.

The skin is constantly receiving bacteria from the air or from objects with which it has come in contact, but the majority of these do not grow on it because of the absence of suitable growth conditions. Few are capable even of surviving on the skin for more than a very short time because of substances which are bactericidal for them. The most constant of the normal flora of the skin are anaerobic diphtheroids—which belong to the genus *Propionibacterium*—and non-pathogenic staphylococci. These organisms are found mostly in the sebaceous glands; their numbers are little affected by washing.

The organisms most frequently and most

constantly found in the nose are non-pathogenic corynebacteria. The nose is also the natural home of *Staph. aureus* and this organism can be isolated from about 50% of nasal swabs taken from normal persons. The occasional *Staph. aureus* on the skin is usually secondary to nasal carriage. About the throat, viridans streptococci, *N. catarrhalis* and staphylococci commonly occur while avirulent strains of pneumococci are not infrequent. The deeper portions of the respiratory tract, the finer bronchioles and the alveoli of the lung are normally sterile.

The mouth has a very mixed bacterial flora, the abundant moisture and the constant presence of small food particles providing an ideal environment for bacterial growth. Both aerobic and anaerobic types are found. The most important of the aerobic organisms are non-haemolytic streptococci and neisseriae, but miscellaneous fungi and actinomycetes are frequently present. Some of the last are believed to be of importance in the formation of dental tartar. Prior to the eruption of the teeth and in the edentulous adult, the flora of the mouth consists almost exclusively of aerobic organisms. Following the eruption of the teeth a number of anaerobic types—spirochaetes, anaerobic vibrios, various anaerobic cocci, and not infrequently *Actinomyces israelii*—are able to establish themselves. These organisms find ideal sites for anaerobic multiplication in the pockets between the gums and the teeth.

The contents of the healthy stomach are practically sterile, as are also those of the duodenum, owing to the presence of hydrochloric acid in the gastric secretion, but the intestine contains an enormous number and variety of bacteria. It has been calculated that an adult excretes in the faeces 3×10^{13} bacteria daily, the majority of which are dead.

The intestinal flora of the breast-fed infant is almost exclusively lactobacillary in type. With the introduction of bottle feeding the lactobacilli became less numerous and, finally, on substitution of solid food and adult diet, the normal mixed but predominantly Gram-negative bacillary intestinal flora becomes established. The most numerous of these are the bacteroids—a miscellaneous group of anaerobic Gram-negative bacilli—and the coliform bacilli, particularly *E. coli.* A variety of other organisms is also found although to a lesser extent, viz.: *Str. faecalis,* clostridia notably *Cl. perfringens* and *Bacillus, Lactobacillus, Proteus* and *Pseudomonas* species. *Staph. aureus* can also usually be isolated from faeces.

In diarrhoea as a result of the rapid movements of the intestinal contents the flora usually undergoes considerable change. Organisms which are normally found in small numbers may become very numerous. The composition of the intestinal flora may be considerably altered in persons receiving wide-spectrum antibiotics; the normal sensitive organisms are inhibited and may be replaced by resistant organisms.

About the external genitals, in both sexes, various Gram-positive cocci, *E. coli,* smegma bacilli and spirochaetes are found. Few bacteria exist in the urethra and the urine in the bladder is sterile in health. The main inhabitants of the adult vagina are the group of lactobacilli known as Döderlein's bacilli. These organisms break down the glycogen produced by the vaginal epithelium with the production of acid, of which they are highly tolerant. Fungi which are highly tolerant of acid are also usually present, but other organisms are found irregularly and in small numbers. The accumulation of glycogen in the vaginal wall is due to ovarian activity. Glycogen is not present before puberty or after the menopause; consequently at these times the flora is mixed rather than predominantly lactobacillary.

The relationship between the commensal bacteria and their animal host is a symbiotic one; the advantages of this relationship to the bacteria are obvious. But there are, possibly in no less degree, also advantages to the host:

1. The bacteria serve as scavengers assisting in the disposal of waste material.
2. In the case of the intestinal bacteria they undoubtedly play a part in the nutrition of the host. Many of the intestinal bacteria can synthesize the major B group vitamins and some can also synthesize vitamins E and K. These syntheses are believed to make a significant contribution to the vitamin requirements of the host.
3. By their presence they tend to exclude pathogenic bacteria and in this way serve to protect the host against disease. This exclusion may be by simple competition with the pathogen for nutrition or it may be by the production of substances that are inhibitory to the growth of the pathogen. Thus, the acid produced by the lactobacilli of the adult vagina protects the vagina

from infection with the gonococcus and other pyogenic organisms. Certain of the mouth streptococci produce hydrogen peroxide which may play a part in the exclusion from the mouth of peroxide sensitive pathogens. Many strains of *E. coli* produce colicines which may protect the intestinal tract from other pathogenic enterobacteriaceae such as the shigella which are sensitive to them.

THE BASIS OF BACTERIAL PATHOGENICITY

The capacity of an organism to produce disease is known as its *pathogenicity*. Pathogenicity is related to a particular host species; organisms which are pathogenic for one way may be quite non-pathogenic for another. The extent to which organisms can inflict disease is known as *virulence*, organisms of high virulence producing infection with much smaller inocula than organisms of low virulence. Pathogenicity is therefore a qualitative term and virulence a quantitative one.

With the exception of the cholera vibrio and the organisms responsible for bacterial food poisoning, bacteria can produce disease only if they invade or gain access to the tissues. In order to do this they must pass through the surface covering of the body—the skin or mucous membranes. It is doubtful whether any bacteria can penetrate the skin unless there has been some degree of trauma, e.g. by wounds, abrasions or insect bites. The mucous membranes are, however, much more permeable and many bacteria appear to be able to pass readily through them.

Invasions of the tissues both through the skin and mucous membranes are of frequent occurrence. The majority of these invasions, however, do not result in disease, as the organisms that have gained entry are rapidly disposed of by the tissue defences. The invading organisms can cause disease only if, after invasion, they can multiply. The capacity to multiply in the tissues is known as *aggressiveness*. The term *invasiveness* is also used, although not strictly correctly, as being roughly synonymous with aggressiveness. Finally, the mere multiplication of bacteria in the tissues does not by itself produce disease unless the bacteria can in some way damage the tissues. The capacity of bacteria to damage the tissues is known as *toxicity*.

Factors contributing to aggressiveness

As soon as a micro-organism gains access to the tissues it is confronted by a mobilization of the defensive machinery of the host. It is obvious, therefore, that in order to establish itself or to multiply it must in some way be able to overcome the host defences. The single most important component of the host's defence against invading bacteria is undoubtedly phagocytosis. Bacteria which are to survive for any time in the tissues must therefore possess some mechanism of defending themselves against this; different bacteria may do so in different ways. They may, as in the case of *Staph. aureus* and *Str. pyogenes*, produce toxins known as leucocidins which actually kill the leucocytes. Only a few bacteria, however, produce such substances. Many possess surface components which confer a resistance to phagocytosis. Important examples of such components are the capsules of the pneumococcus and *Yersinia pestis*, the Vi antigen of *Salmonella typhi* and the type-specific or M proteins of *Str. pyogenes*.

By the liberation of antigens into their environment which can combine with antibody, bacteria can reduce the opsonizing and bactericidal effects of the latter. These soluble antigens are probably the substances described by early workers as *aggressins*.

Some bacteria produce substances which facilitate their rapid spread through the tissues and which thereby contribute to aggressiveness. *Str. pyogenes* secretes streptokinase, a substance which activates a plasma protease precursor. The activated protease in turn breaks down the fibrin formed by the tissues as a method of localizing the infection. *Cl. perfringens*, many strains of *Staph. aureus*, and some strains of *Str. pyogenes* produce an enzyme hyaluronidase which breaks down hyaluronic acid—the intercellular cement substance of the tissues. This results in a reduction in the viscosity of the latter and therefore permits more rapid spread of the organism. *Cl. perfringens* also produces collagenase—an enzyme capable of decomposing collagen. This enzyme plays an important role in the spread of *Cl. perfringens* through infected muscle.

Toxicity

There are numbers of possible mechanisms by which bacteria may cause damage to the tissues.

Action of metabolic end-products

As a result of their growth a number of metabolic end-products, e.g. lactic acid and various protein degradation products, which may have some local toxic effect, are produced. The part played by such substances in the toxaemia of infection is probably small.

Allergic reactions

In tuberculosis the production of significant damage to the tissues depends on the existence of allergy to tuberculin. Allergy to bacterial components probably plays an important part in determining the amount of tissue damage occurring in a number of infections, but its role in this respect is most clear in tuberculosis.

Interference with the nutrition of the host cell

It might be expected that extensive growth of bacteria in the tissues at the expense of nutrient material utilized also by the host cell might result in damage to the latter by interfering with its nutrition. There is reason to believe that this type of damage occurs in the lung in cases of pneumococcal lobar pneumonia.

Bacterial toxins

These are substances produced by or present in bacteria, which have a direct toxic action on tissue cells. Bacterial toxins are conveniently divided into two groups.

The exotoxins. These are toxins which diffuse freely from the bacteria into the surrounding medium and can be obtained in cell-free filtrates.

The endotoxins. These are substances present in the body of the organism and not secreted to a significant extent into the surrounding medium. They may be obtained from the bacterial cells by physically disintegrating them or by various forms of chemical extraction and are liberated to a varying extent under natural conditions by autolysis.

The distinction between exo- and endotoxins is, however, not absolutely sharp. Appreciable exotoxin is released by autolysis of the cell and some endotoxin is usually demonstrable in culture filtrates.

A number of exotoxins have been obtained in a highly purified form and have been found to be proteins; it is probable that all the typical exotoxins are proteins. The organisms producing exotoxin are *Cl. diphtheriae, Cl. tetani, Cl. botulinum, Sh. dysenteriae, Staph. aureus, Str. pyogenes*, and the

Table I.4. Properties of bacterial toxins

Exotoxins	*Endotoxins*
Protein	Function of somatic antigen lipopolysaccharide
Secreted into medium from cell cytoplasm	Present in cell wall and released only on disruption of the cell
Heat-labile	Heat-stable
Converted by H.CHO into toxoid	Cannot be toxoided
Strongly antigenic	Poorly antigenic, if at all
Highly specific for particular tissues, e.g. tetanus toxin for CNS	Non-specific in action
Very high potency (1 mg tetanus toxin = MLD for 1000 tons of guinea pigs)	Low potency (about 1 mg extracted somatic antigen = MLD for 1 mouse)
Produced mainly by Gram-positive organisms, i.e. clostridia, diphtheria bacilli, streptococci, staphylococci	Produced by Gram-negative organisms, i.e. salmonellae, shigellae, brucellae, coliforms

gas gangrene clostridia. Except for *Sh. dysenteriae* (type 1), *V. cholerae* and some strains of *E. coli* exotoxin-producers are Gram-negative.

Endotoxins isolated from the Gram-negative bacilli, i.e. salmonellae, shigellae, *V. cholerae*, coliform bacteria, brucellae and the neisseriae, are present in the organism as components of the somatic or O antigens (p. 243). Endotoxins obtained from *Yersinia pestis* and *Bordetella pertussis* species are not of this type and have properties closely resembling those of the typical exotoxins.

Variations in virulence

As a general rule the virulence of bacteria tends to decrease on prolonged laboratory culture. It is, however, by no means safe for the bacteriologist to assume that culture of pathogenic species which have been maintained for many years in the laboratory can be handled with abandon. In the case of animal pathogenic species the virulence of strains which have been attenuated by artificial culture can usually be increased by passage in an appropriate experimental animal. It is reasonable to assume that in nature, as in the laboratory, bacterial virulence can, in the course of time, be considerably modified in the direction of either enhancement or diminution in virulence. There is reason to believe for example that over the last twenty-five years the virulence of *Str. pyogenes* has been decreasing. Quite apart from the introduction of chemotherapy, infections due to this organism appear to be much less severe now than they once were. On a longer time scale it has been suggested that the spirochaete of syphilis has shown a great diminution in virulence over the last four or five hundred years. It is possible, of course, that in this particular case modern European populations have, as a result of natural selection, somewhat higher resistance to the *Treponema* than their ancestors.

It has frequently been suggested that the development of epidemics may be due to the emergence of organisms possessing an enhanced degree of virulence. Although no laboratory evidence has been produced to support this view, many virologists believe that the great influenza pandemic of 1918–19 was due to the emergence of an influenza virus of enhanced virulence. By analogy with results obtained from experimental animals it might be thought possible that the virulence of bacteria may become enhanced during the course of an epidemic, but so far such speculation has received no laboratory support.

Lesions produced by bacteria in the body

Whether an invasion of the tissues will lead to overt infection depends not only on the virulence and numbers of the invading organisms but also on the resistance of the host. If the resistance of ths host is low, an organism of relatively low virulence may readily establish a progressive infection, whereas if the resistance of the host is high, an organism even of quite a high degree of virulence may fail to do so. The resistance of the host is known as *immunity* and will be considered in more detail in Chapter II.1

The degree of immunity a person possesses against a particular infection determines not only whether he will become infected, but also the extent to which infection progresses once it has become established. If his resistance is high, the infection will pursue a mild course and will frequently be insufficient to produce overt clinical symptoms. The latter is known as *subclinical* or *latent* infection and is of very great importance in stimulating immunity.

Most pathogenic bacteria can give rise to latent or subclinical infections and in communities in which such organisms are endemic overt clinical cases are always accompanied by variable numbers of symptomless infections which can be detected only by serological tests. Some organisms, however, are unable to do so. Outstanding examples of such organisms are *Treponema pallidum* and the gonococcus; these, when present in the body, invariably appear to produce disease.

In most bacterial diseases the main features of the disease are due to the damage produced by the bacteria in the area of the body which is the site of the initial invasion. The characteristics of this local lesion vary with the type of infecting bacterium, on its virulence, on the susceptibility of the host and on the particular tissue of the body affected. Exceptionally, as in gas gangrene, the changes produced may be of a rapidly destructive nature, providing little opportunity for a defensive response on the part of the host. In most cases, however, the damage caused by the bacteria stimulates a defensive response from the host which is known as the *inflammatory reaction*.

Some species of bacteria stimulate an *acute* inflammatory reaction, which is characterized in its early stages by vascular dilatation, a marked exudation of plasma into the tissues and the accumulation of large numbers of polymorphonuclear leucocytes. At a later stage in such reactions pus accumulates; because of this the organisms responsible for this type of lesion, e.g. *Staph. aureus* and *Str. pyogenes*, are described as *pyogenic* organisms. In addition to living and dead polymorphonuclear leucocytes and bacteria, pus contains a considerable amount of desoxyribonucleoprotein derived from the nuclei of the dead leucocytes. It is this substance which gives it its viscous consistency.

If the acute inflammatory reactions fails to check the spread of the infection, the lesion may extend locally in the tissues producing cellulitis. The lymph channels may be involved with the production of lymphangitis and, on reaching the local glands, lymphadenitis; if the infection is not arrested at this stage, the bacteria gain access to the blood stream and may as a result be widely spread throughout the body. This last condition is known as *septicaemia*.

The degree of local damage caused by the invading bacteria may not be sufficiently great to stimulate an acute inflammatory reaction and in this case the reaction is of chronic type, which may be exemplified by the response to the tubercle bacillus. In infections with such organisms the stage of leucocyte accumulation is of short duration and the inflammatory cells present are predominantly of mononuclear type.

When infection is due to organisms which produce potent exotoxins these may cause considerable damage at sites remote from the initial lesion without very much evidence of local reaction. This occurs *par excellence* in tetanus where the essential feature of the disease is the spread of toxin from an infected wound to the central nervous system.

A limited number of organisms, e.g. those of typhoid fever, relapsing fever, undulant fever and Weil's disease, produce little or no lesion at the initial site of invasion. Under these conditions there is, following invasion, bacteraemia with secondary involvement of various organs and tissues. Haematogenous spread with secondary multiplication remote from the site of initial invasion is a common feature of virus infections.

EPIDEMICS

A disease which is continuously and normally present in the community is referred to as *endemic*. A disease which only occurs when introduced from outside the community is referred to as *exotic*. A significant increase in the frequency of overt infection with an endemic disease, or the introduction of an exotic disease, constitutes an *epidemic*. Epidemics on a world-wide scale are referred to as *pandemics*. The terms endemic and epidemic should not be used in respect of animal populations for which the corresponding terms are *enzootic* and *epizootic*.

On first principles, the development of epidemics of diseases normally endemic in a population may be due to one or other of three factors.

1. A change in the resistance of the population of such an extent that the endemic pathogen can initiate an epidemic. The classic work of Topley on the production of experimental epidemics with *Salm. typhimurium* in mice showed clearly the existence of a critical density of susceptibles as a prior condition for the development of an epidemic wave. The immunity of the population as a whole is known as *herd immunity*. A high level of herd immunity for a number of diseases may be achieved by active immunization (Chapter II.8).
2. Occasionally, it is possible that an epidemic may occur without any change in the immunity status of the population, as a result of an increase in the virulence of the organisms responsible. On the whole this is probably a factor of minor importance in the initiation of epidemics.
3. The abnormal dissemination of an endemic pathogen in a susceptible population is responsible, for example, for epidemics of enteric fever, bacterial food poisoning, cholera, plague and yellow fever. Under modern conditions, epidemics of these diseases have been the result of a breakdown in public health control measures.

Seasonal factors are of considerable importance. As a general rule diseases spread by the respiratory route tend to be commoner in the colder months—increased contact indoors is probably an important factor in this but there may also, during the colder months, be some lowering in the local resistance of the respiratory tract. Diseases spread by the gastrointestinal route tend, per contra, to be more prevalent during the summer months—probably because of the greater capacity of the parasite to survive or multiply at higher ambient temperatures. Diseases spread by arthropod vectors occur maximally at a time determined by the life history of the responsible vector and/or its normal host population. However, there are considerable and unexplained differences in the seasonal prevalence of different diseases. Diphtheria epidemics when prevalent tended to occur mainly in the autumn and winter; measles shows a maximal incidence in the spring, poliomyelitis in the late summer and autumn, pertussis in the spring and early summer, and in Europe and Australia bacillary dysentery, although a bacterial gastrointestinal disease, occurs regularly throughout the year.

SOURCES OF BACTERIAL INFECTIONS

Bacterial infections can be classified as *endogenous*, due to organisms of the normal flora, and *exogenous*, due to organisms derived from a source outside the body.

Endogenous infections

These are sometimes referred to as autoinfections. The organisms of the normal flora are normally innocuous, and as we have seen, appear to be of distinct value to the host. Occasionally, however, they may give rise to infection. The most important autoinfections are from *E. coli* and *Str. faecalis* causing infections of the urinary tract, and subacute bacterial endocarditis due to the viridans streptococci. In both these examples two conditions are fulfilled: the organism initiating the infection does so in an area of the body remote from its normal habitat and the infection develops only where there is some tissue abnormality causing a lowering of local tissue resistance. Thus urinary tract infections are commonly associated with local renal abnormalities such as the presence of stone, congenital malformation, tuberculosis or back pressure through ureteric or prostatic obstruction. Subacute bacterial endocarditis occurs only in hearts damaged by previous congenital defect or previous rheumatic heart disease.

Exogenous infections

The great majority of infections are exogenous in origin. In a few instances the primary source of infection is the soil. This is the main source of infection in tetanus and gas gangrene although it is possible that the ultimate source of soil clostridia may be the intestinal tracts of animals in which these organisms are normally present. The soil is also the source of certain fungal infections (Chapter III.22).

Lower animals, either directly or indirectly, are the source of a number of infections. As might be expected, animals which live in close association with man are the greatest danger—cattle being of particular importance. Wild animals, however, especially wild rodents and monkeys, are the source of certain human diseases. They are of particular importance as constituting reservoirs of infection which are extremely difficult to control. Many of the diseases spread from wild animals to man are transmitted by insect vectors.

Man himself, however, is by far the most important source of infection for his fellows. The most obvious human source of infection is the patient suffering from overt or clinically apparent infection. Infection is also frequently transmitted by *carriers*, a carrier being defined as a person who harbours a pathogenic organism in his body but who is not suffering from overt disease. Carriers usually excrete fewer organisms than patients but since they are more in contact with the normal uninfected population and since in many infections they are more numerous than overt cases they are for many diseases equally if not more important as sources of infection. The carrier state is usual during the period of convalescence from most bacterial infections but is not normally of more than a few weeks duration. In certain instances, however, of which typhoid is an outstanding example, the carrier state may be very persistent, continuing for many years and in some cases for life. Persons who carry pathogenic organisms for long periods are known as *chronic carriers*. The term *precocious carrier* is applied to people who carry the causative organism of a disease during the incubation period of the disease before they themselves develop a clinical attack. During the course of epidemics of many infectious diseases, e.g. poliomyelitis, many become infected and carry the causative organism without developing clinical symptoms. These infections, known as *latent* or *subclinical* infections, can as a rule only be recognized by showing that the patient has developed antibody against the causative organism. Carriers of this type are frequently known as *contact carriers*.

TRANSMISSION OF INFECTION

For the spread of infectious disease from one person to another to occur three conditions must be fulfilled.

1. The organism must leave the body of the infected host. In the case of organisms responsible for lesions of the skin or those which produce infections of surface wounds this occurs very readily. Organisms infecting or present in the mouth and respiratory tract are disseminated in the environment of the patient by droplets of salivary, bronchial and nasopharyngeal secretion. The most extensive dissemination occurs through violent expulsive efforts, e.g. in coughing and sneezing. Sneezing is particularly dangerous, and it is estimated that a vigorous sneeze may expel as many as 10^5 to 10^6 bacteria-laden particles. As might be expected, sneezing is important in the dissemination of upper respiratory tract infections, while coughing is particularly important in the spread of lower respiratory tract infections. Contamination occurring from normal conversational efforts, although there are exceptions to every rule, appears to be negligible.

In diseases of the intestinal tract notably the enteric fevers, cholera, bacillary dysentery, poliomyelitis and infectious hepatitis, the causative organisms are excreted in the faeces. Some organisms—typhoid bacilli, *Leptospira icterohaemorrhagiae* and possibly a number of viruses causing systemic virus infection—are excreted in the urine. The urine, however, is not a common source of infection. In certain diseases, however, the organisms are deeply situated in the tissues and cannot spontaneously reach the exterior; most of theses diseases are spread by insect vectors.

2. The organisms must be transferred by some agency to the victim. In some cases this may be by direct physical contact. In the majority, however, some *vehicle* of infection is required. This may be the air, water or food or inanimate objects, e.g. clothing, with which the patient is in close contact and which have been contaminated by him. The latter in their role as agents of infection are referred to as *fomites*.

3. The organisms must gain entry to the body of the new host. There are four ways in which this may occur; across the placenta from the maternal to the fetal circulation, through the respiratory tract, through the intestinal tract or through the surface epithelium, viz. skin, conjunctiva or genital mucosa.

Spread across the placenta

This is not a common route of infection. It is, however, fairly frequent in syphilis, giving rise to the congenital variety of the disease, but has not been clearly established for any other bacterial infection. However, a number of virus infections, notably rubella and cytomegalic inclusion disease, are spread by this route.

Respiratory route

This is firstly the route of infection for bacteria and viruses which cause infections of the respiratory tract itself, but it is also of importance as the route of infection for certain viruses, e.g. smallpox, measles and mumps, which cause generalized or systemic infections. Except in the case of smallpox, in which infection may occur from virus present in skin lesions, these infections are caused by material expelled from the respiratory tract of a patient or carrier. Dust and dried secretions present on clothing or bed clothes or other environmental objects contaminated by the patient appear to be the main sources of infection for bacterial respiratory pathogens. Droplet nuclei, which are dried up small droplets expelled from the respiratory tract, are almost certainly of considerable importance, however, in the spread of virus infections.

Alimentary route

This is the route of infection for diseases affecting the intestinal tract—enteric fevers, bacterial food poisoning, dysentery, cholera and intestinal tuberculosis. It also appears to be the primary route of infection for the enteroviruses and for the virus of infective hepatitis. The sources of these diseases are the faeces of patients or of carriers. Alimentary infection may occur in three ways:

By the contamination of water supplies with excreta. This was at one time much the commonest method of spread of the enteric fevers and cholera. Nowadays however water-borne epidemics of intestinal diseases are very uncommon except in underdeveloped communities. Their elimination has been due to the introduction of satisfactory systems of excretal disposal, particularly by water carriage, and of modern methods of water purification.

2. By the contamination of food. Most epidemics of the enteric fevers occurring in modern urban communities are now spread in this way—the source of infection usually being a carrier involved in the handling or preparation of the food. Under more primitive conditions where there is not a satisfactory system of excretal disposal the food may be infected by flies which feed first on excreta and then transfer the organisms to the food.

3. By the contamination of environmental objects by cases or carriers either directly or by the agency of the hands. It is difficult to eliminate this type of contamination completely but it can be considerably reduced by punctilious personal cleanliness and hygiene. It is probably the method of contamination mainly responsible for the spread of poliomyelitis and of dysentery.

Infection through the body surface

Diseases in which the organism gains access through the outersurface of the body may conveniently be classified according to the precise way in which the infection is transmitted.

Direct physical contact

This group, to which the term *contagious* is applied, is best exemplified by the venereal diseases—syphilis, gonorrhoea and lymphogranuloma venereum; in these the organism is transmitted mainly by genital contact. Other important examples of contagious diseases are trachoma, herpes simplex, impetigo, the common wart, yaws and leprosy. Contagious diseases may also be transmitted, though with less certainty, by indirect contact, i.e. through the intermediary of inanimate objects contaminated by the infected individual. Infection by indirect contact is, however, extremely rare in the common venereal diseases—syphilis and gonorrhoea—because of the low viability of the causative organisms outside the body.

Blood-sucking arthropods

This mode of transmission is more frequent in viral and rickettsial than in bacterial infections. Mosquitoes transmit infections due to the arboviruses. Ticks transmit some arbovirus infections, some rickettsial infections and the tick-borne variety of relapsing fever. Lice transmit epidemic typhus, trench fever and louse-borne relapsing fever. Fleas transmit bubonic plague and murine typhus. Mites transmit scrub typhus and rickettsial pox. Sandflies transmit sandfly fever and Oroya fever.

Many arthropod-transmitted infections occur primarily in nature as infections of animals and man is only an incidental host. As a rule man is infected by the vector responsible for transmission of the disease amongst animals. Occasionally, however, a second vector is required, as in the African variety of jungle yellow fever which is transmitted amongst animals by *Aedes africanus* and from animals to man by *Aedes simpsonii*. For most insect-borne infections of this kind man is epidemiologically a dead end and further transmission amongst the human population does not occur owing to the absence of a vector parasitic on man and capable of transmitting the infection. In the case of yellow fever, however, such a vector is available and the disease once having been initiated in man is spread amongst the human population by *Aedes aegypti*. In a limited number of insect-transmitted infections man is the primary host and there is no animal reservoir of infection. This is so of epidemic typhus, louse-borne relapsing fever, trench fever and sandfly fever.

Insects normally become infected by biting a human or animal host in whose blood the causative organism is present. After this there is in most cases an interval, known as the extrinsic incubation period, during which the insect is incapable of transmitting the infection. During this time the organisms are multiplying in the insect's body. The duration of the extrinsic incubation period depends on temperature, being as a rule shorter at high than at low temperature. Ticks which transmit certain rickettsial and arbovirus infections (p. 394), and trombiculid mites which transmit scrub typhus are unusual in that the infective agent can be transmitted from one generation of insect to the next through the ovum. This is known as transovarial transmission. Whether transovarial transmission is a fortuitous by-product of the anatomy of the insect or whether it indicates that diseases which can be transmitted in this way may at one time have been solely diseases of the insect vector is a matter for speculation. In relation to the latter hypothesis the animal reservoir population has been described as an *amplifier* population. There is little doubt, however, that transovarial transmission is much less important for the perpetuation of the parasite than is the presence of an animal reservoir.

Arthropods may transmit infection in four ways:

1. The infective agent gains access to the salivary glands of the insect and the wound is directly contaminated during feeding. This is apparently the mechanism mainly responsible for the transmission of viral insect-borne infections.
2. The infective agent multiplies in the intestinal tract of the insect and during feeding is regurgitated into the wound. This is the method involved in the transmission of bubonic plague by the flea.
3. The agent multiplies in the intestinal tract and is excreted in the faeces; these are deposited by the insect beside the wound when it bites and are inadvertently scratched by the victim into the wound. This mode of infection is common in the rickettsial diseases.
4. The agent multiplies in the coelomic cavity of the insect and infection is due to contamination of the bite wound with the coelomic fluid of an insect that has been crushed in its vicinity. This type of infection occurs in louse-borne relapsing fever.

The prevalence of arthropod-borne diseases is markedly dependent on climatic conditions; these operate mainly through their effect on the size of the vector population. Climatic factors are of particular importance in relation to diseases transmitted by mosquitoes, which are virtually restricted to tropical countries. Climatic factors will also affect movements of the animal reservoir population in relation to the human population. This effect is particularly clearly seen in the case of Murray Valley encephalitis.

Contamination of wounds

The most serious of these are rabies which is due to the bite of a rabid animal, usually a dog, and gas gangrene and tetanus which are the result of the contamination of severe wounds by clostridia, usually derived from the soil. Of greater practical importance, because they are more frequent, are infections of superficial wounds and burns by staphylococci, streptococci and other pyogenic organisms. The infection of wounds—particularly of operation wounds and burns—during the course of hospital treatment has become a very serious problem since the introduction of antibiotics because the organisms responsible are frequently antibiotic resistant. In some of these cases the infection is due to air-borne spread from the infected wounds of other patients, but in many it is caused by organisms present in the nose or throat of the patient himself or of attendant nurses and doctors. *Staph. aureus* is particularly important as a cause of hospital wound infection. Its role in this type of infection is considered in more detail in Chapter III.2.

Artificial transmission by injection

It is probable that a considerable number of diseases could be spread by injection, but in practice this mode of transmission is only a risk with syphilis and homologous serum jaundice. The organisms responsible for these conditions may be inadvertently introduced by the injection of plasma, serum or whole blood obtained from an infected patient. The risk of transmitting syphilis in this way is virtually eliminated if the plasma, serum or blood is stored for at least 24 hours before it is administered. This will not, however, prevent transmission of homologous serum jaundice. Numerous epidemics of serum jaundice have also occurred through improper sterilization of syringes (p. 444).

METHODS USED TO CONTROL THE SPREAD OF INFECTIOUS DISEASE

Isolating or eliminating the source of infection

In the case of diseases spread from animals to man it may be possible to control these by destroying the infected animals, particularly if this can be followed by adequate quarantine measures to prevent reintroduction of the disease. These procedures have been successful, for example, in almost eliminating rabies from the British Isles.

Isolation of human cases of infectious diseases has long been practised as a method of controlling spread. Its value is greatest in relation to those organisms which usually give rise to overt infection.

Organisms which frequently give rise to subclinical or latent infection are less effectively controlled by isolation procedures. In certain cases, e.g. smallpox and diphtheria, a quarantine may also be imposed on contacts as an additional precaution.

Interrupting the route of spread

In all cases of infectious disease, infectious exudates and excreta should be disinfected and disposed of. For the control of diseases spread by the intestinal route the essential preventive measures are: a satisfactory system of excretal disposal, wherever possible by water carriage; the purification and protection of water supplies destined for human consumption; the pasteurization of milk; and the hygienic preparation and manipulation of food. In relation to the latter it is of particular importance that carriers of organisms which can be spread by food should not be allowed to take any part in food preparation. The control of air-borne infection is more difficult. Good ventilation which will serve to dilute out the organisms present in the air is a simple measure which might be expected to be valuable in offices, lecture halls, theatres or wherever else large numbers of people may congregate. It is probably wise to avoid crowded rooms at times when respiratory diseases are prevalent but there is obviously a limit to the extent to which social contact can be restricted in this way. Overcrowding in dormitories and hospital wards should be avoided. Ultra-violet irradiation and the spraying or vaporization of aerosols—disinfectants such as sodium hypochlorite, hexylresorcinol and triethylene glycol—have been used for the disinfection of air in offices, hospitals and doctors' consulting rooms, but their value in the control of respiratory infection is still in doubt.

The prevention of cross-infection in hospital, particularly the infection of clean surgical wounds, is a problem to which a great deal of attention has been given in recent years. These infections are known as nosocomial infections. The organisms responsible are sometimes described as opportunists. This term denotes organisms which normally live a commensal existence, but assume a pathogenic role when the opportunity arises, e.g. by the infliction of a wound, the development of an abnormality or the lowering of resistance to infection either spontaneously or through the administration of drugs such as immunosuppressives or corticosteroids.

The essential principles in the control of hospital infection are:

1. The exclusion from contact with the patient, both in the operating theatre and in the ward, of all persons with septic lesions of the skin, sore throats or who are known to be carriers of *Staph. aureus* or *Str. pyogenes*; this applies particularly to surgeons and nurses and others directly attendant on the patient. Nasal creams or sprays containing antibiotics not generally used systemically appear to be of considerable value in controlling the staphylococcal carrier state. Particularly good results have been reported for a spray containing framycetin and gramicidin. Patients with established wound infection or other infections, e.g. staphylococcal pneumonia, from which wounds might possibly become infected, should wherever possible be isolated from non-infected patients.

2. The maximum control of environmental contamination in wards and theatres. In the operating theatre this has been traditionally achieved in the first place by insisting that everyone entering the theatre should wear sterile gowns, caps and masks. Ideally, all persons permitted access to the operating theatre should be required to change into clean theatre uniforms. Unsterilized bedding should be excluded from the theatre suite and all movement into or out of the theatre, and in fact within the theatre, should be reduced to a minimum. If possible, positive pressure ventilation using filtered outdoor air should be employed to reduce the risk of aerial contamination. In hospital wards the control of dust is of particular importance. A considerable amount of the dust in a hospital ward is undoubtedly contributed by the bed clothes; the vigorous shaking of bed clothes should therefore

be avoided. At one time the oiling of blankets to prevent dissemination of dust was widely practised. According to Rubbo, however, this is not a satisfactory procedure, because, although it undoubtedly diminishes the extent of dissemination of dust and therefore of the micro-organisms contained therein, it concentrates these organisms at the surface of the blanket and increases the risk of contact infection.

3. Impeccable efficiency in, and continual surveillance of, the procedures for the sterilization of instruments and dressings. In many hospitals most of the sterilization is now carried out in a central sterile supply department.

4. The employment of a completely rigid aseptic technique both during operations and during the dressing of wounds.

5. A full investigation of all cases of hospital infection, with the object of determining the source of the infection and, if possible, how the patient became infected. In the case of staphylococcal infection, phage typing (p. 204) is particularly valuable for this purpose. The administrative complexity of the control of hospital infection is such that it is now recognized that it can only be satisfactorily achieved if it is made the responsibility of a hospital epidemiologist or infection control officer.

Increasing the resistance of the individual

It is possible to increase the resistance of the individual against a number of diseases by various immunization procedures.

In certain circumstances antimicrobial drugs may be given as prophylactics against infection, a procedure known as *chemoprophylaxis*. The administration of chloroquine to give protection against malaria is a particularly important example. The antibiotics and sulphonamides have a definite, though limited, value as prophylactics Their general use as such, however, is not justified because of the danger of increasing the prevalence of drug-resistant strains and of producing allergic sensitization.

At the time of writing the major indications for systemic chemoprophylaxis are for the protection against streptococcal infections of patients who have had acute rheumatism (p. 218), in those with severely traumatized wounds (p. 290), and as cover for certain operative procedures.

I.5

Antibiotics and Other Antibacterial Agents used in Chemotherapy

The term chemotherapy was introduced by Ehrlich to indicate the treatment of microbial disease by the administration of a drug which has a lethal or inhibitory effect on the microbe responsible, but which in therapeutic concentrations has little or no toxic action on the tissues. He described such a drug as a 'magic bullet' which, when introduced into the body, would destroy only the bacteria at which it was aimed. This requirement is not met by the disinfectants and antiseptics which were considered in Chapter I.3. These substances, although highly active against bacteria, have considerable toxic action on tissue cells.

The substances used in chemotherapy are of very diverse chemical structure. For convenience they may be divided into two categories:

1. Relatively simple compounds which were obtained in the first place by laboratory synthesis. The more important compounds of this type in general use are the sulphonamides, isoniazid (INAH), PAS and trimethoprim.
2. Antibiotics. The term antibiotic is applied to any substance produced by a living organism and which is active against other living organisms. Most are produced by soil actinomycetes.

Many of the substances now used in chemotherapy, however, are semisynthetic in that they are derived by chemical manipulation of naturally occurring antibiotics produced in the first place by fermentation processes. The object of this manipulation has been to produce compounds which are more active than the original antibiotic, which act against a wider range of organisms or which have more acceptable pharmacological properties.

The various antibacterial agents differ considerably in their range of action, that is in the number and variety of species against which they are effective. The range of action of the agents currently used is considered in some detail at the end of this chapter. As a general rule Gram-positive organisms are more susceptible than Gram-negatives; the only drug in current clinical use which is active against Gram-negatives only is polymyxin. The neisseriae share the general susceptibility of the Gram-positives. Of the Gram-negative bacilli *Ps. aeruginosa* and *Proteus* species are distinguished by their high resistance to most agents. Tubercle bacilli are also highly resistant to chemotherapeutic agents; of the drugs commonly used against other organisms they are sensitive only to streptomycin. They are, however, susceptible to the action of some drugs, e.g. PAS and INAH, which have no effect on other organisms. Some antibacterial drugs are active also against chlamydiae and rickettsiae but none have any activity against viruses. The specific problems of antiviral chemotherapy are considered in Chapter IV.3.

In respect of the type of action they exhibit against bacteria, antibacterial agents are commonly divided into two classes—bacteriostatic and bactericidal. Bacteriostatic drugs are drugs which in the concentrations attainable in the body merely inhibit bacterial growth. When the growth of bacteria is inhibited, however, they eventually die and to this extent the bacteriostatic drugs may be described as having a bactericidal effect; this effect is, however, secondary to that of growth inhibition. In contrast to the bacteriostatic drugs the bactericidal drugs have a rapid lethal action. Of the drugs in more general use chloramphenicol, the sulphonamides and the tetracyclines belong to the bacteriostatic group and the penicillins, cephalosporins, aminoglycosides, fucidin and nalidixic acid to the bactericidal. As a general rule bactericidal drugs are more effective chemotherapeutic agents than bacteriostatic drugs since they are able to sterilize the tissues by themselves. Because of this, and other things being equal, they are usually preferred. Bacteriostatic drugs are nevertheless very effective chemotherapeutic agents, though necessarily somewhat slower in action. Bacteria whose growth has been inhibited by them eventually die. But of greater importance is the fact that a bacterial population which has been controlled by the use of a bacteriostatic drug can be more readily disposed of by the tissue defences.

MODE OF ACTION

In recent years considerable advances have been made in our knowledge of the mode of action of

Table I.5. Synthesis of cell wall peptidoglycan of *Staphylococcus aureus*

1. UDP-N.Ac glucosamine

   ↓ phosphoenol pyruvate

2. UDP-N.Ac. muramic acid

   ↓ L-alanine, D-glutamic acid, L-lysine, D-alanine, ATP

3. UDP-N.Ac. muramyl-L-ala-D-glu-L-lys-D-ala-D-ala (UDP-N.Ac. muramyl pentapeptide). Formation of D-alanine and D-alanine pentapeptide—step sensitive to *CYCLOSERINE*

   ↓ P-C55 lipid, Mg^{++}

4. P-P-lipid—N.Ac. muramyl pentapeptide

   ↓ UDP-N.Ac. glucosamine

5. P-P-lipid—disaccharide pentapeptide

   ↓ glycl ~ sRNA

6. P-P-lipid disaccharide-L-ala-D-glu-L-lys-D-ala-D-ala (N on lys, with side chain gly–gly–gly–gly–gly)

   (P-P-lipid disaccharide pentapeptide-pentaglycine)

   ↓ cell wall peptidoglycan acceptor

7. P-P-lipid + (disaccharide-pentapeptide-pentaglycine)-(disaccharide-tetrapeptide-pentaglycine)$_n$. Polymerization stage, susceptible to *VANCOMYCIN* and *RISTOCETIN*. Liberated P-P-lipid cleaved to Pi + P-lipid. The reaction is inhibited by *BACITRACIN*.

8. Cross-linking of terminal glycine to penultimate D-alanine in side chain of neighbouring peptidoglycan through peptide bond formation between free glycine NH_2 and penultimate D-alanine COOH. Terminal D-alanine eliminated. This provides energy for creation of peptide bond. Cross-linking is step sensitive to *PENICILLINS* and *CEPHALOSPORINS*.

antibacterial drugs. In most cases, however, it is still far from being formulated in precise chemical terms. The problem can be considered from two aspects: identification of the site of action of the drug and its precise mechanism of action.

Site of action

With a particular drug more than one biochemical effect may frequently be demonstrated. In such cases it is difficult to determine which is the primary effect and which effects are secondary to this primary change. It may be assumed, however, that a biochemical lesion will be responsible for the death of the cell, or for growth inhibition, only if it is demonstrable in the same concentrations of drug as those which exercise an antibacterial effect. With this qualification we can distinguish the following four major loci of action.

1. Inhibition of synthesis of cell wall peptidoglycan. This is the point of action of penicillins, cephalosporins, cycloserine, vancomycin, ristocetin and bacitracin. The precise steps in synthesis with which these interfere are shown in Table I.5.

The membrane lipid carrier involved in the transfer of peptidoglycan fragments to the wall peptidoglycan is a C55 isoprene alcohol-P. During synthesis of the wall it acquires a further P group from UDP (Stage 4). Bacitracin acts by preventing release of the second P group following transfer of the peptidoglycan subunit to the growing wall, thereby preventing participation of the carrier in further rounds of biosynthesis.

2. Damage to the permeability of the cytoplasmic membrane. Such an effect can be demonstrated with tyrocidin, gramicidin, the polymyxins and the antifungal polyene antibiotics.

3. Inhibition of protein synthesis. The aminoglycosides and tetracyclines bind to and inhibit the functions of the 30S ribosome. The aminoglycosides have two effects. They cause errors in translation of certain messenger RNAs; this is demonstrable in vitro but has not been clearly established as occurring in vivo. In addition they prevent binding of amino acyl transfer RNA to the A site. The tetracyclines have a similar effect. Chloramphenicol, the macrolides and the lincomycins bind to and inhibit functions associated with the 50S ribosome. Chloramphenicol inhibits peptide bond formation. The macrolides inhibit translocation of peptidyl tRNA from the A to the P site. The precise action of the lincomycins is not clear.

4. Inhibition of nucleic acid synthesis. A number of antibiotics appear to act by inhibiting the synthesis of nucleic acids. Only two are of importance in chemotherapy. Rifampicin inhibits the synthesis of messenger RNA by its action on the RNA polymerase. Nalidixic acid inhibits DNA replication. Synthesis of nucleic acids is indirectly inhibited by the sulphonamides and trimethoprim.

Mechanism of action

Three general mechanisms by which antimicrobial drugs act may be identified:

1. Competition with a natural substrate for the active site of an enzyme. This is known as nutritional antagonism. The drug may be incorporated instead of the natural compound in a biosynthetic reaction or it may combine so firmly with the enzyme as to block further synthesis.

The classic model of nutritional antagonism is the action of sulphonamides, shown by Woods, in 1940, to interfere competitively with the utilization of *para*-aminobenzoic acid (PABA), of which they are structural analogues. Woods demonstrated that the relationship between PABA and sulphonamide is a competitive one, the inhibitory action of the sulphonamide being reversible by the addition of PABA. It was subsequently shown that, together with glutamate and pteridine residues, PABA is an essential component of the molecule of folic acid. The synthesis of folic acid from its components, and the inhibition of this synthesis by sulphonamide, have been clearly demonstrated in bacterial extracts. It is of interest, however, that while preventing the incorporation of PABA into the corresponding pteroic acid, the sulphonamide is incorporated into a similar but non-functional product. The role of folic acid in the cell has been established as that of providing derivatives which can function as carriers of one-carbon units in the synthesis of various compounds—particularly purine bases, thymine and the amino acids serine and methionine. As might be expected, these end products, as well as folic acid, are able to inhibit sulphonamide activity non-competitively.

A second analogue of *para*-aminobenzoic acid

with appreciable antibacterial properties is *para*-aminosalicyclic acid (PAS). This compound presumably acts in the same way as the sulphonamides but its spectrum of action is surprisingly different, being active only against the tubercle bacillus. Conversely, the growth of tubercle bacilli is not inhibited by sulphonamides. The precise basis for these differences remains unexplained but, presumably, the configuration of the enzyme of the tubercle bacillus responsible for PABA incorporation is such that it combines with PAS and not with the sulphonamides. Per contra, organisms whose growth is inhibited by sulphonamide presumably possess enzymes whose configuration does not permit combination with PAS. On this basis it seems possible that the insusceptibility of many bacteria to sulphonamides is due to a lack of affinity of the homologous enzyme for sulphonamide.

The lack of toxicity of sulphonamides for tissue cells is due to the fact that tissue cells require preformed folic acid, the utilization of which—unlike the incorporation of PABA—is not sulphonamide-sensitive. Appreciable amounts of folic acid and folic acid derivatives are present in animal tissues. These compounds cannot, however, be used directly by the great majority of pathogenic bacterial species. They will not substitute for the PABA requirement of the latter and therefore will not antagonize sulphonamide action.

Attempts have been made to develop other chemotherapeutic agents on the sulphonamide model, which would antagonize the action of various essential growth factors because of structural resemblance to them. A number of these growth factor analogues, e.g. pantoyl taurine and pyridine-3-sulphonic acid, have considerable in vitro antibacterial activity, but none so far has proved to be of any value as a therapeutic agent. In some cases at least this appears to be due to the fact that the antagonized growth factors have been essential factors for the host's cells as well as for the bacteria. The analogues are therefore competitively inhibited by growth factor present in the tissues and if this inhibition is overcome interfere with the metabolism of the host's cells.

Amongst the antibiotics nutritional antagonism has been shown to be the basis of the action of cycloserine, penicillin and the cephalosporins. Cycloserine is a structural analogue of D-alanine. It inhibits the formation of D-alanine from L-alanine and also the synthesis of the D-alanine-D-alanine dipeptide required in the synthesis of peptidoglycan. Penicillin has been shown by Tipper and Strominger to resemble the D-ala-D-ala portion of the peptide side chain of the peptidoglycan, the –CO–N– bond of the β lactam ring corresponding to the –CO–N– bond between the two D-alanines. As a result penicillin combines with the transpeptidase enzyme responsible for cross-linking and blocks its action. The combination of the drug with the enzyme is in this case firm and irreversible. This is also the basis of the action of the cephalosporins, which also possess the β lactam ring.

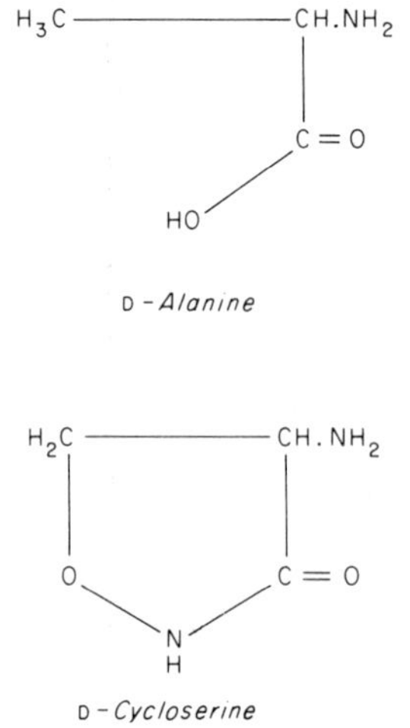

Fig. 1.11. The structural resemblance between D-alanine and D-cycloserine.

2. Combination with an enzyme at a site sufficiently close to the active site as to interfere with its enzymic function. This is possibly the basis of the action of the antibiotics vancomycin, ristocetin and bacitracin, which have not been shown to exhibit any structural analogy with any cell wall component or any intermediate in cell wall synthesis.

3. Combination with non-enzymic structural components. Into the group fall most inhibitors of protein synthesis and the drugs which act by damaging the cytoplasmic membrane.

DRUG RESISTANCE

Acquired resistance

There are two main ways in which bacteria

initially sensitive to a chemotherapeutic agent may become resistant; these are by mutation and by infective transfer.

Mutation

Resistant mutants are readily selected in the laboratory from initially sensitive strains by growing them in progressively increasing but sub-inhibitory concentrations of drug. This may be achieved as follows: a series of dilutions of the drug are made in broth, and inoculated with a sensitive organism. After incubation subcultures are made from the tube containing the highest concentration of drug permitting growth into a second similar series of tubes. A subculture is then made from the highest subinhibitory concentration of this series into a further series of tubes and so on.

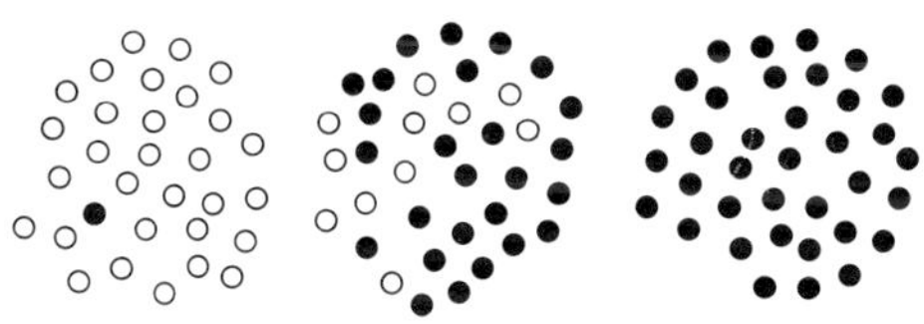

Fig. I.12. A schematic representation of the action of an antibacterial drug in selecting drug-resistant mutants. A, Original sensitive strain with a single drug-resistant mutant. B, In the presence of the drug, sensitive organisms are inhibited and resistant organisms multiply. C, Finally, resistant organisms will completely replace sensitive organisms.

When the technique described above is used, two contrasting patterns in the development of resistance are observed.

1. A high level of resistance may suddenly develop at any stage in subculture; it may even appear at the first subculture. This pattern is characteristic of the development of resistance to streptomycin and, in the case of tubercle bacilli, to INAH.

2. With penicillin, chloramphenicol, the tetracyclines and the sulphonamides, on the other hand, resistance develops slowly and a high level of resistance can be achieved only by a considerable series of transfer in progressively increasing concentrations of drug. This type of resistance appears to develop in a stepwise fashion by a small increment each time. A pattern of resistance development intermediate between these two types is shown by erythromycin and novobiocin.

There is little doubt that, in such cases, the development of drug resistance is due to mutation. These mutations occur spontaneously in cultures and independently of the presence of drug at a frequency of about one in 10^7 to 10^{10} cell divisions. The mutants have a relatively low survival value which is less than that of the parent sensitive strain; they are therefore unable to replace the parent sensitive organisms in cultures not exposed to drug. If the drug is present, however, the parent sensitive is inhibited and the drug-resistant mutants can grow out. The action of the drug is therefore not to stimulate the development of drug-resistant forms but to select them. On prolonged growth in the absence of drug, strains that have been made resistant by selection in the laboratory normally revert to drug sensitivity. This change is due to the occurrence of back-mutations. As a general rule reversion, when it occurs, is incomplete; the organism rarely, if ever, regains its original sensitivity. Different variants show considerable differences in their capacity to revert, some in fact failing to do so under prolonged observation.

There is so far little precise information as to the mechanisms responsible for drug resistance acquired by mutation. The following types of mechanism could be involved:

1. Acquisition by the organism of a capacity to destroy the drug. Some instances have been described of coliform bacteria possessing initially a low level of penicillinase activity acquiring by mutation a high level of activity. No instance has been described, however, of an organism acquiring by mutation the capacity to destroy a drug when it had initially lacked the capacity to do so.

2. Decreased permeability to the drug. There is some evidence that acquired resistance to chloramphenicol and the tetracyclines may be of this type.

3. Diminished affinity of the drug-sensitive site. This probably operates in some cases of laboratory-acquired resistance to streptomycin and penicillin.

4. Increased production of a drug inhibitor. Some organisms resistant to the sulphonamides have been described as having an increased capacity to synthesize PABA; this, however, is not

a general mechanism of sulphonamide resistance.

5. The acquisition by the organism of an altered metabolism in which the drug-sensitive reaction is bypassed. No clear example of this mechanism has been observed.

6. Increased production of the inhibited enzyme. This mechanism has been observed in mutants resistant to cycloserine in which there is increased production of D-alanine racemase and of the D-alanine-D-alanine synthetase.

Organisms may also be isolated which are not only resistant to a drug but which are dependent upon it and which will therefore grow only in its presence; such dependence has been observed in relation to the sulphonamides and streptomycin. As might be expected the growth of sulphonamide-dependent organisms is inhibited by PABA.

Infective transfer

This mechanism was first postulated in 1957, by Akiba in Japan, to explain the considerable increase in the numbers of strains of shigellae and of coliform bacteria possessing multiple resistance to chloramphenicol, streptomycin, tetracycline and the sulphonamides. The resistance exhibited by these organisms to chloramphenicol, tetracycline and sulphonamides was in each case considerably greater than that which could be expected as a result of mutation. Akiba and Watanabe subsequently showed that the increase was due to the infective transmission from resistant to sensitive strains of a genetic determinant for resistance known as a *resistance* or R *factor*. Like sex factors (p. 20) and certain colicinogeny determinants (p. 474), R factors can be transmitted by growing a resistant strain possessing them, with a sensitive recipient. Resistance is transferred much more efficiently by strains that have recently been infected with an R factor than by strains in which an R factor has been firmly established. As in bacterial conjugation and colicinogeny transfer the factor is not demonstrable in the cell-free state and can only be transferred by a conjugation-like mechanism involving cell to cell contact.

R factors consist of two parts, one containing the genes required for transfer (RTF) and the second the genes (r) which determine resistance. The two parts may be permanently integrated to give a composite molecule—the R factor—when they are transferable together in conjugation and transduction. This is normally the situation in *E. coli*. Alternatively they may occur as separate units, each organism possessing one or several resistance plasmids together with a transfer factor. In this case the transfer factor can mobilize the resistance plasmids for transfer. Some organisms possess resistance plasmids alone while others possess transfer factor alone. This can be demonstrated only by infecting a strain possessing resistance plasmids with transfer factor and transferring both to a third strain sensitive to a drug whose resistance is determined by one of the plasmids.

The following biochemical mechanisms of R factor determined resistance have been identified:

1. Enzymic breakdown of the drug as in the hydrolysis of the penicillins by β lactamases.
2. Addition of various groups—acetyl, phosphoryl, adenyl—to the drug which renders it inactive, e.g. acetylation of chloramphenicol and kanamycin, and adenylylation or phosphorylation of various aminoglycosides.
3. Diminution in permeability to the drug. An inducible resistance to tetracycline due to the production of a protein which is inserted into the membrane diminishing its permeability has been described.
4. Production of an altered enzyme with reduced affinity for the drug has been found in cases of resistance to the sulphonamides and trimethoprim.
5. Alteration of drug target. Methylation of 23S ribosomal RNA has been identified as the basis of erythromycin resistance in *Staph. aureus*.

Resistance determined by R factors has been described in respect of most, if not all, of the drugs in common use. R factors are widely distributed among the Gram-negative enteric bacteria: *Escherichia, Klebsiella, Proteus, Salmonella, Shigella*, etc. Transmission has also been reported to *V. cholerae* and *Yersinia pestis*. They cannot be transmitted to Gram-positive bacteria. However, similar plasmids controlling resistance to a variety of drugs are found in staphylococci, neisseriae and *Streptococcus faecalis*. Staphylococcal plasmids are not spontaneously transmissible but are transmissible by transduction by certain bacteriophages. Recently several determinants have been found to be translocatable, i.e. capable of insertion into the bacterial chromosome or into the DNA of other plasmids.

The dissemination of bacterial strains carrying R factors is obviously favoured by the large scale clinical use of antibacterial agents. In this connection, the recovery by Anderson of strains of *Salm. typhimurium* carrying R factors from calves and pigs, is of the greatest importance. Control of the dissemination of these agents must therefore involve not only close attention to drug policy in respect of the human population, but also for domestic animals which constitute a reservoir of such agents, extremely difficult, if not impossible, to control.

The persister state

The name *persister* was originally applied by Bigger to staphylococci which had temporarily acquired resistance to penicillin. It was found that if staphylococci were mixed with high concentrations of penicillin most of the organisms were killed within 24 hours, but a few survived; when the penicillin present was destroyed by penicillinase the survivors sooner or later multiplied. Persisters are believed to be organisms in a dormant or non-metabolizing state in which they are insusceptible to the killing action of the drug. This state of dormancy may be the result of the bacteriostatic action of penicillin. Alternatively, a small proportion of any penicillin-sensitive bacterial population may, at any given time, be in the persister state. The duration of the period of dormancy varies greatly from one organism to another. In some it may last only a few hours, in others for days, in a few for more than a week. As soon as a dormant form resumes activity it becomes sensitive to penicillin and is killed. Penicillin in relation to persisters acts in the same way as a trained police dog in relation to a fugitive. So long as he remains quiet he is not attacked, if he displays any activity it attacks him. Persisters are probably of greatest importance in the chemotherapy of tuberculosis.

Drug resistance as a clinical phenomenon

Since drug resistance can be produced in the test tube, it is to be expected that it will occur also in the bodies of patients who are under treatment. As might be expected it occurs very readily in relation to streptomycin and INAH. In patients under treatment with these drugs organisms may rapidly become completely resistant, due to the selection by the drug of mutants having a high level of resistance. As in the laboratory, resistance may develop regardless of the amount of drug used. With penicillin and other drugs exhibiting the multi-step resistance pattern, however, resistance rarely appears to develop during the course of treatment in organisms originally sensitive provided that adequate amounts of drug are administered. This probably applies also to other drugs showing the same type of in vitro resistance pattern.

It is probable that many cases in which resistance appears to develop during the course of chemotherapy, particularly with drugs showing the multi-step resistance pattern, are due not to the acquisition of resistance by an originally sensitive organism but to replacement or substitution infection. This is a process in which the sensitive organisms initially responsible for the infection are inhibited but the lesion becomes reinfected by resistant organisms present in the environment. It is most likely to occur with surface lesions such as wounds and burns which are directly exposed to environmental organisms. The risk is greatest with patients treated in a hospital environment where resistant organisms are lurking in abundance.

When a patient is treated with an antibacterial drug the drug acts not only on the organisms responsible for his infection but also on the normal flora of his respiratory and intestinal tracts. If this action is sufficient to cause a substantial reduction in the normal flora a bacteriological vacuum is created which tends to become populated with drug-resistant strains. Usually this is without serious consequence for the patient although it may render him a danger to others since he becomes a focus from which resistant organisms may be spread. Occasionally, however, resistant organisms with pathogenic propensities may become established and may initiate an infection. This process is known as superinfection. The most serious infection of this type is the acute enterocolitis which may complicate treatment with the wide-spectrum antibiotics. This is not a common complication of antibiotic therapy but is an extremely grave one. It is most frequently caused by *Staph. aureus*, but *Proteus, Pseudomonas* and other Gram-negative bacilli are sometimes involved. The organisms responsible are frequently resistant not only to the drug being administered but also a number of other antibacterial drugs as well. In many but not in all cases this complication has followed abdominal operation.

Another type of superinfection that has assumed clinical importance is that due to overgrowth of *Candida albicans*—a yeast-like fungus which is a normal inhabitant of the respiratory and alimentary tracts. Following the administration of wide-spectrum antibiotics, especially the tetracyclines, extensive overgrowth of *C. albicans* has on occasion been found in the intestinal, respiratory and urogenital tracts. In exceptional cases the fungus invades the blood stream with the development of metastatic abscesses in the kidney, liver and brain. This complication has also been reported following the administration of penicillin, a drug which appears in fact to have a direct stimulating effect on the growth of *Candida. Candida* infections, as a result of antibiotic treatment in hospital, are being increasingly reported.

Since antibacterial agents can transform sensitive strains into resistant ones and can select for resistant strains, however produced, it is not surprising that, when a drug has been used in a community for some time, the relative proportion of organisms resistant to it should increase. This has occurred in relation to all the agents now in common use, but not to the extent that was at first feared—the increased prevalence of resistant strains having fortunately affected only a few species. The problem is greatest with *Staph. aureus*, many strains of which are now resistant to penicillin, streptomycin and the tetracyclines—the antibiotics which have been most used—and with the enterobacteriaceae as a result of the spread of R factors. It is reported that a small proportion of tubercle bacilli isolated from fresh cases of tuberculosis are by laboratory tests appreciably resistant to streptomycin or INAH and clinical evidence suggests that there has been some increase in the resistance of viridans streptococci and of gonococci to penicillin. There is no evidence of any significant increase in the resistance of other pathogenic species. The fact that the pathogenic staphylococci, organisms which are frequently present in the upper respiratory tracts of healthy individuals, and coliform bacteria, which are intestinal commensals, are the main offenders in this respect suggests that exposure of the respiratory and intestinal flora to drug, an exposure which occurs to some extent every time an antibiotic is administered, is of special importance as a factor in the emergence of resistant strains.

The control of drug resistance

In the control of drug resistance three distinct problems have to be considered:

1. The prevention of the development of resistance in initially sensitive organisms.
2. The prevention of replacement infection with resistant strains present in the patient's environment.
3. The reduction of the apparently inevitable increases in the prevalence in the population as a whole of resistant strains of *Staph. aureus* and coliform bacteria.

In order to prevent the development of resistance in organisms originally sensitive the first principle of chemotherapy is to ensure that an adequate amount of drug is administered. No procedure is more calculated to favour the development of resistance, particularly of the multi-step penicillin type, than half-hearted inadequate treatment. Although the maintenance of an adequate drug concentration in the tissues may be expected to minimize the chance of resistance developing to drugs showing the multi-step resistance it will not do so in the case of drugs showing the single-step resistance pattern especially if treatment with these drugs is prolonged. With these compounds, however, the development of resistance may be diminished by the use of combined therapy. The rationale of combined therapy is that if any organisms are to survive combined treatment they must become resistant to both drugs employed and the probability that they would do so is very much less than the probability that they would become resistant to either drug if given alone. Theoretically the probability of the development of resistance to two drugs given at the same time should be the product of the independent probabilities for each of them. The value of combined therapy in preventing the development of resistance has been clearly shown in the treatment of tuberculosis, in which resistance develops much less frequently if patients are given a combination of two antituberculous agents, e.g. streptomycin and INAH, than if they are given either alone.

The problem of preventing replacement infection in surface lesions, e.g. wounds and burns, is essentially that of the control of cross-infection in general. In cases, e.g. urinary tract infection, in which there is an underlying lesion—stone, congenital defect, prostatic obstruction—which predisposes to infection, this lesion must be dealt with. If this is not done, when one organism has been eliminated by chemotherapy the patient may simply become reinfected with another organism resistant to the drug previously used.

Ultimately the most important of the problems in connection with bacterial resistance is the more general one of preventing increases in the prevalence of resistant strains. Since the prevalence of resistant strains is a direct result of the widespread use of antibacterial drugs, this can be achieved only by ensuring that antibiotic and chemotherapeutic agents are used as little as possible. They should not be given for trivial infections from which the patient would certainly recover. They should not be casually administered as prophylactics (p. 58). They should not, except in emergency, be prescribed blindly in the absence of a bacteriological diagnosis or of a reasonable expectation that the organisms are sensitive. Surface lesions should be treated, if possible, by topical application of drugs not normally used systemically, e.g. bacitracin and neomycin, and only in severe surface infections should systemic treatment be employed.

ANTAGONISM AND SYNERGISM

Antibacterial drugs are frequently given to patients in various combinations not only with the object of preventing the development of resistance, as in the treatment of tuberculosis, but also to achieve a synergistic effect, that is, an effect greater than that which could be achieved with either drug alone.

It is generally agreed that the use of combined therapy is indicated in the circumstances mentioned below:

1. In the case of infections due to organisms of relatively low susceptibility to any of the available antibiotics, e.g. the treatment of infections due to *Str. faecalis* with a combination of penicillin and streptomycin or of *Ps. aeruginosa* infections with a combination of polymixin and carbenicillin. In these circumstances combined therapy should be undertaken only when its potential value is indicated by laboratory sensitivity tests.

2. In mixed infections due to organisms of differing sensitivities, e.g. in combined infection with a streptococcus and a Gram-negative bacillus.

3. In grave undiagnosed infection where combined therapy may be necessary as a life-saving measure.

4. In certain empirically determined conditions in which it has been shown that the combined therapy gives the most satisfactory therapeutic result, e.g. in the treatment of subacute bacterial endocarditis due to viridans streptococci (Chapter III.3).

5. To prevent the development of resistance when antibiotics showing the single step high level resistance pattern are employed.

6. Where a specific combination is markedly synergistic. This is exemplified by the combination of trimethoprim and sulphamethoxazole to form co-trimoxazole.

Under in vitro conditions, however, it has been found that certain antibiotics may antagonize the actions of others and this has naturally given rise to the fear that a similar antagonism might occur during the course of treatment. The antagonisms observed in vitro have, however, been of limited nature. Thus the mainly bacteriostatic drugs, in particular chloramphenicol and the tetracyclines, have been found to inhibit the bactericidal action of penicillin and streptomycin. This antagonism has been found only against organisms highly sensitive to the bactericidal drug and only with certain drug concentrations. The mainly bacteriostatic drugs do not inhibit each other, nor do they affect the bacteriostatic action of penicillin and streptomycin. Penicillin and streptomycin have not been found to antagonize each other and may in fact show synergism. The antagonism of bactericidal by bacteriostatic drugs has been observed not only in the test tube but also in the treatment of certain experimental infections in animals. There is, however, insufficient evidence to indicate to what extent antagonism occurs as a result of combined therapy under clinical conditions.

SENSITIVITY TESTS

These are tests in which the bacteriologist attempts to determine whether an organism is sufficiently sensitive to an antimicrobial drug to permit the use of that drug in treatment. In the tests normally used for this purpose it is the bacteriostatic effect of the drug which is investigated. We determine the extent to which the growth of the organism is inhibited by the drug and on the result of this examination we base an opinion as to whether the drug can be effectively used in treatment.

Although sensitivity tests are of great value in the control of treatment the decision as to which drug or drugs to use should be based in the first place on clinical diagnosis supplemented, where possible, by simple diagnostic laboratory tests. In most types of infection the establishment of a bacteriological diagnosis immediately indicates the appropriate chemotherapeutic agent to use, and sensitivity tests on the causative organism are not normally required. However, in pyogenic infections due to species such as *Staph. aureus*, coliforms, etc., which are not uniformly sensitive, sensitivity tests are essential, although treatment should normally be begun without waiting for the result of these tests.

For clinical purposes sensitivity tests are usually carried out by the agar plate diffusion technique. When carried out on organisms already isolated in pure culture the following technique may be used. The surface of a blood agar plate is flooded with a broth culture or with a suspension in broth of the organism (or with a dilution of these). The excess fluid is pipetted off and the surface of the medium allowed to dry. Discs or tablets impregnated with antibiotic are then placed on it and the plates incubated with the lid uppermost. Multidiscs, in which discs containing individual antibacterials are attached by a stem to a common central disc containing no antibacterial, are now widely employed for sensitivity tests. After incubation the diameter of the zones of inhibition surrounding the discs or tablets is measured. By determining the size of the zones obtained with known sensitive organisms the bacteriologist can fix standards by reference to which the sensitivity of any organism can be graded. The data obtained do not sufficiently closely parallel clinical results to warrant attaching any great significance to relatively small differences in zone diameter and a comparatively simple grading of organisms as into 'sensitive', 'moderately resistant' and 'resistant' is all that is required. The diameter of the inhibition zone depends not only on the sensitivity of the organism but also on the concentration of antibiotic in the disc or tablet used and on its rate of diffusion. It is consequently not permissible to base any opinion on the relative efficacy of different drugs against a particular organism on differences in zone size except in so far as these permit a sensitivity grading such as that indicated above. When organisms are reported as being within an apparently sensitive range the clinical choice of drug should be made on other grounds.

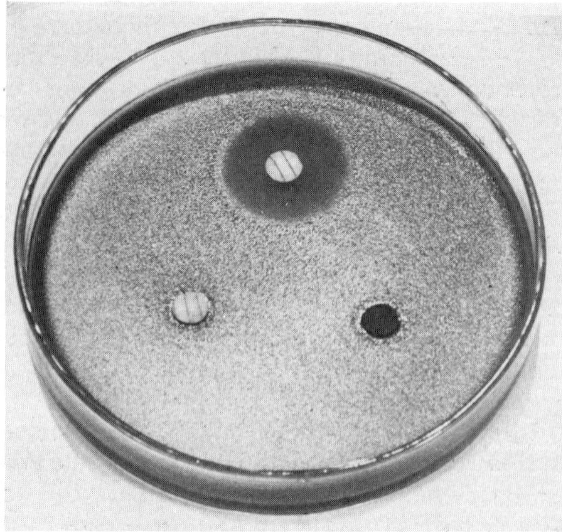

Fig. I.13. An antibiotic sensitivity test using impregnated tablets. Sensitivity is indicated by a zone of inhibition round the tablet.

Sensitivity tests on organisms that have first been isolated in pure culture may not, however, give a completely reliable picture of the sensitivity of the organisms in the original material. This is because the culture tested is obtained by random picking of a single colony from the isolation plate and this colony may differ in sensitivity from the majority of the organisms present. In addition indirect tests are slow, since they require prior isolation of the organism. More rapid and therefore clinically more useful results are obtained if the tests are carried out directly, i.e. using the clinical material (e.g. pus) as an inoculum. Direct tests also give a more reliable picture of the sensitivity of the organisms present, since sampling errors are avoided; they have the disadvantage, however, that they are appreciably less accurate.

The sensitivity of tubercle bacilli to streptomycin, INAH and PAS may be determined by inoculation on to slopes of Löwenstein medium containing various concentrations of drug. The sensitivity is commonly expressed as a ratio of the concentration inhibiting growth to that inhibiting growth of a known sensitive strain, e.g. H37Rv.

If the bacteriologist is required to investigate the possible antagonistic action of one drug on another in relation to a particular organism, tests indicating bactericidal rather than bacteriostatic activity must be employed. In one type of test used for this purpose dilutions of the drugs under test, singly and in various combinations, are made in broth and inoculated with the organism. After overnight incubation subcultures are made on to blood agar from each of the tubes inoculated to determine whether the organisms introduced into it have been killed. If one of the drugs has an antagonistic effect on the other, their combination should be less efficient in killing the organism than the more actively bactericidal of the drugs by itself. Tests of this type are cumbersome and ideally require the testing of a large number of antibiotic concentrations. Whether the information gained is of significant clinical value is doubtful.

Some indication of the existence of synergism between two antibiotics can be obtained by the use of comparatively simple plate tests. For this purpose discs containing the antibiotics are placed on the surface of a seeded plate at a distance apart approximately equal to that of the sum of the diameters of the zones of inhibition produced by each antibiotic independently. If one of the antibiotics potentiates the action of the other the zone of inhibition produced by the latter will be increased. Antagonistic effects are much less readily demonstrable by this technique.

ASSAY OF ANTIBIOTICS

Methods similar to those that have been described for sensitivity testing can be used for assay of the concentration of an antibacterial drug present in various clinical fluids: blood, cerebrospinal fluid, urine, etc. As in sensitivity testing, plate diffusion techniques, although less precise than broth dilution techniques, are technically simpler. The filter-paper disc technique may be used but the 'cup' method is more frequently employed. In this the dilutions of the antibiotic and of the fluid being tested are pipetted into shallow glass or porcelain cylinders placed on the surface of a plate which has been seeded with a known sensitive organism. In a simple but quite satisfactory modification of this technique the dilutions of fluid and antibiotic are pipetted into small holes cut out of the surface of the agar with a cork borer. From the inhibition zones obtained with the known antibiotic dilutions a curve, which relates inhibition zone diameter to antibiotic concentration, can be drawn. If the inhibition zone produced by one of the dilutions of the fluid being examined falls on this curve, the antibiotic concentration present in that dilution can be read off immediately.

LABORATORY USES OF ANTIBIOTICS

Antibiotics have been used in the laboratory for a variety of purposes. Of these the undermentioned are the most important for diagnostic medical bacteriology:

1. They may be incorporated as selective agents in culture media. Thus penicillin may be used to facilitate the isolation of *H. influenzae* from material taken from the upper respiratory tract since it inhibits most of the Gram-positive organisms and the neisseriae present. Neomycin, as in Willis and Hobb's medium, is of value for the isolation of clostridia, particularly of *Cl. perfringens*, from material, e.g. faeces or food, contaminated with other organisms.

2. Antibiotics are universally employed for the control of bacterial contamination in tissue cultures used for virus isolation. The antibiotics normally used are penicillin and streptomycin for the control of bacterial contamination and nystatin for the control of fungal contamination.

3. The pattern of sensitivity of an organism to a battery of antibiotics constitutes a simple method of typing which has been found to be of considerable epidemiological value. It may, for example, be used as a preliminary to phage typing in the investigation of epidemics of staphylococcal infection. The strain isolated from the putative

source of the infection should have the same antibiotic sensitivity pattern as that isolated from the infection. The pattern of sensitivity of an organism to different antibiotics is sometimes known as an *antibiogram.*

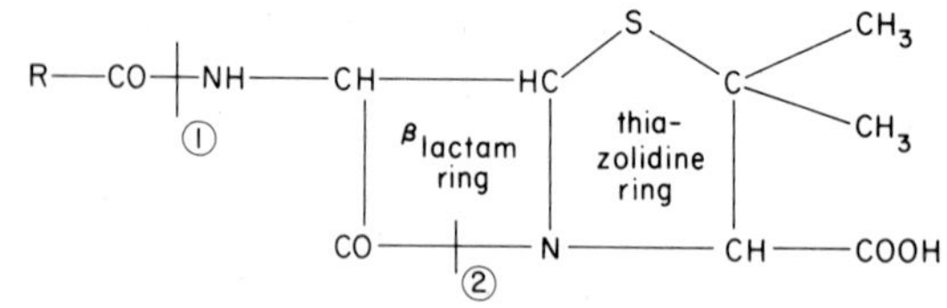

Fig. I.14. The structure of the penicillins. The peptide bond (1) is hydrolysed by penicillin amidase (acylase) to yield 6-amino-penicillinase (6-APA) which has little antibacterial effect. Penicillin amidase, produced by some Gram-negative bacteria, is used in the production of semisynthetic penicillins. It plays little if any part in bacterial response to penicillin. The peptide bond (2) is hydrolysed by penicillinase to yield penicilloic acid. Susceptibility depends on the nature of the side group. Penicillinase is the basis of penicillin resistance in *Staph. aureus* and many Gram-negative bacilli.

SOME ANTIBIOTICS IN CURRENT USE

Antibiotics may be conveniently classified on the basis of the range of bacteria against which they are active.

1. Antibiotics mainly or exclusively active against Gram-positive organisms. The antibiotics falling into this group are penicillin G and V, methicillin, cloxacillin, erythromycin, novobiocin, vanomycin, bacitracin and fucidin. Except for vancomycin these compounds are also highly active against the pathogenic neisseriae—the gonococcus and meningococcus—and, to a lesser extent, *H. influenzae* and *Bordetella pertussis*. Of the Gram-positive group the enterococci are the most resistant, particularly to the less active penicillins—methicillin and cloxacillin.

2. Antibiotics mainly or exclusively active against Gram-negative organisms. The important compounds in this group are the aminoglycosides and polymyxin.

3. Antibiotics which are highly active against both Gram-positive and Gram-negative organisms. This group includes the so-called 'wide-spectrum' antibiotics—the tetracyclines, chloramphenicol, ampicillin and the cephalosporins.

4. Antibiotics active against fungi. These are considered on p. 324.

The penicillins

The penicillins constitute a family of antibiotics, the first member of which was discovered by Fleming in 1929. It was not at that time possible, however, to obtain it in a sufficiently concentrated and purified form for clinical use. This was subsequently achieved by Chain, Florey and their associates in 1941.

Penicillins consist of a nucleus of 6-aminopenicillanic acid (6-APA) to which are attached various side-chains in amide linkage. Penicillin G and penicillin V are produced entirely by fermentation processes using a mutant strain of *Penicillium chrysogenum* which is capable of producing large amounts of penicillin in submerged culture. At an early stage it was found that the *Penicillium* mould was capable of incorporating a variety of substituted acetic acids into the side-chains of the penicillin molecule. The addition of such compounds therefore became standard practice in penicillin production, penicillin G being produced in large amounts following addition of phenylacetic acid and phenoxymethyl penicillin in the presence of phenoxyacetic acid. The next major advance was the discovery by Chain, Bachelor et al. that the penicillin nucleus, 6-APA, was produced in appreciable amounts in fermentation media in the absence of added precursor. From this it was but a short step to the semisynthetic penicillins—methicillin, ampicillin and the isoxazolyl penicillins—in which various side-groups are incorporated by chemical manipulation. In the production of these, however, it has been found more convenient to obtain the parent compound, 6-APA, by treating benzyl penicillin with penicillin amidases (or acylases) of bacterial origin.

Penicillin G fulfils many of the requirements of an ideal chemotherapeutic agent. Active against sensitive organisms in extremely low concentration, it is bactericidal and appears to be without toxicity in the highest concentrations attainable in the tissues. Its defects are its limited range of action, its lability to acid so that it cannot be administered orally, its high susceptibility to bacterial penicillinases and its capacity to induce allergic sensitization. It is these defects that the more recent pencillin derivatives have attempted to rectify. Two of the newer penicillins, methicillin and cloxacillin, are resistant to staphylococcal penicillinase. Ampicillin and cloxacillin are acid-resistant and can therefore be administered orally, and ampicillin and carbenicillin exhibit a wide spectrum of activity embracing many Gram-negative bacteria. Of the five compounds shown in Table I.6, benzyl penicillin and ampicillin possess much the same order of activity against Gram-positive organisms. Methicillin and cloxacillin are appreciably less active but have the enormous advantage that they can be used against penicillinase-producing staphylococci.

Penicillinase

Penicillin is broken down by two types of bacterial enzyme: (*a*) penicillinase, or more correctly β-lactamase, and (*b*) penicillin amidase or acylase. The action of these enzymes on penicillin is shown on p. 10.

Penicillinases are produced by naturally resistant strains of *Staph. aureus*, by certain *Bacillus* species and by a variety of enterobacteriaceae, particularly of the coliform and *Proteus* groups. It is usually demonstrated by a biological assay procedure which determines the capacity of a bacterial culture to destroy the inhibitory activity of penicillin against a known sensitive organism,

Table I.6. Some important penicillins

Name	*Side chain (R)*	*Acid stability*	*Resistance to staphylococcal penicillinase*	*Active against Gram-negative bacteria*
Benzyl penicillin (penicillin G)	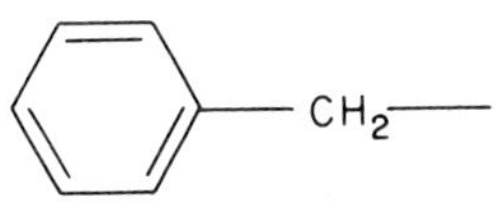	–	–	–
Methicillin (Celbenin)	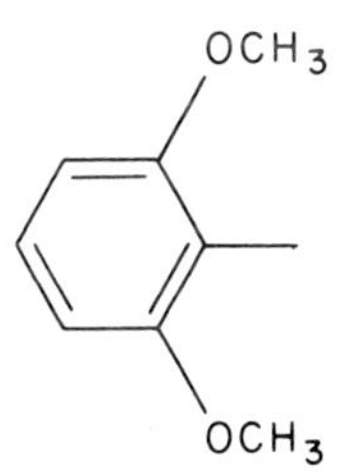	–	+	–
Ampicillin (Penbritin)	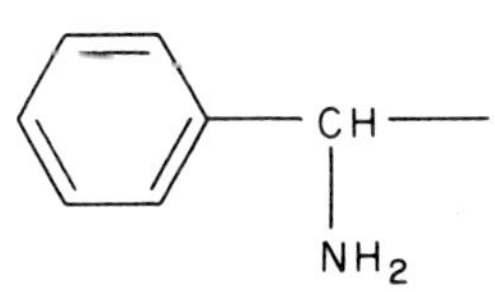	+	–	+
Cloxacillin (Orbenin)	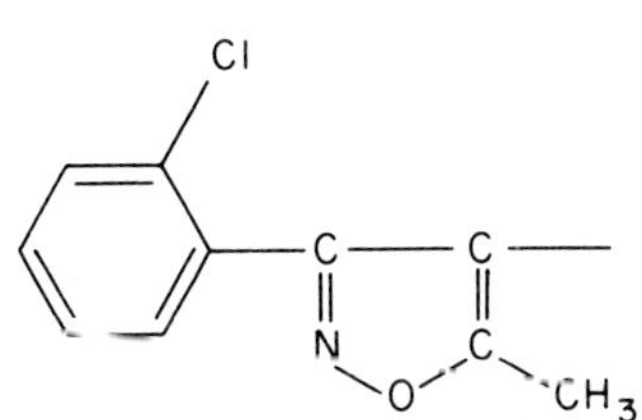	+	+	–
Carbenicillin (Pyopen)	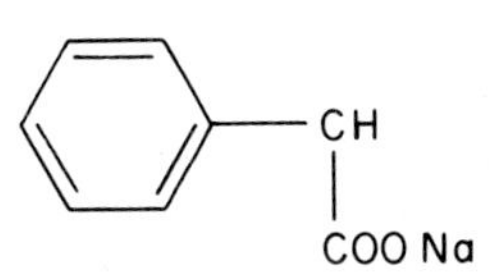	–	–	+ Active against *Pseudomonas*

e.g. the Oxford staphylococcus. As normally carried out, such bioassays usually reflect β-lactamase activity but decisive proof that the agent responsible is a β-lactamase rather than a penicillin amidase requires the demonstration of penicillin acid production, e.g. by chromatography or by the iodometric technique of Perret.

From the clinical point of view the most important of the β-lactamases is that produced by *Staph. aureus*. It is produced by all naturally resistant strains of *Staph. aureus* and is responsible for their resistance. Without penicillinase such strains are intrinsically sensitive to penicillin. This may be demonstrated in sensitivity tests with dilute inocula, when the sensitivity of penicillinase-producing strains approximates that of non-penicillinase-producing strains. When, however, large inocula are employed the organisms appear resistant because of the presence of larger amounts of penicillinase sufficient to destroy the penicillin present.

Staphylococcal penicillinase is inducible, normal wild-type resistant strains producing it in significant amounts only in the presence of penicillin. Non-inducible mutants have however been isolated in the laboratory, and these produce considerable amounts of the enzyme constitutively. Induction of the enzyme is currently explained on the Jacob-Monod hypothesis (p. 24). Inducing capacity is a property of 6-APA and all its known derivatives, including compounds such as methicillin and the isoxazolyl penicillins, which are highly resistant to penicillinase action. It seems very unlikely therefore that it would be possible to develop a penicillin derivative with high antibacterial activity but having little capacity to induce penicillinase.

As previously indicated, the genes controlling penicillinase production in *Staph. aureus* are carried on extrachromosomal genetic elements or plasmids. These may be transduced by bacteriophages from one cell to another, but differ from the resistance transfer factors of the enterobacteriaceae in not being capable of spontaneous transfer.

The penicillinases of the Gram-negative bacteria are a diverse group of enzymes. Those of the coliform bacteria differ from staphylococcal penicillinase in being non-inducible, and in having a more specifically intracellular location. They differ amongst themselves in their activity towards different penicillins, most having much the same order of activity against both penicillin and ampicillin, some being more active against penicillin than against ampicillin, and others, especially in *Klebsiella* strains—being more active against ampicillin than against penicillin. Some Gram-negative bacteria produce penicillinase active against the isoxazolyl penicillins, e.g. cloxacillin, which are highly resistant to staphylococcal penicillinase.

Since on the whole penicillin- and cephalosporin-producing fungi and penicillinase-producing bacteria share the same natural environments, it is reasonable to presume that the capacity to produce penicillinase has evolved in penicillin-sensitive bacteria as a form of protection against penicillins and cephalosporins produced in their environment. That this evolution has not been a recent event is indicated by the observation of Pollock that spores of *Bacillus licheniformis* which had remained dormant on dried plants held in the British Museum since 1689 germinated into vegetative organisms which were active producers of penicillinase. Pollock found also that the penicillinase they produced showed biochemical and immunological properties identical with those of penicillinase produced by *Bacillus* species of the present day. He suggests that penicillinase production may have risen by a mutation affecting the cross-linking enzyme involved in the synthesis of the cell wall (p. 60) whose activity is inhibited by penicillin.

Penicillin resistance of Gram-negative bacteria

Although possessing a peptidoglycan component in their cell walls, Gram-negative bacilli are intrinsically much less susceptible to benzyl penicillin than are Gram-positive bacteria. This could be due to a diminished affinity of the transpeptidation enzyme (p. 60) for penicillin, or to interference by lipid and/or other non-peptidoglycan cell wall components with access of penicillin to the enzyme site. A possible role for the lipid is suggested by the finding that the sensitivity of various Gram-positive bacteria is decreased under conditions leading to accumulation of lipids. The introduction of an amino group into benzyl penicillin, to yield ampicillin, results in considerably increased activity against Gram-negatives. In the case of non-penicillinase-producers this increase must presumably be achieved by an increase in the capacity of the drug to reach or to adsorb to the transpeptidation enzyme. Many coliform bacteria and other enterobacteriaceae, however, are resistant to ampicillin. In most cases this appears to be associated with a β-lactamase active against ampicillin. But there is also evidence that certain Gram-negative bacteria may have an intrinsically high level of ampicillin resistance, quite apart from penicillinase production. The role of penicillin amidase in the resistance of Gram-negatives to the penicillins is in doubt. Some Gram-negatives produce this enzyme, which does not, however, appear to make a significant contribution to resistance under normal growth conditions.

The cephalosporins

The first member of this family of antibiotics, which possess the β-lactam ring structure of the penicillins and which apparently have an identical mode of action, cephalosporin C, was identified by Newton and Abraham in 1953 as one of a number of antibiotics produced by the mould *Cephalosporium acremonium*. This mould had been originally isolated by Brotzu from a sewage outfall in Sardinia in 1945, in a deliberate attempt to isolate an organism producing an antibiotic active against the typhoid bacillus. In addition to cephalosporin C, the mould produces two other components of interest—cephalosporin P, an acidic steroid structurally related to fucidin, and cephalosporin N. Cephalosporin N is in fact a penicillin with an α-amino-adipic acid side-chain. It shows appreciable activity against various Gram-negative bacteria but is considerably less active than ampicillin. Cephalosporin C consists of a nucleus of 7-amino-cephalosporanic acid (7-ACA), which consists of fused β-lactam and dihydrothiazine rings, and an α-amino-adipic acid side-chain. Its potential value was indicated by its resistance to staphylococcal penicillinase and its activity against various Gram-negative organisms, but its antibacterial action was in general much less than that of other penicillins.

Clinical interest has centred mainly on various semi-synthetic derivatives of 7-amino-cephalosporanic acid. These include cephalothin and cephaloridine which are not absorbed from the alimentary tract and must be given by injection. Cephalexin and cephaloglycin are derivatives which can be administered orally. Cephalosporins are highly resistant to staphylococcal penicillinase but susceptible to the penicillinases of most Gram-negative bacteria—though apparently much less sensitive to some than benzyl penicillin. On the other hand, certain *Proteus* and *Klebsiella* strains have been found to produce β-lactamases which are highly specific for the cephalosporins. Whether β-lactamase and cephalosporinase activity is, however, the function of distinct enzymes, is not at the moment clear.

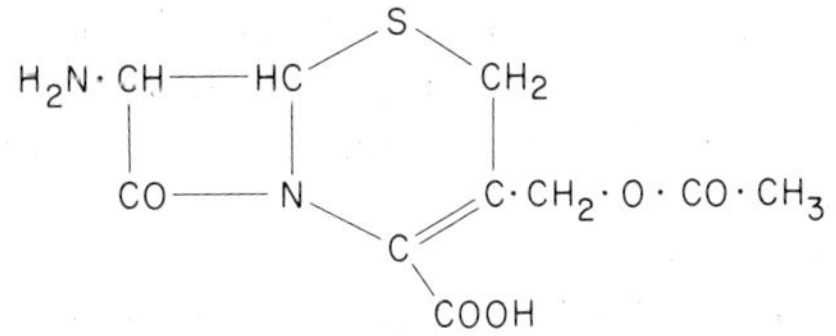

Fig. I.15. The structure of 7-amino-cephalosporanic acid.

A particularly valuable feature of the cephalosporins is that they can be used safely in cases of allergy to penicillin. Whether they can themselves induce sensitization comparable with that produced by penicillin remains to be seen.

The aminoglycosides

Streptomycin

Streptomycin produced by *Streptomyces griseus* was discovered by Waksman in 1944 as a result of an intensive search, prompted by the discovery of penicillin, for chemotherapeutic agents of fungal origin. It is a highly basic compound consisting of the base streptidine, an unusual sugar, streptose and N-methyl-L-glucosamine.

Streptomycin is highly active against the common Gram-negative bacillary pathogens with the exception of *Proteus* species and *Ps. aeruginosa*—organisms which possess a high natural resistance to all the major antibiotics active against Gram-negative bacilli. It is also active against *Staph. aureus, B. anthracis, Actinomyces israelii* and the neisseriae. It has little or no activity against streptococci, pneumococci, clostridia, spirochaetes, rickettsiae or chlamydiae. Undoubtedly the most important and most valuable feature of streptomycin is its high degree of activity against the tubercle bacillus.

Streptomycin is not absorbed from the alimentary tract. It has little power of penetration into the cerebrospinal fluid; its capacity to penetrate into serous cavities is about the same as that of penicillin.

A major difficulty in the use of streptomycin is the readiness with which organisms become resistant to it during the course of treatment. The more prolonged the treatment the greater is this risk. Consequently when streptomycin is used in the treatment of tuberculosis, for which treatment is necessarily prolonged, it must be combined with some other antituberculous drug. The development of resistance may not be a serious difficulty in the treatment of acute coccal and bacillary infections if the course of treatment is of short duration. A further disadvantage in the use of streptomycin is its toxic effect on the eighth nerve.

Neomycin group

This group includes four antibiotics structurally related to streptomycin—neomycin, kanamycin, paromomycin and framycetin. They have a similar range of action but differ in that bacterial resistance to them develops more slowly and in their greater toxicity for the eighth nerve and the kidney. Only kanamycin possesses sufficient marginal safety for systemic use.

Gentamicin

Gentamicin differs from other aminoglycosides in its high activity against *Ps. aeruginosa*. It is sometimes used in the treatment of *Pseudomonas* infections in combination with carbenicillin. It has appreciable toxicity.

Chloramphenicol

This compound, first isolated by Ehrlich and his associates—from an actinomycete, *Streptomyces venzuelae*—in 1947, when it was known as chloromycetin, is now produced by laboratory synthesis. A number of derivatives have been prepared by chemical manipulation but none has shown any improvement in antibacterial action or has remedied the main defect of chloramphenicol, its toxicity to haemopoietic tissues. Chloramphenicol has a wide spectrum of action against both Gram-positive and Gram-negative bacterial species, although it appears to be somewhat less active against Gram-positives than the tetracyclines. Its use as a chemotherapeutic agent has been considerably restricted because of its capacity to cause aplastic anaemia, a complication which has nevertheless been observed only in a small number of cases. As a result of this, it is generally recommended only for the treatment of typhoid fever and *H. influenzae* meningitis, conditions for which it is currently accepted as the drug of choice. Toxicity apart, chloramphenicol possesses many of the properties of an ideal chemotherapeutic agent and is readily absorbed from the intestinal tract. Although some coliform bacteria possess an enzyme *nitroreductase* which reduces the nitro group, this alteration does not appear to affect the antibacterial action. The drug readily crosses tissue barriers and gains easy access to the cerebrospinal fluid. In addition it penetrates tissue cells, a factor which must play a considerable part in determining its efficacy in the treatment of typhoid fever. Because of the hazard of aplastic anaemia it is doubtful if chloramphenicol should ever be given when an acceptable alternative is available. When its use is unavoidable, administration should be controlled by repeated blood examination—a fall in reticulocyte count apparently being demonstrable at a stage at which the damage caused by the drug is still reversible.

The tetracyclines

The tetracyclines are a family of antibiotics produced by various species of *Streptomyces*. The important members of the group are tetracycline, chlortetracycline, oxytetracycline and demethylchlortetracycline. After the penicillins the tetracyclines are undoubtedly the most widely used antibiotics. This may be attributed to their wide antibacterial spectrum, their virtual absence of toxicity except in very high doses, their ease of administration—they are readily absorbed from the intestinal tract—and their very low tendency to induce allergic sensitization. Like chloramphenicol, their action is primarily bacteriostatic; because of this they should preferably not be given in combination with penicillin unless laboratory confirmation of the value of the combination has been obtained. Being amphoteric compounds the tetracyclines are used mainly as the hydrochlorides. Their

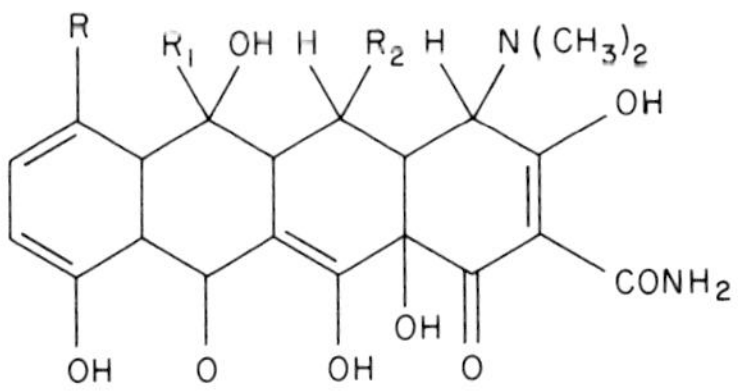

	R	R_1	R_2
Tetracycline	H	CH_3	H
Chlortetracycline	Cl	CH_3	H
Oxytetracycline	H	CH_3	OH

Fig. 1.16. The structure of tetracyclines.

action is optimal at an acid pH.

In recent years a large number of derivatives has been prepared by laboratory synthesis. The most important of these is minocycline which is much more active and better absorbed than the older tetracyclines. It is reported to be highly active against staphylococci resistant to the latter.

The only significant toxic effect of the tetracyclines is on the liver, and then only when very large doses have been given by the intravenous route. In practice the most serious complication of tetracycline therapy is that, by creating a bacterial vacuum in the alimentary tract, intestinal superinfection by *C. albicans, Proteus, Pseudomonas* and, most seriously of all, resistant strains of *Staph. aureus* may develop (p. 207).

The macrolides

The macrolides constitute a group of antibiotics consisting of a macrocyclic lactone ring, linked glycosidically to various sugars. They are basic in character and are most active under alkaline conditions. A number of individual members, differing mainly in their sugar residues, have been described: erythromycin, oleandomycin, carbomycin and spiramycin, of which erythromycin appears to be generally the most effective. The range of action of erythromycin is very similar to that of penicillin but its action is mainly bacteriostatic. For a time, erythromycin was used extensively in the treatment of infections due to penicillinase-producing staphylococci but has now been largely replaced by the new penicillins and cephalosporins. The drug may, however, still be of value for the treatment of streptococcal and pneumococcal infections, in patients exhibiting penicillin hypersensitivity.

Although laboratory mutants show a high degree of cross-resistance to the different macrolides erythromycin-resistant strains of staphylococci isolated from patients are usually sensitive to the other members of the group. In the presence of low concentrations of erythromycin they become resistant. This is apparently due to blocking by erythromycin of their combining site on the 50S ribosome.

Novobiocin

Novobiocin is an acidic antibiotic consisting of a phenolic residue—a substituted coumarin which confers on it a resemblance to dicoumarin—and an unusual sugar, noviose. It is active against most Gram-positive organisms, *H. influenzae,* the neisseriae and, somewhat surprisingly, some strains of *Proteus*, and it is well absorbed from the intestinal tract.

Resistance to novobiocin develops rapidly and, though of low toxicity, the drug has a considerable tendency to produce allergic sensitization. Currently it has only a limited place in chemotherapy.

Vancomycin and ristocetin

These compounds which have considerable toxicity are only active against Gram-positive organisms. Vancomycin has been successfully used for the treatment of staphylococcal and streptococcal septicaemias when it must be given by the intravenous route. Ristocetin has some activity against the tubercle bacillus but currently has no role in treatment.

Cycloserine

Cycloserine is a relatively weak antibiotic active against the tubercle bacillus and a wide range of Gram-positive and Gram-negative organisms. It has been used in the treatment of tuberculosis resistant to the major antituberculous agents and is claimed to be of special value in lesions of chronic type. Cycloserine has considerable neurotoxicity and has given rise to convulsions in a number of patients.

Fucidin

This steroid antibiotic produced by *Fusidium coccinium* is the sodium salt of fusidic acid. Fucidin has a limited range of action, being highly active against staphylococci, neisseriae, corynebacteria and clostridia, but of low activity against streptococci. Resistance develops rapidly in vitro and combination with other antibiotics is therefore advised if prolonged administration is required. For this purpose erythromycin and novobiocin have been recommended. Fucidin is readily absorbed from the intestinal tract. It possesses little if any toxicity.

Lincomycin

This antibiotic shows an antibacterial spectrum very similar to that of erythromycin with which it may show some cross-resistance. Resistance however develops much less readily to lincomycin. It does not possess a macrolide structure. Lincomycin is readily absorbed from the intestinal tract. It may play a minor role in the therapy of resistant staphylococcal infections. Clindamycin is a synthetic derivative of lincomycin which has a higher degree of activity and is more readily absorbable from the alimentary tract.

The rifamycins

This is a group of antibiotics derived from *Str. mediterranei*. Many hundred derivatives have been prepared by laboratory synthesis, the most important of which is rifampicin derived from rifamycin B. Rifampicin is active against a variety of Gram-positive organisms but its most important clinical application is in the treatment of tuberculosis. Since resistance to it develops readily it should always be given in combination with another chemotherapeutic agent.

Polypeptide antibiotics of bacterial origin

A number of highly potent bactericidal antibiotics have been isolated from members of the *Bacillus* genus. They include *gramicidin* and *tyrocidin* which were isolated by Dubos from a soil bacillus—*Bacillus brevis*—and are of historical interest as the first antibiotics discovered as a result of a planned search for such agents; because of their toxicity, however, they are suitable only for local application. *Bacitracin*, produced by *B. subtilis*, is also too toxic for systemic use but is of theoretical interest since it appears to act by blocking some stage in cell wall synthesis.

The polymyxins

These are the only polypeptide antibiotics with a place in current therapy. Five polymyxins—A, B, C, D, and E—have been identified. The commercially available compounds are polymyxin B and colistin or polymyxin E, both supplied as the sulphate or sulphomethyl derivatives. Polymyxins A, C and D are too nephrotoxic for clinical use. Polymyxins B and E are, however, relatively non-toxic and are undoubtedly the antibiotics of choice in the treatment of infections due to *Ps. aeruginosa*. Resistance to the polymyxins is acquired only with difficulty in the laboratory and resistant strains of coliform bacteria and *Pseudomonas* in clinical material are very rare. There is considerable evidence that the polymyxins act by damaging the cytoplasmic membrane. Initially, a rapid leakage of low molecular weight constituents occurs from the cell and if the membrane damage becomes more intense the cell undergoes autolysis. Unlike penicillin lysis, the lysis produced by polymyxin is not dependent on cell growth, and can be produced in protoplasts.

SYNTHETIC CHEMOTHERAPEUTIC AGENTS

The sulphonamides

Prontosil (sulphonamido-crysoidin), introduced by Domagk in 1935, was the first chemotherapeutic agent active against bacteria. In spite of its therapeutic effectiveness, however, it had no in vitro antibacterial activity. Its in vivo activity was shown, by Tréfouel, to be due to the fact that in the tissues it is broken down to yield sulphanilamide or *para*-aminobenzene sulphonamide—the parent drug of the sulphonamide series and from which all the currently used sulphonamides have been derived by substitution for one of the hydrogen atoms of the sulphonamide—SO_2NH_2—group. The most active sulphonamides are those in which the

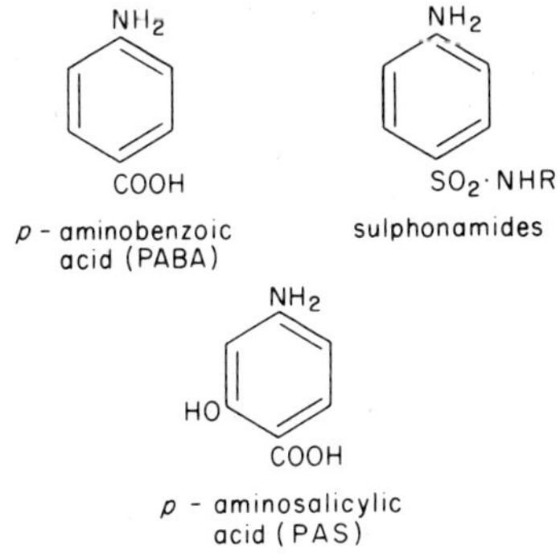

Fig. I.17. The structure of the sulphonamides.

substituent is a heterocyclic ring. In most cases substitution of the *para*-amino group has led to loss of activity, as a result of loss of structural resemblance to PABA (p. 61). One such compound, however, *para*-aminoethyl benzene sulphonamide (Marfanil) has appreciable antibacterial activity. This activity is not of sulphonamide type as it fails to be neutralized by PABA.

To a considerable extent, the antibacterial action of the sulphonamides can be correlated with the electronegativity of the SO_2 group. At pH 7 the electronegativity of this group and, as a result, the degree of dissociation of the compound may be increased by introducing substituent groups of increased electron-attracting power. This process can reach a limit however, the maximum increase in activity being achieved with sulphadiazine which is about 70% dissociated at pH 7. The failure of the more highly dissociated compounds to exert a greater antibacterial effect—one which might be expected from the virtually complete dissociation of PABA—is a result of the diminishing similarity between the electronic structure of the sulphonamide molecule and PABA as substitution progress.

The sulphonamides have a comparatively limited range of action. They are active against the neisseriae—although many strains of gonococci are now resistant—and against *Str. pyogenes*, pneumococci, *B. anthracis* and the gas gangrene clostridia. They are sufficiently active against *H. ducreyi, H. influenzae, Yersinia pestis* and the shigellae to be of therapeutic value in infections caused by these organisms. They are also of value in the treatment of coliform infections of the urinary tract because, although not highly active against the coliform group, they are excreted in effective concentration in the urine. The sulphonamides penetrate readily into the cerebrospinal fluid and are therefore of considerable value in the treatment of meningitis, particularly that due to the meningococcus.

Trimethoprim

This compound has a wide range of action against both Gram-negative and Gram-positive bacteria. It acts by combining with the enzyme dihydrofolate reductase, thereby blocking the synthesis of tetrahydrofolic acid. It has a high selectivity for the bacterial enzyme—about 50 000 times greater than for mammalian enzyme. It is normally administered with a sulphonamide, the combination exhibiting a markedly synergistic effect; the most effective sulphonamide for this purpose and that with which it is normally combined is sulphamethoxazole, in the compound co-trimoxazole.

Nalidixic acid

This compound, one of a series of naphthyridine derivatives, is active against a variety of Gram-negative bacteria but is without effect against staphylococci, streptococci or pneumococci. It is well absorbed from the gastrointestinal tract and is excreted mainly in the urine. There is some evidence that the drug is concentrated in renal tissues. Excellent results have been reported for nalidixic acid in the treatment of urinary infections due to the coliform and *Proteus* groups.

Nitrofurantoin

This compound, a nitrofuran derivative, has been used extensively by the oral route for Gram-negative bacillary infections of the urinary tract. It has in addition appreciable activity against *Str. faecalis*. It is the most active at an acid pH.

para-Aminosalicylic acid (PAS)

As previously considered, *para*-aminosalicyclic acid is a PABA antagonist but is distinguished from the sulphonamides by the fact that its action is exclusively demonstrable against the tubercle bacillus. As an antituberculous agent it is merely bacteriostatic, and as a result, is much less effective than either streptomycin or INAH. As PAS is rapidly excreted it must be administered in high doses, in which, however, it causes gastrointestinal disturbance. No modification of the drug so far achieved has resulted in any improvement in its antibacterial efficacy or in its capacity to persist in the tissues.

Isoniazid (INAH)

Although discovered in 1912, isoniazid was first reported to have activity against the tubercle bacillus in 1952. The β-isomeric form of the drug is inactive, suggesting that it acts as a competitive antagonist. The most obvious metabolite which it might antagonize is nicotinamide. However, the

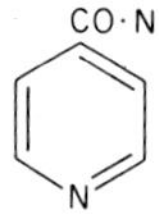

Fig. I.18. The structure of isoniazid (INAH).

drug is without inhibitory effect on NAD-requiring dehydrogenases, nor is its inhibitory action overcome by nicotinic acid or nicotinamide. It is clear therefore that INAH does not act by nutritional antagonism of nicotinamide. An observation which may have some relevance to its mode of action is the fact that tubercle bacilli rendered resistant to it lose the power to synthesize catalase and peroxidase. It has therefore been suggested that peroxidase may convert the drug into an active form. However, the fact that atypical mycobacteria have normal catalase and peroxidase activity but show considerable resistance to INAH, makes this possibility less likely. INAH is an active chelating agent but there is no evidence for a correspondence between chelating activity of this and similar compounds, and their antitubercular action. Recent evidence tends to implicate INAH as having an inhibitory role in the synthesis of lipid. In this way it could cause damage to the cytoplasmic membrane and to intracellular lipid-associated enzyme systems.

INAH is well absorbed from the intestinal tract and has marked powers of penetration across membranes and into cells. It has little, if any, toxic effect, much of which appears to be reduced by concomitant administriation of pyridoxine. Considerable differences have been shown between different individuals in the rapidity with which they metabolize INAH. In some, inactivation of the drug is slow, in others rapid, and there is evidence that these differences are genetically controlled. They are not of significance in chemotherapy provided that adequate doses of drug are administered.

Pyrazinamide

The discovery by Chlorine, in 1945, that nicotinamide possessed some activity against the tubercle bacillus—an activity not possessed by nicotinic acid—prompted the examination of other pyridine derivatives for similar activity. This led to the introduction of the amide,

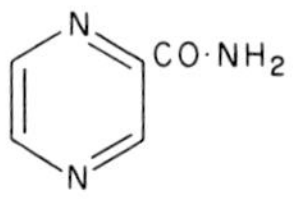

Fig. I.19. The structure of pyrazinamide.

pyrazinamide, which, though not as active an antituberculous agent as INAH is appreciably more active than PAS. Unfortunately, it has a marked toxic effect on the liver and is therefore not an acceptable alternative to PAS in the standard tuberculosis therapeutic regimen. Nevertheless it is of possible value as a second-line drug which can be used in cases showing resistance to the primary drugs—streptomycin, INAH and PAS (p. 275).

Thiosemicarbazones

The thiosemicarbazones have been used extensively in Germany for the chemotherapy of tuberculosis. They are active only against mycobacteria. The most completely investigated member of the group—*para*-acetamidobenzyl thiosemicarbazone, or thiacetazone, has considerable activity against the tubercle bacillus but is, unfortunately, highly toxic. Its degree of toxicity makes it unacceptable except when other antituberculous drugs have proved ineffective.

Ethionamide (α-ethyl thio-isonicotinamide)

This pyridine derivative, because of its possession of the S:C.NH-group, shows a structural

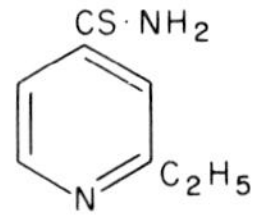

Fig. 1.20. The structure of ethionamide.

resemblance to the thiosemicarbazones. Because of this relationship there is complete cross-resistance between ethionamide and thiacetazone—the most important of the thiosemicarbazones. Ethionamide is readily absorbed from the intestinal tract. The isopropyl derivative, prothionamide, is now regarded by many authorities as the first choice when resistance to one or other of the major antituberculous agents has developed.

The sulphones

These drugs, the most important of which is diaminodiphenylsulphone (DDS), are analogues of PABA. They are of historical interest in that they were the first chemotherapeutic agents shown to be effective in the experimental treatment of tuberculosis. They have appreciable toxicity but are nevertheless the drugs of choice in the treatment of leprosy against which, even though the disease is caused by a mycobacterium, INAH is quite ineffective.

Clofazimine

This is a phenazine derivative with marked antimycobacterial properties which has been found to be of considerable value in the treatment of lepromatous leprosy. It has a high affinity for serum lipoproteins, the complexes being taken up by macrophages in which they are broken down, liberating clofazimine in crystalline form.

Part II

IMMUNITY

II.1

General Consideration of Immune Responses

As a science immunology initially developed in parallel with, and virtually as a branch of, microbiology; its primary concern being with the mechanisms involved in the development of resistance by the body to infectious disease. That clinical attacks of certain diseases could confer a specific immunity or protection against a future attack of the disease has been known from antiquity. This knowledge was first specifically applied to the prevention of disease in the practice of variolation and later of vaccination (p. 414) in the prophylaxis of smallpox.

Subsequent to the development of techniques for the isolation of pathogenic bacteria a major step in the progress of immunology was the demonstration by Pasteur (1878) that it was possible to immunize fowl against chicken cholera by the injection of artificially attenuated cultures of the causative organism, a principle he subsequently applied with dramatic success to the prevention of anthrax in animals and of rabies in man. The first clue as to the nature of the mechanisms responsible for antimicrobial immunity was the demonstration by Von Behring and Kitasato (1890) that the injection of small doses of tetanus toxin into animals stimulated the production of a substance—antibody—which when mixed with it was capable of neutralizing the lethal effects of the toxin. This discovery marks the real beginning of immunology as an experimental science, permitting as it did the possibility of exploring immune responses under in vitro conditions in the laboratory.

From an early date it was apparent that animals could develop antibodies not just against microbes but also against substances from other sources. Thus Ehrlich (1897) showed that animals injected with the haemolytic toxin ricin developed a strong antibody response against the toxin. Since then antibody responses have been demonstrated against a great variety of non-microbial substances, the response being highly specific for the substance which elicited it. Of particular importance for the demonstration of the specificity of the immune response was the finding of Landsteiner (1917) that antibodies could be produced against small molecules or determinant groups introduced into protein carriers. This work laid the foundation for the elucidation of the fundamental role played by chemical structure in immunological specificity.

In the early development of immunology it became apparent that the production of antibodies could on occasion have potentially harmful results. Such potentially harmful results are classified under the heading of hypersensitivities or allergies. The first work on this theme was that of Portier and Richet who, in 1902, showed that dogs which had recovered from an injection of an extract of sea anemones had acquired an altered reactivity so that subsequent injection of the extract resulted almost immediately in their death in a state of acute shock. For this phenomenon the term anaphylaxis, i.e. 'anti-prophylaxis', was applied. Nevertheless it is a characteristic of these reactions that they are responses of the body to contact with various types of material which it recognizes as foreign and which it 'tries' to eliminate; the term immune response is therefore applicable to them. In this wider application the immune response can be regarded teleologically as a response which, though not specifically concerned with immunity against microbial infection, is nevertheless a response which has a protective function, namely the preservation of the constancy of the milieu interieur.

Early experimental work in immunology was

concerned primarily with the production of antibody and this is not surprisingly the aspect of the immune response on which we have most information. A new dimension to immunological research was established with the demonstration by Chase in 1942 of the cellular basis of the tuberculin reaction. This reaction had first been demonstrated by Koch in 1890 (p. 271) and, although possessing a high degree of specificity, was clearly not due to humoral antibodies. The essential clue to its nature was provided by the finding that though it could not be passively transferred, as could anaphylaxis and protective immunity, by the injection of the serum of immunized animals, it could be transferred by the injection of lymphocytic cells obtained from previously sensitized individuals or animals. Subsequently the rejection of tissue transplants derived from other individuals of the same species was shown by the work of Medawar, Mitchison and Gowans to be due to a similar cellular mechanism.

TYPES OF IMMUNE RESPONSE

At this stage therefore two distinct types of immune response had been distinguished: (1) humoral responses resulting in antibody production; the antibody producing cell being identified as a plasma cell, and (2) cellular responses leading to the production of specifically reactive small lymphocytes which behaved as if they carried antibody-like molecules as integral components of their surfaces. Plasma cells which produce antibody and lymphocytes which mediate cellular immune responses are, in fact, the end-products of a process of multiplication and differentiation of antigen-sensitive precursor cells triggered by contact with antigen.

The antigen-sensitive precursor cells of both types of immune response appear to be morphologically small lymphocytes. The available evidence indicates that they belong to two distinct races: the thymus-dependent or T cells which carry the potential for cellular immune responses, and the non-thymus-dependent or B cells which carry the potential for development into antibody producing cells. These two types of cell have a different location in lymphoreticular tissues. In the lymph nodes, which are major sites of immune responses, T cells are found in the paracortical areas and during cellular immune responses there is marked proliferative activity in the cells in these areas of the lymph node. The thymus-dependent small lymphocytes of the paracortical areas are capable of free migration into the tissues and blood stream and represent the bulk of the small lymphocytes found in the peripheral blood and in the thoracic duct. By contrast the non-thymus-dependent antigen-sensitive precursor cells appear to be located primarily in the lymphoid follicles of the lymph node cortex. Unlike the thymus-dependent cells they are confined to this location and show little, if any, tendency to migrate or circulate through the tissues.

MANIFESTATIONS OF THE IMMUNE RESPONSE

Before dealing with the detailed mechanisms involved it would be appropriate to look briefly at the main ways in which immune responses are manifested in vivo in the intact animal. Six main types of situation may be distinguished.

Immunity against microbial infection

This involves the operation of both cellular and humoral immune mechanisms. Humoral responses are the main defence against exotoxin-producing bacteria and against most extracellular microbial pathogens. Humoral responses also play an important part in immunity against certain viral infections. Cellular immune responses are of importance against two types of infection: (*a*) by intracellular pathogens, in particular the DNA-containing viruses and, amongst bacteria, the mycobacteria and brucellae; and (*b*) by fungi which attack the skin. It seems likely, however, that cellular immune responses play some part in immunity against most pathogens. Thus evidence has recently been presented for a substantial role for cellular responses in immunity against staphylococcal infection. Cellular immunity against microbial infection, though determined by the small lymphocytes, appears to be mediated mainly through the macrophages attracted by lymphokine liberated from the small lymphocytes following exposure to antigen.

Hypersensitivity or allergy

These reactions may conveniently be divided into (*a*) the antibody-mediated or so-called immediate-type hypersensitivities and (*b*) the cell-mediated or delayed-type hypersensitivities. Among the immediate-type hypersensitivities we may distinguish anaphylactic hypersensitivity, which is due to a special form of antibody, homocytotropic antibody, with a particular combining affinity for mast cells, and immune complex type hypersensitivity, which is due to the deposition of antigen–antibody complexes in the tissues, particularly in and around the blood vessels. Amongst the cell-mediated hypersensitivities it is convenient to distinguish between sensitivity to microbial antigens, as represented for example by the tuberculin reaction, and contact sensitivity induced by contact of a variety of sensitizing materials, sometimes of very simple structure, with the skin.

Allograft rejection

Animals will accept grafts of tissues from themselves, from their identical twins and from animals genetically identical (syngeneic) as a result of repeated inbreeding. On the other hand, they reject grafts of tissues from all other animals of the same species—these are known as homografts or allografts—and from animals of other species, which are variously known as heterografts or xenografts. In recent years the mechanism and control of allograft rejection has assumed particular importance in the context of tissue and organ transplantation.

Allograft rejection appears to be due mainly to cellular immune mechanisms. In contrast to the role of cellular immune responses in antimicrobial immunity and delayed-type hypersensitivity lymphocytes appear to mediate the allograft rejection by a direct toxic action on the tissues. The evolution of such a mechanism is of considerable interest since transplantation is a phenomenon, which to say the least of it, appeared very late in evolutionary development, while the capacity for allograft rejection appears possibly to have been the earliest immune response to have developed. To explain this anomaly it has been proposed that the allograft rejection mechanism may have developed as a surveillance mechanism concerned with the elimination of malignant cells. This role is supported by the finding that various procedures, e.g. thymectomy and the administration of antilymphocyte globulin, which depress cellular immune responses, are associated with an increased incidence of malignancy.

Autoimmune diseases

Animals do not normally develop antibodies against antigens present in their own tissues. In fact a general tendency to do so would not be compatible with life, a concept early entrenched in immunological thinking and given formal statement by Ehrlich in the expression *horror autotoxicus*. Experimentally, however, it has been found that it is possible to immunize animals with their own tissue antigens particularly when these are injected together with Freund adjuvant (p. 99). In man autoantibodies can arise spontaneously and are frequently, though not invariably, associated with disease in the organ containing the immunizing antigen. In certain cases the autoimmune antibodies may be the result rather than the cause of the disease but in a number a causal role for antibody has been generally accepted. In most of these cases the antigen is one against which tolerance (p. 175) would normally be developed. Such are antigens which are present in the body during the development of the immune system. Why there should be an apparent breakdown of the tolerance mechanism in relation to a particular 'self' antigen is not clear.

Iatrogenic cytotoxic reactions

This designation covers instances of cytotoxic immunological reactions against substances artificially introduced. The principal example is the haemolytic transfusion reaction against blood of an incompatible group inadvertently introduced, e.g. Group A or B blood in a Group O recipient. Another similar reaction is purpura produced following administration of drugs, e.g. apronal (Sedormid), capable of combining with platelets when an immune response develops to the drug–platelet complex. Complement mediated lysis of the platelets results in thrombocytopenic purpura.

Maternal–fetal interactions

Antibody produced by the mother passes across

the placenta and causes cytotoxic damage to fetal cells. The most important example of this type of reaction is haemolytic disease of the newborn, in which mothers are immunized against fetal red cell antigens derived from the father and absent from the mother. IgG antibodies produced in this way cross the placenta and combine with the fetal red cell antigens, leading to complement-mediated lysis and also facilitating destruction of the red cells in the spleen. Other examples are purpura and leucopenia in the fetus due to the action of antibodies against paternal histocompatibility antigens present on the fetal cells and fetal thyrotoxicosis due to autoimmune antithyroid antibodies produced in the mother. In the latter case the antibodies have a stimulatory effect on the fetal thyroid.

Amplifiers

Under in vivo conditions the manifestations of the immune response requires the participation of tissue components other than antibodies or sensitized lymphocytes. Components which are in this way necessary for the expression of the immune response have been described as *amplifiers*. They include (*a*) complement which is necessary for the killing and lysis of certain bacteria and appears to play an important role in phagocytosis, (*b*) polymorphonuclear leucocytes and neutrophils which are actively phagocytic cells vital to the defence of the body against the highly invasive organisms of acute pyogenic infections, (*c*) basophils which play an essential role in immediate-type hypersensitivities and (*d*) macrophages which are vital to the body in its defence against intracellular pathogens.

SPECIFICITY OF THE IMMUNE RESPONSE

The immune response is highly specific. This specificity, which was shown first by Landsteiner in his work on conjugated antigens, is related to the chemical structure of the antigen. Combination of antibody with antigen depends on a complementarity between the structure of the antibody molecule on the one hand and that of the antigen on the other. It is because of this requirement that an antibody will react only with the antigen stimulating its production or with one closely related to it chemically. At the cellular level this is paralleled, although this has not been so extensively documented, by a smaller specificity in the relationship between antigens and the surface receptors of antigen-sensitive precursor cells

Clonal selection

At one time it was thought that the initiation of the immune response leading to antibody production was due to the fact that the antigen in some way instructed a potential antibody-producing cell to make the corresponding antibody. In this view the specificity of the antibody-producing cell arises *de novo* as a result of antigenic stimulation. This hypothesis can also be extended to explain the production of a new surface configuration in sensitized thymus-dependent small lymphocytes. The consensus is now against this view. Instead it is believed that the role of antigen is to stimulate the multiplication and differentiation of cells which, prior to exposure to the antigen, have an in-built genetically determined capacity to react with it. This theory, which is known as the *clonal selection* theory, was first explicitly proposed by Burnet, though the concept of a selective as opposed to an instructive process was first suggested by Ehrlich in 1897. Since the animal body is capable of responding to a great variety of natural and synthetic antigens it must contain, prior to antigenic stimulation, a population of antigen-sensitive cells possessing a specificity for a very large number of antigenic patterns. It has been suggested that the number of distinct specificities required—and this estimate must be to a large extent inspired guesswork—may be of the order of 10^5 to 10^6. All the available evidence indicates that each individual antigen-sensitive cell is capable of responding to only one antigenic pattern. The alternative would be to assume that each cell carries in its genome determinants for all possible antigenic patterns and this would seem to impose on each antigen-sensitive cell a genetic load not consistent with the general economy of nature. This postulated restriction on the capability of individual antigen-sensitive cells in turn implies that since the total population of antigen-sensitive cells in the body is limited there can only at any one time be a small number of cells capable of reacting with a particular antigen.

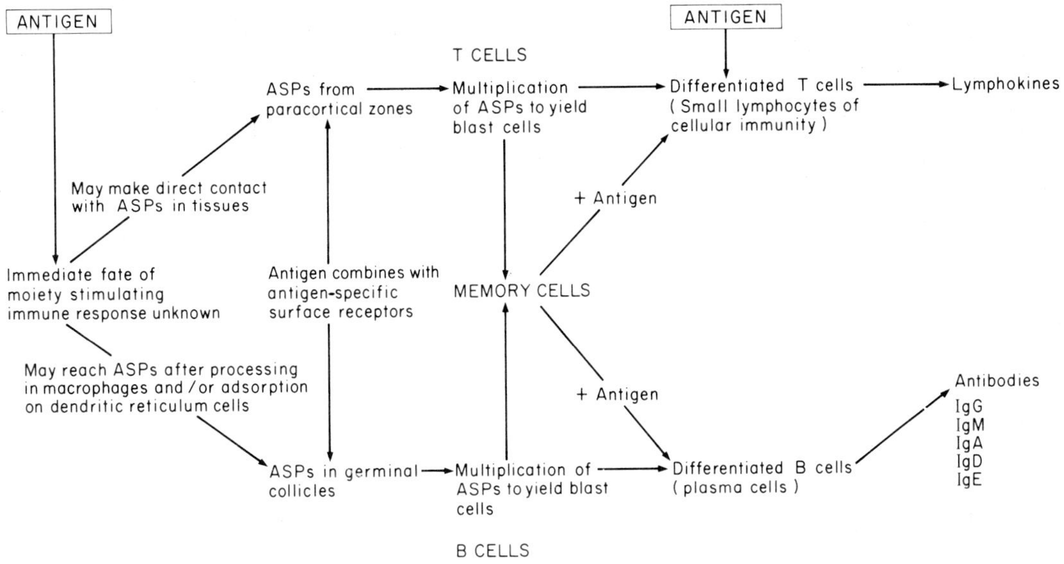

Fig. II.1. The immune response. ASP=antigen-sensitive precursor cell.

It would be surprising if, in the absence of antigenic stimulation, there were the same number of cells capable of responding to each antigenic pattern. It seems more likely that as a result of evolutionary pressures some patterns may have preferential treatment in that there may be *ab initio* a relatively large population of cells capable of reacting with them. It is thought, for example, that the histocompatability antigens involved in allograft rejection may in this sense be privileged antigens, at least for cellular immune responses. A similar evolutionary trend may explain why protein antigens are better antigens than polysaccharides.

According to the clonal selection theory the immune response is due to the expansion, as a result of antigenic stimulation, of a small population of pre-existing antigen-sensitive precursor cells capable of responding, because of their possession of genetically determined surface receptors for a particular antigen, in combination with that antigen, to a large population of cells capable of antibody production and/or to a large population of cells capable by direct interaction with antigen of initiating the sequence of phenomena characteristic of the cellular immune response.

Primary and secondary response

It is a fundamental characteristic of both types of response that an animal which has had a previous experience of an antigen subsequently responds to that antigen more rapidly, more easily, and to a quantitatively greater extent than to a first exposure. The first response to an antigen is known as the primary response and the second, more rapid and quantitatively greater response, as the secondary response. The secondary response is readily explained on the basis of an expanded population of antigen-sensitive cells capable of reacting with the antigen and responding to this stimulation by further proliferation and differentiation. The cells initiating the secondary response are known as *memory cells* and they are the carriers of immunological memory. The end cells of immune responses—the plasma cells and the sensitized small lymphocytes—appear incapable of further differentiation and multiplication. Memory cells, therefore, must presumably have derived from the original antigen-sensitive precursor cells during their antigen-stimulated differentiation and multiplication. A model for this differentiation of antigen-sensitive precursor cells with the production of memory cells is shown in Fig. II.1. In each pathway three distinct cell types are distinguished:

1. Antigen-sensitive precursor cells.
2. Blast cells which result from the antigen-stimulated multiplication of the original antigen-sensitive precursor cells. Cells of this type are readily demonstrable in lymphopoietic tissues undergoing antigenic stimulation. They have a strongly basophilic cytoplasm due to the presence of a high concentration of RNA. Since they stain red with pyronin they are often referred to as pyroninophilic cells.
3. The final products are the plasma cells, the distinctive feature of which is a well developed endoplasmic reticulum, and the fully differentiated T lymphocytes.

According to this schema the memory cells are presumed to arise through an alternative route of differentiation and multiplication of the blast cells. Memory cells are assumed to have a long life in normal circumstances. In their capacity to respond to antigen the memory cells resemble the original antigen-sensitive cells. Consequently it may not be necessary to assume any functional difference between them, other than that the memory cells have arisen as a result of antigenic stimulation. There is, however, some evidence that they may be more resistant to X-irradiation and to some immunosuppressive drugs than are antigen-sensitive precursor cells and it is possible that future work will reveal substantial differences between them.

Choice of response

Both types of immune response, humoral and cellular, may occur to the same antigen. Sometimes, however, the response to the antigen is purely humoral and at other times purely cellular. The precise reason for this selectivity in response has not been fully clarified. There is reason to think, however, that an important factor, if not the decisive factor, is the location of the antigen during the induction of the response. The introduction of soluble antigens into the animal body normally leads to a humoral response. It is probable that such antigens readily gain access via lymph and blood to the lymphoreticular tissues and there stimulate the non-thymus-dependent lymphoid cells in situ. On the other hand, cellular immune responses are readily induced to the cells of transplanted tissues or tumours or to soluble antigens which are injected into the skin or which are injected into the subdermal tissues together with substances such as Freund adjuvant (p. 99) which stimulate the formation of a granuloma in their vicinity, thereby tending to localize them. These observations suggest that cellular immune responses are initiated by antigens which are held locally in the tissues. In this position they may be expected to make contact with the circulating thymus-dependent lymphocytes, which, as we have seen, are the antigen-sensitive cells of the cellular immune response. Since cellular immune responses are associated with hyperplasia in the paracortical areas of lymph nodes it would seem likely that in this event the peripherally stimulated thymus-dependent lymphocytes return to the paracortical area where they undergo differentiation and multiplication. This process of homing to the lymph nodes has in fact been shown to occur in the case of passively transferred thymus-dependent lymphocytes capable of adoptively transferring allograft rejection.

THE LYMPHORETICULAR SYSTEM

As we have seen the small lymphocyte can be regarded as the key cell of the immune response. It is the cell which on antigenic stimulation differentiates to plasma cells and other antibody producers on the one hand and to the effector cells of cellular immune responses on the other. The primary location and source of lymphocytes is the lymphoreticular system, so called because its principal components are lymphocytes and other lymphocyte like or lymphoid cells together with cells known as reticulum cells having long cytoplasmic processes which generate a mesh of reticulin fibres. Though morphologically different the latter are probably derived from macrophages or their immediate precursors.

Three functional compartments of the lymphoreticular system may be distinguished: the stem cell compartment; the primary lymphoid organs; and the peripheral lymphoid organs.

Stem cell compartment

The lymphocytes of the lymphoid organs proper are derived from precursor cells or stem cells which are found in early fetal life in the blood islets of the yolk sac and in the fetal liver and at a later stage in postnatal life in the bone marrow. The seeding of stem cells from these areas to the primary lymphoid organs has been established by the use of chromosomal marker and parabiotic techniques. The morphology of the stem cells has not been identified but as the bone marrow contains a considerable population of small lymphocytes they are thought likely to be small lymphocytes. It is thought also that the stem cells are multipotential and that, depending on local inductive influences in the environment in which they ultimately find themselves, they can develop along one or other of several pathways to form lymphocytes, erythrocytes and possibly granulocytes and macrophages.

Primary lymphoid organs

The central role of the thymus in immunological function was first shown in 1961 by Miller, who found that mice which had been thymectomized on the first day of life showed considerable immunological deficiencies:

1. They were unable to reject allografts.
2. There was a considerable reduction in their circulating small lymphocytes.
3. They rapidly succumbed to microbial infection.

Subsequent work has clarified the precise nature of the immunological defect in thymectomized animals. They have been found to be deficient in the capacity to mount all types of cellular immune response, e.g. allograft rejection, graft-versus-host reactions and delayed-type hypersensitivity, and also to be unable to mount certain humoral responses, e.g. against sheep red cells, diphtheria and tetanus toxoids and certain other antigens. Thymectomy in adult life has much less dramatic results but a reduction in cellular and humoral immune capability is demonstrable after some months. In addition adult thymectomized animals show a delay in recovery from immunological tolerance (p. 170) and if X-irradiated and injected with syngeneic bone marrow do not show the recovery in function found in irradiated bone marrow injected non-thymectomized animals. The role of the thymus has been made clearer by the discovery that certain humoral responses require the cooperation of thymus-dependent helper cells which have a specificity for carrier molecules or groups associated with haptenic or antibody-stimulating antigenic determinants.

The thymus is a lympho-epithelial structure. On section it appears to consist of small discrete lobules 1 to 2 mm across. These are in reality components of a branching coral-like structure. The lobules consist of interdigitating epithelial cells between which are large numbers of lymphocytes; the latter are predominant at the cortex of lobules while epithelial cells are predominant in the medulla. At one time it was thought that the thymus lymphocytes, which are known as thymocytes, were derived from the epithelial cells. It is now apparent that they are derived from lymphoid cells seeded from the stem cell compartments. In X-irradiated animals the thymus may be repopulated with lymphocytes following the injection of syngeneic bone marrow cells. This repopulation cannot be achieved by injection of differentiated lymphocytes derived from the peripheral compartment. Obviously a selective mechanism is at work in the thymus which allows the ingress of stem cells but excludes that of more differentiated cells.

In the young animal thymus lymphocytes show great mitotic activity with an estimated division time of 6 to 8 hours. The majority of these proliferating cells (up to 95%) die in the thymus. The minority (5%) are seeded to the peripheral lymphoid tissues where they are found in the thymus-dependent areas. From here there is no return to the thymus. Thymus lymphocytes differ from those of the peripheral lymphoid tissues in undergoing multiplication in the absence of antigenic stimulation. Nevertheless they are antigenically immunologically competent since germinal centre formation has been observed following direct injection of antigen into the thymus. The thymus lymphocytes of many strains of mice possess an iso-antigenic surface marker, θ, which exists in two allelic forms, θ AKR and θ C3H. This marker is absent from stem cells but persists in thymus lymphocytes after these have been seeded to the periphery. They have consequently been of great value in clarifying migration and other properties of thymus-dependent cells.

The thymus is not simply a nursery for lymphocyte multiplication. Immunological function may be restored to thymectomized animals with the grafting of syngeneic thymus tissue. Some recovery of function also occurs following the implantation of thymus fragments in millipore membranes in the peritoneal cavities of thymectomized animals. This finding clearly indicates the production by the thymus, presumably by the epithelial cells, of a diffusible substance which plays a key role in stimulating the multiplication and differentiation of stem cell precursors. In this context it is significant that in the intact thymus mitoses are most frequent in lymphocytes in close proximity to the epithelial cells. The postulated hormone—thymosine—has not so far been isolated.

Sometimes thymectomy in animals has not been found to produce significant immunological defect. This has particularly been the case with sheep in which thymectomy, even in utero, has been reported not to be associated with any particular immunological definciency. There are a number of possible explanations for such findings:

1. At the time of thymectomy, seeding of thymus lymphocytes may already have occurred to the periphery.
2. It is possible that there may be accessory thymic tissue which has not been removed by the operation.
3. There is some evidence that in some cases the functions of the thymus may be taken over by other tissues. Direct evidence for the role of the thymus from the effects of thymectomy has not been possible in man. Nevertheless in one condition in man—the di George syndrome—nature has itself carried out a thymectomy in that the thymus does not develop and in this condition there has been the anticipated range of immunological deficiency. A similar congenital thymic defect has been observed in mice, one strain of which is hairless and is therefore referred to as nude (nu) mice. Nude mice have been of considerable value as a source of pure B cell lymphocyte populations.

A complementary function in relation to humoral immunological capability has been demonstrated in fowl for the bursa of Fabricius—a lympho-epithelial organ in the hind-gut. A hormonal bursectomy can be achieved in fowl by the administration of testosterone. This results in the loss of the capacity to produce antibody, a loss which can be restored by the grafting or implantation of bursal tissue. These findings indicate a function for the bursa of Fabricius in the control of humoral immune responses analogous to that of the thymus in the control of cellular immune responses. The role of the bursa in fowl raises the question as to whether any organ or organs have a similar function in animals; it has been proposed that this function is carried out by the gut-associated lymphoid tissue (GALT). There is suggestive, but by no means conclusive, evidence in support of this view derived from surgical extirpation experiments in rabbits. Against it, however, is the fact that multiplication of lymphocytes associated with the alimentary tract does not appear to occur in the absence of antigenic stimulation. It is, therefore, possible that in animals the bursal function is achieved in the local environment of the peripheral lymphoid organs.

Peripheral lymphoid organs

These include the lymph nodes, spleen, Peyer's patches of the small intestine, tonsils, adenoids and omental milk spots.

Lymph nodes

The lymph nodes are largely responsible for the immune responses to antigens which have been introduced into or have gained access to the tissues. The spleen appears to be particularly responsible for immune responses, especially the production of antibody, against antigens that have gained access to the blood stream. A schematic representation of the structure of a lymph node is shown in Fig. II.2.

Lymph nodes are essentially filters interposed at intervals in the pathway of return of lymph from the tissues to the blood. The lymph is collected from the tissues by lymph capillaries which, being more permeable than blood capillaries and venules, are the main channel for the return of protein and cells from the tissues. It also collects for subsequent phagocytosis foreign particulate material which has gained access to the tissues. The capillaries join to form larger vessels or trunks which drain as afferent vessels to the neighbouring lymph node. The lymph passes through the node and emerges through the efferent vessels; these ultimately join to form the large lymphatic ducts which discharge directly

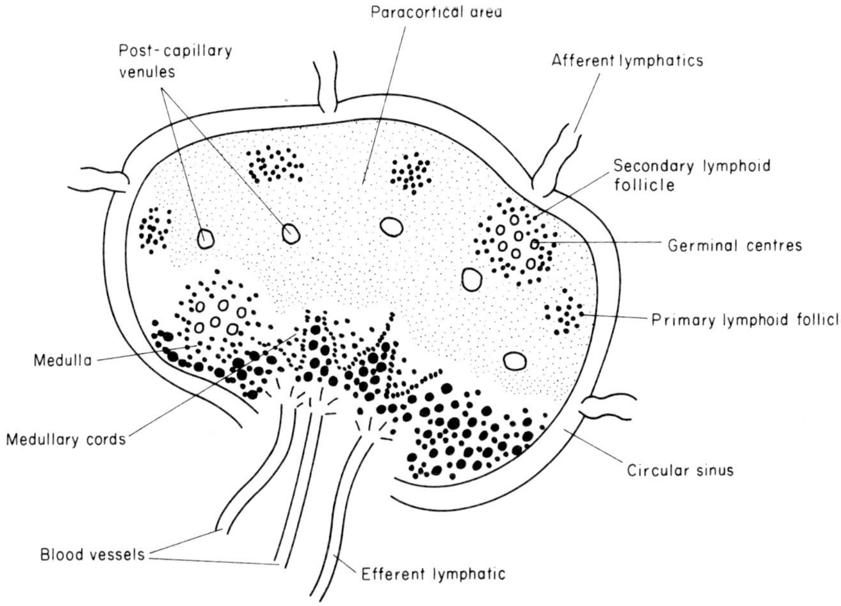

Fig. II.2. A schematic representation of the structure of a lymph node.

into the great veins. Small lymphocytes are the predominant cells in the lymph, constituting about 95% of its cell population. They measure from 6 to 8 μm in diameter, with a relatively large nucleus surrounded by a narrow ring of cytoplasm which shows no significant evidence of functional activity. The following are the principal structural features of the lymph node:

A circular sinus. This is lined on the outside by endothelial cells and on the inside by a variety of cell types. The inner lining of the sinus is freely permeable.

Lymphoid follicles. These are found in the cortex of the node and may be divided into primary and secondary. Primary follicles consist of an accumulation of small lymphocytes and dendritic reticulum cells which form a layer or cap in the peripheral surface of the follicle. The dendritic reticulum cells possess long convoluted processes formed by extensive folding of the membrane of the cell. They are probably a specialized form of macrophage. The secondary lymphoid follicles are larger than the primary and are characterized by the possession of germinal centres consisting of rapidly multiplying pyroninophilic blast cells surrounded by a rim of small lymphocytes. Secondary follicles appear to occur only in response to antigenic stimulation. They are absent from the lymph nodes of germ-free animals. The follicles also show the presence of tingible macrophages which contain a great deal of cellular debris. They are particularly noticeable when the secondary follicle is regressing after antigenic stimulation. They are clearly concerned with the disposal of cells no longer required.

Paracortical zone. This zone is sometimes known as the lymphocytic field and is an area filled with small lymphocytes lying between the follicles and extending into the base of the medullary cords. A prominent feature of the paracortical zone is the presence of post capillary venules which are lined with a high cuboidal endothelium. These are of major importance in the migration of lymphocytes from the blood stream into the paracortical areas.

Medullary cords. These are finger-like processes extending from the paracortical region into the medullary sinuses. In non-stimulated nodes they contain a few plasma cells but in stimulated nodes they contain large numbers of plasma cells together with pyroninophilic blast cells. In sections the latter appear to be migrating from the germinal centres to the medullary cords.

Medullary sinuses. These sinuses, which possess a highly permeable lining consisting mainly of macrophages, drain into the efferent lymphatics of the node. The characteristic medullary cell is the macrophage but after antigenic stimulation the sinuses contain a considerable number of immature plasma cells and antibody-containing blast cells. Medullary sinuses are the principal filter area of the nodes. Antigens, especially particulate antigens, gaining access to them are phagocytosed by the macrophages. During acute inflammation of areas drained by the node there is also an accumulation of polymorphonuclear leucocytes. These block up the outlet to the node like a 'log jam in a mill pond' (Wood) preventing the exit of cells and bacteria and facilitating the phagocytosis of the latter.

Spleen

The spleen differs from the lymph nodes in possessing no draining afferent or efferent lymphatics, access to the organ being via the blood stream. Topographical areas of the spleen known as the white pulp and the red pulp may be distinguished. The white pulp consists of islands of lymphoid tissue around a central arteriole and surrounded by the sinuses and cords of the red pulp.

The red pulp sinuses are concerned with non-immunological functions, e.g. the disposal of red blood cells and the production of red blood cells and granulocytes. The white pulp contains lymphoid follicles and around the central arteriole a periarteriolar lymphocytic sheath corresponding to the paracortical zone of the lymph node. These lymphoid areas are surrounded by a *marginal zone* containing numerous reticular cells and separating the white pulp from the red pulp. The marginal zone is the principal antigen trapping area of the spleen. The red pulp cords project from the marginal zone into the red pulp and contain plasma cells and other antibody-producing cells. They are analogous to the medullary cords of lymph nodes. There is no structure in the spleen corresponding to the medullary sinuses of the lymph nodes.

Migration patterns of lymphoid cells

There is now overwhelming evidence that the paracortical areas in the lymph nodes and the peri-arteriolar lymphocyte sheaths of the spleen are populated by cells originally seeded from the thymus:

1. The lymphocytes in these areas are depleted after thymectomy and in nude athymic mice.

2. Thymic cells labelled in situ with a radioactive isotope migrate to the paracortical areas.

3. Injected T cells localize in the paracortical areas. The thymus cells which localize in the paracortical zones are morphologically small lymphocytes. They enter the zones through the postcapillary venules. There has been considerable controversy as to whether they enter through or between the endothelial cells. Electron micrographs suggest migration through the cells, but recent evidence is more in favour of migration between the cells. Because of their relation to the thymus the paracortical zones of lymph nodes and the peri-arteriolar lymphocyte sheaths of the spleen are known as thymus-dependent areas. These areas show marked hyperplasia on antigenic stimulation leading to cellular immune responses in animals, e.g. following the application of oxazolone to the skin.

The cells of the paracortical areas are actively migratory and constitute a pool of recirculating lymphocytes. This is shown by the fact that these areas are depleted following thoracic duct drainage. Because of this they are sometimes referred to as the 'traffic areas' of the nodes. Lymphocytes leave the node through the efferent lymphatics, eventually gaining access to the blood stream through the great lymph ducts from which they are redistributed through the postcapillary venules to other lymph nodes and also to the tissues. The passage of T cells through the postcapillary venules is clearly a highly specific process which may be based on the presence of specific receptors. The recirculating lymphocytes undergo a very rapid turnover. In the rabbit, for example, it has been estimated that there is a complete turnover of these cells ten times a day.

As indicated above, the traffic from the thymus is in one direction only, only about 5% of thymus lymphocytes contributing to the circulating lymphocyte pool. This, however, is sufficient to maintain the peripheral T cell population since the circulating lymphocytes are extremely long lived. Experimental evidence in support of this has been obtained from labelling experiments. Thus, in rats injected with tritiated thymidine for

12 hours, only 1% of the thoracic duct lymphocytes are labelled, indicating that during this time only 1% of the lymphocytes had divided. Injection of tritiated thymidine for a period of 200 days was necessary in order to label 90% of the cells. In man evidence for the longevity of these cells has been obtained by the demonstration of chromosomal abnormalities after PHA (p. 144) stimulation in patients previously exposed to X-rays. It has been found that small lymphocytes with chromosomal defects induced by X-irradiation may be present for up to ten years after irradiation.

The lymphoid follicles in X-irradiated animals may be reconstructed by injection of suspensions of bone marrow cells. Injection of thymus lymphocytes will reconstitute the paracortical zone in X-irradiated animals but not the lymphoid follicles. It is a reasonable assumption, though this has not been proved formally, that under normal conditions primary follicles have been derived from bone marrow cells and that they are the antecedents of the secondary follicles. The existence and possible role for a bursa equivalent is not clear.

It is reasonable to presume that cells in the germinal centres after antigenic stimulation migrate to the medullary cords of the lymph nodes and to the red pulp cords of the spleen. The evidence in favour of this is:

1. Thymectomy has no effect on germinal centre formation or on the cell composition of the medullary cords or red pulp cords.

2. The demonstration of blast cells apparently migrating from the follicles to the medullary cords of the lymph nodes and red pulp cords of the spleen after antigenic stimulation. Satisfactory marker techniques are not, however, available to supply decisive confirmation of this migration route. It has been clear, however, that significant numbers of plasma cells do not normally migrate from the medullary cords. On the other hand, more primitive antibody-producing cells are found in the medullary sinuses of lymph nodes and in the efferent lymph. These apparently eventually reach the lamina propria of the alimentary tract where they mature into fully differentiated plasma cells. Their precise immunological role is not, however, clear.

IMMUNE DEFICIENCY SYNDROMES

Certain individuals and animals are defective in their capacity to produce immune responses or to translate immune response into an effective clearance function. It is outside the scope of this book to consider these defects in detail. Suffice it to say that there may be a failure at virtually any stage in the development and function of the lymphoreticular system and that these defects are associated with a diminished resistance to microbial infection. In fact immune deficiency syndromes frequently first come to light because of this increased susceptibility to infection.

The more important types of immune deficiency in man are:

1. Reticular dysgenesis, in which there is a failure of stem cell development.

2. The di George syndrome, in which there is defective development of the thymus.

3. Agammaglobulinaemia, in which there is deficient production of immunoglobulins. This may affect all the immunoglobulin classes or may be restricted to a single class. Some cases of agammaglobulinaemia—the so-called Swiss type and ataxia telangiectasia—are associated also with cellular immune deficiency.

4. Tumours affecting antibody-producing cells in which there is excessive production of plasma cells producing a single variety of immunoglobulin. Such tumours are known as myelomas. Excessive production of IgM immunoglobulins not associated with obvious tumours is known as Waldenstrom's disease.

5. Defects in granulocyte production (agranulocytosis).

6. Defects in granulocyte function as in chronic granulomatous disease in which there is defective production of H_2O_2 (p. 152), Chediak-Higashi disease in which there is neutrophil lysosomal abnormality and the lazy leucocyte syndrome in which there is defective mobilization of leucocytes at sites of infection.

II.2

Antigens and Immunogenicity

Substances which on introduction into the animal body can stimulate immune responses are known as antigens. The stimulation of an immune response results from the combination of antigen with specific receptors on the surface of antigen-sensitive cells. As a result these cells undergo multiplication and differentiation, according to their ancestry, into lymphocytes capable of mediating cellular immune responses or into antibody-producing plasma cells. Both of these end products of the immune response are specific for the antigen stimulating their production.

In the past antigens have been defined and their properties examined in respect of their capacity to stimulate the production of antibodies. In this context it is possible to distinguish substances which, as first shown by Landsteiner in his studies on conjugated antigens, are not capable of stimulating antibody production on their own but can do so when combined with protein. Such substances are known as *haptens* and the associated proteins are known as *carriers*. There is now substantial evidence that carrier protein complexes stimulate anti-hapten antibody only if at least two hapten groups are present per carrier.

For many natural antigens it is possible to make a similar distinction within the antigen molecule between haptenic groups or antigenic determinants with which the antibody produced reacts and the rest of the molecule which does not react with the antibody but which is necessary for immunogenicity. This distinction between haptenic and carrier functions of an antigen has assumed considerable importance as a result of the demonstration that production of antibody against haptenic or determinant groups requires in many cases the recognition by non-antibody-producing helper cells of other groups in the carrier portion of the molecule. The distinction between haptenic and carrier groups is, however, not absolute, since groups which behave as haptenic or determinant groups on one occasion may behave as carrier groups on another occasion.

ANTIGENIC MOLECULES

Naturally occurring antigens

Only two groups of naturally occurring compound are clearly immunogenic *per se*—proteins and polysaccharides. Proteins are in general considerably more effective in stimulating antibody production than are polysaccharides. Proteins are highly effective in stimulating cellular immune responses; there is doubt as to whether polysaccharides can do so. Lipids and nucleic acids do not stimulate antibody production on their own but will do so when combined with protein. They are therefore naturally occurring haptens. Bacterial teichoic acids, against which antibody is produced following immunization with intact bacteria, are probably also haptenic rather than immunogenic.

Conjugated antigens

This term is applied to all antigens formed by the artificial complexing of haptens with carrier proteins. The conjugated antigens used by Landsteiner in his classical studies on antigenic specificity were prepared by attaching various groups to an aromatic amine such as aniline; then, following treatment with nitrous acid, a diazo compound is formed which can be coupled to the tyrosine residues of a protein. These complexes are sometimes referred to as azoproteins. Since Landsteiner's work a variety of other conjugated proteins have been prepared. Of particular importance in recent work have been dinitro phenyl (DNP) and 4-hydroxy-5-iodo-3-nitro-phenacetyl (NIP) conjugates. Other haptenic residues incorporated into proteins against which specific antibodies may be produced include amino acids and peptides, sugars, steroids, purines, pyrimidines and penicilloic and penicillenic acids.

Synthetic antigens

In recent years the immunological properties of a variety of synthetic polypeptides have been extensively studied. Antigens of this type are of particular importance in immunological research since their composition and structure is completely known, and can be varied at will. The synthetic polypeptides fall into three groups:

1. Linear polymers of one or more amino acids.
2. Multi-chain polymers with linear side-chains attached to a polypeptide back bone. The most studied of these have been polymers with a poly-L-lysine back bone to which are attached side chains containing DL-alanine, tyrosine and glutamic acid.
3. Conjugated polypeptides. These are polypeptides into which, as with conjugated proteins, various determinant groups have been introduced. The most extensively studied has been DNP-poly-L-lysine. The synthetic polypeptides have been of particular value in providing information on the nature of the determinant groups of proteins, on the conditions for immunogenicity and on the genetic control of immune responses.

REQUIREMENTS FOR IMMUNOGENICITY

As previously discussed the immune response operates as a mechanism for the elimination of foreign material from the animal body. It is therefore a first requirement of an antigen that it should be recognized as foreign by the animal producing an immune response to it. In what does foreigness consist? This is easy to define for compounds which contain determinant groups or residues not represented in the animal's body. DNP, NIP and D-amino acids are obviously foreign and are therefore potential antigenic determinants. Protein antigens, however, are composed of building blocks, namely amino acids, which also occur in animal tissues. In this case foreigness must consist in their occurrence in combinations and arrangements different from those found in the tissues.

Antigens are without exception substances of high molecular weight. When substances of very low molecular weight, e.g. metallic ions such as mercury, nickel, beryllium (p. 196), appear to stimulate antibody production they do so not by a direct action but indirectly by complexing with tissue components, thereby altering the antigenic properties of the latter. The smallest naturally occurring immunogens so far described are the pancreatic hormone glucagon, bacitracin and the hormone angiotensin, with molecular weights of approximately 3800, 1930 and 1030 respectively. Most protein antigens are much larger than this, with molecular weights in the range of 20 000 to 1 million or more. Amongst the synthetic linear polypeptides which have so far been investigated the minimum molecular weight consistent with immunogenicity diminishes with increase in the number of the different amino acids present. Thus for rabbits a 2-amino acid polymer of glutamic acid and lysine with a molecular weight of 38 000 is weakly antigenic while a 3-amino acid polymer of glutamic acid, lysine and tyrosine of molecular weight 13 000 is strongly antigenic. By contrast a homo-polymer of glutamic acid with a molecular weight of 80 000 is non-antigenic. The smallest synthetic antigen so far described is *p*-azobenzene arsonate, with a molecular weight of only 450 daltons.

The extent to which a minimum molecular weight *per se* is an intrinsic requirement for antigenicity is not clear. It is possible that a minimum size may be an essential requirement for the effective operation of an antigen-trapping mechanism. Alternatively a minimum size may be necessary so that other conditions of immunogenicity may be fulfilled. Thus in the case of antigens, e.g. proteins containing building blocks present in the animal's own tissues, the higher the molecular weight the greater will be the number of residues present and the greater will be the chance that these occur in groupings differing from those in the animal's own proteins. This interpretation is supported by the lower molecular weight requirements for linear polymers containing three amino acids than for those containing only two amino acids. In addition antigenicity might require the presence of more than one haptenic group as well as one or more effective carrier patterns. The greater the number of such essential groups present in the molecule the greater the possibility that the antigen will encounter and combine with the appropriate antigen-sensitive lymphocytes. It is also possible that large sections of protein molecules may be immunologically silent even though they possess patterns different from those of the animal's tissues, because the animal's lymphocytes may lack the complementary receptor pattern. Some of these considerations clearly do not apply in the case of molecules containing residues, e.g. DNP, not present in the animal's tissues. In such cases a lower molecular weight might be expected to be sufficient for immunogenicity. This would seem to be the case in respect of DNP octa-L-lysine which is strongly immunogenic while homologous DNP polymers with less than eight lysine residues are, however, non-immunogenic. This particular size is thought to be dictated by the fact that the 8-lysine compound is the smallest of the series which has appreciable rigidity due to its capacity to form stable hydrogen bonds.

THE CHEMICAL BASIS OF ANTIGENIC SPECIFICITY

The specificity of the relationship between antigen and antibody is based on the chemical structure of the antigen, as was first shown by Landsteiner in his studies on conjugated azoprotein antigens. When such antigens are used for immunization, antibodies to both the introduced group and the protein bearer are produced. By using one type of protein for the preparation of the conjugated antigen used in immunization and a second type of protein which did not cross-react serologically with the first for the preparation of the antigen used for testing, Landsteiner was able to investigate the specificity of the antibody directed against the introduced haptenic group. His results showed a very close dependence of serological specificity on chemical structure. Thus serum prepared against a conjugated protein containing a *p*-aminobenzoic acid hapten gave no cross-reactions with conjugated proteins in which the haptens were aniline, *p*-aminobenzene sulphonic acid or *p*-aminophenyl arsonic acid. Sera prepared against azoproteins in which any one of these substances was the introduced hapten gave little or no cross-reactions with any of the others. Each of these compounds is therefore serologically distinct.

The spatial distribution of the radicals incorporated in the molecule was also found to play an important part in determining specificity. Thus *ortho*-, *meta*- and *para*-aminobenzoic acids were found to be completely distinct antigenically. Landsteiner found also that antigenic specificity extended to stereoisomers. Thus, e.g. serum prepared against L-tartaric acid gave negligible reactions with meso- and D-tartaric acids. Similarly Avery and Goebel found it was possible to distinguish by serological reactions between α- and β-glycosidic linkages.

Identification of antigenic determinants in naturally occurring proteins is much more difficult. Firstly an individual protein consists of a much larger number of differing residues and these do not, except to a very slight extent, exhibit a regularly repeating pattern. Secondly the precise composition and arrangement of antigenic determinants depends on the tertiary structure of the protein. Consequently a particular determinant may contain a number of non-sequential amino acid residues. The dependence of antigenic specificity on tertiary structure is shown by the fact that when this structure is altered by denaturation the protein acquires a new antigenic specificity. It has nevertheless been possible in some cases to isolate peptides which inhibit the reactions of antibodies with the original protein. In such cases the larger the peptide the more efficient the inhibition. In a number of immunoglobulin allotypes the antigenic specificity of the protein has been related to the presence of a particular amino acid. Thus the InV 1,2 human IgG allotype has leucine at position 191 in the κ light chain while in the InV 3 allotype this position is occupied by valine. The adjacent residues in this region are known but the precise composition and extent of the antigenic determinant has not been established.

Though a variety of synthetic polypeptides have now been produced which are themselves fully antigenic, none so far has been found to possess an antigenic specificity identical with that of the determinant group of any known protein. This is perhaps not surprising in view of the important part played by conformation in antigenic specificity. An antigenic determinant containing three to five amino acid residues has been proposed for sperm whale myoglobin. When, however, one takes into account that 20 amino acids commonly occur in protein it is clear that even with determinant groups of 3 to 5 residues an enormous number of antigenically distinct proteins is possible.

At one time it was thought that aromatic amino acids were of particular importance in determining antigenic specificity. This view is derived from the fact that the introduction of tyrosine residues into gelatine rendered the latter antigenic. Tyrosine is unlikely to be of particular importance in determining antigenic specificity, as is indicated by the findings that a synthetic polypeptide of glutamic acid, lysine and tyrosine of molecular weight 94 000 is less antigenic than a synthetic polypeptide of glutamic acid, lysine and alanine with a molecular weight of 62 000. Similarly diamino and dicarboxy amino acids do not appear to be of special importance since the free amino and free carboxyl groups of bovine serum albumin (BSA) can be blocked without a significant effect on antigenic specificity. It seems likely that, for proteins at least, the specificity of antigenic determinants is largely a function of their shape.

The determinant groups or *epitopes* of proteins which are involved in stimulating antibody production would appear to be located on the surface of the molecule. Thus by various methods of fractionation it is possible to reveal determinants which are masked in the native protein. The importance of position on the immunogenic activity of the determinant groups is clearly shown in the case of two multichain polypeptides consisting of a poly-L-lysine backbone and side-chains composed of DL-alanine, tyrosine and glutamic acid. In one polymer the tyrosine and glutamic acid residues were at the ends of the chain. In the other these positions were occupied by the DL-alanine residues. Only the polymer with tyrosine and glutamic acid at the ends of the chain were immunogenic for rabbits. Determinant groups may, however, be present on the surface of an otherwise immunogenic molecule but fail to stimulate antibody production. Thus it has been shown that egg white lysozyme, after disulphide bond reduction, can stimulate antibody to determinants which are serologically reactive but non-immunogenic in the native molecule. This suggests that the production of antibody to these determinants requires the cooperation of helper cells which combine with carrier groupings not available on the surface of the native molecule.

THE PHYSICAL STATE OF THE ANTIGEN

The physical state in which an antigen is presented to an animal is of considerable importance in determining its immunogenicity. Antigens delivered in particulate form are in general more potent than if delivered in soluble form. Thus it has been shown by Dresser that mice could not be immunized by injection of bovine gammaglobulin which had been deprived of all aggregated material by centrifugation at 20 000 to 30 000 *g*. The aggregate-free preparations were in fact found to be capable of inducing immunological tolerance when injected in saline solution. If, however, they were injected with Freund adjuvant (p. 99) they became immunogenic.

Poorly immunogenic proteins may in some cases be rendered highly immunogenic by intermolecular cross-linking or by combination with molecules of opposite charge, e.g. with methylated serum albumen. In the latter case antibody is also produced to determinants in the carrier. Immunogenicity of soluble antigens may also be increased by adsorption of the antigen on to particulate materials, such as collodion, charcoal, red cells or bacteria. Possibly the earliest demonstration of this effect was by Ramon who showed that it was possible to produce particularly high levels of antibody to diphtheria toxoid by adsorption of the latter onto tapioca particles.

Immunogenicity is not necessarily enhanced, however, by the presentation of antigenic determinants in a particulate form. Thus Battisto and Borek found that guinea-pigs inoculated intraperitoneally with a picryl hapten coupled to a polyurethane sponge failed to respond to the picryl hapten but instead became tolerant to it (p. 174).

In some cases adsorption of antigen to particulate material may lead to an alteration in the structure of the antigen which is reflected in the nature of the antibodies produced to the adsorbed antigen. Thus it was shown by Kochwa that, although adsorption of human gammaglobulin to polystyrene particles increased its immunogenicity for rabbits, the presence of a large number of adsorbed antigen molecules per particle caused a diminution in the amount of anti-light-chain antibody produced.

The enhanced immunogenicity found on exposure of the animal to the antigen in a particulate form is probably due to the fact that aggregated antigens are taken up and ingested by macrophages more readily than soluble antigens and are in general more easily trapped in the peripheral lymphoid tissues. This enhanced trapping has been directly demonstrated for a variety of antigens.

ADJUVANTS

The term adjuvant was used initially to describe substances which when injected together with antigen increased antibody production and this is the sense in which it will be used here. A variety of substances with diverse chemical composition have been found to have an adjuvant effect. The most important of these and the most widely used are alum and other aluminium salts, water-in-oil suspensions with or without killed mycobacteria, endotoxins of Gram-negative bacteria and low concentrations of antibody. Other substances with adjuvant effect are quaternary ammonium salts, vitamin A and its derivatives and bacteria, notably *Corynebacterium parvum* and *Bordetella pertussis*.

For none of these is it possible to define precisely the mechanism of the adjuvant effect and for many there may be more than one mechanism. They may affect any stage in the immune response from the initial localization of antigen to the final release of antibody. The principal mechanisms which may be involved are as follows:

1. The adjuvant may facilitate the local retention of the antigen, thus preventing its catabolism and thereby maintaining it as a persistent antigenic stimulus. This is thought to be the main effect of alum and aluminium salts and of water-in-oil suspensions.

2. The adjuvant may facilitate trapping of the antigen. This is possibly the basis of the adjuvant action of antibody. Antibody has been shown to increase localization of the antigen in the peripheral lymphoid tissues and has a particularly marked effect on antigen localization on dendritic reticulum cells (p. 91). Increased trapping by macrophages through stimulation of phagocytosis may play a part in the adjuvant effect of endotoxins and of bacterial suspensions.

3. The labilizing of lysosomal enzymes as a preliminary to possible antigen processing by macrophages. This is postulated as the principal effect of surface active agents such as the quaternary ammonium salts. Compounds with surface activity might be expected to damage the plasma membrane and therefore to interfere generally with cellular functions. An adjuvant effect would, therefore, necessarily have to be associated with a specific action on the lysosomal membrane. It is, however, doubtful if there are any compounds which have such a specific effect. Surface active agents which form aggregates with antigen might, however, after ingestion by phagocytosis, be incorporated into phagolysosomes (p. 152) and there exert an effect which would appear to be specific for the lysosomal membrane.

4. Substances on the surface of which antigen is adsorbed, e.g. oil droplets in water-in-oil suspensions, might present or orient the antigen in a suitable physical configuration to antigen-reactive cells. Such a function would be analogous to that postulated by Mitchison for T cells in T–B cell cooperation.

5. Direct stimulation of B cells. The lipo-saccharides of Gram-negative bacteria, which have long been known to have a specific mitogenic effect on B cells, have recently been described by Sjöberg as being capable of substituting for helper T cells in T-dependent humoral responses in vitro.

6. Stimulation of T cell cooperation in T-dependent humoral responses. Allison and Davies reported that complete and incomplete Freund adjuvant, *Bordetella pertussis* and Gram-negative lipopolysaccharides have an adjuvant affect on the production of antibody to the T-dependent antigen BSA in mice only if T cells are present. The adjuvants were ineffective in thymectomized animals to which their effectiveness could be restored by thymus grafting. As a result of these experiments it was concluded that adjuvants acted

by stimulation of T cells either directly or by inducing the liberation of a non-specific T cell stimulant from macrophages.

Freund adjuvant

The adjuvant which has been most studied and most used in experimental work is the Freund adjuvant which consists of a mineral oil, an emulsifier and killed tubercle bacilli. This is known as the complete Freund adjuvant. Without tubercle bacilli it is referred to as incomplete Freund adjuvant. Freund adjuvant has a marked stimulatory effect on antibody production, increasing the amount of antibody produced several-fold and diminishing the amount of antigen exposure required for a demonstrable immune response. It also has the effect of transforming the tolerogenic effect of low doses of antigens which can induce low-dose tolerance into an immunogenic effect and of diminishing the amount of antigen required for the production of high-dose tolerance (p. 175). As well as its quantitative effect on antibody production the complete Freund adjuvant has a qualitative or directive effect in inducing T cell responses as manifested by the development of cutaneous delayed-type hypersensitivity to antigens which normally elicit only humoral responses.

Freund adjuvant is frequently used in experimental work on auto-immunization (p. 85). When administered with normal body components it will break self tolerance, thereby inducing autoimmune responses which may result in the development of symptoms of autoimmune diseases. The best example of such an experimentally induced autoimmune disease is probably the experimental allergic encephalomyelitis (EAE) produced by injecting homologous or autologous brain tissue together with the adjuvant.

The mechanism of action of Freund adjuvant, though extensively studied, is still in dispute. The most obvious histological effect is the production of a granulomatous reaction with the presence of large numbers of macrophages at the inoculation site and in the adjacent lymphatic glands. In addition there is evidence of widespread stimulation of macrophages in the lymphoreticular system. Because of the severity of the gramulomatous reaction Freund adjuvant is not used in human immunization. For the production of this extensive granulomatous reaction the tubercle bacilli are essential. The bacilli may be replaced by the purified wax D component of the bacilli, the principal constituent of which is a peptidoglycolipid which has a structural role in tubercle bacilli analogous to that of peptidoglycan in other bacteria. Highly purified preparations of the peptidoglycolipid are effective. Other mycobacteria may be substituted for tubercle bacilli but their wax D fractions lack the peptide component and are inactive as adjuvants. The situation is further confused by the fact that though in fowl killed tubercle bacilli have a definite adjuvant effect, the wax D fraction derived from them has not.

EFFECT OF HORMONES

As has been previously considered the thymus has a central role in the control of many immune responses and the capacity of thymus fragments after implantation to regenerate thymic function in thymectomized animals indicates that this function is mediated through a secretory product—thymosin—presumably of the thymus epithelial cells. Other endocrine secretions also affect immune responses. Corticosteroids have a lympholytic effect on shortlived lymphocytes (B cells) and the major non-recirculating lymphocyte population of the thymus. Corticoids have been variously reported to have a stimulating and a depressing effect on antibody production. It would appear, however, that the reported stimulatory effect is the result of the release of preformed antibody through lymphocyte destruction. The main effect of corticoids, through their lytic action on B cells, is antibody suppression. Inhibition of antibody production has also been demonstrated for androgens. This is particularly the case in fowl, in which hormonal bursectomy can be carried out through the administration of testosterone. On the other hand oestrogens have been reported to enhance antibody production. There is, however, no evidence, as might be expected from this finding, that females are better antibody producers than males or have a greater resistance than the latter to microbial infection.

Enhancement of antibody production is also reported for the growth hormone—

somatotropin—of the anterior pituitary and for thyroxine.

The effect of hormones on cellular immune reactions is less well documented. Corticoids have a suppressive effect on the expression of delayed-type hypersensitivity but do not affect its induction. Delayed allograft rejection has been reported in hypothyroid rats but the effect of thyroxine on cellular immune responses in general is not clear.

EFFECT OF DIET

The effect to which immune responses are affected by diet is controversial. Increased susceptibility to a variety of bacterial infections has been demonstrated by Dubos in mice maintained on a low protein diet. In general, however, antibody production appears to be maintained at normal levels in the face of considerable depletion of body proteins. Vitamin A deficiency is reported as having a suppressive effect on antibody production. This is consistent with the observation that vitamin A administered together with antigen has an adjuvant effect in mice. This effect was observed both when the vitamin and the antigen were inoculated together and also when the antigen was introduced parenterally and the vitamin by the oral route. Pyridoxine deficiency has been reported as having a suppressive effect on antibody production in rats. No specific effect has been attributed to vitamin C deficiency but in scorbutic animals the expression of cutaneous anaphylactic reactions is reported to be diminished.

NATURAL OCCURRENCE OF ANTIGENS AND HAPTENS

Antigenic and haptenic substances are present in all living organisms, occurring either as structural components or as free compounds in the extracellular fluids. Although all species possess serologically distinct antigens, some antigens are widely distributed, the same compound occurring in a variety of species. Antigens of this type are known as *heterophile* antigens. The most studied of the heterophile antigens is the Forssman antigen which is found in the tissues of a variety of animal species as well as in many bacteria. The Forssman antigen is chemically a glycolipid owing its specificity to amino sugar residues.

The naturally occurring antigens and haptens of most importance to medicine are those found in human tissues and cells and those present in and produced by bacteria and other micro-organisms.

Human tissue antigens

Of the antigens occurring in human tissues those present in the red cells, especially those responsible for the ABO blood groups, have been most studied. These antigens are polysaccharide in nature, possibly occurring in the cells as glycolipid, and owe their specificity to sugar and amino sugar residues. There is now considerable evidence that the ABO antigens are all derived from a basic structure resembling that of the pneumococcus type 14 polysaccharide, by addition of appropriate determinant groups. In this process the antigen responsible for group O specificity appears to be an intermediate, from which, in the presence of the appropriate genes either A or B substances may be derived.

The increased importance of tissue transplantation and the recognition of diseases associated with autoimmune responses (p. 85) have in recent years led to an intensive study of tissue antigens. Of particular importance are those which stimulate allograft rejection. These are known as histocompatibility antigens. In man at least three linked genetic loci controlling these antigens have been identified. The antigens are present on leucocytes and are known as HL-A antigens. They appear to be peptides, each possessing one heavy and one light chain. Recent work has established an association between certain non-microbial diseases and various HL-A antigens.

Bacterial antigens

Bacterial antigens are either excreted by bacteria into their environment or present as structural components of the bacterial cell. The antigens excreted by bacteria are the exotoxins and various extracellular enzymes; these are protein in nature. The bacterial cellular antigens are:

Capsular antigens and haptens. In most cases these are complex mucopolysaccharides which in the pure state are poorly if at all antigenic but which are strongly antigenic in the intact cell. At least one capsular polysaccharide—hyaluronic acid—found in the capsules of group A streptococci, is non-antigenic even in the intact cell. This is possibly because of a tolerance (p. 174) induced by its wide distribution in animal tissues. The *Bacillus* genus is exceptional in that certain species, including the anthrax bacillus, possess capsular antigens composed of a glutamyl polypeptide. Some bacterial species possess surface antigens distributed in a layer too thin to be seen as a visible capsule and which are consequently referred to as microcapsular antigens (p. 6). Such are the M proteins of *Streptococcus pyogenes* which can be inactivated by tryptic digestion and the Vi antigen of *Salmonella typhi* and certain other salmonellae.

Somatic antigens. By definition somatic antigens are antigens present in the body of the organism as opposed to antigens present in flagella and capsules. In practice the term is used principally for antigens present in the cell walls of Gram-negative bacteria, particularly the enterobacteriaceae. These are polysaccharide phospholipid protein complexes the serological specificity of which resides in the polysaccharide component. The most superficial determinants of the somatic antigen complexes are known as O antigens. These may be lost by mutation. In this case other deeper lying determinants, whose reactivity is masked by the normal surface determinants, are revealed. This antigenic loss is associated with a change in colony appearance from smooth to rough and the variation is therefore known as S → R variation.

Polysaccharides are also of considerable importance as antigenic or haptenic components of the cell walls of Gram-positive organisms. Thus in the streptococci they constitute the Lancefield group determining or C substances. In Gram-positive organisms the cell wall teichoic acids (p. 5) also possess serological specificity. The differences in serological behaviour between different teichoic acids appears to be determined by the nature of the sugar residues which they contain. The glycerol teichoic acids also appear to possess a specificity due to the presence of glycerophosphate and alanine residues.

Flagellar antigens. All flagellar antigens which have been investigated chemically have been found to be protein but appear to be incomplete proteins in the sense that they do not contain a full complement of amino acids. Flagellar antigens are frequently referred to as H antigens. Occasionally flagellated bacteria lose their flagella by mutation—a loss often referred to as H → O variation. The variants are non-motile and inagglutinable by O antisera. No other properties of the bacteria are affected.

Fimbrial antigens. The composition of these antigens does not appear to have been studied in detail. Since the normal fimbrial antigens of the enterobacteriaceae are heat labile they are probably protein in nature. The specific F antigen on the sex pili of *E. coli* K12 is, however, probably carbohydrate since the capacity of F plus strains to adsorb male-specific bacteriophage is eliminated by exposure to periodate.

II.3

Antibodies and their Properties

Chemically antibodies are globulins, i.e. proteins possessing a low solubility in water and which may be precipitated by various salting out procedures, e.g. by half-saturation with ammonium sulphate. All have been found to contain some sugar residues and are therefore strictly glycoproteins. Highly purified concentrations of antibodies may be obtained by Cohn's technique of precipitation with ethanol in the cold at controlled pH values. Such preparations are widely used for prophylactic purposes. The discovery in 1938 by Tiselius and Kabat that on electrophoresis antibodies migrated in the slowest moving or γ globulin fraction of the plasma proteins laid the foundation for all subsequent work on their characterization. It was soon apparent, however, that γ globulins as defined by electrophoresis were not a homogeneous group. By analysis in the ultracentrifuge two types of antibody molecule could be identified, differing in sedimentation behaviour, with sedimentation coefficients of approximately 7S and 19S respectively. Further evidence of heterogeneity was obtained by the application of the technique of immunoelectrophoresis introduced by Grabar and Williams (p. 139). Three of the molecular species identified by this technique, now known as IgG, IgA and IgM, were shown to have demonstrable antibody activity. IgG and IgA fell into the low (7S) molecular weight fraction defined by ultracentrifugation and IgM into the high (19S) molecular weight fraction. In 1964 it was agreed that the term *immunoglobulin* should be used as a generic term for antibodies and other compounds of related structure and that each immunoglobulin variety should be designated by the symbol Ig followed by a large letter identifying the class to which it belongs.

ANTIGENIC VARIETIES OF IMMUNOGLOBULINS

To date five immunoglobulin classes, IgG, IgA, IgM, IgD and IgE, have been described. All possess or consist of the same type of structural unit made up of two light and two heavy polypeptide chains. Each class possesses a class-specific antigen which is located in the heavy polypeptide chains. The heavy chains of the different classes are designated by Greek letters γ, α, μ, δ, ϵ. The light chains are of two distinct antigenic varieties or types known as kappa (κ) and lambda (λ), both varieties occurring in each class of immunoglobulin. They were initially identified by the use of antisera against Bence-Jones proteins obtained from cases of multiple myeloma, these proteins being in fact the light chains of the myeloma immunoglobulin (p. 93). The light chains of each immunoglobulin molecule are both of the same antigenic type either κ or λ and each molecule contains only one class of heavy chain. In human serum κ light chains are the more numerous, in total comprising about 60% of all light chains. The proportion of κ to λ chains differs in the different subclasses being highest (6·0) in IgG4 and lowest (1·0) in IgG2. By the use of the symbols indicated above for the light and heavy chains each class of immunoglobulin is assigned a molecular formula. Thus IgG with κ type light chains is $\gamma_2\kappa_2$ and IgE with λ chains is $\epsilon_2\lambda_2$.

Differences in antigenic composition are also demonstrable between different individuals of the same species. These differences define *allotypic* specificity. They are due to antigenic determinants present on the immunoglobulin which are controlled by genetic loci at each of which a

Table II.1. Physical properties of human immunoglobulins

	IgG	*IgA*	*IgM*	*IgD*	*IgE*
Physical state	Monomer	Monomer and polymer (serum); dimer (secretions)	Pentamer and hexamer	Monomer	Monomer
Molecular weight	150 000	160 000 (serum) 370 000 (secretions)	900 000	170 000	185 000
Sedimentation coefficient (S)	7	7 (monomer) 11 (dimer)	19 (pentamer)	7	8
Concentration in serum (mg/ml)	13·2	1·6	0·9	0·1	0·000 33
Carbohydrate (%)	2·9	9·9 (serum) 11·7 (secretions)	11·8	13	11·6
J chain	–	+ in dimers and polymers	+	–	–
Secretory component	–	+	–	–	–
Sublcasses	4	2	1	1	1

Table II.2 Biological properties of human immunoglobulins

	IgG1	*IgG2*	*IgG3*	*IgG4*	*IgA1*	*IgM*	*IgD*	*IgE*
Half-life (days)	23	23	16	24	6	5	3	2
Antibody activity	+	+	+	+	+	+	+	+
Placental transfer	+	+	+	+	–	–	–	–
Complement fixation	++	+	++	–	–	++	–	–
Macrophage binding	+	±	+	±	–	–	–	–
Neutrophil binding	+	+	+	+	+	–	–	–
P–K reaction	–	–	–	–	–	–	–	+

number of alleles is possible. Consequently one individual may have one set of allotypic determinants and another individual a different set. The antigenic types defined in this way—Gm, InV, Am and IsF (p. 111)—are known as *allotypes*. They may be thought of as being analogous to blood groups. A third variety of immunoglobulin antigens is known as *idiotypic*. These are specific for antibody of a single antigen combining specificity produced by a single individual. They are therefore characteristic of a single clone of immunoglobulin secreting cells.

Subclasses and subtypes

The classes are divided into subclasses identified by antigens present like the class-specific antigens in the constant regions (p. 108) of their heavy chains. There are four subclasses of IgG (IgG_1, IgG_2, IgG_3 and IgG_4) the subclass present in higher concentration being IgG1 (60–70% of all IgG), and two subclasses of IgA (IgA1 and IgA2). Two antigenic varieties of IgD and IgM have also been described but whether these are subclasses, V region subgroups (p. 109) or allotypes is not at the moment clear. In addition subtypes of light chains Oz+ and Oz– and Kern+ and Kern– are defined by amino acid differences at positions 190 and 153 respectively.

The varieties of immunoglobulins defined by H chain classes and subclasses and light chain types and subtypes, determined by non-allelic genes, and all of which are present in the sera of each individual of a species, are known as *isotypes*. Further isotypic specificities have been identified in IgG but have not so far made the basis of immunoglobulin classification; they include (*a*) subclass-specific isotypes relating to the CH_2 and CH_3 homology regions of Fc; (*b*) subclass-specific isotypes associated with κ chains in the Fab regions of IgG1, IgG2, and IgG4 proteins, which are probably due to a conformation imposed on the κ chain by its association with the Fd (p. 106) region of the heavy chain subclass; and (*c*) isotypes associated with the Fc region and shared by two or more subclasses.

DETECTION OF IMMUNOGLOBULIN ANTIGENS

Isotypic antigens are detected by the use of antisera raised in animals and may be demonstrated by precipitin reactions using the gel diffusion technique or by immunoelectrophoresis. Allotypic antigens were first defined by their capacity to inhibit agglutination of red cells coated with incomplete anti-Rh antibodies by sera containing the 19S immunoglobulin antibodies produced by patients with rheumatoid arthritis (RAGG reagents). Nowadays they are most often determined by using red cells, in some cases chemically coated with immunoglobulin allotypic protein, and as agglutinating sera antisera produced by artificial immunization of the sera of certain normal individuals which contain antibodies either as a result of transfusion or through immunization across the placenta by maternal antigens (SNAGG reagents).

Individual allotypes are normally determined by showing that the serum under test will inhibit agglutination of the coated red cells by antisera containing antibody of known allotypic specificity. Idiotypic antigens have been most studied in rabbits. They were first described by Oudin. Rabbits (donors) were immunized with *Salmonella* antigens and their sera after immunization used to agglutinate a suspension of the organisms used for imunization. The agglutinated organisms were washed and injected into a further series of rabbits (recipients) which developed antibodies to the immunoglobulins of the donors. The antisera of the recipients were specific for the anti-*Salmonella* antibodies of the donor rabbits. Further investigation has shown that anti-idiotypic antibodies react with the antigen combining site of the immunoglobulin. Antibodies to human idiotypes have also been demonstrated by immunization of rabbits with monospecific human antibody, e.g. to dextran.

STRUCTURE OF IMMUNOGLOBULINS

Before detailed examination of their structure can be carried out immunoglobulins must be obtained in purified form. The normal starting material for such purification is serum. Serum from patients suffering from multiple myelomas or from macroglobulinaemia (Waldenstrom's disease)—conditions in which there is a malignant proliferation of immunoglobulin secreting cells—possess very high concentrations of the homologous immunoglobulins, up to 10 g/100 ml. They have the additional advantage that the excess immunoglobulin is restricted to one particular class or subclass, i.e. they are of the same isotype. The urine of patients with multiple myelomas is also of value for analysis since it contains Bence-Jones proteins which are the light chains of the immunoglobulin produced by the myeloma. Bence-Jones proteins have been of particular value in the elucidation of light chain structure.

Purification of immunoglobulins requires the combined use of a number of fractionation procedures. Isoelectric precipitation with ammonium sulphate or sodium sulphate will separate the globulins from other serum proteins. IgM can be separated from other immunoglobulins by density gradient centrifugation and by gel filtration on Sephadex colums. Separation of the lower molecular weight immunoglobulins can be achieved by chromatography on ion exchange resins such as DEAE cellulose with gradient elution in buffers of various pH values and by polyacrylamide gel electrophoresis. Solutions of immunoglobulins may be concentrated by negative pressure dialysis, freeze drying or dialysis against high molecular weight compounds such as polyethylene glycol. The methods described above may be used for the purification of the different immunoglobulin classes but the immunoglobulins recovered will, of course, comprise a range of molecules of different affinities. For the purification of antibody with a specific combining affinity, i.e. capable of combining with a specific antigen, however, the antibody must be recovered from specific antibody–antigen complexes. These may be obtained as precipitates or by treatment of the serum with an immuno-adsorbent. This is a solid supporting medium such as cellulose or an ion exchange resin to which the antigen has been chemically linked. Finally the antibody is eluted from the precipitate or immuno-adsorbent by one or other of various methods, heating, adjustment of pH, treatment with salts and enzymic destruction of the antigen.

IgG

The first major step in elucidating the structure of the immunoglobulins was the discovery by Porter that the molecule could be broken into three fractions by digestion of rabbit IgG with papain in the presence of cysteine. Two of these, with a molecular weight of 45 000, were of identical amino acid composition. Each contained a combining site for antigen, because of which each is known as an Fab piece. The third fragment had a molecular weight of approximately 50 000 and was of different amino acid composition. It contained most of the carbohydrate associated with the antibody molecule and lacked an antigen combining site. Since it was crystallizable it was designated Fc. The Fc fragment was found to possess the isotypic class and subclass-specific antigenic determinants.

The next major advance was the demonstration by Edelmann and Poulik that on treatment with mercaptoethanol in the presence of 6 M urea the IgG molecule could be broken down into smaller fragments. The fragments produced by this procedure were insoluble in aqueous media and lacked immunological function. However, it was found by Fleischman and Porter that soluble fragments which retained antigen binding capacity could be obtained if the SH groups present in the reduced fragment were blocked by alkylation and the fractions separated by dialysis against weak acid. The fragments could be obtained in pure form by gel filtration and it was shown that each consisted of two polypeptide chains—one with a molecular weight of about 20 000 and the other with a molecular weight of about 50 000. These findings led to the four chain model of IgG (Fig. II.3) in which the light and heavy chains are bound together by –S–S bonds extending from the terminus of the light chains to the hinge region or the N terminal end of the C

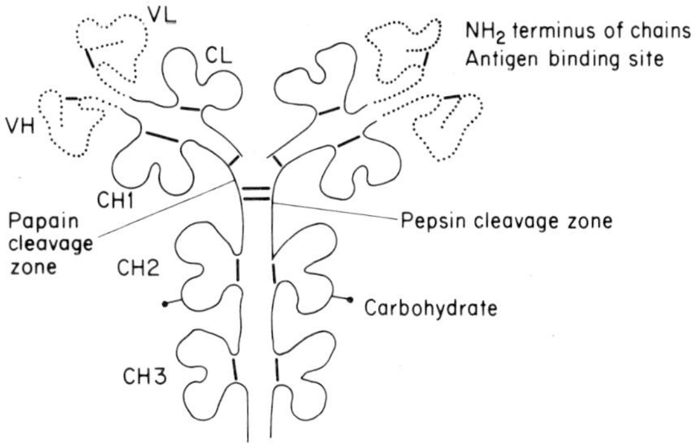

Fig. II.3. The structure of an IgGl molecule. Inter- and intra-chain disulphide bonds are shown as heavy black lines and variable regions are indicated by dotted lines. Homology regions are indicated for the left-hand half of the molecule. Not drawn to scale.

region of the heavy chains and the heavy chains bound together by –S–S bonds in the 'hinge' region of the latter.

The immunoglobulin molecule can be fragmented in other ways. Treatment with pepsin or trypsin will hydrolyse peptide bonds on the C terminal side of the papain site. This results in the liberation of a large fragment $(Fab')_2$ which consists essentially of two Fab-like fragments held together by a disulphide bond or bonds. Since it is divalent, this fragment can, unlike the Fab fragment, cause agglutination and precipitation. Following pepsin treatment a pFc fragment largely consisting of the CH_3 homology region is obtained (p. 111). Cyanogen bromide will hydrolyse peptide bonds adjacent to methionine residues. It has been particularly useful as a reagent for the determination of amino acid sequences.

Since chain separation followed the reduction of disulphide bonds it was apparent that these bonds were of particular importance in holding the chains together in the intact molecule. In the IgG1 molecule one disulphide bond has been identified between the light and heavy chains and two in the hinge region between the heavy chains. The other IgG subclasses differ from G1 in the number of inter heavy chain disulphide bonds. Reduction of disulphide bonds will not itself cause chain separation because the chains are also held together by non-covalent bonds—hydrogen bonds and hydrophobic interactions. These are labilized by treatment with acid, urea or guanidine hydrochloride.

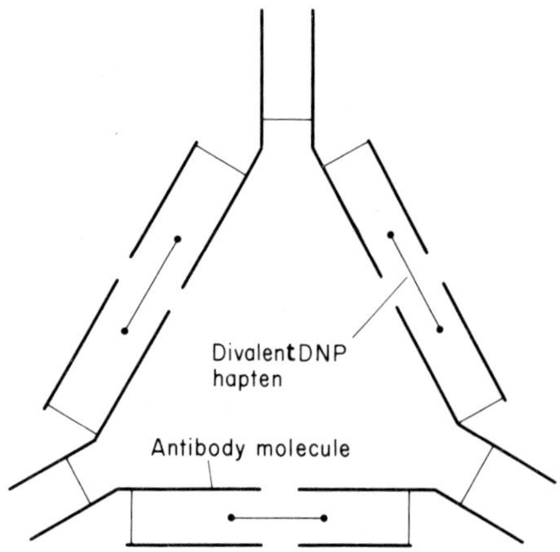

Fig. II.4. The formation of a trimer by the combination of antibody with a divalent DNP hapten.

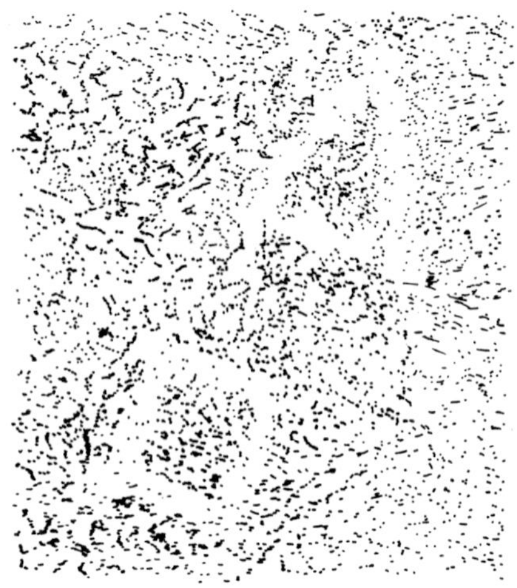

Fig. II.5. Trimer and tetramer complexes formed with antibody and divalent DNA hapten and digested with pepsin. (*Drawn from an electron micrograph of Valentine and Green.*)

In addition to interchain disulphide bonds, there are also intrachain disulphide bonds. Two such bonds have been identified in G1 light chains and four in G1 heavy chains. Each disulphide bond creates a loop containing approximately 60 amino acids. It has recently been found that, though not as susceptible as the hinge region, the region between the loops has a high susceptibility to proteolytic enzymes. This suggests that the loops form folded compact structures or domains with short linear stretches between the loops, the whole comprising a unit of approximately 110 amino acid residues. Two of these domains are in the Fc portion of the molecule and the third in the Fd region, i.e. the constant region of the heavy chain found in the Fab fragment. Initially the domains were envisaged as globular in shape, but recent X-ray studies appear to indicate that they consist of two parallel sheets with a hydrophobic core.

Evidence on the structure of the immunoglobulins has also been obtained from the electron microscopy of antigen–antibody complexes formed with the divalent DNP hapten, DNP-NH-(CH2)8-NH-DNP. The most striking feature of these electron micrographs is the presence of three- and four-sided structures, as shown in Fig. II.5. Dimers and pentamers were also seen. Projections from the sides of these structures are broken down by peptic digestion and clearly represent the Fc portions of the antibody molecules. From the evidence of electron micrographs the IgG molecule when combined with antigen appears as a Y-shaped structure, the arms of the Y corresponding to the Fab region and its stem to the Fc region. The Fab and Fc regions of the combined molecule are about 70Å in length.

The region of the heavy chain joining the Fab and Fc moieties, and which also contains the inter heavy chains disulphide bonds, is known as the hinge region. It is the region which hinges to allow the Fab and Fc regions to assume a variable angle in relation to one another which will permit combination with antigen, leading to the formation of antibody–antigen complexes. Evidence obtained from X-ray scattering experiments indicates that in the uncombined state the molecule has a T configuration with the Fab and Fc regions at right angles to each other. From fluorescent depolarization studies it would appear that there is not, as might be expected, completely free mobility at the hinge region. Electron microscopy with divalent DNP haptens indicates, however, that the Fab regions can rotate through an angle of up to 60° from their position in the uncombined T conformation.

By controlled reduction it is possible to split the inter heavy chain disulphide bond(s) but without affecting the other disulphide bonds. This results in the release of two half-molecules, each consisting of one light and one heavy chain. If the half molecules are held at an acid pH and mixed with half-molecules of antibodies of different specificity prepared in the same way they combine randomly to form molecules with antibody sites of two different specificities as well as molecules with antibody sites of identical specificity. Antibodies with mixed specificities do not, however, occur under natural conditions, although they have been of considerable value in research work. Hybridization is due to covalent (disulphide) bond formation and non-covalent bond formation between paired heavy chains. Heavy chains of one immunoglobulin class have a highly specific tendency to pair with other heavy chains of the same class but a low tendency to pair with heavy chains of a different class. Thus, Gray, Abel and Zimmerman found that in mixtures of IgG and IgA heavy chains 95% of the pairs formed between IgG chains and only 5% between an IgG chain and an IgA chain.

IgM

On mild reduction with mercaptoethanol IgM dissociates into 7S units with the same four-chain structure as IgG. From its molecular weight IgM normally occurs as a pentamer of five 7S subunits. Occasional hexamers have been found in electron micrographs. Electron microscopy shows the subunits in the uncombined molecule, and often also in the combined molecule, to be radially disposed in a stellate formation with the arms facing out. In company with IgE, IgM has been found to possess four homologous loops or domains in its heavy chain, three of which are in the Fc region, and the fourth in the Fd region.

IgM, as well as dimeric IgA, possesses an additional polypeptide chain which is known as the J chain with a molecular weight of approximately 24 000 daltons. The J chains of polymeric IgM and IgA have been found to be identical both antigenically and in amino acid composition. They have no sequence homology with any of the Ig regions or with the secretory component of IgA. They appear to have a function in joining the subunits together by binding to the C terminal ends of the H chains, apparently through the formation of disulphide bonds. Only one J chain is found per polymer molecule. Each contains four half cystines—two of which form intra-J chain bonds and two of which link the J chain to adjacent immunoglobulin heavy chains. It has been suggested that IgM J chains have two important functions: they are necessary for the initiation of polymerization and they act as a signal for the IgM to IgG switch (p. 116). J chains are produced along with immunoglobulin in plasma cells.

IgM has a considerably higher carbohydrate content than IgG. At least five carbohydrate residues have been identified in each IgM subunit, two being located in the Fc region, one in the Fd region and two in the hinge region. There is also one carbohydrate residue associated with the J chain.

The valency of IgM is in doubt. There is some evidence that it may be 10 for small antigens but only five for various large antigens, possibly as a result of steric hindrance by the antigen. There is also evidence that the two Fab regions of each monomer may differ in combining affinity, one having a relatively high affinity for antigen and the other having a low affinity. Each monomer may, therefore, possess one 'strong' and one 'weak' arm.

Electron micrographs of IgM in combination with polymerized flagellin of *Salmonella typhimurium* and with polio viruses have provided evidence on the configuration of the molecule in the combined state. Three major types of arrangement have been observed:

1. A stellate arrangement with a radial disposition of the monomers in a single plane.
2. A 'staple-like' arrangement in which the Fc portion is in one plane and the Fab regions are disposed at an angle to the Fc portion but all pointing in the same direction, the molecule having a crab-like appearance.
3. The Fab portions are arranged at an angle to the Fc region but different Fab regions point in opposite directions. From these appearances it would seem that the Fab regions have considerable rotatory mobility in relation to the Fc region.

IgA

In human serum IgA appears as a monomer but in secretions as a dimer in which the subunits are complexed with a fucose containing glycoprotein known as the secretory component (SC) which has a molecular weight of approximately 67 000. Secretory IgA constitutes the bulk of the IgA produced by the body. The interchain disulphide bonding of the two subclasses of IgA (IgA1 and IgA2) is different. In IgA1, disulphide bonding is similar to that in IgG molecules, disulphide bonds being formed between adjacent heavy chains and between heavy chains and light chains. In IgA2 molecules with Am allotypic specificity, however, there are no disulphide bonds between the heavy chains and the light chains. Instead there are disulphide bonds between the two light chains and between the two heavy chains, in each case occurring at the C terminus of the chain—the light and heavy chains being non-covalently bonded. Although in man serum IgA occurs in monomeric form, in various animal species, dimeric serum IgA is found.

The secretory component is apparently synthesized in mucosal epithelial cells rather than in plasma cells and is conjugated through disulphide linkages with IgA dimers formed from cells in the lamina propria of the gut after secretion. It is possible to demonstrate combination of the secretory component with IgA monomers under in vitro conditions. Under these conditions, however, the secretory component is also capable of complexing with IgG and IgM but has a very much higher affinity for IgA. It is possible that the secretory component serves to protect IgA from the destructive action of proteolytic enzymes present in body secretions, particularly of the alimentary tract. In addition to the secretory component each molecule of dimeric IgA possesses, like IgM, a J chain which is bonded to the heavy chains by disulphide linkages and binds the monomers together, or contributes to the binding. There is some evidence that the J chain is necessary for the binding of SC to IgA monomers.

IgE

IgE was first identified by Ishizaka in the serum of atopic patients who showed immediate-type hypersensitivity to ragweed and subsequently by Johannsen and Brannich in the serum of a patient with multiple myeloma. Its role in immediate anaphylactic type hypersensitivity in man has now been conclusively established (p. 187). IgE differs from other immunoglobulin classes in its heat sensitivity, being rapidly inactivated at 56°C, and in its capacity for fixation through its Fc portion to basophilic polymorphonuclear leucocytes (mast cells) of the same species. Because of this it is referred to as homocytotropic in contrast to other immunoglobulins which are known as heterocytotropic since they fail to combine with basophils of the producing species but combine with basophils of other animal species. IgE is unable to cross the placenta from the maternal to the fetal circulation or to fix complement but can agglutinate red cells to which its homologous antigen has been chemically coupled. Like IgM the constant region of the heavy chain has four disulphide bonded loops. The concentration of IgE in normal serum is very low but much higher concentrations (up to 1 μg/ml) may be found in the sera of atopic individuals and of patients with nematode infestations. IgE has also been demonstrated in nasal and bronchial secretions but, unlike secretory IgA, secretory IgE has not been found to be associated with a secretory component. It is possible, therefore, that IgE associated with mucosal surfaces may be rapidly inactivated by bacterial and other proteases.

IgD

This immunoglobulin was first recognized in a myeloma protein. It occurs in very low concentrations in normal serum. It possesses antigen-binding capacity but has not so far been extensively investigated. Recent work has identified IgD in association with IgM on the surface of B lymphocytes. It has been found in the absence of IgM on umbilical cord lymphocytes.

SEQUENCE DATA

In recent years an enormous amount of information has been accumulated on the amino acid sequences of immunoglobulins. As sources of immunoglobulin for this purpose, myeloma proteins have been of particular value since, being derived from single clones, they are chemically homogeneous. In addition individuals with myelomas as a rule excrete the light chains of the myeloma protein in the urine; these have long been known in clinical medicine as Bence-Jones proteins.

Examination of the sequences of the light chains of two myeloma proteins by Hilschman and Craig revealed virtually identical amino acid sequences in the C terminal half of the chains with marked differences in sequences in the N terminal half. Further study has confirmed and extended these findings and it is now established that both light and heavy chains of all immunoglobulins have an N terminal variable region (V) of approximately 110 residues and a constant C terminal region (C).

The C terminal regions are constant in sequence for each class and/or subclass and within each of the κ and λ types, apart from the allotypic differences which in the human have been demonstrated in κ light chains and in IgG and IgA heavy chains, i.e. InV, Gm and Am respectively, and the isotypic differences which define the Oz and Kern subtypes of λ chains.

In man amino acid variability has been found which corresponds with some of the known allotypes (p. 112). The most clearly defined are a leucine/valine interchange at position 191 on κ chains which defines the InV allotypes. InV 1, 2 has a leucine residue in this position while InV 3 has a valine residue. The designation InV 1, 2 indicates reactivity with antisera possessing two different specificities. These differences may reflect differences in the specificity of the antisera for differences in orientation in the region of the molecule containing residue 191. Allotype-related sequence differences have also been found in human heavy chains, e.g. an arginine/lysine interchange at position 214 on Gamma 1 chains defines the Gm 4 allotype. Three amino acid interchanges Asn–Glu–Ileu/Glu–Gln–Met at positions 356, 357, 358 on Gamma 1 chains are associated with the Gm1 allotype and a Phe/Tyr interchange near the C terminus of the Gamma 3 chain define the Gm5 and Gm21 allotypes. Allotype associated changes determined by the d and e loci of rabbit immunoglobulins have been identified at positions 214 and 309 on the heavy chain. The Kern and Oz subtypes of λ chains are defined by a serine/glycine and lysine/arginine interchange at positions 153 and 191 respectively.

The variability found in the V region of each chain is fundamental to the recognition function of the immunoglobulins. Nevertheless the V regions show a considerable degree of sequence regularity. There is firstly a regularity derived from the overall resemblance of individual chains by virtue of which it is possible to distinguish VK, VH and VL sequences. These three sets of sequences can be distinguished from each other and the individual members of each set show considerable resemblence to each other. For each set it is possible to identify a basic sequence defined by the residue occurring most frequently at each position. In determining these residues it is necessary, in order to secure maximal homology, to make adjustments in alignment at positions at which deletions or insertions are presumed to have occurred. The basic sequence defined in this way has not been found to occur in any immunoglobulin so far examined. It may be regarded as a sequence from which all variable regions have developed in the progress of diversification. Each V family, VL, VK and VH, is associated, as indicated by its name, exclusively with the corresponding C region sequence in the intact immunoglobulin molecule. In the formation of complete immunoglobulin molecules the VL sequences combine only with CL, VK with CK and VH with CH regions. In the V regions of the λ, κ and heavy chain families a number of subgroups can be distinguished, the individual members of each subgroup showing a very high degree of resemblance to each other. Three such subgroups are recognizable in κ light chains, and five in λ light chains and three (or four) in heavy chains.

In each V region subgroup the variability which occurs at different positions is not completely random. For some positions exhibiting a low variability only one alternative amino acid appears to occur. There are, however, certain regions known as 'hot spots' or 'hypervariable' regions with a high degree of variability at which four or five amino acids may occur. In a human IgG protein hypervariable regions have been described by Cohn at positions 24–34, 50–55 and 88–97 in light chains and at positions 30–35, 50–65, 81–88 and 95–102 in heavy chains. Since in this protein intrachain disulphide bonds unite half cystines at positions 23 and 88 in the light chains and at positions 22 and 92 in the heavy chains, two of the light chain and three of the heavy chain hypervariable regions are brought close together in the intact molecule. These juxtaposed regions are ideally suited to constitute the antibody site and the evidence from affinity labelling (see below) strongly indicates that they do so. On this basis the residues at the N terminal end of the light and heavy chains do not, as might be imagined, contribute to the binding site. In fact the degree of variability found from positions 1 to 23 of both light and heavy chains is quite insufficient to constitute a basis for antibody diversity. In Porter's original work the antigen-binding site was found to be located in the Fab fragment prepared by papain hydrolysis. More recently it has been shown that a fragment produced by pepsin hydrolysis of Fab which contained light and heavy chain variable regions and little if any of the C regions possessed antibody activity against the DNP hapten. This is conclusive evidence, the first obtained that the binding site is located in the variable region.

The role of hypervariable regions in the construction of the antigen combining site is supported by the results of affinity labelling experiments. The procedure adopted in these experiments is to react antibodies prepared against hapten protein conjugates with hapten containing a substituted diazonium or bromacetyl group. The substituted hapten then binds covalently to certain amino acid residues, particularly tyrosine, lysine and histidine residues, in or near the antigen combining site. The heavy and light chains are then prepared from the antibody molecule and the amount and location of the labelling hapten identified by sequence analysis combined with specific tests for the hapten. By the use of this technique the labels were found in one case to attach to amino acids in position 32, 33, 54 and 94 of the light chains, all of which residues are within the hypervariable regions. Labelling in heavy chain regions of high variability has also been found but the heavy chains are considerably less labelled—one-third to one-quarter—than light chains. On the other hand, separated heavy chains have been reported to bind more antigen than light chains. The reason for these different findings is not known but both indicate participation of both light and heavy chains in the combining site.

Another technique which has been used in analysis of the combining site is that of site modification. In this technique the antibodies are exposed to reagents which selectively modify certain amino acids and in so doing block combination of hapten with its specific antibody. The amino acids most susceptible to specific modification are lysine, arginine and tryptophane and each has been shown to be involved in the reactions with certain haptens. Reference has been made (p. 103) to the occurrence of idiotypic antigens which are specific for immunoglobulins of a particular combining specificity. Evidence that these antigens are located in the combining site has been obtained by a number of workers in the findings that the combination of antibodies with anti-idiotype sera can be blocked by prior exposure of the antibody to the homologous hapten. Other workers have found; however, that the same idiotypic specificity may be present in antibodies of different combining specificity. It is possible therefore that idiotype specificity at times depends on residues in the combining site which are not directly involved in combination with antigens.

The size of the combining site has been estimated by determining the number of haptenic determinants which give maximum binding to antibody or which most efficiently inhibit binding of the homologous antigen. Hexasaccharides and hexapeptides were found to give maximal binding and/or inhibition. These figures are similar to those obtained for the number of molecules of substrates binding to the active site of the enzyme lysozyme for which it has been estimated that the site is composed of from 15 to 20 amino acid residues. It is probable that the same number of amino acids could be contributed by the juxtaposed hypervariable regions of the light and heavy chains of immunoglobulins. From the data obtained by X-ray analysis the binding site constitutes a cavity between the heavy regions, the cavity having a different size and shape in different molecules. There is some evidence that the size of the site may be increased by prolonged immunization. This would explain the increase in antibody affinity associated in prolonged immunization with broadening of its specificity.

INVARIANT RESIDUES OF VARIABLE (V) REGIONS

A striking feature of the invariant residues is the frequency with which the amino acid glycine has been identified in close relation to the hypervariable regions. Glycine is of fundamental importance in three-dimensional structure of immunoglobulins and this allows very free rotation around two bond angles ϕ and ψ. This confers a high degree of flexibility on the coiling of the heavy chain, thereby facilitating the association of the site regions of homologous light and heavy chains. The appreciable number of invariant residues other than glycine and the considerable number of locations at which only limited amino acid substitution is possible confers, on the other hand, a stable background for V region variation; a background which appears to have remained constant far back into evolutionary time. Such stability clearly provides the necessary structural basis for light and heavy chain association. A similar function can be assigned to the constancy of the C region. In addition the stability of the C regions permits them to carry out non-specific biological functions, namely complement fixation, placental transfer, skin fixation, control of metabolic decay, polymerization and binding to phagocytic cells as a preliminary to phagocytosis, attributed to these regions.

HOMOLOGY REGIONS

Sequence homologies, suggesting origin from a common gene which has duplicated itself during the course of evolution, have been identified within both polypeptide chains. Thus a κ type IgG_1 myeloma protein which has been extensively examined by Edelman and Poulik has been found to show the following areas of homology:

1. Between VL and VH.
2. Between CL and three sections of the C region of the heavy chain, each corresponding to approximately one third of the chain which have been designated CH 1, CH 2, and CH 3.

No significant homologies have been identified between V and C regions.

In relation to their possible evolution from a single gene it is significant that each homology area is characterized and limited by the presence of a single disulphide bond. Homology has also been demonstrated between human light chains and those of other animals, notably between λ light chains and light chains of rhesus monkeys, baboons, guinea-pigs, mice and dogs and between κ light chains and light chains of rhesus monkeys, baboons, sheep, cows, rabbits, pigs, guinea-pigs, mice and rats. All these homology data point to a primordial sequence of about 107 amino acids from which the entire set of immunoglobulin genes has been generated by gene reduplication of a 100–117 residue unit.

In this context the discovery by Berggård and Bearn, in the urine of patients suffering from a variety of disease states in which renal tubular reabsorption was affected, of a peptide β2m of molecular weight 12 000 containing a loop forming domain defined by a single disulphide bond is of considerable interest. In its amino acid sequence, when aligned for maximum similarity, it bore considerable resemblance (21–7%) to the homology regions of IgG. The degree of resemblance was of the same order as that of the various homology regions to each other. The peptide was later found to be identical to the light chains of the HL-A histocompatibility antigens, to the heavy chains of which it appeared to be bound by non-covalent bonds. It did not carry any of the antigenic specificity of the HL-A antigens. A considerable amount of the peptide present on the surface of the lymphocytes appeared to be in a free form. Its function is unknown but its existence suggests the possibility that the histocompatibility antigens and the immunoglobulins have evolved from the same precursor.

POLYSACCHARIDE

The amount of carbohydrate associated with the molecule is different for the different classes of immunoglobulin, being minimal with IgG and maximal with IgM and IgE. The carbohydrate appears to be almost invariably restricted to the heavy chains. In IgG, there is a single polysaccharide molecule attached to the CH_2 domain. In an IgM protein five molecules have been described at positions 170, 342, 405, 412 and 574. The sugars found include fucose, mannose, galactose, N-acetyl glucosamine, N-acetyl galactosamine and N-acetyl neuraminic acid. The function of the polysaccharide residues is unknown but their location suggests that they may play some part in the biological functions of the constant regions.

ALLOTYPES

In addition to *isotypic* antigens which are present on the immunoglobulins of all individuals of a particular species, there are *allotypic* antigens which differ in different individuals of the same species. Allotypic antigens are controlled by autosomal genes which obey classical Mendelian laws of segregation. Allotypes are of considerable importance in genetic analysis. In particular, linkage relationships between genes coding for immunoglobulin structure can only be established by showing that genetic markers for the corresponding immunoglobulin or immunoglobulin regions are transmitted together. In man the determination of immunglobulin allotype is, like blood grouping, of considerable potential practical value in cases of doubtful paternity.

Human allotypes

The existence of human immunoglobulin allotypes was first recognized by Grubb and Laurell who in 1956 identified the first Gm allotype, Gml, which they found to be present on the immunoglobulins of approximately 50% of Caucasians. Four independent systems of human immunoglobulin allotypes are now recognized. These are Gm on γ chains, Am on α chains, InV on κ light chains and Isf which is associated with IgG1 heavy chains. A number of Gm markers and the InV markers have been shown to be correlated with differences in amino acid sequences in the relevant chains. It is presumed that all the Gm markers and the single Am marker will eventually be shown to be due to a sequence difference also. The status of the Isf marker is not clear since though present on IgG1 chains it has not been found to exhibit linkage relationship with any of the Gm markers. It may be associated with the carbohydrate residue attached to the G1 chain. Isf is unusual in that its frequency increases with increasing age in Caucasians but not in Negro populations.

Gm markers. Two systems of nomenclature have been used to designate human Gm markers—by small letters of the alphabet and by Arabic numerals; the latter system will be adopted here. The different subclasses of IgG carry different Gm markers (Table II.3). All those whose location has been identified so far have been found to be associated with the C regions of the heavy chains. Several have been assigned to a particular homology region, some to various Fc fragments obtained by proteolytic digestion; a few have been shown to be due to amino acid interchanges in particular positions while others are associated with differences in sequence over a stretch of the chain. Thus Gm3 and Gm17, whose determinants are true alleles (homo-alleles), have lysine and arginine respectively at position 217 while Gm1 is associated with sequence differences from positions 355 to 358.

Table II.3. Subclass distribution of Gm allotypes

Heavy chain subclass	*Gm allotypic markers*
G1	1 2 3 17 18 20
G2	23
G3	5 6 11 10 13 14 15 16 21 24 25
G4	–

Non-markers. This term is applied to antigens which are allotypic markers for one chain but which occur in other chains as non-allotypic markers. Thus non-1 is a homo-allele to Gm1 on IgG1 chains but occurs on all G2 and G3 chains on which it is therefore a non-allelic marker. A number of such non-markers which are allelic for one chain and which are non-allelic for others have been identified, i.e. non-5, non-11 and non-21 which occur on all G1 and G2 chains. A slightly different type of non-allelism is shown by the determinants (4a) and (4b) which are allelic on G4 chains; 4a, however, is present on all G1 and G3 chains and 4b on all G2 chains.

The Gm factors appear to have arisen comparatively recently in mammalian evolution, not being found in the lower primates. Polymorphism in a number of Gm factors is, however, demonstrable in chimpanzees and orang-utans. It has been suggested that Gm1 and Gm-1 sequences have both been derived from a precursor sequence present in old world monkeys.

Am. Only one marker, known as Am1, has been identified on IgA proteins. It occurs in some proteins of IgA2 subclass. Antisera specific for the allelic alternative have not been found but it is usually referred to as Am-1.

The genes controlling the Gm and Am markers present in any one individual are closely linked, though occasional cross-overs resulting in reassortment have been identified in family studies of their inheritance. Each individual possesses two sets of genes coding for their allotypic markers. Each set comprises a complex allele or haplotype, only one such haplotype being expressed at the level of the immunoglobulin producing cell. This phenomenon is known as allelic exclusion, since the alleles are codominants some cells express one haplotype and some cells the other. Individual immunoglobulin molecules express only one haplotype but the totality of immunoglobulin molecules each individual possesses comprise all the markers represented in his genotype.

InV markers occur on κ light chains. Two allotypes are distinguished, InV 1, 2 and InV 3. κ chains lacking InV determinants are known but are very rare. Though present on light chains full expression of InV allotypes is only found in intact immunoglobulin molecules or in heavy light chain combinations. Isolated light chains have only about one-fiftieth of the activity of the light heavy combination. InV 1, 2 proteins differ from InV 3 proteins in a leucine/valine interchange at position 191. Some sera react with all InV 1, 2 proteins, others only with some of them, allowing the distinction to be drawn between InV 1 and InV 1, 2. A specificity for InV 2 has not, however, been defined. It is possible that the differences in reactivity of sera used for the detection of InV reflect differences in some other site of the molecule which induces a conformational change at position 191.

Rabbit allotypes

Genetic markers have been identified in rabbit light chains of both κ and λ types and on IgG, IgM and IgA heavy chains. The major rabbit allotypes and their locations in the immunoglobulin molecule are shown in Table II.4. Rabbits are unusual in that some markers, those determined by locus *a*, are present on the heavy chains. The discovery of heavy chain markers was the first indication that V and C regions may be coded for by separate genes.

Investigation of the mode of inheritance in rabbit allotypes has shown that (*a*) genes coding for κ and λ light chain determinants segregate independently of each other and also independently of the genes for the heavy chain determinants and (*b*) the V region heavy chain markers and C region heavy chain markers behave as a single linkage group.

GENETIC CONTROL OF IMMUNOGLOBULIN STRUCTURE

As we have seen, each animal is capable of producing a very large number of different types of immunoglobulin molecule, differing in respect of class and subclass, as defined by amino acid sequences in the C region, and in antigen combining specificity, as determined by amino acid sequences in the V region. Undoubtedly the simplest explanation for this heterogeneity is that genes for each distinct type of immunoglobulin molecule are carried in the germ line. But since there exist only a limited number of C region sequences it is quite unnecessary to postulate duplication of C region genes for each V region sequence. The most satisfactory model would be one in which the V and C regions were coded for by separate genes and this is the model now generally accepted. The number of C region genes required on the basis of present knowledge would be one for κ, four for λ and 9 or 10 for heavy chains. No further somatic diversification of any of these genes is required. The concept of separate genes coding for the constant and variable regions of immunoglobulin chains is supported by a number of lines of evidence. Firstly sequence data show that (*a*) each of the heavy chain constant region classes and subclasses may be associated with V regions of any combining specificity and (*b*) random association of C subclasses with the corresponding V subgroup sequences. Secondly it is

Table II.4. Major rabbit allotypes

Locus	*Allotypes*	*Location chain*	*Region*	*Sequence difference*
a	a1, a2, a3	H	V	Multiple
b	b4, b5, b6, b9	κ	Probably C	Multiple
c	c7, c21	λ	Probably C	
d	d11, d12	γ	C	Met/Thr at 225
e	e14, e15	γ	C	Thr/Ala at 309
f	f71, f72, f73	α	C	
g	g74, g75	α	C	
Ms	Ms2, Ms4, Ms5, Ms6	μ	Probably C	

Allotype Ms1 is probably a VH marker and Ms3 a CK marker expressed only in the presence of μ chains.

supported by genetic evidence: (*a*) in the rabbit the VH region allotypes controlled by the a locus (p. 113) are found in association with IgG, IgA, IgM and IgE constant regions; (*b*) the discovery of a patient with two myeloma proteins of IgM and IgG2 classes which shared VH regions of the same idiotype and of apparently identical amino acid sequences; and (*c*) the demonstration in the rabbit of IgG and IgM anti-*Salmonella typhi* antibodies sharing the same idiotype and with identical V region sequences.

Linkage relationships

These can be conclusively established only by examination of the segregation patterns of genetic markers. The evidence is incomplete but all points in the same direction. Firstly, in the rabbit, linkage has been established between V and C region markers for heavy chains. In man linkage has been established between Gm markers on IgG and the Am marker of IgA and a lack of linkage between Gm markers and the κC region markers InV 1, 2 and InV 3 (p. 112). In the rabbit, the κ, λ and heavy chain C region markers are unlinked. These findings are most readily explained by postulating three unlinked families of genes, for κ light chains, for λ light chains and for heavy chains. This model is supported by the association shown by sequence analysis of κ, λ and heavy V region subgroups exclusively with κ, λ and heavy C region sequences respectively.

Genetic basis of V region variability

Although there is general agreement on the broad outline of the scheme presented above there is not general agreement on the minimum number of V region genes required. There are two contrasting views:

1. Each V region specificity is coded for by a separate gene and a somatic mechanism for the generation of diversity is not required. This is the germ line hypothesis. It postulates that all information for the synthesis of individual V regions is carried in the germ line.
2. The number of V region genes is limited but is expanded prior to contact with antigen by a somatic mechanism.

At the moment it is not possible to decide which of these two contrasting views is correct. Attempts to determine the number of V line genes in the mouse have given values from 40 to 500. The upper limit would be consistent with the germ line theory, the lower limit with the somatic variation theory. The germ line theory has the inherent attractiveness that it does not postulate any novel type of genetic mechanism for the generation of diversity. This would occur as a result of random germ line mutations in the course of evolution with selection by normal environmental pressures. The recent finding that in rabbits and mice idiotypic specificities which are associated with the combining site exhibit classical Mendelian inheritance is a particularly strong argument in favour of the germ line theory.

If we postulate that there are 500 VL and 500 VH genes in the germ line then, assuming that there is completely random association of light and heavy regions (the evidence so far obtained suggests that there is considerable but not complete freedom of association), the total number of specificities which could be generated is 250 000. If the humoral immune mechanism is regarded as having being derived in the first instance for defence against invading micro-organisms, such a figure could well be in excess of the number of combining specificities required. It would also be within the capability of the mammalian genome requiring at most 0·0025% of the total DNA. If, however, only about 10% of light chains and heavy chains could combine the demands for 250 000 combined specificities would require 0·025% of the genome. The germ line theory must therefore be regarded, from the point of view of the demands it makes on chromosomal DNA, as distinctly within the bounds of possibility, but only if the requirements for V region sequences and the restrictions on light heavy chain association are not significantly greater than those indicated above.

The principal difficulty in the germ line hypothesis is that it does not explain the maintenance, for possibly some millions of years, of genes determining antibodies against antigens which the organism is unlikely ever to encounter, such as the range of synthetic antigens to which it is exposed by human intervention. In this context it has been suggested that they may be a much greater sharing of specificities than was at one time thought, so that certain naturally occurring antigens may share antigenic patterns with synthetic antigens. A further difficulty in the hypothesis is that it presents us with the problem of explaining the maintenance of V region allotypes in the rabbit (p. 113) and of V region subgroups in man and other animals. It is difficult to see how these differences could have been preserved in the course of evolution, since they would be expected to be eliminated by occasional recombination.

The somatic variation theory proposes that a limited number of V region genes is present in the germ line and that these are expanded by some somatic mechanism during ontogeny to yield molecules with a large number of individual specificities. The finding of a large number of invariant sequences in the V regions (24 were identified in VHI, VHII and VHIII regions in two mouse myelomas and in rabbit and guinea-pig pools by Capra and Kehoe) constitutes strong evidence for a limited number of V region genes diversified by some somatic process. The principal difficulty in the somatic variation theory is that it requires a very extensive variation occurring during a short period of embryological development.

Expansion of V region genes could occur as follows:

1. By random somatic mutation. This is the most widely accepted possible mechanism and was that first proposed by Burnet as a corollary of his clonal selection hypothesis. Somatic mutation, however, other than that which may occur in malignant cells, is a novel genetic mechanism for which there is no factual evidence and no parallel in other genetic systems.

2. A second possible mechanism proposed by Jerne is that each individual possesses two subsets of genes: a set (A) which determines antibodies which react with the individual's own histocompatibility antigens and a set (B) coding for antibodies reacting with the histocompatibility antigens of other members of the species. Jerne proposed that genes of subset A are driven by a self-histocompatibility antigens, with which the antibodies they control react at a very early stage of embryonic development, so that they produce antibodies with random non-self specificities. This ingenious theory which appears in the first instance to have been proposed because of the close genetic relationship between histocompatibility antigens and immune response genes in mice is lacking firm experimental support. Such support might be obtained from an examination of the range of responses produced by animals possessing a wide range of self-histocompatibility antigens, e.g. heterozygous, allophenic mice as compared with mice homozygous for one set of parental strain histocompatibility antigens.

3. The third possible origin for somatic variation is by recombination between limited sets of V genes. Evidence for such recombination as a somatic event has not been obtained and it has not been possible to explain the generation of any set of immunoglobulin V region sequences from known or possible prototype sequences by this method.

Integration of V and C regions

If, as seems likely, the V and C regions are coded for by separate genes, it becomes necessary to postulate some mechanism for the association of the V or C genes or their products in the antibody producing cell. The available evidence rules out the likelihood that integration occurs at the level of the polypeptide chain. Pulse-labelling experiments with labelled amino acids have shown a gradient in label concentration from the C terminus to the N terminus of the chain indicating the presence of a single growing point starting from the N terminus in the polypeptide. If there were more than one growing point the label concentration would show discontinuity throughout the chain. The size of the polysomes synthesizing each chain agrees with the chain length data. Thus the L chains of mouse myeloma protein are synthetized in 190S polysomes with a 9–11S messenger RNA and the heavy chains are synthesized on 270 polysomes with 14–16S messenger RNA. The size of the RNA molecules involved, which have been isolated and used to effect chain synthesis in vitro, is somewhat in excess of that required for the corresponding chain. This, however, is a standard feature of eucaryotic translational systems. The complete stability in composition which has been shown in successive generations of myeloma proteins argues against integration at the level of the polypeptide chain or at the level of messenger RNA since in both these processes mistakes might occasionally be expected to occur.

It seems most likely that integration occurs at the DNA level. This implies some form of translocation or juxtaposition during the course of differentiation from germ cells to producer cells of a V region gene to the C region, or vice versa, and their combination to form a new gene. A plasmid-like mechanism has been proposed by Edelmann and Gally. This envisages:

1. Excision of a V gene by pairing and recombination between the homologous regions of adjacent V genes.

2. Translocation of the excised V gene to the C region.

3. Integration of the V gene into the C region by pairing and recombination. The entire process is analogous to plasmid integration and release in bacteria.

This attractive model explains gene variability and translocation by the same mechanism. In addition it is suggested that new nucleotide sequences generated by recombination between the V and C genomes would provide a stimulus to transcription. The principal difficulty in the model is that it does not explain the restricted association of V and C genes of the same family. Since homology has been demonstrated between the CH1, CH2, CH3 and CL polypeptide regions it might be expected that if during translocation the excised gene is free of the chromosome VH might associate with CL as well as with CH. A model which would explain this restricted association would be by the formation of a loop in the genome in such a way as to bring the appropriate V region gene into proximity with the C region gene of the immunoglobulin to be synthesized. The loop hypothesis provides a neat explanation of the IgM to IgG switch in immunoglobulin production in single cells for which there is now considerable evidence.

Phenotypic expression

Although each individual must carry genes coding for different immunoglobulin classes, subclasses or subgroups in the germ line, at the level of the producer cell, only genes for one type of light chain and one type of heavy chain are expressed in the immunoglobulin product. This involves two types of restriction in the antibody producing cell, on the expression of the germ cell genome:

1. Only one of each pair of chromosomes carrying immunoglobulin determinants is expressed. This is known as allelic exclusion. It is demonstrated by the finding that in animals heterozygous for allotypic markers only one marker per region is found in the product of a particular cell.

2. On each chromosome only one V gene and one C gene of either κ or λ clusters and one VH and one CH gene are expressed. All the other immunoglobulin genes on the chromosome are excluded. This is known as class exclusion.

As a result of these exclusion processes each immunoglobulin molecule produced possesses two identical light chains and two identical heavy chains. The molecule has, therefore, two identical combining specificities. The biological objective of the exclusion process must therefore be to secure this identity, which is an essential requirement for the expression of antibody function. At the moment we are in the dark as to the mechanisms involved. It could be due either to the deletion or to the repression of immunoglobulin genes other than those immediately required by the producing cell. There is reason to think that it may occur at a comparatively late stage in the differentiation of the producer cell since it has been shown that in the earlier stages of immune response, the surface receptors of individual lymphocytes may be of more than one immunoglobulin class or subclass and of more than one allotypic specificity.

IgM to IgG switch

In the early stages of the immune response both IgM and IgG antibodies are produced; agglutination test systems usually show a rise in IgM before there is demonstrable IgG but this may depend to some extent on the high sensitivity of the agglutination reaction for the estimation of IgM. As the response progresses IgM antibody is no longer produced in significant amounts but since it is rapidly metabolized its concentration in serum begins to fall. On the other hand the concentrations of IgG antibody continues to increase so that in the later stages of the response the immunoglobulin produced may be virtually entirely IgG. This progression suggests that cells which were producing IgM in the early stages of the response might be switched to IgG production in the later stages. The occurrence of this postulated IgM to IgG switch in individual cells is supported by considerable though not conclusive evidence:

1. Morphologically IgM producing cells appear to have a more blast like character than IgG producing plasma cells.

2. Chicks which have been exposed to anti-IgM in ovo subsequently fail to produce IgM or IgG on exposure to antigen.

3. The demonstration in rabbits and in man of IgM and IgG molecules of identical idiotype. Idiotypic specificities are characteristic of the V regions of an individual clone of immunoglobulin producing cells.

4. The finding by Fudenberg in the same human patient of both IgM and IgG myeloma proteins possessing identical L chains and identical amino acid sequences in their VH regions. The identity of V regions in both proteins is taken as conclusive evidence that they are produced by the cells of the same clone, indicating that in the same cell a specific V region gene can associate with μ and γ chains. This finding also provides strong support for the hypothesis of separate V and C region genes.

II.4

Antibody Production

The clonal selection theory of antibody production postulates that the role of antigen in the humoral immune response is to stimulate cells genetically predetermined for production of the homologous antibody. It is virtually a prerequisite of such stimulation that the lymphocytes should possess specific receptors capable of combining with the specific antigen. The receptors should, therefore, have a specificity and a structure corresponding to that of the antibody produced or capable of being produced by the cell. The concept of the close relationship or possible structural identity of the cell receptors of the antibody producing cells with the antibody they produced was first formulated by Ehrlich. During the period of dominance of instructive theories of antibody production it was, however, regarded as irrelevant. It was restated in the context of a clonal selection theory of antibody production in 1955 by Jerne and since then has received a substantial body of experimental support. Recent work has also indicated that in many cases antibody production is a result of a cooperative effort between two cell lines, T cells and B cells, the B cells actually producing antibody and the T cells functioning as helper cells which also possess a specificity for the antigen. It is therefore necessary to assume the presence of specific antigen receptors on both T cells and B cells.

The evidence for the presence of antigen-specific receptors on the surface of lymphocytes is as follows:

1. The capacity of lymphocyte suspensions to produce antibody against an antigen may be specifically removed by passage or filtration of the suspension through an immuno-adsorbent column containing the antigen while the capacity of the suspension to produce antibody to an unrelated antigen is unaffected. The capacity of lymphocyte suspensions used in such experiments to produce antibody may be demonstrated by injecting them into syngeneic X-irradiated animals together with the antigen or in vitro by the plaque (PFC) technique. The cells removed by the antigen in this way are almost certainly B cells. The cells required as helper cells for the production of antihapten antibodies in response to hapten carrier conjugates do not adsorb to carrier-coated immuno-adsorbent columns. It has been shown, however, that mouse T cells reacting with histocompatibility antigens would adsorb to tissue cultures with cells possessing the corresponding antigen, the tissue culture behaving in effect as immuno-adsorbent.

2. If lymphocytes of an animal that has been immunized against heterologous red cells are mixed with a suspension of red cells, the red cells are adsorbed to the lymphocyte surface forming a ring or rosette around it. Antibody-producing cells also form rosettes but these are several cell layers deep, in contrast to the rosettes formed by non-antibody-producing cells which are composed simply of a single layer of cells. Rosette-forming cells are also present in non-immune animals but in much smaller numbers than in immune animals. That both T and B cells can form rosettes is shown by the fact that the number of rosette formers is reduced following treatment of the suspension with complement and anti-θ or anti-MBLA sera. If rosette-forming cells are removed from a lymphocyte suspension by centrifugation the capacity of the suspension to produce homologous antibody is eliminated.

3. The adsorption of labelled antigen to lymphocytes may be demonstrated by autoradiography when the label can be seen attached to the surface membrane of the cell.

Autoradiographic studies have shown that the antigen receptors are distributed in discrete clusters over the cell membrane.

NATURE OF LYMPHOCYTE RECEPTORS

Evidence for the presence of immunoglobulins on the surfaces of lymphocytes has been obtained by the use of the following procedures:

1. Direct visualization of the determinants by immunofluorescence and by autoradiography.
2. Development of blastogenic transformation following exposure of lymphocytes to antibodies directed against allotypic Fab or class-specific determinants.
3. Opsonic adherence of lymphocytes to macrophages after exposure to anti-immunoglobulin sera.
4. The cytotoxic effect of anti-immunoglobulin sera in the presence of complement.
5. Inhibition of rosette formation or of in vitro plaque formation by immunoglobulin antisera or of blastogenic transformation induced by various antigens.

In mice the cells produced during the early stages of immune response may carry determinants of more than one immunoglobulin class but later in the response individual cells appear to possess determinants of only one class. Class exclusion and also allelic exclusion (p. 116), which have been found to operate in respect of surface immunoglobulin determinants of cells obtained in the later stage of the immune response, provide strong evidence that that immunoglobulin identified on the cell surface by these techniques has not been adsorbed passively from the surrounding fluid. Since passive adsorption would be expected to be random, individual lymphocytes would be expected to possess determinants of more than one class or specificity. A further argument against passive adsorption of immunoglobulins has been the complete failure to sensitize lymphocytes passively with immunoglobulins in such a way that rosette formation is inhibited by the use of anti-immunoglobulin sera.

Evidence for the occurrence of immunoglobulin determinants on the surface of B cells is provided by the finding that exposure to anti-Fab serum will cause about 80% reduction in the number of plaque-forming cells demonstrable in vitro. In addition it was shown that direct plaque formation can be inhibited by anti-μ and indirect plaque formation by anti-γ serum. It has been estimated that each B cell may carry 10^4 to 10^5 immunoglobulin-like receptors on its surface. There is also specific evidence for the occurrence of immunoglobulin receptors on T cells. Thus the mitogenic response of human peripheral blood lymphocytes to tuberculin is blocked by anti-light-chain serum. In mice the capacity to mount a graft-versus-host reaction and the capacity to transfer delayed-type hypersensitivity passively are both blocked by exposure of the cells to an anti-IgG serum. More cogent evidence has recently been obtained by Marchalonis in the isolation and physical characterization of IgM-like material from human and murine lymphocytes.

Despite the considerable body of indirect evidence for immunoglobulin-like antigen receptors on T lymphocytes, direct microscopic evidence is lacking. Thus Raff found that the proportion of mouse lymphocytes on which immunoglobulin receptors could be demonstrated by immunofluorescence increased sharply after thymectomy. In addition the number of lymphocytes in different organs which were positive by immunofluorescence was inversely related to the proportion of θ-positive lymphocytes present. Similarly Jones and Roitt were unable to demonstrate by autoradiography using labelled anti-light-chain sera immunoglobulin receptors on rabbit and mouse lymphocytes which had undergone blastogenic transformation after stimulation with soluble PHA. These were presumably T lymphocytes since there is considerable evidence that in the mouse only T lymphocytes are stimulated by PHA.

The difficulty of demonstrating immunoglobulin-like receptors on the surface of T lymphocytes raises the possibility that T cell receptors are not immunoglobulin-like at all. They could be gene products of immune response genes (p. 129) or alternatively surface components resembling histocompatibility antigens and coded for by the same region of the chromosome. A second possibility is that T cells may possess much fewer immunoglobulin-like receptors than B cells.

It would appear that in the differentiation of antigen-sensitive precursor cells to plasma cells there is a progressive loss of immunoglobulin surface receptors. In a study using immunofluorescence, Pernis was unable to demonstrate any IgG receptors in IgG-producing plasma cells of the rabbit. Some IgG-producing cells, however, appeared to possess IgM receptors. All IgM-producing cells, on the other hand, were found to possess demonstrable IgM receptors. The significance of the presence of IgM on IgG-producing cells is not clear. It could represent the persistence of IgM-type receptors on which there had been a clonal switch to IgG production.

PRODUCTION OF ANTIBODIES

As we have seen the production of antibody is an event which follows exposure of antigen-sensitive lymphoid cells to antigen. This exposure requires the binding of antigen to the antigen receptor present on the cell and presupposes physical contact of the antigen-reactive cell with the antigen. If the antigen is from a pathogenic micro-organism, this readily occurs following access to and multiplication of the micro-organisms in the tissues. From the tissues the antigen is carried, principally via the lymphatic vessels, to the neighbouring lymph nodes. In the case of the commensal organisms resident in the alimentary and respiratory tracts there is opportunity for direct contact with the lymphoid cells of the lymphoid collections, e.g. Peyer's patches, tonsils and adenoids, of the alimentary tract. Antigens which gain access to lymph nodes stimulate principally the production of IgG and IgM antibodies. Antigens found in the alimentary tract stimulate preferentially the production of IgA antibodies. This is presumably because the lymphoid cells associated with the alimentary tract have principally IgA-producing potential. IgE immunoglobulins are produced particularly in response to various antigenic substances, e.g. dust and pollens, which gain access to the respiratory tract, presumably because there is a preponderance of cells with IgE potential in the respiratory lymphoid tissues. The stimulation of antibody production by organisms of the normal alimentary and respiratory commensal flora is of considerable importance in the production of antimicrobial immunity in the early days of life, since many antigens of commensal micro-organisms cross-react with antigens of more highly pathogenic organisms. Antibody production by the spleen is the result of simulation by antigens which have gained access to the blood stream. This may occur through the bacteraemia sometimes associated with bacterial infections or through the fact that antigenic fragments of micro-organisms have gained access to the blood. By the use of in vitro tissue culture techniques, it has been established that antibodies are also produced by liver and bone marrow cells derived from primed animals. The significance of these organs as sources of antibody production under in vivo conditions is not clear. Antibodies have also been demonstrated in urine in certain cases of urinary infections and in the milk of lactating animals. Whether these are produced by local collections of plasma cells or whether they are secreted from the circulation is not clear.

SUBCELLULAR SITES OF ANTIBODY SYNTHESIS

The subcellular sites of synthesis have been examined by electron microscopic and autoradiographic studies using leucine-^{3}H as precursor for the polypeptide chain and galactose-^{3}H for the carbohydrate moiety. Such studies have shown that the light and heavy chains are synthesized independently on two different classes of ribosome, light chains on polyribosomes of approximately 190S with an associated messenger RNA of 9–11S and heavy chains on larger ribosomes of approximately 270S with a corresponding messenger RNA of 14–16S. The chains are synthesized on polyribosomes bound to the endoplasmic reticulum. Synthesis of the individual chains proceeds from a single growing point, the synthesis of individual light chains being complete in approximately 30 seconds and of individual heavy chains in approximately 60 seconds. Therefore it is not surprising that light chains are formed in considerable excess because of the shorter time required for their synthesis. Because of this it has been proposed that the light chains in some way trigger the release of the heavy chains from the polyribosomes. The peptide chains are synthesized first, most of the intra- and inter-chain disulphide links and the entire incorporation of carbohydrate being accomplished after the chains have been released from the polyribosomes. The immunoglobulin molecules are transferred via the Golgi apparatus, in which hexose transferases necessary for carbohydrate incorporation are found. From the Golgi apparatus they pass into the cisterns of the smooth endoplasmic reticulum in which the formation of the inter-chain disulphide bonds initiated on the ribosomes is continued. Accumulation of antibodies in the cisternae constitutes the Russell bodies which may be seen in haematoxylin and eosin stained plasma cells as hyaline amorphous masses and are readily demonstrable by electron microscopy. They appear to be transmitted to the outside of the cell via pockets of endoplasmic reticulum which constitute the secretory vesicles. These fuse with the surface membrane of the cell, the outer part of the membrane breaking down and the remainder maintaining continuity with the rest of the plasma membrane. At this stage the immunoglobulin, at least in the case of IgG, remains attached to the membrane, apparently through its Fc portion. Eventually the immunoglobulin molecules are detached from the membrane and become completely free. This mechanism of release is described as reversed pinocytosis. It resembles that responsible for the discharge of secretory products from the storage granules of other types of secreting cell.

There are a number of different ways in which the individual light and heavy chains may be associated to form a tetrameric four-chain molecule. The commonest pathway, demonstrated in both mouse and human myeloma proteins, is as follows:

1. H + H = H2.
2. H + L = H2L
3. H2L + L = H2L2.

In contrast to this pathway rabbit lymph node cells have been found to yield HL as the only disulphide-bonded intermediate. In this case the operative pathway must be:

1. H + L = HL
2. HL + HL = H2L2.

PRIMARY AND SECONDARY RESPONSES

The clonal selection theory of immunity requires that, prior to first contact with an antigen, the individual should possess a population of cells specifically reactive with that antigen. Stimulated by antigen this initial population of cells gives rise to the effector cells of the immune response and in addition, to a population of cells which persist in the body and which are available to react with the antigen at a later date. These cells are known as memory cells. In their specific reactivity with antigen they resemble the precursor antigen-reactive cells. There is evidence, however, that they differ qualitatively from the latter in their higher resistance to X-rays and to various immunosuppressive agents.

The response to a first contact with antigen which leads to the production of memory cells is known as the primary response (p. 87). Response to subsequent contact with an antigen of an animal previously primed by exposure to that antigen is known as the secondary response. In the primary response there is a lag period of variable duration before antibody is demonstrable in the serum. The antibody level then rises to a peak concentration which may be maintained for weeks or months depending on the particular antigen or animal involved. After this the antibody concentration falls slowly to the pre-immunization level. If at this time the animal is again exposed to the antigen it develops a typical secondary response. This is characterized by a lowering of the antigen concentration required to trigger the response, by a shorter lag period, by the eventual development of higher antibody levels in the blood and by a much longer persistence of detectable antibody. The differences between the primary and secondary responses are due to the fact that at the beginning of the secondary response there exists a much larger population of antigen reactive cells than at the beginning of the primary response. The rate of antibody increase during both responses is very much the same. It shows an approximately linear relationship between the logarithm of antibody concentration and time after exposure to the antigen. This relationship reflects the fact that the increase in antibody levels is due to a progressive increase through the division of precursors in the number of antibody-producing cells.

The phase of decline is initiated when the rate of degradation of antibody is in excess of the rate of synthesis. That some synthesis occurs throughout a considerable part of the decline phase is clear from the finding that some animals fed during this period with labelled amino acids incorporate some of the labels into their immunoglobulins. The rate at which immunoglobulins are degraded varies with different immunoglobulin classes and with different species. This rate can be measured by the injection of homologous labelled immunoglobulin and is stated as the half-life of the immunoglobulin, that is, the time required for 50% reduction in its concentration in the serum. The half lives of various immunoglobulin classes are shown in Table II.2 (p. 103). Smaller animals, rabbits, guinea-pigs and rats, show overall much shorter immunoglobulin half-lives than do larger animals.

The distinction between primary and secondary responses is readily demonstrable for protein antigens. For polysaccharide antigens, however, the distinction has not been so clear. There are three possible explanations for this:

1. The response to polysaccharide antigens appears to be much less thymus-dependent than the response to protein antigens. The lack of memory could thus be readily explained if it resided entirely in the T cell population. As against this, however, there is good evidence for the existence of memory in both T and B cell populations.

2. The responses to a first injection of polysaccharide antigen is in fact a secondary response because the animal has been previously exposed to the polysaccharide or to an antigen chemically related to it, as a component of organisms of the respiratory or intestinal flora.

3. In some cases the antigen may persist indefinitely in the tissues because the animal possesses no mechanism of breaking it down. This may occur following injection of pneumococcal polysaccharide in mice. Under such circumstances the animal is in a state of continuous antigenic stimulation. If the amount of antigen injected is very large this may be sufficient to induce a state of immunological paralysis.

CHANGES DURING IMMUNIZATION

When antibody concentrations are assayed by agglutination tests the first immunoglobulin class to be detected is IgM. If, however, assays are based on antigen-binding capacity (p. 140) IgM and IgG antibodies appear at much the same time. It seems likely, therefore, that studies which have shown the appearance of IgM before IgG reflect the higher sensitivity of the agglutination technique for the measurement of IgM. The IgM response, however, is more transient than the IgG response. Consequently in the later stages of the response IgG antibodies can be detected in the presence of minimal levels of IgM. This difference is more marked in the secondary than in the primary response, a finding which has given rise to the idea that there is no immunological memory in respect of IgM. That there is immunological memory for IgM responses, however, is shown by the fact that the levels of IgM in the early stages of the secondary response are demonstrably higher than those found in the primary response.

Under the conditions of natural infection a similar greater persistence of IgG than IgM has been demonstrated. Thus in chronic brucellosis and in the typhoid carrier state conditions in which there is repeated exposure to antigen as a result of chronic persistent infection, the antibodies detected are non-saline agglutinating IgG antibodies demonstrable by the antiglobulin technique (p. 138).

Changes in the affinity of antibody for antigen have also been demonstrated during the course of the immune response. Affinity is measured by the association constant. This constant increases up to a maximum with increasing time after immunization and is greater, for any particular interval, in the secondary response than for the same interval after immunization in the primary response. In experimental animals it has been found that when low doses of antigen are used for immunization the affinity of the antibody is greater than when large doses of antigen are used. This is presumed to be due to the fact that with low doses of antigen only cells with high affinity receptors are stimulated whereas with high doses of antigen receptors of both high and low affinity are stimulated. In the same manner it seems likely that during the course of the immune response, as the concentration of antigen falls, there should be a preferential stimulation of cells with high affinity receptors. It would not be surprising, therefore, if cells of such high affinity clones are predominant among the memory cells of the secondary response.

ANTIGEN LOCALIZATION

The sites of antigen localization in the lymph nodes and the spleen have been investigated using immunofluorescence with fluorochrome labelled antisera or autoradiography with antigens directly labelled with radioactive isotopes, usually ^{125}I or ^{131}I. In the lymph nodes two sites of antigen localization have been identified: the medullary macrophages and the cortical dendritic reticulum cells. Antigen is demonstrable in association with medullary macrophages within a short time after its introduction into the tissues. This antigen has been found to be present on the surface membrane and also within the macrophages in which it appears to be sequestered within lysosomal inclusions. Most of the lysosomal antigen is rapidly degraded but a fraction of it is believed to persist in what has been designated by Mitchison as a 'storage compartment'.

Surface binding of antigen to macrophages is particularly characteristic of peritoneal exudate macophages. It is not a significant feature of the medullary macrophages of the lymph nodes. The antigen associated with the dendritic reticulum cells of lymph follicles is, however, present on the surface membrane of the cell. Follicular localization requires the presence of antibody which has been demonstrated on the cell surface together with antigen by immunofluorescence. In animals previously immunized it may occur rapidly within a few hours after injection but in normal non-immune animals follicular localization is delayed until some antibody has been produced. In the spleen there is little or no trapping of antigen by macrophages since these are few in number and there is no area corresponding to the medullary sinuses of the lymph node. The principal trapping area is the marginal zone in which there are large numbers of cells resembling the dendritic reticulum cells of lymph nodes. As with the latter, antigen appears to be trapped on the surface membranes of the cells and to require presence of antibody which is produced shortly after exposure to the antigenic stimulus.

ROLE OF THE MACROPHAGE

There is considerable doubt as to the role of the macrophage in the immune response. That it does have a role is suggested by the close association often observed in lymph nodes between macrophages and lymphocytes. Some workers in fact claim to have shown fusion of macrophage and lymphocyte surface membranes which allows the passage of processed antigenic material from the macrophage to the lymphocyte. In addition macrophage uptake of antigen occurs at a very early stage after injection of the antigen and might, therefore, be expected to play an important role in the early production of antibody. Quite apart from deliberate prior immunization, however, particulate antigens appear to be particularly taken up by macrophages. Amongst soluble antigens some, e.g. ferritin, are more readily adsorbed on to reticulum cells. This may be because of the prior existence of antibody capable of binding with ferritin.

A positive role for the macrophages is strongly indicated by a variety of in vivo transfer experiments. Thus it has been found that lymphocytes which had been exposed to macrophages incubated with sheep lysed red cells readily produced antibody against sheep red cells on injection with the latter into an X-irradiated host. Exposure of lymphocytes alone to sheep red cells prior to injection was without effect.

There is abundant evidence that macrophages do not produce antibody. Their putative role must, therefore, relate to some sort of antigen processing necessary for the humoral immune response. The following are the main possibilities:

1. Macrophages are concerned with elimination of excess of antigen which might otherwise have a tolerizing affect. This is unlikely to be the only role of macrophages, as is indicated by experimental evidence for a potentiating effect of the macrophage cited above.
2. Concentration of antigen on the macrophage surface may in some way facilitate lymphocyte stimulation. This concentration does not, however, appear to occur to any significant extent with lymph node medullary macrophages.
3. Antigen fixed to the macrophage surface or persisting in a storage compartment may be readily transported via the macrophages to lymphoreticular tissues throughout the body and in this way facilitate wide diffusion of the antigenic stimulus. This is a conceivable role for the macrophage under in vivo conditions.
4. Macrophages are responsible for the

breakdown of particulate antigens into smaller fragments which are effective in inducing immune responses. In vitro experiments of Nossal and Ada suggest that this may be an important in vivo function. These workers found that in an in vitro lymphocyte tissue culture system macrophages were required for synthesis of antibody to sheep red blood cells but not to flagellin. Microparticulate and soluble red blood cell fractions could stimulate antibody production in vitro without macrophage cooperation.

5. Macrophages may play a significant part in the T-B cell cooperation required for the production of certain antibodies.

6. Fishman and his co-workers have reported that RNA extracted from macrophages exposed to the particulate antigen phage T_2 could stimulate antibody production by lymphocytes maintained in in vitro culture or in diffusion chambers in an X-irradiated host. With a high concentration of antigen, two waves of antibody production were observed, the first consisting of IgM and the second consisting mainly of IgG. They identified two RNA fractions as being responsible: a 19S ribonuclease-sensitive fraction which was not precipitable by specific antibody and which was, therefore, presumed not to contain antigen and a 23–28S ribonuclease-resistant fraction which was precipitated by antibody and, therefore, contained antigen.

They found the antigen-free fraction to be responsible for initial IgM production and the antigen-containing fraction to be responsible for IgG production. When the macrophages and lymphocytes were obtained from animals of different allotypic specificities it was found that the IgG antibody produced had the allotypic specificity of the recipient but the IgM antibody had the allotypic specificity of the macrophage donor. It was therefore postulated that the RNA responsible for IgM production was an informational macromolecule which was transferred from the macrophages to the lymphocytes and in the latter coded in some way for the production of IgM. There are, however, a number of difficulties in the acceptance of this hypothesis: (*a*) Two species of RNA would be required—one for the heavy and one for the light chains—and because of the considerable degree of specificity of the association of heavy and light chains, both species must enter the same recipient lymphocyte. (*b*) The RNA extracted from the exposed macrophages has not been found to possess a different base composition from that of normal macrophage RNA. (*c*) It is doubtful whether the RNA extracts which stimulated IgM production were completely free of antigen. Fishman's hypothesis is stimulating but so far has not received experimental confirmation.

7. The formation in macrophages of a highly immunogenic RNA antigen conjugate. There is considerable evidence that RNA from various sources form firmly bound complexes with antigen and that the antigen bound in this way is a much more effective immunogen than unbound antigen. It is doubtful, however, whether such complexes are formed within macrophages by a biological process and evidence has been presented by Roelants that they are, in fact, artefacts produced by various extraction procedures. In addition no significant relationship has been established between the capacity of an antigen to form complexes with RNA—which is a function primarily of the negative charge on the antigen—and its immunogenicity.

THE ROLE OF DENDRITIC RETICULUM CELLS

It seems likely that dendritic reticulum cells have an important role in antibody production. Since, however, localization of antigen to reticulum cells requires the presence of antibody, this role would be exercised in the later stages of the primary response and in the secondary response. In mice a clear role for these cells has been defined experimentally by Hanna. Old mice were found to give very much poorer 7S antibody responses than young mice. The response of the older mice could, however, be considerably enhanced by inserting into their spleens plugs of reticulum cells derived from young mice. On the other hand, dendritic reticulum cells are not essential for the immune response since they are absent from tissue cultures of lymphoid cells in which responses leading to 7S antibody production can be produced in vitro.

COOPERATION BETWEEN T CELLS AND B CELLS

The cooperation between T and B lymphocytes in

antibody production has been demonstrated for a variety of antigens both in vivo and in vitro and for both primary and secondary antibody responses. By the use of chromosomal markers it was shown by Mitchell and Miller that under conditions of cooperation the antibody-producing cells were the B cells. Not all antigens require T–B cell cooperation for antibody production. Non-T-dependent antigens are uniformly antigens with numerous identical determinant groups or epitopes, e.g. polysaccharides and the synthetic antigen polyvinyl pyrrolidone. T-dependent antigens, on the other hand, are those with only a few determinant groups of any particular specificity. The phenomenon has been most studied in respect of protein–hapten conjugates, the T cells combining with the carrier protein, in which context they are referred to as helper cells, and the B cells responding to the conjugate with antigen production.

The precise mechanism of T–B cell cooperation has not been determined. The main possibilities are:

1. T cells are required for the correct physical orientation of the antigen at the surface of the B cells. In this case the antigen constitutes a bridge between the T cells and the B cells.
2. On antigenic stimulation the T cell releases an antigen receptor which bonds non-specifically to macrophages, the latter presenting the antigen to the B cells.

In either of these cases it is possible that some non-specific substance or factor is liberated from the T cells or the macrophages which has a short range of action but which is necessary as a second stimulus to the B cell.

There is a considerable amount of evidence in support of the latter hypothesis:

1. In an in vitro system of cultured spleen cells it was found by Hartman that T cells from animals primed with sheep red blood cells enhanced the stimulation of B cells by horse red blood cells, if sheep red blood cells were present at the same time.
2. The demonstration by Katz et al. That graft-versus-host reactions, which are believed to be due to an attack on T cells, can stimulate antibody production to T-dependent antigens both in vivo and in vitro.
3. The finding of Sjöberg that polysaccharide from *E. coli*, a non-specific mitogen for B cells, could confer responsiveness on thymectomized mice to sheep red blood cells. Even if in certain cases B cells are stimulated non-specifically by a T cell product, the available evidence supports the view that the initial reaction between T cells and antigen is specific. This is a necessary consequence of the fact that both tolerance and memory can be induced in T cells and that T cell helper activity is abolished if the cells are exposed to an antigen carrying a radioactive label whose disintegration causes their destruction.

ONTOGENY OF THE IMMUNE RESPONSE

It is frequently stated that the lymphoreticular system is immature at birth and that, in humans at least, full responsiveness does not develop until the end of the first year of life. This postulated immaturity is often used as a reason for delaying vaccination procedures until such time as a satisfactory immune response might be expected to occur. The argument, however, is of doubtful validity. It is true that only low levels of immunoglobulins are demonstrable at birth and in the immediate postnatal period, the immunoglobulins present being passively transmitted either, as in man, across the placenta or, in the case of certain other species, by absorption from the colostrum and the milk in the first few days of life. The absence of active immunoglobulin production by the fetus can, however, be attributed largely to a lack of antigenic stimulation. After birth the neonate is stimulated by antigens released from organisms which establish themselves as normal commensals in his alimentary tract. In gnotobiotic animals, i.e. animals raised in a germ-free environment, normal antibody levels are attained very much more slowly than they are in animals raised in a normal bacteria-laden environment. It is, of course, true that the response of neonates to a variety of antigens, e.g. to polio virus vaccine, is much poorer than that of older animals, full responsiveness to these antigens only being achieved at the end of the first year. However, this diminished responsiveness can probably be attributed to the inhibitory effect on the immune response of antibody transmitted from the mother rather than through any major inadequacy in the infant's response capability. This is shown by the fact that when maternal antibody is not present, the response of the neonate to a variety of antigens, e.g. diphtheria toxoid, is as great as that of older children. The neonate is defective only in the amount and variety of antibodies which it produces and that its response to most antigens must necessarily be of primary rather than of a secondary type.

Not only has the neonatal animal a developed capacity to respond to antigenic stimulation but so also has the fetus. This has been shown by a variety of lines of evidence:

1. Plasma cells which are only produced in normal animals as a result of the antigenic stimulation of antigen reactive precursors have been demonstrated in human fetuses at five months' gestation.
2. Children suffering from the rubella syndrome due to intrauterine infection from the mother during the first three months of pregnancy (p. 399). have been found to exhibit, when examined at a later date, considerably higher antibody levels than cohort controls at ages of up to 10 years, at which time all maternally transmitted antibody must have long since disappeared.
3. 19S antibody to the protozoon *Toxoplasma gondii*, infection with which, like rubella, is acquired in utero, has been identified in fetal serum at the twenty-eighth week of gestation. Since 19S antibodies are incapable of passing through the placenta this 19S antibody must have been produced by the fetus.

The investigation of immune responses in the fetus is clearly difficult since it involves multiple intrauterine manipulations. Some data are available however, particularly those obtained by Silverstein relating to the responses of sheep. Silverstein has transplanted homografts (allografts) to sheep fetuses at about the middle of their gestation period of 150 days and has found the homografts to be rejected with the same efficiency, with the same time-sequence and with the same histological reactions as they are in the adult sheep. The production of antibodies to bacteriophage antigens—in this case X174—was demonstable at 40 days' gestation. The demonstration of antibodies at earlier stages than this was not technically feasible. Consequently it was not possible to identify the earliest stage at which the fetus could produce antibody but it was possible to show that after 40 days the capacity to mount an immunological response to different antigens increased in step-wise fashion and in such a way that an antigenic hierarchy could be established, immunological responsiveness to the various antigens in the hierarchy being achieved sequentially. On the basis of Silverstein's work it would appear that the order of antigens in the hierarchy may be different for different animal species. The development of responsiveness was an all-or-none phenomenon, the capacity to respond to a particular antigen, when it appeared, being the same in the fetus as in the adult. The existence of an antigenic hierarchy of this sort is difficult to explain on the germ-line theory of the immune response (p. 114). It would, however, be more consistent with somatic mutation theories, but in this case it would be necessary to postulate a non-random component in the somatic mechanism for the generation of antibody diversity. Alternatively sequential responsiveness might be due to a progressive development either in antigen-handling capacity or in the synthesis of histocompatibility antigens linked to components which play a part in genetic differences in immunological responsiveness.

There is evidence that the capacity of the individual animal to produce IgM antibodies develops prior to the capacity to produce IgG antibodies. IgM antibody, which must have been produced by the fetus since it does not cross the placenta, is demonstrable at birth. On the other hand it would appear that IgG antibody production only begins to occur at or about the time of delivery. This conclusion is based on estimates derived from the rate of increase of the infant's IgG antibodies and the rate of decline of IgG antibodies derived from the mother, which have a half-life of about 30 days. Adult levels of IgM production are attained at about a year, at which time IgG has only reached about 50% of the adult level. At birth the concentration of IgG antibody derived from the mother is usually fully equal to and sometimes in excess of adult levels. This maternally transmitted IgG antibody is of great importance in the protection of the neonate during the early months of life and some of it may persist in the infant's circulation up to the age of one year.

Original antigenic sin

This phenomenon was first described by Francis in individuals who had been immunized with Type A influenza virus vaccines (p. 420). Francis found that such individuals developed antibody not only against the antigens of the vaccine strain but also against related but distinct antigens of the strain of virus with which, it could be presumed from epidemiological data, they had been infected on a previous occasion. The phenomenon was made the basis of an exploration, first by Francis and Davenport and later by other workers, of the antigenic composition of influenza viruses to which individuals of different age groups in various populations had first been exposed. In this way it was possible for them to suggest the possible antigenic composition of viruses responsible for epidemics for which no laboratory data were available. Thus they postulated that the influenza virus responsible for the 1918–9 pandemic was antigenically related to the swine influenza virus. On similar grounds the virus responsible for the 1957 pandemic, the so-called Asian virus, is believed to resemble antigenically the virus responsible for the 1890 pandemic, since individuals who were old enough to experience the 1890 pandemic were found to have in their sera appreciable antibody against the 1957 Asian strain.

The phenomenon of original antigenic sin has been extensively documented in experimental animals primed and subsequently challenged with various hapten carrier conjugates. Thus rabbits primed with dinitrophenyl gammaglobulin conjugate and subsequently challenged with a trinitrophenyl gammaglobulin conjugate have been found to produce antibody with greater avidity for the dinitrophenyl hapten than for the trinitrophenyl hapten.

ANTIGENIC COMPETITION

It has been known for a long time that simultaneous immunization of human subjects with two or more antigens may lead to a lower antibody response to one or other of the antigens than if the antigens are given singly. The phenomenon is demonstrable in experimental animals with a variety of defined antigens. Thus rabbits readily produce antibody specific for the introduced hapten against protein substituted with poly D-ala side-chains or with poly L-ala side-chains if either of these substituted proteins is given separately. However, if both are given simultaneously, or if the animal is immunized with a protein substituted with both poly D-ala and poly L-ala side-chains, only antibody to the D-ala hapten is produced. Similar competition was demonstrated between the azobenzene arsonate and dinitrophenyl haptens. Rabbits were capable of developing antibody to both hapten carrier conjugates when these were given independently but if they were immunized with carriers substituted with comparable amounts of both hapten the response to the azobenzene arsonate hapten was depressed. Since the haptens are non-cross-reacting competition is not possible at the B cell level. On the other hand, since the same carrier was involved competition could have readily occurred at the T cell level.

Antigenic competition does not appear to be a serious practical problem in the immunization of humans with the multiple component vaccines, e.g. triple vaccine (p. 169). With the triple vaccine in fact the pertussis component appears to have a distinct potentiating or adjuvant affect (p. 98) on the immune response to diphtheria and tetanus toxoids. Competition may, however, occur to a significant extent in individuals given a mixture of antigens to one of which they have been previously immunized. In this case much of the response may be directed preferentially towards that antigen. Antigenic competition has also been demonstrated between complex proteins with no demonstrable common antigenic determinants, e.g. between haemocyanin and BSA in mice. In this case the competition occurs in both directions and is greater if one antigen is given first rather than if both are given simultaneously. Competition has also been demonstrated between different types of cells, e.g. inhibition of allograph rejection with prior administration of haemocyanin and red cells.

EFFECT OF SIZE OF ANTIGEN DOSE

The effect of antigen dose is complex and differs with different antigens and in different animals. For each antigen and animal it is, however, possible to define a minimum dose below which there is no detectable response on either primary or secondary stimulation. Although there are no absolute limits it can be said that in general proteins are better antigens than polysaccharides and that high molecular weight proteins are more immunogenic than low molecular weight proteins. Above the minimum dose necessary to produce response, the response increases with increasing dose up to a maximum above which it may be possible to demonstrate reduction in the amount or avidity of the antibody, due apparently to immunological tolerance.

Low doses of antigen in the immunizing range combine preferentially with high affinity cells and as a result lead to the production of antibody with high affinity. On progressive immunization with low doses of antigen, high-affinity antibodies may be progressively selected and as a result ultimately a high level of high affinity antibody may be produced. With large doses of antigen, cells of low as well as of high combining affinity are stimulated. In prophylaxis against microbial infections, however, immunization, leading to a high concentration of low affinity antibody may have a greater protective effect than appreciably lower concentration of high affinity antibody.

Virtually no systematic investigation of the effect of different immunization schedules and doses with a range of antigens on the levels and avidity of the antibody produced and of the extent and duration of the associated memory has been carried out in humans. In practice immunization schedules are chosen for their convenience rather than for their capacity to elicit optimum responses. Thus in experimental animals immunization courses often consist of 6 to 12 injections at 2 to 3 day intervals, the animals being bled from one to two weeks after completion of the course. The courses of immunization normally used in humans against non-living microbial antigens usually consists of 2 or 3 injections at intervals of two to eight weeks.

PHYLOGENY OF THE IMMUNE RESPONSE

It is frequently stated that invertebrates are incapable of immune responses. This is certainly true in relation to antibody production. In many invertebrates, however, non-specific bactericidal substances and haemagglutinins have been recognized and in certain cases have been shown to increase in concentration in response to antigenic stimulation. This type of response, however, is regarded as being more analogous to enzyme induction than to immune responses in animals. None of the humoral substances identified possesses the specificity of antibodies or resembles the latter in chemical structure. Cellular recognition and *quasi*-immune phenomena are more in evidence. The sponges possess a clearly defined cell recognition mechanism. Thus if two species of sponge are disaggregated into single cells and allowed to reaggregate, the reaggregation is specific. Hybrid forms are not found. This process could forseeably be a forerunner of immune recognition.

All invertebrates so far examined possess cells capable of phagocytic function. It was in fact the demonstration by Metchnikoff that mesodermal phagocytes in the water flea *Daphnia* protected it against invasion by pathogenic fungi which lead him to postulate, as we now know erroneously, that the immunity to infection exhibited by the animal body was due entirely to the activity of phagocytic cells. Phagocytosis in invertebrates is enhanced, in a manner analogous to opsonization (p. 156), by various non-specific antimicrobial substances and haemagglutinins which they produce. The closest approach to specific immune responses found in invertebrates is the cellular allograft rejection in higher species. This coelomycytes observed in annelid worms. The process has the essential characteristics of homograft rejection in higher species. This finding points strongly to cellular immune responses as being phylogenetically prior to humoral immune responses.

All vertebrates so far examined possess well defined cellular and humoral immune response mechanisms. All possess circulating lymphocytes. All exhibit cellular graft rejection of a specific character and all produce antigen-specific antibody-like substances. In the most primitive vertebrates, the hagfish, there are no discrete collections of lymphoid cells although hagfish can produce an IgM-like immunoglobulin. In the more advanced cyclostomes, the lampreys, there is evidence of a primitive thymus consisting of epithelial cells and focal collections of lymphoid cells. Clearly defined thymuses, lymphopoietic spleens and plasma cells are first seen in higher elasmobranchs, e.g. sharks. True lymph nodes are found only in mammals. Fowl are unique in possessing as a primary lymphoreticular structure, the bursa of Fabricius (p. 111), for which no precise equivalent has been found in mammalian species.

Evolution of immunoglobulins

Considerable information on the evolution of immunoglobulins has been obtained from amino acid sequence data of immunoglobulins from a primitive gene coding for approximately 110 amino acid residues, i.e. possessing the dimensions of one light or heavy chain domain. The evidence would suggest that the earliest evolutionary divergence of this primordial gene was into primitive V and C region genes. The emergence of heavy and light chain C regions and V regions subgroups were later events. The most primitive of the heavy chains appears to have been the μ chain. μ-like immunoglobulins have been found in cyclostomes. γ-like molecules have been found in the elasmobranchs. It has been proposed that δ and ϵ heavy chains appeared first in the higher fish, the chondrostei and teleostei, and that the α chains first appeared in amphibia. The emergence of γ and α subclasses is believed to have first occurred in the mammals. With certain assumptions about the rates of mutation it has been estimated that the entire evolution of the immunoglobulins covered a period of some 400 million years.

THE GENETICS OF IMMUNE RESPONSES

So far we have considered the possible genetic basis for the capacity of an animal to respond with specific antibody production to a great variety of chemically dissimilar antigens. It has now been firmly established, however, that different animals of the same species may show marked differences in their capacity to mount immune responses, either of humoral or cellular types, to the same antigen. Such differences were first shown in 1938 by Gorer and Schutze who observed a correlation in strains of mice that were genetically resistant of *Salmonella* infection between resistance and the capacity to produce anti-*Salmonella* antibodies. Mice with a high level of resistance were good antibody producers and mice with a low level of resistance were poor antibody producers. Later work by other workers with a variety of complex antigens clearly indicated that such differences in immune response had a genetic basis. Thus Schiebel in 1943 showed that it was possible, by selective breeding from a colony of random-bred guinea-pigs containing both good and poor responders to diphtheria toxoid, to produce a strain of high response capability by mating good responders and a strain of low response capability by mating poor responders. The difference in response in each of the selectively bred populations was specific for diphtheria toxoid.

Although experiments such as those described above indicated that animals could differ markedly in their capacity to respond to a particular antigen and that this capacity had a genetic basis they did not permit a precise genetic analysis. This lack of precision was due in the first instance to the fact that the complex antigens employed possessed a number of distinct antigenic determinants, the response to the whole antigen representing the summation of individual responses to a variable number of these determinants. For precise analysis experiments in which the response to a single antigenic determinant could be examined were essential. This requirement has been met in three ways:

1. By the use of a variety of synthetic polypeptide antigens made up of repeating units composed of one or a limited number of amino acids and of hapten conjugates of these polypeptides.
2. By the use of potent multideterminant antigens of natural origin administered in low concentrations. Under these conditions the immune response is directed mainly or exclusively to one antigenic determinant. The antigens employed in this way have included bovine and human serum albumin, hen ovalbumin and bovine gammaglobulin.
3. By the use of native antigens known to stimulate weak immune responses. These have been antigens derived from other animals of the same species (allogeneic antigens) which differ only slightly in composition from those of the animals being immunized. They include histocompatibility antigens responsible for transplantation immunity (p. 85) and erythrocyte antigens.

The principal questions which must be answered

in respect of immune responses to antigens towards which differences in immune response capability exist are:

1. Are the differences genetic in origin?
2. If they are, is the capacity to respond due to a single gene?
3. If so, is the gene dominant or recessive?
4. Does it show linkage to other genetic markers?
5. If linkage is established, an attempt is made by the examination of recombinants to map the location of the gene on the relevant chromosome or linkage group.
6. What is the nature of the gene product? Is it, for example, the presence of specific receptors on the surface of the lymphocytes of the responding animal?
7. What is the precise function, at the cellular level, of the putative gene product in controlling immune response capability?

To answer the genetic question proposed it is essential to have highly inbred strains of animals, all the members of which are genetically identical, apart from sex differences, and homozygous at all major loci, and which give strain-specific responses to the antigen employed. Many strains of laboratory mice with well documented genetic markers are available and these animals' responses have consequently been particularly examined. Less extensive but important work has been carried out with random-bred strains of guinea-pig and rats. In these various species most information has been derived from an examination of the responses of the parent strains and of hybrids derived from them to synthetic polypeptides and polypeptide hapten conjugates and the discussion will therefore be concerned mainly with the investigation of responses to these antigens.

Synthetic polypeptides and hapten polypeptide conjugates

Poly-L-lysine and DNP poly-L-lysine

The responses to these compounds were originally observed by Benacceraf and his colleagues. The simplest synthetic polypeptide antigens are linear polymers of a single amino acid. No such homopolymers are immunogenic in mice but two-poly-L-lysine (PLL) and poly-L-alanine (PLA), when given in complete Freund adjuvant (p. 99), will stimulate cellular immune responses in inbred strain 2 guinea-pigs but will not elicit responses in strain 13 guinea-pigs. Strain 2 guinea-pigs differ from strain 13 guinea-pigs in that neither will accept a tissue graft or skin graft from the other because of differences in their major histocompatibility antigens. PLL and PLA will also stimulate an immune response in 30% of the Hartley random-bred guinea-pig strain. Strain 2 guinea-pigs and the responding 30% of Hartley strain guinea-pigs, when immunized with DNP-PLL conjugate, produced antibody to DNP and delayed-type hypersensitivity (p. 193) to DNP and PLL. Strain 13 guinea-pigs and non-PLL-responding Hartley strain guinea-pigs showed no response to DNP-PLL. DNP-PLL responders also produced antibody directed to the hapten component when immunized with benzyl penicilloyl–PLL and *p*-toluene sulphonyl-PLL while non-responders did not. The latter finding would identify the defect in responders as being in their capacity to produce a response to the carrier PLL rather than to the hapten. This is confirmed by the observation that non-responders produce anti-DNP antibodies when immunized with a conjugate of DNP–PLL coupled to bovine serum albumin (BSA). Animals which were previously rendered tolerant to BSA, on the other hand, produced no anti-DNP antibodies.

The gene involved in these responses is known as the PLL gene and, although it controls the response to PLL, it has been found not to be specific for PLL since PLL-responders produced anti-DNP antibodies when immunized with DNP-poly-L-alanine (DNP–PLA), DNP-protamine and a conjugate of DNP with a copolymer of glutamic acid and lysine DNP–GL. The PLL gene, therefore, controls the recognition of PLL, PLA, protamine and GL.

In hybridization experiments between strain 2 and strain 13 animals the progeny were all found to be responders to DNP–PLL, indicating that the responses are controlled by a dominant autosomal gene. The finding that the Hartley strain responders and the responders of the progeny of strain 2 × strain 13 F_1 animals with parental strain 13 animals all showed the presence of the strain 2 major histocompatibility antigen as detected by the mixed lymphocyte reaction (MLR) and the lymphocytotoxicity test while non-responders lacked the strain 2 antigen indicated linkage of the gene determining the latter with the PLL gene.

Responsiveness was found by Green and Benacerraf to be transferable to lethally X-irradiated strain 13 animals by the injection of suspensions of lymph nodes and spleen from strain 2 × strain 13 F_1 hybrids. It was necessary to use strain 2 × strain 13 F_1 hybrid cells as donors since strain 2 cells would cause severe graft-versus-host reactions in strain 13 recipients. The suspensions used in these experiments were described as containing about 95% of small lymphocytes. Since macrophages from responders and non-responders were found to be capable of phagocytosing the conjugates used for immunization to the same extent and since the conjugates were found to be broken down equally well in vivo by both responders and non-responders the lymphocytes were clearly implicated as the likely carriers of the gene product conferring responsiveness.

Mechanism of action. From these results it seems reasonable to exclude not only the macrophages but also the antibody-producing cell precursors (B cells) as the determinants of responder status. The concurrence of responsiveness to DNP–PLL, BP–PLL and *p*TS–PLL clearly implies that responsiveness depends on carrier recognition. The production of anti-DNP antibody on stimulation with DNP–PLL–BSA conjugates indicates that non-responders possess the appropriate antibody-producing cell precursor. The responsible cell would therefore appear to be the T cell, in its role as the cell controlling carrier recognition. This identification is supported by the finding that responders do and non-responders do not develop delayed-type hypersensitivity to PLL *per se*.

As to the nature of the T cell component which determines responsiveness there are three possibilities:

1. It might be an immunoglobulin-like surface determinant responsible for antigen recognition. Evidence for the occurrence of such determinants on T cells has already been discussed (p. 119). Although it is possible, and perhaps in the present state of our knowledge probable, that T cell recognition is mediated through immunoglobulin-like receptors, such receptors are unlikely to be the product of the PLL immune response gene since, in view of the fact that this gene controls responsiveness to PLL, PLA, GL and protamine, the gene product would need to possess a broadness of specificity quite distinct from that characteristic of the immunoglobulins (p. 96). The specificity of the cell determinant would seem likely to be related primarily or exclusively to the strong positive charge carried by the antigens involved.

2. The component might be a special class of T cell antigen receptor unlike the postulated immunoglobulin-like receptors on B cells. If antigen recognition by T cells is due to a distinct type of T cell receptor, the number of such receptors required would appear to be much less than the number of immunoglobulin receptors needed for B cells because of (*a*) their potentially wider reactivity and (*b*) the lack of necessity to provide for a variety of V region subgroups (p. 109) and of immunoglobulins of varying affinity for each antigen.

3. The determinant might be a histocompatibility antigen or other membrane component which controls immune responsiveness by permitting or preventing approximation of antigen to the appropriate T cell receptor. This it might do by altering the charge or configuration of the surface of cell membrane. At present this appears the most plausible working hypothesis for PLL responses.

Linear copolymers

Strain 2 guinea-pigs can and strain 13 guinea-pigs cannot produce immune responses to linear copolymers of glutamic acid and alanine (GA) while strain 13 animals can and strain 2 animals cannot develop immune responses to copolymers of glutamic acid and tyrosine (GT). The response to GA is closely linked to the PLL gene but, since some Hartley strain guinea-pigs which respond to PLL do not respond to GA, the GA and PLL genes must be distinct. Both the GA and the GT genes are autosomal dominant genes.

Mice do not develop immune responses to polymers of two amino acids but can do so to certain polymers of three amino acids, e.g. glutamic acid, alanine and tyrosine (GAT) or glutamic acid, lysine and alanine (GLA). The responses are controlled by dominant autosomal genes located in the chromosomal region containing the major mouse H2 histocompatibility antigens. As in the case of guinea-pig responses to PLL responsiveness is of the all-or-none type.

Branched chain copolymers

Mice are capable of mounting immune responses to branched chain amino acid polymers made up of a poly-L-lysine backbone to which are attached side chains of DL-alanine with glutamic acid and either phenylalanine, histidine or tyrosine at the tips of the chains. The corresponding compounds are designated PheGAL, HGAL and TGAL. These compounds were first used in studies of immune responses by McDevitt and Sela who found that different inbred strains of mice showed quantitative differences in the amount of antibody produced to them, some animals being characterized by a high response, some by a low response and some by an intermediate response. Apart from the fact that PheGAL was the strongest antigen for all strains the responses of each strain to the individual polymers were quite independent of each other, a particular strain being capable of a high response to any one polymer and a low response to others. The specificities of the antibodies produced was found to be directed to the two terminal residues of the amino acid chains. There are also qualitative differences in that high responders produced both 19S and 7S antibodies and low responders only 19S antibodies.

By hybridization experiments it was shown that the responses were determined by dominant autosomal genes. Thus the F_1 hybrids produced by crossing the CBA strain (a low responder strain to TGAL) with C57 (a high responder strain to TGAL) were all high responders. When the F_1 hybrids were crossed with parental C57 animals the progeny were high responders; when crossed with CBA animals 50% of the progeny were high and 50% low responders. The inheritance data obtained indicated either a single locus with three closely linked genes or a single locus with multiple alleles each of which determined a specific pattern of response to the three antigens. The locus which is known as Ir-1 has been found to be closely linked to alleles at the major mouse histocompatibility region—H2. Examination of the responses of animals bearing recombinant H2 alleles have located the Ir-1 locus in the middle of this region close to a serum substance locus. That the Ir-1 gene product is present on T lymphocytes with specific reactivity was shown by experiments of McDevitt et al. in which thymectomized lethally irradiated C57 × DBA F_1 hybrid mice which are normally high responders to both TGAL and HGAL were given fetal liver cells as a source of B cells and a thymus implant as a source of T cells from CBA mice, which are low responders to TGAL and high responders to HGAL. When challenged with TGAL no TG-specific antibody was produced but on subsequent challenge with HGAL, HG-specific antibody was produced.

The responses of mice to a further group of branched chain amino acid polymers with PLL as a backbone and side-chains of proline with TG and PheG as end groups has also been examined. The capacity to produce anti-TG and anti-PheG antibodies in response to these compounds depends on the presence of the appropriate Ir-1 locus allele. Independently of their Ir-1 locus content however certain mice were found to produce antibody to the backbone pro-L. Hybridization experiments have identified the gene responsible as an autosomal dominant gene designated Ir-3 which does not show linkage to the H2 locus.

Mechanism of Ir-1 gene responses. Because of the serological specificity of the responses controlled by the Ir-1 gene the possibility that the defect in low responders is related to the genes controlling immunoglobulin variable regions obviously requires consideration. This would be difficult to reconcile with the findings that low responders produce 19S antibodies of the same specificity and are capable of producing 7S antibodies though at a low level in response to the antigens complexed with methylated BSA as carrier. In addition the defect cannot be due to a genetically determined deficiency of V region genes in low responders, since no linkage has been demonstrable between the Ir-1 gene and the genes determining heavy chain allotypes. Such a defect might also be expected to affect all the relevant V region genes. It could, however, be due to the development of a form of split tolerance affecting the extent of IgG responses generated by exposure to determinants common to the different polypeptide antigens and histocompatibility or other surface antigens present on the cell membranes of the cells of low responders. There are a number of arguments against this hypothesis:

1. The high responding state is dominant in the F_1 hybrids of high responders and low responders in which, if the tolerance hypothesis is correct, it should be suppressed by the presence of the tolerizing antigen, which should, if it is controlled by a dominant gene, like all known histocompatibility antigens, be expressed in the hybrid. The possibility cannot, however, be excluded that the tolerizing antigen may differ from known histocompatibility antigens in being determined by a recessive gene. This difficulty would appear to be partially met by experiments in which allophenic (tetraparental) mice constructed by fusion of good and low responding to TGAL embryos were good responders.

2. It has not been possible to demonstrate any cross-reactivity between the polypeptides and the histocompatibility antigen of low responders. C57 mice carrying histocompatibility allele H2b are high responders to TGAL and low responders to HGAL while C3H mice carrying the allele H2k are high responders to HGAL and low responders to TGAL. TGAL antisera do not cross-react with H2k tissue antigen nor do H2k antisera cross-react with TGAL antigen. Similarly HGAL antisera do not cross-react with H2b tissue antigens and H2b antisera do not cross-react with HGAL.

As in the case of the PLL gene in guinea-pigs the evidence points to T cells as being the cells carrying the Ir-1 gene product. Non-responsiveness affects only 7S responses which are T cell dependent. 19S responses which are not T cell dependent are not affected. Conjugation with methylated BSA will elicit 7S responses from low responding strains. Thymectomized irradiated TGAL high responding mice reconstituted with syngeneic bone marrow show the same level of 19S antibody production as low responders and no 7S response on challenge with TGAL.

As to the nature of the gene product the main possibilities are those which have already been discussed in connection with guinea-pig PLL responses. In this case, however, much more precise information on the chromosomal location of the controlling locus is available. Its location within the H2 chromosomal region raises more acutely the possibility that Ir-1 genes or alleles may determine histocompatibility antigens which play an important part in antigen recognition by T cells. So far, however, there is no evidence that the Ir-1 gene product is a known histocompatibility antigen though it could, of course, be a histocompatibility antigen not so far recognized. The location of the Ir-1 gene within the H2 region does not, however, necessarily mean that its product is a histocompatibility antigen. In this connection it is significant that the gene is located close to genes controlling the synthesis of a serum protein which is unlikely to be in any way related to a histocompatibility antigen. Indirect evidence against the identification of the Ir gene product with a histocompatibility antigen is the failure to establish any linkage between other immune response genes, e.g. Ir-3, and genes controlling any known histocompatibility antigens.

The hypothesis that Ir-1 gene product is a surface component which controls access of antigen to T cells faces the difficulty that in contrast to the PLL situation in the guinea-pig the product must have a high degree of specificity since it can distinguish between compounds closely resembling one another in chemical structure. In the case of molecules such as TGAL, HGAL, and PheGAL one would have expected by analogy with PLL that the carrier function would be related to the poly-L-lysine backbone. This does not appear to be the case. If it were, a lack of response to one of these compounds would be associated with a lack of response to the others.

Multi-determinant natural antigens

In guinea-pigs the response to low doses of bovine serum albumin (BSA) or human serum albumin (HSA) is controlled by an autosomal dominant gene linked to the major histocompatibility antigen locus. Responsiveness which is of the all-or-none type is found in strain 2 but not in strain 13 animals.

In mice autosomal dominant genes linked to the H2 locus are involved in the responses to ovalbumen, ovmucoid and bovine gammaglobulin. The response to ovmucoid and bovine gammaglobulin are probably due to the same gene which is found in mice carrying the $H2^a$ and $H2^k$ alleles. The response to ovalbumen is found in mice bearing the alleles $H2^b$, $H2^d$ and $H2^q$ and is probably due to a distinct immune response gene.

The response of inbred strains of mice to the red blood cell antigens EA^a and EA^b found only in certain wild-type mice is due to an autosomal recessive gene linked to the minor histocompatibility antigen loci H3 and H6. In this case it is possible, since it is controlled by a recessive gene, that non-responsiveness could be somatically induced by tolerance to a histocompatibility antigen. The response of mice to keyhole limpet haemocyanin and to a bacteriophage phage Fd are also controlled by autosomal dominant genes but no linkage to any histocompatibility antigen locus has been demonstrated.

BIOLOGICAL FUNCTION OF IMMUNE RESPONSE GENES

At present one can only speculate on the possible biological advantages to an animal of the possession of immune response genes and the identification of the relevant gene products is essential to a precise definition of the problem. Nevertheless considerations of possible biological advantage of different models might provide important indications of the possible genetic and cellular mechanisms involved. At the outset it seems clear that if all individuals of a population or species possess the same potential, at the genetic level, for immune responses to the antigens they are likely to encounter in their environment then the presence in individuals of the species of additional mechanisms whose effect would be to limit the expression of their immune response potential would be disadvantageous rather than advantageous unless some compensating advantage were to result from the presence of the mechanism responsible for the limitation. This might occur if the limitation were imposed by histocompatibility antigens or other surface components which affected only incidentally the reaction between antigen and the receptors of antigen sensitive cells.

From what has been said it seems very unlikely that differences in responsiveness are due to differences in V gene content. If the immune response gene product is a non-immunological type of antigen receptor intraspecies differences between individuals in their possession of T cell receptors of different specificities could be an advantage by permitting a larger genetic pool for sharing between the individual members of the species, a pool which could be drawn on for defence, e.g. against microbial antigens, when environmental demands are made on it. In this case a mechanism for the maintenance of heterozygosity at the immune loci would obviously be of considerable importance.

II.5

Antigen–Antibody Reactions

In this chapter we shall be concerned primarily with those antigen–antibody reactions which are demonstrated under in vitro conditions.

VALENCY OF ANTIBODIES AND ANTIGENS

As we have seen all monomeric antibodies have two antigen combining sites; the antibodies are therefore divalent. Polymeric antibodies, e.g. IgA and IgM, have higher valencies corresponding to the number of monomeric units in the polymer. If the antigen or hapten has only a single combining group it will combine with only a single molecule of antibody. Under these circumstances it is impossible for an antigen–antibody lattice to be built up. If, however, the antigen is multivalent, three-dimensional lattices can be built up. The result of this process is precipitation in the case of soluble antigens and agglutination in the case of insoluble antigens, e.g. bacteria. In both cases, provided sufficient antibody and antigen have combined, the complexes are visible to the naked eye. The principal naturally occurring multivalent antigens are polysaccharides and proteins. In polysaccharides multivalency is determined by repetition of identical sugar residues. In many proteins, however, combination of residues to yield a specific determinant group may occur only once in the protein. Nevertheless proteins are multivalent because of their possession of a number of different determinant groups each of which combines with a different antibody.

The number of determinant groups is not, however, the only factor controlling the reaction. Neither agglutination nor precipitation will occur in the absence of electrolyte. This is believed to be because, for the effective formation of a lattice, the antibody and antigen molecules should be able to approach one another so closely that combination can occur. Under normal conditions this approximation is inhibited by mutual repulsion between the negatively charged antigen and antibody molecules. When this negative charge is reduced by adsorption of the cations of electrolytes, usually Na^+ of NaCl, adequate approximation for antigen–antibody combination can occur. The correctness of this view is borne out by experiments in which antibody molecules are acetylated. This increases the negative charge on the molecules and their consequent hydrophilic character; as a result the antibodies are incapable of agglutination and precipitation because of their electrostatic repulsion of the antigen and their heightened affinity for water.

AFFINITY

A basic requirement for antigen–antibody combination is that the antigen and antibody should have an affinity for one another. Affinity may be measured objectively for monovalent haptens by the technique of equilibrium dialysis. In this the antibody and antigen are placed initially in different compartments separated by a dialysis membrane which is permeable to the antigen but impermeable to the antibody. After diffusion has been allowed to occur antigen molecules in the antibody compartment combine with the antibody and are trapped in this compartment, in which there is consequently an excess of antigen which is represented by the antigen bound to antibody. The extent of this excess is a measure of the affinity of the antibody for the antigen.

Most populations of antibody molecules from

the same animal are heterogeneous in their binding affinity. The affinity revealed by this technique is therefore the average of a heterogeneous collection of molecules of different affinities. Antibody molecules which are completely homogeneous in respect of their affinity for a particular antigen have been found to be produced by immunization of animals with certain streptococcal and pneumococcal polysaccharides. Another source of homogeneous antibodies are myeloma proteins.

BONDING

Antigen–antibody complexes have a high degree of stability. This stability is created by a number of factors which include every type of intermolecular reaction. Of these undoubtedly the most important are the non-specific intermolecular forces between instantaneous dipole moments, known as Van der Waals' forces. The latter only become effective when the molecules are very close together, their strength varying inversely with the sixth power of the intermolecular distance. Other intermolecular forces which also take part in antigen–antibody combinations are electrostatic or Coulomb forces, hydrogen bonds and hydrophobic interactions between apolar groups on the antigen–antibody surfaces. As a result of the operation of these various forces the antigen–antibody union is a very firm one. Under certain conditions, however, dissociation may occur, e.g. on dilution or on addition of fresh antigen to an antigen–antibody complex. In the latter case some of the added antigen will substitute for the antigen already built into the complex.

SPECIFICITY

As previously indicated the immune response is highly specific; an antibody combines maximally or only with the antigen used to elicit its production. The most important determinant of this specificity is the complementarity between the shape of the antibody combining site and that of the determinant group of the antigen so that the antigen fits snugly into the combining site on the antibody. The relation has appropriately been described as that between a lock and its key. Cross-reactions, however, frequently occur, the same antibody reacting with more than one antigen. There are two reasons for such cross-reactivity:

1. The existence in complex antigens of identical determinant groups. In this case the corresponding antibody can be absorbed out of a serum with an antigen which contains it while leaving behind antibodies to other determinants on the eliciting antigen.
2. On similarity between determinants in different molecules. In this case there will not be a close fit between antibodies and antigen; the union will be correspondingly loose and readily dissociable and the affinity will therefore be low.

AVIDITY

When an antiserum containing a heterogeneous population of antibody molecules is confronted with an antigen, e.g. a protein possessing a heterogeneous group of antigenic determinants, the multiplicity of intermolecular combination which ensues creates a firmness of binding known as avidity. In such a case avidity is the summation of the affinity of individual antibodies with their specific determinants. In practice it has been found, however, that the firmness of the union of antigen and hapten molecules bound together by two antibodies is more than double that of each antibody–antigen complex formed by a single molecule with consequent enhancement of avidity.

THE AGGLUTINATION REACTION

Agglutination is the clumping that occurs when particulate antigens, e.g. bacteria and red cells, are mixed with homologous antiserum. It can only occur if the latter contains antibody to antigens located on the surface of the particle, and only in the presence of electrolyte. Agglutination can only be demonstrated satisfactorily if the antigens used are stable in suspension prior to addition of the antiserum.

Agglutination tests are of two types.

Direct agglutination tests

The antigens involved in these tests are structural components of the surfaces of cells, e.g. of bacteria or red cells.

Direct bacterial agglutination tests may be carried out by *slide* or *tube* methods. Slide methods are merely qualitative or at most semi-quantitative procedures. They are usually carried out with living organisms taken from colonies on plates or from the growth on an agar slope and are of considerable value in the preliminary identification of certain bacteria, notably *Salmonella, Shigella, Vibrio* and *Brucella* species. Tube tests are more precise and are normally carried out as quantitative procedures. Killed bacterial suspensions are generally used, the method of killing depending on the antigen required for the test. Serial dilutions of the serum are made and the highest dilution capable of agglutinating the organism under standard defined conditions is determined; this dilution is known as the *titre* of the serum. Tube agglutination tests are of great importance in confirming the results of slide agglutination tests for the purpose of bacterial identification, and in the estimation of the antibody content of a patient's serum.

Agglutination tests carried out by serial dilutions of antiserum frequently show prozones, that is the absence of agglutination in higher concentrations of antibody. This is often followed by tubes, containing lower concentrations of antibody, which show positive agglutination. Occasionally the prozone may extend throughout the range of the agglutination test. By the use of various techniques, notably the antiglobulin technique, it has been found that such prozones are due to non-agglutinating or blocking antibodies. These are particularly common in *Brucella* agglutination tests but occur in other systems as well. The nature of the blocking effect has not been explained. Possibly the most plausible explanation is that instead of forming links with antigenic determinants on different cells the antibodies combine only with adjacent identical determinants on the same cell. Other possibilities are that the antibody molecules are too short to span the distance between different bacteria or that one of the antigen combining sites is for some reason non-functional and the antibody molecules are therefore univalent in practice.

Passive agglutination tests

In these the antigens are adsorbed or otherwise attached to the surfaces of carrier particles which are then agglutinable by antibody to the adsorbed or attached antigen. The most generally useful carrier particles are undoubtedly red blood cells, human and sheep red cells having been particularly used. Many bacterial polysaccharide antigens readily adsorb to untreated cells. Adsorption is more effective at 37°C than at lower temperatures and is often considerably enhanced following pretreatment of the antigen by boiling or by exposure to dilute alkali. Several serologically distinct polysaccharide antigens can be adsorbed to the same red cells, apparently without mutual interference, and adsorbing, presumably, to different receptors on the cell surface. Adsorption of bacterial antigen may on occasion mask certain of the surface groupings of the red cell. Thus adsorption of *Salm. typhi* Vi antigen to human red cells reduces agglutinability by anti-A and anti-B antibodies, and adsorption of certain *Klebsiella* capsular antigens blocks the adsorption of influenza virus.

Passive haemagglutination tests have been found to be of considerable value for the estimation of antibody to Gram-negative bacillary somatic antigens (p. 240), and in the identification of a variety of streptococcal antigens which have been found to be capable of causing red cell sensitization.

Generally, protein antigens adsorb poorly to untreated red cells, but as shown by Boyden, strong adsorption occurs if the cells are first treated with low concentrations of tannic acid. The precise nature of the change that tannic acid produces in red cells is unknown. Protein antigens may also be coupled to red cells by chemical means, e.g. through the intermediary of *bis*-diazotized benzidine which links the protein to a substrate, possibly another protein, on the red cell surface. Coombs has described an ingenious coupling technique in which the protein is first linked by *bis*-diazotized benzidine to a non-agglutinating anti-D serum and the latter adsorbed by antibody antigen union to D-positive red cells.

A number of antigens will adsorb to particles of polystyrene latex and of bentonite. So far, however, these carriers have been used mainly for the demonstration of the rheumatoid factor—found in the sera of patients with rheumatoid arthritis—which behaves as if it were an antibody to human gammaglobulin.

Antiglobulin or Coombs's technique

As previously indicated, some antibodies, because of an unknown structural property, are unable to cause agglutination in saline media. They may be demonstrated by Coombs's antiglobulin technique, in which sensitized cells are exposed to antiserum against the corresponding immunoglobulin. This technique was first introduced for the demonstration of non-agglutinating anti-Rh antibodies in cases of human Rhesus immunization but it has also been used to demonstrate non-agglutinating antibody to *Salmonella, Shigella* and *Brucella* antigens in human sera. The relative concentrations of agglutinating and non-agglutinating antibodies in human sera have been found to vary independently. Some immune sera may contain only low titres of agglutinating antibody but very high titres of non-agglutinating antibody.

THE PRECIPITIN REACTION

When the antigen is in solution and provided antibody and antigen are present in suitable proportions, the antigen–antibody complex is precipitated. Like agglutination, precipitation can occur only in the presence of electrolyte.

The precipitin reaction may be applied to the detection of either antibody or antigen, but is not as extensively used in bacteriology for either of these purposes as the agglutination reaction. Although of relatively limited practical value it is, however, of great theoretical importance and its investigation has thrown considerable light on the nature of antigen–antibody reactions in general. Its main use in the detection of bacterial antigens is in the Lancefield grouping of streptococci and in the identification of antigenic components, e.g. of *Bacillus anthracis* and *Yersinia pestis* in the tissues of infected animals. Its only important application in the demonstration of antibody is in the Kahn and other flocculation tests used in the diagnosis of syphilis.

The precipitin reaction is more dependent than the agglutination reaction on the relative amounts of antibody and antigen present. Inhibition of precipitation may occur through the presence of excess antigen or of excess antibody. In both cases soluble non-precipitating antigen–antibody complexes are formed. Inhibition of precipitation in the presence of excess antigen is the greater difficulty and is shown by all precipitating antisera. Inhibition of precipitation because of antibody excess is shown only by certain sera, notably those prepared by immunization of horses with protein antigens.

Precipitation techniques

Ring test

This is the method normally used when the precipitin reaction is employed for the detection of antigen. A small amount of antiserum is placed in the bottom of a precipitin tube (a very small test tube) and the antigen solution carefully layered on top of it. The precipitate forms as a narrow ring at the junction of the two fluids. The ring test is not significantly affected by inhibition due to antigen excess. This is because at the zone of junction the two fluids diffuse into each other and at some point in this zone the optimal relative concentrations of antibody and antigen necessary for precipitation will be present.

Simple mixture

In this procedure the antigen solution and serum are mixed, incubated and then observed for precipitation. Since inhibition of precipitation in the region of antigen excess is frequently found, it is usual to set up a number of tubes each containing different dilutions of antigen but a constant amount of serum. The simple mixture technique is employed in the Kahn and other flocculation tests used for the diagnosis of syphilis. It may also be used for the estimation of the amount of antibody in serum by a serial dilution procedure analogous to that employed in agglutination tests; for this purpose a low concentration of antigen must be employed.

Optimal proportions technique

If a series of tubes are set up each containing the same amount of antibody but varying amounts of antigen it will be found that a precipitate forms first with one of the dilutions of antigen. The ratio of antigen to antibody present in this tube is known as the *constant antibody optimum ratio*. This optimum ratio is the same for all dilutions of antibody. This procedure can readily be applied to the estimation of antibody and will allow a comparison of the amount of antibody in an unknown serum with that in a standard serum. The method can also be adapted for the estimation of antigen.

With certain sera, a *constant antigen optimum ratio* can also be demonstrated. If varying amounts of antibody are mixed with a constant amount of antigen, the antigen–antibody ratio giving optimal flocculation is the same for all dilutions of antigen. This procedure is used commercially in the preliminary assay of diphtheria antitoxin. The constant antigen optimum ratio can be determined only for sera which show an inhibition of precipitation in the presence of antibody excess.

The constant antibody and constant antigen optimum ratios are not identical. In the constant antibody titration at the optimum ratio all antibody and antigen is precipitated. In the constant antigen titration on the other hand optimum precipitation occurs in the presence of antibody excess.

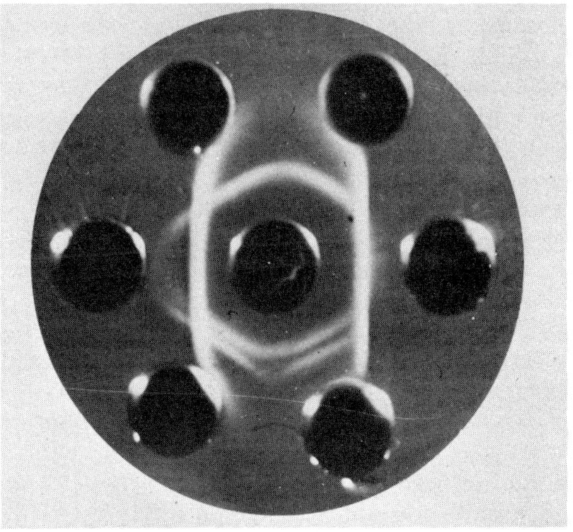

Fig. II.6. The Ouchterlony agar gel diffusion technique for the precipitin reaction. The centre well contains antiserum and the peripheral wells contain antigen solutions. The antiserum possesses three distinct antibodies. Note the reactions of identity (fusion of lines) and non-identity (crossing of lines).

Absolute quantitative methods

When any of the methods so far described are used for the estimation of antibody only a relative indication of the concentration of antibody is obtained. The results therefore give no indication of the amount of antibody protein present and only permit a comparison of the antibody content of an unknown serum with that of a known serum. An absolute estimation of the antibody content of a serum in terms of antibody protein can be obtained by precipitating all the antibody with antigen and determining the amount of antibody nitrogen present in the precipitate by chemical methods. The results obtained by this method in the estimation of both antibody and antigen have a degree of accuracy unattainable with other procedures.

Gel diffusion techniques

These are modifications of the precipitin technique in which the reactions are allowed to occur in agar. In the Ouchterlony technique antigen and antibody are allowed to diffuse towards one another from wells cut in an agar plate, a line of precipitate which usually takes 24 hours or longer to develop forming in the zone between the wells. If more than one antigen reacting with the antiserum is present each will give rise to a separate precipitate line and the method can therefore be used to show the minimum number of antigens present in a solution. The Ouchterlony method may also be used to determine the relationship between the antigens in different antigen solutions precipitating with the same antiserum or the relationship between the antibodies present in different antisera precipitating with the same antigen solution. If, in the first case, the antigens in each solution are identical they will give lines of precipitate which fuse together; this is known as the *reaction of identity*. If the antigens are different they will give lines of precipitate which intersect each other; this is known as the *reaction of non-identity*. In the reaction of *partial identity* the two precipitate lines join but one forms a 'spur' projecting beyond the other; this occurs when the antigens have a common structure but that forming the spur contains an additional antigenic determinant.

Immuno-electrophoresis

The status of electrophoresis as one of the most powerful tools of biochemical research, permitting the separation of macromolecules carrying different electrical charges is now well established. By combination with the agar gel double diffusion precipitin technique, the sensitivity and discrimination of the procedure are enormously increased, allowing the differentiation of antigens on the basis of both serological specificity and electrophoretic mobility. The reaction is usually performed in thin layers of agar gel on glass slides and is carried out in two stages. In the first, the antigen mixture is introduced into a small well cut into the agar and electrophoresed, the different components of the mixture separating according to their electrophoretic mobility. When the electrophoretic run is complete a gutter is cut along the length of the slide and antiserum containing antibody to the antigens present in the mixture pipetted into it. The slides are then kept at room temperature in a closed container to minimize evaporation. After about 24 hours precipitin lines develop in the agar, the centre of each coinciding with the area of maximum concentration of the corresponding antigen. This technique has proved especially valuable for the identification and differentiation of serum proteins, some 75 of which have been distinguished by its use. It has been of particular importance in the differentiation of the immunoglobulins and of the components of complement, and has been widely used for the separation of a variety of bacterial and viral antigens.

Single radial immunodiffusion

This technique is frequently employed for estimating the concentration of protein in a solution and has been particularly applied to estimating the concentration of different immunoglobulin classes in a serum. Antibody to the protein is incorporated in agar on a glass slide or a plastic tray and the antigen allowed to diffuse outwards from wells punched in the agar. Immediately adjacent to the well the antigen concentration is high and soluble antigen–antibody complexes form; further out at optimum antibody–antigen ratios a ring of precipitate is formed, the diameter of which is directly proportional to the concentration of the antigen. By plotting a calibration curve for a particular antigen with a given serum the concentration of antigen in an unknown solution can immediately be read off.

ANTIGEN-BINDING TECHNIQUES

In these the concentration or antigin-binding affinity of an antiserum is determined by estimating the amount of radioactive antigen which would bind with a given volume of the serum when the antigen is in excess. In the Farr procedure the antibody–antigen complexes are precipitated with 40 to 50% ammonium sulphate. The amount of label in the precipitate is then determined by scintillation counting. The method is only applicable when the antigen employed is soluble in ammonium sulphate. Alternatively the antibody may be precipitated by the use of an antiglobulin serum. By the employment of immunoglobulin class-specific antisera the concentration of the different immunoglobulin classes reacting with the antigen may be determined.

If the antigen is rendered insoluble by cross-linking with *bis*-diazotized benzidine or glutaraldehyde or by coupling to an immuno-adsorbent column and exposed to antibody the antibody can be eluted by lowering the pH and its concentration estimated by, for example, the single radial immunodiffusion technique.

Radioimmunoassay

This is probably the most sensitive method for the estimation of antigen. It has been extensively applied to the assay of various hormones, e.g. insulin. The procedure employed is first to mix a known amount of isotopically labelled antigen, together with unlabelled antigen, with specific antibody. Then either the antigen–antibody complex is precipitated with ammonium sulphate or antiglobulin sera or the free uncombined antigen is removed, e.g. by adsorption onto charcoal; the ratio of free to bound radioactivity is then estimated. The labelled and unlabelled antigens will compete with each other for combining sites on the antibody in proportion to their relative concentration. From the ratio of free to bound radioactive antigen the concentration of unlabelled antigen can then be calculated. This method can only be used in the presence of antigen excess but is capable of measuring as little as 10^{-12} g/ml of antigen.

REACTIONS INVOLVING COMPLEMENT

The nature and functions of complement will be considered in detail in Chapter II.7.

Immune haemolysis and complement fixation tests

As we shall see, red cells which have been sensitized by homologous antibody are lysed in the presence of complement. This phenomenon is of great practical importance as an indicator in complement fixation tests for the detection and estimation of both antibodies and antigens. Complement fixation tests have been most used for the detection of antibody particularly in the diagnosis of syphilis (p. 299) and various viral diseases.

The principle of the tests is as follows. If a serum containing antibody is mixed and incubated with its homologous antigen in the presence of complement (which is usually obtained from the guinea-pig) the complement is fixed. If to this mixture is added an indicator system consisting of a suspension of red cells (sheep cells being most frequently used) and antibody to the red cells no lysis of the cells occurs since there is no free complement to produce it. This is a positive result. If, however, the serum employed in the first stage of the test does not contain antibody to the antigen, the complement added is not fixed and is free to cause lysis of the red cells in the second stage of the test. This is a negative reaction. Complement fixation tests can be rendered quantitative for the amount of antibody present by varying the amount of complement or by using a standard amount of complement and varying the concentration of the antiserum.

Immune adherence

Particulate antigens sensitized by their homologous antibodies in the presence of complement will adhere to primate red cells, the platelets of various species and macrophages and polymorphonuclear leucocytes. Adherence occurs through the activated C3b component of complement. Immune adherence is of importance in phagocytosis, but, using red cells or platelets, can be made the basis of a sensitive procedure for antibody or antigen estimation.

Conglutinating complement fixation

As considered below (p. 162) the sera of many animals, especially bovines, contain a substance—conglutinin—which binds with the activated C3b component of complement fixed on to sensitized cells. If the complement employed is deficient, as is horse complement, in the later components of the complement complex, haemolysis does not occur, but instead the cells are agglutinated by the conglutinin. Like immune haemolysis conglutination can be used as an indicator for complement fixation; fixation of a horse complement by an antigen–antibody complex is shown by demonstrating its non-availability for the sensitization of red cells to conglutination.

MICROSCOPIC TECHNIQUES

Immunofluorescence

As first shown by Coons antibodies can be coupled to dyes which fluoresce in ultra-violet light thereby making it possible to recognize localization of the antibody in the tissues. By the use of antibodies specific for the antigens of particular tissue components the distribution of the antigen in the tissues can be visualized. The dyes commonly used, which are known as fluorochromes, are fluorescein isothiocyanate which gives a green fluorescence and lissamine–rhodamine which gives an orange fluorescence. The fluorescence can only be seen when the preparations are examined in a dark field microscope system.

Immunofluorescence may be used for the detection of both antigens and antibodies. Three procedures are employed:

1. Direct staining, when the coupled antibody is specific for the antigen under investigation. The antigen may be a microbial as well as a tissue antigen.

2. Indirect staining. The material under examination is first exposed to an uncoupled antiserum, washed and exposed to fluorochrome-coupled antiserum specifically reacting with the

gammaglobulin of the species from which the antiserum used in the first stage of the technique was obtained. This procedure has the advantage that only a single coupled antiserum is required for the detection of a range of antigens. The procedure can also be used for the detection of antibody, e.g. in autoimmune and immune complex diseases, or of bacterial antibody as in the fluorescent treponemal antibody test employed in the diagnosis of syphilis.

3. The sandwich technique. This is a test for antibody localized in tissues. The tissue is first exposed to a preparation of the antigen against which the antibody is directed and then to coupled antiserum prepared against the antigen.

Ferritin-labelled antibodies

Antibodies can be coupled to the protein ferritin which has a high electron density because of the large amount of iron it contains. Ferritin-coupled antibodies can consequently be used for demonstration of antigen by electron microscopy. It is particularly suitable for the demonstration of surface tissue components and of viral antigens which adhere to the cell surface. In immunology it has been of special value for the demonstration of immunoglobulin receptors on the surfaces of B cells.

CELLULAR TECHNIQUES

Jerne plaque technique

This technique was introduced for the demonstration of antibody-producing cells in various lymphoreticular tissues and has been of great value in the investigation of problems connected with antibody production. In its simplest form it is employed to detect cells producing antibody against red cells, e.g. sheep red cells. For this a suspension of lymphoid cells from animals immunized against sheep red cells is prepared, mixed with a suspension of sheep red cells, and the mixture incorporated in agar. The mixture is incubated and complement added when lysis of the red cells in the vicinity of an antibody-producing lymphocyte occurs. This area of lysis is known as a plaque. When the test is performed as described above only IgM-producing cells are detected. IgG-producing cells can be detected by the addition of anti-IgG serum; in this case cells with surface IgG will lyse on subsequent addition of complement. The plaque technique may also be used for the detection of cells producing antibodies which can be adsorbed to or chemically coupled to red cells if the latter are incorporated in the agar cell mixture.

Rosette formation

The object of this technique is to demonstrate immunoglobulin receptors on the surfaces of lymphocytes. When lymphocytes from animals which have been immunized with red cells are mixed in suspension with the latter, the red cells form a ring of cells around each immune lymphocyte, the whole complex forming a 'rosette'. If the lymphocyte is actually producing antibody against the red cells the rosette will be several layers thick, whereas if the lymphocyte possesses receptors only but is not producing antibody the red cells will form a single layer. If sheep red cells are used they will form rosettes non-specifically around T cells. This reaction is sufficiently specific for T cells in non-immune cell populations that rosette formation with sheep cells has been extensively used for the identification of T cells in human peripheral blood.

NEUTRALIZATION

When an antigen has a demonstrable in vivo or in vitro effect, this effect can often be neutralized by antibody. Tests in which this is demonstrated are known as neutralization tests.

Toxin neutralization

Bacterial exotoxins are highly antigenic and their activity may be completely neutralized by appropriate concentrations of antibody. Antibody to bacterial exotoxins is usually referred to without qualifications as *antitoxin*. Bacterial endotoxins, at least of the type found amongst the enterobacteriaceae, are poorly if at all antigenic and their toxicity is incompletely neutralized by antibody (p. 243). Such antibody, whose specificity is directed against the polysaccharide component of the somatic antigen complex, is not usually referred to as antitoxin.

The neutralizing capacity of an antitoxin can be assayed by neutralization tests in which mixtures of toxin and antitoxin are injected into susceptible animals. The most widely used indicator of toxin activity in such tests is the lethal effect of the toxin on the animal, the concentration of an antitoxin being estimated by determining its capacity to neutralize this lethal effect. With toxins, such as the diphtheria toxin, which produce reactions in the skin, these can be used as indicators of toxin activity. This phenomenon is the basis of the Schick test used for the demonstration of immunity against diphtheria (p. 231) and can also be applied to the assay of diphtheria antitoxin, using the guinea-pig as test animal.

If a toxin has a demonstrable in vitro effect, this effect is neutralized by antitoxin. Thus antitoxins to haemolytic toxins neutralize the haemolytic action of the toxin. This neutralization can be made the basis of an in vitro method of assaying the antitoxin. The only procedure of this type of clinical importance is the estimation of the antibody to streptolysin O (antistreptolysin O). This antibody is significantly increased in patients with acute streptococcal infection and in those suffering from acute rheumatism (Chapter III.3).

The specific neutralization of toxin by antitoxin is utilized in the Nagler reaction used for the identification of *Clostridium perfringens*. The test depends on the capacity of antitoxin to neutralize the lecithinase activity of *Cl. perfringens* and is usually carried out in plate cultures (Chapter III.18).

Toxin is not destroyed by combination with antitoxin. By certain procedures, for example by treatment with dilute acids, by freezing and thawing or even by simple dilution, toxin may be dissociated from combination with antitoxin; the liberated toxin is fully active. Toxin–antitoxin mixtures are therefore quite unsuitable as immunizing agents.

Neutralization of enzyme activity

Antibodies to enzymes can inhibit the enzymatic activity of the enzyme. As in the case of toxins the enzyme may be dissociated from the enzyme–antibody complex in a fully active form. The mechanism of neutralization is not fully understood, but may be due either to neutralization of the active group of the enzyme by the antibody or to combination of antibody with determinants on the enzyme distinct from the active group. Such combination may interfere with the effective binding of the enzyme to its substrate by a steric effect or by inducing a conformational change in the surface of the enzyme. In some cases, however, the activity of an enzyme may increase by combination with antibody; this appears to occur particularly when the enzyme combines with low molecular weight substrates. An interesting example of such enhancement was the demonstration by Pollock that the activity of a penicillinase, produced by *Bacillus licheniformis*, against benzyl penicillin and cephalosporin C was inhibited by antibody to the enzyme, but that its activity against methicillin was enhanced. These findings lend support to the view that the primary effect of antibody is the production of conformational changes in the enzyme surface. These changes may diminish the affinity of the enzyme for one substrate while increasing its affinity for others.

II.6

Lymphocytes and Lymphokines

LYMPHOCYTE TRANSFORMATION

It is a basic requirement of the immune response that, on contact with a specific antigen, lymphocytes carrying homologous receptors are stimulated to proliferate. In the whole animal this proliferation is reflected in the expanded population of specifically antigen-sensitive lymphocytes and in the biological effects of their interaction with antigen.

Under in vitro conditions the stimulation of lymphocytes which eventually leads to their proliferation may be demonstrated in a variety of ways. Thus, if a suspension of immune lymphocytes derived from lymph nodes, spleen, peripheral blood or peritoneal exudates are exposed to their homologous antigen, the lymphocytes undergo blast transformation in which they show nuclear enlargement and the development of mitotic changes. This blast transformation is readily detected by conventional staining techniques. Lymphocyte transformation is, as might be expected, associated with DNA synthesis, which may be detected by the incorporation by stimulated cells of ^{3}H or ^{14}C thymidine. Thymidine incorporation occurs only in cells showing morphological transformation. The latter, therefore, can be detected by autoradiography. In general, however, the number of cells showing morphological blast transformation is greater than the number showing thymidine incorporation. This difference may be due to technical factors or it could indicate that the morphological changes may occur in the absence of DNA synthesis.

Blast transformation and DNA synthesis can also be stimulated non-specifically, i.e. other than by exposure to the specific antigen. It may be induced by reacting lymphocytes with antisera containing antibodies to various immunoglobulin determinants. In fact it was the demonstration of transformation following exposure to anti-immunoglobulin sera by Sell and Gell in 1965 that first established the presence on the lymphocyte surface of immunoglobulin-like receptors. Transformation may also be induced by exposure to plant mitogens, e.g. phytohaemagglutinin (PHA), concanavalin A (Con A) and pokeweed mitogen (PWM), and by various bacterial products, the most important of which are the Gram-negative endotoxins and subtances present in staphylococcal and streptococcal culture filtrates. The number of cells which are transformed by these various agents is, as might be expected, considerably greater than the number of cells specifically transformed by homologous antigen.

Non-specific mitogens appear to exhibit some degree of specificity towards the lymphocyte sub-population stimulated. Thus soluble PHA and concanavalin A are highly specific T cell mitogens. Gram-negative endotoxins and aggregated tuberculin are described as being highly specific B cell mitogens. Insoluble plant mitogens and staphylococcal enterotoxins are mitogenic for both T and B cells. Phytohaemagglutinin, as its name implies, also has the capacity to agglutinate red cells but this capacity is due to components distinct from that responsible for mitogenic activity and haemagglutinin-free and non-mitogenic haemagglutinin PHA preparations derived from crude PHA are commercially available.

PHA transformation has been extensively used in the investigation of immune deficiency states in man. PHA-induced transformation of peripheral blood lymphocytes is depressed in Hodgkin's disease, the di George syndrome, Swiss-type agammaglobulinaemia and ataxia telangiectasia, all of which are conditions with T cell deficiencies. Depression of PHA transformation has also been reported in non-lymphoid malignancies and would lend some support to the postulated surveillance function of cellular immune responses in the suppression of malignant cells. The demonstration of impairment of PHA transformation of peripheral blood lymphocytes has also a potential field of application in the monitoring of immunosuppressive therapy.

Small lymphocytes are highly refractory to virus infection. Lymphocyte transformation has, however, been shown to be associated with the acquisition of the capacity to support the growth of certain RNA viruses. This phenomenon has been extensively investigated by Bloom, particularly in respect of infection with the vesicular stomatitis virus and Newcastle disease viruses. The acquisition of this capacity was shown by plating virus-infected lymphocytes over an appropriate target cell monolayer (p. 367), the cells of which acquire the capacity to sustain virus growth and to produce plaques on the target monolayer. An unexplained finding reported by Bloom is the fact that not all transformed cells support virus growth. He has suggested that the latter capacity may possibly be restricted to the effector cells of cellular immune responses.

Mixed lymphocyte reaction

If the lymphocytes of two genetically distinct animals of the same species are cultured together the cells of each animal will stimulate those of the other to undergo blastogenic transformation. The reaction may be made a one-way reaction if the cells of one animal are treated with X-rays or mitomycin C to inhibit DNA synthesis. The determinants which induce transformation are known as LADs (lymphocyte activating determinants) and in both mice and men they are controlled by genes closely linked to the genes which determine the antigens responsible for allograft rejection. The transformed cells are cytotoxic for lymphocytes which possess different histocompatibility antigens and are known as 'killer' cells.

LYMPHOKINES

The term lymphokine was introduced by Dumonde for biologically active substances which may be demonstrated by various in vitro and in vivo techniques in the cell-free supernatants of transformed lymphocyte cultures. A variety of such substances has been described by various workers, each being defined operationally in relation to the particular techniques used for its demonstration. It is possible that a number of the techniques used may in fact identify the same substance. Thus it seems probable that a single substance is defined by the macrophage migration inhibition technique, by macrophage aggregation and by macrophage spreading inhibition and it remains to be decided whether this substance is distinct from the lymphotoxin demonstrated by cytotoxicity tests. On the other hand, a particular technique may reflect the activity of more than one substance. Thus it seems likely that inflammatory changes in the skin ascribed to the skin reactive factor are due in fact to the combined effect of a number of substances. It is also not clear to what extent the various substances identified or produced in the process of or as a result of cell transformation are normal lymphocyte products produced by untransformed cells but whose release is triggered by cell transformation. That this may be the case is suggested by the finding of substances with mitogenic cytotoxic and migration inhibitory properties in the supernatant of lymphoid cell culture lines.

The production of lymphokines is generally thought of in the context of the cellular immune response from which it might be concluded that lymphokine production is characteristically a property of T cells. In some instances, for example in the production of the migration inhibition effect, this certainly appears to be the case and the lymphokine may be thought of as a T cell product analogous to the B cell product, immunoglobulin. It is probable, however, that lymphokine production is not an exclusive property of T cells and that some biologically active substances other than immunoglobulin may be produced by B cells.

It is usually assumed that lymphokines produced in response to antigenic stimulation of specifically sensitized cells are the product of the cells initially stimulated by the antigen. This is not necessarily the case. It may well be, as suggested by Pick and Turk for the production of migration inhibition factor, that antigenic stimulation results in the release of an antibody-like substance which complexes with further antigen molecules and stimulates the release of lymphokine from bystander cells. Such a sequence of events constitutes an effective amplifying mechanism by virtue of which a comparatively small population of sensitized cells could produce a substantial biological effect.

Migration inhibition factor (MIF)

This factor was discovered independently by Bloom and Bennett and by David in 1966. These workers showed that the outward migration of cells from capillary tube cultures of peritoneal exudate cells from guinea-pigs sensitized with various antigens, and which had developed delayed-type hypersensitivity to these antigens, could be inhibited by exposure of the cells to the homologous sensitizing antigen. As a practical test migration inhibition appears to be of considerable value for the detection of a cellular immune response to antigens which give ambiguous delayed skin reactions, for example antigens which themselves produce a non-specific reaction or in animals that have developed a concomitant Arthus-type reactivity (p. 191).

A dissection of the migration inhibition phenomenon into its component parts revealed that the cell whose outgrowth was inhibited was the macrophage and the inhibition of migration of macrophages was due to a substance liberated from sensitized small lymphocytes on exposure to antigen. This can be demonstrated in tests employing mixtures of lymphocytes from animals sensitized by a particular antigen and macrophages from unsensitized animals. Under these conditions migration inhibition is demonstrable on exposure to the antigen when the sensitized small lymphocytes constitute as little as 1% of the entire cell population. The phenomenon can be demonstrated with lymphocytes from any lymphoid source but lymphocytes from peripheral blood and from lymph nodes have been most used. That the production of migration inhibition factor, in man at least, is the property of T cells is supported by the finding that it can be produced by the lymphocytes of individuals with agammaglobulinanaemia and is not produced by the lymphocytes of individuals with the di George syndrome.

The migration inhibition factor, as demonstrable in the supernatants of transformed lymphocytes, has a molecular weight estimated to be in the region of 35 000 to 55 000. On electrophoresis it has approximately the mobility of serum albumen. The migration of macrophages exposed to migration inhibition factor (MIF) is not antigen-dependent. This finding, together with the physical properties of MIF, would indicate that it is not a cytophilic antibody, although in certain instances cytophilic antibodies have been shown to be capable of sensitizing macrophages to inhibition by antigen. It is possible, however, as has been suggested by Pick and Turk, that sensitized lymphocytes on antigenic stimulation release an antibody-like substance which, on complexing with antigen, stimulates lymphocytes standing by in the neighbourhood to produce MIF. In support of this hypothesis they have demonstrated that lymphocytes of non-sensitized animals will produce MIF after exposure to lymphocytes of animals immunized against red cell antigens, and subsequently stimulated by these antigens or to the supernatants of cultures of such immune antigen-stimulated lymphocytes or by exposure to antigen–antibody complexes.

The precise mechanism by which migration inhibition factor acts on macrophages has not been established. Evidence has been obtained, however, that exposure to MIF increases the stickiness or viscosity of the macrophage. The nature of the surface change responsible for this increase in viscosity has not been identified but evidence has recently been presented by Pick that MIF may induce activation of the adenyl cyclase system, with resulting increase in intracellular cyclic AMP concentration. This in turn might activate an enzymic system involved in increased viscosity of the cell surface.

Macrophage migration inhibition is of considerable potential biological importance. It seems likely that it is responsible for the predominantly mononuclear character of the exudates in delayed-type hypersensitivity. This could be explained by the fact that macrophages which have been extravasated from the blood vessels as a result of the inflammatory stimulus are unable to return to the blood stream and, therefore, accumulate in the tissues. Macrophage migration inhibition would also explain the early observation of Rich and Lewis that following injection of tuberculin into the peritoneal cavities of tuberculous guinea-pigs the macrophage content of the free peritoneal fluid was considerably reduced but that the missing macrophages could be recovered from the surface of the peritoneum to which they were found to be adherent.

MIF can also be assayed or demonstrated by its capacity to inhibit the migration of cells from cultured fragments of spleen. MIF-containing supernatants have been found to inhibit the spreading of macrophages on glass surfaces. This inhibition, and also the aggregation of macrophages in cultures, is almost certainly due to the migration inhibition factor.

Mitogenic factors

The supernatants of transformed lymphocytes contain substances which are able to induce blast transformation and thymidine uptake in normal lymphocytes. That mitogenic factor is distinct from migration inhibition factor is shown by the lack of parallelism in mitogenic and migration inhibitory activity, e.g. as demonstrated by Flad and Hochapfel in the supernatants of lymphocyte cultures from the blood of Mantoux-positive donors. As in the case of MIF it is possible that mitogenic factor produced as a result of antigenic stimulation may not be the direct product of the stimulated cell but produced from attendant cells after stimulation of the latter by a complex of antigen and antibody-like substance derived from the former. This would accord readily with the observation that mitogenic activity may be demonstrable in supernatants before obvious blastogenic cell transformation has occurred.

Lymphotoxin

Although the rejection of allografts is almost certainly due to the direct cytotoxic and cyto-aggressive action of lymphocytes, there is evidence for the presence of a non-specific cytotoxic substance in the supernatant of transformed lymphocytes. Although this substance probably does not play a significant role in allograft rejection it may be of importance in the immunological surveillance of neoplastic cells.

The cytotoxic effect of transformed cell supernatants may be demonstrated by a variety of techniques all of which measure the degeneration or death of non-lymphoid cells in tissue culture. The principal methods employed are as follows:

1. Enumeration of the proportion of cells which may be detached from surface cultures by treatment with trypsin. The detachable cells are non-viable, having been killed by lymphotoxin.
2. The release of ^{51}Cr from labelled cells. This probably measures damage to cell surface membrane.
3. The release of ^{14}C thymidine from cells grown in the presence of this label. ^{14}C thymidine release requires the breakdown of cellular DNA.

Different types of tissue culture have been found to show considerable differences in their susceptibility to lymphotoxin; this susceptibility depends on the concentration of the lymphotoxin to which they are exposed. The highest degree of susceptibility so far recorded has been shown by tissue cultures of L cells. Lymphotoxin-containing supernatants also possess demonstrable growth-inhibitory properties. Thus they will prevent the development of clones of Hela cells and will inhibit the incorporation of tritiated thymidine and of ^{14}C-labelled amino acids. These inhibitory effects are presumed to be due to the lymphotoxin components responsible for cell death. Lymphotoxin is also reported to damage fowl red cell membranes, damage which can be detected by the release of incorporated ^{51}Cr label.

Chemotactic factors

Chemotaxis may be studied in vitro by the use of a simple cylindrical two-compartment chamber, the compartments of which are separated by millipore membrane filters. The potentially chemotactic substance is placed in the lower compartment and the cell suspension the mobility of which is being measured in the upper. Migration is detected by microscopic examination of the membrane and is quantitated by estimating the numbers of cells migrating through it. Chemotaxis of macrophages is determined by using millipore filters with an average pore diameter of 5 μm and chemotaxis of polymorphonuclear leucocytes by filters with an average pore diameter of 0·65 μm. Suspensions of polymorphonuclears suitable for these tests can be readily obtained by intraperitoneal injection of rabbits with glycogen. With this technique the same rabbit can be used on a number of occasions. Suspensions of mononuclear cells may be obtained by injection of rabbit peritoneal cavities with mineral oil. This procedure has the disadvantage that the rabbit can only be used as a source of cells on one occasion.

Supernatants of transformed lymphocytes possess both polymorphonuclear and macrophage chemotactic factors. These have been shown by Ward to differ in electrophoretic mobility from one another and, also, from migration inhibition factor. Migration inhibition factor has no effect on the response of macrophages to the macrophage chemotactic factor. It would therefore appear that the mechanism involved in the random movement of macrophages is different from that found in the directional movement observed in chemotaxis.

Interferon

Interferon-like substances have been found to be liberated from transformed lymphocytes. Such substances are presumed to be of considerable importance in the contribution of cellular immune responses to antiviral immunity (p. 375). The relationship between the interferons produced from transformed cells and those produced by virus infected cells or cells stimulated to produce interferon with double stranded RNA (p. 364) has not been determined.

Skin reactive factors

The presence of substances producing inflammatory changes in the skin of guinea-pigs was first demonstrated by Bennett and Bloom in 1968 using lymph node lymphocytes stimulated by PPD. Similar substances were later shown by Pick and Turk to be liberated from peripheral blood lymphocytes and lymph node lymphocytes not only by antigen but also by plant mitogens. As described by Bennett and Bloom, skin reactive factor produces an indurated inflammatory change in guinea-pig skin reaching a peak 6 or 12 hours after injection. On histological examination there is a predominantly mononuclear response. The active agent or agents were found on Sephadex filtration to have a molecular weight in the region of 35 000 to 85 000. The reactions observed are clearly analogous to those produced by cutaneous injection of antigens in animals exhibiting delayed type hypersensitivity.

It seems possible that skin reactions produced by lymphocyte culture supernatants are due to the combined action of a number of different components, in particular migration inhibition factor and polymorphonuclear leucocyte and macrophage chemotactic factors. These observations are clearly related to the demonstration by Willoughby and Spector in 1962 of a lymph node permeability factor which could be extracted from leucocytes by various physical procedures, e.g. sonication, and which were found by these workers to be present at sites of hypersensitive reactions induced by injection of tuberculin into sensitized animals. Lymph node permeability factor was described as increasing vascular permeability, as being chemotactic and as leading to the deposition of fibrinoid material in the tissues. It could be differentiated from other potential pharmacological mediators of inflammatory reactions by its lack of effect on smooth muscle and its differential susceptibility to various drugs.

II.7

Antimicrobial Immunity

In order to survive in a world of pathogenic micro-organisms the animal body must possess a resistance against these organisms which will prevent them from invading the tissues and which, if they have invaded, will rapidly dispose of them and neutralize their toxic products. For such resistance it is customary to use the term immunity. In the past it has been usual to distinguish innate or *basic immunity*, which is defined as the immunity that an individual possesses by virtue of his genetic or constitutional make-up, and *acquired immunity*, which the animal acquires during the course of its life. Since, however, acquired immunity is based on the capacity of an animal to mount an immune response which has as its basis an expansion of a population of immunologically competent cells, the capacity of which is genetically determined, this differentiation though apparently useful is not semantically satisfactory. It is more appropriate to distinguish non-specific and specific mechanisms of immunity, the distinction being that specific mechanisms are the expression of the immune response and are directed against specific antigenic components or products of the invading micro-organism. Although such a distinction is convenient, specific and non-specific immunity do not exist in completely water-tight compartments since non-specific factors, e.g. complement and phagocytic cells, greatly increase the efficiency of the immune response as it affects defence mechanisms. In this context they have been described as amplifiers of immunity.

NON-SPECIFIC IMMUNE MECHANISMS

It is convenient to consider non-specific immune mechanisms on an anatomical basis. First there are the surface defence mechanisms which protect the tissues from invasion and secondly the tissue defence mechanisms which serve to eliminate the organisms which have invaded the tissues.

Surface defences

The first line of defence is the integrity of the surface layers of the body, namely the skin and mucous membranes. The skin constitutes a greater barrier against bacterial invasion than do the mucous membranes, since in some cases bacteria appear to be capable of passing directly through the mucous membrane, whereas it is extremely doubtful if any can ever pass through the intact skin. Mechanical breaks in the surface layers, therefore, facilitate tissue invasion, a fact which is readily attested by the readiness with which invasion and infection may occur through wounds and abrasions of the skin.

In addition to their mechanical integrity the skin and mucous membrane are bathed in various secretions and excretions, namely sweat, tears, mucus, saliva and gastrointestinal juices. These act mechanically by washing off bacteria which have adhered to the body surfaces. This effect is enhanced by the beating action of the cilia of ciliated epithelium which creates currents serving to dislodge bacteria from the deeper structures and, in the case of the mucous glands of the alimentary tract and the epithelium of the external ear, by outgrowth of cells from the depth of the structure to the surface. Many of the body secretions contain antibacterial substances which have a distinct microbicidal affect. Lysozyme is present in high concentration in tears; fatty acids are found in the secretions of the sweat and sebaceous glands; hydrochloric acid is present in high concentration in the gastric fluid.

An important role in the protection of the body surface must also be assigned to the commensal bacteria, many of which produce substances which are inhibitory to other organisms and which in addition by their mere presence prevent pathogenic organisms from gaining a foothold. Examples of such substances are hydrogen peroxide produced by salivary streptococci, fatty acids produced by the resident flora of the skin and by vaginal lactobacilli and the bacteriocines produced by many organisms of the normal flora, particularly the coliform bacteria of the alimentary tract. Because of the defensive role of antimicrobial substances produced by bacteria the application of stringent sterilization procedures to the skin with the object of reducing the population of resident skin flora, e.g. prior to surgical operations, must often be regarded as ill considered.

Tissue defences

Tissue defences fall into two categories, humoral and cellular. Apart from antibody, which is specific, and complement, which though non-specific acts only on micro-organisms sensitized by antibody (p. 162), a variety of non-specific antibacterial substances have been described in the tissues. These include lysozyme, haematin, β lysin, properdin (p. 163) and various basic peptides. The presence of these various substances in the tissues appears to be a secondary phenomenon resulting from the disintegration of various types of cell in which they are produced. Thus lysozyme and basic peptides are derived from leucocytes, haematin from red cells and the β lysins which are active only against Gram-positive bacteria from platelets. Tissue fluids also contain the iron-transporting compound transferrin, which exerts an antibacterial effect by diminishing the iron concentration in the local environment of the organism and therefore has an indirect antimicrobial action.

Phagocytosis and subsequent intracellular killing is undoubtedly the most important component of the tissue defences. Phagocytosis is in essence an extension of pinocytosis, the mechanism by which all cells imbibe nutrient material. Consequently phagocytosis can to some degree be regarded as a general cellular function. The most important of the phagocytic cells—the professional phagocytes—however, are the polymorphonuclear leucocytes and the macrophages. This is so because of their greatly enhanced capacity for phagocytosing particles which have combined with their homologous antibodies. The most important polymorphonuclear leucocytes are the neutrophils; the eosinophils have slight phagocytic activity and the basophils none. Basophils are, however, important in the reactions of anaphylactic hypersensitivity (p. 182).

Opsonization

The phagocytosis of bacteria, particularly pathogenic bacteria, is greatly enhanced by the presence of antibody which combines with the bacteria; this process is known as opsonization. The role of antibody in this process is often referred to picturesquely as rendering the bacteria more palatable to the phagocyte. Although this is the end result of opsonization, such a description does not indicate how this result is achieved. One way in which it may be achieved is that combination with antibody, which at physiological pH has a low negative charge, reduces the net negative charge on the bacterial surface. This charge is maximal for those bacteria that possess strongly acidic groups on the cell surface, e.g. the uronic acids of the polysaccharides of the bacterial capsules and microcapsular layers. Recent work has shown, however, that other factors are involved. Thus there is a direct binding of the antibody via its Fc portion to Fc receptors on the surface membrane of the phagocyte. Such receptors have been demonstrated for polymorphonuclear leucocytes and for macrophages. In addition bacteria which have been sensitized by IgM and IgG antibodies will fix complement. This fixation will trigger the immune adherence phenomenon mediated by the activated C3 component of complement (p. 163), as a result of which the cells should be more readily phagocytosed.

Antibody appears to be of much greater importance for the phagocytosis of virulent than of non-virulent bacteria. This could be due in part to the neutralization of substances toxic to phagocytes and in part to the fact that virulent bacteria seem, in some cases at least, to have a higher negative surface charge than avirulent bacteria; but even virulent bacteria may be substantially phagocytosed in the absence of antibody by the mechanism described by Wood as surface phagocytosis. Wood has shown that considerable phagocytosis of virulent bacteria occurs if the leucocytes are brought into contact with the bacteria not in fluid medium but on a solid surface, e.g. on the surface of filter paper or of fibrin. Wood postulated that under these circumstances the phagocyte can pin the bacteria against the solid support; he believed that this mechanism was of considerable importance in the tissues where tissue cells and fibrin produced in the inflammatory reaction may supply the solid support required.

Polymorphonuclear leucocytes

The polymorphonuclear leucocyte is an end cell with a short life in the tissues of probably not more than 24 hours. It is derived from bone marrow precursors which are also believed to be the stem cells for the monocytes. In the production of polymorphonuclear leucocytes the stem cell undergoes a number of divisions, producing the morphologically distinguishable stages of myeloblast, promyelocyte and myelocyte. The latter differentiate through the metamyelocyte form to the mature leucocyte, readily recognizable by its multilobular nucleus and by the presence of cytoplasmic granules of basophil, eosinophil or neutrophil type. Neutrophil polymorphonuclear leucocytes play a primary role in the phagocytosis and killing of ingested micro-organisms.

Two types of neutrophil granules are differentiated, differing in their size, in the stage at which they develop and in their composition. The larger or azurophil granules appear at the promyelocyte stage. They are essentially lysosomes and contain a battery of hydrolytic enzymes, the enzyme peroxidase and various antibacterial basic peptides to which the general name phagocytin has been applied. The small granules appear at the myelocyte stage and have a less obviously lysosomal character. They are deficient in acid hydrolases but contain considerable amounts of an antibacterial protein, lactoferrin and the enzyme lysozyme (p. 5). In addition to these substances present in their granules polymorphonuclear leucocytes contain hydrogen peroxide and a protamine-like protein, leukin, apparently derived from the nuclear fraction of the leucocytes and which has been shown to have antibacterial activity against certain Gram-positive bacteria.

The total number of neutrophils present in the blood stream is about 6×10^{11}. This population is supported by a large reserve of mature cells in the bone marrow, about 3×10^{12}. The latter can be discharged rapidly into the blood stream when required. The number of neutrophils normally present in the tissues is not known but it has been estimated that about 10^{11} cells are lost from the tissues each day. It is believed that most of these are lost by migration through the epithelium of the respiratory and intestinal tracts.

Macrophages

The macrophages are conventionally divided into two classes: the free or wandering macrophages of the tissue spaces and the fixed or sessile macrophages which are found as lining cells of the blood sinusoids of the liver and spleen and the lymph sinuses of lymph nodes; the former constitute the cells referred to by histologists as Kuppfer cells.

Macrophages are derived from precursor stem cells in the bone marrow. Their distribution through the body occurs via the blood stream, in which they appear as monocytes. The fixed macrophages of the blood and the lymph sinusoids appear to have a predilection for endothelial cells to which they become attached. The mode of this attachment is not known. Tissue macrophages possess large granules which contain acid hydrolases and therefore may be classified as lysosomes. These do not contain any of the antibacterial substances found in the granules of neutrophil polymorphonuclear leucocytes. The macrophages of the tissues, in contrast to the neutrophils, appear to have a very long life. Like the neutrophils, however, they are to all intents and purposes end cells but they appear to be capable of dividing in response to certain stimuli, although they do not normally do so. They can, however, undergo marked changes in morphology and activity (p. 137). By fusion they generate the giant cells which are characteristic of the histological foreign body reaction and of the granulomas found in tuberculosis. Giant cells are also formed, possibly in some cases by nuclear replication without cell division in response to certain viral infections. It seems reasonable to expect that the defensive function of the macrophages is enhanced by giant cell formation.

Phagocytosis

Phagocytic cells, both neutrophils and macrophages, will undoubtedly encounter invading micro-organisms by random collision in the tissues but this random process is not sufficient for the defence of the body. There is clear evidence that phagocytic cells are attracted to invading micro-organisms by a process of directional motility known as chemotaxis. This name indicates movement of the phagocyte towards the micro-organism along a concentration gradient of a chemotactic substance. Among the more important neutrophil chemotactic agents are substances liberated during the complement cascade (p. 161), C5a, C3a and C567. Others are kallikrein, denatured and conjugated proteins, staphylococcal and streptococcal products, chemotactic lymphokines and cyclic AMP.

On microscopic examination neutrophils moving in response to a chemotactic stimulus show a *quasi*-amoeboid motion. A membranous protrusion or lamellipodium first forms and adheres to the substratum; then the cytoplasm streams into the protrusion. In this process the nucleus at first lags behind but eventually moves forward after the terminal lamellipodia which do not contain cytoplasm have become detached. The distal portion of the cell appears to be projected forward by some type of contractile mechanism; the precise biochemical basis for this is still unexplained. It has been suggested that the chemotactic agent combines with stereospecific receptors on the surface of the leucocyte which in the uncombined state block the action of a membrane localized esterase. Activation of this esterase is then thought to lead to an increase in cyclic AMP, an increase which is the immediate trigger to the contractile process.

Ingestion of the particle is preceded by close contact of the particle with the surface of the cell. The particle is then enclosed in a phagocytic vacuole formed by invagination of the cytoplasmic membrane in contact with it and a surrounding of the particle by membranous pseudopods on its external surface. The phagocytic vacuole is finally formed and the external surface of the cell regenerated by membrane fusion. The process of ingestion is energy-requiring, the energy requirement being obtained principally from anaerobic glycolysis. As the phagocytic vacuole is being formed the neutrophil granules fuse with it and discharge their contents into its interior. Microscopically this process is marked by degranulation of the cytoplasm.

Killing of ingested cells usually occurs within 5 to 10 minutes after formation of the phagosome. Some bacteria can, however, persist within the cell and some can even kill the phagocyte. A variety of mechanisms has been described as being involved in intracellular killing. Substances released from the granules, e.g. basic peptides and lysozyme, have a direct bactericidal affect against various bacteria. Phagocytosis results in a stimulation of anaerobic glycolysis with the production of a high intracellular concentration of lactic acid which also has a bactericidal affect. Phagocytosis also activates the hexose monophosphate pathway, leading to the production of high intracellular concentrations of hydrogen peroxide. It has been proposed that this pathway is enhanced in the presence of vitamin C. Hydrogen peroxide has a direct killing action. Its effect is enhanced by peroxidase liberated from the granules, in the presence of which there is a halogenation, in particular iodination, of bacterial proteins. A further antibacterial effect is attributed to lactoferrin, which interferes with bacterial metabolism by lowering the local concentration of iron.

Bacteria that have been killed by the mechanisms described above are then digested by the various hydrolytic enzymes of the phagolysosomes. Digestion is usually slow, taking some 30 to 60 minutes for completion. The end-products of digestion are variously disposed of; some are excreted through the membrane, others may escape by reverse pinocytosis. It is likely, however, that the major event in the removal of end-products is the disposal or excretion of the neutrophils through the mucous membranes of the gastrointestinal and respiratory tracts. Under normal conditions it has been estimated that about 10^{11} cells disappear from the tissues by these routes each day.

The importance of phagocytosis by neutrophils in the defence of the body against bacterial infection is shown by the high susceptibility to such infections occurring in individuals with agranulocytosis produced by various drugs, e.g. the sulphonamides, in animals depleted of neutrophils by administration of benzol and in individuals with defective neutrophil function.

Phagocytosis by macrophages is also of considerable importance in protection against infection. Macrophages are susceptible to chemotaxis by substances liberated in bacterial infections and also by macrophage-specific lymphokines. Macrophages are highly efficient in killing bacteria but the mechanism by which they do so is not clear since, although they possess lysosomal hydrolytic enzymes, bactericidal substances similar to those found in neutrophils have not been identified in macrophage granules. The macrophages adhering to the endothelial cells of the liver and splenic sinusoids are mainly responsible for the removal of bacteria and other particles which have gained access to or have been introduced directly into the blood.

The inflammatory reaction

When bacteria gain access to the tissues in significant numbers they trigger a series of events which results in mobilization of the tissue defences and which are known collectively as the inflammatory reaction. The inflammatory reaction may be acute or chronic. In both there is an increase in vascular permeability leading to the passage of fluid and of phagocytic cells through the walls of the blood vessels, particularly the capillaries, into the tissues. The visible signs of inflammation are swelling due to fluid extravasation, redness due to hyperaemia and increased local temperature due to increases in local metabolism. The acute response is rapid in onset, more intense than the chronic inflammatory response and characterized by the accumulation of large numbers of neutrophils. Since it is easier to observe and greater in intensity it has been more extensively studied than the chronic inflammatory reaction.

The initial event in acute inflammation is an increase in vascular permeability due primarily to a widening of the intracellular gaps between the endothelial cells. This change may be produced not only by bacterial infection but also by trauma and by the accumulation of vasoactive amines such as histamine and serotonin. The precise mechanism by which bacterial infection causes these changes is not known but since they are most marked in the capillary walls proximal to the bacteria they are presumed to be due to the action of substances liberated during the infection both directly from the bacteria themselves and from the tissues. These substances probably act in the first instance by inducing a contraction of the endothelial cells; when the stimulus becomes more intense there may be actual degeneration of the capillary wall. Concomitant with this vascular change there is a slowing of the rate of blood flow, possibly due to an early loss of fluid from the blood vessels. When this occurs leucocytes move from the centre to the periphery of the capillary and begin to adhere to the endothelial cells. This process is known as margination. Eventually the marginated cells work their way through the widened intercellular gaps, ending up outside the blood vessel. During this process of emergence they can be shown to undergo considerable distortion. Once in the tissues they are drawn by chemotaxis towards the site of the invading organisms.

Acute inflammation is accompanied by an increase in the number of neutrophils present in the circulating blood (known as leucocytosis), the degree of leucocytosis paralleling the extent of the inflammatory reaction. Leucocytosis is due to the liberation of leucocytes from the bone marrow reserves. It is possibly evoked by the stimulus of substances released from the tissue leucocytes as a result of bacterial damage during the inflammatory reaction.

Another general effect of the inflammatory response, particularly of the acute inflammatory response, is fever or pyrexia. The mechanism of the pyrexia which occurs during the course of inflammation has been considerably clarified by the work of Wood in rabbits. The body temperature is regulated by hypothalamic centres and the increased temperature in inflammation is due to the action of endogenous pyrogen liberated from the leucocytes which affects the hypothalamic centres in such a way as to increase the level at which temperature-controlling homeostatic mechanisms become effective. This effect is analogous to a change in the setting of a thermostat. Endogenous pyrogen is demonstrable in the serum and plasma of animals exhibiting a pyrexial reaction. Its origin from polymorphonuclear leucocytes is shown by the failure of exogenous pyrogens, such as Gram-negative endotoxins, to elicit a pyrexial reaction in animals depleted of polymorphonuclear leucocytes by treatment with nitrogen mustard and by the fact that it can be extracted from neutrophils by a variety of procedures, e.g. by ultrasonication. Purified preparations of the endogenous pyrogen have been obtained and it has been identified as a glycoprotein with a molecular weight of about 14 000. Its mode of action is obscure but its release from neutrophils appears to be related to loss of potassium ions from the cell. Thus the addition of potassium will prevent the release of pyrogen from neutrophils exposed to exogenous pyrogens such as endotoxins. This inhibition is blocked by substances, such as ouabain, which prevent cellular uptake of potassium. The release of pyrogen is an energy-requiring process, the energy being derived principally from anaerobic glycolysis. It is, therefore, a component of the overall stimulatory effect of phagocytosis on the metabolism of the cell.

By prior exposure to endotoxin, animals may be rendered tolerant or refractory to its pyrogenic effect. Animals rendered tolerant in this way are, however, fully susceptible to the pyrogenic effect of endogenous pyrogens. Except in the case of highly temperature-sensitive organisms such as *Treponema pallidum* it is unlikely that pyrexia itself has a directly bactericidal effect. Its effect is presumed to be mediated through thermal acceleration of the rate of tissue bactericidal reactions.

The increased volume of fluid in the tissues during inflammation results in distension of the lymph capillaries which leads in turn to an increase in lymph flow to the local lymph node. It is not surprising, therefore, that during inflammation these should also be the sites of an inflammatory reaction, readily demonstrable clinically by an enlargement and tenderness of the nodes draining an infection site. The increased concentration of cells in the medulla of the nodes results in blocking of the hilar outlets of the medullary sinuses. This blocking has been compared by Wood to a 'log jam at the outlet of a mill pond'. It serves to prevent the outflow of bacteria from the gland, thus giving the phagocytic cells an enhanced opportunity for contact with the bacteria. Bacteria which do not stimulate a marked cellular inflammatory reaction, e.g. streptococci, may, since medullary blockage is minimal, pass rapidly through the nodes and thereby gain access to the circulation, through which they are carried throughout the body, giving rise to foci of metastatic infection characteristic of the septicaemic state.

Another component of the inflammatory reaction is the local deposition of fibrin. In a non-immune animal this can be attributed largely to the liberation of thrombokinase from damaged tissue cells. In the immune animal thrombokinase may be liberated from platelet aggregates generated by the operation of the immune adherence phenomenon (p. 141) and through a direct activation of the Hageman coagulation factor by antigen–antibody complexes. A very important role has been ascribed by Menken to fibrin deposition as a method of walling off and preventing the spread of microbial infection. It seems likely that such a role is achieved not so much by the local trapping of bacteria within a fibrin barrier as by creating solid surfaces which permit the operation of surface phagocytosis (p. 151).

Although the primary result of the inflammatory reaction can be attributed to the fact that it brings the phagocytic mechanisms of the body fully into action, it also has an effect on humoral antibacterial mechanisms by increasing local concentrations of antibody and of complement. It may, therefore, be presumed to potentiate complement bactericidal and bacteriolytic effects (p. 162). This effect is believed to be minimal in infections with organisms liberating ammonia, which has an anticomplementary effect, particularly infections of the urinary tract with urea-splitting organisms. The accumulation of leucocytes increases tissue concentrations of lactic acid and the bactericidal effect of this is increased by the increased carbon dioxide tension associated with inflammation. It is reasonable to suppose that a significant antibacterial effect is exercised in the tissues by various substances—basic peptides, hydrogen peroxide, peroxidase, lactoferrin—liberated from disintegrating leucocytes. When the inflammatory reaction is severe, leading to the death of large numbers of neutrophils, it results in the formation of pus which is a mixture of living and dead leucocytes together with intracellular material liberated from the latter. The characteristic viscosity of pus is due to the desoxyribonucleoprotein liberated from dead leucocytes. As the inflammatory reaction subsides the dead leucocytes and tissue cells and their extravasated contents are broken down by tissue proteases and other hydrolytic enzymes. The resulting fluid, which has a high protein content, and the remaining intact cells are removed through the lymphatic system.

The very great importance of the inflammatory reaction in the defence of the animal body against bacterial infection is shown by the increased susceptibility to septicaemic invasion following the local injection of adrenaline which has the effect of preventing capillary dilatation and also by the increased risk of bacterial tissue infections associated with the administration of steroids which, by some mechanism not yet explained, have an appreciable blocking effect on the inflammatory reaction.

Certain bacteria, notably mycobacteria, actinomycetes and *Treponema pallidum,* when they have invaded the tissues, do not stimulate an acute inflammatory reaction but instead produce a chronic inflammatory reaction characterized by the presence of macrophages and lymphocytes and which, in its later stages, becomes walled off by fibrous tissue. Why such bacteria produce a chronic rather than acute inflammatory reaction is not fully explained. It is possible, however, that their failure to induce an acute reaction is due to the fact that they themselves inflict relatively slight damage on the tissues and, as a result, substances which might increase vascular permeability or have a chemotactic effect are not produced. In this case the mononuclear inflammatory response may be analogous to that produced by inert foreign bodies, the mononuclear cells being retained at the site through the action of MIF (p. 146) released locally.

SPECIFIC IMMUNITY

Specific immunity is immunity directed against antigenic components or products of microorganisms to which the animal is capable of developing a specific immune response. Consequently we may distinguish two types of specific immunity, humoral and cellular. Humoral immunity is due to the presence of antibodies which are the products of B cells; cellular immunity is due to T lymphocytes which possess surface components or receptors capable of specifically recognizing antigen. In the defence of the body the specific cellular and humoral immune systems are not complete in themselves, in that for the exercise of their protective function they require the cooperation and participation of certain of the components considered under non-specific immunity, notably polymorphonuclear leucocytes, macrophages and the complement system.

As we have seen, exposure to antigen induces a proliferation of antigen-sensitive or immunocompetent cells of both B cell and T cell lines. This proliferation leads to an expanded population of end cells and of memory cells; the latter outnumber the original antigen-sensitive cells and are subject to simulation on future contact with antigen. Following a primary contact with an organism an animal will, as a result of its expanded population of memory cells, have a much greater specific immunity than that of the unexposed animal. The difference between the exposed and the unexposed animal is, therefore, basically quantitative. However, when considered in relation to the outcome of infection and the survival of the animal the difference is qualitative. In this context the unexposed animal may, therefore, be designated as non-immune and the exposed animal as immune.

Effector mechanisms of specific immunity

Under this heading will be considered the various ways in which antimicrobial antibodies and T lymphocytes specific for microbial antigens determine immunity against microbial infections. In the translation of immune potential into the reality of protection against infection specific immune mechanisms interact with the more important components of non-specific immunity, the polymorphonuclear leucocytes, macrophages and the complement system, which are designated as amplifiers (p. 86) of specific immune mechanisms.

Humoral immunity

Specific antibodies directed against microbial antigens are important determinants of immunity against virtually all microbial infection with the exception of mycobacterial and superficial fungal infections. They are the sole determinants of specific immunity against toxin-producing bacteria, namely diphtheria bacilli and the clostridia, and they are mainly and possibly in some cases exclusively responsible for a specific immunity against infections produced by other bacteria. The role of antibody in antimicrobial immunity is shown by a number of lines of evidence:

1. Individuals with agammaglobulinaemia or with defects in gammaglobulin production such as are found in the various dysgammaglobulinaemias and in multiple myeloma exhibit a high degree of susceptibility to bacterial infection.
2. In cases in which the antigens responsible for the stimulation of protective antibody have been identified, there is a parallelism between antibody production and resistance to infection.

It is convenient to distinguish the environments in which antimicrobial antibodies act since it would appear that different environments are protected by different immunoglobulin classes.

Blood stream. Bacteria which have gained access to the blood are exposed to all the immunoglobulin classes but particularly to IgM and IgG which are present in highest concentration. Since IgM antibodies do not pass as readily from the blood to the tissues as IgG antibody, it would be expected that they have a particular role in the clearance of bacteria from the blood.

Tissues. IgG antibodies readily pass from the blood stream to the tissues and, therefore, must be particularly responsible for the defence of the latter.

Mucous membranes. The principal antibody class associated with the mucous membranes is IgA which may be both locally and centrally produced. The presence of high concentrations of IgA in association with mucous membranes may be the result of a resistance conferred by the secretory component against proteolytic digestion. It is possible that IgE antibodies may also play a part in the defence of the mucous membranes (p. 108). They appear to be specifically implicated in immunity against nematode parasites.

The precise mechanisms by which antibodies confer immunity against microbial infection may be classified as follows:

1. Neutralization of bacterial toxins. Immunity to diseases, e.g. tetanus and diphtheria, the manifestations of which are due to the action of bacterial exotoxins, depends on neutralization of the toxin. In diphtheria the local invasion of the tissues by the organisms appears to be a sequel to mucosal damage produced by the local action of the toxin. Consequently local invasion as well as general toxaemia is prevented by antitoxin. Antibody to the erythrogenic toxin of *Streptococcus pyogenes* will prevent the development of the scarlet fever symptom complex though it will not prevent local tissue invasion by the organism. Toxin neutralization probably plays a major role in immunity against anthrax, an organism which is a potent toxin-producer under in vivo but not under in vitro conditions. It may be presumed that antibodies to leucocidins play a part in controlling infection due to organisms, e.g. *Staphylococcus aureus* and *Streptococcus pyogenes,* which produce extracellular leucocidins. The mode of action of antitoxic antibodies has not being fully clarified. It could be due to a direct combination of the antibody with toxic groupings, to a combination with non-toxic determinants of the toxin molecule in such a way as to block the activity of the toxic groups or to facilitating phagocytosis and/or enzymic breakdown of the toxin.
2. Combination of antibody with antigenic groups on the surface of micro-organisms renders the latter susceptible to phagocytosis.
3. Sensitization to the bactericidal and bacteriolytic actions of complement. Most Gram-negative bacteria are susceptible to the bactericidal action of complement fixed to sensitized cells; the complement reaction sequence is identical to that involved in the lysis of red cells. The bactericidal effect of complement appears to be due to lesions produced in the cytoplasmic membrane which may be demonstrated by electron microscopy. The precise mechanism of production of the lesions is, however, somewhat different from that responsible for the production of lesions in the red cell membrane, in that the antibody which sensitizes to complement lysis combines with antigens present in the bacterial cell wall and not those present in the cell membrane. Lysis, however, is due to membrane damage. The K antigens of the klebsiellae and the Vi antigens of the salmonellae have been found to possess considerable anticomplementary effect. This is probably due more to their physical location than to their chemical structure since by their presence they serve to separate any complement which may be fixed from its lytic substrate in the membrane. It is presumably because of this separation that rough bacterial

variants are more susceptible to complement lysis than smooth forms. Lysozyme has been reported to enhance the bactericidal effect of complement. It may possibly do so through breaking down the peptidoglycan layer of the cell wall, thereby exposing the complement-susceptible sites on the cytoplasmic membrane. There is also evidence that cells which have been killed by complement are more readily transformed into spheroblasts by lysozyme.

4. Virus neutralization. Viruses which have been combined with homologous antibodies are unable to infect cells. In some cases combination with antibody prevents cellular invasion. In others, though invasion is not prevented, the antibody appears to interfere with an early stage, e.g. uncoating, in virus replication.

5. It has been suggested that IgE antibodies, through their capacity to mediate the release of histamine from basophils, may play an important part in the expulsion of parasites, e.g. helminths, by inducing the contraction of the smooth muscle of the alimentary tract. It is conceivable that IgE antibodies may exercise this function in respect of a variety of antigenic particles which may be found in both the alimentary and the respiratory tracts.

Cellular immunity

Since cellular immune responses as demonstrated by cutaneous delayed-type hypersensitivity develop during the course of a great variety of microbial infections it is reasonable to assume that they have a role of some importance in antimicrobial immunity. In most bacterial infections cell-mediated immunity, that is, immunity determined by specifically reactive T cells, is clearly secondary in importance to humoral immunity. In immunity against intracellular pathogens, namely viruses and, amongst the bacteria infecting man, the mycobacteria and the brucellae, cellular immune reactions are of major importance. This is shown by the fact that the development of specific immunity to such organisms is unimpaired in individuals suffering from hypogammaglobulinaemia.

The role of cellular immunity is most clearly defined in respect of infection with the tubercle bacillus. There is abundant evidence that infection with the tubercle bacillus leads to a specific immunity. This has been shown in a number of studies establishing that in communities in which tuberculosis is widely present individuals who have had a tuberculous infection in early life, as demonstrated by a positive tuberculin test, are much less likely to develop tuberculosis in later life than are tuberculin-negative individuals who have not had such prior exposure. The immunity developing in the course of tuberculosis is manifestly not due to antibody production. Killed vaccines which will stimulate antibody production do not produce immunity. In experimental animals immunity cannot be passively transferred to normal animals by injection of antibody in the serum of an infected or immunized animal. The cellular basis of the immunity has been established by the demonstration that susceptible animals may be protected against infection by the adoptive transfer of lymphoid cells from immune animals. The identification of the cells responsible as T cells is shown by the greatly enhanced susceptibility to infection of thymectomized animals.

The precise mechanisms involved in cell-mediated immunity have been extensively investigated by Mackaness using as in vivo models infections of mice with a variety of intracellular bacteria, BCG, *Listeria* and *Brucella*. Mackaness found that animals suffering an active infection with any one of these pathogens was resistant to the others. He attributed this cross-resistance to the presence in infected animals of activated macrophages. These could be recovered from the peritoneal cavities of infected mice and could be shown by a variety of criteria to differ markedly from normal peritoneal macrophages. Morphologically they show an increase in mitochondria and lysosomal granules. They spread more actively on glass surfaces, have a greatly enhanced phagocytic activity and are more actively microbicidal for intracellular bacteria. These activated macrophages have been appropriately described as 'angry' macrophages.

Using *Listeria* infection in mice as a model for the study of the mechanism of immunity to intracellular pathogens Mackaness found that substantial immunity, which could be estimated by assay of the number of organisms surviving, after injection, in the spleen and liver of infected animals, could be transferred by intravenous injection of splenic lymphocytes from *Listeria*-infected donors. It could also be transferred by injection of lymphocytes from BCG-infected donors but only if the recipient animals were inoculated with BCG at the same time. No resistance was transferable by injection of lymphocytes from normal donors. Similarly, immunity to intracellular parasites may be conferred by the adoptive transfer of lymphocytes from animals immunized with bovine gammaglobulin if, and only if, bovine gammaglobulin is injected at the same time.

The most likely interpretation of these findings is that antigen-sensitive small lymphocytes when exposed to their specific antigen liberate a substance which induces macrophage activation, transforming them into 'angry' macrophages. As previously considered lymphocytes on exposure to their specific antigen under in vitro conditions liberate substances active on macrophages, e.g. the migration inhibition factor and the macrophage chemotactic factor. So far, however, a macrophage activating factor capable of producing angry macrophages has not been demonstrated under in vitro conditions. In fact, paradoxically, the migration inhibition factor would appear to have the opposite effect, in so far as it inhibits not only the migration of macrophages but also the spreading of macrophage on surfaces (p. 147). This paradox has not been resolved. It is possible that macrophage activation can occur only under in vivo conditions. However this may be, there is little doubt that the interaction of lymphocytes with homologous antigen leads to activation of macrophages in such a way that the latter acquire enhanced microbicidal activity for intracellular parasites. This process involves at least two stages, the first of which, the interaction of the lymphocyte with the antigen, is specific for the homologous antigen. The second stage, that of macrophage activation, is non-specific in that activated macrophages are capable of destroying a range of intracellular parasites antigenically distinct from those responsible for lymphocyte stimulation.

It has been shown that the resistance of mice to intracellular parasites is considerably reduced in animals which have been subjected to neonatal thymectomy or treated with antilymphocyte serum. These findings, together with the observation that immunity to such organisms runs parallel to demonstrable cutaneous delayed-type hypersensitivity, clearly implicate the T cells as the lymphocytes involved in these reactions. T-cell mediated immunity also plays an important part in immunity against viral infection but the precise mechanism involved differs somewhat from that involved in antituberculous immunity. There may be an increased uptake and intracellular inactivation of viruses by macrophages but a more important part may be played by the release of interferon-like substances, which act by interfering with virus replication, following contact of T lymphocytes with antigen.

IMMUNIZATION AGAINST MICROBIAL INFECTION

The term immunization is used to designate the procedures and circumstances which lead to the acquisition by an animal of enhanced specific antimicrobial immunity. Immunization may be *natural,* i.e. occurring spontaneously during the course of the life of the animal, or *artificial*; each of these varieties may be divided into *active* and *passive*. Active immunization is the stimulation of an immune response in an animal by exposure to microbial antigens. Passive immunization is the introduction into the body of an individual or animal of antibodies from an actively immune person or animal. Artificial immunization will be discussed in Chapter II.8.

Natural immunization

Active

Natural active immunization is the development of immunity as a result of a spontaneous contact during the normal course of life with the antigens of pathogenic micro-organisms. There are three ways in which this may occur:

1. As a result of overt infection with the organism. It is well known that an attack of many diseases produces a long-lasting and solid immunity. The extent of this immunity depends on a number of factors which may differ considerably between different diseases. It is more solid and persistent following infections in which there has been substantial invasion of the tissues than after infections in which the organisms are mainly confined to the body surfaces. Thus immunity produced by the common cold and other viruses causing respiratory infections is in general of relatively short duration. On the other

hand the immunity produced by infections with substantial tissue invasion, e.g. yellow fever, smallpox, typhoid fever and plague, is of long duration and may persist to a greater or lesser extent for the life time of the individual.

A form of dissociated immunity is shown in the case of streptococcal infections. A patient may be susceptible to a particular serotype of *Streptococcus pyogenes* but resistant as a result of prior infection with a toxin-producing strain of different serotype to the erythrogenic toxin. If the infecting strain is a producer of erythrogenic toxin the individual will be immune to the symptom complex of scarlet fever though susceptible to the throat infection caused by the organism. Other individuals infected with the same strain who do not possess antitoxin will, however, develop the scarlet fever complex. In this case immunity as a result of a previous attack, though it has not prevented infection, has considerably modified its course.

2. Immunity may develop as a result of contact with a pathogenic organism in which only trivial or minor invasion of the tissue occurs but which may nevertheless be sufficient to stimulate an immune response. Minor invasions of this type are known as latent or subclinical infections. The majority of pathogenic organisms can cause latent infections. Outstanding examples are the enteroviruses, the arboviruses and tubercle bacilli, in which the number of subclinical infections greatly outnumber the overt clinical cases. For poliomyelitis the ratio of subclinical infections to cases in non-immunized communities is of the order of 20 to 1 and in the case of certain arbovirus infections (p 391) the ratio is reported to be as high as 1000 to 1.

Another example for which quantitative data is available is diphtheria. In London in 1942 it was estimated that the number of children with latent infections occurring during any one year, as estimated by the development of negative Schick test, outnumbered overt infections by about 10 to 1. The occurrence of latent infection in the community is sometimes referred to as latent epidemization. It might be expected that the immunity resulting from latent infection would not be as substantial as that resulting from an actual attack of the disease but unfortunately no data are available from which a conclusion may be drawn on this point.

3. Exposure to commensal organisms of the normal flora. When animals are maintained in a germ-free state from birth onwards (these are known as gnotobiotic animals) their serum gammaglobulin levels are much lower than are those of normal animals and there is an associated hypoplasia of the lymporeticular tissues. When, however, gnotobiotic animals are exposed to the normal germ-laden environment, their alimentary tracts rapidly become colonized by bacteria and there is a concomitant increase in gammaglobulin levels and an associated hyperplasia of the lymphoreticular tissues. The low but significant gammaglobulin levels found in gnotobiotic animals are attributed to the stimulation by antigens, some possibly of microbial origin, present in the food. Although there is no direct evidence in support of this hypothesis it is reasonable to conclude that antibodies produced in response to stimulation by the antigens of commensal organisms may, if these antigens are related to those of pathogenic organisms, have a protective value.

Passive

In man, other primates and various rodents transfer of antibodies occurs mainly as an antenatal event, across the placenta from the maternal to the fetal circulation. In rabbits and guinea-pigs antenatal transmission of antibodies also occurs, but principally by secretion into the uterine lumen followed by absorption via the yolk sac. In ruminants, horses and pigs there is little if any antenatal transmission of antibody, the placenta in these species being impermeable to antibody, probably because of its thickness, as it has five or six membranous layers. Consequently antibody transmission is mainly a postnatal event via the milk or the colostrum. The colostrum secreted in the first few days after delivery contains a high concentration of antibody and during this period the alimentary mucosa of the newborn animal is highly permeable to the ingested antibody. This postnatal alimentary route is generally assumed to be of little importance in man but it can hardly be without significance that human colostrum has a considerable immunoglobulin content, particularly of IgA. It has been suggested that this antibody is not absorbed from the alimentary tract but is of considerable local value there in preventing the establishment of viruses, e.g. the enteroviruses, which naturally infect the alimentary tract. If this is the case it would provide considerable justification for the continuation of breast feeding in communities in which such viruses are prevalent as causes of neonatal disease.

The placental route of transfer is highly selective, permitting the passage of IgG antibody only. The capacity of IgG antibodies to traverse the placenta is due to the presence of a placental transfer site on the Fc portion of the molecule. Because of the failure of IgE antibodies to cross the placenta, immediate-type allergic sensitivity is not transmitted from the mother to the fetus. The barrier to the passage of IgM antibody also has the advantage that it prevents the transfer of isoagglutinins, which in cases of ABO blood group incompatibility would otherwise have a damaging effect on the fetal red cells.

After birth there is a progressive fall in the concentration of maternally derived antibodies. The maternal antibody would appear to have a half-life in the infant's circulation of about 60 days. This period has been confirmed by a number of lines of evidence:

1. The persistence of immunoglobulin in the circulation of hypogammaglobulinaemic children born of normal mothers.
2. The persistence of maternal immunoglobulin identifiable by the possession of an allotypic specificity distinct from that of the infant.
3. The persistence of isotopically labelled gammaglobulin after intravenous injection.

On the basis of these findings most of the maternal antibody should have disappeared from the infant's circulation by about six months of age. Nevertheless some maternal antibody may persist for longer than this. Thus early observations indicated that in some cases diphtheria antitoxin could be demonstrated in low concentrations for up to one year after delivery in children born of immunized mothers but which themselves did not appear to have been exposed to antigenic stimulation and measles antibody has been demonstrated up to nine months after delivery. It would seem likely, therefore, that sufficient maternal antibody may persist in the fetal circulation to provide protection against many commonly recurring diseases up to the age of one year.

COMPLEMENT

The term complement is applied to a system of components present in the serum of man and other animals which can be activated by antigen–antibody complexes and which in the activated state can produce a damaging effect on cell membranes. The complement system consists of 11 different proteins: Cl, a complex of three components q, r and s, C2, C3, C4, C5, C6, C7, C8 and C9. The proteins differ in electrophoretic mobility, ranging from the α to the γ 2 regions, and in molecular weight, having sedimentation constants ranging from 4 to 11S. These differences permit their separation by physical methods. The components C1 and C2 are rapidly inactivated at 56°C. Complement as a whole is therefore heat-labile, although many of the components are relatively heat-stable. The amount of complement present in an animal's serum is not increased by immunization. However, complement is biologically of considerable importance as an amplifier of immune reactions involving humoral antibodies and is believed to play an important role in the defence of the body against microbial infection.

The activity and properties of the complement system are most readily studied in relation to its capacity to fix to and cause the lysis of red blood cells which have been sensitized with homologous antibody. This type of haemolysis is known as immune haemolysis. It is usually examined in a system consisting of sheep red cells, guinea-pig serum as a source of complement and the serum of a rabbit immunized with sheep red cells as a source of antibody. In the sheep red cell system the antigen involved is present in the red cell membrane. Complement, however, will also induce lysis of red cells passively sensitized with extraneous antigens, e.g. bacterial polysaccharides, in the presence of their homologous antibodies.

Reaction sequence in immune haemolysis

The individual complement components are bound or fixed to the cell–antibody complex in a sequential series of reactions in which the binding of each component triggers the activation and binding of components acting later in the sequence. Once the reaction series has been initiated it proceeds with almost complete invariability to its end, resembling in this way a variety of other biological cascade phenomena. In the sequence which follows E=erythrocyte, A=antibody and Cl etc.=complement components. Bold letters indicate an enzymically active component.

Stage 1

$$E + A \rightarrow EA$$

Only IgG and IgM antibodies are capable of sensitizing red cells to complement lysis, since they alone among the immunoglobulins possess the complement binding site. This is located on the CH_2 homology region of the Fc portion of the molecule and is not exposed when the antibodies are in the uncombined state. On combination of antibody with antigen, however, the complement site is exposed. It has been estimated that a single molecule of IgM will sensitize red cells to complement lysis. On the other hand, sensitization by IgG requires that at least two IgG molecules should be bound at adjacent sites on the red cell surface. If the IgG binding sites are too far apart, as in the Rh system, complement binding and lysis cannot occur. The requirement for two adjacent IgG molecules has not been fully explained. It may be that two adjacent IgG Fc regions must interact to create the active site or that complement possesses two receptors for antibody, stable complexes being formed only if both receptors can bind the antibody. In the case of the IgM molecule the adjacent Fc regions required are provided within the molecule itself. Since many molecules of IgG may have to be bound to a red cell to provide two adjacent Fc regions, IgM, even on a weight basis, is considerably more efficient in sensitizing to complement lysis than is IgG.

There are considerable differences in the effectiveness with which the complement of any one species will be bound by complexes formed by antibodies of other species. Thus guinea-pig complement reacts well with complexes formed with guinea-pig, rabbit or human antibodies, but very poorly with complexes formed with dog antibody and not at all with complexes formed from chicken antibody.

Stage 2

$$EA \xrightarrow[Ca^{++}]{Clqrs} EA\ \mathbf{C1}$$

Ca^{++} is necessary for the formation of the C1 complex from its individual components. Of these C1q is the component which binds directly to antibody; C1q will also bind to aggregated gammaglobulin and will precipitate soluble antigen–antibody complexes. Binding of C1q activates **C1r**, which is an enzyme, permitting it to act on and activate C1s. Activated **C1s** is an esterase, the activity of which can be demonstrated by hydrolysis of synthetic esters such as TAME (tosyl-arginine-methyl-ester) or ATEE (acetyl tyrosine ethyl ester). Its activity is inhibited by esterase inhibitors, e.g. difluorophenyl phosphate (DFP).

Stage 3

$$EAC1 \xrightarrow{C4} EA\mathbf{C1\ 4b} + 4a$$

Much of the activated **4b** remains in the solution where it degenerates to an inactive form, C4i. C4b appears to fix to both the cell membrane and the antibody.

Stage 4

$$EAC14b \xrightarrow[Mg^{++}]{C2} EA\mathbf{C1\ 4b\ 2a} + C2b$$

The binding of C2a yields an enzymically active complex **C4b2a**, known as C3 convertase. At this stage C1 is irrelevant and may be eluted without affecting the reaction sequence. The binding of C2a is unstable and is only stabilized by the addition of further components.

Stage 5

$$EAC14b2a \xrightarrow{C3} EA\mathbf{C14b2a3b} + C3a$$

The moiety C3b is the major component of complement which is fixed in the reaction sequence. C3b appears to attach mainly to the cell membrane. The new complex has peptidase activity through which it appears to act on C5, splitting it to C5a and C5b. C5b forms complexes with C6 and C7.

Stage 6

$$EAC14b2b3b \xrightarrow{C5\ C6\ C7} \mathbf{C14b2a3b5b67} + C5a + \mathbf{C5b67}$$

The activated **C567** complex released into solution will fix to the membranes of nearby cells on which it will trigger the terminal stages of the sequence, leading to the lysis of the cell. This type of lysis is known as reactive lysis.

Stage 7. Fixation of **C567** allows binding of C8. This by itself will cause lysis of the cell, but lysis is considerably enhanced by binding of C9. Similar enhancement can be produced by various chelating agents. In immune haemolysis C8 probably functions as a phospholipase.

Inhibitors

Two inhibitory proteins have been clearly identified which can control to some extent the progression of the complement cascade.

C1 inhibitor acts on fixed **C1,** inhibiting the enzymic activity of **C1s**. It acts slowly and does not significantly affect the rate of immune haemolysis. However, it considerably inhibits the activity, and possibly also the activation, of any **C1s** which may be present in solution. Deficiency of **C1** inhibitor is found in hereditary angioneurotic oedema. The oedema in this condition is probably due to a direct effect of activated **C1** on vascular permeability. Why the symptoms of this condition should be intermittent while the inhibitor deficiency is permanent has not been explained.

C3b inhibitor is an enzyme which acts on fixed C3b, exposing on the latter the site for bovine conglutinin. An alternative designation is KAF or conglutinin activating factor. By a complex series of reaction, C3b can generate a C3 convertase from a serum glycoprotein. This reaction is controlled by the plasma KAF. In individuals deficient in KAF there is an excessive consumption of C3 nd an increased susceptibility to pyogenic infections.

Biological effects of complement

Cytolysis

The end result of the complement reaction sequence is the lysis of the cell; the immediate cause of this is the production in the cell membrane of lesions appearing as holes or pits on electron microscopy. These lesions are possibly produced as a result of enzymic action of C8 on a phospholipid component of the membrane.

When complement-mediated membrane damage occurs in tissue cells the cells are killed; this cytocidal reaction is of considerable importance in histocompatibility testing. In the killing of tissue cells IgG is considerably more effective than IgM. The reason for this difference is not known. Tissue cells are more difficult to lyse than red cells since they have considerable capacity to repair the lesions produced by complement. Complement-mediated damage also leads to the killing of a variety of Gram-negative bacteria. In sensitizing to a bactericidal effect IgM antibodies are more effective than IgG. Bacteria that have been killed as a result of the action of complement may be converted into sphaeroplasts in the presence of lysozyme and then undergo cytolysis. Vibrios are particularly susceptible to this lytic effect. It has recently been shown that complement causes dissolution of the membranes of enveloped oncogenic RNA viruses. This is not surprising since the membrane of these viruses is derived largely from that of the host cell (p. 345).

Conglutination

Since the turn of the century it has been known that fresh unheated serum of certain animals, notably the horse, can render human and other red cells agglutinable by normal bovine serum. This agglutination is due to a combination of a substance known as conglutinin, which is present only in the serum of bovines, with fixed C3b which has been modified by the conglutinin activating factor KAF. The reaction requires the presence of Ca^{++} ions. Conglutinin is a normal serum component and is not an immunoglobulin. The receptor for conglutinin has been shown to be a mannose containing glycopeptide which is present also in yeast cell walls, its susceptibility apparently being due to N-acetyl D-glucosamine residues. The sera of man and of animals other than bovines do not possess conglutinin but they do possess autoantibodies known as immunoconglutinins (IKs). Human immunoconglutinin is an IgM antibody which reacts with fixed C3b. The C3b receptor for IK is different from that which binds to conglutinin and the binding does not require the presence of calcium ions. The level of immunoconglutinin is increased by infections and by immunization. Its presence in normal serum is believed to be due to the antigenic stimulus of complement fixed to red cells sensitized by adsorption of cold agglutinins. Recently an immunoconglutinin has been identified in saliva. It differs from serum immunoconglutinin in reacting with **C567** as well as with C3b.

Immune adherence

Cells or complexes which have fixed C3b have the capacity to adhere to primate red cells and non-primate platelets. Adherence is through a receptor site on C3b which differs from that involved in haemolysis and conglutination. Adherence of complexes to red cells in this way is believed to considerably facilitate phagocytosis. Fixed C3b may, however, have a more direct role in phagocytosis by mediating, in a similar manner, adherence of immune complexes to macrophages and polymorphonuclear leucocytes.

Virus neutralization

Evidence has recently been obtained that fixation of C1 and C4 has an enhancing affect on the neutralization of herpes virus by antibody. Whether a similar phenomenon occurs with other animal viruses is not yet known.

Indirect effects due to components released during the reaction sequence

Chemotaxis. Three complement derivatives C3a, C5a and activated **C567**, which are released into solution during complement fixation, have been found to be chemotactic for polymorphonuclear leucocytes.

Anaphylatoxin formation. Exposure of serum to immune complexes has long been known to lead to the generation of a substance, anaphylatoxin, which can stimulate the release of histamine from mast cells (p. 182). This effect appears to be mainly due to C3a and C5a.

Alternative pathways of complement activation

In recent years considerable attention has been given to alternative routes of complement activation. Of particular importance are those which can occur under in vivo conditions; mechanisms which would allow bypassing of the earlier complement components are of special interest since they may permit the operation in vivo of a complement reaction sequence in animals deficient in these components and in the absence of antigen–antibody reactions. Alternative routes which probably operate under in vivo conditions include the activation of C1s by plasmin and other proteases, of C3 by plasmin and of C5 by lysosomal enzymes.

An important pathway involves participation of a 5S serum globulin—properdin—first described by Pillemer. Properdin, which is lacking in immunoglobulin determinants, appeared to have a non-specific bactericidal effect on various Gram-negative bacteria. There has been considerable confusion as to the status and significance of properdin and many workers have thought the functions ascribed to it were in effect those of low-titred relatively non-specific natural antibodies. Recent work, however, indicates that properdin has an important biological role. Certain polysaccharides, especially Gram-negative endotoxins, and aggregated IgA, can form complexes with properdin which, in the presence of serum factors resembling but not identical with C4 and C2, are capable of activating C3 through the production of a C3 convertase. A similar convertase is also produced through the interaction of C3b with serum components. The latter reaction is inhibited by KAF. C3 can also be activated by a complex formed by cobra venom and a serum cofactor. This activation has been widely used for demonstrating the presence of later components of the reaction sequence in various animal species. Cobra and other snake venoms have long been known to have haemolytic properties. These properties, however, appear to be due only to a slight extent to complement activation.

Role of complement in protection against microbial infection

Because of its diverse biological effects it would be reasonable to expect that complement should have an appreciable protective value against microbial infection. Conclusive formal evidence that this is the case has, however, been difficult to obtain. In this context data on the susceptibility to microbial infection of animals genetically deficient in the capacity to synthesize various complement components would be of particular value. A variety of such genetic defects has been described. Deficiency in the production of C4, C5 and C6 has been observed in guinea-pigs, mice and rabbits respectively. In none of these cases, however, has there been a clear demonstration of increased susceptibility to infection. In man deficiency in C2 production has been recorded, also without any concomitant increase in susceptibility. Increased susceptibility has, however, been reported in an individual showing hypercatabolism of C3 due to deficiency of C3b inhibitor and in infants showing a dysfunction in C5.

Biosynthesis of complement

It is only in recent years that it has been possible, as a result of the development of techniques for the accurate identification of complement components, to investigate the sites of complement biosynthesis. So far the information obtained is very incomplete but the findings indicate that different components are produced in different organs. Thus the gastrointestinal tract appears to be the main source of C1 and it seems likely that the individual C1 components are synthesized in different types of cell. Evidence has been presented that C2 and C4 are synthesized mainly by macrophages and C3, C6 and C9 and C1 inhibitor mainly in the liver. C8 appears to be synthesized mainly in the spleen.

The phylogeny of complement

By the use of the cobra venom activator system certain arthropods, horseshoe crabs and sipunculid worms, have been found to synthesize the components in the reaction sequence from C3 through to C9. These species do not appear to produce C1, C4 or C2. Amongst vertebrates no complement components have been found in lampreys even though they are capable of producimmunoglobulin. Functional complement is produced from the elasmobranchs onwards. Since arthropods, which do not produce immunoglobulins, nevertheless produce components of the complement system it seems likely that complement did not in fact evolve as an amplifier of immune responses, though it may now function as such, but that it had a more general biological function.

II.8

Artificial Immunization

Like natural immunization, artificial immunization may be either active or passive. *Active* immunization is the exposure of an individual to the antigens of a pathogenic organism which are capable of stimulating the production of antibody which will protect the individual against the disease produced by that organism. Such antigens are sometimes referred to as protective antigens. The term *passive* immunization, as normally used, is applied to the introduction into an individual of antibody derived from an animal or from another individual actively immunized by either natural or artificial exposure. This more precisely is passive humoral immunization. Passive cellular immunization is also possible by the injection of lymphocytes from actively immune animals. This is known as adoptive immunization. In humans adoptive immunization by lymphocytes derived from the peripheral compartment (p. 90) is not carried out because of the danger of graft-versus-host reactions. The grafting of thymus tissue has, however, been successfully used for the correction of certain types of cellular immune deficiency.

Since the development of immunity following active immunization involves the proliferation and differentiation of antigen-reactive lymphocytes it is a slow process requiring, in the case of non-living antigen preparations, at least a few weeks to reach significant levels. When it has developed, however, it is persistent, lasting in some cases, particularly after the inoculation of live vaccines, for many years. In general artificial active immunization produces a less persistent immunity than does an attack of the actual disease. In contrast to active immunity, passive immunity is rapidly established, the individual possessing a maximum level of immunity immediately after injection of the antibody. The immunity produced, however, is of short duration, persisting only for as long as effective concentrations of antibody remain in the blood and tissue fluids. Because of these differences passive immunization is an emergency procedure carried out when it is necessary to establish immunity quickly, as in individuals who have been in contact with patients suffering from serious microbial infection. Active immunization, on the other hand, is an elective procedure carried out for the long-term protection of individuals who might at some future date be in contact with the disease.

PASSIVE IMMUNIZATION

Passive immunization is used for protection against certain diseases caused by viruses and against bacterial diseases in which the lesions can be attributed mainly or entirely to the action of bacterial exotoxins. From time to time there have been claims, e.g. in pertussis, that immunity can be produced against various bacterial diseases by the injection of antibodies against intrinsic microbial antigens. There is, however, no instance in which this can be said to have been satisfactorily established.

Antitoxins

Antitoxins are used almost exclusively for protection against two diseases, tetanus and diphtheria. The commercially available antitoxins are prepared by immunization of horses. The suitability of horses is due to their large size and their general docility on manipulation rather than to any special capability for mounting an immune response. Horse antitoxins, however, have a major disadvantage in the readiness with which they produce allergic reactions. Frequently this is a comparatively mild serum sickness syndrome; exceptionally the injection of horse antitoxin may produce fatal anaphylactic shock. In the early days of passive immunization with antitoxin crude horse serum was used and reactions were frequent. With modern refined antitoxins the incidence of reactions has been very considerably reduced.

For the preparation of antitoxin the horses are first immunized with a series of injections of toxoid. These are followed by a series of injections of toxin. The purified antitoxin is prepared from immune horse serum by iso-electric precipitation or with organic solvents. In the final stages of preparation the antitoxin is concentrated by pressure dialysis and further refined by treatment with pepsin. Peptic digestion yields an $F(ab')_2$ fragment which lacks the placental transfer site present on the Fc portion of the molecule (p. 105). Consequently the fetus is not protected by preparations of enzyme-refined antitoxin injected into the mother before delivery.

Although the incidence of reactions has been considerably reduced following the use of refined antitoxins, reactions do occasionally occur and, because these may be of the fatal anaphylactic type, many physicians consider that where possible, horse antitoxin should not be used in prophylaxis. A satisfactory alternative which has been used for the prophylaxis of tetanus is antitoxin derived from actively immune human subjects. The availability of this preparation, however, is very limited.

For the therapy of the serious exotoxic diseases, diphtheria and tetanus, administration of antitoxin, and this will usually be horse antitoxin, is imperative. Antitoxin administered during the course of such diseases has no capacity to reverse any tissue damage already produced by the toxin, nor does it appear to be able to dissociate toxin already bound to tissue cells. Its value lies in its capacity to neutralize unbound toxin and thereby to prevent further tissue damage.

The duration and degree of immunity produced following the administration of antitoxin prepared in another species, e.g. the horse, will depend on the amount of antitoxin administered. With the antitoxin dosages currently in use immunity has a maximum duration of four to six weeks. After this immunity has disappeared the individual is fully susceptible and therefore requires active immunization. As a matter of practical politics passive immunization carried out as an emergency procedure should always be followed by active immunization. If, however, an interval of four to six weeks is permitted before active immunization has commenced, many patients may not return for further immunization. As a compromise active immunization may be carried out routinely 10 to 14 days after passive immunization. In practice, regimens following this procedure have been found to be eminently satisfactory, permitting the establishment of an adequate level of active immunity and at the same time ensuring a good return rate of patients.

Passive immunization against viral diseases

The antibodies used for prophylaxis of viral diseases, excluding rabies (p. 405), are derived from human sera. Because of the risk of producing serum jaundice (p. 444), crude human sera, though at one time employed, are no longer used. The preparations now employed are gammaglobulin concentrates obtained by precipitation with salt or with organic solvents. They may be dispensed in a lyophilized form which is dissolved before use or in concentrated solutions with an immunoglobulin content of from 15 to 20 g/100 ml which represents a 10–20-fold concentration of the antibody in the original serum. With the purification procedures currently used the immunoglobulin is almost entirely of the IgG class.

Two varieties of human immunoglobulin preparation may be distinguished:

1. Immunoglobulin prepared from pools of sera derived from individuals whose immune status is unknown or has not been determined. Formerly such preparations were referred to simply as gammaglobulin but they are now designated by WHO as *human normal immunoglobulin*. Human normal immunoglobulin has been found to contain antibody to a variety of viruses (poliomyelitis, measles, herpes simplex, influenza, hepatitis and rubella and other viruses depending on their prevalence). Human normal immunoglobulin is of most value in the

prophylaxis of measles in young or debilitated children and of hepatitis, against both of which it can provide a high degree of protection. It has also been used to a limited extent for the prophylaxis of chickenpox in premature and debilitated infants.

2. Immunoglobulin from individuals known to be immune to specific diseases against which protection is required. This is known as *human specific immunoglobulin.* It is derived variously from the sera of convalescents or of individuals recently recovered from infection. Human specific immunoglobulin has been used successfully for the prophylaxis of mumps in adults (p. 430), in this case being prepared from the serum of convalescents. Immunoglobulin from recently vaccinated individuals is of value in the protection of contacts of smallpox cases who have been seen at too late a date for effective immunization by vaccination. It is reported that it is of some value also in the treatment of some of the complications of smallpox vaccination (p. 415).

Repair of immune deficiency in antibody-deficient syndromes

Patients suffering from syndromes characterized by hypogammaglobulinaemia (p. 93) have a high susceptibility to bacterial infection, particularly to infections by Gram-positive organisms. Such infections are usually the cause of death in these patients. Hypogammaglobulinaemic patients can be given considerable protection against bacterial infections by the injection of human normal immunoglobulin. A major difficulty in this procedure, however, is that the administration of the immunoglobulin must be continued throughout the life of the patient. Where the immunoglobulin is derived from an individual with an allotypic specificity different from that of the recipient it would be expected that repeated administration would stimulate the production of anti-allotype antibodies which would result in the elimination of the immunoglobulin. Too little experience has been obtained in this field, however, to permit an assessment of the extent to which this may occur.

ACTIVE IMMUNIZATION

Active immunization may be achieved by exposure of individuals to living or killed organisms, to microbial toxins normally administered as toxoids or to fractions or extracts derived from intact organisms known to contain the components responsible for the stimulation of protective antibody. In the last resort the effectiveness of these preparations in protecting against disease can only be determined by trials in human subjects which will permit adequate statistical analysis. Such trials would not be feasible during the development of a potentially immunogenic preparation. For this purpose the efficiency of agents chosen in active immunization is monitored initially by determining their capacity to protect laboratory animals. This raises, however, two problems of particular difficulty:

1. Diseases against which it is desired to produce protection, e.g. measles and rubella, may not be capable of producing infection in laboratory animals. In such cases the monitoring of vaccine efficacy can only be carried out in trials in human subjects.

2. Many organisms against which it is desired to produce protection, though they may produce infection in animals by parenteral injection, are not capable of doing so by the natural route through which infection occurs in the human subject. Thus the efficacy of pertussis vaccines is normally determined by their capacity to protect against intracerebral infections in mice. The extent to which efficacy in protecting against natural human infection can be assessed from the results of such animal experiments is doubtful.

Living vaccines

Immunization with fully virulent pathogenic organisms capable of multiplying in the body is obviously not a feasible prophylactic procedure. Many organisms, however, have now been attenuated by various techniques of laboratory passage in such a way that their virulence for man is enormously decreased and these attenuated strains can be safely used for immunization. The possibility of attenuating micro-organisms so that they may be safely used for immunization was first demonstrated by Pasteur, who found that cultures of the chicken cholera bacillus, *Yersinia multocida,* which had been maintained in serial subculture in the laboratory, had become non-pathogenic and that fowl inoculated with these cultures were immune to infection with virulent chicken cholera bacilli. Subsequently he applied this principle to the development of attenuated vaccines for immunization against anthrax (p. 228) and rabies (p. 405). At present attentuated vaccines are widely used against poliomyelitis, yellow fever, measles, rubella and tuberculosis. In animals attenuated vaccines are used for protection against anthrax and brucella.

The attentuation of virulent organisms is in fact a process of selection of mutants of low virulence. It is achieved by long periods of growth under abnormal conditions outside the normal animal host. Illustrative examples are BCG vaccine attentuated by prolonged growth on a bile potato medium, the Sabin polio vaccine attentuated by serial subculture in kidney tissue cultures and the 17D yellow fever vaccine attentuated by passage in mouse tissue cultures and chick embryos.

The immunity produced by the inoculation of attenuated organisms approximates that produced by natural infection and is generally greater than that produced by inoculation of killed organisms. In fact against certain diseases (smallpox, yellow fever, measles, rubella and tuberculosis) it is not possible to produce a satisfactory level of immunity by inoculation of killed vaccine. The possible reasons for this are:

1. Inoculation of living organisms results in local multiplication of the organism and consequently exposure of the host to a larger mass of immunizing antigen.

2. Antigens which stimulate the production of protective antibody may be produced as a result of the growth of micro-organisms in the environment of the tissues which are not produced under the conditions of in vitro culture.

3. In certain cases, notably in virus diseases and tuberculosis, intracellular multiplication of the organism, possibly by ensuring peripheral localization, facilitates the development of a cellular immune response (p. 84) by increasing the possibility that the relevant antigens will make contact with T lymphocytes.

The attenuation of micro-organisms for vaccine production presents certain problems:

1. The degree of attenuation should be such that in normal subjects the organism cannot cause a progressive infection. Otherwise the hazards of immunization may be as great or greater than those of the overt disease.

2. On the other hand the organism must not be over-attenuated. In some cases attenuation with the maintenance of immunizing potency may not be possible because, in the attenuation process, the antigens responsible for the stimulation of protective antibody are major determinants of virulence and may be lost. This is certainly the case in respect of avirulent variants of *Streptococcus pyogenes* lacking the M antigen, rough variants of pneumococci lacking the capsular polysaccharide antigens and, though possibly to a lesser extent, variants of *Salmonella typhi* lacking the Vi antigen. In other cases, e.g. measles, attenuation, though not resulting in loss of immunizing antigen, may for some reason lower the capacity of the vaccine to multiply in the tissues to such an extent that it is insufficiently immunogenic.

3. The attenuated variant should be stable, that is it should have no significant tendency to revert to its original virulent form. This stability should be maintained even if the vaccine is transmissible from one individual to another in a vaccinated population. This requirement has received particular attention in the development of attenuated poliomyelitis vaccines.

4. In certain cases attenuation is not a practicable proposition because of the antigenic diversity and lability of the organisms involved. In such cases any change in the antigenic composition of the infecting strain would render obsolete the attenuated variants previously used.

Non-living vaccines

As a general rule, with the exception of the toxoids, non-living vaccines produce a much less effective and less persistent immunity than live vaccines.

Killed vaccines

The possibility that vaccines of killed organisms might be effective immunizing agents was first demonstrated in 1897 by Wright, who showed that killed suspensions of typhoid bacteria could give substantial protection against typhoid. Killed vaccines now have an established role in immunization against typhoid, cholera, plague, pertussis, poliomyelitis (though for the immunization of children killed poliomyelitis vaccine has been largely replaced by live vaccine), epidemic typhus, influenza and rabies. Other effective killed vaccines, though of less widespread applicability, are arbovirus vaccines and vaccines against various rickettsial infections. Autogenous vaccines prepared from organisms isolated from the patient have been used in the prophylaxis of recurrent staphylococcal infection and of various rheumatic diseases and bronchitic conditions. Satisfactory evidence for their effectiveness has never, however, been obtained.

In the preparation of killed vaccines the micro-organisms may be killed in a number of ways, e.g. by heating, treatment with formalin, phenol or alcohol, the precise method employed differing with different organisms. In general the method used is that which produces least damage to the immunizing antigen. The vaccines are preserved with an antiseptic, e.g. dilute phenol, merthiolate or alcohol.

Toxoids

Toxoids are bacterial toxins which have been treated in such a way as to deprive them of their capacity to damage the tissues without impairing their capacity to stimulate the production of antitoxin. Only protein exotoxins can be converted into toxoids. Some protein exotoxins cannot be toxoided since the treatment required to deprive them of their toxicity at the same time deprives them of their antigenicity. This is the case with the erythrogenic toxin of *Streptococcus pyogenes*. Some degree of spontaneous toxoiding occurs in preparations which have been allowed to stand for some time but the method used for commercial preparations of toxoid is by treatment with formalin. The effectiveness of formalin appears to depend on its capacity to combine with free amino groups. Only two toxoid preparations are currently used for active immunization, those prepared from the diphtheria and tetanus toxins. The preparations currently used are highly purified in order to diminish the possibility of adverse reactions in vaccinated individuals. In the case of diphtheria, however, highly purified toxoids may be appreciably less immunogenic than cruder preparations. This might be due to the presence of bacterial components which have an adjuvant action or of bacterial antigens other than toxins which may make some contribution to active immunity.

Component vaccines

Theoretically it should be possible to produce active immunity with fractions or extracts of organisms containing the antigenic components required for the stimulation of protective antibody. So far, however, only one such preparation is generally available, the influenza virus subunit vaccine which contains the virus haemagglutinins. The vaccine is usually prepared by treatment of the virus with a detergent, sodium deoxycholate, which ruptures and disaggregates the external haemagglutinin-containing virus envelope (p. 421). It has advantages over whole virus vaccines in that it stimulates less reactions while still being a highly effective immunizing agent.

Combined vaccines

Combined prophylactic preparations containing more than one immunizing agent are universally employed for the elective immunization of young children, as they have the enormous advantage of diminishing the number of injections which have to be given for an adequate course of immunization. The most widely used combined preparation is the triple vaccine containing diphtheria and tetanus toxoids and killed *Bordetella pertussis*. Vaccines containing *Bord. pertussis* are potent immunizing agents since the organism has a considerable adjuvant effect (p. 98). It has been thought that in the administration of combined vaccines there might be the danger of antigenic competition which might reduce the immune response to the most immunogenic of the components. This, however, appears to be a problem only in persons who have previously been immunized with that component.

ADMINISTRATION OF PROPHYLACTICS

Indications

All members of the community should be immunized in infancy or early life against the more serious diseases prevalent in the community. The conditions for which a programme of general immunization is required naturally vary from one community to another and also in the same community from time to time. Thus when a disease which was once widely prevalent has been brought under effective control by active immunization and/or other procedures it may be possible to modify the immunization programme. This appears now to be the position in relation to tuberculosis in Scandinavian countries and possibly in other western European countries as well. In western Europe diseases for which general immunization of the community is currently required, and against which children should therefore be routinely immunized, are diphtheria, tetanus, whooping cough, poliomyelitis, measles and rubella. Immunization against smallpox and tuberculosis is currently optional but in Afro-Asian countries immunization against these two diseases should be also carried out as a routine.

Individuals travelling to countries where they may encounter diseases not prevalent in their own country. Important examples are yellow fever, cholera, plague and various rickettsial infections.

In the face of an impending epidemic. Thus in a community in which smallpox does not occur as an endemic disease, mass vaccination may be required if the disease is introduced into the community from outside and the immediate measures to contain it have been unsuccessful.

Conditions of special risk. Under this heading we may include the use of influenza vaccine for elderly people and patients with cardiovascular insufficiency, immunization against rabies of individuals believed to have been bitten by rabid animals and immunization against tetanus of patients sustaining severe wounds.

Route of administration

This differs with different vaccines. The following are the routes normally employed:

1. Toxoids, triple vaccine and pertussis and measles vaccines by the intramuscular or subcutaneous routes.
2. Cholera, influenza, rubella, typhus, yellow fever vaccines by the subcutaneous route.
3. Typhoid and paratyphoid vaccine (TAB) by the subcutaneous or intradermal route.
4. Poliomyelitis vaccine by the oral route.
5. Smallpox and BCG vaccines by intradermal inoculation; effective vaccination is manifested by the development of a localized cutaneous lesion.

As previously considered the mucosal surfaces are protected largely by IgA antibody, much of which appears to be produced in local lymphoid tissue in the respiratory and alimentary tracts. It is therefore possible that effective immunization against organisms which gain access by these routes may be achieved by direct introduction of antigen into the respiratory tract, e.g. by intranasal spray, or into the alimentary tract by the oral route. The effectiveness of oral administration has been clearly shown in the use of the live polio vaccines. The immunizing efficiency of respiratory or oral administration of vaccines other than polio vaccines has, however, still to be assessed.

Dose

It might seem that the maximum doses of vaccine which can be introduced without producing a toxic reaction would be desirable. In this context, however, the important criterion of dose is whether the amount of vaccine used is sufficient to stimulate an adequate immunity and there would seem to no reason to exceed this. Furthermore, it has been shown that exposure to quite small doses of antigen which do not themselves stimulate a significant primary response can, nevertheless, condition the animal to an appreciable secondary response. There are acceptable arguments against the use of large doses of vaccine. The possibility of inducing immunological tolerance (p. 175) could well be a serious problem in the immunization of children but there is insufficient information available on the basis of which to assess its reality. In addition there is appreciable experimental evidence that low doses of antigen produce preferentially a high-avidity antibody. Consequently the higher concentration antibody produced in response to exposure to a large dose of antigen might, because of its avidity, be of no greater protective value than the lower concentration of highly avid antibody produced in response to a low dose of antigen.

Number of inoculations. In the case of live vaccines introduced intradermally or subcutaneously, a single inoculation constitutes sufficient exposure for initial immunization. However, for immunization with live polio vaccine by the oral route, three doses are usually given (p. 384). This is to ensure an adequate 'take' of the vaccine (p. 385). With non-living vaccines and toxoids effective immunization requires two or more injections of the antigen.

Duration of immunity

With the passage of time the immunity produced by active immunization inevitably diminishes, as does the immunity produced by natural active immunization. The rate of decline is much greater in some cases than in others. Thus immunity is very persistent following smallpox and yellow fever vaccination or immunization with tetanus toxoid, but quite transient following immunization with TAB vaccine.

Not only does the duration of immunity vary with the antigen employed, but with the same antigen it can differ considerably from one individual to another. Some individuals are good responders, others are poor responders. A waning immunity can be reinforced, however, by periodic booster injections. Where the initial immunity has been transient, as following immunization with TAB vaccine, booster inoculation would be required at frequent intervals, in the case of those who continue at risk. In respect of diseases with which an individual may periodically come in contact appreciable natural boosting of immunity will occur under natural conditions. Although insufficient to protect completely against infection, a waning immunity may nevertheless be of considerable value in preventing a serious overt attack of the disease, permitting instead a latent subclinical infection. The latter may in turn effect a considerable restimulation of the antibody-producing machinery. This natural boosting of immunity is believed to be of considerable importance in the maintenance of immunity following a single administration of BCG vaccine, one inoculation of which therefore, in a community in which tuberculosis is prevalent, may give rise to a long-lasting immunity.

Hazards

Active immunization is not without its hazards. It is essential, therefore, that immunization programmes should only be adopted when the hazard associated with immunization is less than the hazard of infection by natural exposure. The nature of this problem is best illustrated with reference to smallpox. In European countries and North America, smallpox is now an exotic disease and smallpox surveillance to prevent importation of the disease and its subsequent spread is highly effective (p. 415). Under these conditions the risks of a community-wide programme of smallpox immunization are appreciably greater than the risk of natural infection. In communities where it is endemic, however, community-wide immunization is imperative. In general one may say that measures which have been responsible for the elimination of a disease should not necessarily be continued to prevent its reintroduction.

The hazards of immunization may be classified as follows:

Toxic reactions. Local reactions at the site of inoculation are to be expected to a greater or lesser extent following the introduction of any vaccine, living or dead. General reactions manifested by malaise and pyrexia are also not uncommon. With killed vaccines general reactions are particularly associated with typhoid and paratyphoid vaccine. Different individuals in the same population differ markedly in their susceptibility to these reactions, differences which are not demonstrably related to an active immunity to the corresponding antigens.

Sepsis. With the stringent security precautions now mandatory in the commercial preparation of prophylactics and provided a satisfactory aseptic technique has been used in inoculation the chances of septic complication except after smallpox vaccination (p. 416) are now extremely remote.

Progressive infection. When live vaccines are used in individuals with immune deficiency there may be progressive multiplication and possibly dissemination of the vaccine strain. The most important examples of such progressive infection are progressive vaccinia following smallpox vaccination (p. 416) and generalized mycobacterial infection following BCG vaccination (p. 276). Because of the possibility of this hazard, live vaccines should not be given to individuals with immunological defect, particularly individuals with defects in cellular immune response.

Encephalopathies. Under this heading are included

postvaccinal encephalitis, which is a particular hazard after smallpox vaccination of older individuals, encephalitis which may follow rabies immunization and convulsions which sometimes occur in infants after immunization with pertussis or triple vaccine containing a pertussis component. The pathogenesis of these conditions has not been elucidated. However, there is some evidence that postvaccinal and postrabies encephalitis may have an autoimmune aetiology (p. 85). A similar condition has been reported recently following inoculation of certain commercial brands of measles vaccine but has been very much milder than the encephalopathies following smallpox and rabies vaccination. There appears to be an increased risk of postvaccinal encephalitis when smallpox vaccination is combined with yellow fever immunization. It is possible that this risk may also be increased with combinations of smallpox vaccination with other live vaccines. It is, therefore, advisable that smallpox vaccination should not be given within three weeks before or after the administration of another live vaccine.

Infection of the fetus. Because of the capacity of the rubella virus to infect the fetus (p. 399) live rubella vaccine should not be given within the first three months of pregnancy. The same restriction is now considered desirable for all other live viral vaccines, with the possible exception of polio vaccine.

Provocation poliomyelitis. Before the introduction of general campaigns of polio immunization paralytic poliomyelitis was observed to occur, though infrequently, following the injection of vaccines containing a pertussis component. The pathogenesis of this condition is obscure but it is thought that the injection of the pertussis component renders the nerve axons more accessible to virus invasion. Provocation poliomyelitis does not appear to be a risk if polio immunization is carried out concomitantly with or before pertussis immunization.

Allergic reactions. Anaphylactic reactions could conceivably occur after the introduction of any vaccine. In practice, however, they appear only to be of practical importance following immunization with toxoid preparations or with viral vaccines prepared by propagation in eggs. Anaphylactic reactions to toxoid preparations appear to be mainly reactions to non-toxoid components of the culture medium in which the organism was grown and are of negligible importance following administration of modern highly purified toxoids. In immunization against diphtheria an indication of the possibility of such a reaction is provided by the development of a marked pseudo-positive reaction to the Schick test (p. 231). Anaphylactic reactions to vaccines prepared in eggs may be prevented by prior inquiry into a history of egg-hypersensitivity. If such a history is elicited the vaccine is probably best not given, but if its administration is a matter of urgency it should be given cautiously and in small doses. Anaphylactic reactions, if detected in time, may be aborted by the injection of dilute adrenaline solution. A solution containing 1 in 1000 adrenaline should be at hand when any immunization, either active or passive, is attempted.

If BCG vaccine is given to tuberculous subjects allergic reactions of delayed type are a potential hazard and may be sufficiently severe at the site of infection to precipitate progression of the lesion. It is prevented by the performance of tuberculin tests prior to immunization; only negative reactors should be immunized. In community-wide programmes in developing countries, tuberculin testing prior to BCG immunization may, however, not be practicable. Under these circumstances, it is accepted as being quite permissible to proceed with the immunization without prior tuberculin testing. If in exceptional circumstances there should be activation of a latent lesion or stimulation of an active lesion in a tuberculous subject the infection is still amenable to control by chemotherapy.

Negative phase. Following the administration of TAB vaccine a 'negative' phase may occur during which the bactericidal power of the blood against *Salmonella typhi* is considerably reduced; the phenomenon was first described by Wright in 1901. The reality of this phase has been the subject of controversy. Evidence has, however, been presented by Raettig on both experimental and epidemiological grounds that such a phase may occur and that immunization during the incubation period of typhoid may considerably increase the severity of the disease in those who contract it in spite of having being immunized. Because of this it would seem advisable not to vaccinate with typhoid vaccine during the course of an epidemic of typhoid fever. No similar effect has been demonstrated for live vaccines; in fact there is evidence that the administration of live polio vaccine may, by interfering with the growth of the infecting strain, tend to abort an epidemic.

II.9

Immunosuppression

Immunosuppression may be specific or non-specific.

SPECIFIC IMMUNOSUPPRESSION

Specific immunosuppression may be achieved by the administration of antibody or antigen. Suppression by antigen is known as immunological tolerance. Suppression by antibody may be due simply to the neutralization of antigen by the antibody, thereby preventing it from gaining access to the lymphoreticular tissues, provided sufficient antibody is administered. There is some evidence that low doses of antibody may in fact enhance the immunogenicity of an antigen. Antibody can also suppress immune responses to antigen in a manner analogous to feedback inhibition of enzymic activity. This mechanism is probably responsible for the inhibition of antibody production achieved when anti-Rh antibody is given in the terminal stages of pregnancy to Rhesus-negative mothers who are believed to be carrying Rhesus-positive fetuses. It probably also explains the fall in IgM antibodies associated with increase in IgG antibodies during the course of immunization with most antigens.

Glenny and Hopkins first demonstrated in 1942 that animals could be rendered non-responsive by exposure to antigen. They found that rabbits injected repeatedly with large amounts of diphtheria toxoid produced progressively less and less antitoxin to successive injections. In the 1940s it was shown by Felton and his colleagues that mice injected with a single large dose of pneumococcal polysaccharide were immunologically paralysed in that they failed to produce antibody to the polysaccharide and were at the same time not protected against pneumococcal infection on subsequent challenge. If, however, they were injected with a smaller dose of polysaccharide they did develop antibody and were protected against infection. The paralysis induced by the large injection persisted for most, or all, of the life time of the animal. This persistence was shown to be related to the fact that the polysaccharide was not broken down or excreted by the animal but was retained in the tissues. Felton believed that the polysaccharide had induced a paralysis of antibody-producing cells. Alternatively it was suggested that antibody was produced but that it was soaked up by the retained antigen so that none was free in the circulation. That this was not the explanation was readily shown, however, by the use of the fluorescent antibody technique, with which it was possible to establish that antibody-producing cells were absent from the tissues of the tolerant animal. The development of tolerance under natural conditions was first demonstrated by Owen in 1945. Owen showed that each of non-identical cattle twins which had shared the same placenta—a phenomenon common in cattle and known as synchorial twinning—was a chimaera in that it possessed functional haemopoietic tissues characteristic of the other twin. Sometime later a similar phenomenon was observed by Dunsford in a woman of blood group O who, during embryonic development, had acquired erythropoietic tissue derived from her non-identical twin of blood group A. As a result she possessed a mixture of groups O and A red cells in her circulation and at the same time lacked an alpha isoagglutinin. The far-reaching implications of these observations were first explicitly recognized in 1949 by Burnet and Fenner, who saw in them a vital clue to the mechanism by which animals may become non-reactive to antigens present in their own tissues. They postulated that if an animal is exposed to an antigen during the course of embryological development before the effective establishment of its immunological machinery, its capacity to respond to that antigen in later life might be suppressed. Decisive experimental support for this hypothesis was subsequently obtained by Medawar and his associates who showed that each member of the pair of synchorial cattle twins was capable of accepting a skin graft from the other twin and that if embryonic or newborn mice were injected with tissues of mice from an unrelated strain they were able in later life to accept skin grafts from that strain.

From the work of Medawar et al. it seemed likely that there was a critical period in early life during which exposure to antigen would lead to the development of tolerance. That this was not the case soon became apparent from the findings that antibody production to certain antigens could be demonstrated in the newborn animal and in the fetus and that tolerance could be produced in adult animals but was much more difficult to produce than in the fetus or the newborn.

The mechanism by which tolerance is established has still to be fully explained. The simplest model is that originally proposed by Burnet. This postulates that the tolerance-inducing exposure results in the killing of specifically reacting clones of antigen-sensitive cells. Although the antigen-sensitive cells are killed, recovery from tolerance can still take place through the differentiation, in the course of time, of a new population of antigen-sensitive cells. This could only occur in the absence of antigen. In the presence of antigen any such cells would in turn be eliminated.

From the examples of tolerance considered above it can be seen that tolerance can be established in respect of both cellular immune responses and humoral immune responses. Of the two, tolerance affecting humoral immune responses is the more susceptible to experimental study. Tolerance to proteins and protein–hapten complexes has been particularly examined. Tolerance may also be induced to a variety of synthetic polypeptides—in recent years this model has been increasingly used—and to a variety of polysaccharide antigens. Tolerance to erythrocyte antigens is of particular interest since it permits the use of Jerne's plaque-forming centre technique (p. 142) for the investigation of the phenomenon at the cellular level. It might be expected that tolerance would also develop to microbial antigens as a result of massive infections in the adult or as a result of infections acquired in utero or in very early life. In general, however, this does not appear to be the case. One example of the development of tolerance against a micro-organism as a result of infection in utero has, however, been well documented, namely that developed against lymphocytic choriomeningitis virus in mice. This virus is of low intrinsic virulence for mice and in infected mouse colonies causes a persistent asymptomatic infection. Mice readily become infected from their mothers in utero or during the neonatal period. Following such infections the virus is demonstrable in the brain, blood and other tissues. Despite this, the mice do not make antibody against or develop hypersensitivity towards the virus.

The cellular immune responses in respect of which tolerance may be established and which have been most studied are those against homologous tissue transplantation antigens. This type of tolerance is of particular importance in that it could possibly have considerable significance in the future for the management and control of allograft rejection. In general, the ease with which tolerance can be established against transplantation antigens is related to the degree of genetic and antigenic relationship between the donor and recipient. Where there is a large genetic barrier the establishment of tolerance is difficult; where the genetic barrier is slight the establishment of tolerance is relatively easy.

As previously indicated tolerance is much more readily established in young animals than in older ones. In adults it may, in fact, not be possible to establish tolerance at all against certain antigens against which tolerance can be readily developed in the young animal. Tolerance may, however, be established in the adult under two conditions; either following the administration of a very large amount of antigen or by the administration of antigen combined with immunosuppressive agents. When given together with immunosuppressive agents it is generally the case that the tolerizing dose of antigen is the same for the adult as for the young animal. This suggests that the difficulty of establishing tolerance in the adult is not due to the presence of a larger population of antigen-sensitive cells or to an intrinsically greater susceptibility to tolerance induction but rather to the fact that in the adult a positive immune response is for some reason more readily achieved.

For most antigens tolerance can be induced, both in the adult and the young animal, only with doses of antigen considerably greater than those necessary to stimulate an immune response. As a general rule the minimum tolerizing dose is about 10^4 times the minimum immunizing dose. In recent years, however, considerable interest has been aroused by the demonstration by Mitchison and Dresser that with certain weak antigens tolerance can also be induced by doses of antigen lower than those required for immunization. Thus 0·01 mg of bovine gammaglobulin is sufficient to paralyse an adult CBA mouse. This dose of antigen will immunize the animal, however, when injected together with Freund adjuvant. Certain substances do not produce low zone tolerance when injected into normal untreated animals but will produce it if administered along with immunosuppressive therapy. Low dose tolerance can also be produced by a number of 'non-immunogenic' antigens. These are antigens which are not immunogenic alone but which are immunogenic when injected with Freund adjuvant. The most studied of such antigens are various poly-D amino acids, e.g. capsular poly-D glutamic acid of *Bacillus anthracis,* which are highly tolerogenic in mice. These findings support the view that potentially antigenic substances will produce tolerance if the animal is exposed to them under conditions and in concentrations which will not evoke an immune response. This is consistent with the finding that the exposure of animals to non-immunogenic poly-DL-alanine will produce tolerance to poly-DL alanyl protein, a complex in response to which the animal would normally produce antibody reacting with the DL amino acid residues.

There is also evidence that the physical state of the antigen is of considerable importance in tolerance induction. Thus immunizing concentrations of bovine gammaglobulin have been found to lose their immunizing capacity and become paralysing if centrifuged at a speed sufficiently high to remove all aggregated material from the solution. A somewhat similar situation has been observed with a *Salmonella* flagellin. Polymeric flagellin is a potent immunogen and will not produce low zone tolerance. If, however, the flagellin monomer or a cleavage product A is injected, low zone tolerance is produced.

Loss of tolerance

The maintenance of a tolerant state requires the persistence of the antigen in the tissues. In the case of grafted tissues, tolerance persists indefinitely since the animal continues to be exposed indefinitely to the tolerizing antigen. With the exception of pneumococcal polysaccharides in mice, which are not catabolized, soluble antigens are eventually eliminated and as a result tolerance spontaneously terminates. Termination of tolerance does not, however, occur immediately. Tolerance to red cells has been observed to persist for months after the disappearance of foreign red cells from the circulation. In the case of protein antigens, the period for which tolerance may be maintained has no apparent simple relationship to the rate of catabolism and elimination of the antigen. It is possible that the persistence of tolerance after the disappearance of antigen from the circulation is due to the retention of some portion of the antigen within certain tissue cells, possibly macrophages. Studies with protein antigens, e.g. bovine serum albumin and flagellin labelled with radioactive iodine, in which it was possible to detect only a few molecules of antigen have, however, failed to reveal intracellular retention.

There is evidence that when recovery from tolerance occurs it is, for some time at least, only partial. Thus in recovery from tolerance to protein antigens the antibodies appearing react with only a limited number of determinants of the tolerizing antigen. Consequently the cells responsible for recovery must possess only a limited range of antigenic reactivity. This would be completely consistent with the recruitment hypothesis on which it might be expected that the regeneration of a full range of cells reacting with all the determinants of a complex antigen such as a protein might take a considerable time.

Of particular interest is the finding that tolerance may be abrogated by immunization with a cross-reactive antigen. This has been achieved in three ways:

1. Tolerance induced by conjugates of a hapten

with one protein can be broken by conjugates of the same hapten with a different protein; the antibody produced in response to the latter is hapten-specific. Thus it has been shown by Weigle that tolerance to picryl bovine serum albumin conjugates may be broken by conjugates of picryl chloride with other carriers.

2. Tolerance induced to a protein antigen may be broken by protein hapten conjugates. Thus tolerance to bovine serum albumin can be readily broken by immunization by bovine serum albumin substituted with mixed sulphonyl and arsonyl groups.

3. Tolerance induced to one protein may be broken by exposure to a cross-reacting protein. The degree of cross-reactivity is critical. Thus tolerance to bovine serum albumin may be broken by horse serum albumin which shows 15% cross-reaction with bovine serum albumin but not by sheep serum albumin which shows 75% cross-reaction.

In the last two examples it is possible that the antibodies produced reacting with the original tolerance-inducing protein may not be directed towards the determinant groups of the protein against which tolerance was induced. Weigle has shown, however, that after abrogation of BSA tolerance by a cross-reacting protein the animals will respond to a further injection of BSA-producing antibody indistinguishable from that normally produced in response to BSA in non-tolerant animals.

The phenomenon of tolerance abrogation by cross-reacting antigen may be of significance in relation to the development of autoimmune disease since it would seem to provide a mechanism by which immune responses to self antigens, to which tolerance had been established, may occur. Thus modification of a tissue antigen to yield antigen with the right degree of cross-reactivity to break tolerance might occur as a result of a combination with microbial products, structural alteration by microbial enzymes, somatic mutation or degenerative changes occurring spontaneously during the process of ageing. Alternatively autoimmunization might occur as a result of exposure to an antigen similar to a tissue antigen and possessing the right degree of cross-reactivity or to an antigen identical with the tissue antigen but associated with a different carrier. Such a mechanism would support the proposal that rheumatic carditis is due to the stimulation of autoimmunity to a heart muscle antigen by exposure to a cross-reacting streptococcal antigen (p. 216).

Mechanism of tolerance induction

There is a convincing body of evidence that immunological tolerance is the result of a change in the lymphoid cell population. It could conceivably be due to an alteration in the macrophages so that these lose the capacity to handle or process the antigen. No such alteration has, however, been demonstrated. In fact the macrophages of a tolerant animal ingest the tolerizing antigen quite normally. It would at any rate be difficult to imagine how such a specific interference with macrophage function could be established. There are two main lines of evidence for the postulated change in the lymphocyte population:

1. Lymphoid cells from tolerant animals lack reactivity to the tolerizing antigen. (*a*) They will not restore immunological function towards the antigen when introduced into the X irradiated animal. (*b*) They do not undergo antigen-stimulated proliferation either in vivo or in vitro (*c*) Lymphoid cells from animals rendered tolerant to soluble antigens do not produce antibody in vitro. (*d*) Lymphocytes from animals rendered tolerant to tissue transplantation antigens fail to show the mixed lymphocyte interaction (p. 145) shown by normal cells.

2. Tolerance may be broken by the adoptive transfer of lymphocytes from a normal non-tolerant donor. There are, however, a number of reports of failure to restore competence by adoptive transfer. In such cases it is possible that the tolerant cells have a suppressive effect on the normal non-tolerant cells.

The simplest explanation of the change in the lymphoid cell population of the tolerant animal is that the antigen-sensitive cells have been either eliminated or irreversibly inactivated. Elimination could occur either, as originally suggested by Burnet, by the death of the cells following the tolerizing exposure to the antigen or by differentiation of the antigen-sensitive cells to the end cell state without multiplication. In either case it would be necessary to explain why the antigen under immunizing conditions could stimulate multiplication and differentiation of the antigen-sensitive cells and under tolerance-inducing conditions would have a completely different effect. It has been suggested by Mitchison that the difference, at least in the case of tolerance induced in respect of humoral responses, is in the mode of presentation of the antigen to the antigen-sensitive cell. The normal route of antigen presentation in responses leading to antibody production may involve preliminary stages of processing in macrophages and/or attachment of the antigen to dendritic reticulum cells. If on the other hand this route were bypassed and the antigen were to gain direct access to lymphocytes it might kill them, possibly by allowing the fixation and subsequent action of complement, or induce end cell differentiation without reproduction.

The second major possibility—that antigen-sensitive cells can be irreversibly inactivated while still remaining viable—has at the moment little support. One possible way in which such inactivation could occur is by a permanent blocking of surface receptors by antigen. It seems unlikely, however, that the binding of the antigen to the cell surface could be sufficiently strong to maintain paralysis for months or years unless, as appears to be the case for pneumococcal polysaccharides, there was a large excess of unmetabolizing antigen in the circulation.

The finding that tolerance to an antigen can be abrogated by exposure of the animal to a cross-reactive antigen would seem to constitute *prima facie* evidence against the hypothesis that tolerance is due to elimination of antigen-sensitive cells since it would appear to require that cells capable of producing antibody against the tolerizing antigen persist in the body. There are two possible ways out of this difficulty. First, it is possible that the antigen-sensitive cells are heterogeneous in their affinity for the antigen and that cells with low affinity for the original antigen have a high affinity for the cross-reactive antigen. These cells, therefore, fail to react with the primary antigen but could react with the cross-reacting antigen. A second possibility is that, in the case of thymus-dependent antigens, tolerance affects not the primary non-thymus-dependent cells which produce antibody but thymus-dependent helper cells. If the helper cells for the relevant determinants on the inducing antigen were eliminated while those for different but equally effective determinants of the cross-reacting challenge antigen were retained, the challenge antigen could be mobilized or focused for appropriate stimulation of the antibody producing cells.

It has been suggested by Hellstrom and Hellstrom that tolerance to allografts may be due to an antibody which protects the cells of the allograft from the reactive lymphocytes of the recipient. These workers consider that there is, therefore, a close resemblance between allograft tolerance and antibody-mediated enhancement of tumour growth. It is possible that some cases of allograft persistence could be due to a mechanism of this sort but there is no reason to think that it could provide a general explanation for allograft tolerance in respect of which the evidence for a central mechanism affecting the lymphoid cell population is now considerable.

NON-SPECIFIC IMMUNOSUPPRESSION

Surgical removal of lymphoreticular tissues

Thymectomy, particularly neonatal thymectomy, has a profound effect on the capacity for immune response. Cellular immune responses are particularly affected. In fowls humoral responses are eliminated by extirpation either surgically or chemically of the bursa of Fabricius. Splenectomy appears to have some effect on humoral responses to antigens administered intravenously. It is reported that this effect is largely on 19S antibody production. Thoracic duct drainage by removal of circulating lymphocytes has a marked effect on cellular immune responses and on T cell dependent humoral immune responses.

X-irradiation

Sublethal doses of X-rays have a markedly destructive effect on lymphoid cells. Ultimately, the precise time schedule depends on the animal and the radiation dose, but on average in a matter of three to four weeks the lymphoid tissues are regenerated and appear histologically normal. Irradiation acts almost entirely on the primary response and affects antibody production much more than cellular immune responses. Antibody production, once initiated, however, is not significantly affected by radiation, the effect of which is therefore in the inductive phase of the response. The selective effect of radiation on the primary response is *prima facie* evidence for a difference between the antigen-sensitive cells of the primary response and the memory cells of the secondary response. The timing of irradiation in relation to antigen exposure is critical. In experimental animals inhibition of response is maximal when irradiation precedes antigen administration by 24 to 48 hours. A lesser but appreciable reduction in response is found when irradiation is given approximately 50 days before the primary stimulus. If the animal is irradiated within the first three days after antigenic stimulation the antibody levels may be above normal. After this period irradiation has little effect. X-irradiation was used in the early days of transplantation surgery to control allograft rejection but is no longer employed for this purpose.

Cytotoxic drugs

A rational classification of immunosuppressive drugs would be based on the precise stage of the immune response at which they act. At present, however, our knowledge is too incomplete to make such a classification feasible. They must therefore be classified on the basis of chemical structure. In contrast to X-rays, immunosuppressive drugs tend on the whole to be more active against cellular than against humoral immune responses. Like X-rays, however, they have little effect on the production of antibody once this has been initiated and, with the exception of antilymphocyte serum, they have a relatively slight effect on memory cells. In general their principal target appears to be the small lymphocyte. Because of their activity against actively proliferating cells cytotoxic drugs have had considerable application in the chemotherapy of malignant disease. They may also possibly have some degree of potential toxicity for normal tissue cells and this limits their use in treatment. In recent years drug-induced immunosuppression has been of considerable importance in transplantation surgery for the control of allograft rejection. They are also used in the treatment of autoimmune diseases.

Alkylating agents. The most important of these are nitrogen mustard, busulphan (Myleran), chlorambucil and cyclophosphamide. The primary action of these drugs is the methylation of guanine residues. Some can form cross-links between opposing DNA strands. Because of the indirect effect on DNA they are sometimes described as radiomimetic. These changes can be effected in resting cells and the main effect of these drugs appears to be to prevent the transformation of undifferentiated lymphocyte precursor antigen-sensitive cells to blast cells. Cyclophosphamide is the agent which has been most used. It is the least toxic and differs from other alkylating drugs in that it is effective when administered after as well as prior to antigenic stimulation.

Antimetabolites. These include folic acid analogues which inhibit the action of dihydrofolate reductase, the most important of which is methotrexate (amethopterin), and purine and pyrimidine analogues. The most important of the latter group is azathioprine, which is an imidazole derivative of 6-mercaptopurine. Azathioprine has a marked effect on cellular immune responses, though antibody production is also reduced. It is currently the drug of first choice for the control of allograft rejection. It is appreciably less toxic than either 6-mercaptopurine or methotrexate.

Corticoids. These are also extensively used for the control of allograft rejection in man together with azathioprine, the corticoid most used being prednisolone. The corticoids have a lytic effect on lymphocytes in vitro but it is not clear whether they exhibit this action in vivo. They have a suppressive effect on both humoral and cellular immune responses. It has been suggested that this may be due to their capacity to stabilize macrophage lysosomes and so interfere with the processing of antigen by macrophages which may play some part in the initiation of immune responses. By their depressing effect on inflammatory reactions corticoids also interefere with the efferent limb of the cellular immune responses. This could probably be their main role in the control of allograft rejection.

Antilymphocyte serum. The immunosuppressive effect of antilymphocyte serum was first demonstrated in 1963 by Woodruff and Anderson, who obtained prolongation of skin allograft survival in rats by treating them with an antiserum prepared by immunization of rabbits with rat thoracic duct lymphocytes. At present antilymphocyte serum, or more correctly its gammaglobulin fraction, obtained from horses immunized with human lymphocytes, is widely used as a supplementary immunosuppressive measure in the control of allograft rejection following organ transplantation in man.

Antilymphocyte serum has a suppressive effect on both cellular and humoral immune responses, its effect on cellular immune responses being much the more marked. It is effective against the memory cells of the cellular immune response, but has relatively little effect on secondary humoral responses. It appears to act by combining directly with the lymphocyte, as shown by the fact that lymphocytes exposed in vitro exhibit diminished graft-versus-host reactions. It has also been found to diminish the inflammatory reaction to certain pyogenic bacteria, e.g. *Staphylococcus aureus*.

Shortly after injection of antilymphocyte serum there is a considerable drop in the lymphocyte count with depletion of the paracortical areas of the lymph nodes. Under in vitro conditions it has a cytotoxic effect on lymphocytes in the presence of complement. In the absence of complement it stimulates lymphocytes to blast cell transformation. Lymphocyte transformation has also been demonstrated after prolonged in vivo administration. On withdrawal of the serum there is a recovery of the suppressed immunological responsiveness. This recovery has been found to be thymus-dependent, a finding which supports the proposal that the main target of the serum is the thymus-dependent lymphocyte. The mechanism of action of antilymphocyte serum has not been fully explained. Although the serum has a cytotoxic effect on lymphocytes in vitro in the presence of complement and produces lymphopenia in vivo there does not seem to be a clear correlation between cytotoxic and immunosuppressive effect. Possibly the simplest explanation of its action is that the serum blindfolds the lymphocyte so that the latter is incapable of responding to antigenic stimulation. Against this explanation, however, is the finding that lymphocytes from animals treated with ALS, when cultured in vitro retain their non-responsiveness for some generations after exposure.

The administration of antilymphocyte serum to patients is frequently associated with pain at the site of injection and occasionally with anaphylactoid reactions. These complications are considerably diminished by the use of antilymphocyte globulin. A more serious hazard of immunosuppression by antilymphocyte serum is the occasional development of malignancy, presumably due to an interference with the surveillance function of thymus-dependent lymphocytes.

For the preparation of antilymphocyte sera for experimental use lymphocytes have been variously obtained from the thoracic duct, lymph nodes, spleen, thymus and peripheral blood. For human immunization the principal lymphocyte sources have been spleen and thoracic duct. Antilymphocyte sera have a high degree of species specificity but this is by no means absolute and there is some evidence of sharing of antigens between lymphocytes and macrophages. The sera also contain red cell agglutinins which must be removed by absorption with red cells before use and antineutrophil antibodies which appear to be the antibodies responsible for the depression of inflammatory reactions to pyogenic bacteria which has been observed to follow administration of the serum.

Miscellaneous compounds

A variety of substances has been described as having immunosuppressive effects under experimental conditions. At the moment these are of theoretical rather than of practical importance.

Colloidal dyes, e.g. trypan blue. These have received little attention in recent years. It is possible that they may act by overloading phagocytic cells.

Antibiotics. The most interesting of these is ovalicin, and antibiotic from *Pseuderotium ovalis.* Ovalicin is described as inhibiting antibody production and as having a marked suppressive effect on the development of experimental allergic encephalomyelitis, presumably due to interference with the cellular immune reaction. Its practical utility in man is limited by its appreciable toxicity.

Plant mitogens. The most studied of these are phytohaemagglutinin (PHA) and pokeweed mitogen. These stimulate mitosis and DNA synthesis and the resultant transformation of small lymphocytes into blast cells. They are glycoprotein in nature and are lethal in high concentration. They also cause agglutination of red cells. Their immunosuppressive effect is believed to be due to the masking of antigen receptors on lymphocytes. These differ from the receptors involved in haemagglutination.

Carcinogens. The most studied of these is the potent chemical carcinogen methylcholanthrene. This compound has a marked inhibitory effect on the development of antibodies as measured by response to red cells and on cellular immune responses. Chemical carcinogens produce tumours which carry new tumour-specific antigens. These antigens should render them susceptible to cellular immune surveillance. It is, therefore, possible that their oncogenic effect is a result of their capacity to depress such surveillance.

Viruses. Infection with the measles virus has long been known to suppress tuberculin sensitivity and to enhance susceptibility to tuberculosis. The rubella virus diminishes the susceptibility of small lymphocytes to transformation by PHA but appears to have no immunosuppressive action as such. A number of oncogenic viruses, notably the Friend and Moloney viruses which produce leukaemias in mice, have a considerable suppressive effect in the preleukaemic phase on the capacity of mice to produce antibody against red cell antigens. These viruses are enveloped, RNA viruses resembling myxovirus and paramyxovirus groups in this respect. So far, however, human RNA viruses have not been shown to have either immunosuppresive or oncogenic activity.

Gram-negative endotoxins. Following injection of the endotoxins of various Gram-negative bacteria, experimental animals show a phase of increased susceptibility to a variety of bacterial infections. This phase is then followed by a phase of increased resistance. It is possible that these changes are the result of an action on phagocytic cells. Endotoxins are quite good adjuvants but some workers have found that they may depress antibody responses to various antigens. In contrast to their erratic effect on humoral responses they appear to depress consistently cellular immune responses such as the acquisition of tuberculin hypersensitivity. They have also been found to depress the normal lymphocyte transfer reaction and to be mitogenic for B cells.

Ribonucleases. A ribonuclease distinct from the known ribonucleases is present in normal serum and was shown by Mowbray to have a distinct immunosuppressive effect. This is obviously of interest in relation to Fishman's theory of antibody production (p. 124).

II.10

Hypersensitivity

By definition hypersensitivity is a state in which there is an increased reactivity of the body to external stimulation. In this sense it is essentially a statistical concept and applies to responses irrespective of whether or not they have an immunological basis. However, the term hypersensitivity is applied by immunologists to states in which the exaggerated response has a demonstrable immunological basis and as currently used is synonymous with the term *allergy,* which as originally defined by Von Pircquet who introduced it means altered reactivity, but is in practice applied to states of increased reactivity.

It has been customary to distinguish two types of hypersensitivity: *immediate* hypersensitivity in which the response appears within a few seconds to some hours after exposure to the stimulus and *delayed* hypersensitivity in which the response is delayed for several hours. This difference is most readily demonstrable when the hypersensitive reaction is elicited by skin testing. If, however, we consider reactivity in tissues other than skin the distinction is less clear, tissue responses associated with immediate skin reactivity frequently being delayed for some time after contact with the relevant antigen. When the mechanism underlying the two types of reactivity had been clarified it was apparent that classical immediate-type reactions, typified by weal and flare responses in the skin on the introduction of the antigen, were due to humoral antibody production and were in some way the result of combination of antigen with humoral antibodies. By contrast classical delayed-type reactions, e.g. the tuberculin reaction, were found to be cell-mediated immune responses due to reaction between antigen and homologous receptors on T lymphocytes.

IMMEDIATE OR ANTIBODY-MEDIATED HYPERSENSITIVITY

Two types of antibody-mediated hypersensitivity may be distinguished: anaphylactic and immune complex hypersensitivity.

Anaphylactic hypersensitivity

This form of hypersensitivity is due to the release of vasoactive amines (histamine and serotonin) as a result of the combination of antigen with antibodies, known in this context as homocytotropic antibodies, which have become fixed to basophilic polymorphonuclear leucocytes. Other substances may also be released from cells or otherwise produced during the course of anaphylactic reactions and contribute to their symptomatology. Thus when the reaction occurs in lung tissues as in asthma, slow-reacting substance (SRS-A) is produced and is at least as important as vasoactive amines in the clinical manifestations of this condition. Other substances which have been suggested as being produced or released in anaphylactic reactions are bradykinin, acetylcholine and the prostaglandins, but the evidence that these substances play any role in anaphylactic syndromes in man or guinea-pigs is quite inconclusive and any role they might play is almost certainly a very minor one.

Substances such as those mentioned above, which are released on antigenic stimulation and which are responsible for the symptomatology of anaphylaxis, are known as mediators. Undoubtedly the most important of these mediators is histamine, which can both stimulate smooth muscle contraction and increase the permeability of capillaries or venules. The clinical symptoms of anaphylaxis will depend on the organ in which these changes occur. This is usually the organ in which the antigen–antibody reaction occurs. The organ which is the seat of an anaphylactic reaction is known as the *shock* organ.

Although the combination of antigen with antibody fixed to mast cells is the principal method of histamine release, histamine may also be released by the generation of anaphylatoxin from the C3a and C5a components of complement formed as a result of antigen-antibody reactions occurring in the fluid phase. Anaphylatoxin may also be formed in the bypass reactions leading directly to C3 activation. It seems likely that these mechanisms of histamine release may be of importance in the inflammatory reaction following exposure to microbial antigens and in tissue reactions of the immune complex type (see below). Anaphylatoxin formation would not, however, appear to play a significant part in anaphylactic hypersensitivity.

The evidence that the mast cell is the principal cell involved in anaphylactic reactions is that (*a*) mast cells are the principal source of histamine and serotonin in most if not all animal species; (*b*) mast cell degranulation occurs during the course of anaphylactic reaction; (*c*) mast cells may be passively sensitized by homocytotropic antibodies. On subsequent exposure to antigen they release histamine.

The capacity of homocytotropic antibodies to combine with mast cell receptors is clearly due to a specific configuration on the Fc portion of the molecule. Unlike complement activation cell fixation of homocytotropic antibody is not dependent on prior combination with antigen. Except in the case of fixation of human homocytotropic antibody to rat mast cells and to the mast cells of non human primates, there is so far no evidence that homocytotropic antibodies can combine with the cells of species other than that in which they were produced.

Anaphylactic-type reactions may occur in two clinical forms:

1. A generalized shock reaction known as systemic anaphylaxis or anaphylactic shock.
2. A localized reaction known as atopy.

Systemic anaphylaxis

Systemic anaphylaxis is most readily induced by the intravenous injection of antigen into sensitized animals but it may also be induced, though less readily, by injection of antigen by the subcutaneous or intramuscular route or in highly sensitive subjects by injection into the skin. All animals may be sensitized to produce systemic anaphylaxis on contact with antigen. Guinea-pigs are particularly susceptible and protein antigens are particularly potent in inducing anaphylactic sensitization. Doses of as little as 0·0001 ml of foreign protein, e.g. horse serum, may be sufficient to sensitize a guinea-pig. This is known as the sensitizing dose. The reactions of shock can be produced by injection of a large amount of the sensitizing antigen some 10 to 21 days after the sensitizing injection. This is known as the shock dose. In the guinea-pig death occurs within a few minutes of the shock injection due to suffocation from contraction of the smooth muscle in the walls of the small bronchioles. That histamine plays a major role in systemic anaphylaxis in the guinea-pig has been established by a number of lines of evidence:

1. Guinea-pig bronchiolar tissue is highly susceptible to histamine and the symptoms of shock can be mimicked by the injection of histamine.
2. Shock is associated with an increase in blood histamine levels.
3. Histamine is released from perfused or chopped lung tissue or from the basophils of sensitized guinea-pigs following exposure to the sensitizing antigen.
4. The development of the shock syndrome can be substantially, though not completely, prevented by the administration of adrenaline and the antihistamines. The failure of antihistamines to prevent the development of the syndrome completely is taken as evidence of the participation of other pharmacological mediators.

In rats the picture of anaphylactic shock is somewhat different, the main features being circulatory collapse due to dilatation of the splanchnic blood vessels with draining of the circulation from the periphery and increased intestinal peristaltic activity due to contraction of the smooth muscle of the alimentary tract. Rats are considerably less susceptible to histamine than guinea-pigs and it is possible that the clinical features of shock, which are little affected by antihistamines, are due largely to the action of serotonin. In the rabbit death is associated with extreme dilatation of the right side of the heart caused apparently by pulmonary hypertension produced by constriction of the pulmonary arterioles—a histamine effect.

Animals can be passively sensitized to anaphylactic shock by the injection of antiserum obtained from a sensitive animal of the same species. Guinea-pigs are unique amongst experimental animals in that they may be sensitized passively by the injection of antibodies from certain other species, notably human and rabbit antibodies. In this context these antibodies are known as *heterocytotropic* antibodies.

Anaphylactically sensitized animals may be *desensitized* if they are given a series of injections of antigen, at intervals, over a period of some weeks. In many cases this procedure stimulates the production of a precipitating antibody which will combine with the injected antigen and will prevent the latter from reaching the fixed homocytotropic antibody. This precipitating antibody is sometimes referred to as a blocking antibody. Precipitating blocking antibodies are also produced if the animal is initially given one or more of a series of large doses of antigen instead of small doses of antigen. Non-reactivity due to the production of blocking antibodies is sometimes referred to as anti-anaphylaxis. In some cases, however, it is believed that the refractory state is due to the fact that the animal has been rendered tolerant to the sensitizing antigen.

Experimental correlates of the anaphylactic state in sensitized animals

Organ anaphylaxis. If the uterus, ileum or a bronchial segment from a sensitized animal is exposed to antigen it will contract as a result of smooth muscle spasm. Thereafter it will be refractory to antigenic stimulation. On the other hand it retains its sensitivity to histamine and serotonin. The initial contraction is due to the fact that a sufficient number of mast cells are present in the sensitized tissue. On exposure to antigen, however, all the histamine is released and while the tissue retains its histamine sensitivity there is no further supply of histamine in the mast cells to stimulate it. Organ anaphylaxis may be induced passively in the isolated organs of normal animals by exposing these to serum from sensitized animals. After washing, the smooth muscle of the passively sensitized tissue will contract on exposure to antigen.

Active cutaneous anaphylaxis. Injection of a small amount of antigen into the skin of a sensitized animal leads to the development of a weal and flare reaction as a result of increased vascular permeability. This permeability is more clearly demonstrated if the animal has been injected intravenously, some time prior to the test dose of antigen, with a dye such as Evans' blue which does not normally escape from the small blood vessels. If the permeability of the latter is increased, however, the dye does escape and causes a visible bluish colouration of the tissues around the site of antigen injection. The blueing reaction is positive even in the absence of a visible weal and flare.

Passive cutaneous anaphylaxis. The serum of a sensitized animal is injected intradermally into the skin of a normal animal. From four to 24 hours later antigen together with a dye such as Evans' blue is injected intravenously. If the serum contains anaphylactic sensitizing antibody blueing occurs around the site of antigen injection. This reaction can be used to quantitate the amount of sensitizing antibody by the estimation of the blue area of the skin measured on the under side of the skin after killing the animal. This technique is relatively easy to carry out and has been of particular value in the elucidation of the mechanism of anaphylactic hypersensitivity. It is analogous to and in principle derived from the Prausnitz-Kuestner reaction used for the detection of human reaginic antibodies (see below).

Histamine release from sensitized cells. The release of histamine from the blood leucocytes (subsequently identified as basophils) of human allergic subjects, on exposure to antigen was first shown by Katz and Cohen in 1941 and has been increasingly used as an experimental procedure. Histamine release from basophils is associated with degranulation. Histamine is also released, after exposure to antigen, from basophils which have been passively sensitized with homocytotropic antibodies. The release of histamine from chopped lung tissues of actively sensitized guinea-pigs as first demonstrated by Mongar and Schild in 1951 has also been extensively used. In this system many of the important features controlling histamine release by antigen were first clarified.

Systemic anaphylactic-type hypersensitivity in man

Man develops systemic anaphylactic hypersensitivity much less readily than the guinea-pig. Systemic anaphylaxis has been described under the following conditions:

1. After injection of diphtheria or tetanus antitoxins. The majority of people affected have had a history of spontaneous local hypersensitivity, particularly of asthmatic type (see below), to horse dander. Less frequently, sensitization is the result of previous administration of horse serum.

2. Following the administration of drugs, penicillin—to which many people have now been rendered hypersensitive—undoubtedly being the worst offender. Many cases of fatal anaphylaxis in man, following the injection of penicillin, have occurred as a result of self-administration. In persons highly sensitive to penicillin it has been estimated that as little as 3×10^{-6} units of benzyl penicillin can give rise to a systemic reaction.

3. Following insect bites, particularly bee or wasp stings. It has been estimated that from six to 12 people per year die in the British Isles in this way. Persons with an allergy to insect bites appear to acquire first a delayed hypersensitivity, which is then followed by an immediate skin reactivity. After further bites this cutaneous hypersensitivity usually disappears, but in a few individuals it may become intensified, ultimately to form the basis of a systemic anaphylactic reaction.

4. Following the injection of various prophylactics, e.g. tetanus toxoid (p. 289), prepared from media containing components to which very exceptionally an individual may exhibit a spontaneous hypersensitivity.

5. Following intradermal injection of pollen extracts for the identification of allergens in atopic, particularly asthmatic subjects.

6. Following miscellaneous intravenous injections, e.g. of contrast media in intravenous pyelography.

7. Following spontaneous rupture or, more frequently, surgical incision of a hydatid cyst.

The commonest clinical pattern in systemic anaphylaxis in man is that of severe respiratory distress due, as in the guinea-pig, to bronchiolar constriction, together with the signs of vascular collapse consequent on vasodilatation. Laryngeal oedema, which may itself be sufficient to cause death, is an important additional component. As in the guinea-pig, the main pharmacological mediator appears to be histamine.

Local anaphylaxis (atopy)

Under this heading will be considered local anaphylactic reactions which occur spontaneously in response to substances encountered during everyday life. In man these local anaphylactic states are frequently referred to as idiosyncrasies or collectively as atopy. The different varieties of atopy are normally classified on the basis of the organ in which they occur:

Allergic rhinitis in which there is considerable exudation of fluid in the nasal passages and conjunctiva. This condition is commonly known as hay fever since in rural areas in which it is common the condition is particularly associated with exposure to allergens present in hay. A variety of pollens and dusts may, however, act as sensitizing agents. The fluid exudate in hay fever is attributed to increased vascular permeability and the responsiveness of the condition to antihistamines supports the role of histamine as the principal mediator.

Asthma. In this condition there is contraction of the smooth muscles of the bronchioles. This leads during an attack to difficulty in breathing, particularly in expiration. This gives rise to the 'wheezing' which is the classical symptom of the condition. It is probable that SRS-A plays a major role in asthma since antihistamines and serotonin antagonists have little if any effect on the clinical symptoms.

Gastrointestinal allergies are associated with vomiting and diarrhoea due to contraction of the smooth muscle of the alimentary tract. Histamine and serotonin are the presumed mediators.

Allergic reactions involving the skin. These include *eczema,* which as an allergic manifestation primarily if not exclusively affects infants, and *urticaria,* so called because it resembles nettle rash. Urticarial skin lesions are localized areas of oedema which are directly due to increased vascular permeability. When the oedema involves the subcutaneous tissues as well it is known as giant urticaria or angioneurotic oedema. Although these various types of skin reaction may have an allergic basis, very frequently they do not. Hay fever and asthma, on the other hand, appear invariably to have an allergic aetiology.

In no case has it been possible to identify the determinant groups of the allergens responsible for the various clinical types of atopy. There is little doubt, however, that the principal allergens responsible for the respiratory allergies hay fever and asthma—are substances present in plant pollens and in various vegetable and animal dusts. Recently evidence has been presented that allergic reactions due to house dust are induced by components or excreta of the house mite *Dermatophagoides pteronyssinus.* In both types of respiratory allergy the allergic reaction is induced by contact of the allergen with basophils located in the appropriate region of the respiratory tract. Gastrointestinal allergies are normally due to foods, the most important of which are nuts, eggs and fish, to which the individual is hypersensitive. Allergic reactions in the gastrointestinal tract are normally triggered by ingestion of the allergen but may occasionally be triggered by inhalation. Similarly, asthmatic symptoms may be induced by ingestion. Why, under certain circumstances, the shock organ should be remote from the organ through which immediate contact with the allergen is made has not been explained. Eczema in children appears to be due most usually to ingestion of the allergen, eggs being particularly potent in this respect. Urticarial skin reactions may also be due to ingestion, notably of various drugs when the skin lesions frequently occur as components of the serum sickness syndrome (p. 192). Urticarial reactions of papular type appear often to be allergic reactions to insect bites.

Urticarial reactions of particular interest are those produced as result of exposure to cold and to light; the nature of the allergen responsible presents an obvious immunological problem. It has been suggested that allergy induced by light is directed against a photoproduct of some normal tissue component. There is no evidence, however, that an analogous product is produced by exposure to cold. In this case it is possible that the allergen is a tissue component which cross-reacts with some extrinsic antigen to which the individual has been sensitized and that the triggering of the reaction by cold depends on some peculiar property of the homocytotropic antibody involved. This cannot, however, be the capacity to fix to basophils since reaginic antibodies mediating cold hypersensitivity have been found to fix efficiently to basophils at 37 °C.

Experimental procedures used in connection with human atopy

The procedures used are analogous to those employed in experimental anaphylactic hypersensitivity in animals:

Skin tests with the appropriate antigen. These tests which are usually carried out by a needle prick are not without danger since even very small amounts of certain antigens, e.g. penicillin, may induce systemic anaphylactic reactions.

The Prausnitz–Kuestner reaction. This test was historically the basis of the procedure for demonstrating passive cutaneous anaphylaxis in animals. Kuestner had a gastrointestinal allergy to fish. The test was devised by Prausnitz who injected some of Kuestner's serum into the skin of a volunteer subject and sometime afterwards an extract of the offending fish, when a weal and flare reaction developed at the site of the injection. The P–K reaction is rarely if ever carried out nowadays because of the risk of transmitting homologous serum jaundice (p. 444). It is nevertheless of considerable importance not only historically but also as the basis for the definition of reaginic antibodies as the responsible antibodies were originally called. Recently it has been found that positive reagin tests may be elicited by injection of human atopic serum into monkey skin followed by injection of the antigen. This procedure though potentially useful is obviously limited by the availability of monkeys as test animals.

Passive organ anaphylaxis. Such tests can be carried out on human tissues, notably appendix and bronchial rings removed at operation. Human reagins will also sensitize monkey tissues, e.g. monkey ileum.

Demonstration of histamine release, on exposure to antigen, from actively or passively sensitized human or monkey mast cells. A recent development of considerable potential has been the demonstration that human reagins may sensitize rat mast cells. The value and limitations of this procedure for the investigation of atopic states has still to be fully explored.

Degranulation of both actively and passively sensitized basophils after exposure to antigen. Degranulation is easier to measure than histamine release, particularly if rat basophils can be used, and may well prove to be a satisfactory routine clinical procedure for the estimation of reagins.

A conspicuous feature of human anaphylactic states is the presence of a blood eosinophilia and the presence of eosinophils in exudates, e.g. in the mucous secretions of hay fever. The role of eosinophils in allergic reactions is so far unknown. In the guinea-pig it has been shown that heat-stable reaginic antibodies (see below) can mediate the release of a specific eosinophil chemotactic factor. This factor does not appear to be released in the course of passive cutaneous anaphylaxis mediated by classical heat-labile reaginic antibodies.

Identification of reagins with IgE immunoglobulin

The implication of a new immunoglobulin class-IgE in human atopic hypersensitivity was discovered by Ishizaka. This antibody is present in very low concentration in serum but is found in appreciable concentration in relation to the lymphoid tissue of the respiratory and alimentary tracts. This distribution raises the possibility that IgE antibodies have a role in the protection of the respiratory and alimentary tracts, e.g. of the respiratory tract from pollens and dusts and of the alimentary tract from parasitic worms.

The following lines of evidence identify IgE as the principal if not the sole reaginic antibody in man:

1. The serum levels of IgE are considerably higher in atopic than in normal individuals.
2. Purified IgE from individuals with various types of atopy give positive PK reactions. This capacity is eliminated by precipitation of the serum with specific anti-IgE serum or by passage through immunoadsorbent columns to which anti-IgE antibody has been coupled.
3. A parallelism has been demonstrated in ragweed hypersensitivity between reagin concentration as estimated by the P–K reaction and the concentration of antigen-specific IgE in the serum of atopic patients.
4. Cutaneous weal and flare reactions and histamine release from sensitized cells can be induced by anti-IgE antibodies but not by antibodies to other immunoglobulin classes.
5. P–K reactions elicited in human and monkey skin can be blocked by prior injection of non-antigen-specific IgE from IgE myelomas.
6. IgE antibodies show the same characteristics in respect of heat and sulphydryl reagent lability, inability to cross the placenta and incapacity to fix complement as do reaginic antibodies.

Properties of reagins

Reaginic antibodies as demonstrated by the P–K test have the following properties:

1. They are markedly heat-labile, being rapidly inactivated at 56°C. In general heating appears to affect only the sensitizing capacity of reaginic antibodies, the capacity of the antibodies to combine with antigen remaining unchanged.
2. High lability to sulphydryl reagents. This lability presumably indicates that the cell-binding site of the reaginic antibody contains at least one –S–S– linkage.
3. Incapacity to fix complement. This indicates either the absence of the complement combining site or its masking in the combined reagin molecule. The failure to fix complement would indicate that complement reactivity plays no part in anaphylactic reactions.
4. Inability to pass across the placenta from the maternal to the fetal circulation.
5. Capacity for long fixation in the skin in P–K reactions. Positive reactions can be obtained for up to 2 weeks after injection of reagin. Since immunoglobulins of other classes persist for at most a few hours sufficient time must be allowed to elapse to permit them to diffuse away from the site of introduction before the injection of antigen. Because of this it has been thought that reaginic antibodies fix slowly to mast cells. This, however, does not appear to be the case, fixation being virtually instantaneous. Consequently the so-called latent period required before weal and flare reactions are demonstrable is simply the time required for diffusion of non-reaginic antibodies away from the test site.

Reaginic antibodies with properties very similar to those of human IgE have been demonstrated in experimental anaphylactic hypersensitivity in various animals, monkeys, rabbits, guinea-pigs, rats and mice. Monkey IgE shows a close antigenic similarity to human IgE. Unlike IgA, IgE antibodies present in secretions appear to have properties similar to those of IgE in serum. In guinea-pigs, rats and mice heat-stable homocytotropic antibodies of IgG class have also been identified. These heat-stable homocytotropic antibodies are of slower electrophoretic mobility than the heat-labile reaginic antibodies. The heat-stable antibodies of guinea-pigs and mice are also stable to sulphydryl reagents and lack ability to fix complement. In biological properties all differ from heat-labile antibodies in persisting for at most a few days at passively prepared skin sites and in having a much shorter latent period for the development of passive cutaneous anaphylactic sensitization. Heat-stable guinea-pig and mouse homocytotropic antibodies have been shown to be capable of crossing the placenta from the maternal to the fetal circulation. Results suggesting the presence of heat-stable non-IgE homocytotropic antibodies have been obtained in humans in cases of anaphylactic hypersensitivity to certain protein antigens, but the position requires further clarification.

Mechanism of sensitization

The specific binding of IgE antibody to mast cells has been shown by the findings that:

1. Amongst human leucocytes only the basophils will fix human ^{125}I-labelled IgE myeloma protein and will alone react demonstrably with ^{125}I-labelled anti-IgE antibodies. In both these cases the presence of the labelled protein on the basophils has been demonstrated by autoradiography.
2. After exposure to labelled anti-IgE, histamine was shown by Ishizaka to be released and to be associated with basophil degranulation. Under these circumstances histamine release and degranulation could be prevented by EDTA. In this case virtually all the labelled cells present were found to possess basophilic granules. By contrast in the absence of EDTA a high proportion of the labelled cells did not possess granules, this proportion showing a parallel relationship to the percentage of histamine released.

The capacity of reaginic antibodies to combine both with mast cells and with antigen constitutes *prima facie* evidence that the combination with mast cells occurs through a receptor on the Fc portion of the molecule. Decisive evidence in support of this has been obtained by Ishizaka in the findings that:

1. Fc fragments of IgE myeloma proteins obtained by papain digestion are capable of blocking the sensitization of human and monkey skin with reaginic antibodies whereas the Fab fragment obtained by papain digestion and the $F(ab')_2$ fragment obtained by peptic digestion failed to do so.
2. The release of histamine from human leucocytes following exposure to anti-IgE was greatly increased following prior exposure of the cells to the Fc fragments of myeloma IgE. Similarly it was found that chopped monkey lung exposed to the Fc fragments released histamine on subsequent exposure to anti-IgE whereas unsensitized lung tissue did not. The reason why unsensitized lung tissue does not release any demonstrable histamine on exposure to anti-IgE is not clear. It may be because of the low concentration of mast cells in normal lung tissue as a result of which the histamine concentration released was below the level of detectability. The precise nature of the IgE receptor on mast cells has not been determined but the demonstration by Bach et al. That the receptor on rat peritoneal mast cells is inactivated by phospholipase C suggests that it is a phospholipid.

Both histamine and serotonin are present in platelets and it is possible that platelets may make some contribution to the total anaphylactic reaction. That vasoactive amines may be released from platelets following the formation of antigen–antibody complexes has been clearly demonstrated. This mechanism requires complement and is not clearly related to IgE fixation; it is unlikely, therefore, to play a significant role in the anaphylactic syndrome. A complement-independent mechanism for the release of histamine from platelets has been described in rabbits. It is claimed that the combination of antigen with IgE antibody on the surface of circulating basophils resulted in the liberation of a factor (PAF) which could induce histamine release from normal rabbit platelets.

The identity of the reacting antibody with IgE was shown by the fact that PAF release on exposure to antigen could be inhibited by prior treatment of the basophils with specific anti-IgE antibody. The occurrence of this mechanism in man has not so far been demonstrated but it is clearly a way in which the anaphylactic response may be amplified.

Release of histamine from mast cells

The composition of mast cell granules has been determined for granules released by osmotic lysis and harvested by density gradient centrifugation. The granules have the following composition: protein, most of which is a basic polypeptide, 60%; heparin, about 30%; histamine, about 10%. The concentration of 5-hydroxytryptamine (serotonin) is quite low, usually less than 1%. The matrix of the granule appears to consist of a complex between the basic polypeptide and the acidic heparin, histamine and presumably also 5-hydroxytryptamine being bound through their amine groups to the carboxyl groups of the polypeptide. As long as the granules are suspended in deionized water they retain their histamine and 5-hydroxytryptamine content. However, if they are exposed to sodium or other cations the amines are immediately released. On washing of the depleted granules in deionized water the cation is removed and the granules are again capable of taking up histamine. The granules therefore function as ion-exchange resins from which there can be an exchange of cations for amines. Since, after exposure to the extracellular environment, they are not surrounded by a membrane, this exchange resulting in the release of the amines occurs rapidly following their release from the cell.

The release of the granule from the cell is a complex process which is not yet fully understood. It is not dependent on any humoral factor such as complement and is, therefore, entirely the result of combination of antigen with the cell-fixed reaginic antibody. There is now virtually conclusive evidence that this release can only occur if the antigen is divalent so that it can combine with two immunoglobulin molecules located close together on the cell surface membrane. Thus in animals sensitized to both DNP and benzyl penicilloyl (BPO) haptens, histamine release did not occur following exposure of the cells to protein hapten conjugates each molecule of which contained only one DNP or one BPO group. Histamine release could occur, however, following exposure to conjugates each molecule of which contained at least one DNP and one BPO group.

It has been customary to postulate that the release of granules from basophils is the result of damage to the cell membrane. This, however, is far from the case. Release requires the presence of cations, is dependent on temperature, requires a source of energy which may be from glycolysis or from oxidative phosphorylation and is dependent on the activity of a di-isopropyl fluorophosphate (DFP) sensitive esterase.

In the intact cell, the granules are surrounded by a perigranular membrane. The first stages of release of the granules consist in the fusion of the perigranular membrane of peripheral granules with the limiting membrane of the cell. This leads to the formation of localized pores through which cations and water pass rapidly, resulting in rupture and extrusion of the granule and permitting subsequent ion exchange.

Following contact of antigen with sensitized cells two stages can be distinguished in the release process. The first stage is temperature-dependent, is independent of calcium and leads to an activated cell. The biochemistry of this stage is complex. It is triggered by antigen and requires the activity of the DFP-sensitive esterase. There is now accumulating evidence that cyclic AMP plays an important modulating role in the first stage. It is inhibited by exposure to dibutryl cyclic AMP which increases intracellular cAMP concentration directly, by catecholamines which increase intracellular cAMP concentration by stimulation of adenyl cyclase and also by methyl xanthines which inhibit the phosphodiesterase responsible for intracellular cAMP breakdown. At present the most likely role for cAMP appears to be that, analogous to its action on glycogen phosphorylase, of modulating the activity of the DFP-sensitive esterase. The second stage is that responsible for granule release; as described above this stage is antigen-independent but calcium-dependent. It also occurs optimally at 37°C. The membrane fusion which is associated with the second stage does not have a lethal effect on the cells since cells are found to be actively motile for several hours after degranulation.

Immune complex hypersensitivity

In recent years it has been established that the deposition of immune complexes within blood vessel walls and in the tissues can trigger tissue damage. This has been shown in experiments on actively or passively sensitized animals subsequently exposed to antigen and by the direct injection of preformed immune complexes. The effect of immune complexes is closely dependent on the relative proportions of the antibody and antigen they contain. Complexes formed in the equivalence zone and in the presence of antibody excess are precipitated and the aggregates thus formed are readily removed from the site of deposition by polymorphonuclear leucocytes and by macrophages. The complexes formed in the presence of antigen excess, particularly in moderate antigen excess, are soluble and are not readily disposed of. These, being persistent, are much more dangerous.

Tissue events initiated by immune complex formation

These may be conveniently divided into primary and secondary events.

Primary events

1. The formation of pharmacologically active substances liberated during the course of the complement cascade (p. 161). These include the fragments C3a and C5a which have both anaphylactogenic and chemotactic activity for neutrophil leucocytes, and the complex **C5b67** which is neutrophil chemotactic but not anaphylactogenic.

2. Activation of the Hageman factor. This factor plays the key initial role in the blood clotting mechanism. That Hageman factor can be activated by antigen antibody complexes has undoubtedly been demonstrated. What is not clear, however, is the extent to which this activation and its subsequent numerous results are of importance under in vivo conditions. The possible role of activated Hageman factor may be summarized briefly as follows: (*a*) In the presence of a cofactor, the activated factor generates the formation of plasmin from plasminogen. (*b*) As a result of its proteolytic capacity plasmin can induce fibrinolysis and can transform kininogens into kinins. The most important kinin is the nonapeptide bradykinin which has the capacity to induce contraction of smooth muscle and to enhance capillary permeability. (*c*) Plasmin releases a fragment from the activated Hageman factor which catalyses the generation of the enzyme kallikrein from its inactive precursor pre-kallikrein which is present in the plasma. Kallikrein is probably the principal stimulus to bradykinin formation. Kallikrein activation can also be induced by activated Hageman factor directly.

The important biological results of the activation of the Hageman factor are, therefore, clotting, fibrinolysis and bradykinin formation. In this case it is not surprising that there should exist homeostatic mechanisms controlling the operation of the various biochemical intermediates in the tissues. Of particular importance in this respect is the activity of an inhibitor which acts on C1 esterase, on plasmin and on kallikrein. This is the inhibitor whose deficiency is responsible for the condition of hereditary angioneurotic oedema.

3. Immune complex formation in small blood vessels results in platelet aggregation. This is probably a complement-mediated phenomenon as described under the immune adherence reaction (p. 141).

4. It is possible, though this has not been proved, that immune complexes formed in the tissues may directly produce some degree of damage to tissue cells and may also damage vascular endothelium.

Secondary events

The most important of the secondary results of immune complex formation in the tissues are those due to the chemotaxis of neutrophil polymorphonuclear leucocytes. The importance of the leucocyte is indicated by the fact that a significant degree of immune complex damage does not occur in animals depleted of polymorphonuclear leucocytes by treatment with, for example, nitrogen mustards. By itself the accumulation of neutrophils would not be expected to initiate tissue damage. This appears to be essentially the result of liberation from the neutrophils of lysosomal and other enzymes which are the immediate causes of the damage. This release may be triggered by ingestion of the antigen–

antibody complexes. One of the most important of the lysosomal enzymes liberated appears to be an elastase which has the capacity of digesting elastic tissue in general and in particular the basement membranes of small blood vessels. There is also some evidence that substances liberated from leucocytes may activate the kinin system and may themselves have a direct chemotactic effect on leucocytes. Histamine is released during the course of immune complex reactions firstly by the action of anaphylatoxins generated in the complement activation sequence and also as a result of platelet aggregation and will obviously contribute to inflammatory reactions.

Immune complex syndromes in experimental animals

There are a variety of circumstances in which cellular damage may be produced in experimental animals as a result of the injection of antibody directed against components of normal tissues, e.g. Masugi-type nephritis produced by the injection of antibody to glomerular basement membrane; syndromes caused by the injection of antibody into animals, e.g. guinea-pigs, directed against Forssmann antigens; and purpura and anaemia produced by injection of antibodies against platelets and red cells respectively. In each of these cases the antibody is directed against a normal tissue component and the experimental syndrome is in this respect a potential model of antibody-mediated autoimmune disease.

Here, however, we are concerned primarily with conditions in which the immune complex is formed by combination of antibody whether actively induced or passively injected with a foreign antigen. Two contrasting situations may be distinguished:

1. The Arthus phenomenon in which the tissue damaging complexes are aggregates formed in the presence of considerable antibody excess.
2. The experimental serum sickness syndrome in which the damaging complexes are soluble complexes formed in the region of moderate antigen excess.

The Arthus phenomenon

The Arthus phenomenon was first described in 1903 by Maurice Arthus who found that rabbits which had been immunized with horse serum responded on subsequent subcutaneous injection of horse serum with a severe haemorrhagic oedematous reaction. The phenomenon is now usually studied in primed animals given subsequent intradermal injections of purified homogeneous proteins. The haemorrhagic oedematous lesions only develop in animals possessing precipitating antibodies to the injected antigen and the severity of the reaction is related to the titre of the precipitating antibody. In very severe reactions there may be necrosis of the injected skin. Histologically, polymorph aggregates are found inside the small blood vessels and there is an accumulation of polymorphs in the walls of the blood vessels, particularly those of the small venules. There is also extravascular accumulation of polymorphs in the tissues which are demonstrably oedematous. A striking feature of the histological picture is the necrosis of the blood vessel walls which is the direct cause of haemorrhages in the tissues.

The localization of antigen–antibody complexes in the tissues has been studied by both immunofluorescence and autoradiographic techniques. Shortly after injection of the antigen the complexes are found extravascularly but very soon afterwards can be seen to be predominantly located in the walls of the small blood vessels in which they appear to be formed by reaction between antibody diffusing out of the blood vessels and antigen diffusing inwards from the tissues. The complexes, both within the vessel walls and in the tissues, are rapidly ingested by polymorphonuclear leucocytes. In many cases the ingested complexes are digested but during this process the leucocytes are not infrequently killed. At a later stage in the reaction undigested or incompletely digested complexes from dead leucocytes are found within macrophages.

The sequence of events in a typical Arthus phenomenon would appear to be as follows:

1. The generation of leucocyte chemotactic substances from complement components. The key role of polymorphonuclear leucocytes is shown by the fact that Arthus reactions do not occur in animals depleted of leucocytes by treatment with nitrogen mustards. The central importance of complement is shown by the failure of these reactions to occur in C3-deficient rabbits and that in guinea-pigs passive reactions can be

produced only by gamma 2 antibodies which fix complement but not by gamma 1 antibodies which are incapable of doing so.

2. Following the accumulation of polymorphs there is a release of various enzymes of which possibly the most important is the elastin digesting protease. The mechanisms involved in enzyme release and/or death of the polymorphs is so far unexplained.

The experimental serum sickness syndrome

Since serum contains a large number of immunologically different proteins a precise analysis of the role played by complexes of antibodies with these proteins is not feasible. Such an analysis can, however, be achieved in experimental animals injected with purified homogeneous foreign proteins. That a variety of vascular lesions could be produced by such purified preparations was first demonstrated by Hawn and Janeway. Subsequent work by Germuth and by Dixon et al. using isotopically labelled antigens established that practically all the animals injected developed some degree of glomerulonephritis, a lesser proportion showing generalized arteritis resembling polyarteritis nodosa in man and carditis resembling the carditis of acute rheumatism. The time of development of these lesions was closely related to the immune phase of antigen elimination being first demonstrable at the time of demonstrable inflexion of the antigen concentration curve, indicating the beginning of immune elimination and reaching a maximum at or shortly after the disappearance of free antigen and the appearance of free antibody in the circulation. In general the lesions had completely resolved within 14 to 21 days after the introduction of antigen.

Immune complex syndromes in man

Arthus-type reactions

Although haemorrhagic inflammatory Arthus-type reactions were occasionally seen when diphtheria and tetanus antitoxins were used for treatment or prophylaxis of diphtheria and tetanus respectively, they are now rarely encountered.

Serum sickness syndrome

In contrast to Arthus-type reactions which only occur in individuals who have been given a series of injections of serum, the serum sickness syndrome follows a single injection of serum, usually after an interval of seven to 14 days. In individuals who have at some previous time had a prior injection of serum the interval is appreciably shorter. The serum sickness syndrome is sharply distinguished from acute anaphylactic reactions due to homocytotropic antibodies which appear within a few minutes to a few hours after the administration of serum and which though relatively uncommon are much more dangerous since they may lead to the death of the patient in anaphylactic shock. Serum sickness is much less common following the use of refined modern antitoxins than after the administration of the crude serum preparations employed in the early era of passive immunization.

The principal features of the condition are pyrexia, glandular enlargement, particularly of the glands draining the injection site, pain and swelling of the joints and the development of an urticarial rash. Occasionally there is a transient nephritis and rarely symptoms of meningitis and carditis. Symptoms usually disappear in from three to five days after the injection of serum. Shortly after the appearance of symptoms precipitins for the antigen responsible are demonstrable in the patient's serum. The role of immune complexes in the aetiology of serum sickness is supported by the delay in the development of symptoms, the invariable development of precipitins and the general resemblances between the condition and experimentally produced immune complex disease in animals. The analogy with experimental disease in animals has been made particularly clear by the studies of Kendall on the effects of injecting bovine serum albumin into human subjects. In one subject who developed the serum sickness syndrome the onset of symptoms coincided as in animals with the immune inflexion in the curve of antigen concentration in the serum. In another patient in whom serum sickness did not occur there was no such evidence of immune elimination.

The serum sickness syndrome may also occur following the administration of various drugs, particularly penicillin. As in the case of anaphylaxis induced by penicillin sensitization it is believed to be due to a combination of degradation products of the drug with tissue protein. It has been suggested that allergy to drugs such as penicillin, which are not in themselves antigenic, may in fact be directed towards antigenic impurities present in the drug preparations. So far, however, there is no clear evidence that, although possible, this is ever the case.

Spontaneous immune complex disease in man

There is evidence suggesting that certain spontaneously occurring diseases in man may be due to immune complex formation. The following are of particular importance.

Post-streptococcal glomerulonephritis. Renal biopsy studies using the immunofluorescence technique have shown the presence in the glomeruli of irregular deposits of immunoglobulin and complement together with in certain cases the streptococcal type specific antigen M protein, usually type 12 (p. 215). The distribution of the complexes and the nature of the associated histological changes closely resemble those found in the kidneys of rabbits with the experimental serum sickness syndrome. Similar irregularly distributed immune complexes consisting in this case of DNA and antinuclear antibodies have been demonstrated in the glomeruli of patients with nephritis occurring in the course of the systemic lupus erythematosus syndrome.

Polyarteritis nodosa. This condition is characterized by necrotic and inflammatory lesions in the walls of small blood vessels resembling those seen in experimental animals with Arthus-type reactions. The evidence that polyarteritis nodosa is an immune complex disease is strengthened by the occasional occurrence of a similar syndrome in patients receiving penicillin, sulphonamides and various other drugs.

DELAYED-TYPE HYPERSENSITIVITY

As previously indicated the term delayed-type hypersensitivity is applied to responses whose immunological basis is the interaction of antigen with T cells of immune subjects. They are therefore reactions elicited in the course of cellular as opposed to humoral immune responses. This aetiology is reflected in the application to them of the term cell-mediated hypersensitivity. The skin reactions observed differ from the weal and flare reaction of anaphylactic type (p. 184) in the following characteristics:

1. In their slowness of onset. They take on average 24 to 48 hours to produce a macroscopically visible result.
2. The reactions are inflammatory in character, the dominant cells present being small lymphocytes and macrophages.

The elucidation of the mechanisms involved in these cutaneous delayed reactions has been considerably facilitated by the fact that they can be produced in experimental animals. These experiments have in particular defined the nature of the antigens responsible for production of delayed cutaneous reactivity. The antigens responsible are, somewhat surprisingly, considerably more restricted in chemical class than those capable of stimulating humoral antibody production. Thus delayed reactions can be stimulated by proteins, protein–hapten complexes, synthetic polypeptides and mycobacterial peptidoglycolipids but not by polysaccharides, e.g. the capsular polysaccharides of the pneumococcus and the lipopolysaccharides of the enterobacterioceae, which are non-thymus-dependent. Why certain antigens, e.g. polysaccharides, are apparently not capable of stimulating T cells is not clear. It has been suggested that it is associated with a repetitive determinant pattern in the antigen structure leading to an increased avidity of the antigen for the B cell which would allow B cell stimulation to occur in the absence of T cells. This does not, however, explain its failure to stimulate T cells unless the corresponding receptor pattern were absent from the latter. It might well be the case that in the course of evolution such cells have not emerged simply because they were not needed as helper cells in antibody production.

That cutaneous delayed-type hypersensitivity has an immunological basis is shown by its antigenic specificity, by its susceptibility to suppression by immunosuppressive agents and by the fact that it does not occur in animals which have been rendered immunologically tolerant to the homologous antigen. The reactivity is not, however, passively transmissible to normal animals by injection of serum from a sensitive donor but can be transmitted by the adoptive transfer of lymphocytes from such donors. That it is not due to humoral antibodies is shown by a variety of lines of evidence, the most important being that it does not occur in hypogammaglobulinaemic patients or in fowl that had previously been subjected to bursectomy (p. 90). The possibility has been entertained that it may be due to a cytophilic antibody which has the capacity for adsorption to macrophages since these antibodies undoubtedly occur and macrophages are important components of the cell population at sites of delayed reactions. There is, however, no real evidence to support this possibility. On the contrary, macrophages sensitized with macrophage cytophilic antibody have been found to be incapable of conferring cutaneous hypersensitivity. All the evidence points to the T lymphocytes as being the cells involved. Thus sensitization does not occur under conditions of T cell suppression or defect, e.g. following thymectomy, in cases of natural thymus deficiency, in animals as in nude mice, in man in the syndromes of reticular dysgenesis, the di George syndrome and ataxia telangiectasia or following the administration of antilymphocytic globulin. The identification of T cells as the operative cell is supported by the finding that in conditions leading to cutaneous hypersensitivity, there is considerable proliferation in the paracortical areas of the nodes draining the site of introduction of the antigen; in certain experimental syndromes the paracortical areas may be the sole site of proliferation in the nodes.

The T cells responsible for delayed-type cutaneous reaction are not necessarily identical with the T cells responsible for the helper effect in antibody production (p. 125) but so far no differentiation between helper T cells and the T cells involved in delayed cutaneous hypersensitivity has been possible. If there is such a difference it is likely that the T cells of delayed hypersensitivity are identical with cells which liberate lymphokines (p. 145) on exposure to antigen, since it seems probable that the inflammatory reaction elicited in the skin is in fact due to the activity of lymphokines. A provisional model for the production of the skin reaction envisages, as a first stage, contact of the antigen with specifically reacting small lymphocytes resulting in the liberation of lymphokines. Chemotactic lymphokines liberated in this way induce the accumulation of polymorphonuclear leucocytes and of macrophages at the reaction site. Mitogenic factor could induce small lymphocyte proliferation. This in turn might lead non-specifically to the release of lymph node permeability factor from non-antigen specific lymphocytes. This might in turn facilitate migration of inflammatory cells into the tissues by causing increased vascular permeability. The paucity of polymorphonuclear leucocytes in developed lesions, although these cells may be numerous in the early stages of the reaction, may be explained either by their local death or by their drainage from the reaction site by the lymphatics. Macrophages, on the other hand, remain at the site for a considerable period; this could be explained by the action of macrophage inhibition factor (MIF) which in in vitro migration inhibition experiments prevents their random movement possibly by affecting the surface viscosity of the macrophage.

Delayed hypersensitivity to microbial antigens developing during the course of infection

Delayed cutaneous reactivity as a sequel to prior infection can be demonstrated to a variety of microbial antigens. The paradigm of these reactions is the tuberculin reaction which is demonstrable in individuals and animals who have at one time been infected with tubercle bacilli. The reaction is elicited by a protein component of the bacilli known as tuberculoprotein or tuberculin, now available for testing in highly purified form as the purified protein derivative (p. 272). Although the altered tissue reactivity on subsequent contact with the tubercle bacillus which results from infection and the effect of intradermal injection of tubercle bacilli into tuberculous animals was described by Koch in 1891 (p. 271) the cutaneous reactivity to the intradermal inoculation of tuberculin was first used for the purpose of determining the incidence of tuberculosis in a human population by Von Pirquet in 1907. In Von Pirquet's investigations the tuberculin was introduced by a scratch technique. This was later replaced for the investigation of tuberculin hypersensitivity in man by the Mantoux technique in which graded doses of tuberculin are injected intradermally and later still by the Heaf multiple puncture technique.

Like other proteins tuberculin will not, if injected by itself, induce cutaneous sensitivity though it is capable of stimulating the production of antibody. The combination of tuberculin with the wax D component of the bacillus, however (p. 271), will induce skin reactivity. The cellular immune response produced in this manner is in some unexplained way insufficient to protect against infection with living tubercle bacilli. That tubercle bacilli possess some component which has a directive effect on the cellular immune response was first shown by Dienes in 1926. Dienes found that various proteins which stimulated antibody production but did not elicit cutaneous hypersensitivity could nevertheless do so if injected directly into a tuberculous focus. The identification of the substance responsible as part of the lipid component of the bacilli was later demonstrated by Raffel and its characterization as a peptidoglycolipid was due to the work of Lederer and of White.

The histological changes in the skin reaction sites have been extensively investigated, especially in guinea-pigs. In this animal perivascular accumulations of cells are apparent in four to six hours after intradermal injection of the antigen. At this stage the exudate consists of about 60% polymorphonuclear leucocytes and 20% each of lymphocytes and macrophages. Over the next two to three hours the proportion of polymorphonuclear leucocytes falls rapidly to about 25%, remaining constant at this level up to about 24 hours after injection. The proportion of lymphocytes present in the lesion progressively increases throughout the reaction, eventually constituting from 50 to 60% of the cellular content at 48 hours. Macrophages, the other important component of the exudate, reach a peak at about eight hours, when they constitute about 50% of the cells present, falling with increase in the small lymphocyte population to about 25% at 48 hours. It will be seen, therefore, that the skin reaction site exhibits a dynamically changing cell population. In man inadequate data are available for the early stages of the reaction. At 24 and 48 hours, however, about 90% of the cells present are small lymphocytes, the remainder being macrophages with a virtually complete absence of polymorphonuclear leucocytes.

Passive transfer

The passive transfer of tuberculin cutaneous hypersensitivity was first demonstrated by Chase by the intraperitoneal or intravenous injection of live peritoneal exudate cells from an infected guinea-pig as donor into a normal animal as recipient. After intravenous injection of the cells cutaneous hypersensitivity was demonstrable in 24 to 48 hours. Sensitivity persisted only for about seven days in the absence of further exposure to antigen but after repeated skin testing positive reactions were obtained for up to four months. Subsequently, following the development of the inbred 2 and 13 strains of guinea-pigs (p. 130), it was found that if donor and recipient belong to the same inbred strain, sensitivity persisted up to about 500 days. In this situation there is clearly colonization of the recipient's tissues by the donor's cells—a state known as adoptive transfer. Passive transfer was achieved in man by Lawrence by the intradermal injection of blood leucocytes of tuberculin-positive donors. Positive reactions could be elicited in the recipients by challenge at the injected site at about 18 hours. A generalized cutaneous sensitivity was not, however, demonstrable until about six days after the initial cell transfer. It was found to persist for two to three months.

Transfer factor

The transfer of tuberculin sensitivity in man by the intradermal injection of extracts of leucocytes obtained from sensitive donors prepared either by osmotic lysis or by freezing and thawing was reported by Lawrence in 1954. Cutaneous sensitivity appeared in four to 18 hours after injection and was found to be capable of persisting for up to two years. In addition serial transfer of cutaneous reactivity was possible by the same technique. Thus, following injection of cell extracts of donor A into recipient B, extracts of cells from recipient B were capable of inducing sensitivity in recipient C. The hypersensitivity transferred was immunologically specific. Thus in addition to tuberculin sensitivity, specific sensitivity to diphtheria toxoid, coccidioidin and streptococcal antigen and tissue compatibility antigens could also be transferred by injection of the appropriate cells, the cutaneous reactivity of the recipient being specific for the antigen to which the donor had been sensitized. Transfer factor is species-specific, human transfer factor being incapable of transferring sensitivity between man and animals. Similar factors have not so far been described in species other than man.

The transfer factor present in leucocyte extracts has been found to be resistant to DNAse and RNAse and trypsin and to be heat-labile, being inactivated by heating at 56°C for 30 minutes; it is excluded by Sephadex G25 and therefore has a molecular weight of less than 10 000 and is non-antigenic and fails to react with immunoglobulin antisera. According to the results of Lawrence it has both polynucleotide and polypeptide as components. In addition to transferring cutaneous hypersensitivity in vivo transfer factor will sensitize lymphocytes in vitro so that they will respond by DNA synthesis as indicated by thymidine uptake on response to stimulation by the homologous antigen. Like in vivo cutaneous hypersensitivity the in vitro sensitization is antigen specific.

Transfer factor and/or some similar substance is also liberated from leucocytes following stimulation and exposure to the homologous antigen. Antigen-liberated transfer factor has been found to be much more effective than the dialysable transfer factor of cell extracts in in vitro sensitization of leucocytes. Thus the dialysable transfer factor can sensitize at most about 3% of non sensitive lymphocytes both in vitro and in vivo whereas the material liberated from leucocytes after exposure to antigen will sensitize some 15 to 20% of lymphocytes. The simplest explanation for these differences is that the material from lymphocytes exposed to antigen contains mitogenic factor (Mit) or some similar factor which potentiates the stimulation of lymphocytes initially transformed by the dialysable factor.

Although Lawrence's results are fully reproducible, there is at the moment no explanation of the mechanism of action of the transfer factor. The principal possibilities are:

1. The transfer factor contains an informational macromolecule which is incorporated into the recipient's cells. Lawrence does not consider this a likely hypothesis since the polynucleotide has the physical characteristic of a double-stranded RNA rather than DNA and the amount present in dialysed preparations would appear to be insufficient to code for the differences in specificity required.

2. The second possibility which cannot be excluded is that the transfer factor preparations contain the antigen in the form in which it finally stimulates T cells. In this case, however, the antigenic fragment would require to be of quite remarkable potency to explain the serial in vivo transferability of sensitivity.

Other microbial antigens

Delayed-type cutaneous reactivity has been demonstrated to a variety of other microbial antigens. Of particular importance in this context are the brucellae, leprosy bacilli and chlamydiae, notably lymphogranuloma venereum. Delayed skin reactivity has also been reported to streptococcal and staphylococcal antigens and to diphtheria toxoid. The delayed hypersensitivity to diphtheria toxoid is in fact responsible for the pseudoreaction in the Schick test (p. 231). Delayed hypersensitivity frequently develops in fungal infections particularly in fungal infections which involve the skin. Delayed-type reactions have been demonstrated in many virus infections, the so-called reaction of immunity which develops to vaccination against smallpox in immune persons being an outstanding example. It has also been suggested that delayed cutaneous hypersensitivity to viral antigens is the basis of the skin rashes found in these conditions, outstanding examples being the rashes of measles and rubella. It is not surprising that many of these reactions have been made the basis of skin tests used for diagnostic and epidemiological purposes.

Contact sensitivity

This denotes a hypersensitivity that develops from the contact of potentially sensitizing material with the skin. Spontaneous hypersensitivity of this type is quite common in man, when it is known as contact dermatitis. It can be produced as a result of exposure of the skin to a variety of substances: sulphonamides, organic arsenicals, antibiotics, formalin, paraphenylene-diamine (a substance used in the dyeing of furs), active agents present in poison ivy, primula and other plants, laundry marking-ink derived from the nut of the ral tree, picric acid and various ions including mercury, iodine, nickel, cobalt, zirconium and beryllium.

In contact sensitivity it may be possible to

identify the substance responsible for the sensitivity by patch tests in which the suspected substance is applied to the skin. In such cases positive reactions may be obtained not only from the area initially sensitized, but from any part of the skin surface.

Contact sensitivity can be readily produced in the guinea-pig and this model has been extremely useful in the study of delayed-type hypersensitivity in general. Most work has been carried out using picryl chloride and nitro- and halo-substituted benzenes, e.g. 2,4-dinitrochlorobenzene and 2,4-dinitrofluorobenzene. These compounds have been found to form complexes readily with the amino and sulphydryl groups of proteins. Some substances which are incapable of combining with protein themselves, however, give rise to degradation products which can combine with protein. Thus it is probable that penicillin hypersensitivity of anaphylactic as well as of delayed type is due to the formation of complexes of protein with benzyl penicilloic and benzyl penicillenic acid. As in the human subject, positive reactions may be elicited from any part of the skin surface. In spite of this, the skin does not itself appear to be sensitized since a skin graft from a sensitive animal to a normal non-sensitive animal does not retain the sensitivity. A skin graft from a non-sensitive animal to a sensitive animal, however, does acquire sensitivity. It is clear that the sensitivity is due to the migration of sensitized cells or, possibly, of some soluble substances capable of effecting sensitization into the skin.

THE SHWARTZMAN REACTIONS

These are reactions which may be produced in various animals, but most readily in rabbits, by the endotoxins of a variety of Gram-negative bacteria and by certain other macromolecular substances. Although they do not appear to have an immunological basis they exhibit a superficial resemblance to certain hypersensitivity states. Two types of reaction have been described—local and general. In the local reaction a small amount of endotoxin is injected into the abdominal skin of the rabbit and then after 24 hours a large dose of endotoxin is injected intravenously. Within a few hours after the second injection the prepared area of skin shows the development of petechial haemorrhages which coalesce to form a large circular haemorrhagic area. In the production of the generalized reaction both injections are given intravenously. In this case in from 12 to 24 hours after the second injection many of the injected animals die exhibiting at post mortem various haemorrhagic lesions, of which the most prominent is a bilateral haemorrhagic cortical necrosis of the kidneys. There is no evidence that Shwartzman-type reactions occur during the course of any bacterial infections but it has been suggested that certain of the biological effects associated with the injection of the endotoxins of the Gram-negative bacteria may in fact be due to reactions of this type occurring in naturally sensitized animals.

The mechanism of the Shwartzman reactions is still in dispute. Recent evidence suggests that in the generalized reaction the first injection causes an alteration in fibrinogen so that it becomes precipitable by a substance or substances released from leucocytes by the second injection and that the precipitation of the altered fibrinogen in the small blood vessels is the immediate cause of the vascular damage observed.

Part III

SYSTEMATIC BACTERIOLOGY

III.1

The Classification of Bacteria

In Chapter I.1 bacteria were subdivided on purely morphological grounds into a number of large groups. Such a classification is, however, inadequate for practical purposes for which a greater degree of subdivision is needed. There is general agreement that bacteria should be classified as far as possible in the same way as plants and animals. On this basis each distinct kind of bacterium is designated as a species and species with a number of common characters are grouped into genera.

Each species receives two Latin or Latinized names, e.g. *Staphylococcus aureus,* both of which should be italicized. The first name is the genus name; it is a noun in the singular and is always written with a capital initial letter. The second name is the species name; it is usually either a noun in the genitive qualifying the genus name, e.g. *Salmonella typhi,* or an adjective agreeing with the latter in gender, e.g. *Proteus vulgaris.* The species name is always written with a small initial letter.

The genus name may be used in the plural for descriptive purposes; in this case it is written with a small initial letter and is not italicized. We may for example describe the properties of the clostridia or of the neisseriae.

For practical purposes many species are subdivided into *types*; these are not regarded as being sufficiently distinct from each other as to warrant being given species rank. Subdivision into types is usually based on antigenic differences but occasionally other criteria are employed (Chapter I.2). The term *group* is also used to indicate a subdivision of a species, but loosely and without any generally agreed application. It is usually employed for a number of types with sufficient resemblance to each other to justify their being grouped together. The ultimate unit in descriptive bacteriology is the *strain.* The term is applied to all the descendants derived by single colony isolation from a single source.

The application to bacteria of a system of classification similar to that employed for plants and animals, however, raises two major problems:

1. The concept of species is based on sexuality, a species being a group of organisms that regularly interbreeds. This concept cannot, however, be applied directly to bacterial taxonomy since bacteria are normally asexual and sexual equivalents in bacteria are of too restricted a distribution to constitute a basis for a comprehensive classification. Consequently species in bacteria must be defined on the basis of other criteria. For this purpose a variety of properties is used, i.e. structure (at both the anatomical and the antigenic level), staining reactions, cultural characteristics and biochemical capabilities. In the application of these criteria to taxonomic definition two approaches are possible. (*a*) Certain characters may be regarded as being of particular importance, e.g. coagulase production for the definition of *Staph. aureus* or the production of the Lancefield Group A hapten for the definition of *Str. pyogenes.* In this way species are defined on the basis of a limited number of 'important' properties. This is the conventional and traditional approach and for the day-to-day identification of individual bacteria the only feasible one. Its main disadvantage is that the importance of particular characters may be determined by a variety of adventitious factors—the techniques available, the fashions of the day and the particular interests of the taxonomist. The selection of differential criteria must therefore, to some considerable degree, be arbitrary. (*b*) The second approach involves the

abandonment of the weighting of characteristics and the acceptance of all characteristics as of equal importance. This is the basis of modern numerical or neo-Adansonian taxonomy. The data obtained are fed into a computer which provides association indices, on the basis of which taxonomic groups or 'clusters' of individual organisms may be defined.

Although sexual equivalents do not provide a practicable basis for bacterial classification, the concept of sexuality has in recent years been reintroduced at a biochemical level in the determination of DNA base composition. Wide differences in base composition, as shown by guanine-plus-cytosine content, have been found amongst bacteria. So far, data available for several hundred bacterial species have shown on the whole very substantial consistency within individual species, which are clearly defined by current procedures. Thus strains of *Staph. aureus* have a guanine-plus-cytosine content of 30 to 40%, while tubercle bacilli have a guanine-plus-cytosine content of 60 to 70%. Such findings clearly indicate that DNA base composition must in future be taken into account in species definition.

A more sophisticated method for determining strain relationships is by the technique of DNA–DNA hybridization. The essence of the technique is to mix heat-denatured fragments of single-stranded DNA for two different organisms under conditions which will allow pairing to take place. If one of the strains has been grown in a medium containing tritiated thymidine a calculation of the amount of radioactivity present in the reassociated hybrid molecules will indicate the extent to which hybridization has occurred and this will be a measure of the extent to which there are similar or homologous sequences in both strains. The extent of hybridization that can occur between the DNA and RNA of different bacteria can be determined by a somewhat similar method. Although an immense amount of information has been obtained by the use of hybridization techniques, in the great majority of cases confirming existing associations, the techniques are too complex for purposes other than research.

2. The second problem is that of classifying species and genera into higher groups in a hierarchical arrangement of families, tribes, orders and classes. This has been the classical approach of earlier editions of *Bergey's Manual of Determinative Bacteriology,* which has presented the

Table III.1. Groups and genera of medical importance

Gram-positive cocci	Gram-negative anaerobic bacteria	Spiral and curved bacteria
Staphylococcus	*Bacterioides*	*Spirillum*
Streptococcus	*Fusobacterium*	*Bdellovibrio*
Pneumococcus		
Peptostreptococcus	Gram-negative facultatively anaerobic bacteria	Spirochaetes
		Treponema
Gram-negative cocci	Enterobacteriaceae	*Borrellia*
Neisseria	*Escherichia*	*Leptospira*
	Edwardsiella	
Gram-negative aerobic rods	*Citrobacter*	Mycoplasmas
Pseudomonas	*Salmonella*	*Mycoplasma*
Bordetella	*Shigella*	
Francisella	*Klebsiella*	Actinomycetes and related organisms
Brucella	*Enterobacter*	*Cornyebacterium*
	Hafnia	*Actinomyces*
Gram-positive endospore-forming rods	*Proteus*	*Nocardia*
Bacillus	*Yersinia*	*Mycobacterium*
Clostridium	Vibrionaceae	
	Vibrio	Rickettsias
Gram-positive non-spore-forming rods	Unassigned	*Rickettsia*
Lactobacillus	*Haemophilus*	*Rochlimaea*
Listeria	*Pasteurella*	*Coxiella*
Erysipelothrix	*Streptobacillus*	*Chlamydia*
	Calymmotobacterium	*Bartonella*

system of classification most widely adopted by bacteriologists. The ultimate object of such an arrangement is to attempt to define evolutionary or genetic relationships between bacteria. Most bacterial taxonomists, however, now feel that the drawing up of such a rigid hierarchical system is not justified, since it must at best be speculative and is therefore largely a waste of time. They consider instead that the object of classification should be to present a determinative key which will facilitate the identification of individual bacteria. This is the point of view taken in Cowan and Steel's *Manual for the Identification of Medical Bacteria.* Bergey's recent eighth edition adopts a compromise position in which groups of bacteria defined by a few readily determined criteria are considered in 19 separate parts, each of which is then to a greater or lesser extent subdivided in a hierarchical fashion. The groups, as defined by Bergey, of medical importance are shown in Table III.1.

III.2

The Staphylococci

Staphylococci are Gram-positive, non-capsulated, non-sporing cocci which occur characteristically in irregular grape-like clusters. This arrangement is due to the way in which the organisms divide, successive divisions tending to occur in different planes. Their average size is from 0·7 to 1 μm. Staphylococci are facultative anaerobes but grow best under aerobic conditions.

Staphylococcus aureus (Staphylococcus pyogenes)

After overnight incubation *Staph. aureus* produces a colony about 2 mm in diameter, with a smooth, highly refractile surface and an entire edge.

The colonies of most strains have a characteristic golden-yellow colour due to the carotenoids—δ carotene and rubizanthine. On blood agar the colonies are usually surrounded by a narrow zone of haemolysis. *Staph. aureus* is highly tolerant of salt and media containing a high concentration of salt are therefore of value for isolation from contaminated sources.

Most stains produce gelatinase and ferment a variety of sugars—of which mannitol is of some differential importance—to give acid but no gas. All strains are strongly catalase-positive. About 95% of strains produce lysozyme to which they themselves are insensitive.

Antigenic structure

A species-specific antigen demonstrable by precipitation tests has been identified in the cell wall, from which it can be readily extracted. The antigen appears to be a ribitol teichoic acid which owes its serological specificity to N-acetyl glucosamine residues. A non-species-specific red cell sensitizing antigen is demonstrable in cell extracts and culture supernatants. A similar antigen which is produced by a variety of Gram-positive organisms is glycerol teichoic acid.

Using slide agglutination tests with living 5-hour cultures Oeding has been able to distinguish at least 26 antigenic components. These antigens occur in different combinations in different strains and on the basis of their distribution it is possible to define a large number of agglutination types. The main limitation of this method as an epidemiological tool is that the majority of strains normally encountered appear to belong to a limited number of the possible types.

At least 90% of strains produce a low-molecular-weight basic protein—the A protein; this is present on the surface of the cell, is antiphagocytic and functions as an agglutinogen. It has the unusual property of combining non-specifically with the Fc region of IgG, the mixture being capable of fixing complement.

Bacteriophage types

On the basis of their susceptibility to a series of symbiotic bacteriophages *Staph. aureus* strains may be classified into a number of types (p. 473). Most strains are susceptible to lysis by more than one typing phage. In such cases the type of the strain is defined by its pattern of susceptibility to different phages. With the currently standard set of 24 typing phages, several hundred bacteriophage types or patterns can be distinguished. The majority of strains isolated from human sources are susceptible to one or other of the typing phages, but a minority are untypable. The types fall into four broad groups—I to IV—the strains within each group being as a rule susceptible only to phages active within that group.

As for *Salm. typhi* (p. 242), the bacteriophage type of a particular strain may be modified by

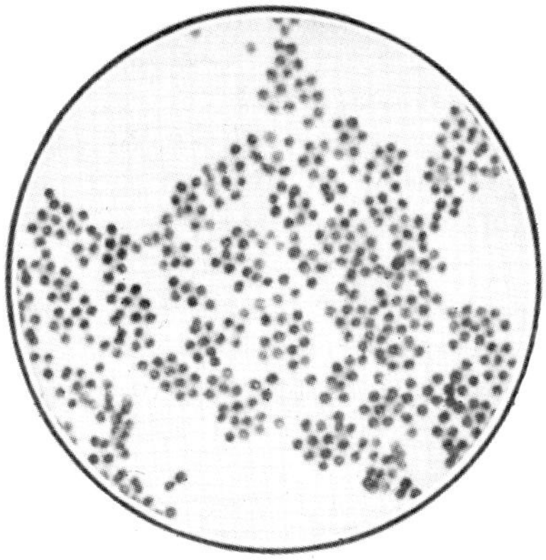

Fig. III.1. *Staphylococcus aureus* from an agar culture. ×950.

infecting it with a different symbiotic bacteriophage. This transformation is believed to have occurred under natural conditions and to have been responsible for the emergence of the 80, 81, 52, 52A type. As most, if not all, staphylococci are lysogenic, it would seem possible that the type-specificity of lysogenic strains might depend on the nature of the symbiotic phage they carry.

Phage typing has been of particular value in epidemiological investigations, especially in the investigation of sources of infection in outbreaks of bacterial food poisoning and in tracing the spread of staphylococcal cross-infection in hospital. Because of the larger number of types it reveals—several hundreds have now been encountered—phage typing is potentially a more useful epidemiological tool than serological typing.

Pathogenicity

At least four antigenically distinct haemolysins, α, β, γ and δ, are found in cell-free filtrates. The α, γ and δ haemolysins are produced by the great majority of strains irrespective of source. The β haemolysin is produced mainly by strains from animal sources.

Of the four the α haemolysin is undoubtedly the most important in pathogenicity. It can be transferred to non-toxigenic strains by lysogenization. Its haemolytic range is narrow, being lytic for rabbit, to a lesser extent for sheep and only slightly for human red cells, but it is toxic for a variety of cells in tissue culture. It has a rapidly lethal effect on experimental animals following intravenous inoculation and produces necrosis of the skin on intradermal injection.

The toxin has been obtained in a highly purified form as a protein with a molecular weight of about 44 000. It shows a peculiar behaviour towards heat, crude preparations being inactivated rapidly at 60°C. Their activity is to some extent restored if they are subsequently heated at temperatures between 80 and 100°C. This paradoxical behaviour has been attributed to combination of the toxin with a heat-labile inhibitor at 60°C, at higher temperatures the inhibitor being destroyed and the toxin released. On further heating, however, the toxin is inactivated.

The β haemolysin is lytic for sheep but not for human or rabbit red cells. The haemolytic process, initiated by incubation of cells with the haemolysin at 37°C, may also be completed by exposure to a variety of substances—glycerol, low concentrations of δ lysin, supernatants of cultures of *Str. agalactiae* (p. 216) and of various other organisms. The β lysin has not been obtained in purified form but appears to be a phospholipase, hydrolysing various sphingomyelins, the distribution of which would explain the specificity of the lysin for the red cells of different animal species. It is of doubtful toxicity for experimental animals and tissue cultures and is unlikely to play any role in pathogenicity.

The γ haemolysin is lytic for red cells from a variety of species. It is lethal for rabbits and produces oedematous skin lesions in rabbits and guinea-pigs. It can be transferred to non-toxigenic strains by lysogenization.

The δ lysin is lytic for sheep, guinea-pig, human and rabbit cells. It has been isolated as a crystalline protein with a molecular weight of about 70 000. Like the β haemolysin, it is much less toxic for experimental animals than is the α lysin. It can release phosphorus from various glycerophosphatides and is particularly active on phosphatidylinositol.

Staphylococcal filtrates have considerable toxic action on human white cells. The substance responsible, known as *leucocidin,* is a complex of two serologically distinct components. These components will adsorb to red cells treated with tannic acid. Antibody to each of them can consequently be estimated independently by passive haemagglutination tests. Leucocidin appears to have a direct action on the leucocyte surface and, following entry into the cell, on the lysosomal membranes.

Some strains of *Staph. aureus* produce an enterotoxin which is responsible for a variety of bacterial food poisoning. The enterotoxin is heat-stable and can withstand boiling for 30 minutes. Consequently it may be present in food which has been heated sufficiently to kill all the staphylococci present. The existence of toxin was first recognized by feeding experiments in human volunteers and in monkeys and unfortunately these are still the only satisfactory test animals. Its effect appears to be due to a direct action on the central nervous system. The enterotoxin has been purified by Dack, who found it to be a trypsin-resistant protein. Five distinctive antigenic types of enterotoxin (A, B, C, D and E, of which A is the most important) have been identified. The toxins may be demonstrated by the Outcherlony agar gel diffusion technique, using antisera against purified toxins. A reversed passive haemagglutination test is reported to be very sensitive.

All strains of *Staph. aureus* have the property of clotting human and rabbit plasma by inducing the transformation of fibrinogen into fibrin. This is due to the secretion by the staphylococcus of a substance known as coagulase. Coagulase, however, is not itself the active agent of clotting but reacts with a second substance, variously referred to as *coagulase-reacting factor* or *coagulase activator,* present in plasma and in tissue fluids, to produce the active agent; the latter has properties very similar to those of thrombin, both splitting off identical peptides from fibrinogen and both being inhibited by the esterase inhibitor di-isopropyl-fluorophosphate (DFP) and by heparin. Because of this resemblance it is referred to as coagulase–thrombin.

Coagulase is antigenic; immune sera can be prepared against it in animals, and antibodies to it are demonstrable in the sera of patients suffering from chronic staphylococcal infections. Seven distinct antigenic varieties of coagulase have now been described. The majority of human strains of *Staph. aureus* produce coagulase A.

Coagulase is an extremely important component of the offensive armoury of *Staph. aureus.* By laying down a fibrin barrier which is impermeable to leucocytes in the immediate vicinity and on the surface of the organism, it gives the latter considerable protection from phagocytosis. Since under in vitro conditions coagulase is produced in the lag phase and in the early log phase, it can be presumed to protect the organism from phagocytosis at a time when such protection is most needed, i.e. during the early stages of infection. In addition there is evidence that coagulase protects the staphylococcus against bactericidal substances present in the tissue fluids. How this is achieved is not known.

For the demonstration of coagulase 0·1 ml of an overnight broth culture of the organism and 0·9 ml of a 1 in 10 dilution of human or rabbit plasma are pipetted into a Kahn tube. The mixture is incubated at 37°C and the result read after two hours. If the organism produces coagulase, the plasma will form a clot which is usually sufficiently firm to remain adherent to the sides of the tube when the latter is inverted. It is important that incubation should not be prolonged for more than two hours, since on prolonged incubation some organisms are capable of utilizing the citrate employed as anticoagulant and, as a result, spontaneous clotting of the plasma may occur.

When staphylococci are mixed on a slide with the plasma of certain animal species the staphylococci are rapidly clumped or agglutinated. It was first assumed that this agglutination was due to coagulase activity. It was shown by Duthie, however, that this was not the case, the clumping being due to the combination of fibrinogen in the plasma with a receptor, known in this context as *clumping factor,* present on the surface of the organism.

Staphylococcus aureus also shows some fibrinolytic activity, but not as actively as *Str. pyogenes.* Fibrinolysin is produced by the action of a staphylococcal product, *staphylokinase,* on plasminogen—a protease precursor which is a normal component of plasma. Staphylokinase is produced only during the later stages of growth and its contribution to staphylococcal pathogenicity is therefore in doubt. It may, however, play a part in the production of staphylococcal septicaemia, by dislodging infected intravascular thrombi. Most strains of *Staph. aureus* also produce hyaluronidase and desoxyribonuclease.

Staphylococci have been shown to produce a number of lipolytic enzymes at least one of which, described by Alder and Gillespie, can produce opacity in egg yolk media and may have some relation to pathogenicity. Its substrate appears to be the lipoprotein, *lipovitellin*. This enzyme is produced by most strains isolated from boils and carbuncles and is thought to play a part in the invasion of staphylococci through the intact skin.

A toxic complex capable of impairing mitochondrial respiration has been described in 70% of strains of *Staph. aureus* by Lominski et al. It is known as the succinic oxidase factor, since it has a particularly marked effect on the oxidation of succinate. Since it is active on lipids extracted from mitochondria it is thought to act by disorganizing the electron transport chain. It probably does not pass through intact cell membranes but may pass through membranes which have been damaged by the α toxin.

Staphylococcal infections

The nose is the natural home of *Staph. aureus* and the organism can be isolated from the anterior nares of about 50% of normal persons. In many of these the carrier state is of short duration, but in some it is more persistent. From 5 to 10% of people are skin carriers. Staphylococci may be isolated from the faeces of about 20% of normal persons. The perineum, particularly in infants, is also an important carrier site.

Staphylococcus aureus has a marked predilection for the skin. It produces pustules, boils and carbuncles—the organisms responsible in many cases being derived from the nose—and is a frequent cause of sepsis in wounds and burns. About 10% of people with cutaneous staphylococcal lesions are subject to recurrent infections, which are particularly frequent in persons infected with so-called hospital staphylococci (p. 208). The capacity to produce boils and other cutaneous staphylococcal infections is apparently not possessed equally by all strains of *Staph. aureus.* Many strains capable of causing hospital wound sepsis do not, appparently, cause boils. *Staph. aureus* appears to be the causative organism of most cases of impetigo contagiosa, a condition which may also be caused by *Str. pyogenes* (p. 215). Most of the strains of *Staph. aureus* isolated from cases of impetigo belong to a single bacteriophage type, type 71, and are strong producers of hyaluronidase. The latter property may explain the epidermal character of the lesion in impetigo, which contrasts with the subepidermal type of infection found in boils and carbuncles. A closely similar condition in babies, pemphigus neonatorum, is also caused by the staphylococcus.

Recently outbreaks of staphylococcal toxic epidermal necrolysis (TEN), with extensive splitting of the epidermis and the production of a lesion resembling that produced by scalding, have been reported. The effect is due to a newly described toxic product for which the name exfoliation has been proposed.

Staphylococcus aureus is the causative agent in about 90% of cases of acute osteomyelitis. This condition, which occurs mostly in children, is the result of haematogenous spread from a superficial site of infection such as a boil or furuncle. Staphylococcal osteomyelitis may also occur as a result of infection of a compound fracture.

Staphylococcus aureus is responsible for a relatively uncommon but severe type of pneumonia. In the adult it is usually secondary to some other infection of the respiratory tract and is particularly important as a complication of influenza. The pulmonary lesions in staphylococcal pneumonia are of bronchopneumonic distribution and frequently progress to abscess formation.

Reference has already been made to the severe enterocolitis which may complicate treatment with antibiotics, especially the tetracyclines, which are highly active against the normal Gram-negative flora of the intestinal tract. This condition is most frequently caused by *Staph. aureus.* As a rule it occurs sporadically but epidemics have been described. It is thought that only certain strains of staphylococci have the capacity to produce enterocolitis.

Staphylococcus aureus is an important cause of bacterial food poisoning. This condition is an intoxication and not an infection. It is due to the ingestion of enterotoxin formed as a result of multiplication of the organisms in food. The foods responsible are usually made up meat dishes and foods containing a high concentration of milk or cream. Food poisoning is most likely to occur if the food has been held under warm conditions for some time prior to eating, thereby permitting bacterial multiplication. The food is usually infected in the course of preparation by a nasal carrier or by an individual with a superficial staphylococcal lesion, e.g. furuncle, boil or impetigo.

Staphylococcal food poisoning is usually rapid in onset, the symptoms, the most prominent of which are vomiting and abdominal pain, appearing within four to six hours of ingestion of the food. It is of short duration, usually resolving in a further 24 hours, has a negligible mortality and is a much less serious condition than salmonella food poisoning. The lack of fever and the prominence of vomiting as a symptom help to differentiate it clinically from the latter condition.

Hospital infections

In recent years considerable attention has been directed to the problem of the infection, in hospital, of clean surgical wounds by pyogenic organisms. *Staph. aureus* is the single most important cause of such infections. Thus in a survey of wound sepsis carried out by the Public Health Laboratory Service staphylococci were isolated alone in 45% of cases and in association with coliform bacteria in a further 15%. There is a widespread belief that these infections have increased in frequency. Reliable evidence on this point has, however, been difficult to obtain, since the incidence of sepsis is not only different in different hospitals but it is also different in different types of operation. As a rule wound infection merely results in a delay in healing, though this, by prolonging the patient's stay in hospital is of considerable economic importance. Occasionally, however, it may be the starting point of septicaemia which not infrequently leads to the death of the patient.

Though in some cases wound infection is due to a strain of *Staphylococcus* which the patient was carrying on his admission to hospital, it is much more frequently due to organisms which he has acquired from the hospital environment—the so-called hospital staphylococci. The application of the phage-typing technique has shown that the majority of strains of staphylococci responsible for hospital infections belong to a quite limited number of bacteriophage types. This finding has created the concept of epidemic or hospital strains of enhanced virulence and infectivity. The existence of such strains—although earlier suggested by paediatricians as being involved in neonatal infections—was first clearly demonstrated by the identification of the notorious phage type 80 as being responsible for a high proportion of epidemics of hospital sepsis in Australia in 1953 to 1954 and, shortly afterwards, as an important cause of hospital cross-infection in other parts of the world. In 1957 strains giving the pattern 52, 52A, 80, 81 started to appear. This pattern was shown to be the result of lysogenization, with type determining phages, of 80/81 strains. The phages 52, 52A, 80 and 81 appear in fact to determine a closely interrelated complex of strains which may give all possible combinations of reactions with these typing phages. From 1957 onwards group III strains of type 83A, which is quite unrelated to the important 80/81 complex, appeared as important causes of epidemics.

It is not known which properties of epidemic staphylococci are ultimately responsible for their high infectivity and their capacity to persist in the hospital environment. There is some evidence that they may have a higher degree of viability outside the body than non-epidemic strains. There is, in addition, evidence that they not only possess a higher infectivity than non-epidemic strains but also are more pathogenic. Thus Barber found that highly drug-resistant strains more frequently give rise to generalized, as opposed to localized, infections than drug-sensitive strains. Moreover, carriers of epidemic strains more frequently develop infections caused by the carried strains than do carriers of non-epidemic strains.

Although they may be isolated from floors, hangings and other inanimate objects, the great reservoirs of staphylococci in the hospital environment are patients and members of the hospital staff who are carriers or who are suffering from staphylococcal infections. As sources of staphylococcal infection *people* are much more important than *things* and patients suffering from overt infections—of which the most dangerous in this context are exposed skin lesions, enterocolitis and pneumonia, from which the environment is readily contaminated—are much more important than carriers.

Occasionally a hospital experiences a sharp increase in the incidence of staphyloccal wound infection. The more serious epidemics appear in most cases to be produced by strains of staphylococci, e.g. the notorious phage type 80, with an exceptional capacity to spread and to produce superficial cutaneous lesions. Epidemics may on occasion be due to a breakdown at some point in the sterilizing, operative or wound dressing procedures or to a nurse or surgeon who has a superficial staphylococcal lesion or who is a persistent nasal carrier.

A high incidence of staphylococcal infection has also been reported in newborn infants in maternity wards. Cutaneous lesions and blepharitis are of frequent occurrence and staphylococcal pneumonia is not uncommon. Infants are particularly susceptible to colonization with *Staph. aureus* and within a short period after birth from 80 to 100% of those born in hospital become first skin and then nasal carriers. Other frequent sites of colonization are the umbilical stump and the perineum. Newborn infants frequently transmit the staphylococci they acquire in hospital to their mothers during breast feeding. This often results in the production in the mother of a staphylococcal mastitis which may not become clinically apparent until after she leaves hospital.

Diagnosis

Colonies of *Staph. aureus* are usually readily recognizable on blood agar, although occasional strains of the organism are encountered which do not produce the characteristic golden pigment. For normal diagnostic purposes the identity of the organism is confirmed by coagulase tests.

Small colony (G) variants may from time to time be isolated from clinical material. They appear to be fairly readily selected on media containing aminoglycoside antibiotics. Small colony variants have also been described which show a high degree of resistance to methicillin. It is thought that the genetic defect in the production of these variants involves a permeability block. Since many of the variants described have been resistant to linoleic acid it has been postulated that they are more suited than the parent staphylococci to survival on the skin. The clinical significance of these variants has, however, to be fully assessed.

Capsulated strains which produce diffuse mucoid colonies are occasionally isolated from clinical material. They have been estimated to represent about 4% of all isolates. Though coagulase-positive they are insusceptible to clumping by fibrinogen, presumably because the capsule obscures the fibrinogen-combining site on the cell wall. These strains are also insusceptible to typing phages and are inagglutinable by typing sera. Non-capsulated variants derived from them have been found to be much less virulent for mice than the parent strains. A high proportion of capsulated strains are non-haemolytic.

In recent years considerable interest has been taken in the possibility of using serological tests for the diagnosis of staphylococcal infections. In most cases serological tests are quite unnecessary, but in certain occult infections, e.g. osteomyelitis of the spine and hip, a reliable serological test would be of considerable value. Relatively simple techniques have been developed for the estimation of antibodies to α haemolysin, leucocidin, coagulases and staphylokinase. Of these, the tests for antileucocidin—carried out by the haemagglutination technique—appears to be the most generally useful.

Treatment

Of all bacterial species *Staph. aureus* has shown the greatest tendency to develop resistance to antibiotics to which, when first introduced, the species as a whole appeared uniformly sensitive. Because of this, the choice of antibiotic in any particular case should, where possible, be based on the results of sensitivity tests. Antibiotic-resistant strains of staphylococci are much more numerous in the hospital environment than in the general population. Nevertheless, the number of resistant strains isolated from sources outside hospital has, though to a much lesser extent, increased as well. Systemic antibiotic therapy should not be used for mild cutaneous staphylococcal infections which can be effectively treated by topical applications of various antimicrobials which are not suitable for systemic administration, e.g. bacitracin, neomycin, chlorhexidine, framycetin and gramicidin, given singly or in various combinations.

Penicillinase plasmids

Naturally occurring penicillin-resistant strains of *Staph. aureus* owe their resistance to the production of penicillinase (p. 71). Staphylococcal penicillinase is inducible and is excreted by the organism into the medium. The amount excreted, however, varies considerably in different strains. Three antigenic varieties of penicillinase have been distinguished: A, B and C. Type B only occurs in phage group II strains, and types A and C in strains of groups I and III. The type B enzyme is much less active and has a lower substrate affinity than types A and C. This may explain the lower frequency as causes of human infections of phage group II strains compared with strains of groups I and III.

Penicillinase production is controlled by extra-chromosomal genetic determinants known as *plasmids* (p. 20). These resemble R factors but are not spontaneously transmissible. They may, however, be transferred from one strain to another by transduction, for which phage 80 has been most frequently used. The plasmids may be spontaneously lost and in some cases the rate of loss is higher than 1 in 1000 cell divisions—a frequency much greater than that of spontaneous mutations. Loss of plasmid is accelerated by exposure of the cell to acridine dyes. In addition to penicillinase production, penicillinase plasmids have been found to carry determinants for other bacterial characters, viz. resistance to mercuric chloride and to erythromycin, and a colonial morphology marker. They may be divided into two groups, *pla* I and *pla* II, depending on their compatibility in the same cell. On transduction of a plasmid of one compatibility group into a cell already possessing a plasmid of another compatibility group, both plasmids can coexist in the cell as separate entities and each may be lost from the cell independently. This coexistence is referred to as the *polyplasmid* state. If, however, a plasmid of one compatibility group is transduced into a strain already harbouring a member of this group, the polyplasmid state is not produced but recombination between the plasmids may occur. On the basis of their compatibility group, the antigenic type of penicillinase produced, and the non-penicillinase markers they carry, a variety of types of plasmid have so far been distinguished. The properties of the first six plasmids to have been described are shown in Table II.2.

Many workers have reported the isolation of penicillinase-producing staphylococci which are also resistant to methicillin. Methicillin-resistant staphylococci are also resistant to the isoxazolyl penicillins and the cephalosporins. This resistance is not due to the production of a modified penicillinase and is not, as are the penicillin plasmids, transducible. Its genetic basis is unknown. Methicillin-resistant cultures show marked heterogeneity in the resistance of individual clones. Consequently in the ordinary type of disc test, particularly with discs containing low concentrations of methicillin, cultures with a considerable number of highly resistant organisms could be classed as sensitive. Phenotypic expression of resistance is greatly increased in cultures grown at 30°C or in cultures grown at 37°C in the presence of a variety of salts of which sodium chloride has been most used.

Many strains of staphylococci resistant to penicillin are resistant to mercury. These strains are also usually resistant to a variety of other antibiotics and to cadmium and arsenate and produce considerable amounts of A type penicillinase. The genetic determinants for the control of resistance to these compounds appear to be invariably located on a penicillinase plasmid. On the other hand, mercury-sensitive cadmium- and arsenate-resistant strains which produce C type penicillinase appear to carry chromosomal resistance determinants. Strains with similar metal ion resistance patterns, but producing A type penicillinase, carry the resistance determinant on a plasmid.

Table III.2. Penicillinase plasmids in *Staphylococcus aureus*

	Plasmid type					
	α	β	γ	δ	ϵ	ζ
Phage groups of carrier strain	I/III	I/III	I/III	II	I/III	I/III
Erythromycin resistance	−	−	+	−	−	−
Penicillinase type*	A	C	A	B	A	C
Resistance to $HgCl_2$	+	+	+	+	−	−
Compatibility group†						
pla I	+	−	+	?	+	+
pla II	−	+	−	?	−	−

After M. H. Richmond (1965) *Br. med. Bull.*, *21*, 260.

* Types differ serologically: B is less active against penicillin than A and C.

† Plasmids in the same compatibility group cannot coexist in one cell as separate components, but may undergo recombination.

Prophylaxis

The control of hospital infections is now a problem of such magnitude as to require planning at the administrative level. The general principles which should be adopted are:

1. The elimination of all reservoirs of infection. This involves the institution of measures designed to reduce dust, the sterilization of bedclothes, the control of nasal carriers (for which topical application of antiseptic creams or sprays has been favourably reported on) and the exclusion of all persons with superficial septic lesions from contact with the patient, either in the operating theatre or in the ward.

2. The employment of a fully aseptic technique in the operating theatre and during the dressing of wounds.

3. The recording and investigation of all cases of wound sepsis. Phage typing should be employed in the investigation of all major outbreaks of infection. If several phage types are found the presumption is that the outbreak is due to a breakdown at some point in aseptic technique. The isolation of a single phage type from the majority of cases would indicate on the other hand the likelihood of infection from a single source or the wide dissemination of a rapidly spreading epidemic type.

Staphylococcal vaccines have been extensively used in both the treatment and the prophylaxis of recurrent cutaneous staphylococcal infections and have occasionally been found to be of value. Better results have been claimed in these conditions with a staphylococcal toxoid, prepared by treatment of a filtrate containing the α toxin with formalin. The immunity produced by the toxoid is of short duration and frequent boosting doses are required. The toxoid is sometimes of value in aborting a superficial staphylococcal infection such as a boil if given at the first sign of development of the lesion.

Staphylococcus epidermidis

Staphylococcus epidermidis is a commensal of low pathogenic potential. In the past it has been principally isolated from stitch abscesses but in recent years it has assumed greater importance through its implication in postoperative endocarditis following cardiac surgery for valve prosthesis. It has also been responsible for infections in hydrocephalic children in whom Spitzholter valves have been inserted to drain the ventricles. In both these circumstances infection has developed several months after the operation, indicating that it is almost certainly due to bacteraemic spread. Most strains isolated have been penicillin-resistant.

III.3

The Streptococci

Cocci which occur in chains are known as streptococci. Streptococci are Gram-positive, non-motile and do not form spores. They grow poorly on simple media but growth is greatly enhanced by the addition of fermentable carbohydrate (e.g. glucose), serum or blood.

CLASSIFICATION

Most of the streptococci from human sources are aerobic and facultatively anaerobic. The classification of the aerobic streptococci has, however, given rise to considerable difficulty. The criterion which has so far been of most value in classification is the presence of polysaccharide haptens which can be extracted from the organisms by a variety of methods and which are demonstrable by precipitin tests. On the basis of these haptens 18 groups of streptococci (A, B, C, D, E, F, G, H, K, L, M, N, O. P, Q, R, S and T) may be distinguished, all the strains in a particular group possessing an antigenically identical polysaccharide. Since they were first recognized by Lancefield these divisions are known as Lancefield groups.

In the majority of the Lancefield groups the group polysaccharide appears to be bound firmly to the cell wall mucopeptide, presumably through covalent bonds, but not in sufficiently superficial a location as to be demonstrable by agglutination tests.

Most of the organisms which possess Lancefield group haptens produce a wide zone of haemolysis on blood agar which is usually referred to as β haemolysis. This is an area of complete clearing of from 2 to 5 mm in diameter surrounding the colony. Since the Lancefield group haptens were first recognized amongst the β haemolytic strains, organisms possessing them are frequently referred to as haemolytic streptococci. The term is a convenient one but it should be noted that it is by no means a correct description of the group since many are non-haemolytic or may produce the viridans change.

In addition to being classified into groups on the basis of their possession of Lancefield haptens many streptococci have been assigned species rank. In some instances, e.g. groups A, B and N, the species is co-terminous with a Lancefield group. In some, e.g. groups C and D, the group is subdivided into species distinguished by various physiological characteristics. In others, species are defined by physiological characteristics only, without reference to serological properties, e.g. *Streptococcus sanguis* and *Streptococcus salivarius*. Many streptococci, however, including all those of groups E, L, M, O, P, Q, R, S and T, are classified simply by serological group and have not been given a species designation.

Other criteria that have been suggested as a basis of classification, but whose ultimate value for this purpose has still to be determined, are the chemical composition of the cell wall, particularly the presence or absence of rhamnose as a cell wall component, and the production of various red cell sensitizing antigens which are secreted by the organism into its environment but which would nevertheless appear to have a close association with the cell wall.

Streptococcus pyogenes

The most important of the haemolytic streptococci as causes of human infection are those of Lancefield group A. Group A streptococci form a homogeneous group which has received species rank as *Streptococcus pyogenes*.

Morphology

Streptococcus pyogenes is capsulated, the capsules consisting of hyaluronic acid. Capsules are demonstrable, however, only in young cultures examined after two or three hours' incubation. In the later stages of growth the capsular hyaluronic acid is secreted into the medium and in some cases broken down by the action of hyaluronidase.

Cultural properties

On blood agar *Str. pyogenes* produces small colonies about 1 mm in diameter which have a finely granular or matt surface. The colonies are surrounded by an area of clear β haemolysis which may be as much as 5 mm in diameter. The haemolysis on aerobic plates is due to the oxygen-stable haemolysin. A small number of strains are deficient in this haemolysin and show haemolysis only on anaerobic or deep culture. Occasionally strains are encountered which synthesize large amounts of hyaluronic acid and which give rise to large mucoid colonies measuring about 4–5 mm in diameter. Glossy colonies which lack hyaluronic acid and sometimes also M protein are occasionally encountered.

Antigenic structure

The species may be subdivided, as first shown by Griffith using the slide-agglutination technique, into a number of distinct antigenic types. Type specificity depends in most cases on a group of antigenic trypsin-sensitive alcohol-soluble proteins—M proteins—which are also, because of their antiphagocytic properties, essential for virulence. The M proteins may be extracted from the cells with hot acid and typing is now normally carried out by precipitin tests with acid extracts. Over 50 M antigens have been identified.

Most strains possess one of a second group of acid-labile trypsin-resistant antigens known as T antigens. These are demonstrable in agglutination tests and have a distribution which is independent of that of the M protein.

The hyaluronic acid of the capsule is non-antigenic, presumably because of its structural similarity to the hyaluronic acid of animal tissues.

Pathogenicity

Streptococcus pyogenes produces a number of substances which play a part in pathogenicity. The most important of these are the *O* and *S haemolysins, streptokinase, desoxyribonnuclease* and the *erythrogenic* or *Dick* toxin.

The O haemolysin is an oxygen-labile exotoxin which readily undergoes oxidation to an inactive form. It may be obtained in an active state by growing the organism in broth containing a reducing agent such as sodium hydrosulphite. The O lysin is lethal on injection into animals and has in particular a highly toxic effect on the heart. It is strongly antigenic and antibody to it is found in the sera of many persons during the course of, or during convalescence from, streptococcal infection. This antibody—antistreptolysin O—may be demonstrated in vitro by its capacity to inhibit the haemolytic action of the toxin. Streptolysin O is strongly inhibited by cholesterol to which it is apparently adsorbed and which is probably its normal substrate. Like complement it causes lysis by the production of holes in the cell membrane. Similar lysins are also produced by pneumococci, *Bacillus cereus* and various species of *Clostridium.*

In contrast to streptolysin O, streptolysin S is oxygen-stable. It is more closely bound to the cell than streptolysin O but active cell-free preparations of it can be obtained by extraction with serum, albumin, RNA and detergents, to which it is transferred from a cellular carrier, probably a component of the cell membrane. There is evidence that the various extracting substances also stimulate the synthesis of the lysin. Streptolysin S appears to be a protein but is non-antigenic. Its haemolytic action is antagonized by a lipoprotein present in normal serum. The concentration of this inhibitor is diminished in the course of streptococcal infection. Streptolysin S has been shown to be capable of producing necrosis of the convoluted tubules of the kidneys of mice when implanted intraperitoneally. It is highly toxic for white cells and is probably the leucocidin of the organism.

The erythrogenic toxin is so called because it produces a local erythema when injected into the skin. It is responsible for the rash of scarlet fever. Human skin is exquisitely sensitive to its action, reactions being produced by as little as 10^{-6} µg of the toxin. It will produce pyrexia in rabbits and skin reactions in previously sensitized guinea-pigs and it enormously enhances the susceptibility of rabbits to endotoxic shock. The toxin is a relatively heat-stable protein. It is produced by only certain strains of *Str. pyogenes,* and it is only these strains which can cause scarlet fever. Its mode of action is unknown. At least three different antigenic types of Dick toxin have been identified; only one of the types is commonly found, being that produced by about 80% of toxin-producing strains. The capacity of *Str. pyogenes* to produce erythrogenic toxin depends on lysogenization by a specific temperate bacteriophage. Toxin production only occurs when the phage is induced (p. 466) to give rise to mature phage particles.

Culture filtrates of *Str. pyogenes* are actively fibrinolytic for human fibrin. The fibrinolytic agent is a plasma protease (plasmin) produced by activation of plasminogen by a plasma kinase which is formed from a plasma precursor by the action of the streptococcal agent—*streptokinase.* Streptokinase almost certainly plays a part in the pathogenicity of *Str. pyogenes* by breaking down the fibrin barrier formed by the tissues in response to infection. In contrast to staphylokinase it is produced maximally in the early stages of growth and can therefore be presumed to be produced at an early stage of streptococcal infection. Streptokinase is antigenic and increase in the antistreptokinase activity of serum is frequently found in the course of streptococcal infections. Two distinct antigenic varieties of streptokinase (A and B) have been described.

Four antigenically distinct desoxyribonucleases (streptodornases)—A, B, C and D—have been described; of these B is the most antigenic in man. One or more of these is produced by all strains of *Str. pyogenes.* Since DNA may be present in considerable concentration in inflammatory exudates and accounts for much of their viscosity, these enzymes are probably the substances mainly responsible for the serous character of streptococcal pus. Mixtures of streptokinase and streptodornase have been used therapeutically for breaking down blood clots and fibrinous exudates in closed spaces such as the joints or pleural cavity. Like streptokinase, desoxyribonuclease is produced in the early stages of streptococcal growth. It seems likely, therefore, that together they may contribute appreciably to the spreading character of streptococcal infection.

Some strains produce *hyaluronidase,* an enzyme which breaks down the hyaluronic acid of the tissues. In most cases the amount of hyaluronidase produced in vitro is small and can be detected only by very sensitive tests. In the course of streptococcal infection, however, most patients develop antibody against hyaluronidase, suggesting that the enzyme may be produced in large amounts in the tissues. Hyaluronidase also breaks down the hyaluronic acid of the streptococcal capsule; consequently strains which produce a large amount of hyaluronidase produce little hyaluronic acid. Hyaluronidase might be expected to play a part in the virulence of *Str. pyogenes* by facilitating its spread through the tissues, but whether, or to what extent, this effect is counterbalanced by its capacity to break down the hyaluronic acid capsule is not clear.

Another substance of dubious significance in pathogenicity is a protease, produced in cultures grown at 37°C but not at 32°C, which is capable of breaking down the M type specific protein. Streptococcal protease also inhibits the production of streptokinase. Since it is produced only at an acid pH it may not be present during the early stages of infection.

Infections due to Streptococcus pyogenes

Streptococcus pyogenes is intrinsically a much more dangerous organism than *Staph. aureus,* and has a much greater tendency to spread in the tissues, therefore being more likely to give rise to septicaemia, the most serious complication of pyogenic infection. The last few decades, however, have seen a considerable fall in the incidence of streptococcal infection, as well as a marked diminution in its severity.

The main site of streptococcal infection is the throat, where a purulent tonsillitis is the most typical lesion. This condition is found characteristically in older children and in adults. In younger children the infection is more diffuse—pharyngitis rather than tonsillitis—and less acute in character. The more acute localized lesion of the older person is believed to depend on a previously acquired allergy.

When a throat infection is caused by a strain which produces the erythrogenic toxin the resultant symptom complex is known as scarlet fever. During the course of outbreaks of scarlet fever many people develop a sore throat but without the appearance of a skin rash. This is because they are immune to the erythrogenic

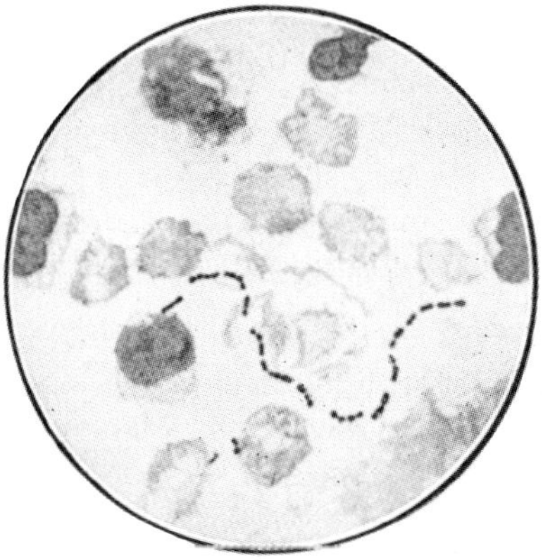

Fig. III.2. *Streptococcus pyogenes* in pus. × 950.

toxin owing to antitoxin in their blood, developed as a result of a previous infection with a toxin-producing strain of *Str. pyogenes* of a different serotype from that responsible for the present infection. Though immune to scarlet fever they can nevertheless develop a streptococcal throat infection because their blood does not contain type-specific antibody for the currently infecting strain.

From the throat the infection may spread locally, particularly to the neighbouring lymphatic glands and to the middle ear. Occasionally a peritonsillar or retropharngeal abscess is produced and, infrequently, diffuse cellutis of the floor of the mouth, known as Ludwig's angina. From the throat the organism may also invade the lungs, causing pneumonia of severe type. Streptococcal pneumonia occurs infrequently as a primary condition and is usually a complication of other infections involving the respiratory tract such as influenza.

Streptococcus pyogenes is an important cause of superficial infections. Infections of wounds, abrasions and burns are of frequent occurrence. In recent years, however, the staphylococcus has complely ousted *Str. pyogenes* as the primary cause of wound infection in hospitals. Local extensions of the infection to subcutaneous tissues may give rise to cellulitis and, if unchecked, to septicaemia.

Erysipelas is a diffuse streptococcal skin infection involving the superficial lymphatic vessels. The skin is swollen with a brawny induration and the area of inflammation shows a characteristic sharp edge. *Str. pyogenes* also appears to be responsible for some cases of impetigo.

Streptococcus pyogenes is the most important cause of puerperal sepsis; in this condition the organisms are almost invariably derived from attendant nurses or doctors or from contaminated instruments. In the pre-antibiotic era puerperal sepsis progressing to septicaemia was a major cause of maternal mortality.

A small proportion of normal individuals carry *Str. pyogenes* in their throats. The carrier state appears to be most frequent during the winter and spring. Some persons are nasal carriers. Both types, but particularly the nasal carrier, are of great importance in the spread of streptococcal infection, the mechanism of environmental con tamination being essentially the same as that already considered for the staphylococcus.

Sequelae of infection

There is a considerable body of evidence that *Str. pyogenes* is in some way the cause of acute rheumatism and of most cases of acute glomerulonephritis in man. These conditions differ, however, from the suppurative infections in that they only appear between the first and third week after infection. Many cases are not preceded by overt streptococcal infection but in many of these, high titres of antibodies to streptococcal extracellular antigens, particularly to streptolysin O, are frequently demonstrable. Streptococci are not directly demonstrable in the lesions in either condition and it is clear that they are not due to direct invasion of the tissues by the organism. Apart from the differences in the organs involved, there are certain differences in the mode of occurrence of these conditions which must be taken into account in any proposed explanation of their pathogenesis:

1. When streptococcal in origin, acute glomerulonephritis is usually associated with infection due to particular serotypes of *Str. pyogenes,* particularly type 12. No specific serological types have so far been associated with acute rheumatism. Since streptococcal immunity is type-specific these findings presumably explain the fact that acute nephritis is a non-recurrent condition whereas acute rheumatism recurs with each subsequent streptococcal infection.

2. There is some evidence that susceptibility to acute rheumatism but not to acute nephritis is genetically determined. It appears to occur most

frequently in non-secretors of blood groups other than O. In addition, persons who suffer from rheumatic fever develop higher antibody responses to streptococcal antigen and appear to have a higher susceptibility to recurrent streptococcal infection than the normal subject.

There are two main theories to explain the role of the streptococcus in producing these conditions:

1. They may be due to the direct toxic action of some streptococcal product. The association of acute nephritis with only a few serological types of *Str. pyogenes* suggests that these strains may in fact produce some substance that is toxic to the kidneys. Claims to this effect have not, however, been adequately documented. In the case of acute rheumatism the presumed toxic component would have to be produced by all the serological types of streptococci. The most likely substance would appear to be streptolysin O but the susceptibility of the rheumatic subject to recurrent attacks of acute rheumatism in the presence of a high titre of antistreptolysin O renders this hypothesis unlikely.
2. They may have an immunological basis. This could be an allergy to some streptococcal component—the allergic reaction occurring in the tissues giving rise to tissue damage—or it could be an autoimmune process resulting in the development of antibody reacting with a normal tissue antigen. An immunological basis could provide a satisfactory explanation for the delay in the development of the lesions.

The evidence for an immunological basis is clearer in acute nephritis than in rheumatic fever. Glomerular lesions resembling those occurring in acute nephritis in man have been produced in rabbits by the injection of protein antigens. When a single large dose of protein was injected the nephritis was transient. When the protein was injected in a series of small doses the renal lesions were progressive and permanent. Nephritis may also be produced in rabbits by the injection of antigen–antibody complexes. It is possible therefore that in man acute nephritis may be due to the production of Arthus-type reactions in the glomeruli, as a result of the formation or retention of antigen–antibody complexes. The most likely antigen would in this case be the M protein. This possibility is consistent with the demonstration, by the fluorescent antibody technique, of the presence of gammaglobulin and complement in glomeruli from cases of acute nephritis and with the lowering of serum complement levels found in the condition. These complexes have a characteristic, irregular 'lumpy bumpy' distribution different from the linear distribution found in experimental autoimmune nephritis.

Cardiac lesions have been produced in rabbits as a component of the serum sickness syndrome, following the injection of large amounts of protein. The lesions found, however, have consisted primarily of a periarteritis and were quite different from the lesions observed in acute rheumatism in man. Lesions more closely resembling those of rheumatic carditis were observed (by Murphy) in rabbits, following a series of intradermal injections of streptococci of different serological types. In man the presence of gammaglobulin adsorbed to heart muscle cells has been described and also the presence in the blood of antibodies to human heart tissue.

The latter have been found, however, in conditions other than rheumatic fever and may be a result rather than the cause of the disease. The occurrence in streptococci of an antigen which will stimulate the development of antibodies which react with heart muscle cells and which can be absorbed out with streptococcal cell walls has been reported. It is conceivable that antibody to this antigen might be responsible for the cardiac lesions in acute rheumatism. An alternative possibility is that a streptococcal product might confer antigenicity on an antigen present in the heart, thereby allowing it to bypass the tolerance mechanism (p. 85).

Erythema nodosum is also in some cases associated with streptococcal infection. It may be an allergic reaction but is more probably due to deposition of call wall fragments in the skin, since similar lesions can be produced in animals by such fragments. The damaging component in the latter has been identified by Schwab as the peptidoglycan. A number of cases of this condition, however, are undoubtedly associated with tuberculosis and various fungal infections.

Other haemolytic streptococci

Streptococci of Lancefield group B constitute the species *Str. agalactiae*. This organism is the main cause of bovine mastitis and is only very slightly if at all pathogenic for man. It may occur as a commensal in the nose, throat and vagina.

Streptococci of groups C and G can produce lesions very similar to but less severe than those of *Str. pyogenes*. Many strains of groups C and G produce streptokinase and streptolysin O and a number also produce erythrogenic toxin. Other points of resemblance between groups A, C and G are the findings that strains of all three groups are lysed by an enzyme produced when group C streptococci are infected with a specific bacteriophage and that strains of groups C and G possess an R antigen serologically identical with that found in group A. Group G streptococci occur not infrequently as throat commensals. Group C streptococci are primarily animal pathogens, being responsible for mastitis in cows and for strangles and contagious bronchopneumonia in horses. On the basis of their biochemical characteristics they have been subdivided into four species, the species responsible for human infections sometimes being known as *Str. equisimilis*.

Group D strains rank next in importance to group A strains as causes of human infection. Since their natural home appears to be the intestinal tract of man and other animals group D streptococci have been grouped together as 'enterococci'. Two species of enterococci occur as human commensals, *Str. faecalis* and *Str. faecium*. *Str. faecalis* is the species most frequently encountered.

Although normally non-pathogenic, enterococci may invade the tissues and produce pyogenic lesions. The lesions most frequently caused are urinary infections and wound sepsis. The enterococci are second only to coliform bacilli as a cause of acute pyelitis and are occasional causes of puerperal sepsis and of subacute bacterial endocarditis, usually of severe type. Many of these infections have occurred after genitourinary manipulation.

Streptococcus faecalis is of considerable value as an indicator organism in the bacteriological examination of water. Its presence in a water sample is presumptive evidence of faecal contamination and since it dies fairly quickly in water is an indication of recent contamination.

Strains of groups F, H and O occur as commensals in the throat. They may be either β haemolytic or of viridans type. Amongst the viridans strains of group F are a group of strains which have been assigned species rank as *Streptococcus MG*. They form a serologically homogeneous group, and produce colonies which fluoresce in ultra-violet light. Their importance can be attributed to the fact that antibody to them, demonstrable by the agglutination test, is present in the sera of cases of primary atypical pneumonia, a condition with which, however, the streptococcus has no causal relationship (p. 283).

For some strains of group H, which have the distinctive property of producing, when grown in sucrose broth, a dextran precipitated by antiserum to the capsular polysaccharide of the type 2 pneumococcus, the species *Str. sanguis* has been proposed. Many strains of *Str. sanguis* lack the group H hapten, however, and the species, as currently defined, is clearly heterogeneous. *Str. sanguis* strains appear to be of importance as causes of subacute bacterial endocarditis.

Streptococci of other Lancefield groups are primarily of animal origin; though some have occasionally been isolated from human lesions, their pathogenicity for man is doubtful.

The viridans streptococci

Many streptococci isolated from human sources do not possess Lancefield polysaccharide haptens. Most of these produce a greenish discoloration of the blood in the immediate vicinity of the colony—the greenish zone being frequently surrounded by a very narrow rim of almost complete haemolysis. This type of change is usually referred to as α haemolysis. It is to these organisms that the designation viridans streptococci is primarily applied by bacteriologists. The term is, however, a confusing one since many viridans-type streptococci isolated from human sources possess identifiable Lancefield group haptens. The majority, however, do not. They are a heterogeneous group and their classification presents particular difficulties because of the paucity of criteria which will permit satisfactory species definition. Viridans streptococci occur as commensals of the upper respiratory tract. They are of particular importance as causes of subacute bacterial endocarditis.

Subacute bacterial endocarditis

Although viridans streptococci are responsible for

the great majority of cases, subacute bacterial endocarditis is occasionally due to *H. influenzae* and *H. parainfluenzae* (p. 261) and, very rarely, to *Coxiella burneti* (p. 319).

It occurs usually in valves which have been damaged by a previous rheumatic endocarditis or which are the seat of a congenital defect. Bicuspid aortic valves are particularly prone to involvement. The organisms gain entrance to the valves from the blood stream. Other important precipitating causes are tonsillectomy and manipulation of the cervix uteri. In the case of group D streptococci, the infection frequently occurs after genitourinary manipulation. In group D infections, previous cardiac damage does not appear to be so essential a predisposing condition as in viridans infections.

Anaerobic streptococci

These organisms are found as normal inhabitants of mucous membranes, but can occasionally invade the tissues, producing pyogenic lesions. They can cause infection in any organ but are most commonly found as invaders of wounds, particularly those that have been extensively damaged or contused. They are of special importance as causes of a severe type of puerperal sepsis. They have marked proteolytic properties and usually produce large amounts of hydrogen sulphide from sulphur-containing amino acids. Because of this, pus from anaerobic streptococcal lesions has a unpleasant putrefactive smell. Anerobic streptococci are assigned by Bergey to a separate genus, *Peptostreptococcus,* in which a number of species are defined.

Diagnosis

Streptococci are usually isolated from clinical material by aerobic culture on blood agar. For the isolation of *Str. pyogenes,* however, anaerobic cultures give more satisfactory results particularly for the recognition of the occasional strain which is deficient in the S lysin and which does not therefore produce haemolysis on surface aerobic culture.

When a haemolytic streptococcus has been isolated on culture its Lancefield group should be determined. For this purpose the group hapten is extracted and its identity determined by a tube precipitin test (p. 138) with appropriate grouping sera.

An alternative method of grouping, which has been used increasingly in recent years and which does not require prior extraction of the antigen, is by the fluorescent antibody technique. It is highly specific but occasionally gives false-positive reactions with some strains of group C. The bacitracin sensitivity test, which has been recommended as a substitute for grouping in the identification of group A strains, has a low degree of specificity and is not reliable.

The diagnosis of subacute bacterial endocarditis is made by isolating the organism from the blood. If the patient is under treatment with penicillin, penicillinase should be incorporated in the broth used for culture. Since many of the streptococci responsible grow slowly, blood culture bottles should not be discarded as negative for at least four weeks.

Estimation of antistreptolysin O titres is sometimes carried out for the diagnosis of infection with *Str. pyogenes,* and particularly to indicate recent episodes of infection in children suffering from juvenile rheumatism. It should be noted that antistreptolysin titres when significant are indicative only of streptococcal infection and are not specifically diagnostic of rheumatic fever. It is claimed that rheumatic fever subjects show an unusual response to streptolysin O—the antibody appearing later than in uncomplicated streptococcal infection, but attaining higher levels and persisting for much longer.

Treatment

Streptococcal throat infections should be energetically treated with penicillin, and the drug is usually continued for some time after the disappearance of symptoms in order to minimize the risk of the development of acute rheumatism as a sequela. Penicillin has given excellent results in the treatment of subacute bacterial endocarditis due to viridans streptococci, although some viridans strains are reported to have appreciable resistance to penicillin. In such cases and in enterococcal infections, combined therapy with penicillin and streptomycin has proved satisfactory. Erythromycin is a satisfactory alternative to penicillin in the treatment of streptococcal infections in general.

Penicillin and the sulphonamides have both been used successfully in the prophylaxis of streptococcal infection in persons who have at one time suffered from acute rheumatism. Penicillin, because of its capacity to sterilize the tissues, has now been accepted as the agent of choice for this purpose. In the case of children it is recommended that prophylaxis should be continued at least until adult life. Penicillin has also been used as a prophylactic for children living in closed communities in the face of an epidemic of streptococcal throat infection but is not generally used for this purpose because of the dangers of sensitization and because it interferes too much with natural active immunization.

III.4

The Pneumococci

Diplococcus pneumoniae (Pneumococcus)

The pneumococcus has much in common with the streptococci but is assigned to a distinct genus, *Diplococcus,* of which it is the only member.

Morphology

The pneumococcus is a non-motile, non-sporing Gram-positive coccus. In material taken from the body it occurs characteristically in pairs of flame-shaped cocci about 1 μm in length, the rounded ends of the cocci being adjacent to each other. This typical morphology is not often seen in laboratory cultures where the appearance and arrangement is similar to that of the streptococci. The pneumococcus is capsulated, the capsule being best seen in material taken directly from the tissues.

Cultural properties

On blood agar normally capsulated strains produce a highly characteristic, smooth, flat colony frequently showing umbilication and resembling a draught-board piece in appearance. Many of the strains found as normal inhabitants of the nasopharynx are non-capsulated on primary isolation and as a result give rise to rough colonies. The colony on blood agar is surrounded by a zone of greenish discoloration of the medium and often by an outer zone of haemolysis—the changes being similar to those associated with the viridans streptococci. Type 3 pneumococci produce large, highly mucoid colonies of characteristic appearance.

Pneumococci readily undergo autolysis in culture, the flat appearance of the colony in fact being due to these autolytic changes. The enzyme system responsible for autolysis may be activated by a number of agents such as bile salts, saponin, soap and alkali. This is the basis of the bile solubility test, which is of considerable value in the differentiation of pneumococci from viridans streptococci.

Pneumococci are highly sensitive to ethylhydrocupreine (Optochin), a property which is also valuable in differentiating them from viridans streptococci.

Antigenic structure

In their antigenic composition pneumococci resemble the streptococci. They possess a species-specific polysaccharide hapten, which is chemically, although not serologically, similar to the Lancefield group haptens. This antigen is precipitated by a globulin which is found in high concentration in the sera of patients with various acute diseases. The most important of the pneumococcal antigens are the capsular polysaccharides on the basis of which the species is subdivided into serologic types, some 80 of which have so far been described. Serologic typing may be carried out by the capsule-swelling or by the agglutination reactions. The purified polysaccharides, although fully antigenic for man, are non-antigenic in the rabbit. Pneumococci also possess type-specific M proteins whose distribution closely parallels that of the capsular polysaccharides. The reactivity of the type-specific proteins is masked by the presence of the capsule. They do not have an antiphagocytic effect.

Rough pneumococci derived from capsulated organisms of one type can be permanently transformed into capsulated organisms of a different type by treatment with DNA derived from the latter (p. 23). This transformation is of considerable historical interest, since its discovery by Griffith was the first indication that properties could be transferred from one micro-organism to another. The extent to which this type of change occurs under natural conditions is unknown, but it is likely to be infrequent since the in vitro transformation requires special conditions and occurs at a low rate.

Immunity to pneumococcal infection depends on the presence of antibody to the capsular polysaccharides and is, therefore, type-specific. Consequently an individual can undergo a number of infections, each infection being due to an organism of different serologic type.

Pathogenicity

The pneumococcus produces an oxygen-labile haemolysin resembling streptolysin O. Pneumococcal cultures which have undergone autolysis also contain a substance which can produce purpuric lesions in experimental animals. Like many viridans streptococci, pneumococci produce large amounts of a neuraminidase. None of the substances referred to above, however, appears to play any significant part in pathogenicity. Consequently it has been suggested that the severe toxaemia which is so characteristic a feature of pneumococcal pneumonia is due to the fact that the organisms growing in the lung deprive the tissue cells of adequate nutriment. It is significant in this connection that the dramatic improvement in cases of lobar pneumonia, known clinically as the crisis, coincides with an outpouring of antibody and cessation of growth of the organism. The improvement in the patient at the time of the crisis is certainly not the result of any marked change in the lesion in the tissues, which remains histologically precisely the same for some time after the crisis; it may therefore well be the result of the mere inhibition of bacterial growth.

The capsule of the pneumococcus is essential for virulence, its role being to protect the organism from phagocytosis. This protective function is lost on combination with specific antiserum. The role of the capsule in virulence is clearly shown in an interesting experiment carried out by Dubos, who found that it was possible to remove the capsule from type 3 organisms without affecting their viability by exposing them to the action of a polysaccharide-splitting enzyme which he had isolated from a soil bacillus. The decapsulated organisms were avirulent.

Pneumococcal infections

Pneumococci are widely distributed as commensals of the upper respiratory tract. Many of the commensal pneumococci are, however, non-capsulated and avirulent. The majority of pneumococcal infections are due to a limited number of capsular types, the most important of which are types 1, 2 and 3. These appear to be much more invasive than the usual commensal types; in addition they are more likely to show epidemic spread in closed communities.

Pneumonia is the most important disease caused by the pneumococcus. Although occurring as a primary condition, pneumococcal pneumonia is more frequently secondary to some damage to the respiratory tract, especially that caused by virus infections, e.g. influenza, which render the lungs more liable to bacterial invasion.

Pneumococcal pneumonia occurs clinically in two forms, bronchopneumonia and lobar pneumonia, classified according to the distribution of the area of consolidation in the lung. The two types of pneumonia show important differences in incidence and pathogenesis. Broncopneumonia is characteristically a disease of young children and of the older adult, i.e. over 50 years. It is essentially a sporadic infection and is usually due to one of the less commonly pathogenic types which occur as commensals in the upper respiratory tract. In contrast lobar pneumonia is almost exclusively a disease of the age group 10 to 50 years. Lobar pneumonia occurs both as a sporadic and as an epidemic disease. Epidemics of the disease occur almost exclusively in closed communities, e.g. barracks and asylums. In such communities an epidemic type when once introduced spreads rapidly. It has been suggested that the typical lobar distribution of pneumococcal pneumonia in the adult may be the result of a previously acquired allergy. In the allergic individual contact with the organism might lead to an outpouring of fluid by which invading organisms are swept throughout an entire lobe.

The pneumococcus is occasionally responsible for acute episodes of infection in individuals suffering from chronic bronchitis but is of less importance in this respect than *H. influenzae*. It is probably the organism most frequently found in pyogenic infections of the air sinuses. It is a frequent cause of otitis media and is occasionally responsible for a primary peritonitis occurring in young girls.

The pneumococcus is one of the major causes of acute pyogenic meningitis. This condition is commonest in children, in whom it frequently occurs as a complication of otitis media. It may also develop as a sequel to fractures of the skull. Occasionally pneumococcal meningitis assumes a very acute and rapidly fatal form. In these cases the appearance of the cerebrospinal fluid is highly characteristic showing the presence of large numbers of pneumococci with comparatively few pus cells.

Diagnosis

The diagnosis of pneumococcal meningitis can usually be made with reasonable certainty from the examination of Gram-stained films of a centrifuged deposit from the cerebrospinal fluids. For the diagnosis of pneumococcal pneumonia, however, culture is essential.

When Gram-positive cocci of typical colonial morphology are found on blood agar they can usually be safely identified as pneumococci. The typical colony appearance is, however, not always observed and the organisms may then be indistinguishable from viridans streptococci. Their differentiation from the latter can be made either by the bile solubility test or by determining their susceptibility to ethylhydrocupreine. For the latter test a subculture is made to a blood agar plate from a suspicious colony and on the inoculum is placed a filter paper disc impregnated with 1/1000 ethylhydrocupreine. If the organism is a pneumococcus, an appreciable area of inhibition of growth will appear round the disc. Viridans streptococci are not inhibited, or are so only slightly.

Treatment

Pneumococci are sensitive to sulphonamides, penicillin and the wide-spectrum antibiotics. Penicillin is the agent of choice; in the treatment of meningitis it is claimed to give best results when combined with a sulphonamide and should be given by the intrathecal as well as by the intramuscular route.

III.5

The Neisseriae

The neisseriae are aerobic Gram-negative cocci which characteristically occur in pairs. The cocci have a highly distinctive bean or kidney shape. In the diplococcal form they are arranged with their long axes parallel and with their flattened sides adjacent to each other. They are non-motile and non-sporing and are devoid of morphological capsules. All the neisseriae give strong positive oxidase (i.e. indophenol oxidase) reactions.

The pathogenic members of the genus, *Neisseria meningitidis* and *Neisseria gonorrhoeae,* are extremely delicate organisms and are among the more difficult to grow in the laboratory. They die out rapidly outside the body and in the laboratory, if not preserved in the dry state, must be frequently subcultured. Though Gram-negative their antibiotic susceptibility pattern is very similar to that of the streptococci. They are strictly human parasites and on inoculation possess a very low pathogenicity for experimental animals.

In agglutination, precipitin and complement fixation tests, both species appear to be antigenically heterogeneous and show in addition considerable antigenic overlap with *Neisseria catarrhalis* and other non-pathogenic neisseriae.

Neisseria meningitidis (Meningococcus)

The diplococcal form of *N. meningitidis* is most readily seen in films prepared from the cerebrospinal fluid from cases of meningococcal meningitis. Capsules are not demonstrable by ordinary microscopic techniques; they have, however, been demonstrated by the use of the capsule swelling reaction with group-specific serum.

On blood agar the meningococcus produces, after about 24 hours' growth, a smooth, translucent and rather viscid colony from 1 to 2 mm in diameter, with a shining surface and an entire edge. It is non-haemolytic. The meningococcus ferments glucose and maltose without gas production and has no action on saccharose. Fermentation tests are best carried out on agar media containing added rabbit serum.

Seven serological groups showing some degree of antigenic overlap have been distinguished. Many strains, however, cannot be serologically grouped.

Meningococcal infections

Meningococci are present in the nasopharnyx of from 2 to 5% of normal people. These commensal strains are most frequently of group B. Group A strains, which are those mainly responsible for epidemics, rarely occur as commensals.

The meningococcus is the most important cause of acute pyogenic meningitis. Meningococcal meningitis occurs most commonly, although not exclusively, in children. Mortality from the condition is maximal at the extremes of life. Meningitis, however, is the final and most dramatic phase of meningococcal infection, the evolution of which can be divided into three stages. In the first stage the organism appears in the nasopharynx, where it may cause pharyngitis. In the second stage the blood is invaded. The characteristic lesion of this stage is the development of petechial skin lesions, due to foci of infection in the capillaries. In the third stage the organism localizes in the meninges producing an acute pyogenic meningitis. The infection may abort at either of the earlier stages without the production of meningitis.

In some cases of meningococcal infection bacteraemia may become a chronic condition without evidence of meningeal localization. In a

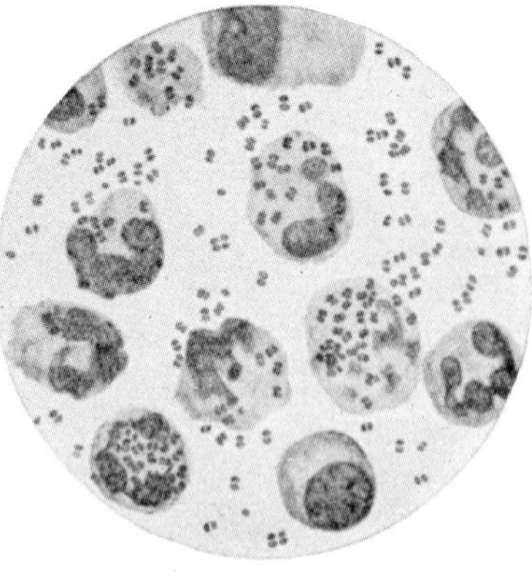

Fig. III.3. Meningococci in purulent cerebrospinal fluid.

small proportion of cases the bacteraemic stage appears as a fulminating septicaemia which may be fatal in a few hours. In this condition, known as the Waterhouse–Friderichsen syndrome, the patient dies in a state of severe shock; recent evidence identifies this syndrome as a manifestation of the diffuse intravascular coagulating syndrome found in other high-grade endotoxaemias (p. 236).

Meningococcal meningitis may occur either as a sporadic or as an epidemic infection. The development of epidemics is favoured by conditions of overcrowding and they are particularly liable to occur in institutions with a susceptible young population, e.g. schools and barracks. Under such conditions carriage rates of the order of 90% have been found in the absence of overt infection.

Metastatic meningococcal lesions also sometimes occur in the heart, where they give rise to endocarditis, or in the joints, where they cause a purulent arthritis.

Diagnosis

In meningitis the organisms may be demonstrable in Gram films prepared from the centrifuged deposit of the cerebrospinal fluid. In some cases, however, even when there is a considerable amount of pus, they may be so few in number that this is not possible. Since the meningococcus dies rapidly outside the body, it is imperative that cultures should be made as soon as possible, blood agar normally being used. The plates are best incubated in an atmosphere containing 5 to 10% carbon dioxide. The cerebrospinal fluid should also be incubated directly, or better still with the addition of an equal volume of glucose broth, and, after incubation, subcultures made on to blood agar. This procedure may yield a growth of the organism when immediate plating has been negative. A precipitin reaction using the ring technique for the demonstration of meningococcal antigens in the cerebrospinal fluid is employed by some workers as a rapid diagnostic method; it has been of particular value when organisms cannot be demonstrated in stained films. In early cases it may be possible to isolate the organism from the blood. The blood should be inoculated into a blood culture bottle and should also be spread directly on to a blood agar plate.

The identification of *N. meningitidis* in the cerebrospinal fluid presents no difficulty. In cultures taken from the nasopharynx, considerable trouble is experienced, for a number of other Gram-negative diplococci are frequently found in that situation. The chief one of these likely to cause confusion is *N. catarrhalis,* which may be distinguished from the meningococcus by its capacity to grow on nutrient agar at room temperature, by its lack of fermentative actions on sugars, by its colonies which are larger and white and by the fact that it is usually spontaneously agglutinable in saline.

Treatment

Penicillin is the drug of choice in the treatment of meningococcal meningitis.

Neisseria gonorrhoeae (Gonococcus)

Neisseria gonorrhoeae resembles *N. meningitidis* very closely in microscopic appearance. In stained films of the pus obtained from the urethra in acute gonorrhoea, the great majority of the cocci are seen to lie within leucocytes. As many as one hundred cocci may occasionally be seen within a single pus cell.

The gonococcus is even more difficult to grow than the meningococcus. A satisfactory medium contains 20% of citrated blood with an agar concentration sufficiently low to produce a distinctly soft medium. The colonies, which usually appear within 24 hours, very closely resemble those of the meningococcus. The gonococcus ferments glucose with the production of acid only. It does not ferment maltose or saccharose.

Four biotypes, 1 to 4, differing in colonial morphology and auto-agglutinability, are distinguished. Serological classification is not satisfactory.

Gonococcal infections

Man is the only animal naturally infected by gonococci. Infection is almost always direct from individual to individual and almost exclusively by venereal contact. Only types 1 and 2 are pathogenic.

The gonococcus attacks chiefly the urethra in both the male and the female, producing an acute urethritis. From the urethra the disease, if untreated, tends to spread locally. In the male spread occurs to the prostate, testes, seminal vesicles, epididymis and sometimes the periurethral tissues. In the female the cervix uteri is invaded early; later the infection may spread to Bartholin's glands, to the uterus, producing endometritis, and to the Fallopian tubes, ovaries and pelvic peritoneum. In adult women the vagina usually escapes because of the acid pH of the vaginal secretions but severe vulvovaginitis can occur in young girls. Outbreaks of the latter condition were at one time not uncommon in children's hospitals and schools.

Gonococcal conjunctivitis occurring in the newborn is now a very uncommon condition because of the universal practice of instilling sulphonamide or silver nitrate into the newborn infant's eyes immediately after delivery. The condition is due to infection of the child's eyes during delivery from gonococci present in the genital tract of the mother.

Diagnosis

In the male a diagnosis of gonorrhoea can usually be made in the acute stage by the demonstration of typical intracellular Gram-negative diplococci. In chronic cases, and in the female, a microscopic diagnosis is rarely possible and isolation of the organism is essential. Where possible cultures should be made directly from the patient on to the medium.

In chronic cases, in which microscopic and cultural examination may be equivocal, and in late complications of the disease, such as arthritis, the gonococcal complement fixation test may be of value.

Treatment

The treatment of gonorrhoea was revolutionized by the introduction of the sulphonamides, but unfortunately after a few years' use of these drugs a high proportion of strains isolated were found to be sulphonamide-resistant.

Penicillin has for many years been the treatment of choice. Recent reports, however, show an increasing number of strains with a substantial resistance to penicillin. Fortunately, gonococci appear also to have regained, to a very considerable extent, their original sensitivity to sulphonamides.

Gonorrhoeal patients not infrequently have a coexistent syphilitic infection contracted at the same time. The administration of penicillin or any other antibiotic active against *Treponema pallidum* to such patients may delay or in some cases completely inhibit the development of the primary syphilitic lesion. It is therefore important to carry out serologic tests for syphilis over a period of some months after such treatment.

III.6

Bacillus

Organisms of the *Bacillus* genus are aerobic, Gram-positive, sporing rods. Only one species, *Bacillus anthracis,* is pathogenic for man.

Bacillus anthracis

Morphology

Bacillus anthracis is a large non-motile bacillus measuring 5 to 10 μm × 1 to 2 μm. In the tissues it is found singly, in pairs and in short chains and is surrounded by a capsule which is continuous over the whole chain. In culture long chains of bacilli may be found.

Spores are produced only in the presence of oxygen. Consequently they are found in culture only when the organism is grown under aerobic conditions; they are never seen in material taken directly from the tissues. Characteristically the spores are centrally situated, of the same width as the bacillus and oval in shape.

Cultural properties

The anthrax bacillus is aerobic and facultatively anaerobic. The organism grows well on all ordinary media at temperatures between 19 and 40°C. Growth is not improved by the presence of blood. After overnight incubation the colonies on agar are highly characteristic. They are of a greyish white or cream colour, opaque, with an uneven surface and an irregular wavy edge, showing the so-called Medusa head or 'barrister's wig' appearance. On microscopic examination the end of the colony can be seen to consist of long, interlacing strands comprising parallel chains of bacilli. When grown under conditions which favour capsule development, e.g. on serum agar or in the presence of a high concentration of carbon dioxide, the colony has a smooth, mucoid character. Normally the organism is non-haemolytic on blood agar, but a few strains may show a narrow zone of haemolysis.

In gelatin stab-culture *B. anthracis* produces a highly characteristic type of liquefaction in which lines of growth radiate from a central stem, the whole resembling an inverted fir tree with delicate filamentous branches.

Antigenic structure

The main component of the capsule is a polypeptide of D-glutamic acid. Similar glutamyl polypeptides are found in other *Bacillus* species. Although essential for virulence there is some doubt as to whether the polypeptide is antigenic. A polysaccharide which cross-reacts serologically with the type 14 pneumococcal capsular polysaccharide is the principal antigenic component of the cell wall. Antibody to this antigen nevertheless plays no part in protection against infection. An antigen which stimulates the production of protective antibody is demonstrable, however, in the oedema fluid produced in response to the inoculation of living anthrax bacilli. This antigen is not produced in the conventional type of laboratory culture but is produced when the organism is grown in the presence of a high concentration of bicarbonate.

Pathogenicity

The anthrax bacillus produces a specific toxin which is lethal and oedema-producing on injection into experimental animals. The toxin, which is not demonstrable under ordinary cultural conditions, was first identified in the

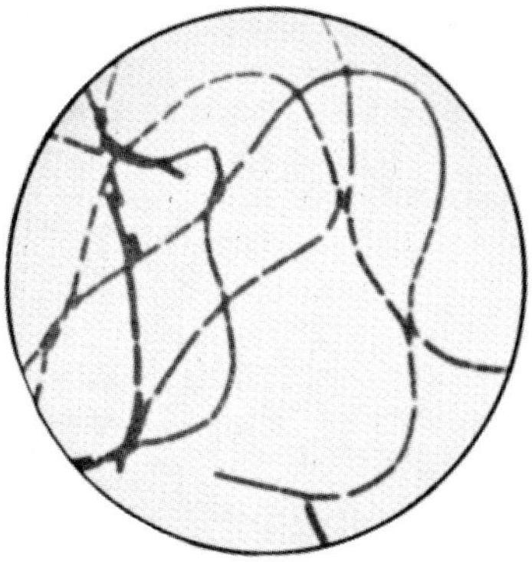

Fig. III.4. *Bacillus anthracis* from an agar culture. × 750.

plasma of animals dying of anthrax infection. It was subsequently found in very young cultures of the organism growing in heparinized or defibrinated guinea-pig blood. As shown by Smith, the toxin is a complex of the three protein components I, II and III, each of which by itself is non-toxic. Component II is identical with the protective antigen referred to above. It is the only component which is antigenic *per se*. A mixture of components I and II has marked oedema-producing capacity but relatively low lethality. The addition of component III considerably enhances lethality while reducing oedema-producing capacity. The lethal effect of the toxin appears to be the result of shock but the precise nature of the biochemical lesion involved is not known.

Anthrax

The natural disease affects herbivora, chiefly sheep and cattle, in which the mortality may be as high as 80%. Horses, goats and pigs are less commonly attacked. In animals, anthrax is a septicaemic disease which may be acute or subacute in type. The most marked post-mortem finding is the great enlargement of the spleen, from which fact the name 'splenic fever' has been given to the disease.

There is little doubt that infection in animals is chiefly by the ingestion of spores present in their food. Since, however, laboratory animals can only rarely be infected by feeding, it is probable that some damage to the mucous membranes, such as might be inflicted by thorns, is essential. Infected animals excrete large numbers of anthrax bacilli in discharges from the mouth, nose and rectum and in this way contaminate their pasturage, where the spores may survive for years and constitute a source of infection for other animals.

The disease is uncommon in animals in Britain, most infections being caused by spores present in imported foods. Consequently the majority of cases occur in winter when imported foodstuffs are generally used. In countries in which it is enzootic the disease occurs most frequently during the late summer and autumn.

Anthrax is not a common disease in man. In countries where the disease is prevalent in animals infection occurs mainly in persons whose work brings them into close contact with animals: veterinary surgeons, butchers and farmers. In others it is found mainly as an industrial disease, infection occurring almost entirely in workers whose occupation involves the handling of imported hair, hides or wool. Infection of members of the general population may occur through the use of shaving brushes, the bristles of which have come from infected animals.

In man the disease assumes three clinical forms: cutaneous, respiratory and intestinal. In the cutaneous form, which is the commonest variety of the disease, the spores or bacilli gain entrance through cuts or abrasions or possibly by way of the hair follicles. The face and neck, hands, arms or back are the usual sites for the characteristic lesion, which is known as the 'malignant pustule'. From one to three days after infection a small papule, usually painful, occurs. This soon becomes vesicular, the contained fluid being clear or blood-stained The centre becomes necrotic and black in colour and around it a ring of vesicles forms: the whole area is congested and oedematous. The cutaneous lesion usually resolves spontaneously but in a minority of cases, if untreated, the blood stream is invaded, and the outcome is, invariably, fatal.

The respiratory form of anthrax, known as 'woolsorter's disease', occurs almost exclusively in the wool industry, as a result of the inhalation of dust from infected wool. Respiratory anthrax has a high mortality rate and the organisms appear to multiply mainly in the mediastinal glands from which the blood stream is rapidly invaded. Though it is the normal route of infection in animals, infection by the intestinal tract is rare in man. As in the respiratory form the condition carries a grave prognosis.

Diagnosis

In the early stages of cutaneous anthrax the organisms are usually readily demonstrable in the skin lesions; in the later stages, however, this may not be possible. Whatever the source of the material examined, the organism should be isolated on culture.

The cultural features distinguishing the anthrax bacillus from other sporing bacilli are the characteristic colony, absence of haemolysis on sheep blood agar, absence of motility, typical fir-tree growth in gelatin, susceptibility to specific bacteriophages, sensitivity to penicillin and production of morphological capsules on serum or bicarbonate agar.

Treatment

Penicillin is the treatment of choice. Although very effective in treatment, penicillin does not modify the evolution of the skin lesion since it has no effect on the anthrax toxin. The treatment of respiratory and intestinal anthrax is, however, unsatisfactory.

Prophylaxis

A considerable degree of immunity can be produced against anthrax in animals by the inoculation of vaccines prepared from living attenuated strains. This was first demonstrated by Pasteur, using a culture attenuated by prolonged growth at 42°C, in his classic experiment at Pouilly-le-Fort in 1881.

Because of the difficulty of ensuring the stability of the attenuated strains prepared by Pasteur's method his method of vaccine preparation has been abandoned. The most satisfactory of the vaccines now in use for immunization of animals appears to be that of Mazzuchi. This consists of a spore suspension of the stable attenuated Carbazoo strain in 2% saponin. The immunity produced by any form of anthrax vaccine is of short duration and animals require annual reinoculation. Vaccines are not used for human immunization but preparations of the protective antigen may be valuable for persons exposed through their occupation.

Prophylaxis, so far as man is concerned, is chiefly a matter of industrial hygiene, adequate exhaust ventilation to remove dust and the sterilization of wool, hides and hair. In Britain the control of industrial anthrax has been greatly facilitated by restricting the importation of potentially infected hides and hair to a single port—Liverpool. Since most infections occur through the skin, the hands of the workers should be protected as much as possible. Abrasions should be treated immediately with iodine.

Other bacilli

Non-pathogenic aerobic sporing bacilli are widely distributed in nature. They are of interest in that several produce inducible extracellular penicillinases and some, e.g. *B. polymyxa,* produce polypeptide antibiotics. *B. cereus* is the species most likely to be confused with the anthrax bacillus. It has been implicated as the cause of outbreaks of bacterial food poisoning.

III.7

The Corynebacteria

The corynebacteria are irregularly staining, Gram-positive, non-sporing, non-motile and non-capsulated rods. They frequently show club-shaped swellings, a characteristic which is responsible for the name of the genus. Only one species, *Corynebacterium diphtheriae,* is pathogenic for man. Other species—*Corynebacterium ovis, Corynebacterium pyogenes* and *Corynebacterium equi*—are important animal pathogens.

Corynebacterium diphtheriae

Morphology

Diphtheria bacilli vary considerably in morphology. The bacilli may appear as short, more or less uniformly stained organisms, but more often, as long ($5 \times 0{\cdot}4$ μm), slender, slightly curved, irregularly stained bacilli. The most striking morphological feature is the presence of the metachromatic granules—intensely staining round or oval bodies usually located at the ends of the bacilli—which consist of highly polymerized metaphosphate. In films made from cultures, diphtheria bacilli, particularly the longer forms, tend to be grouped in pairs with the individual organisms lying at an acute angle. The pairs frequently occur in groups, giving rise to the 'Chinese letter' appearance.

Cultural properties

The diphtheria bacillus grows best aerobically; under anaerobic conditions growth is poor. It will grow on nutrient agar but growth is better on media containing blood or serum. For isolation, Löffler's serum medium and media containing potassium tellurite, which are highly selective for the diphtheria bacillus and on which it produces a black colony, are the media of choice.

Three cultural types of diphtheria bacillus, distinguished by their colonial morphology on a special blood tellurite medium, were described by McLeod. These types are known as *gravis, intermedius* and *mitis,* the names indicating the severity of clinical disease with which each type was found to be most frequently associated. This association is, however, no longer considered to be of general validity.

Diphtheria bacilli ferment glucose and maltose, producing acid but no gas. Except for a few very rare strains they do not ferment saccharose.

A large number of serologic types have been defined by agglutination tests. Serological typing is of some epidemiological value but, though the antigens responsible appear to be restricted to diphtheria bacilli, is not carried out as a routine procedure.

Pathogenicity

The diphtheria bacillus produces a potent exotoxin which has been isolated as a crystalline protein with molecular weight of approximately 64 000. It is extremely potent and is lethal for most animals in concentrations of the order of 100 to 150 mg/kg body weight.

The toxin has a cytopathic effect on most animal cells in tissue culture. Under these conditions it has been shown to inhibit protein synthesis without affecting cellular respiration. Its action has been defined as inactivating the translocation factor EF_2, responsible for transfer of peptidyl tRNA from the A to the P site of the ribosome. For this inactivation NAD is required, the toxin apparently catalysing the formation of an inactive ADP ribose EF_2 complex.

Toxin is produced only by strains which are lysogenic for certain symbiotic bacteriophages, of which the phage β and its variants are the most important. Lysogenization of non-toxigenic strains with β phage usually, though not invariably, converts them into toxigenic strains. The gene *tox,* which controls toxin production, has been shown to be located on the bacteriophage chromosome. The lysogenic state is extremely stable and it is doubtful if non-lysogenic, non-toxigenic cured variants occur under natural conditions. The production of toxin requires phage induction and this occurs spontaneously in a small proportion of the organisms of a lysogenic strain. Most of the toxins appear to be released, however, before the cell undergoes lysis.

The production of toxin is also related to the iron concentration of the medium. Maximum yields have been found in concentration of iron considerably below those required for optimum growth. The precise role of iron limitation in toxin production is not clear. In low iron concentration diphtheria bacilli have a prolonged generation time, show a low lipid content and possess a demonstrably thinner cell wall. These findings suggest that a low iron concentration may increase the permeability of the cell to the toxin, thereby permitting its release. The highest levels of toxin production, some 10 to 12 times greater than for other strains, have been reported for the Park Williams No. 8 strain. This strain is consequently the standard strain used throughout the world for toxin production. Its high level of toxin production may be due to the fact that it is capable of growth at levels of iron concentration which would be inhibitory for other strains.

Diphtheria

The natural disease in man affects chiefly the upper part of the respiratory tract, especially the fauces, the toxin of the organism producing necrotic changes in the mucous membrane of the affected area. This stimulates a local inflammatory reaction as a result of which is formed a 'false membrane', composed of laminae of fibrin, in which are entangled leucocytes, red cells, epithelial cells and, in the more superficial layers, diphtheria bacilli. The membrane is most conspicuous in the tonsillar area but, as the condition progresses, spreads over the palate around the pharyngeal wall and into the larynx. In the latter situation it can cause respiratory obstruction leading, unless relieved by tracheotomy, to the death of the patient. In addition to membrane formation there occurs a general toxaemia as a result of spread of toxin in the blood stream from the local lesion. This produces remote toxic effects, particularly on the adrenals, which cause a fall in blood pressure; on the heart muscle, which may be sufficient to produce cardiac failure; and on the kidneys, leading to albuminuria. The acute stage of the infection is not infrequently followed by peripheral neuritis, usually of a transient character. This condition is possibly due to allergic sensitization, rather than to the direct action of the toxin. In some cases, particularly in young children, the nasopharynx is mainly involved. A false membrane may be found in other parts of the body, such as the ear, conjunctiva, vulva and vagina, but much more rarely than in the respiratory tract. Infection of wounds or abrasions with the production of an indolent ulcer is not uncommon in the tropics but is rare in temperate zones. This type of infection is associated with relatively little systemic disturbance.

Diagnosis

Early diagnosis of diphtheria is of the greatest importance. Films made from the lesion are quite unsuitable for a microscopic diagnosis and culture is therefore essential.

A throat swab is charged from the area involved, care being taken to include a portion of any membrane present. The swab is inoculated on to a Löffler serum slope and on to a blood tellurite plate. The swab should also be inoculated on to blood agar for the detection of *Str. pyogenes.* A preliminary diagnosis may be made from smears prepared from overnight cultures. In all cases, however, biochemical tests and tests for toxigenicity should be carried out.

Virulence tests may be carried out by injecting the growth from a young serum culture subcutaneously into each of two guinea-pigs, one of which has been protected by 250 units of antitoxin, injected the previous day. If the organism is a toxigenic diphtheria bacillus the unprotected animal will die, usually within five days, showing at post mortem the characteristic signs of diphtheritic toxaemia while the protected animal will survive.

Fig. III.5. The plate virulence tests for *Corynebacterium diphtheriae*. Two strains have been streaked vertically and a strip of filter paper has been laid horizontally across the plate. The strain on the left has yielded four lines of toxin–antitoxin precipitate and is therefore toxigenic; the strain on the right is non-toxigenic.

The performance of virulence tests has been made very much simpler by the plate technique introduced by Elek. In this method a strip of filter paper soaked in antitoxin is laid across a plate of serum agar. The strains to be tested are streaked at right angles to the strip and the plate incubated. If the strain is toxigenic, a line of precipitate is observed in the angle between the streak of growth and the filter paper strip. The test has been found to give very good correlation with in vivo toxigenicity tests.

Treatment

Since the efficacy of antitoxin depends on the stage of the disease at which it is given, it is of the utmost importance that the antitoxin should be given as early as possible, *without waiting for the bacteriologist's report.* A dose of 5000 to 100 000 units of antitoxin is recommended, depending on the severity of the case and the stage at which it is seen. In severe cases, and in cases seen at a late stage, antitoxin should be administered by the intravenous route. In other cases the intramuscular route is to be preferred to the subcutaneous.

The diphtheria bacillus is sensitive to penicillin and penicillin has been found to be a useful adjunct in treatment. Although it does not affect the toxaemia it assists in the sterilization of the tissues, thereby shortening the course of the disease and preventing the development of a convalescent carrier state.

Immunity

Immunity against diphtheria can be produced both by an attack of the disease and by a subclinical infection, as well as by artificial immunization and is due to antitoxin in the blood of the immune person. This can be detected by the Schick test, in which a small amount of toxin in a total volume of 0·2 ml is injected into the skin of one forearm and a similar amount of heated toxin, which acts as a control, into the other. Readings should be made twice, the first at 48 hours, the second from five to seven days after the test. A positive reaction is shown by the development in 24 to 48 hours at the site of injection of the toxin, of a red flush, from 1 to 5 cm in diameter. The colour usually begins to fade after about a week, to be succeeded by a brownish discoloration which, in exceptional cases, may persist for some months. The blood antitoxin concentration required to give a negative reaction with the test concentrations of toxin usually employed is on an average 0·005 unit/ml. This is usually but not always sufficient to protect against diphtheria. Since some subjects may give non-specific allergic reactions to components of the toxic filtrate other than the toxin it is necessary, in order to detect these 'pseudo'-reactions, to carry out a control test by injecting into the other arm the same amount of toxic filtrate which has been heated sufficiently to destroy its toxin. Such reactions consist of a transient erythema which develops as a rule within 24 hours and usually disappears completely in from five to six days, at which time the reaction to the unheated toxin is still present. Schick tests are mainly used for detection of immune status in older children (over 10 years) and adults as a preliminary to immunization.

Prophylaxis

Contacts may be passively immunized by the injection of antitoxin, 500 units being the dose normally used. The immunity produced is of short duration, lasting at most for a few weeks.

Passive immunization is therefore only an emergency measure but, since it is rapidly established, is a very effective one. To be effective, antitoxin must be given as soon as possible after exposure and preferably within 24 hours. Antitoxin may be combined with active immunization for children who have not been previously immunized, but it appears to interfere to some extent with the efficacy of active immunization.

For the production of active immunity the following preparations are used:

Formol toxoid (FT). This preparation consists of toxin which has been treated with formalin to deprive it of its toxicity. It may be used by itself or, more frequently, in combination with various other antigens, e.g. tetanus toxoid and pertussis vaccine. By itself, formol toxoid is a relatively poor antigen but in combination with pertussis vaccine its antigenicity is considerably enhanced. Pertussis-containing vaccines are suitable only for the vaccination of young children.

Adjuvant vaccines (APT and PTAP). The antigenicity of formol toxoid can be greatly enhanced, as in APT and PTAP by combination with adjuvants. APT is a suspension of the precipitate formed by the addition of aluminium hydroxide to formol toxoid. PTAP is prepared by the adsorption of a purified toxoid on to aluminium phosphate. Although highly efficient as prophylactics, adjuvant preparations have the disadvantage that they give rise to a higher proportion of reactions—particularly in older children and adults—than formol toxoid. Consequently they are only recommended for immunization of children of less than 10 years of age.

Toxoid antitoxin floccules (TAF). This prophylactic is a suspension of the precipitate formed by the addition of antitoxin to toxoid in optimal proportions. It is the purest of all the diphtheria prophylactics and appears to give the least reactions on injection. It has about the same immunizing potency as formol toxoid. It has been used mainly for the immunization of older children and adults in whom the alum-containing preparations would be more liable to produce reactions.

Ideally, a primary course of immunization should be completed during the first year of life. For this purpose the triple vaccine (ATP), which contains diphtheria and tetanus toxoids and killed *Bordetella pertussis* vaccine, is the prophylactic of choice. A booster dose should be given at the age of 18 months and another at four to five years, just prior to school entry.

Other corynebacteria

The term 'diphtheroid' is loosely applied to other species of corynebacteria. Some of these resemble the diphtheria bacillus closely in appearance and may cause difficulty in identification. Non-pathogenic corynebacteria are frequently found as commensals in the upper respiratory tract and are fairly constantly present in the nose. They can usually be differentiated from *C. diphtheriae* by their fermentation reactions, most strains either failing to ferment glucose or fermenting saccharose.

The named species most frequently found as commensals are *Corynebacterium pseudodiphtheriticum* (Hofmann's bacillus), which is often present in the nose and less frequently in the throat, and *Cornybacterium xerosis,* which is sometimes found in the conjunctiva. *C. pseudodiphtheriticum* is readily distinguishable morphologically from the diphtheria bacillus. It is shorter, does not possess metachromatic granules and usually exhibits an unstained central bar. It does not ferment any sugars and is not pathogenic for guinea-pigs. *C. xerosis* is morphologically similar to the *mitis* type of diphtheria bacillus but ferments saccharose and is non-pathogenic for guinea-pigs.

III.8

Commensal Intestinal Bacteria

In the next three chapters we shall be concerned with organisms of the family enterobacteriaceae, so called because their normal habitat is the intestinal tract of man and other animals. The enterobacteriaceae are Gram-negative non-sporing bacilli which grow readily on ordinary laboratory media under both aerobic and anaerobic conditions. All ferment glucose and most ferment a variety of other sugars as well. The enterobacteriaceae have been extremely difficult to classify precisely since they include a large number of complexly interrelated bacterial types. This difficulty is particularly acute in the classification of the coliform group. Consequently the definition of genera has in many cases been quite arbitrary and has tended to change with the acceptance of new criteria for taxonomic characterization. From the medical point of view, for which their pathogenic potential is of a primary importance, they may be subdivided into two large groups:

1. Organisms occurring primarily, or exclusively, as intestinal pathogens. This group comprises the genera *Salmonella, Shigella* and *Arizona* which will be considered separately in subsequent chapters.
2. Organisms occurring primarily as intestinal commensals. These fall into two groups: the *coliform* group and the *Proteus* group.

For the differentiation of the enterobacteriaceae the capacity to ferment lactose, though not accepted as a primary taxonomic criterion, is of considerable practical importance. Lactose is fermented with the production of acid and gas by most of the eschericheae. It is not fermented by any of the salmonellae, shigellae—with the exception of *Shigella sonnei* which is a late lactose fermenter—or proteae.

THE COLIFORM GROUP

In this group six genera may be distinguished: *Escherichia, Citrobacter, Klebsiella, Enterobacter, Serratia* and *Hafnia.* Of these *Escherichia* and *Klebsiella* are the most frequently isolated from human infections.

Escherichia is represented by a single species, *E. coli,* and *Klebsiella* by three species, *Kl. pneumoniae, Kl. rhinoscleromatis* and *Kl. ozaenae.* The features distinguishing *E. coli* and the typical, and most important, *Klebsiella* species, *Kl. pneumoniae,* are shown in Table III.3.

Table III.3. Typical reactions of *Escherichia coli* and *Klebsiella pneumoniae*

Reaction	*E. coli*	*K. pneumoniae*
Indole	+	–
Methyl red	+	–
Voges Proskauer	–	+
Growth in citrate as sole carbon source	–	+
Adonitol and inositol fermentation	–	+
Decomposition of urea*	–	+

* This reaction occurs slowly, taking 3–4 days to be detectable as a rule. Decomposition of urea by *Proteus* is much more rapid, being detectable within 24 hours.

Escherichia coli

Escherichia coli is the typical coliform organism of the intestinal tract and is present in large numbers in the faeces of man and other animals. It is non-capsulated and usually motile, although in some cases motility may be extremely difficult to detect. It usually occurs as a short rod (0·5 × 2 to 5 μm in size) but short coccobacillary and long filamentous forms are occasionally observed, the latter particularly in urine. It exhibits no special arrangement. On solid media it produces relatively large colonies which are usually smooth and glistening. Variants are, however, occasionally isolated from pyogenic infections which give rise to very minute colonies. These small colony variants produce the normal type of colony if they are grown on media to which various organic sulphur compounds, e.g. cystine, have been added. On blood agar *E. coli* is usually non-haemolytic but haemolytic strains are occasionally encountered.

The great majority of *E. coli* strains ferment lactose promptly, i.e. within 24 hours, with the production of acid and gas. Glucose, maltose and mannitol are also fermented, usually with gas production. Fermentation of dulcitol and saccharose is variable. *E. coli* does not liquefy gelatin or produce hydrogen sulphide and will not grow in media containing potassium cyanide.

Because of its invariable occurrence in faeces *E. coli* is of considerable value as an indicator organism in the bacteriological examination of water supplies since if it is present it can be presumed that the water has been contaminated from a faecal source.

Citrobacter

Organisms of this genus show considerable resemblance to *E. coli.* The genus as now defined includes all the lactose-fermenting Bethesda-ballerup strains formerly classified as paracolon bacilli. *Citrobacter* strains occur as intestinal commensals and are commonly isolated from water when they are classified by water bacteriologists as intermediates. They have not so far been shown to be pathogenic.

Klebsiella pneumoniae

In contrast to *E. coli, Kl. pneumoniae* is capsulated and non-motile. The capsule is large and conspicuous and is often several times the size of the bacillus. *Kl. pneumoniae* is usually shorter and plumper in appearance than *E. coli* and when present in the body frequently occurs as a coccobacillus (0·5 to 1·5 × 1 to 2 µm), which may be mistaken for a coccus. On solid media it produces a large white highly mucoid colony which is very characteristic.

Klebsiella pneumoniae ferments lactose, glucose, maltose, mannitol, saccharose, adonitol and inositol. Dulcitol may or may not be fermented. From sugars other than glucose *Kl. pneumoniae* is a less active gas producer than *E. coli.* Gelatin is not liquefied and hydrogen sulphide not produced.

Klebsiella pneumoniae occurs in smaller numbers in the intestinal tract than *E. coli.* Certain serotypes—these are the strains previously known as *Bact. pneumoniae* or colloquially as Friedländer's bacilli—are not infrequently found both as commensals and as pathogens in the respiratory tract. In spite of its name, *Kl. pneumoniae,* like *E. coli,* may cause a variety of pyogenic lesions throughout the body and is frequently isolated from infections of the urinary tract. Strains with identical properties are also widely distributed in nature being found on grasses, grains and decaying vegetation and in soil and water. These strains, which were formerly known as *Bact. aerogenes,* are included by Kaufmann in the species *Kl. pneumoniae* but are assigned by Bergey to a separate genus *Aerobacter* as *Aerobacter aerogenes.*

Other klebsiellae

Also included in the genus *Klebsiella* are two atypical species, *Kl. rhinoscleromatis* and *Kl. ozaenae,* isolated from cases of rhinoscleroma and ozaena respectively. These species are capsulated and non-motile, do not produce acetyl methyl carbinol and are of more restricted fermentative capacity than *Kl. pneumoniae.*

Enterobacter, Serratia and Hafnia

Organisms assigned by Bergey to the genera *Enterobacter, Serratia* and *Hafnia* show considerable resemblance to *Klebsiella.* They exhibit a similar, though much lower, pathogenicity. Many strains of *Serratia* produce a reddish colony due to the pigment prodigiosin.

Antigenic structure of coliform bacteria

Four distinct types of antigen are found in the coliform group:

1. Thermostable somatic antigens chemically resembling those of the salmonellae.
2. H or flagellar antigens.
3. K or surface antigens. These are surface

antigens which, when present, mask the agglutinability of the organism by O antiserum. In *E. coli* strains most of the K antigens are non-capsular, surface antigens and are relatively heat-labile; in these respects they resemble the Vi antigens of the salmonellae. They are divided by Kaufmann on the basis of their physical properties into L and B antigens; L antigens are totally destroyed—both in respect of agglutinability and capacity to combine with antibody—by exposure to 100°C for one hour. The agglutinability of the B antigens is also destroyed by this degree of heat treatment but their capacity to combine with antibody is not. Individual strains possess at the most one K antigen. In *Kl. pneumoniae* strains the K antigens are true capsular antigens; they are stable at 100°C but their agglutinability is usually destroyed by some hours exposure to 120°C. They are referred to by Kaufmann as A antigens.

4. Fimbrial antigens. These are heat-labile, being destroyed in one hour at 100°C. Fimbrial antigens are not usually found in cultures grown on solid media. They are not of any importance in classification.

On the basis of their antigenic composition both *E. coli* and *Kl. pneumoniae* can be subdivided into a large number of serological types. In *E. coli* strains 150 serologically distinct O antigens, 90 K antigens and 50 H antigens have so far been described. These antigens may occur in various combinations and consequently permit the differentiation of a very large number of distinct serotypes. Pathogenic strains of *E. coli* usually possess heat-labile K antigens and usually belong to a limited number of serological types. Strains commensal in the intestinal tract belong to a larger number of serotypes and a higher proportion of such strains lack K antigens.

In *Kl. pneumoniae* over 70 distinct capsular antigens have been described. Because of the technical difficulties of examining the *Klebsiella* O antigens antigenic typing of *Klebsiella* strains is based entirely on capsular antigen composition. Of the capsular antigenic types, types 1, 2 and 3 are those formerly known as Friedländer's bacilli or *Bact. pneumoniae. Kl. rhinoscleromatis* possesses the capsular antigen 3; capsular antigens 4, 5 and 6 have been identified in strains of *Kl. ozaenae.*

Infections due to coliform bacteria

While confined to the intestinal tract the coliform bacteria lead a saprophytic existence and may, in fact, by the synthesis of certain vitamins, contribute to the nutrition of the host. In certain circumstances, however, they gain access to the tissues and give rise to pyogenic lesions.

Coliform bacteria are the most frequent causes of urinary infections. These infections are usually associated with some underlying predisposing factor in the urinary tract, such as the presence of stone, congenital defect, urinary retention or tuberculosis. Such underlying conditions must therefore always be sought in acute urinary tract infections, particularly if these are of a recurrent character. Infection may also follow interference with the genitourinary tract, e.g. by catheterization—an indwelling catheter being particularly hazardous—or operations on the prostate and bladder.

In most cases of renal infection the infection is comparatively superficial, but in severe cases these may be extensive invasion of the renal tissue, with the development of pyelonephritis. This complication appears particularly liable to occur in young children in whom urinary tract infections must be vigorously treated, since they can cause serious damage to the kidney. There is appreciable evidence that in older children and in adults deeper invasion of the kidney, with the development of pyelonephritis, may be associated with high antibody levels to the infecting strains. Serological tests may, therefore, be of considerable value in controlling therapy and evaluating progress.

E. coli is an important cause of postoperative wound infection though its role in this condition ranks second to that of *Staph. aureus* (p. 208). *E. coli* has frequently been isolated from the lesions of cholecystitis, appendicitis and peritonitis, but in these conditions other organisms are usually also present and the precise contribution made by *E. coli* to the disease picture is difficult to assess. It is possible that it may simply play the part of a secondary invader.

Certain serological types of *E. coli* have been shown to be responsible for some epidemics of infantile gastroenteritis. So far 18 distinct antigenic types, distinguished by their somatic antigen composition, have been implicated in this condition. The pathogenicity of a number of these serotypes has been confirmed by feeding experiments in both animals and children. Not all strains of the enteropathogenic serotypes, however, appear capable of causing gastroenteritis. The capacity to cause gastroenteritis seems almost certain to be due to the production of an enterotoxin. Living cultures of enteropathogenic strains have been shown to cause accumulation of fluid in ligated segments of rabbit small intestine. Enterotoxic activity has been found in cell-free filtrates of cultures by some workers. No damage, however, was observed in the gut epithelium and it is likely that the toxin has a mode of action similar to that of cholera toxin (p. 250). In certain cases of gastroenteritis invasion of epithelial cells and of the subepithelial tissues, however, undoubtedly occurs. Enterotoxins produced by strains infecting pigs have been shown to be controlled by a plasmid. This may also be the case with human strains.

The condition is much more common in bottle-fed than in breast-fed infants, possibly because with bottle feeding there is a greater opportunity for the child to become infected. Infants with gastroenteritis cause heavy contamination of their environment. This derives not only from the faeces which may be a virtually pure culture of *E. coli* but also from the vomit in which the organisms are usually present in considerable numbers.

Klebsiella pneumoniae type strains are responsible for about 1% of bacterial pneumonias. The pneumonia produced closely resembles in its early stages that produced by the pneumococcus; at a later stage a considerable amount of destruction of lung tissue is found, frequently with the formation of multiple abscesses.

In recent years it has been recognized that absorption and dissemination by the blood stream of endotoxins of *E. coli* and other Gram-negative bacteria, even in the absence of bacteraemia, can be accompanied by severe clinical symptoms and a high mortality rate. Endotoxaemia is associated with various coagulation abnormalities which may be sufficiently severe to cause the condition of disseminated intravascular coagulation, low blood fibrinogen levels, bleeding, particularly into the gastrointestinal tract, hypotension and haemorrhagic or septic pneumonia. In more severe forms the patient succumbs with a syndrome of acute shock.

Endotoxaemia is due to abnormal levels of endotoxin in the blood. This may be because of massive loading of the blood with endotoxins from some septic focus or on occasion to absorption from the gastrointestinal flora. It is frequently associated with some interference with the efficient functioning of the reticuloendothelial system, as for example by the administration of cytotoxic drugs. The overall mortality in Gram-negative endotoxaemia is about 60%; the mortality in cases with bacteraemia and toxaemia is about 70%.

Gram-negative endotoxins have the capacity in very low concentrations of producing a visible gel in lysates of washed amebocytes of the horse crab *Limulus* prepared by freezing and thawing. Endotoxin concentration is readily measured by this method since increase in concentration of the endotoxin results in increased rates of gelling of the lysate. The method is capable of detecting as little as 0·0005 μg/ml of endotoxin in plasma and serum. The *Limulus* test is of considerable prognostic value.

Diagnosis

For the diagnosis of urinary tract infections, stained and unstained preparations of the centrifuged deposit should be examined. Acute infections usually show a considerable number of leucocytes in the urine; this, however, is not always the case. At the same time a loopful of the uncentrifuged and well-mixed urine should be inoculated onto blood agar and MacConkey plates. The number and variety of organisms growing on culture should be noted. If a standard 5 mm loop has been employed, the presence of 500 coliform bacteria per plate can be accepted as presumptive evidence of infection, even in the absence of clinical symptoms or of excess leucocytes in the urine. A quantitative estimate of this type is of value in distinguishing between an actual infection and contamination occurring during the taking of the urine. Such a distinction could also be made by the examination of a catheter specimen. In the case of children, however, for whom urinary examination is most frequently required, catheterization is not now considered justified.

In the identification of enteropathogenic *E. coli* from cases of gastroenteritis the identity of the strain as an enteropathogenic type must be confirmed serologically. Since these strains do not differ colonially from commensal *E. coli* at least 10 colonies from the primary plate should be examined by the slide agglutination technique with polyvalent sera. If positive they should be tested with appropriate type-specific sera, their final identification being confirmed by tube agglutination tests with heat-killed suspensions.

THE PROTEUS GROUP

This group comprises two genera, *Proteus* and *Providencia.*

Proteus

Proteus are actively motile, non-capsulated, non-lactose-fermenting enterobacteriaceae. They are highly pleomorphic, varying in length from short coccobacillary forms to long filaments which are frequently slightly curved. They differ from other enterobacteriaceae in their capacity to break down phenylalanine to phenylpyruvic acid—the PPA reaction—and in their active urease production. The most immediately recognizable property of *Proteus* strains is their capacity to swarm on solid media. If inoculated in the centre of an agar plate the growth spreads rapidly over the surface of the medium, frequently in concentric rings and forming a thin transparent layer which may be easily overlooked. Since this swarming growth may result from the presence of a single organism on a plate, *Proteus* may be erroneously regarded as the causative organism of a pyogenic lesion.

As first shown by Dienes, some strains of *Proteus,* when inoculated at different points on an agar plate, produce a mixed swarming growth, while with others no mixing occurs, a line of demarcation showing no growth appearing between the strains. The failure to mix appears to occur when the strains differ in antigenic structure and particularly when they differ in their H antigens.

The genus comprises four species, *Proteus vulgaris, Pr. mirabilis, Pr. morganii* and *Pr. rettgeri,* differentiated by their biochemical characteristics. Of these *Pr. mirabilis* is that most frequently encountered in diagnostic material. With the exception of some strains of *Pr. rettgeri* all species produce gas in fermentation reactions.

Antigenic structure

Proteus strains are antigenically heterogeneous. Three varieties known as *Proteus OX19, OX2* and *OXK* are of importance since they may be agglutinated by the sera of patients suffering from certain rickettsial infections. These three types are differentiated by their O antigen composition. In their biochemical properties *OX19* and *OX2* strains resemble *Pr. vulgaris* and *OXK* strains resemble *Pr. mirabilis.* So far on the basis of their H and O antigen composition over 100 serotypes have been described in *Pr. vulgaris* and *Pr. mirabilis* strains.

Pathogenicity

Proteus is of low primary pathogenicity. In recent years, however, because of their high antibiotic resistance, the frequency with which *Proteus* strains have been isolated as causes of cross-infection in hospitals, particularly of operation wounds and of the urinary tract, has considerably increased. Though the species as a whole has a high natural resistance to antibiotics sensitivity tests will in many cases reveal an antibiotic to which a particular strain is susceptible.

Providencia

This group of strains closely resembles *Proteus* biochemically and is classified by Bergey as *Proteus inconstans.* Like *Proteus* they give a positive PPA reaction. Providence strains, however, do not swarm and rarely decompose urea. To date at least 150 serotypes based on O and H antigen composition have been described. Strains of this type have been isolated from urinary tract infections and may be present as commensals in the intestinal tract.

ANAEROBIC GRAM-NEGATIVE BACILLI (BACTEROIDACEAE)

A considerable number of species of non-sporing anaerobic or micro-aerophilic Gram-negative bacilli have been identified as normal parasites of the alimentary tract. Two genera have been distinguished: *Bacteroides* and *Fusobacterium.* They have been found in association with inflammatory lesions of the intestinal tract such as appendicitis and cholecystitis and in gangrenous lesions both in the intestinal tract and elsewhere in the body. The bacilli are considerably more pathogenic than was previously thought. They have been estimated as being responsible for about 5% of all Gram-negative septicaemias. Lincomycin and its derivative clindamycin have been reported to give good results in the treatment of *Bacteroides* infections.

III.9

The Salmonellae

Salmonellae are Gram-negative, motile, non-sporing, non-capsulated bacilli morphologically indistinguishable from *E. coli*. They do not produce indole and do not hydrolyse urea. They are non-lactose-fermenters but ferment glucose and various other sugars, most species producing acid and gas. Salmonellae exist in nature primarily as parasites of the intestinal tract of man and other animals.

The nomenclature of the salmonellae has undergone several changes. Some of the species earlier described were named after the disease with which they were associated and these names, such as *Salm. choleraesuis, Salm. typhi, Salm. paratyphi* and *Salm. enteritidis,* have for the most part been retained. A few species have been called after the persons who discovered or worked with them (e.g. *Salm. schottmuelleri*) or of the patient from whose body they were isolated (e.g. *Salm. thompson*). Many have been named after the locality in which they were first isolated, e.g. *Salm. dublin, Salm. london, Salm. poona, Salm. aberdeen, Salm. minnesota,* etc. Since the number of species recognized is steadily increasing—over 1000 have now been described—it has been suggested that the practice of giving a separate name to each new species should be abolished and new species designated by their antigenic formulae. This is a more rational procedure since the great majority of species differ only in antigenic composition.

On the basis of their pathogenicity the salmonellae may be divided into two groups:

1. The enteric fever group comprises the organisms capable of causing the enteric fevers in man, *Salm. typhi* and the paratyphoid bacilli. These species are found only in the intestinal tract of man for whom they have a high degree of pathogenicity and in whom they frequently cause invasive disease.

2. The food poisoning group are essentially parasites of animals from whom man is occasionally infected. Their pathogenicity for man is relatively low, the usual result of infection being the production of gastroenteritis. In this condition they penetrate to the subepithelial tissues but rarely invade the blood stream.

Cultural properties

For the isolation of the salmonellae, various selective media are employed, the most important being desoxycholate citrate agar on which they produce translucent, colourless non-lactose-fermenting colonies.

With the exception of *Salm. typhi* the salmonellae ferment maltose, glucose, mannitol and usually dulcitol producing acid and gas. None ferment lactose or saccharose. In the case of the commonly isolated aerogenic species, the amount of gas produced is often small and naturally occurring variant strains which produce no trace of gas have been described. *Salm. typhi* ferments maltose, glucose and mannitol, producing acid but not gas. Some species can be subdivided on the basis of certain fermentation reactions into biochemical types which tend to show a characteristic geographical distribution.

Antigenic structure

The antigenic structure of the salmonellae is complex. The antigens of importance in identification are the H or flagellar antigens, the O or somatic antigens and the Vi antigens.

H antigens

These are protein in nature and are destroyed by heat, e.g. by boiling for a few minutes, and by exposure to alcohol. They are not destroyed by formalin, and H suspensions may therefore be prepared by the addition of formalin to broth cultures. Agglutination of H suspensions by H antiserum occurs rapidly, the organisms aggregating into characteristic loose, open floccules. Both in rabbits and man the H antigens are much more highly antigenic than the O antigens. About 90 distinct H antigens are known; these are considered below under diphasic flagellar variation.

O antigens

These are phospholipid protein polysaccharide complexes, the antigenic specificity of which resides in the polysaccharide component. They are stable to heat, withstanding a temperature of 100°C for several hours, and are unaffected by exposure to alcohol. O suspensions may therefore be obtained from motile organisms by heat or by alcohol treatment, both of which procedures destroy the H antigens.

About 60 serologically distinct O antigens have been recognized. Most species possess more than one O antigen, consequently many species share the same O antigens. On the basis of their O antigen composition the species can be classified into some 40 serological groups, each group being characterized by the possession of an O antigen not present in other groups. Thus the antigen 2 is characteristic of group A, the antigen 9 of group D and so on. The organisms most commonly encountered fall into the first five of these A, B, C, D and E.

The chemistry of the polysaccharide has been considerably clarified in recent years by the work of Westphal and his associates. Its general structure is shown below. It possesses a backbone of heptose and 2-keto-3-deoxyoctonic acid. Anatomically, the backbone is the innermost layer of the polysaccharide, and in the complete antigen is complexed with the lipid and/or protein components. To the backbone are attached side chains which may be divided into (*a*) a core section of N-acetyl glucosamine, glucose and galactose common to all *Salmonella* O antigens and (*b*) repeat units which carry the O serological specificity. The structure of the repeat antigens, 4,12 and 9,12 is shown in Table III.5 It will be seen that the sugars tyvelose and abequose, which are 3,6-dideoxy-hexoses, are the components determining group specificity. Thus tyvelose determines the specificity of group antigen 9 and abequose the specificity of group antigen 4. That of group antigen 12 is determined by the remaining sugars of the repeat unit. Other important dideoxyhexoses are paratose, which carries group antigen 2 specificity, and colitose, which is shared by some less common salmonellae and certain coliform bacteria.

Vi antigens

Recently isolated strains of *Salm. typhi* possess an additional surface antigen which is assumed to be

Table III.4. Antigenic structure of certain Salmonellae

Group	*Species*	*Phase variation*	*O antigens*	*H antigens* Phase 1	Phase 2
A	*Salm. paratyphi* A	Monophasic	(1), 2, 12	a	
B	*Salm. paratyphi* B	Diphasic	(1), 4, (5), 12	b ⟷	1, 2
	Salm. typhimurium	Diphasic	(1), 4, (5), 12	i ⟷	1, 2, 3
C	*Salm. paratyphi* C	Diphasic	6, 7, (Vi)	c ⟷	1, 5
	Salm. oslo	Diphasic	6,7	a ⟷	e, n, x
	Salm. newport	Diphasic	6, 8	e, h ⟷	1, 2, 3
	Salm. tennessee	Monophasic	6, 7	z_{29}	
D	*Salm. typhi*	Monophasic	9, 12, (Vi)	d	
	Salm. enteritidis	Monophasic	(1), 9, 12	g, m	
	Salm. dublin	Monophasic	1, 9, 12	g, p	

( ) indicates that the antigens may not be present.
⟷ indicates phage variation.

present in a layer, covering or masking the O antigen since it prevents the agglutination of the organism by O antiserum. Because of its presumed association with virulence it is known as the Vi antigen and may be regarded as being analogous to the non-capsular K antigens of the coliform group. The agglutinability of suspensions of *Salm. typhi* by Vi antiserum is markedly heat-labile, being lost after exposure to 60°C for one hour. Vi antigens serologically similar to those of *Salm. typhi* are found in *Salm. paratyphi* C and in certain coliform bacteria, but in these species the antigens do not render the organisms O inagglutinable, presumably because they are present in much lower concentrations on the cell surface.

Antigenic variation

A number of types of antigenic variation occur in the salmonellae.

H → O variation

This is a loss of flagellar antigens due to a loss of flagella; as a consequence of this flagellar loss, the organisms are non-motile.

Table III.5. Structure of Salmonella somatic antigens

Each *Salmonella* species possesses a number of O antigen specificities each of which is described as a separate O antigen and designated by an arabic numeral. The different O antigens are not, however, different molecules but *different determinants* on the *same* molecule

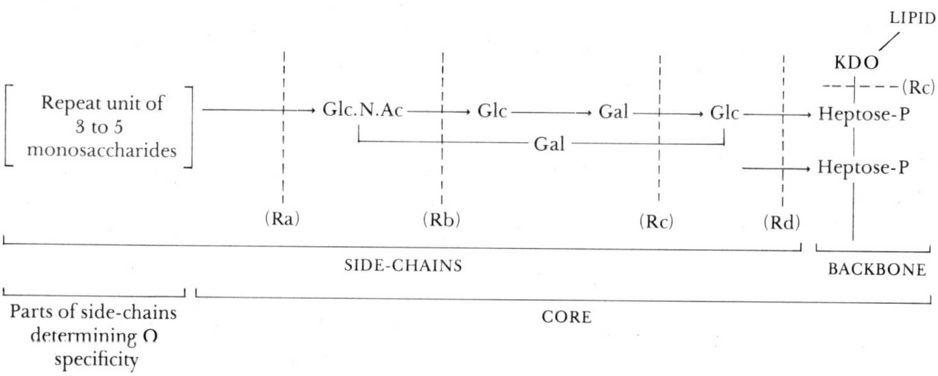

Rough (R) mutants arise by failure to synthesize the entire polysaccharide. Different mutants are distinguishable by the sites at which the synthetic block occurs. These are indicated on the above structure by interrupted lines and the designation of the R mutant—(Ra), (Rb) etc.

Typical repeat units

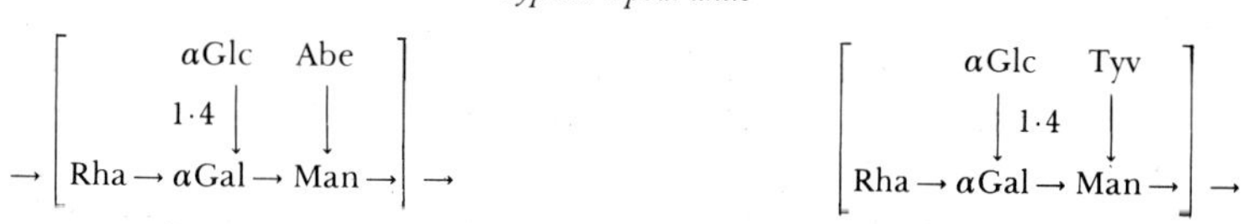

Group B. Antigens 4,12
e.g. *Salm. typhimurium*

Group D. Antigens 9, 12
e.g. *Salm. typhi*

Antigen 5 by O-acetylation of galactose

Abe—abequose
Gal—galactose
Glc—glucose
Glc N.Ac—N-acetyl glucosamine
KDO—2-keto-3-deoxyoctonic acid
Man—mannose
P—phosphate
Rha—rhamnose
Tyv—tyvelose

S→R variation

As a result of mutation the organism may lose, to a greater or lesser extent, the capacity to synthesize the polysaccharide of the somatic antigen. Types of variant have been recognized, differing in the extent to which this capacity has been lost. These antigenic mutants are in fact biochemical mutants, lacking the capacity to synthesize the appropriate sugar or sugar nucleotide required in transfer reactions or deficient in specific transferases. The most extreme defect is that in which there is a failure to assemble the polysaccharide backbone because of an inability to synthesize the heptose component. The commonest is that involving failure to synthesize the repeat units of the side-chains. In this case antigenic specificity is dictated by the core of the side-chain, particularly by the terminal N-acetyl glucosamine residues. Since the core is common to all *Salmonella* antigens, organisms in which this form of variation has occurred are antigenically indistinguishable. The loss is associated with a change in colony appearance from smooth to rough, because of which it is known as S→R variation. Rough organisms are also spontaneously agglutinable in saline—apparently because of a loss of hydrophilic properties conferred by the specific portion of the side-chains—and are avirulent.

S→R variation occurs infrequently under natural conditions but more frequently in strains which are maintained for long periods in laboratory cultures, and particularly when the medium contains a considerable amount of carbohydrate.

Diphasic flagellar variation

Many *Salmonella* species are *diphasic* in relation to their flagellar antigens. By this is meant that the organism may have one or other of two sets of flagellar antigens; these sets are known as phase 1 and phase 2 antigens. Although a particular bacterium can have flagellar antigens only in one phase, it can give rise to descendants with flagellar antigens in the second phase. In a mass culture of diphasic species one or other of the two phases is usually predominant so that the culture is agglutinated by only one of the phase antisera. In most cases the predominant phase is phase 1. Some *Salmonella* species, e.g. *Salm. typhi,* are *monophasic,* i.e. the flagellar antigens can exist in only one phase.

The first phase 1 antigens identified were designated by small letters a, b, c, etc., but, as more and more were discovered, the alphabet was soon exhausted and later antigens were therefore designated by z with a suffix number. At present at least 60 phase 1 antigens are recognized, the majority of *Salmonella* species possessing only one such antigen. Phase 1 antigens are therefore much more specific than O antigens. Phase 2 antigens, on the other hand, are less specific than either phase 1 H or O antigens. The majority of *Salmonella* species possess phase 2 antigens. Those first described were designated by Arabic numerals 1 to 7. When the phase 2 antigens of an organism belong to this series the antigen 1 is always present, usually with at least one other antigen of the series. None of this series of phase 2 antigens has so far been recognized as phase 1 antigens in any species. A number of species, however, possess, as phase 2 antigens, antigens which occur as phase 1 antigens of other species. The most important of these are e, n, x and z_{15}.

V→W variation

This is a variation involving loss of the Vi antigen. Loss of Vi antigen by *Salm. typhi* may be partial and the organism is then agglutinable by both O and Vi antisera. These intermediate variants are referred to as VW forms. Partial or complete loss of the Vi antigen renders *Salm. typhi* insusceptible to the action of the Vi bacteriophages used in bacteriophage typing.

Bacteriophage types

A number of *Salmonella* species can be subdivided into types on the basis of their susceptibility to bacteriophage. These are considered in detail in Chapter IV.13. Phage typing is of most importance in relation to *Salm. typhi* in which a total of 80 types and subtypes have been defined. The phages used for *Salm. typhi* are specific for the Vi form and are active only against organisms possessing their full quota of Vi antigen. Since the types exhibit high degree of stability under natural conditions, phage typing is of considerable value in epidemiological investigations, and it has been widely and successfully used to trace the source of outbreaks of typhoid fever. Thus, if a carrier is responsible for an epidemic, he should harbour the phage type found in that epidemic; if he does so a *prima facie* case for his responsibility is established.

Pathogenicity

The toxicity of the salmonellae is due to the somatic antigen complex. The basis of the toxicity for the somatic antigen has not been completely determined but clearly depends on the presence of lipid A—a minor phospholipid component of the complete antigen (p. 240). The smallest toxic molecule is a lipopolysaccharide obtained by phenol extraction and containing the somatic polysaccharide and an intact lipid A component. When the latter is modified by deacylation the phospholipid loses its toxicity. The evidence suggests that lipid A is the major determinant of toxicity but it is not at the moment possible to say that it is the exclusive determinant. The participation of lipid A in toxicity provides an explanation for the failure of antibody to the somatic polysaccharide to neutralize toxicity completely. Its capacity to do so at all can be attributed to a steric effect, by virtue of which it interferes either with the fixation or activity of lipid A.

The injection of small amounts of purified somatic antigen intravenously into man and other animals gives rise to pyrexia and leucopenia followed by leucocytosis. The pyrexia appears to be mainly due to the action of endogenous pyrogens, liberated from leucocytes as a result of the action of the endotoxin. When injected intravenously in large doses the endotoxin produces a condition of shock due to the intravascular coagulation syndrome.

Following injection of endotoxin into animals a tolerance to further injections, which lasts for some days, is induced. This tolerance extends also to the chemically related antigens of the shigellae and other Gram-negative bacilli; since it shows no serologic specificity it is not due to antibody production. It is possible that this non-specific tolerance may play a part in recovery from *Salmonella* infection, but there is no evidence on this point.

The somatic antigens of the salmonellae undoubtedly play an important part in determining virulence, since in the course of S→R variation the virulence of the organism is invariably lost, even though the variant rough form is still endotoxic. In *Salm. typhi* the Vi antigen, which is almost invariably present in freshly isolated strains, is believed to be of particular importance. It seems probable that by virtue of a strongly acidic character it may confer on the surface of the organism marked anti-phagocytic properties. Because of this, and its relatively low antigenicity—antibody to it being produced slowly and only in small amounts—it could contribute substantially to rendering Vi-containing cells much less susceptible to the humoral defence mechanisms of the host than non-Vi forms. This in turn would explain why typhoid bacilli are infective for man in much smaller numbers than are other salmonellae.

SALMONELLA INFECTIONS

The salmonellae are strict pathogens and have no habitat other than the human or animal body. The source of human infection is, therefore, a human or animal case or carrier, the organisms being excreted in the faeces or urine and transmitted by food or water which is ingested by another subject. The portal of entry of the organisms is always the mouth, irrespective of the subsequent sequence of events. Three types of clinical disease are associated with *Salmonella* infection: enteric fever, bacterial food poisoning and *Salmonella* septicaemia.

Enteric fever

The term 'enteric fever' should be used to indicate the *clinical* type of disease about to be described, irrespective of the causative *Salmonella* species. The terms 'typhoid fever' and 'paratyphoid fever', often used synonymously with 'enteric fever', imply a bacteriological diagnosis (of infection with *Salm. typhi* or *Salm. paratyphi A, B* or *C* respectively) and should be used only when this has been established. Enteric fever is characterized by fever of continuous type, enlargement of the spleen, usually a 'rose-spot' rash and leucopenia. Headache and constipation are common in the early stages of the disease but diarrhoea with 'pea soup' stools is usual later when ulceration of the bowel occurs. The causative organism can be isolated by blood culture, especially in the early stages of the disease, and later from the faeces and, sometimes, from the urine. The fatality rate of enteric fever normally lies between 10 and 20%.

Enteric fever is most commonly caused, throughout the world, by *Salm. typhi.* Other *Salmonella* species, especially *Salm. paratyphi A* and *B,* may, however, be responsible both for sporadic cases and for epidemics. In Western Europe and North America, *Salm. paratyphi B* infection is common while *Salm. paratyphi A* or *C* are seldom, if ever, isolated. In recent years there has been a considerable reduction in the number of cases of severe fever due to the typhoid and paratyphoid bacilli, with a concomitant relative increase in cases due to salmonellae normally associated with food poisoning.

The source of infection is a human case or carrier who excretes the organism either in the faeces or, less commonly, in the urine. In order that the disease may be spread, the organisms must be conveyed thence in sufficient numbers to the alimentary tracts of other people. There are two main ways in which this may occur:

1. By contaminated water. This epidemiological variety is most likely to occur in communities with inadequate systems of sanitation. It is unlikely to be found in communities with satisfactory methods of sewage disposal and water purification, unless there is some breakdown in these systems due, for example, to lack of adequate control, as in the 1962 Zermatt outbreak. Epidemics of enteric fever due to a contaminated water supply are characteristically explosive in onset and affect a large number of people. An uncommon mode of infection is the consumption of shellfish taken from the estuary of a river contaminated by sewage. A number of milk-borne epidemics have occurred in which the infection was traced to the use of contaminated water for the washing of bottles and other utensils. Canned foods may be infected if the cans are not completely watertight and if, after sterilization, they are cooled and washed in infected water. This appears to have been the mode of infection in the 1963 Aberdeen outbreak.
2. By food which has been handled by carriers. Outbreaks of this type are much less extensive than water-borne epidemics, affecting only the persons who have ingested the contaminated foods.

Enteric fever, despite its name, is not primarily a disease of the alimentary tract, but is rather a generalized infection in which the gall bladder and the lymph follicles of the small intestine are typically involved. It is probable that the bacteria which have entered the body by the mouth and have survived the acidity of the gastric juice do not at first proliferate in the small intestine but leave it by way of the lymph follicles. From there the organisms pass to the mesenteric lymph nodes and thence to the blood stream by way of the thoracic duct. A transitory bacteraemia follows, but the organisms are rapidly removed by the reticuloendothelial cells, especially by those of the spleen, liver and bone marrow which form the principal foci of infection. In them, and in the mesenteric lymph glands, the organisms multiply and produce a secondary bacteraemia which is of greater severity than the first and which marks the onset of the actual disease. At this stage, the bacteria are widely distributed throughout the tissues of the body and may readily be isolated by blood culture.

In the faeces the organisms reach a maximum during the third week and then gradually decline during convalescence. Reinvasion of the bowel occurs through two channels. The first of these is the infected Peyer's patches and solitary lymph follicles of the small intestine which discharge salmonellae into the bowel and which may later become the sites of ulceration, haemorrhage and perforation. The main reservoir from which salmonellae reach the lumen of the gut is the gall bladder which is constantly infected.

In about 25% of cases of typhoid fever, *Salm. typhi* may be found in the urine after the third week. Since the urine is free from pus there is probably no actual infection of the urinary tract. In contrast to this, when localization of infection in the renal system occurs in *Salmonella* septicaemia, the condition is always a purulent one.

Carriers

Following an attack of typhoid fever, the great majority of cases cease to excrete *Salm. typhi* during convalescence. About 5% of cases continue to excrete the organism during 6 months after the attack, while some 3% are still excretors at the end of a year. The majority, if not all, of those who are still excreting typhoid bacilli a year after convalescence become permanent carriers and continue to excrete the organism indefinitely.

The risk of a typhoid patient becoming a carrier increases with age. Thus in an investigation, reported by Ames and Robbins, in New York State, 10% of patients in the 50 to 60 age group became carriers, as against 2·9% at all ages. The risk appears to be appreciably greater in the female than in the male, particularly in the fifth decade. Since the seat of residual infection in faecal carriers is in the gall bladder, it can be reasonably presumed that the higher carrier rate in women is associated with their greater proneness to gall bladder disease. Urinary carriers are much less common than faecal ones and it is possible that urinary carriage only develops when there is some concomitant renal lesion.

Diagnosis

Blood culture. In enteric fever, isolation by blood culture is the method of choice. The earlier it is performed the greater is the chance of success. During the first week, *Salm. typhi* can be isolated from the blood of 90% of cases of typhoid fever, falling to about 40% in the fourth week.

Isolation from faeces or urine. The chances of isolation from the faeces increase from about 50% in the first week to about 80% in the third week. Thereafter the chances of isolation fall off rather sharply.

Cultures of faeces should be made by inoculating a tube of tetrathionate broth or selenite broth and dexoxycholate citrate medium. After overnight incubation, a loopful of the enrichment broth is plated on desoxycholate citrate agar. Pale non-lactose-fermenting colonies on DCA media are picked off and examined biochemically and serologically.

When other methods have failed it may be possible to isolate the organism from the urine. This can rarely be achieved before the end of the third week.

The Widal test. In the course of the enteric fevers, the patient usually develops a high titre of agglutinins for the causative organism. The detection and estimation of these antibodies is carried out by the Widal reaction. Since the diagnostic level of the antibodies to the various antigenic components H, O and Vi is different, antibody to each must be determined independently. For this purpose H, O and Vi suspensions must be used, standardized suspensions being commercially available so that the results obtained in different laboratories are comparable. If possible two specimens of serum, the second taken about 10 days after the first, should be examined to demonstrate a rising antibody titre. If this is not feasible a diagnosis may have to be made on the results of a single serum examination. In either case it is rarely possible to obtain serological evidence establishing or supporting a diagnosis of enteric fever before the beginning of the third week.

Identification of carriers

When large numbers of people are to be examined for the carrier state, a full bacteriological examination, although the ideal, is often impracticable. Under these circumstances examination of the patient's serum for antibody is of considerable value as a screening procedure. It has been generally accepted that the Vi antibody titre is of particular value for this purpose and that Vi antibodies are unlikely to be found in significant amount unless typhoid bacilli are still present in the body.

To establish a diagnosis of the carrier state the organism must be isolated and, since excretion may be very intermittent, this should if necessary be attempted on a number of occasions. If repeated examination of faeces and urine is unsuccessful it may be possible to isolate the bacillus from samples of duodenal contents obtained by intubation. When a carrier has been identified it is essential, in order to establish a presumptive link between the carrier and an epidemic, to show that the phage type of the organism which he carries is the same as that responsible for the epidemic.

Treatment

Chloramphenicol is the antibiotic of choice in the treatment of typhoid fever. It has a marked effect in diminishing the duration of the disease and in lowering the fatality rate. It is most effective in the early stages of the infection. Unfortunately, relapses are of frequent occurrence even after intensive treatment with chloramphenicol, and it does not prevent the development of the carrier state. Chloramphenicol-resistant strains have been isolated in Mexico. As an alternative to chloramphenicol a combination of trimethoprim and sulphamethoxazole has been favourably reported on.

Prophylaxis

The methods used for the prevention of enteric fever fall into two categories. The first and more important aims at preventing the spread of the organism. This is a public health problem, the essential measures being the provision of a good water supply and of satisfactory methods for the disposal of excreta, the isolation of cases and the control of carriers.

Chloramphenicol, which has been so effective in the treatment of the clinical disease, is unfortunately of no value in the treatment of the carrier. Good results have recently been claimed for ampicillin. This drug in fact appears to be more effective in the treatment of the carrier than of the established disease, but it appears to be necessary to prolong the treatment for some months. The carrier state can undoubtedly be cured by cholecystectomy, but this is, however, a somewhat drastic measure even in the interests of public health.

The second method, introduced by Almroth Wright at the end of the last century, is that of specific prophylaxis and aims at protecting the individual by active immunization. TAB vaccine is a suspension of killed *Salm. typhi* and *Salm. paratyphi A* and *B*. *Salm. paratyphi C* is sometimes included and the name TABC vaccine employed. TAB vaccine has been subjected to extensive trials under field conditions during two world wars, when conditions have been especially favourable for the spread of enteric fever, and there can be little doubt that its use has resulted in a marked lowering in the incidence of the disease among populations exposed to the risk of infection. At best this protection is of the order of 75%. Experience gained during World War II, however, strongly suggests that the severity and fatality rate of the disease among those who do contract it in spite of inoculation are no lower than among uninoculated people. Many experienced workers now consider that monovalent *Salm. typhi* vaccine, which gives much less reaction, is to be preferred to the combined vaccine. Vaccination is by two subcutaneous injections with an interval of four to six weeks between them. Booster doses are required every three years.

Immunization is normally considered necessary only for those who are exposed to a special risk of infection. It is used for service personnel, for people travelling to countries where sanitary arrangements are not satisfactory and for persons such as laboratory workers and staffs of infectious disease hospitals who are particularly exposed by virtue of their occupation. Vaccination is not necessary for children under two years of age, since they are relatively insusceptible to infection.

Salmonella septicaemia

The next clinical type of infection to be considered is *Salmonella* septicaemia. This resembles a septicaemia due to the pyogenic cocci, with a spiky temperature of intermittent or remittent type. Occasionally there is an antecedent history of diarrhoea, but this is by no means constant. The causative organism can usually be isolated from the blood during the course of the disease. This is the type of disease most frequently caused by *Salm. paratyphi C* and *Salm. choleraesuis*, but it may be due to any species except *Salm. typhi*.

Salmonella food poisoning

In great Britain the salmonellae are the most important single cause of bacterial food poisoning. In America, however, the salmonellae, though important, appear to take second place to *Staph. aureus* as causes of this condition. In general, *Salm. typhimurium* is the species most frequently encountered, but the frequency with which any species is prevalent varies in different places and at different times.The epidemiological behaviour of *Salm. heidelberg* is of particular interest. This organism was quite unknown as a cause of bacterial food poisoning in the United Kingdom prior to 1951 but since then has been isolated with increasing frequency; other European countries appear to have had a similar experience.

Infection is the result of ingestion of contaminated food, that most frequently responsible being meat, usually in the form of made-up or reheated dishes. Eggs are an important but less frequent source of infection and several outbreaks have been traced to the consumption of milk and sweets. The risk of *Salmonella* infection is greatest with foods which have been inadequately cooked and with food which has been kept warm after cooking, thus permitting bacterial growth. Meat may be infected because it has been derived from an infected animal or it may be infected at any time after slaughter. Cattle, sheep and pigs are frequently carriers of salmonellae. Occasionally the carrier state develops into invasive disease during transit from farm to abattoir. After slaughter meat may be infected in storage, while displayed for sale or in the course of preparation. This type of infection is frequently due to rodents, particularly mice which may infect the food directly if they have access to it, or the food may be indirectly contaminated from their droppings by flies or hands. A number of outbreaks have been traced to *Salmonella* 'virus' preparations, which were at one time popular for the eradication of mice. Cats and dogs may also be infected and might on occasion infect food. Infection may also occasionally occur from human carriers, but this mode of infection is probably not very frequent. Fowls and ducks are frequently infected and may infect their eggs. Infection of the egg may occur either in the oviduct or after laying; the latter is likely to occur if the egg is deposited in a moist environment which damages the surface waxy film, thereby permitting bacteria to pass through the shell. When eggs are pooled as in the preparation of custards, baking powders and spray-dried egg a few infected eggs may contaminate an entire batch. This may result in large-scale epidemics such as occurred following the introduction of American spray-dried egg in Great Britain in World War II.

Salmonella food poisoning is a gastroenteritis; only the alimentary tract is involved and general invasion of the body does not occur. Following ingestion of the contaminated food there is commonly an incubation period of about 24 hours, representing the time necessary for multiplication of the organisms in the intestine. In addition to vomiting and diarrhoea, which may be severe and prolonged over several days, there is usually pyrexia. The organism can frequently be isolated from the stools and vomit as well as from the infected food. In premature or debilitated children, *Salmonella* infection is liable to be particularly serious, on occasion giving rise to severe enterocolitis.

The value of chemotherapy is in doubt; it may actually prolong the period of parasitism with the organism.

Diagnosis

Bacterial food poisoning may also be caused by the following organisms:

1. Shigellae (Chapter III.10).
2. Strains of *Staph. aureus* which produce enterotoxin (Chapter III.1).
3. *Cl. perfringens* (Chapter III.18).
4. *Vibrio parahaemolyticus,* which is the cause of a toxic form of bacterial food poisoning first recognized in Japan (Chapter III.11).

In the diagnosis of bacterial food poisoning, the patient's faeces should be examined and also, if available, a specimen of vomit and some of the incriminated food. A direct film stained by Gram's method may be valuable in revealing gross contamination of the food. The following cultural examinations may be attempted: inoculation on to Leifson's medium and into selenite broth for isolation of salmonellae and shigellae; on to salt agar for isolation of *Staph. aureus*; into heated and unheated Robertson's meat medium; and on to an anaerobic blood agar plate for the isolation of clostridia. Serological tests for the demonstration of antibody are of little value in the diagnosis of bacterial food poisoning due to *Salmonella* species, since the condition is of such short duration. They may, however, permit a retrospective diagnosis, although the antibody response observed is usually poor.

III.10

The Shigellae

The shigellae are non-motile, non-capsulated, Gram-negative rods with the general properties of the enterobacteriaceae. Except for *Shigella sonnei*, which is a late lactose-fermenter, they are non-lactose-fermenters. Like the salmonellae, the shigellae must be finally identified by their serological reactions. They differ from the other enterobacteriaceae so far considered in that they occur exclusively as human pathogens.

Shigella dysenteriae. These are non-mannitol-fermenting shigellae and comprise eight serotypes of which the classical *Shigella shigae* is type 1. Some of the types show characteristic biochemical reactions. All ferment glucose producing acid but not gas.

Shigella flexneri. This species is subdivided into six antigenic types, each of the types possessing an antigen specific for the type. In addition, the different types posess group antigens which occur in each type either singly or in various combinations.

Shigella boydii. These strains biochemically resemble *Sh. flexneri*, but differ from it serologically. They possess type-specific antigens distinct from those of the *flexneri* group and do not possess group antigens. On the basis of their type-specific antigens they have been subdivided into 15 serological types.

Shigella sonnei. This organism differs in a number of respects from *Sh. flexneri*. It is a late lactose- and saccharose-fermenter. Late fermentation of these sugars is due to the appearance of variants in culture. If the organism is plated on to MacConkey's medium, the variants appear after some days' incubation as pink-coloured papillae in the centre of otherwise colourless colonies. Indole is not produced.

Shigella sonnei rapidly undergoes S→R variation. On primary isolation on MacConkey's medium colonies are frequently encountered which are partially rough and partially smooth—the rough organisms being present in a fan-shaped segment, the edge of which extends beyond the margin of the colony; this appearance is said to resemble that of a bomb crater.

On the basis of their capacity to produce various colicines *Sh. sonnei* strains have been assigned to 16 types each type being characterized by the production of a specific colicine. This procedure has been applied with some success as a method of epidemiological typing.

Pathogenicity

The shigellae are endotoxic, the endotoxins, like those of the salmonellae, being associated with the somatic antigens. *S. dysenteriae* type 1 produces in addition a potent enterotoxin demonstrable by the ligated rabbit gut technique. It has been isolated from cell-free filtrates and has a molecular weight of approximately 82 000. The toxin probably has a mode of action similar to that of cholera toxin (p. 251).

Bacillary dysentery

Dysentery is an inflammatory condition of the intestine, mainly the large intestine, characterized in its more severe forms by necrosis, sloughing and ulceration of extensive areas of the mucosa, with blood and mucus in the faeces. It may be divided into two main types: amoebic, due to *Entamoeba histolytica*, and bacillary, due to the shigellae. In the pathogenesis of bacillary dysentery the organism first attaches to the surface of the epithelial cells, which it then invades and in which it multiplies, spreading laterally to adjacent cells. Sloughing of the necrotic epithelium gives rise to the ulcers characteristic of the condition.

Cases of bacillary dysentery differ considerably in severity. In some there is only a mild diarrhoea and, as indicated in the previous chapter, the dysentery bacilli may on occasion give rise to the symptoms of bacterial food poisoning. The most severe type of dysentery is that caused by *Sh. dysenteriae* type 1. In this there is usually considerable toxaemia, sometimes with an effusion into the joints or evidence of myocarditis.

Dysentery occurs both as an epidemic and as an endemic infection. Epidemics of major proportions occur mainly in tropical communities and particularly in those with low standards of sanitation. In such communities the prevalence of dysentery shows a marked seasonal incidence, epidemics occurring at times of maximal density of the fly population. This relationship emphasizes the importance of flies, which act as mechanical vectors, in transferring infection from faeces to food.

In temperate zones dysentery exists primarily as an endemic infection but with a maximal incidence in the late winter and early spring. In European countries and in America, in spite of their advanced standards of sanitation and hygiene, endemic dysentery appears, in the last few decades, to have shown a marked increase in prevalence. In contrast to tropical and sub-tropical dysentery *Sh. sonnei* is the main organism responsible, with *Sh. flexneri* next in frequency. Both of these species have considerable tendency to establish a symptomless, but transient, carrier state. In the British Isles endemic dysentery is primarily an infection of urban communities with a maximum incidence in young children, the main foci of infection being communal nurseries, children's homes and primary schools. From such foci, once they have become established, dysentery bacilli can only be eradicated with the greatest difficulty. Dysentery is also wide-spread in mental institutions; this epidemiological variety, which is known as asylum dysentery, is usually associated with a more severe clinical disease than is the usual endemic type. In the maintenance of endemic dysentery flies are of minor importance, though they must of course be kept under control. The disease is mainly transmitted as a result of the casual contamination by cases and carriers of objects in their environment. The organisms are then transferred from contaminated objects by the hand of the recipient to the mouth. Communally used lavatories appear to be of particular importance in the spread of dysentery under these conditions.

Diagnosis

In the early stages of the disease microscopic examination may be of value in differentiating bacillary dysentery from amoebic dysentery. In bacillary dysentery 90% of the cells in the stools are of polymorphonuclear type while these cells are in a minority in amoebic dysentery. In the latter condition microscopic examination should reveal the presence of *Ent. histolytica*.

Rectal swabs may be used for isolation and, when large numbers of specimens must be examined, are very convenient. They do not, however, give as high a proportion of isolations as faeces. Since shigellae die out rapidly in faeces, culture should be attempted as soon as possible. Failing this, the specimen may be preserved for a few days in buffered glycerol saline. If mucus is present it should be used as an inoculum. The cultural techniques employed are the same as for the isolation of the salmonellae. Serological tests for the demonstration of antibody are not of value.

Treatment

Mild cases of dysentery are probably best not treated with an antimicrobial. For more severe cases the drug used should be chosen on the basis of sensitivity tests.

Prophylaxis

The prevention of dysentery depends on the same principles as does that of the enteric fevers. However, specific prophylaxis by vaccination has not so far been found feasible.

III.11

Vibrio, Spirillum, Pseudomonas

The three genera to be considered in this chapter are assigned by Bergey to the family Pseudomonadaceae which are distinguished from the Eubacteriales by their polar flagellation.

VIBRIO

The genus *Vibrio* comprises a group of non-sporing, non-capsulated, motile, Gram-negative rods with a curved shape and a characteristic 'darting' mobility. They grow readily on ordinary laboratory media. Most species are free-living organisms occurring naturally in soil and water. Some are pathogenic for various animals and only two, *V. cholerae* and *V. parahaemolyticus,* are pathogenic for man. Two biotypes of *V. cholerae,* the classical and the El Tor type, which show marked epidemiological differences, are distinguished.

Vibrio cholerae (Vibrio comma)

Morphology

The cholera vibrio measures from 1·5 µm to 3·0 µm by about 0·5 µm. As seen, usually in enormous numbers, in the dejecta from cases of cholera, it is definitely curved and from its resemblance to a comma it was given the name 'comma bacillus'. The curve, however, is not flat, but is in two planes. Frequently the organisms occur in pairs, giving either a C or, more commonly, an S shape. In the stools the majority of the vibrios appear to lie with their long axes parallel, this appearance being described as resembling 'fish in a stream'.

Cultural properties

Growth is best under slightly alkaline conditions (pH 7·6 to 8) and will occur at a pH of 9 to 9·5, sufficiently high to inhibit the growth of most other pathogenic bacteria. On agar the colonies of freshly isolated strains are usually thin and very transparent but a number of variant types have been described. In peptone water and in broth *V. cholerae* grows rapidly. In fluid media some strains produce a surface pellicle which tends to grow up the wall of the tube.

Indole is produced and nitrates are reduced to nitrites. The 'cholera red reaction'—the production of a pink colour on the addition of a mineral acid to a peptone water culture of the vibrio—is due to the presence of both indole and nitrites in the culture. It is a non-lactose-fermenter. Gelatin and serum are rapidly liquefied.

Antigenic structure

On the basis of their O antigen, the cholera and cholera-like vibrios can be divided into six serological groups. The cholera vibrio belongs to subgroup 1 of this classification. Two distinct antigenic subtypes of *V. cholerae,* the Inaba and the Ogawa varieties, have been distinguished.

Pathogenicity

By the use of the ligated rabbit gut technique, De and Chatterjee have shown that the cholera vibrio secretes an enterotoxin which causes accumulation of fluid in the gut lumen. The toxin has been found to produce diarrhoea in suckling rabbits, dogs and human volunteers when introduced directly into the gut; there is little doubt that it is the chief agent of the vibrio's pathogenicity. The enterotoxin has been classified as a heat-labile trypsin-resistant protein with a molecular weight of approximately 90 000. Its mode of action has not been completely clarified but its ultimate effect is to increase the secretion of fluid from the gut wall into the lumen. The immediate cause of this loss is stimulation of adenyl cyclase activity following combination of the toxin with gangliosides in the gut wall. This leads to an increase in cAMP which plays some as yet unidentified role in the secretion of fluid. Evidence that cAMP concentration is the decisive factor has been obtained in the following findings: (*a*) prostaglandins and theophylline which increase cAMP concentration markedly increase fluid secretion; (*b*) ethacrynic acid, an adenyl cyclase inhibitor, causes reduction in the amount of fluid secreted in response to the toxin; and (*c*) the toxin causes an increase in adenyl cyclase activity in intestinal cells of laboratory animals.

The toxin also has the effect of increasing the permeability of rabbit skin following intradermal inoculation. The skin reaction is extremely sensitive and can be produced by as little as 0·1 mg of purified toxin.

The cholera vibrio produces a neuraminidase (receptor-destroying enzyme) resembling that of the influenza virus (p. 421). No role in pathogenicity has so far been attributed to this enzyme, nor to a mucinase which is demonstrable in tissue cultures and which can break down intestinal glandular mucin.

Cholera

Cholera is an acute disease marked by intense diarrhoea and vomiting. As a result there is a great loss of fluid which may, in severe cases, amount to as much as two-thirds of the plasma volume. The dehydration in turn leads to haemoconcentration, anuria and severe muscular cramps. In the absence of treatment, case mortality rates of the order of 60 to 70% have been frequent. Death is usually due to renal failure. The choleraic stool has a characteristic appearance. It is almost completely fluid with flakes of mucus floating in it, the common term 'rice water stool' being a fairly good description of this appearance. The stool contains large numbers of cholera vibrios, in some cases up to 10^9 organisms per ml having been observed.

The intestinal mucosa is not invaded and there is never a general bacteraemic spread of the organisms through the body. They appear in fact to multiply largely on the luminal surface of the intestinal mucous membrane. Biopsy examination during life has shown that the epithelium is, to all intents and purposes, normal in appearance.

The vibrios are rapidly eliminated, the great majority of cases ceasing to excrete them about three weeks from the onset of the disease. It is doubtful whether a persistent carrier state ever develops; if it does it is extremely rare. Healthy persons who have been in contact with a case may, however, become infected and may excrete the organism for a short time without the development of clinical symptoms. Such subjects, known as contact carriers, are of considerable importance in maintaining cholera as an endemic infection. Occasionally contact carriers after carrying the organism for some time develop the clinical disease, possibly as a result of a dietary indiscretion; in these circumstances they are sometimes referred to as precocious carriers.

Classical cholera is characteristically an epidemic disease. It was virtually unknown outside India before the nineteenth century, but since then many epidemics have occurred throughout the world—the most recent being in Egypt in 1947 and Thailand in 1958. In the Ganges delta in India it appears to have been endemic for centuries and this appears to have been the main source from which most pandemics have originated and from which infection of the rest of the Indian subcontinent has occurred. The magnitude of this endemic infection is shown by the fact that in East Pakistan alone there were approximately a quarter of a million deaths from cholera in the period 1948 to 1959. In India and Pakistan the disease achieves a maximum epidemic prevalence in the months April to June and October to December. A second great focus of endemic infection is in the Yangtse valley in China. Notable water-borne epidemics outside India were the London Broad Street pump epidemic of 1854, in which, for the first time, the role of water in the spread of cholera was brilliantly established by John Snow, and the Hamburg epidemic of 1892 which dramatically demonstrated the effectiveness of sand filtration as a method of controlling water-borne disease.

In endemic areas at non-epidemic periods the disease is probably maintained, like endemic dysentery, by casual and environmental contamination from both cases and carriers. In the transmission of the disease from one place to another the precocious or incubating carrier is of special importance. Major epidemics of cholera appear to be due, almost invariably, to contaminated water supplies in which, particularly if alkaline, the vibrios may survive for a considerable time; this is, apparently, a frequent cause when water is stored in large tanks, as in certain tropical communities.

The El Tor biotype was first isolated in 1905 from pilgrims returning from Mecca who were being held in the El Tor quarantine camp in the Sinai peninsula. Prior to 1960 the organism was identified as a pathogen only in Indonesia where it caused a cholera-like disease of high infectivity. Since 1960, however, the El Tor biotype has shown remarkable capacity for epidemic spread, reaching India in 1963 and Europe in 1970. This may be due to its capacity to survive outside the body, which is greater than that of the classical cholera vibrio. It is more frequently responsible than the latter for mild or asymptomatic infections, more frequently infects children and has a higher capability for establishing a symptomless carrier state.

Diagnosis

A provisional diagnosis of cholera can frequently be made by direct microscopic examination of the stools; the occurrence of large numbers of vibrios in a 'rice-water stool' is virtually conclusive evidence of infection. The vibrios may be sufficiently numerous in the faeces to permit the direct performance of agglutination tests with faecal material.

In isolating the organism it is essential that cultures, for which either faeces or a rectal swab may be used, should be carried out as soon as possible. If delay cannot be avoided, the specimen should be kept in a buffered saline mixture; for this purpose a borate saline mixture has been described as being satisfactory. The specimen should be inoculated into a tube of alkaline peptone water and on to an appropriate solid selective medium. After about six hours the peptone water will usually show an extensive surface growth which can then be subcultured onto the selective medium employed. A variety of selective media has been recommended for the isolation of the cholera vibrio. These include Dieudonné's alkaline blood agar, MacConkey's medium and Monsur's medium which contains tellurite, taurocholate and gelatin. On this the vibrio produces a colony with a black centre and surrounded by a definite halo as a result of gelatin breakdown.

Strict criteria are required for the identification as *V. cholerae* of a vibrio isolated from the faeces of a suspected carrier or from a sample of water. The main differential characteristics of *V. cholerae* are fermentation of saccharose and mannose but not of arabinose—a positive cholera red and a negative Voges–Proskauer reaction. In most cases an organism giving these reactions is found to be the cholera vibrio. The organisms should be agglutinated by an O sub-group 1 serum.

The El Tor biotype was initially distinguished from the classical type by the fact that it produced a haemolysin for goat red cells. Many strains now isolated, however, do not produce haemolysin under normal cultural conditions. The most important distinguishing features are its capacity to cause haemagglutination of fowl red cells, its comparative resistance to polymyxin B and to cholera phages to which classical vibrios are susceptible.

Serological tests on the patient's serum are of little value in diagnosis during the acute stage of the disease. They may, however, be of some value in establishing a retrospective diagnosis and in the detection of carriers. The tests are carried out with a *V. cholerae* O suspension.

Treatment

Though the cholera vibrio is sensitive to sulphonamides and to various antibiotics, chemotherapy appears to have little, if any, effect on the clinical course of the disease. It may, however, diminish the period of convalescent carriage and therefore reduce the risk of the patient being a source of further infection. The most important measures in treatment are those designed to make good the fluid loss. The adoption of such measures has, in recent years, effected a reduction in case mortality from about 70% to about 20%.

Prophylaxis

As in the case of enteric fever, a dramatic reduction in the incidence of cholera followed the introduction of the large-scale purification of water supplies and of sanitary methods for the disposal of excreta. These methods, combined with hygienic handling of food, should control epidemic spread of the disease; it may, however, be difficult to eradicate the disease from the remaining endemic foci and more difficult to eliminate the El Tor than the classical biotype.

A killed vaccine prepared from the Inaba and Ogawa strains is widely used for active immunization. Vaccination is employed for people going to countries where cholera is endemic and also in mass immunization programmes in endemic zones. If possible, two injections should be given with an interval of four weeks between each. In mass immunization programmes only one injection may be feasible. The vaccine is much less toxic than TAB vaccine, with which it may safely be combined. Immunity produced by vaccination is of short duration and booster doses should be given six-monthly for those who continue to be exposed to infection.

Vibrio parahaemolyticus

This is a marine vibrio which is widespread in eastern Asia and is responsible for a toxic variety of food poisoning. It can be isolated from about half the cases of food poisoning during the summer months in Japan and is being increasingly recognized as a cause of this condition throughout the world. On occasion it causes a severe cholera-like disease and it appears to produce a cholera-like enterotoxin. It is isolated on media containing a high proportion of salt.

SPIRILLUM

This is a genus of small, actively motile spiral organisms morphologically resembling the spirochaetes. Spirilla differ from the latter, however, in being rigid and in owing their motility to polar flagella.

Spirillum minus

This is the only pathogenic member of the genus. It is a common parasite of the respiratory tract of rats in various parts of the world. It is a small, actively motile organism—measuring from 2 to 5 μm in length, usually with two or three complete spirals. It has not so far been maintained in culture on inanimate media and is consequently regarded by some workers as being more properly classed with the spirochaetes than with the bacteria.

Spirillum minus is the cause of a variety of rat-bite fever. In this condition after some weeks, the site of the wound, which may have healed, becomes inflamed and breaks down, the drainage glands become swollen, there is pyrexia, sudden in onset and remittent in type, and a purpuric skin rash. The spirilla are present in the blood but are few in numbers and are difficult to demonstrate. The easiest method of diagnosis is to inject blood intraperitoneally into a mouse or guinea-pig, in the blood of which the spirilla can be demonstrated after about 10 days by dark ground illumination.

This variety of rat-bite fever must be distinguished from that due to *Streptobacillus moniliformis* (p. 284). Both varieties respond to treatment with penicillin.

PSEUDOMONAS

Organisms of the genus *Pseudomonas* are widely distributed in nature occurring as free-living organisms in the soil, in water and in decomposing organic material. Bergey lists over 100 species, many of which are plant pathogens. A striking feature of the genus is the great diversity of compounds which can be utilized by different members for energy production.

Pseudomonas aeruginosa (Pseudomonas pyocyanea)

Cultural properties

Pseudomonas aeruginosa is an actively motile, non-sporing, non-capsulated, Gram-negative rod. It is a strict aerobe and grows readily on ordinary media at 37°C. It produces two pigments: pyocyanine, which is blue, and fluorescin which is greenish yellow and fluorescent. The pigments diffuse throughout the medium, giving the latter a bright green colour by which the organism is readily identified. On milk agar a degree of clearing due to caseinogenase is found around the colonies. Some strains, particularly those from chronic lesions, have a mucoid character. Most strains break down glucose which is the only common sugar attacked, with the production of acid but no gas. This is not strictly a fermentation but an oxidation and results in the production of gluconic acid. Gelatin is rapidly liquefied. All strains give a positive oxidase reaction. Cetrimide agar is a useful selective medium for isolation.

Pathogenicity

Pseudomonas aeruginosa appears to possess little power of initiating infection by itself. Not infrequently, however, it occurs as a secondary invader of infected tissues or of tissues which have been traumatized by operation or other manipulation. Children and persons with diminished resistance, e.g. from administration of corticosteroids, are particularly susceptible to infection. As the species *Ps. aeruginosa* is highly resistant to the commonly used antibiotics and because of its capacity to persist and even to multiply in unpromising environments, it has in recent years assumed considerable prominence as a cause of hospital infections, especially of infections of wounds and of the urinary tract. Most *Pseudomonas* infections of the urinary tract occur either as a sequel to operations, following catheterization or complicating chronic genitourinary infections. The capacity to produce specific bacteriocines (pyocines) has been of value in epidemiological investigations. Typing by bacteriophage susceptibility, pyocine susceptibility and serology has also been performed.

Pseudomonas aeruginosa has a special preference for moist environments; it has been frequently isolated from disinfectant solutions—notably chloroxylenol and cationic detergents—to which it is often highly resistant. Its occurrence in eye drops and lotions is a particular hazard in ophthalmology since it may produce a post-operative ophthalmia which may lead ultimately to blindness. Other important sources of infection are bed-pans, urine bottles and water taps.

For the treatment of *Pseudomonas* infections, polymyxin B, E gentamicin and carbenicillin are the drugs of choice.

Pseudomonas mallei (Loefflerella mallei)

Pseudomonas mallei, the causative organism of glanders, is a non-motile, non-capsulated, non-sporing Gram-negative bacillus, which sometimes occurs in the form of long, slightly branched filaments. It is aerobic and grows moderately well, but rather slowly, on ordinary media. For purposes of isolation, blood agar is quite satisfactory.

Guinea-pigs are highly susceptible. In male animals, following intraperitoneal inoculation, marked inflammatory changes are found in the tunica vaginalis of the testis. This is known as the Straus reaction. Guinea-pig inoculation, preferably by the subcutaneous route, may be used as a method of isolation, the organism being recoverable from the regional lymphatic glands.

Glanders

Glanders is a disease essentially of the horse family, man being only occasionally and incidentally infected. The disease in both animals and man is characterized by the occurence in infected tissues of granulomatous nodules composed of leucocytes, epithelioid cells and connective tissue. Glanders is now virtually unknown in horses in the British Isles but still occurs in Eastern Europe and in parts of Asia.

Infection with *Ps. mallei* is rare in man and is found mainly in those whose work involves close contact with horses—the organism usually gaining access through wounds and abrasions. Cultures of the organism are, however, very highly infective and have given rise to a number of laboratory infections.

In man, glanders may be either acute or chronic. The primary lesion is generally in the skin, more rarely in the mucous membrane of the nose, mouth or eye and is associated with inflammation of the draining lymphatic vessels. In acute cases there is pyrexia, very severe prostration and usually a generalized pustular rash. In chronic cases secondary foci occur throughout the body.

Pseudomonas pseudomallei (Loefflerella pseudomallei)

Pseudomonas pseudomallei is a short, Gram-negative bacillus, frequently showing bipolar staining. It resembles *Ps. mallei* closely, differing antigenically and in being motile and in liquefying gelatin. It is pathogenic for guinea-pigs and rats and produces a Straus reaction in male guinea-pigs.

Pseudomonas pseudomallei is a normal parasite of the respiratory tracts of many rodents. In man it

produces melioidosis—a disease, usually of acute character, in which lesions resembling those of glanders occur in the lungs and other organs. Melioidosis is a rare disease and is confined to tropical countries. Tetracycline appears to be the drug of choice for treatment.

III.12

Yersinia, Pasteurella, Francisella

Yersinia pestis

Morphology

Yersinia pestis is found in its most characteristic form in material taken from early plague lesions in man or animals. It is short and thick, 1 to 2 μm by 0·5 to 1 μm in size. It has rounded ends, presenting a boat-shaped appearance, and may frequently appear almost coccal in outline. It occurs in the body either singly or in pairs, short chains being exceptional. In fluid media it tends to grow in long chains. In the body and on serum agar at 37°C it possesses a capsule which is often, however, rather poorly defined. *Y. pestis* is non-motile.

Cultural properties

The bacillus grows on ordinary media but growth is rather slow. The optimum growth temperature is 28 to 30°C. On blood agar after 24 hours' growth the colonies are very small—0·1 to 0·2 mm in diameter—translucent and slightly viscid. It is non haemolytic. If grown in broth on the surface of which a little oil or butter is floated the organism grows in a stalactite formation suspended from the undersurface of the oil.

On the basis of glycerol fermentation and nitrite production three epidemiological types, each of which shows a specific geographic distribution are distinguishable.

Antigenic structure

Yersinia pestis has a complex antigenic structure. Three different types of antigen have been described:

1. A heat-labile capsular protein antigen. This antigen is only produced in significant amounts when the organisms are grown at 37°C. It is the antigen responsible for the stimulation of protective antibody.
2. The plague toxin (see below).
3. A somatic antigen complex. One or more of the components of this complex are identical with somatic antigen components of *Y. pseudotuberculosis*.

Pathogenicity

The plague bacillus produces a protein endotoxin which can be converted by treatment with formalin into toxoid. The toxin will adsorb to tannic acid treated red cells and antitoxin can consequently be titrated by the passive haemagglutination technique. The toxin has a damaging effect on the peripheral blood vessels. On intradermal injection in laboratory animals it causes a local oedema which may be followed by tissue necrosis. The toxin presumably plays a part in the production of the tissue lesions found in plague.

Plague

Plague is primarily an infection of rodents, particularly of the rat, in which it generally takes the form of an acute and rapidly fatal septicaemia. A rat usually becomes infected by the bite of a flea, tick or louse which has previously bitten a rat suffering from plague septicaemia. Man is only incidentally infected and except in pneumonic plague the disease is not transmissible from man to man.

Human plague is a most serious disease. It occurs either in large-scale epidemics in towns or cities, this type being known as *urban plague,* or as sporadic cases in rural areas—the so-called *rural* or *sylvatic plague*.

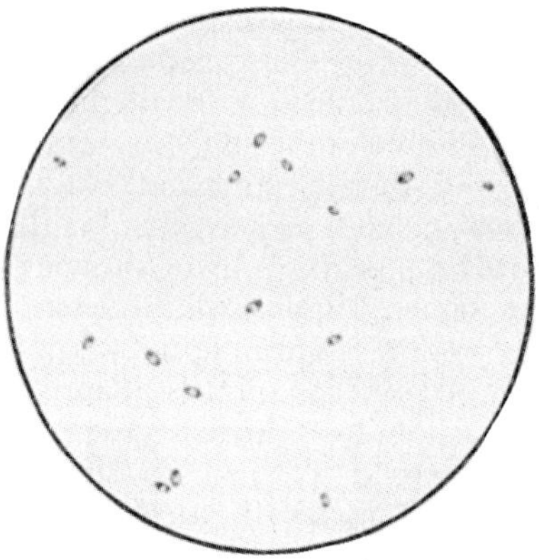

Fig. III.6. *Yersinia pestis* from a young agar culture. × 950.

Urban plague is now mainly a disease of tropical and subtropical countries and is spread to man from infected rats by rat fleas, the most important of these being *Xenopsylla cheopis*. The flea becomes infected when it bites an infected rat; the organisms then multiply in the proventriculus of the flea and are regurgitated into the wound when the flea bites a fresh victim. Infection may also apparently occur on occasion, through minute abrasions of the skin, from bacilli present in dust.

Epidemics of urban plague are always preceded by an outbreak (epizootic) in rats. Grey rats are usually the variety first infected, the disease then spreading to the black rat. Finally, when the rat population is seriously reduced in numbers by the disease, the fleas turn to man as an alternative host. Epidemics of plague occur only when the ambient temperature is in the range 10 to 30°C. At temperatures above 30°C fleas rapidly become non-infective. The human flea *Pulex irritans* appears to be incapable of transmitting the disease.

From time to time pandemics of plague have swept throughout the world and all communities appear at one time or another to have been infected. It is probable that the Black Death of the fourteenth century was a pandemic of plague, at first bubonic and later pneumonic, spread by ships returning from the Middle East. From being one of the great scourges of mankind, plague is now, presumably as a result of control of the rodent and vector populations, of relatively infrequent occurrence. This fall in the incidence of the human disease has unfortunately, however, been associated in certain parts of the world with progressive extension of the prevalence of *Y. pestis* in wild rodent populations.

Sylvatic plague appears as a sporadic disease occurring in rural areas—the sources of infection being various wild rodents, some 200 varieties of which have now been shown to be infected. In different parts of the world the infection predominates in different species. Thus, ground squirrels and voles are of particular importance in the USA and bandicoots in India. Amongst the rodent population the disease is spread by fleas and other biting insects. Human infection, however, more often occurs as a result of handling or being bitten by an infected animal and has been found chiefly in workers in woods, in hunters and in children. In addition to giving rise to occasional human cases the wild rodent reservoir is of considerable importance as the continuing source of infection for the rat population and its existence therefore constitutes a serious menace to man.

In man plague occurs in two clinical forms: *bubonic*, which is much the commonest variety, and *pneumonic*. Bubonic plague is characterized by the appearance of swelling and inflammation in the lymphatic glands draining the site of the bite. Since bites usually occur on the leg the inguinal glands are those most frequently involved. The pathological condition of the swollen glands, which are known as buboes, is one of intense inflammation with haemorrhages, leading eventually to necrosis. In the early stages the glands are packed with enormous numbers of coccoid, polar-staining bacilli and a film prepared from the gland pulp may resemble a pure culture of the organism.

Pneumonic plague is characterized by a rapidly spreading bronchopneumonia of haemorrhagic type. Since the bacilli are present in large numbers in the sputum the disease is extremely infectious. As might be expected of a respiratory disease, epidemics are commonest in the colder months of the year. The disease probably originates in the first place from a patient suffering from bubonic plague in which the infection has localized, as a result of metastatic spread, in the lung. Pneumonic plague does not occur in epidemic form in India; epidemics appear to occur almost exclusively in Manchuria, Northern China and Mongolia.

Diagnosis

The demonstration of typical polar-staining

bacilli in fluid aspirated from a bubo or in the sputum of a case of suspected pneumonic plague is strongly suggestive of infection. Isolation of the bacillus is readily achieved by inoculation on to blood agar, the plates being incubated at 30°C. Since the organism is highly infective, laboratory diagnosis should be attempted only by skilled and immunized personnel. Identification is by demonstration of the morphological and biochemical characteristics of the organism and of its agglutinability by specific antiserum. The susceptibility of *Y. pestis* to a specific bacteriophage is of value in differentiating this species from *Y. pseudotuberculosis*. A particularly detailed and careful examination, which should include a virulence test in guinea-pigs or white rats, is essential for the identification of organisms isolated from rats or from suspected human cases in non-plague areas.

The organism may also be isolated by direct inoculation of a guinea-pig. This should be attempted if few or no organisms are demonstrable in smears prepared from the bubo. If the material for examination has undergone extensive secondary bacterial invasion, some of it should be rubbed on to the shaved skin of the guinea-pig; if *Y. pestis* is present it will penetrate the skin and set up an infection. Infected guinea-pigs usually die within a week after inoculation when the characteristic local and general lesions may be demonstrated at post-mortem examination.

Yersinia pestis can frequently be isolated from the blood in the early stages of the disease. For this purpose a broth containing cystine appears to give best results.

Treatment

Tetracycline is the drug of choice in the treatment of plague.

Prophylaxis

In countries where plague occurs the most important methods in its prevention are: the reduction of the rat population by the use of chemical rodenticides—dicoumarin, sodium fluoroacetate and α naphthyl-thiourea, the control of fleas for which DDT is of great value, and the early diagnosis of plague epizootics in rats.

Killed vaccines have been widely used for active immunization. They appear to give appreciable, if not complete, protection against infection and somewhat reduce the severity of the disease in those who, in spite of inoculation, contract it. The immunity is of short duration and revaccination at six-monthly intervals is necessary for those who continue to be exposed. A living vaccine prepared from a mutant of diminished virulence has also been employed. It has been found to confer much greater protection in animals than heat-killed vaccine. The vaccine, which is given in a single dose, appears to be particularly suited for mass immunization campaigns in native populations.

Other yersiniae

Yersinia pseudotuberculosis is the cause of pseudotuberculosis, a disease which occurs in epizootic and enzootic form in a variety of animals. In the chronic type of the disease, nodular lesions resembling those of tuberculosis occur in lymphatic glands, liver, spleen and lungs. In rodents the disease may occur in an acute form which resembles plague.

Yersinia pseudotuberculosis differs from *Y. pestis* in the following respects: it is non-capsulated, motile in 18-hour cultures at 22°C, urease-positive and insusceptible to *Y. pestis* bacteriophage. Serological differentiation of the two species, which are antigenically related, is not always satisfactory.

At one time human infection with *Y. pseudotuberculosis* was thought to be extremely rare; in recent years, however, an increasing number of infections due to this organism have been reported. These infections may take the form of a severe typhoid-like disease with a high fatality rate or of a relatively mild condition involving the caecum and terminal ileum and their draining lymphatic glands and often simulating acute appendicitis. This latter condition has been observed mainly in children and young adults and is the more frequent type of human infection. The demonstration of agglutinins, using live suspensions of *Y. pseudotuberculosis*, appears to be of value in establishing the diagnosis.

Yersinia enterocolitica closely resembles *Y. pseudotuberculosis*, from which it differs in various biochemical reactions. It is reported to cause gastroenteritis, appendicitis and mesenteric adenitis following the ingestion of food infected from animal sources. However, its precise pathogenic role has yet to be determined.

Pasteurella multocida

Pasteurella multocida is the cause of chicken cholera and of haemorrhagic septicaemia in various mammals. It is capsulated, non-motile at 22°C, produces indole and is unable to grow on desoxycholate citrate medium.

Pasteurella multocida rarely infects man, but the number of diagnosed human infections appears to be increasing. Many of these have been cases of localized wound infection following an animal bite, but generalized infection, in some cases showing localization in the brain or lungs, has been reported.

Francisella tularensis

Francisella tularensis is a very small Gram-negative bacillus which exhibits very marked pleomorphism in culture In the tissues it is found in large numbers inside the tissue cells, where it appears to be capsulated; capsulation has not, however, been demonstrated in culture. It is non-motile and non-sporing.

It will not grow on ordinary media but may be isolated, though with difficulty, by the use of special media, e.g. coagulated egg yolk or a medium containing glucose, rabbit blood and cysteine.

Tularaemia

Tularaemia is a plague-like infection of wild rodents. It appears to be widely distributed in the USA where it was first reported and is present also in Japan and in northern Europe. In most animals the disease appears to be spread mainly by ticks which can transmit the parasite transovarially. Human infection is uncommon and is usually due to the handling of infected animals, particularly rabbits and hares, the infection probably occurring mainly through abrasions of the skin. In the commonest type of infection a small papule which eventually breaks down to form an ulcer appears at the site of introduction of the organism, usually on the finger; the local lymph glands become enlarged and may undergo ulceration. Infection may also occur through the conjunctiva with associated involvement of the pre-auricular and cervical lymphatic glands. A generalized form of the disease also occurs probably as a result of respiratory or intestinal infection. At post-mortem examination there are widespread necrotic lesions in various organs.

Very successful results have been obtained in cases of tularaemia by treatment with streptomycin.

Diagnosis

The best method of isolating the organisms is to inject a guinea-pig intraperitoneally with material from the human lesions and obtain a culture from the necrotic foci which develop in the liver, spleen and lungs of the animal.

The agglutination test is of considerable value—diagnostic titres being usually found in the second week of the disease. The serum of tularaemia patients may also agglutinate *Br. abortus* to a fairly high level. A delayed type of hypersensitivity may be demonstrated by the intradermal injection of a killed suspension of the organism. The skin test is usually positive during the first week of the disease.

III.13

Haemophilus, Bordetella

HAEMOPHILUS

This is a genus of very small non-motile, non-sporing aerobic Gram-negative bacilli of highly parasitic character. All species require one or other or both of two accessory growth substances—X factor or haematin and V factor or phosophopyridine nucleotide. Both of these factors are conveniently supplied in blood.

Haemophilus influenzae

Morphology

The morphology of this organism varies considerably. The type most frequently observed is a short coccobacillus measuring 1·5 × 0·3 μm. Some strains are more definitely bacillary and some are filamentous in type. There is a tendency for coccobacillary organisms to become filamentous on subculture. Fully virulent strains are capsulated; strains occurring as secondary invaders are frequently non-capsulated.

Cultural properties

Haemophilus influenzae is strictly aerobic, little if any growth occurring under anaerobic conditions. It requires both X and V factors for growth. On blood agar, colonies are just visible as very minute, clear, drop-like non-haemolytic dots at the end of 24 hours. Better growth is obtained on media in which the contents of the red cells have been liberated either by heat, as in the so-called 'chocolate' agar and in Levinthal's medium, or by tryptic digestion as in Fildes's medium. The latter two media have the advantage of being quite transparent; when grown on them capsulated strains yield iridescent colonies and non-capsulated strains non-iridescent colonies.

Whether grown deliberately or accidentally in mixed culture with staphylococci, *H. influenzae* show the phenomenon of satellitism; this is a great increase in the size of the colonies in the immediate vicinity of a staphylococcal colony and is due to the liberation of V factor from the staphylococcus.

Haemophilus influenzae is a very delicate organism and dies out rapidly outside the body. It is particularly susceptible to drying. It is important therefore that culture should be carried out as soon as possible after the clinical material has been taken.

Antigenic structure

Capsulated strains can be divided into six serologic types, a to f, the type-specificity of which depends on haptenic polysaccharides present in the capsule. Typing can be carried out either by the agglutination or capsule-swelling reactions. The capsules are readily lost in laboratory culture. Many of the strains isolated from normal throats and some of those obtained from lesions are non-capsulated on primary isolation.

Infections due to Haemophilus influenzae

Haemophilus influenzae owes its name to the belief, at one time widely held, that it was the cause of influenza. This belief arose from the frequency with which it was encountered in cases of influenza in the great pandemics of 1890 and 1918 to 1919. Many of the strains isolated were undoubtedly responsible for a secondary pneumonia complicating the disease, but others were almost certainly non-pathogenic commensals.

Haemophilus influenzae can frequently be isolated from the normal respiratory tract; these strains are usually non-capsulated and avirulent. The organism is frequently a secondary invader of the respiratory tract in influenza and is occasionally responsible for pneumonia complicating the disease. *H. influenzae* plays a role of great importance in the acute exacerbations occurring in the course of chronic bronchitis and is apparently more frequently responsible for these exacerbations than any other organism.

Haemophilus influenzae is an important cause of acute pyogenic meningitis in children, ranking next in importance as a cause of this condition to meningococci and pneumococci. *H. influenzae* meningitis is sporadic in occurrence. The majority of infections occur in children under two years of age. In its pathogenesis the condition closely resembles meningococcal meningitis. The organism first sets up a focus of infection in the pharynx from which it invades the blood stream, meningeal involvement occurring as a result of this haematogenous spread. Occasionally the organism localizes in the joints, where it produces an acute pyogenic arthritis, or in the valves of the heart with the production of subacute bacterial endocarditis.

By direct spread from the nasopharynx, it may give rise to sinusitis, otitis media and, in children, epiglottitis. In the last condition, the oedema and inflammation produced may be of very rapid development and may be so severe as to cause marked respiratory obstruction. *H. influenzae* is an occasional cause of acute conjunctivitis.

Diagnosis

In cases of *H. influenzae* meningitis the organisms can usually be demonstrated in Gram films of the centrifuged deposit from the cerebrospinal fluid and as a rule show the typical coccobacillary morphology. They are characteristically found within the polymorphonuclear leucocytes. It may be possible to detect the presence of type-specific antigen in the cerebrospinal fluid by the ring precipitin technique.

Blood agar may be used for routine cultures from sputum, but when, as in meningitis, *H. influenzae* is sought specifically, its detection is rendered easier if one of the special media such as Levinthal's or Fildes's medium is used; these media, being transparent, permit the ready recognition of the characteristic iridescent colonies of capsulated strains. Their identity may be confirmed by the Neufeld reaction with specific antiserum, if this is available. The majority of infections in children are due to type b strains. If a specific antiserum is not available, the identification of *H. influenzae* may be achieved by demonstration of its specific growth requirements. Primary infections in childhood are frequently associated with bacteraemia and in such cases blood cultures should be carried out.

Treatment

Chloramphenicol has so far been the antibiotic of choice in the treatment of meningitis on account of its ready diffusion into the cerebrospinal fluid. Clinical experience has shown that it is best combined with a sulphonamide. *H. influenzae* is also highly sensitive to ampicillin but this does not readily gain access to the cerebrospinal fluid. Its role in the therapy of *H. influenzae* meningitis has still to be determined. Ampicillin, the tetracyclines and trimethoprim–sulphamethoxazole have been used in the treatment and chemoprophylaxis of chronic bronchitis.

Haemophilus parainfluenzae and Haemophilus haemolyticus

These species, which are frequently found as respiratory commensals, produce colonies similar to those of *H. influenzae*. *H. parainfluenzae* resembles the non-capsulated strains of *H. influenzae* but differs in requiring V factor but not X factor for growth. It occurs as a respiratory commensal but can cause conjunctivitis and subacute bacterial endocarditis. *H. haemolyticus* requires both X and V factors and is haemolytic on blood agar.

Haemophilus ducreyi

Haemophilus ducreyi is difficult to isolate; it requires X factor and is haemolytic. It is the cause of soft chancre or chancroid, a fairly common venereal disease which may be confused with syphilis.

Haemophilus ducreyi shows the same susceptibility as *H. influenzae* to chemotherapeutic agents, but treatment with sulphonamides or streptomycin is preferred by most workers, since these drugs will not mask primary syphilitic lesions as may occur if the condition is treated with a drug active against *Treponema pallidum*.

BORDETELLA

Bordetella pertussis

Morphology

Bordetella pertussis is very similar to *H. influenzae* in morphology but is inclined to be more definitely bacillary and less pleomorphic. Capsules are demonstrable in young cultures.

Cultural properties

Bordetella pertussis is difficult to isolate, requiring specially enriched media such as that of Bordet and Gengou, which is a glycerol potato extract agar containing a high concentration of defibrinated horse blood. The medium may be rendered considerably more selective by the addition, as recommended by Lacey, of a diphenylamine and of penicillin in a concentration of 0·25 units/ml. The colonies after 72 hours incubation are very small, glistening and with a pearly appearance. They are surrounded by a zone of haemolysis which has an ill-defined periphery. Although blood is required for its isolation *Bord. pertussis* can be grown in subculture on agar containing serum, albumen or charcoal. These substances do not supply a nutrient requirement but permit growth by neutralizing traces of oleic acid which are usually present in ordinary media and on glassware, and to which *Bord. pertussis* is very sensitive.

Antigenic structure

Bordetella pertussis is antigenically extremely complex, up to 14 different antigens having been recognized in various strains in agar gel diffusion tests. The most important of these antigens are the capsular and somatic antigens which function as agglutinogens, the haemagglutinin, the heat-labile endotoxin, the histamine-sensitizing factor and the protective antigen. A number of heat-labile capsular antigens have been identified in individual strains; these occur in combinations of two or more and permit the recognition of a number of different serotypes. All the serotypes, however, share a common antigenic component—the type 1 antigen.

Capsular antigenic loss may occur by a reversible process, designated by Lacey as modulation, and by an irreversible process of mutation analogous to S→R variation amongst the salmonellae. As described by Lacey, modulation can be induced by altering the ionic composition of the growth medium.

A heat-labile protein endotoxin has also been identified which appears to occur as an intracellular component. It is produced only by capsulated strains and can be liberated by autolysis and by physical disruption of the cell. Although the toxin is strongly antigenic, antitoxin does not appear to be produced during the course of infection or as a result of artificial immunization. Because of this the role of the toxin in pathogenicity is equivocal.

A substance with the property of increasing sensitivity of mice to histamine occurs as a component of the cell wall. It is markedly heat labile, being destroyed in half an hour at 75°C. Like the toxin, it is found only in fully virulent capsulated organisms.

Immunization with a suitably prepared vaccine will produce a substantial degree of protection against pertussis. The nature of the antigen responsible for the stimulation of protective antibody is, however, still in doubt. It is clearly not identical with the somatic antigens, haemagglutinin or the heat-labile endotoxin. It has recently been suggested by Preston and Evans on the basis of intracerebral mouse protection tests that the capacity to stimulate the production of protective antibody is a function of the capsular antigens.

Whooping cough

Bordetella pertussis is an exclusively human pathogen. It is the cause of pertussis or whooping cough. Though present in large numbers in the respiratory secretions of infected patients it shows no tendency to spread elsewhere in the body. It is rarely recoverable from normal subjects.

Whooping cough is a disease of childhood. Its mortality has decreased considerably during recent years, but it is still one of the most serious of the epidemic childhood diseases. It is particularly dangerous in the very young child, over half the deaths occurring in the first year of life. It is probably more highly infectious than any other bacterial disease of the respiratory tract.

The clinical course of whooping cough can be divided into three stages. The first is a catarrhal stage, lasting about 14 days, during which the child has a mild but progressively increasing cough. The catarrhal stage is followed by the paroxysmal stage which lasts for a further fortnight; the typical paroxysmal cough is the most marked feature of this stage. The final or convalescent stage also lasts about two weeks.

In the early stages of infection, the organisms are mainly found in the trachea and the bronchi. In the later stages they invade the lung itself and may be observed in the alveoli. They may produce localized pneumonic changes in the lung tissue, but of more serious import is their capacity to predispose the lung to secondary invasion with pyogenic organisms. Atelectasis may result from blocking of the bronchioles with mucus and may lead to bronchiectasis, which causes serious and permanent damage to the lung tissue. Anoxaemia resulting from atelectasis and pulmonary infiltration is responsible for the convulsions which occur in a small proportion of cases. Whooping cough is sometimes followed by an encephalitis, the precise pathogenesis of which is unknown.

Diagnosis

Plates of Bordet–Gengou's medium may be inoculated by spreading a specimen of sputum, by allowing the patient to cough on the plate and so distribute droplets containing the organism over its surface (cough plate method) or, best of all, by spreading material collected with a swab from the posterior nasopharyngeal wall. The latter can be obtained by a perinasal swab inserted along the floor of the nose. The organisms are identified on the basis of morphology, colony form and agglutination by specific antiserum. The fluorescent antibody technique has been favourably reported on for the direct demonstration of the organism in nasopharyngeal swabs.

A slide agglutination test with a suspension of living organisms is sometimes used to detect antibody in the patient's serum. It has the limitation of giving a positive result only when the disease is well established—usually not before the third week. A complement fixation test is also used.

Treatment

The wide-spectrum antibiotics—chloramphenicol, the tetracyclines and ampicillin—are reported to be of value if given at an early stage of the disease, but of little value in cases of more than one week's duration.

Prophylaxis

Pertussis immunization with a killed vaccine is usually combined with immunization against tetanus and diphtheria (p. 169). In infants and young children reactions following vaccination are negligible. Severe reaction may, however, occur in children of school age. The development of encephalopathy some hours after injection has been described, but fortunately is extremely rare. In recent years the protective effect of vaccination appears to be less than formerly in the UK. This has been attributed to antigenic variation in the currently prevalent strains.

It is claimed that gamma globulin obtained from adults injected repeatedly with pertussis vaccine is capable of conferring passive protection. The value of this procedure is *sub judice*.

Bordetella parapertussis

This organism, which has been isolated from cases of mild whooping cough, is very similar to *Bord. pertussis* in its cultural properties. Though it possesses some antigenic components in common with *Bord. pertussis* the two species can be distinguished serologically.

III.14

The Brucellae

The brucellae are very minute, Gram-negative, non-motile, non-sporing rods which may be so short as to appear coccal in morphology. Capsules have been described as being present in freshly isolated strains, but if this is correct they are soon lost. Three species pathogenic for man are *Brucella abortus, Br. melitensis* and *Br. suis*.

Cultural properties

The brucellae are aerobic, but primary growth from small inocula is improved by mild reducing conditions. They will grow on ordinary media but growth is slow. For isolation, enriched media—liver extract broth and liver extract agar—have been widely used, but more modern media—serum dextrose agar and Albimi agar—appear to give better results. Growth on artificial media is slow and colonies may not become visible for 48 hours. When freshly isolated the colonies are smooth, circular, convex and translucent, but these organisms readily undergo $S \rightarrow R$ variation to yield a slightly rough, yellowish brown colony.

Conventional fermentation tests are of little value in differentiating the different *Brucella* species. For this purpose the following criteria have been widely used:

1. Requirement for CO_2 on primary isolation. *Br. abortus* will grow on primary culture only in the presence of a high concentration, about 10%, of CO_2.
2. Production of H_2S. When grown on liver extract agar *Br. abortus* and American strains of *Br. suis* produce H_2S for at least four days. *Br. melitensis* and Danish strains of *Br. suis* either do not produce H_2S or produce a small amount detectable for one day only.
3. Sensitivity to certain dyes, of which thionine and basic fuchsin are the most generally useful. The dyes are usually incorporated in the growth medium; the optimum concentration of dye depends to some extent on the medium used but usually lies between 1/30 000 and 1/60 000 for thionine and between 1/25 000 and 1/50 000 for basic fuchsin.

In recent years, however, an increasing number of isolates have been found which have shown atypical cultural and antigenic reactions. For the classification of these two further types of tests have been used:

1. Metabolic tests to determine the capacity of the organisms to oxidize a variety of amino acids and sugars. The substrates of most differential value in these tests are glutamic acid, ornithine, lysine and ribose. Glutamic acid is oxidized by *Br. melitensis, Br. abortus* and most strains of *Br. suis*; ornithine and lysine only by *Br. suis* and ribose by *Br. suis* and *Br. abortus*.
2. Determination of the susceptibility of the organisms to a bacteriophage which is only significantly active against *Br. abortus*.

Antigenic structure

The behaviour of the brucellae in agglutination tests has been explained by Miles on the assumption that they possess two antigens, A and M, which are present in each of the species in different proportions, A being the dominant antigen in *Br. abortus* and *Br. suis* and M the dominant antigen in *Br. melitensis*. Sera prepared against any of the species will contain antibody against both antigens but can be rendered monospecific for either antigen by controlled absorption with a strain in which the other antigen is the major component. Monospecific sera agglutinate only organisms in which the corresponding antigen is present as the major component, but a minority of strains are agglutinable by both antisera. Work in recent years has indicated that the antigenic composition of the brucellae is more diverse than was originally thought, some strains which are biochemically and culturally *Br. abortus,* reacting with anti-M but not with anti-A sera, and strains which are biochemically and culturally *Br. melitensis,* reacting with anti-A and not with anti-M sera.

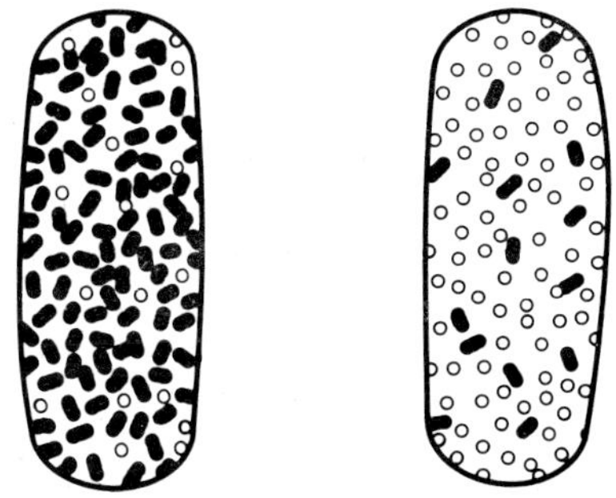

Fig. III.7. The antigenic composition of *Brucella melitensis (left)* and of *Br. abortus (right). Br. suis* resembles *Br. abortus,* with a lower A to M ratio.

Undulant fever

The brucellae are primarily animal pathogens; cattle, goats and pigs are particularly affected and are the main source of human infections. Under natural conditions infection of goats and sheep is mainly due to *Br. melitensis,* infection of cattle to *Br. abortus* and infection of pigs to *Br. suis. Br. abortus* infection of cattle is widespread throughout the world. Infection of goats and sheep with *Br. melitensis* mainly occurs in Mediterranean countries. *Br. suis* infection of pigs though occurring in various parts of the world, mostly in a sporadic form, is found as an epizootic infection mainly in the pig rearing areas of the U.S.

In cattle, goats and sheep the brucellae are responsible, on primary infection, for a form of contagious abortion. When established in a herd, however, they give rise to chronic and latent infection with abortion as an infrequent manifestation. Infected animals excrete the organism freely in the uterine discharges, in the faeces and in the urine. In most cases the organisms localize in the mammary glands and are excreted in the milk; this is of particular importance in the transmission of infection to man. In pigs, however, abortion is infrequent and the mammary glands are rarely involved.

In man the brucellae give rise to three types of infection:

Table III.6. Differentiation of *Brucella* species

	Br. melitensis	*Br. abortus*	*Br. suis*
Usual reservoir	Sheep, goats	Cattle	Pigs
Require 5–10% CO_2	–	(+)	–
H_2S production	–	(+)	(++)
Growth in presence of			
Thionine	(+)	(–)	+
Basic fuchsin	+	(+)	(–)
Predominant antigen	(M)	(A)	A
Sensitivity to specific brucella phage	–	(+)	–
Oxidation of:*			
Glutamic acid	+	+	(+)
Ornithine	–	–	+
Ribose	–	+	+

Adapted from S. S. Elberg (1965).
Signs without brackets indicate invariable reactions. Signs within brackets indicate commonest reactions.
Variations within each species in respect of CO_2 requirement, H_2S production, dye susceptibility and predominant antigen permit differentiation of biotypes. The less common biotypes usually have a restricted geographic distribution.
* Other substrates used: L-lysine, D- and L-alanine, L-arginine, L-asparagine, L-arabinose, D-galactose, D-xylose.

1. Latent infections can be detected only by serological methods. These appear to be common in persons whose occupation brings them into contact with infected animals.

2. Acute infections, the most prominent feature of which is a daily or diurnal remitting fever which is responsible for the name—undulant fever—by which they are commonly known. This type of fever occurs most typically in *Br. melitensis* infections. It has in the past been generally assumed that the fever is due to the liberation of brucellae into the blood stream from their sites of multiplication in the reticuloendothelial system. This view is almost certainly incorrect. It is considerably more likely that it is due to the liberation of endogenous pyrogens from polymorphonuclear leucocytes. In addition to fever and toxaemia there may be a wide variety of symptoms referable to involvement of particular organs. Lymph nodes, liver and spleen—the chief sites of the reticuloendothelial system are frequently enlarged. There may also be evidence of hepatitis, cholecystitis, subacute bacterial endocarditis or osteomyelitis. As a general rule *Br. melitensis* and *Br. suis* cause more serious clinical disease than *Br. abortus*. Pyogenic lesions are relatively frequent in infections caused by *Br. suis*.

3. Chronic low-grade infections usually showing periodic exacerbations. Pyrexia may be slight or absent, the common clinical manifestations being sweating, lassitude and joint pains. In this type of infection, which usually occurs as a sequel to an acute infection, the organisms can only rarely be isolated from the blood. In contrast to its occurrence in animals, abortion appears to be a rare event in man.

Brucellosis is primarily a disease of the reticuloendothelial system in which the organisms multiply with the production of a granulomatous tissue reaction. The brucellae have an unusual predilection for intracellular growth and may be demonstrated inside phagocytic cells, endothelial cells and fibrocytes. Their intracellular location explains the difficulties found clinically in the satisfactory treatment of brucellosis by chemotherapeutic agents and also why the condition can persist for a long time in the presence of a high titre of circulating antibody.

The brucellae appear to spread from the initial site of introduction, via lymphatic channels, to the local lymph glands in the cells of which they multiply. From the lymph glands they are discharged into the blood stream by which they are disseminated to other reticuloendothelial sites throughout the body. They have a particular predilection for the placenta which they appear to reach via the lumen of the uterus. Their predilection for the placenta may possibly be due to the presence of erythritol which has been found to stimulate the growth of brucellae in culture. They multiply within the placental cells and it is generally believed that abortion is due to a necrotizing action of the locally produced endotoxin.

Brucellosis is commonest in individuals, e.g. veterinarians and farmers, who work with infected animals. The organisms may gain entry to the body through abrasions in the skin, through the conjunctiva, through the alimentary tract and possibly also through the respiratory tract. Individuals of the general population may also become infected—almost invariably as a result of the ingestion of infected cow's or goat's milk, consumed in the raw state. In milk products, e.g. butter and cheese, the organisms die out rapidly and these are rarely if ever sources of infection. The incidence of overt brucellosis in persons known to have consumed infected milk is very low. This would seem to indicate a high degree of resistance in man to infection by the alimentary route.

Diagnosis

The most consistently successful method of diagnosis is by the demonstration of antibodies in the patient's serum. In a well-established case titres, in the Felix method, of up to 1 in 5000 are frequently found. In some cases of brucellosis the patient's serum shows the presence of a marked prozone, i.e. inhibition of agglutination in the higher serum concentrations in agglutination tests. In chronic infections agglutination titres may be low or absent. That in such cases the serum may nevertheless contain a high titre of a non-agglutinating antibody may be shown by the use of the antiglobulin test (p. 138). This antibody is mercaptoethanol-resistant, has a marked capacity to fix complement and can be shown in antiglobulin tests with antisera specific for the main immunoglobulin classes (p. 105) to be IgG, though some IgA which does not fix complement is also usually present. By contrast the agglutinating antibodies present in the sera of acute cases are of the IgM class though as can be shown by their mercaptoethanol resistance considerable amounts of IgG are also normally present.

The persistence of antibody after infection is variable. In some cases it falls rapidly to a low level but in others it persists at a high level for some years. It is possible that some of the latter cases are examples of persistent latent infection.

An attempt should also be made to isolate the organism from the blood. It is frequently stated that the chance of obtaining a positive blood culture is greatest if the blood is taken at the height of a pyrexial period. This is largely based on the assumption that the fever is due to the liberation of organisms into the blood from the reticuloendothelial system. There is considerable doubt, however, as to whether this view is correct. It is probably more important to make persistent attempts to isolate the organism than to choose a particular time for doing so. Even then, blood culture is often unsuccessful. Blood culture may be carried out using liver extract broth or glucose serum broth or by Castaneda's method, in which a mixture of blood and broth is poured over the surface of agar in a specially prepared bottle. Cultures should be incubated in an atmosphere of 5 to 10% carbon dioxide and should not be regarded as negative until after six weeks.

It may on occasion be possible to isolate the organism from the bone marrow or from the lymph glands. It has also been isolated from the cerebrospinal fluid, bile and urine.

During the course of infection a delayed-type hypersensitivity develops and may be demonstrated by intradermal injection of killed organisms or of various extracts (brucellin, brucellergin). As diagnostic procedures skin tests have the same limitation as the tuberculin reaction (p. 271), in that they give positive results not only if the patient is suffering from an active infection but also if he has had a prior infection.

Treatment

The best results in the treatment of brucellosis are reported for a combination of streptomycin and tetracycline. Although marked symptomatic improvement is readily achieved it has been found difficult to ensure complete sterilization of the patient's tissues by antibiotic therapy. Relapses are frequent even after prolonged treatment and are probably due to the release of organisms from intracellular sites of multiplication in which they are inaccessible to chemotherapeutic agents.

Prophylaxis

Experience in various parts of the world has shown that it is possible by use of eradication programs to build up *Brucella*-free herds and to maintain them free from infection.

Immunization with a live vaccine prepared from the attenuated S19 strain of *Br. abortus* confers a substantial degree of immunity and has been widely used for the prophylaxis of the disease in cattle.

Brucellosis occurring in members of the general population who are infected by ingestion of infected milk or milk products may be eliminated by the pasteurization of all milk destined for human consumption.

III.15

The Mycobacteria

The mycobacteria are slender, non-capsulated, non-motile, non-sporing rods. They are very difficult to stain but when stained the stain cannot be removed by treatment with strong mineral acid. Because of this they are known as *acid-fast bacilli*.

The genus includes both pathogenic and saprophytic species. The former are more difficult to grow and are in general more acid-fast than the latter. The most important of the pathogens are the tubercle bacilli and the leprosy bacillus, *Mycobacterium leprae*.

Mycobacteria have a much higher lipid content than other organisms; the lipids, mainly phospholipids (or phosphatides) and various waxes, appear to occur primarily as components of the cell wall. On hydrolysis a variety of fatty acids have been obtained, some of which, branched-chain fatty acids—tuberculostearic, phthienoic, mycocerosic and mycolic acids—are apparently unique to the mycobacteria. It is probable that the high lipid content of the mycobacteria is responsible for many of their biological properties.

1. Acid-fastness is believed to be due to the presence of mycolic acid, a component of various wax fractions.

2. The lipids probably confer a low permeability to water-soluble compounds, which might explain the resistance of tubercle bacilli to chemical disinfection.

3. There is considerable evidence that the high lipid content of the tubercle bacillus plays a major part in determining the type of tissue reaction it evokes in the animal body. This renders it difficult to dispose of, as a result of which it stimulates a granulomatous foreign-body type of reaction. In addition the wax D component (p. 271) plays a decisive role in inducing delayed-type hypersensitivity.

TUBERCLE BACILLI

There are five distinct varieties of tubercle bacilli: human, bovine, avian, murine and piscine. Only the human and bovine varieties cause tuberculosis in man. These, though formerly assigned to the same species, are now usually designated as separate species—the human variety as *Myco. tuberculosis,* and the bovine variety as *Myco. bovis.* The avian, murine and piscine varieties produce lesions resembling those of human tuberculosis in birds, voles and fish respectively, and are now classified as separate species, *Myco. avium, Myco. microti* and *Myco. marinum,* respectively.

In the description which follows we will be concerned only with the properties of *Myco. tuberculosis* and *Myco. bovis.*

Mycobacterium tuberculosis and Mycobacterium bovis

Morphology

Mycobacterium tuberculosis is a thin, straight or slightly curved rod measuring from 2 to 5 μm by 0·3 μm. By contrast *Myco. bovis* tends to be somewhat shorter and thicker. The organisms may stain uniformly or present a beaded, irregularly stained appearance. In the tissues the bacilli occur singly, in pairs arranged at an angle, or in clusters of organisms usually lying parallel to one another. In cultures of virulent organisms clusters or 'cords' of parallel bacilli are a characteristic feature.

Tubercule bacilli are both acid- and alcohol-fast and are normally demonstrated microscopically by the Ziehl–Neelsen technique. They may also be demonstrated by fluorescence microscopy. There is some evidence that when growing in the tissues non-acid—fast forms of the organism occur.

Cultural properties

Tubercle bacilli grow slowly, their average generation time being about 24 hours and prolonged incubation of cultures is therefore necessary. On primary culture colonies are not usually visible in less than 10 to 14 days. They are strict aerobes and grow best at 37 °C.

For isolation, coagulated egg media, e.g. Löwenstein's medium, are those most frequently employed. Although special media are required for their isolation, tubercle bacilli are not nutritionally exacting. On subculture they grow well on simple synthetic media incorporating an ammonium salt or amino acid, e.g. asparagine, as nitrogen source, and glucose, provided a large inoculum is employed. They will not grow in such media, however, from small inocula; this is apparently because they are highly susceptible to the toxic action of very small amounts of various substances, particularly of fatty acids, which may occur as impurities in reagents and on glassware. The toxic action of these substances can be neutralized by the addition of serum or serum albumin. A simple fluid medium containing albumin and a synthetic fatty acid ester, known commercially as Tween 80 (sorbitan mono-oleate), is widely used in experimental work and in sensitivity testing. Tween 80 stimulates the growth of the organism and also causes it to grow in a dispersed fashion rather than in the pellicles found in other fluid media.

On primary culture *Myco. tuberculosis* and *Myco. bovis* show characteristic differences in cultural characteristics. On egg media the human type produces a dry wrinkled warty growth. *Myco. bovis* grows much less luxuriantly and yields a growth which is more powdery and less cohesive than that of *Myco. tuberculosis.* This type of growth is referred to as *dysgonic* in contrast to the more profuse growth of *Myco. tuberculosis* which is described as *eugonic.* The types differ also in the way in which they respond to the presence of glycerol. Glycerol in a concentration of 0·75% has a stimulating effect of the growth of the human type but has no such effect on the growth of the bovine type. In fact the growth of some strains of *Myco. bovis* may be slightly inhibited. The species may be differentiated by pathogenicity tests. Both are pathogenic for guinea-pigs, which develop a local caseous lesion and signs of disseminated tuberculosis, but *Myco. bovis* alone is highly pathogenic for rabbits. The niacin test is also of considerable value in species differentiation. This compound is readily demonstrable in cultures of *Myco. tuberculosis* but not, or only to slight extent, in cultures of *Myco. bovis.*

Antigenic structure

A variety of antigenic components, both proteins and polysaccharides, has been identified in tubercle bacilli. Of these the most important are the protein antigens, known collectively as *tuberculoprotein* or *tuberculin,* which are capable of eliciting the allergic reactions demonstrable in the tuberculin test. Slight antigenic differences can be shown between the tuberculins of *Myco. tuberculosis* and *Myco. bovis* but they are nevertheless so similar that they behave virtually identically in skin tests. Protein antigens which cross-react serologically with the tuberculoproteins of these species are also present in other mycobacteria.

Tubercle bacilli have been examined serologically by a variety of methods—agglutination, complement fixation and the Ouchterlony technique. These studies have shown that the mammalian type bacilli form a serologically homogeneous group. Serological tests have not been of practical value, however, in the identification and differentiation of mycobacteria.

Pathogenicity

In the tissues of infected animals and man the tubercle bacillus causes the production of a characteristic and unique lesion—the tubercle. In its typical form the tubercle shows a peripheral zone of fibroblasts and lymphocytes, an inner zone of epithelioid cells and in its centre one or more giant cells. As it increases in size necrosis of the central portion occurs. The necrotic material is not normally digested; consequently it has a cheesy appearance and consistency, on account of which it is known as caseous, i.e. cheesy, material. The necrotic changes found in the tubercle are clearly the result of tuberculoprotein hypersensitivity which develops during the course of infection. They are almost certainly due to the action of toxic lymphokines liberated from sensitive lymphocytes following exposure to tuberculoprotein. Normally, when necrosis of tissue occurs the necrosed tissue is digested by tissue proteolytic enzymes. In caseous lesions, however, this normal proteolysis is inhibited, presumably due to the inhibition of tissue proteases. It has been suggested that the inhibitor is a lipid component of the bacillus.

Although the phosphatide and tuberculoprotein of the tubercle bacillus may account for the microscopic anatomy of tuberculous lesions they are present in attenuated bacilli which are unable to establish a progressive infection. Some other factor or factors must therefore be required for virulence. The nature of this virulence factor is, however, still in dispute. In this connection the isolation by Bloch from virulent bacilli, by extraction with petrol ether, of a lipid claimed to have a toxic action on tissue cells and leucocytes and capable of inhibiting leucocyte migration has aroused considerable interest. Since extraction with petrol ether disrupts the cords invariably found in virulent cultures the lipid has been designated 'cord factor'. Cord factor is a component of the wax fraction of the tubercle bacillus and has been identified as a complex of mycolic acid and trehalose. Since, however, the factor is present in attenuated tubercle bacilli and saprophytic mycobacteria its role in virulence is not clear.

Killed tubercle bacilli have considerable toxicity for experimental animals on injection but a similar toxicity is shown by the saprophytic mycobacteria. This toxicity presumably contributes to the overall effect of tuberculosis infection.

Tuberculosis

Tuberculosis is a disease of world-wide prevalence and is responsible for more serious illness than any other of the infectious diseases. Over the last century, however, its incidence in most countries has diminished enormously. Thus the mortality in England and Wales in 1855 was 3626 per million; by 1960 it had fallen to 75 per million. Some of this fall has undoubtedly been due to an improvement in social conditions, but in recent years the improvement has been mainly due to the introduction of chemotherapy, effective case finding and prophylaxis by BCG vaccination. Nevertheless the World Health Organization estimate that throughout the world there are still some 15 million infectious cases, mostly in the developing countries. Most of these cases can be attributed to exogenous infection, but in communities in which the disease has shown a marked fall in incidence this has occurred particularly in the younger age groups and cases in older age groups are probably due to endogenous reinfection from a latent infection acquired early in life.

In communities where there is a high prevalence of the disease the majority of people make their first contact with the tubercle bacillus in childhood and as a result develop a primary childhood infection, generally of pulmonary type. This type of infection has a high tendency to spontaneous cure. In those who have recovered from it, it leaves an appreciable but by no means complete resistance to reinfection. At the same time it renders the tissues hypersensitive or allergic so that they react quite differently to future contact with the tubercle bacillus. This change in reactivity is demonstrable by the various types of tuberculin test.

Pulmonary tuberculosis is the commonest form of the disease; it is almost invariably due to the human-type bacillus. Infection appears to occur mainly by inhalation of droplets or droplet nuclei from an infected case. Tubercle bacilli remain viable in contaminated dust for a considerable time and it was at one time thought that this was an important, possibly the major, source of infection. Dust particles, however, appear to be almost invariably greater than 5 μm in diameter—the upper limit of size above which particles are normally incapable of gaining access to the alveoli to initiate a pulmonary infection.

The pulmonary lesions in childhood infection show some striking differences from those in the adult. They are usually relatively slight and in most cases undergo fairly rapid resolution. In a small proportion of cases, however, the disease is not arrested and assumes a rapidly progressive form. In the adult, on the other hand, the lesions are much more chronic and with little tendency to spontaneous cure. There is also considerable destruction of tissue with caseation and cavity formation. In childhood there is invariably an involvement of the mediastinal glands; in the adult this mediastinal involvement does not occur.

These differences are usually attributed to the operation in the adult of allergy and immunity resulting from childhood infection. However, when infection occurs in adult life the lesions are still those of classical adult type. This may be due to the fact that immunity and allergy develop more rapidly in the adult than in the child.

Infection with bovine bacilli is almost invariably due to the consumption of milk from an infected cow. In European countries and in America it is now, as a result of the effective control of the disease in bovines, very uncommon. These infections occur mostly in children, and usually present as abdominal tuberculosis with involvement of mesenteric glands.

Age is an important factor in determining the severity of the disease. In children of less than five years, and particularly in infants, the mortality is high. In children of from five to 15 years the mortality is low and asymptomatic latent infection is usual: this is the age group of typical childhood primary tuberculosis. These differences appear to be due entirely to differences in physiological susceptibility; the milder disease in the older child is not due to acquired immunity. After 15 years of age the mortality from the disease again increases.

In experimental animals genetic factors are of considerable importance in determining susceptibility to tuberculosis. Evidence that a genetic factor is also of importance in man is provided by the finding in twins of which one is infected that infection of the second twin is commoner when they are identical than when they are non-identical. There is also epidemiological evidence that resistance to infection is greatly diminished by malnutrition. In Great Britain and Germany there was a considerable increase in the incidence of the disease in both world wars, an incidence which is usually attributed to a lowering of nutritional status. That nutrition affects susceptibility receives considerable support from experiments of Dubos, who found that the susceptibility of mice to experimental infection could be greatly increased by feeding them on a protein-poor diet. Mice in a state of nutritional deficiency may even develop progressive lesions following inoculation of the attenuated BCG strain. A very marked predisposition to tuberculosis is found in patients suffering from silicosis, a disease in which a progressive fibrosis of the lungs occurs as a result of the inhalation of particles of silica.

Allergy in tuberculosis

The development of allergy in the course of tuberculous infection was first demonstrated by Koch and the reaction of the tuberculous animal to reinfection is in consequence known as Koch's phenomenon. When a normal guinea-pig is injected subcutaneously with a suspension of tubercle bacilli a slight local nodule forms within some hours and disappears in a few days. After about 10 to 14 days a hard nodule appears, which breaks down to form an ulcer. The infection spreads to the local lymphatic glands, which become large and caseous, and ultimately becomes disseminated through the body. If, on the other hand, a tuberculous guinea-pig, which has been infected some four to six weeks previously, is injected, a small area of swelling develops within about 48 hours and rapidly progresses to form a shallow ulcer. The ulcer heals within a short time and there is no involvement of the local lymphatic glands. Koch found also that similar local inflammatory reactions could be produced in tuberculous guinea-pigs by the injection of killed bacilli or of small amounts of tuberculin. Following the injection of tuberculin the tuberculous animal develops a severe and often fatal systemic reaction. In addition to this general reaction a local reaction also occurs at the site of the injection and a marked focal reaction around the tuberculous lesions in the tissues, the main feature of which is a gross enlargement of the capillaries.

These reactions of the tuberculous animal are due to allergy to tuberculoprotein which is acquired during the course of the initial infection. Allergy may also be induced by the injection of killed tubercle bacilli but not by the injection of tuberculoprotein *per se*. It can, however, be produced by injection of tuberculoprotein together with the wax D fraction.

In man, as in the guinea-pig, allergy to tuberculoprotein develops as a result of tuberculous infection. After recovery from infection the allergic state persists for many years even in the absence of further contact with the tubercle bacillus. It may be demonstrated in a variety of ways, in all of which tuberculoprotein is applied to or injected into the skin. These procedures are known as tuberculin tests and are widely used both to indicate that an individual has at some time suffered from a tuberculous infection and to a lesser extent for the diagnosis of a current infection (p. 274). A positive reaction is indicated by the development of areas of swelling and erythema around the site of introduction of the tuberculin. The skin reaction is believed to be due to the release of lymphokines (p. 148) from the lymphocytes of sensitive persons. A combination of different lymphokines is probably involved.

For tuberculin testing tuberculoprotein is used either in the form of Old Tuberculin or PPD. The name Old Tuberculin is the name applied to a preparation obtained by the original method of Koch. It consists of a filtered glycerol broth culture of the tubercle bacillus concentrated to one-tenth of its original volume by evaporation. PPD is a highly purified preparation of tuberculoprotein prepared from the growth of the bacillus in a special synthetic medium. Since it is purer than Old Tuberculin and therefore less likely to give rise to non-specific reactions PPD is now the reagent of choice for skin testing.

A false-negative reaction is sometimes given in the tuberculin test. The test is frequently negative in cases of advanced tuberculosis, in miliary tuberculosis, and in children following excessive exposure to sunlight or during the course of one of the exanthemata. A negative reaction may also be obtained in the early stages of infection.

Immunity

There is little doubt that immunity in tuberculosis is due to delayed hypersensitivity. This is shown by the following facts:

1. In communities in which tuberculosis is widely prevalent a positive tuberculin test, as a result of childhood infection, is correlated with subsequent immunity.
2. Immunity can be produced by the inoculation of BCG vaccine which induces hypersensitivity.
3. In man immunity develops normally in hypo-gammaglobulinaemia.
4. In experimental animals immunity cannot be passively transferred by the injection of the serum of immunized animals.

The precise mechanism involved in the production of immunity requires further classification. In essence, however, it is due to the activation of macrophages following stimulation of the lymphocytes of immune animals by tuberculin. Activated or 'angry' macrophages show an increase in mitochondria and lysozymes, in their capacity to spread on surfaces, in phagocytic activity and in bactericidal properties. Their bactericidal activity, as shown in an animal model, is non-specific, extending to other intracellular organisms, e.g. brucellae and *Listeria monocytogenes*. The initial lymphocyte stimulation is a specific event and the link with macrophage activation is almost certainly the production of a macrophage-activating lymphokine.

There are, however, certain difficulties in associating immunity with hypersensitivity which require explanation:

1. It has been shown by Rich that it is possible to desensitze an allergic animal, as demonstrated by loss of cutaneous reactivity, without affecting its immunity. As against this, however, loss of cutaneous reactivity may not indicate total loss of hypersensitivity. Thus animals which have lost their cutaneous reactivity may still be killed by intravenous injection of tuberculin.
2. It has been found by Raffel that it is possible to sensitize an animal by injection of tuberculoprotein together with a wax fraction of the organism, presumably wax D (p. 268), without stimulating immunity. This may have been because the antigenic stimulus was not sufficiently intense.
3. There is no correlation, either in animals or in man, between the extent of the hypersensitive state, as manifested by the strength of cutaneous tuberculin reactions, and the immune state. In fact in the British Medical Research Council trial (1956), it was found that children who gave strongly positive tuberculin reactions were much more likely to develop tuberculosis subsequently than those who gave weakly positive reactions, while children inoculated with vole bacillus vaccine, so that they exhibited a low-grade cutaneous reactivity, were as well protected as children who developed a more intense cutaneous reactivity as a result of BCG vaccination. These findings do not, however, invalidate the association of hypersensitivity with immunity; they could be explained on the grounds that an excessive inflammatory reaction might facilitate wider dissemination of the bacilli by opening up the blood and lymphatic channels.

Diagnosis

Only a brief outline of the various techniques used in the laboratory diagnosis of tuberculosis will be undertaken.

The material to be examined depends on the site of the infection: in suspected tuberculous meningitis the cerebrospinal fluid, in suspected genitourinary infection the urine and in suspected pulmonary tuberculosis the sputum. In certain cases of suspected pulmonary infection, however, it may not be possible to obtain a specimen of sputum. This particularly happens with young children who frequently swallow any sputum produced. In such cases tubercle bacilli may be demonstrable in the faeces or in material obtained by gastric lavage or by the use of a laryngeal swab.

Three methods are used for the demonstration of tubercle bacilli: (*a*) microscopy, (*b*) culture and (*c*) animal inoculation. Ideally all three methods should be used. Although, in general, culture and animal inoculation are more sensitive than microscopy, they may give negative results in patients harbouring bacilli resistant to INAH. Positive microscopic results associated with negative culture may also be found during the course of chemotherapy, just prior to the stage of total elimination of the bacilli. At this stage the bacilli demonstrable are presumably of diminished viability.

Microscopy. Films should be prepared from all materials suspected of containing tubercle bacilli and an attempt made to demonstrate the organisms microscopically. Great care must be taken in microscopic diagnosis because of the danger of confusing saprophytic acid-fast bacilli which may be present in reagents or on glassware with tubercle bacilli. It is essential therefore that all containers used in the collection of specimens should be thoroughly cleaned, new cleaned slides should be used for the preparation of films and all stains should be made up in distilled water. For the collection of specimens of sputum waxed cardboard cartons which can be discarded are strongly recommended.

Two microscopic procedures are available: the Ziehl–Neelsen method and fluorescence microscopy (Chapter I.2). In most laboratories the Ziehl–Neelsen method is the one routinely used but in laboratories carrying out a large number of daily examinations fluorescence microscopy is frequently preferred since it permits a more rapid examination of films. Positive results obtained by fluorescence microscopy can be confirmed by restaining the film by the Ziehl–Neelsen method but experienced workers do not as a rule find this necessary.

Sputum may be examined directly or after concentration for which Petroff's method (see below) may be used. Fluids, e.g. cerebrospinal fluid, pleural fluid and urine, should be centrifuged and films made from the deposit. In the case of cerebrospinal fluid films should also be prepared from the fine web-like clot which is often present.

As a general rule the finding of acid–alcohol-fast bacilli of typical morphology in sputum or in material coming from the interior of the body, e.g. cerebrospinal fluid, pleural fluid or pus from an unopened abscess, can be taken as conclusive evidence of tuberculosis.

Great care must, however, be taken in the microscopic identification of acid–alcohol-fast bacilli found in urine, as tubercle bacilli. If the urine has not been taken by catheter, smegma bacilli may be present and may be morphologically indistinguishable from tubercle bacilli. They should, however, be found only rarely in mid-stream specimens if these have been properly taken; nevertheless if acid-fast bacilli are present, the examination should be repeated on a catheter specimen.

Culture. The medium most widely used for the isolation of tubercle bacilli is the Löwenstein-Jensen medium incorporating 0·75% glycerol. At least two tubes of this medium should be inoculated from each specimen. Many bacteriologists consider it advisable to inoculate one or more tubes of Dorset's egg medium as well on the grounds that the bovine type of tubercle bacillus may be inhibited by the concentration of glycerol in the Löwenstein-Jensen medium. Cultures are examined at intervals and any suspicious colonies stained by the Ziehl–Neelsen method. They should not be discarded as negative in less than eight weeks.

Saprophytic acid-fast bacilli and atypical mycobacteria also grow on egg media and may show considerable colonial and microscopic resemblance to tubercle bacilli. Because of this an organism should not be accepted as a tubercle bacillus until it has been subjected to further tests, the most important of which is guinea-pig inoculation. INAH-resistant strains of tubercle bacilli are, however, usually of low virulence for guinea-pigs, producing at most a local lesion. They require particularly careful examination to differentiate them from atypical myocobacteria (p. 276).

For the isolation of tubercle bacilli from sputum, the sputum is first treated to break down the mucus present, either by incubation with proteolytic enzymes or trisodium phosphate or with agents such as sodium hydroxide or sulphuric acid which, as well as breaking down mucus, also kill commensal bacteria. Of these methods, alkali treatment, as in Petroff's method, is probably the most widely used. The method may also be used for the treatment of pus and deposits from fluids contaminated or infected with other organisms. Deposits from uncontaminated fluids may be inoculated without treatment.

Guinea-pig inoculation. For normal drug-sensitive strains guinea-pig inoculation has much the same sensitivity as culture. The material to be inoculated is treated as for culture and is injected either subcutaneously or intramuscularly into the animal's thigh. For routine diagnostic work the animal is killed after eight weeks and a post-mortem examination is then carried out. In positive cases films should be made from the local caseous lymphatic glands and stained by the Ziehl–Neelsen method. Guinea-pig inoculation should not be regarded as an alternative to culture, since many INAH-resistant strains are avirulent for guinea-pigs.

Other methods. There are no satisfactory serological methods for the diagnosis of tuberculosis. A red cell agglutination test, in which the patient's serum is tested for its capacity to agglutinate cells which have been coated with a polysaccharide antigen present in Old Tuberculin, has been used; it is not reliable, however, since it gives negative results in cases of active tuberculosis and, like the tuberculin test, indicates at most, when positive, that the subject has at one time been infected.

The tuberculin test is primarily used to indicate past infection with the tubercle bacillus—its main practical applications being as a preliminary to BCG vaccination (see below) and in epidemiological surveys to determine the incidence of latent tuberculosis. In certain circumstances, however, the test has considerable diagnostic value. As a general rule, except under the conditions previously noted, a negative reaction can be taken as indicating the absence of tuberculous infection. The value of a positive reaction in indicating current infection depends on the age of the person and the degree of prevalence of tuberculosis in the community. In infants and in young children the test is of considerable diagnostic value, the younger the child the greater the value, since the younger the child the less likely is it to have had a prior infection. A positive reaction in a child of less than two can, in any community, be taken as conclusive evidence of the presence of active tuberculosis. In the adult, however, the position is different. In communities where there is a high prevalence of tuberculosis a positive reaction usually means only that the person was at one time infected. The lower the prevalence of the disease, however, the greater is the probability that a positive reaction indicates current infection.

Treament

A number of chemotherapeutic agents are active against the tubercle bacillus and have given dramatic results in treatment. The most important of the agents used are streptomycin, PAS and INAH. The most effective of these are streptomycin and INAH which are bactericidal; PAS is purely bacteriostatic. Tubercle bacilli, however, readily become resistant to streptomycin and INAH but, as discussed in Chapter I.5, the development of resistance is greatly diminished when a combination of two or more of the drugs is employed. In practice, the best results have been obtained with regimens in which treatment has started with all three drugs. The regimen can be modified subsequently, according to the results of sensitivity tests.

With effective chemotherapy it is now clear that, when the organisms are originally sensitive, it should be possible to obtain a bacteriological cure in virtually every case. This has been shown clearly in a trial, organized by the International Union against Tuberculosis, in which all patients who were maintained on a regimen of streptomycin, INAH and PAS for six months, followed by PAS and INAH for a further six months, became bacteriologically negative. In the past, however, the therapy of tuberculosis has not been as successful as this and the development of resistance, even with combined therapy, has been not infrequent. Thus in a random sample carried out by the British Tuberculosis Association in 1960, it was estimated that, in Britain, there were then 3500 patients infected with tubercle bacilli resistant to one or more of the major drugs and 1800 with bacilli resistant to all three. These results can undoubtedly be attributed to inadequate treatment. It has been suggested that host factors may also be of importance in the genesis of therapeutic failures and in permitting the emergence of resistant strains. Such factors would include the presence of open thick-walled cavities in which bacilli multiply rapidly because of free access of oxygen and into which there may be a slow penetration of drug and, in the case of INAH, rapid metabolism of the drug by rapid inactivators (p. 78). It would appear, however, from data presented by Mitchison, that the presence or absence of cavitation is only of importance with inadequate therapeutic regimens, and provided that sufficient INAH is given, there is no difference in success rate between rapid inactivators of INAH and slow inactivators.

It is a general finding that in the later stages of treatment, on what is obviously an effective chemotherapeutic regimen, patients continue to excrete tubercle bacilli which, on subsequent sensitivity testing, are found to be sensitive. These bacilli could either be located in some focus of infection into which the drugs had failed to penetrate or they could be sensitive organisms which have remained in a dormant or non-metabolizing state—'persisters'—because of which they are insusceptible to the action of the drug. The latter appears to be the more usual alternative. There is some evidence that this problem is minimized when high doses are used in the early stages of treatment.

Other drugs of importance are ethambutol, prothionamide, pyrazinamide, rifampicin, capreomycin and thiacetazone. So far these have been mainly second-line drugs. In recent years, however, ethembutol has tended to replace PAS as a primary drug, since it is more readily tolerated by patients, and a combination of ethambutol with rifampicin and either prothionamide or pyrazinamide has been used as a primary treatment regimen. A combination of thiacetazone and INAH has been used considerably in developing countries because of its low cost.

If at all possible sensitivity tests should be carried out on organisms isolated before the beginning of treatment. At present the majority of organisms isolated from fresh cases of tuberculosis, i.e. cases which have received no previous chemotherapy, are fully sensitive to the three major antituberculous drugs. Nevertheless a proportion of fresh isolates—about 5% in the British Isles—are resistant to one or other of these drugs. The proportion does not seem to be increasing, almost certainly because of the much more rigid and effective therapeutic regimens now adopted. In Africa, however, the incidence of primary resistance is much higher and presumably reflects the less satisfactory conditions under which treatment must often be carried out.

It is essential to carry out sensitivity tests on isolated strains prior to treatment. However, when the organisms are initially sensitive there is little point in carrying out sensitivity tests in the course of treatment. For the control of treatment simple smear examination, using a standardized technique which will permit a valid quantitative assessment of the number of organisms present, and cultural examination provide in most cases all the information required.

Prophylaxis

Neither dead vaccines of tubercle bacilli nor tuberculin will stimulate immunity against tuberculosis. Immunity can, however, be produced by the use of a living vaccine, consisting of a strain of bovine bacilli, *Bacille Calmette-Guérin* (more commonly known as BCG), attenuated by prolonged culture on a bile-containing medium. The BCG strain is not completely avirulent. It will grow to a limited extent in the tissues; in fact its immunizing effectiveness appears to depend on its being able to do so. It is, however, attenuated to the extent that it does not normally initiate a progressive infection. The BCG vaccine strains currently employed differ considerably in their degree of attenuation. Consequently their immunizing potency must be kept continually under review, particularly to ensure that they do not become over-attenuated.

The vaccine is introduced, usually in the deltoid region, by intradermal inoculation or by the multiple puncture method. Following intradermal inoculation a small nodule develops and is usually well established in about three weeks. This local lesion may progress to ulcer formation, but the process is quite benign and the ulcer heals by scar formation. In a few cases there is an associated enlargement of the cervical and axillary lymph nodes which may, although very rarely, progress to abscess formation. A few cases have been recorded in which the BCG strain has given rise to a progressive infection, but the risk of this occurring must be regarded as infinitesimal. BCG vaccination can, in fact, be regarded as being as safe as any other prophylactic measure. The risk against which it protects is currently in most communities immeasurably greater than any risk attendant on the vaccination process.

Contraindications to BCG vaccination are measles, pertussis, eczema or tuberculosis. Moreover, BCG vaccine should not be given to patients under prolonged cortisone treatment, nor at the same time as any other inoculation.

A tuberculin test should normally be carried out as a preliminary to vaccination; only those who are tuberculin negative are inoculated. In the case of young children, particularly those who have been in contact with a case of tuberculosis, it is desirable also to carry out a second tuberculin test six weeks after the first, the child having been removed from all contact with the disease during the intervening period. The double test minimizes the possibility that the child is infected at the time of inoculation. If the child were infected a short time prior to the first test, he might give a negative reaction because the allergic state had not had sufficient time to develop. After a further six weeks, however, the allergic state should have developed sufficiently to give a positive reaction. After a satisfactory intradermal inoculation the reaction should be positive at the end of six to eight weeks. A tuberculin test is not, however, necessary, inspection of the lesion being sufficient to show that the vaccine has taken.

In mass immunization programs in communities where there is a high incidence of the disease the World Health Organization has recommended that BCG vaccination be undertaken without prior tuberculin testing in order to ensure a high rate of acceptance of the vaccine. In such circumstances any focal reactions produced in individuals with active tuberculosis can be controlled by chemotherapy.

We are not yet in a position to assess fully the effects of BCG vaccination, but from the data so far available it is clear that it gives a considerable amount of protection against the disease. Thus in extensive trials which have been carried out by the Medical Research Council in children in industrial areas in England vaccination was found to confer approximately 80% protection over a follow-up period of 15 years. The value of BCG vaccine is greatest in communities which have a high normal incidence of tuberculosis. In communities where, possibly as a result of vaccination, the incidence of the disease is low, there is a considerable case for not attempting general vaccination. In such communities the disease can be controlled by effective case finding and chemotherapy. Under these conditions the tuberculin reaction can be used as a diagnostic test, whereas if there were general vaccination it would have no such value. Whatever the policy towards general immunization persons in special risk categories, nurses, medical students, laboratory workers and children from a tuberculous environment, should certainly be vaccinated.

Chemoprophylaxis with INAH given over a period of one year has been recommended for young children who give a positive tuberculin reaction, in order to prevent dissemination of the disease, and for children in contact with a case of active pulmonary tuberculosis.

Many communities have succeeded in greatly diminishing the incidence of bovine tuberculosis and some in eliminating the disease altogether by improvement in milk hygiene. For this purpose the most important measure is pasteurization which, if properly carried out, can be completely relied upon to kill any tubercle bacilli present in the milk. The safety of milk can be further ensured by the building up and maintenance of tuberculin tested herds.

ATYPICAL MYCOBACTERIA

Four groups are recognized:

1. Photochromogens, which produce a bright orange pigment which is developed in cultures incubated after exposure to light.
2. Scotochromogens, which are pigmented when grown in the dark.
3. Non-pigmented slow-growing mycobacteria.
4. Rapid growers.

For their identification and differentiation from tubercle bacilli a variety of tests are employed.

Atypical mycobacteria are not transmissible from man to man and the principal reservoirs of infection appear to be soil and water. Some, notably the photochromogen *Myco. kansasii,* cause pulmonary lesions resembling tuberculosis, usually in middle-aged men whose lungs have become damaged by chronic bronchitis. Some, especially the scotochromogen *Myco. scrofulaceum,* cause cervical lymphadenitis in children. Atypical mycobacteria, especially the slow-growing non-pigmented *Myco. intracellulare,* which is very similar to *Myco. avium,* are frequent causes of latent infections which give rise to weak positive tuberculin reactions. The rapid grower *Myco. fortuitum* has been isolated from superficial abscesses.

Mycobacterium leprae

In infected tissues, leprosy bacilli appear as acid-fast bacilli, morphologically resembling tubercle bacilli. Because of this they have been assigned to the *Mycobacterium* genus as *Myco. leprae.* The protection against leprosy recently shown to be given by BCG vaccination supports the validity of this classification.

So far no organism with any real claims to be regarded as the leprosy bacillus has been isolated by laboratory culture. It can, however, be propagated in the footpads of mice. Under these conditions the organism has a long generation time of 12 to 13 days, a finding consistent with the chronicity of the lesions characteristic of leprosy in man. No other laboratory animal has been reported as susceptible.

Leprosy

Leprosy is divided into two varieties; lepromatous and tuberculoid, which differ in the type of tissue lesion produced. This in turn depends on differences in the resistance of the patient. In the lepromatous type, in which resistance is low, there is a diffuse inflammatory reaction, the granulation tissue showing extensive infiltration with mononuclear type cells, and the presence of giant cells. Leprosy bacilli are usually present in large numbers, occurring either in groups of parallel bacilli or in large globular clumps known as globi. In the tuberculoid type, in which the resistance of the patient is high, the lesions are characteristically fibrotic with relatively few inflammatory cells and very scarce bacilli. The main lesions in leprosy are in the skin and nerves. In individual patients the disease is often predominantly cutaneous or predominantly neural. The cutaneous lesions occur mainly in exposed parts of the body—the face, hands, arms and legs. In the lepromatous type the lesions are nodular and frequently ulcerate and become secondarily infected. In the tuberculoid type they are raised erythematous macules. The neural lesions affect particularly the sensory nerves, leading to anaesthesia, deformities and trophic changes; they are usually of tuberculoid type. The mucous membrane of the nose is frequently involved, particularly in the lepromatous type, and bacilli may be demonstrable in the nasal secretions. Though mainly a superficial disease, invasion of the blood stream may occur and the organisms may spread to the viscera.

Characteristically, leprosy is a disease of communities in which standards of hygiene are low. Its infectivity is low and close and prolonged contact is necessary for infection. Susceptibility is maximal in childhood and most infections occur before adult life. The extreme chronicity of the disease is matched by the long incubation period, usually from two to four years. Not infrequently infection contracted in childhood is not manifested clinically until adult life. Subclinical infection is not uncommon and the disease can regress spontaneously.

Diagnosis is usually made by the demonstration of bacilli in material obtained by skin biopsy or in scrapings from the nasal mucosa.

Diaminodiphenyl sulphone (DDS) is the drug of choice in the treatment of leprosy. Treatment must be continued for several years sometimes throughout the life of the patient. DDS must be administered with care since it is highly toxic. In clinical trials the phenazine derivative, clofazimine, is reported to be a highly effective chemotherapeutic agent.

Lepromin reactions

Lepromin is an extract of infected leprous tissue used for skin testing in leprosy. A number of methods of preparation have been employed. In the original method tissue was boiled and the larger particles removed by centrifugation. Two types of reaction are demonstrable. The Fernandes reaction, a classic delayed type hypersensitivity, reaches a maximum in two to four days. As the reaction fades it is succeeded by nodular infiltration of the skin which is maximal in from three to four weeks; this is known as the Mitsuda reaction. Although of little use in diagnosis, the lepromin test has considerable prognostic value. It is usually negative in lepromatous infections, indicating a poor prognosis, and positive in tuberculoid infections, indicating a good prognosis.

The precise immunological basis of the Mitsuda reaction is unknown. Since the reaction is only produced by antigens containing particulate material dervived from the tissues, it is possible that it may to some extent be due to altered tissue antigens. In relation to this possibility it is of interest that patients with lepromatous leprosy who exhibit a negative Mitsuda reaction appear to have a considerably diminished capacity for allograft rejection.

Prophylaxis

The finding that persons giving a positive Mitsuda reaction are less likely to develop leprosy than those giving a negative reaction, together with the observation that BCG vaccination can induce lepromin hypersensitivity, has led to the hope that BCG vaccination might have a protective effect against leprosy. A trial in Uganda has confirmed this expectation. Similar results were obtained in New Guinea, but in Burma no benefit appeared to be conferred by BCG vaccination. The value of BCG in prophylaxis is therefore in doubt.

Favourable results have been reported for chemoprophylaxis with DDS in children in contact with cases of leprosy.

Mycobacterium ulcerans

This organism is responsible for a disease characterized by the occurrence of granulomatous ulcers in the skin and subcutaneous tissues, particularly of the arms and legs. Visceral involvement is unknown. The condition, which is uncommon, has been reported from Australia, West Africa and Mexico.

Mycobacterium ulcerans differs from the tubercle bacillus, which in many respects it resembles, in having an optimum growth temperature in the range 25 to 35°C. Like the tubercle bacillus growth is slow. Rats and mice may be infected with the development of ulcerative lesions of the extremities but the organism is non-pathogenic for guinea-pigs. The superficial and peripheral distribution of the lesions can be related to the temperature requirements of the organism, the lesions occurring in those areas of the body whose temperature is normally less than 37°C.

Similar lesions contracted from swimming baths and aquaria contaminated with *Myco. marinum* have been reported from a number of countries.

NON-PATHOGENIC MYCOBACTERIA

Saprophytic non-pathogenic mycobacteria are widely distributed in nature, occurring in the soil, on grass and in water. The occasional presence of organisms of this type on rubber corks and in tap water could cause confusion in the microscopic diagnosis of tuberculosis. The saprophytic mycobacteria resemble the tubercle bacillus morphologically but differ in that they grow well at room

Table III.7. The effects of BCG vaccination in the prophylaxis of leprosy in the Uganda trial 1960–64

Tuberculin status at intake	*Unvaccinated*		*BCG vaccinated*		*Reduction by vaccination*
	Total	Cases*	Total	Cases*	(%)
Negative	2930	20 (0·68%)	2844	4 (0·14%)	79
Weak positive	5141	69 (1·34%)	5247	14 (0·27%)	80
Strong positive	1081	9 (0·83%)	–	–	–

* Cases of leprosy discovered at follow-up two to three years after the beginning of the trial.
Protection against leprosy by BCG vaccination is almost the same as against tuberculosis. The disease was of the *tuberculoid* variety; vaccination might not give the same protection against *lepromatous* leprosy.
After Brown, J. A. K. and Stone, M. M. (1966). *Br. med. J.*, **1**, 1.

temperature and on nutrient agar.

Mycobacterium smegmatis

This non-pathogenic mycobacterium is of greater practical importance than those described above. It occurs in smegma and is occasionally found in urine in which it may be confused with the tubercle bacillus. As a rule *Myco. smegmatis* is less acid-fast than the tubercle bacillus and many strains are decolorized by alcohol, but strongly acid–alcohol-fast strains occur. The organism grows readily on nutrient agar and has an optimum growth temperature of 37 °C.

III.16

The Actinomycetes

The actinomycetes are a group of organisms intermediate in properties between bacteria and fungi. They are Gram-positive, branching filaments, non-sporing, non-motile and non-capsulated. As with the fungi the filaments interlace to produce a tangled mycelial growth. Two genera are distinguished: *Actinomyces* and *Nocardia.*

ACTINOMYCES

The genus *Actinomyces* consists of anaerobic or micro-aerophilic non-acid-fast species of highly parasitic nature. Three species are distinguished: *A. bovis,* the cause of actinomycosis or lumpy jaw in cattle; *A. baudettii,* the cause of actinomycosis in cats and dogs; and *A. israelii,* the cause of actinomycosis in man.

Actinomyces israelii

In the tissues *A. israelii* occurs in the form of yellowish granules which are usually readily visible to the naked eye. These granules, known as sulphur granules, are in fact colonies of the actinomycete. The centre of the granule consists of a dense felted mass of branching Gram-positive filaments each about 0·5 µm in diameter. In older colonies the filaments fragment to yield bacillary and coccal forms. At the edge of the granule they terminate in radially disposed, pear-shaped bodies or clubs, almost certainly formed by the deposition of lipid from the tissues around the end of the filaments. In contrast to the filaments the clubs are Gram-negative.

Actinomyces israelii is difficult to isolate and even more difficult to maintain in laboratory culture. Isolation may be attempted by inoculation of the granules, preferably after washing in sterile water to remove contaminants, onto blood agar or brain–heart infusion agar. Plates are incubated anaerobically, preferably in the presence of 5% CO_2. In addition, the washed granules should be incorporated in a glucose or serum agar stab in which, after a few days' incubation, micro-aerophilic colonies appear in a zone from 0·5 to 2 cm below the surface of the medium. On the surface of solid media *A. israelii* produces irregular, warty colonies adherent to the medium.

Actinomycosis

Actinomyces israelii is a strict parasite. It is frequently present in the mouths of normal individuals, where it is found mainly round the teeth and in the tonsillar crypts. It normally lives a harmless parasitic life in the mouth and, with other mouth actinomycetes, may play a part in the formation of dental tartar. Trauma, e.g. dental extraction and appendicectomy, is an important factor in the development of actinomycosis.

Actinomycosis, although of world-wide distribution, is an uncommon disease. It occurs in three main clinical forms:

1. Cervicofacial. In this form, which appears to be the commonest variety, the lesions occur mainly in the cheek and submaxillary regions.
2. Thoracic, in which the lesions occur in the lungs.
3. Abdominal, where the lesions are usually in the region of the caecum.

From these regions the organisms may pass by the portal vein to the liver, which may as a result become riddled by actinomycotic abscesses. Histologically the lesions of actinomycosis are granulomatous in character, usually with considerable suppuration and the formation of discharging sinuses. The high incidence of the lesions among farmers has not been explained.

Actinomycotic lesions usually contain a variety of anaerobic bacteria, particularly a small Gram-negative bacillus, *Actinobacillus actinomycetemcomitans*. It appears to play no pathogenic role.

Penicillin is the drug of choice in treatment.

NOCARDIA

The genus *Nocardia* comprises a group of aerobic actinomycetes which are widely distributed in the soil where they live a purely saprophytic life. Many are acid-fast.

They grow readily on agar media. Occasionally they are pathogenic to man, in whom they give rise to chronic granulomatous lesions in which, as in actinomycosis, granules consisting of colonies of the organism are frequently demonstrable.

Nocardiosis occurs in two clinical forms: systemic nocardiosis and mycetoma. In systemic nocardiosis, lesions occur mainly in the lungs but there may be haematogenous spread to other organs, particularly the brain. The condition is uncommon and is most frequently due to *N. asteroides,* an acid-fast species which, on agar media, produces an orange star-shaped colony from which the species derives its name. Mycetoma is a chronic granulomatous condition, mainly localized to the region of the foot (Madura foot), which occurs in certain tropical countries. It is most frequently due to *N. madurae,* a non-acid-fast species, but may also be caused by several species of filamentous fungi. Infections due to the latter are usually referred to as maduromycosis.

Sulphonamides have given excellent results in the treatment of *Nocardia* infections, if given early in the disease. In mycetoma, chemotherapy is combined with surgical treatment. Infections due to filamentous fungi are not susceptible to chemotherapy.

III.17

Mycoplasmas and Miscellaneous Bacteria

MYCOPLASMAS

Mycoplasmas are a group of organisms widely distributed in nature as parasites and pathogens of various animal species. Thirty-six species are listed by Bergey, seven of which have been isolated from various human sources. The relationship between the different species and their matural animal hosts is highly host-specific. Only one species *Mycoplasma pneumoniae,* the cause of mycoplasmal pneumonia (primary atypical pneumonia), is of proved pathogenicity for man.

Mycoplasmas differ from other bacteria in not possessing a cell wall, the cell being bounded by a thin cytoplasmic membrane. In spite of this they exhibit considerable stability. Consequently the cell membrane must be considerably stronger than that of bacteria. Morphologically, mycoplasmas are highly pleomorphic. Four morphological types are found:

1. Small coccoid bodies measuring 100 to 250 nm in diameter. These constitute the basic reproductive unit of the organism. The small bodies are bounded by a membrane about 7·5 nm thick. The membrane consists of lipid (mainly phospholipid) with cholesterol interspersed between the phospholipid molecules. It also contains protein, most of which is probably on the external surface of the membrane.
2. Large cells usually ranging from 500 to 1000 nm in diameter which contain a number of particles of subunit size.
3. Very large empty cells about the size of a leucocyte which appear to be degenerative and non-viable.
4. Filamentous forms of various sizes and shapes of 60 to 75 nm in diameter. Filament formation is very marked in some species, notably *M. mycoides,* the cause of contagious bovine pleuropneumonia. The precise relationship between these various morphological forms has not been clearly defined.

Mycoplasmas require enriched media for growth. Media both liquid and solid, containing yeast extract and a high concentration of animal protein as a source particularly of lipids and sterols, are commonly used. Most species are facultative anaerobes but some are microaerophilic. On solid media they produce a very characteristic fried egg type of colony consisting of a central button which grows deeply into the agar and a flat translucent peripheral zone which spreads out over the surface of the medium.

Biochemically, carbohydrate-fermenting and glucose-utilizing and non-carbohydrate-fermenting species are distinguished. The latter utilize fatty acids and amino acids, particularly arginine, as a source of energy (one group of strains—the T strains—utilizes urea).

Antigenically mycoplasmas have been analysed by a variety of serological techniques. The most widely used test for strain characterization is the growth inhibition test. In this, discs containing antiserum are placed on plates inoculated with mycoplasma. A positive reaction is the appearance of a clear zone of growth inhibition around the disc. Mycoplasmas differ from bacteria in that their growth is inhibited by antiserum in the absence of complement, a property which they share with the leptospirae.

It is probable that many of the pathogenic strains produce toxins but the only species for inflamed with the development of a local lymphangitis and lymphadentis and the appearance of septicaemic symptoms. Endocarditis and arthritis may occur as complications. The organism can also gain entry through the gastrointestinal tract, but this mode of infection appears to be very rare.

Penicillin is the drug of choice in treatment.

Isolation is most readily achieved by inoculation of the blood into a good infusion broth to give a final blood concentration of 20%.

Listeria and Erysipelothrix

These genera resemble each other closely. Both are widely distributed in nature as pathogens of a variety of animal species.

Listeria monocytogenes is a Gram-positive, non-sporing aerobic bacillus. It is sluggishly motile, exhibiting a characteristic end-over-end or 'tumbling' motility. It will grow on nutrient agar, but growth is considerably improved by the addition of glucose, blood or serum.

Listeria monocytogenes is responsible for a natural infection occurring in sheep, cattle, rabbits, guinea-pigs and other animals. In similar animals the infection takes the form of a septicaemia which is associated with areas of focal necrosis in the liver. The presence of monocytosis—which gives the organism its name—is a characteristic feature. In cattle, sheep and goats the infection takes the form of meningo-encephalitis; in these animals monocytosis is unusual. In man, as in cattle, *L. monocytogenes* may cause meningo-encephalitis, but the condition is very rare and only a limited number of cases have been reported. In these cases monocytosis was observed in the peripheral blood. The mode of infection has not been discovered. A generalized infection in infants associated with focal necrosis of the liver and spleen has also been described. The disease which is known as granulomatosis infantiseptica has a high mortality rate. The infection is believed to be transmitted from the mother to the fetus in utero. In contrast to the disease in the infant the infection in the mother is usually mild and may be asymptomatic. Penicillin and tetracycline appear to be therapeutically effective.

Erysipelothrix rhusiopathiae is very similar to *L. monocytogenes* in its properties. It differs from *Listeria* antigenically, in certain biochemical reactions and in being non-motile. On primary isolation it tends to be micro-aerophilic.

Erysipelothrix rhusiopathiae is responsible for a natural infection in a variety of animals which is characterized by septicaemia and the presence of focal lesions in the liver. In swine it causes the condition known as swine erysipelas. The organism also occurs as a commensal on the skin and scales of various types of fish.

Human infection is not uncommon. It occurs mainly in those whose work involves the handling of fish, meat or poultry, the organism gaining access through abrasions of the skin. At the site of inoculation a purplish red oedematous area develops. The lesion spreads by peripheral extension while at the same time clearing in the centre. The condition has some resemblance to erysipelas and is on this account known as erysipeloid. The disease is a mild one with a low fatality rate, but relapses are frequent. Isolation of the organism from the skin lesions is difficult and is only likely to be successful with material taken by skin biopsy. Penicillin appears to be the most effective chemotherapeutic agent.

Lactobacillus

Bacteria of the genus *Lactobacillus* are mostly long slender rods which are Gram-positive, non-motile, non-capsulated and non-sporing. They are, for the most part, micro-aerophilic or anaerobic. Although they grow best in media of neutral reaction, they are aciduric, that is, they are capable of growing in media, e.g. broth containing 0·5% glacial acetic acid, sufficiently acid to inhibit the growth of other organisms.

Lactobacilli are the major organisms present in the faeces of breast-fed babies; they are also found in the faeces of adults, but in much smaller numbers. The species usually encountered are *Lactobacillus acidophilus* and *L. bifidus*. The strains of lactobacilli occurring in the adult vagina, known as Döderlein's bacilli, closely resemble *L. acidophilus* and may in fact belong to this species.

Some workers believe that *L. odontolyticus*, as well as certain streptococci, may play a part in the aetiology of dental caries. It is supposed that the organisms break down food residues with the production of lactic acid, which dissolves out the calcium phosphate of the enamel. As a result of this initial breach, the dentine is exposed and is then disintegrated by various proteolytic organisms.

Calymmatobacterium granulomatis (Donovania granulomatis)

This organism is the cause of granuloma inguinale, a chronic granulomatous venereal disease which occurs only in tropical countries. The organism is demonstrable within mononuclear cells obtained from the lesions in which it appears as a Gram-negative capsulated bacillus. It has been grown in the yolk sac of the chick embryo and, under in vitro conditions, in egg yolk. It appears to have some antigenic relationship to *Klebsiella rhinoscleromatis*.

III.18

The Clostridia

The genus *Clostridium* is a genus of Gram-positive, anaerobic, sporing bacilli. The natural habitat of the clostridia is the soil. From the soil they may gain access to the intestinal tracts of animals and man, from which certain species can frequently be isolated. The genus is a very large one, 61 different species being described by Bergey. Only a few of the species described are pathogenic for man.

The clostridia differ morphologically from the aerobic sporing bacilli in that in most cases the spore is normally wider than the body of the bacillus. Some, notably *Cl. sporogenes,* spore readily; with others, particularly *Cl. perfringens,* sporing is infrequent. In general, spore formation is diminished in the presence of fermentable carbohydrate. In the pathogenic species, except for *Cl. tetani,* the spore is central or subterminal in position; *Cl. tetani* possesses a terminal spore. All the pathogenic species are motile with the exception of *Cl. perfringens,* which is also the only species to produce a capsule.

On the basis of their biochemical behaviour most of the pathogenic clostridia can be divided into two contrasting groups:

1. Predominantly *saccharolytic* organisms. These are active fermenters of sugars, do not digest serum, produce a pinkish colour in Robertson's meat medium, which is not decomposed and yield cultures which frequently have a smell reminiscent of rancid butter—due to the presence of higher fatty acids such as butyric acid.
2. Predominantly *proteolytic* organisms. These are not as active sugar fermenters as the first group. They digest serum, produce a black colour in meat which is extensively decomposed and their cultures have a putrefying smell due mainly to the liberation of hydrogen sulphide from proteins.

Clostridium tetani does not fall into either of these groups, being only mildly proteolytic and not at all saccharolytic.

The biochemical activities of most importance in species differentiation are the fermentation of glucose, saccharose and lactose, appearance of the growth in Robertson's meat medium and effect on gelatin, serum and litmus milk. Clostridia may reduce the indicators present in sugar media and consequently false negative fermentation reactions may be obtained. The correct reaction is apparent if further indicator is added to the medium prior to reading the results of the test.

The pathogenic clostridia are very active exotoxin producers and it is to this property that their pathogenicity is mainly or entirely due. The demonstration of the appropriate toxin and its neutralization by specific antitoxin are essential for the identification of the pathogenic species. For this purpose guinea-pigs and mice, which are highly susceptible to the toxins of the clostridia, are commonly employed.

The identification of the pathogenic clostridia usually affords little difficulty. Since, however, clinical material frequently contains more than one species of *Clostridium,* and since some species tend to swarm on solid media, it is often difficult to isolate single species in pure culture. The material should be plated directly on to a well dried blood agar plate which is incubated in a McIntosh and Fildes jar under anaerobic conditions. Willis and Hobb's medium is of considerable value for the isolation of clostridia because of its selective and differential properties. Tubes of Robertson's meat medium should also be inoculated for enrichment purposes. After inoculation one tube is incubated unheated and a second after heating to 80°C for 10 minutes. Subculture from meat medium is often successful in isolating a *Clostridium* when direct plating is

Table III.8. Biochemical reactions of certain Clostridia

Species		*Glucose*	*Lactose*	*Sucrose*	*Milk*	*Gelatin*	*Serum*
Cl. perfringens	Mainly saccharolytic	AG	AG	AG	AG	+	–
Cl. septicum		AG	AG	AG	Ag	+	–
Cl. novyi		AG	–	–	Ag	+	–
Cl. fallax		AG	AG	AG	Ag	–	–
*Cl. tertium**		AG	AG	AG	Ag	–	–
Cl. bifermentans	Mainly proteolytic	AG	–	–	D	+	+
Cl. botulinum†		AG	(AG)	–	D	+	+
Cl. histolyticum		(A)	–	–	D	+	+
Cl. sporogenes		AG	–	–	D	+	+
Cl. tetani		–	–	–	–	+	–

* Grows appreciably under aerobic conditions.
† Changes produced by Type A and most Type B strains.
A=acid; G=large amount of gas; g=small amount of gas; D=digestion; +=liquefaction. ( ) not produced by all strains.

negative. Identification of the non-pathogenic clostridia, particularly of the proteolytic varieties, may be extremely difficult. The demonstration of antibody in the patient's serum is of no value in the diagnosis of clostridial infections.

Clostridium tetani

Morphology

Clostridium tetani is a slender bacillus measuring 2 to 5 μm by 0·4 to 0·5 μm. It produces a round terminal spore which gives the organism a 'drumstick' appearance. Spores are produced in the body as well as in culture. The drumstick appearance is not a completely reliable diagnostic criterion since it is also shown by certain non-pathogenic clostridia, notably *Cl. tetanomorphum.*

Cultural properties

Clostridium tetani produces a delicate irregular colony with fine offshoots. If the medium is moist, the growth is in the form of a thin film which spreads over the surface except in the case of occasional non-motile strains which produce isolated colonies. This swarming capacity is made use of in Fildes's method for the isolation of *Cl. tetani* in which primary cultures in peptic blood broth after heating to kill non-sporing organisms are inoculated into the condensation water of a blood agar slope. It may then be possible to obtain a pure culture from the edge of the swarming film of growth which may have spread a considerable distance up the surface of the slope.

On the basis of its flagellar antigen composition the species can be sub-divided into ten antigenic types. One of the types—type VI—is non-flagellated. All the antigenic types have a common O antigen.

Pathogenicity

Clostridium tetani produces a very powerful exotoxin which has a selective action on the central nervous system. There is a considerable body of evidence that the toxin acts centrally, i.e. on the nerve cells in the brain and spinal cord, rather than on the peripheral nerves. Its central action results in an exaggerated reflex motor response to sensory stimuli. This appears to be due to interference with the control of lower motor neurons by upper motor neurons. How this inhibition is achieved is not known.

The toxin is protein in nature and has been obtained in crystalline form. It is destroyed by heating to 65°C for 5 minutes, by the action of acid and by proteolytic enzymes. It is therefore inactivated in the stomach and the intestinal tract and is consequently without effect if ingested. It is extremely potent, 1 mg of a purified preparation containing around 6×10^6 mouse lethal doses. Culture filtrates differ considerably in potency but an active filtrate may contain as much as 10^5 mouse lethal doses per ml.

Filtrates of *Cl. tetani* also contain an oxygen-labile haemolysin which resembles and is immunologically related to streptolysin O. The haemolysin is produced by strains of the organism which fail to produce the neurotoxin. No role has been assigned to it in pathogenicity.

Tetanus

The tetanus bacillus is widely distributed in the superficial layers of the soil in all parts of the world, but is more often found in richly cultivated land than in virgin soil. It is commonly present in the faeces of herbivora, particularly of horses, and in some countries, e.g. in China, in the faeces of man.

Tetanus is due to infection of a wound —particularly of a deep penetrating wound or one in which there has been considerable destruction of tissue or which has been contaminated with soil—by tetanus spores. Wounds contracted in warfare are particularly liable to infection with tetanus bacilli. Infection of such wounds with pyogenic organisms increases the risk of tetanus still further. These factors predispose to the disease by creating an anaerobic environment in the wound suitable for the growth of the organism. Thus extensive laceration by interfering with the blood supply causes necrosis with resultant anaerobiosis, and soil lowers the oxidation–reduction potential of the tissues, mainly through the action of its calcium salts. Tetanus may, though infrequently, follow a trivial wound—particularly in children, who are more susceptible than adults—or may result from the infection of a burn or of a vaccination site (p. 415). In primitive communities tetanus neonatorum due to infection of the umbilical stump is not uncommon. In the past a number of infections have been traced to the use at operations of improperly sterilized catgut but as reliably sterile catgut is now available commercially this mode of infection should no longer occur. On occasion tetanus following operations has been ascribed to contamination of the operation wound by non-sterile instruments or other materials brought in contact with the wound or by the dust of the operating theatre from which tetanus bacilli have occasionally been isolated.

The bacilli inflict very little local injury, and there is normally no invasion of tissues, although they have, on a few occasions, been isolated from the spleen. The symptoms of tetanus are due entirely to the action of the neurotoxin, produced in the wound, on the central nervous system.

There has been considerable controversy as to the route by which the toxin gains access to the central nervous system. If toxin is injected into the limb of an experimental animal and the nerves supplying the limb are divided, tetanus does not occur. The toxin must therefore pass via the nerves. Both axonal spread and spread by the perineural lymphatics have in the past been strongly supported as the probable route. There is more convincing evidence, however, that the toxin travels in the fluid between the nerve fibres—its upward passage being accelerated by pressure on the nerves from muscular contractions.

The first symptom of tetanus is usually a tingling sensation in the vicinity of the wound. This is followed by the development of the characteristic muscular spasms occurring first in the proximity of the wound, then in the masseter muscles of the jaw and ultimately involving all the voluntary musculature. Spasm of the masseter muscles is frequently the first obvious sign of the disease, and is responsible for its popular designation—'lock-jaw'. Death is due to cardiac and respiratory failure resulting from involvement of vital centres in the brainstem.

Treatment

The results obtained with antitoxin alone in the treatment of tetanus have been disappointing. This is probably due to the fact that by the time symptoms appear the toxin is so firmly bound to the nervous tissue that it cannot be removed by antitoxin. With current therapeutic regimens in which antitoxin and heavy sedation are combined with thorough surgical cleaning of the wound and chemotherapy to control secondary bacterial invasion, the prognosis has, however, greatly improved. In general the chance of success varies with the incubation period. If this is less than five days the patient usually dies; if the incubation period is longer than five days most cases should recover.

Prophylaxis

Active immunization against tetanus is carried out by the subcutaneous or intramuscular injection of toxoid. Reactions are rare and usually mild. A full course of injection consists of two injections each of 1 ml of toxoid with an interval of not less than six weeks between each injection followed six to

18 months later by a third injection. Active immunization has been mainly used for individuals, especially service personnel, who are particularly at risk but as tetanus is now a more frequent cause of death than diphtheria there is a good case for routine immunization for all members of the population in infancy or childhood. For this purpose the toxoid is frequently given as a component of triple vaccine.

It is of the utmost importance that all patients with wounds should receive adequate and immediate prophylactic treatment. There is, however, some doubt as to what is the best such treatment to employ. For many years the procedure followed by the British Army authorities—the value of which was first established in troops wounded on active service in the 1914–18 war—was widely adopted as a basis for the prophylaxis of civilian personnel. This consisted of the administration of tetanus antitoxin (ATS) to non-immune persons, with a booster dose of tetanus toxoid to those who were deemed to be immune. Immune persons were defined as those who had received two doses of tetanus toxoid within six months before the accident, or three doses, i.e. a full course, of toxoid within five years before the accident. In recent years, however, the acceptability of this regimen for civilians has been challenged by a number of workers. The grounds on which this challenge is based are:

1. The high incidence of reactions following the administration of ATS prepared in the horse. Under the conditions of active service the risk of death from an anaphylactic reaction might be much less than that of death following tetanus. This, however, does not necessarily apply to a civilian population.

2. In subjects who have received a previous injection of antitoxin the antitoxin is rapidly eliminated as a result of an immune response. Because of this the patient is protected only for a fairly short time.

3. Many cases of tetanus, particularly in children, occur as a result of minor wounds. These wounds admittedly carry a lower risk of tetanus than severe wounds but are much more numerous in incidence; in Cox and Knowelden's survey, which covered a seven-year period, minor wounds were responsible for the majority of cases of tetanus investigated. This disparity must, to some extent, be due to the fact that only the more serious wounds were reported and adequately treated, while trivial wounds were either not reported or were not regarded as requiring specific tetanus prophylaxis.

On these grounds Cox and Knowelden recommended the abandonment of antitoxin. Instead they proposed, together with the thorough cleaning of the wound, the administration of penicillin and, regardless of the state of immunity of the patient, an injection of tetanus toxoid. This would be followed, for those who were not previously immunized, by a complete course of active immunization. Following the adoption of this regimen they found no significant increase in the incidence of tetanus. Most authorites, however, are reluctant to abandon the use of antitoxin completely and consider it should be given if the wound is thought to carry a particularly high risk of tetanus, especially if there has been an appreciable delay in treatment. The administration of antitoxin is of greater urgency when wounds have been contracted in areas in which there is a high incidence of tetanus, as obtains in many developing countries. As an alternative to horse antitoxin, human gamma globulin obtained from persons who have been immunized with tetanus toxoid could, if available, be employed. If the number of cases to which antitoxin is given were small the general substitution of human antitoxin for horse antitoxin, which would eliminate the risk of an anaphylactic reaction, would be feasible. Human antitoxin has been found by Suri and Rubbo to persist much longer in the circulation than horse antitoxin and since it can be used in low doses it does not interfere with the immunological responses to toxoid administered at the same time.

Clostridium botulinum

Clostridium botulinum, the cause of botulism in man and of similar intoxications in various animals, is a large thick bacillus with oval, central or subterminal spores. Six distinct types of *Cl. botulinum,* A, B, C, D, E and F, are described. The types differ in producing antigenically distinct exotoxins and in certain biochemical properties. *Cl. botulinum* occurs more frequently in virgin than in cultivated soil.

Cultural properties

Different strains of *Cl. botulinum* differ in their temperature optima. Some have an optimum at or about 37°C; this appears to be the case for those most frequently responsible for botulism in man. Others have an optimum in the range of 20 to 30°C. The colonies have a variable appearance but are uniformly haemolytic. On the basis of their biochemical characteristics strains of *Cl. botulinum* are divided into proteolytic (ovolytic) and non-proteolytic (non-ovolytic) groups. All type A and some type B strains belong to the proteolytic group. These strains are classified by Bergey as *Cl. parabotulinum.*

Pathogenicity

Though antigenically distinct the toxins of the various types have apparently an identical pharmacological action. Like the tetanus toxin they act specifically on the central nervous system. They differ from the tetanus toxin, however, in that they act peripherally at the myoneural junction, where it is thought that they interfere with the synthesis or release of acetylcholine.

The toxins are not inactivated by the gastric hydrochloric acid or by the proteolytic enzymes of the intestinal tract. From the intestinal tract they are absorbed into the blood and from the blood gain access to the central nervous system. It is thought that the absorption of toxin through the alimentary mucosa requires a preliminary dissociation of the large toxin molecule into smaller fragments.

The different toxins differ in their lethality for different animal species. This probably explains why, under natural conditions, certain *Cl. botulinum* types are associated with disease of particular animals. The great majority of cases of botulism in man are due to type A strains, a minority to type B and a few to type E.

Type A and type B toxins have been isolated as crystalline proteins. The type A protein has a molecular weight in the undissociated state, of about 10 000. The toxins are fairly resistant to heat, withstanding exposure to 80°C for some time; they are, however, rapidly destroyed at 100°C. They are extremely potent. 1 mg of the purified toxin has been estimated to contain about 20×10^6 mouse minimum lethal doses and would constitute a lethal dose for 1000 tons of guinea-pigs.

Botulism

Botulism is an intoxication not an infection; it is due to the ingestion of toxin preformed in food by the growth of the organism. Since the organism is strictly anaerobic it is most likely to grow in food which has been canned or bottled. Nowadays, however, the risk from commercially canned food is virtually non-existent, since in commercial processing the food is heated sufficiently to ensure the destruction of any spores which might be present. Consequently most cases occurring now are caused by food which has been preserved at home. In highly acid foods, e.g. fruits, with a pH of less than 4·5, germination of the spores is inhibited. The danger of botulism from such foods is therefore negligible. Vegetables are now the greatest hazard because of the readiness with which they may become contaminated by spores present in the soil.

Botulism differs from other forms of food poisoning in that gastrointestinal symptoms are very slight. The action of the toxin on the parasympathetic nervous system is responsible for the most prominent symptoms, involvement of the eye muscles with protrusion of the eyeballs, ptosis, loss of accommodation, dilated pupils, as well as aphonia and dysphagia. Fever is generally absent and consciousness is retained. Symptoms usually appear within 24 hours of eating the food and death, due to respiratory or, more rarely, to cardiac failure, occurs 4 to 8 days later. The severity of the symptoms depends on the amount of food (and therefore toxin) consumed. This fact may explain the differences in the fatality rates in different outbreaks which average about 65%. The toxin can be absorbed through the respiratory tract, and its presence in aerosols constitutes a serious hazard for laboratory workers.

Antitoxin has been used therapeutically in the disease in man in a few cases, but it is of little or no value.

Diagnosis

Diagnosis is usually made on clinical grounds and, from the character of the disease, presents little difficulty. Bacteriological confirmation is not possible by the isolation of the organism from the body but an attempt should be made to demonstrate the organism and its toxin in the suspected food.

Prophylaxis

Prophylaxis is chiefly a matter of the hygienic preparation of foodstuffs. Since the toxin which is the cause of the disease is thermolabile, the heating of any food, which may be suspected, to 100°C for a few minutes immediately before use, will make it quite safe. Anerobic conditions are, of course, most perfect in canned and bottled foods, and these should be rejected if they are not in perfect condition, as judged both by the eye and by the nose. Since gas is produced by the growth of *Cl. botulinum* all canned foods should be discarded if the tin is blown. The ends of blown tins are convex and give a drum-like note on percussion. Blowing of tins is not necessarily, and is nowadays in fact rarely, due to the growth of *Cl. botulinum.* It should be remembered that food in which *Cl. botulinum* has grown may show no obvious signs of decomposition. Because of the very real danger of botulism from home preserved foods, home preservation should be restricted to fruits.

Though of little or no value in treatment antitoxin appears to be of considerable value in prophylaxis and when a case of botulism has occurred should be given to all individuals partaking of the food responsible. Toxoids have been prepared from type A and type B toxins by treatment with formalin and can be used for the active immunization of laboratory personnel working with *Cl. botulinum.*

THE CLOSTRIDIA OF GAS GANGRENE

A great variety of clostridial species has been isolated from cases of gas gangrene. Many of these, however, are purely saprophytic or are present as secondary invaders. The most important species and those primarily responsible for the condition are *Cl. perfringens, Cl. novyi* and *Cl. septicum.* Each of these is an active toxin producer and, although frequently present along with other species, can cause gas gangrene by itself. *Cl. perfringens* is the species most frequently found, being present in about 60% of all cases.

Clostridium perfringens (Cl. welchii)

Morphology

Clostridium perfringens is a large, Gram-positive, non-motile, bacillus with square-cut or slightly rounded ends, usually measuring from 4 to 8 μm by 0·8 to 1·2 μm. Capsules are present in the tissues, but are found in cultures only when the medium contains serum. The organism produces oval subterminal spores which are of smaller diameter than the bacillus. Spores are not seen in the tissues and are not produced in media containing any carbohydrate fermented by the bacillus. Special media have been described, however, e.g. Ellner's medium, which will stimulate spore formation in culture.

Cultural properties

Clostridium perfringens is a strict anaerobe. On blood agar it produces round, smooth, colonies with an entire edge and and is haemolytic. *Cl. perfringens* is actively saccharolytic but exhibits little proteolytic activity. A very characteristic change—the so-called 'stormy fermentation'—is produced in milk cultures. There is a coagulation of the casein into a clot, which floats in an almost clear whey and the clot and layer of cream are torn and riddled as a result of the active evolution of gas.

The species is antigenically heterogeneous.

Pathogenicity

Clostridium perfringens is a very prolific exotoxin producer. In the species as a whole a total of eight toxins, lethal for mice on intravenous inoculation, have been identified. On the basis of the major lethal toxins they produce, strains of *Cl. perfringens* can be assigned to one or other of six subtypes A, B, C, D, E and F. In addition to the lethal toxins a number of extracellular enzymes—collagenase, proteinase, hyaluronidase, desoxyribonuclease and neuraminidase—are also identifiable in culture filtrates. Some of these also tend to show a characteristic type distribution.

Only types A and F are pathogenic for man and only type A is involved in the production of gas gangrene. Type A strains have also been implicated as causes of a variety of bacterial food poisoning. In distinction from normal type A strains, strains isolated from cases of food

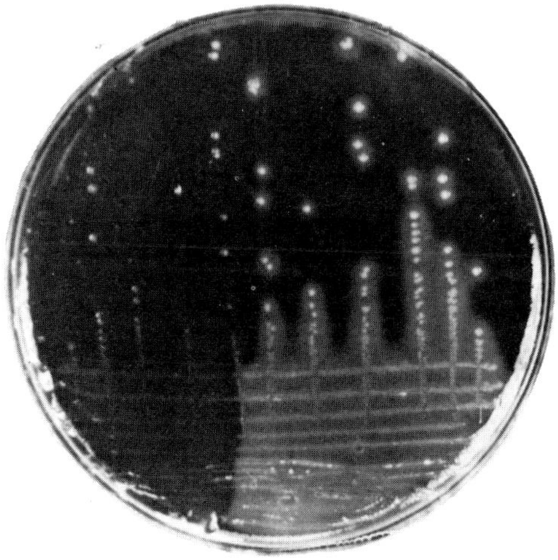

Fig. III.9. *Clostridium perfringens* showing the Nagler reaction on serum agar.

poisoning have been found to be non-haemolytic, only weakly toxigenic and produce spores with an unusually high degree of heat resistance. Meat which has been kept in a warm environment after cooking is the main source of these strains. In fact heating appears to activate spore germination. Type F strains have been found to be responsible for certain cases of human enterotoxaemia; this is a serious but uncommon condition in which there is a severe sloughing enteritis involving both the large and small intestines. Types B, C and D produce similar syndromes in various animals, particularly sheep.

The most important of the toxic substances produced by type A strains is the α toxin. This is a lecithinase which breaks down lecithin to yield a glyceride and phosphoryl choline. To distinguish it from lecithinases from other sources, e.g. snake venom haemolysins, which have a different mode of attack, it is known as lecithinase C. When injected into animals it causes necrosis of tissue cells and is lethal if given in sufficient amounts. It also haemolyses red cells by breaking down the lecithin of the red cell membrane. In its haemolytic action the α toxin shows the hot–cold effect which is also exhibited by the staphylococcal β lysin. Other types of *Cl. perfringens* also produce α toxin, but in much less amount than type A.

The α toxin may be demonstrated by the Nagler reaction. For this the organism is grown on plates containing 20% human serum or egg yolk; as a result of lecithinase activity lipid is deposited around the colony, giving rise to a readily visible zone of opalescence. If antitoxin is spread over half the plate, it inhibits lecithinase activity and therefore prevents the development of opalescence on that half of the plate. Lecithinase activity and its specific inhibition by antitoxin may also be demonstrated by growing the organism in broth containing serum or egg yolk. Calcium and magnesium ions are necessary for lecithinase activity. *Cl. novyi* and *Cl. bifermentans* also produce lecithinase. The lecithinase produced by *Cl. bifermentans* is antigenically identical with that of *Cl. perfringens* and is neutralized by *Cl. perfringens* antitoxin.

The α toxin is the agent of primary importance in the pathogenicity of *Cl. perfringens*. Since lecithin is a normal component of cell membranes, the toxin is obviously in a position to cause grave damage to tissue cells. There is also evidence that it can interfere with cellular metabolism by its action on essential lipoproteins. Furthermore, by its local destructive action the α toxin creates conditions favourable to further growth of the organism. The α toxin may not, however, be solely responsible for the marked toxaemia found in gas gangrene. Patients who fail to respond to treatment with very large amounts of antitoxin can frequently be saved by surgical removal of the affected tissue. This suggests that the organism produces some further but as yet unidentified toxic substance by its growth in the tissues.

Other substances produced by *Cl. perfringens* type A, and which may play some part in its pathogenicity, are the θ toxin—an oxygen-labile haemolysin resembling streptolysin O—hyaluronidase, desoxyribonuclease and collagenase. Collagenase breaks down collagen fibres and therefore presumably contributes to the disintegration of muscle which is the most dramatic feature of *Cl. perfringens* infection.

Clostridium septicum

Clostridium septicum is a straight or slightly curved bacillus, measuring from 3 to 8 µm by 0·4 to 0·8 µm, but very much longer forms are frequently seen, both in the tissues and in culture. It is feebly motile. It produces oval, central or subterminal spores which are, however, rarely seen in infected tissues.

Clostridium septicum is a saccharolytic anaerobe. On blood agar it usually produces a large colony with an irregular surface and with coarse projections at the edge. Some strains produce colonies very similar to those of *Cl. tetani.*

Pathogenicity

Clostridium septicum has considerable invasive powers and has frequently been isolated from the blood stream. It produces an exotoxin which is lethal on injection into animals and is specifically neutralized by antitoxin. Filtrates of *Cl. septicum* have haemolytic properties which are believed to be due to the lethal toxin. *Cl. septicum* also produces hyaluronidase, collagenase and desoxyribonuclease.

Clostridium novyi (Cl. oedematiens)

This bacillus, which measures from 3 to 10 μm by 0·8 to 1·0 μm, resembles *Cl. perfringens* in appearance, but is usually rather longer. Spores, which are oval, thicker than the bacilli and subterminal, are freely produced in culture, even in the presence of fermentable carbohydrate.

Clostridium novyi is a saccharolytic anaerobe. It is one of the strictest of the anaerobes. On solid media it produces a colony similar to that of *Cl. septicum.* Type A strains produce a characteristic iridescence when grown on egg yolk agar.

Pathogenicity

Clostridium novyi is a more active toxin producer than either *Cl. perfringens* or *Cl. septicum,* the mouse minimum lethal dose of a toxic filtrate being about 0·0002 ml. The toxin which has been obtained in a highly purified form is lethal and necrotizing but is not haemolytic; it produces a gelatinous oedema by its action on muscle. Other toxic components identified in culture filtrates of *Cl. novyi* include two distinct lecithinases which differ antigenically from the α toxin of *Cl. perfringens,* an oxygen-labile haemolysin resembling streptolysin O and a lipase. On the basis of the toxins they produce, strains of *Cl. novyi* are divided into four types, A, B, C and D. The lethal toxin is produced by types A and B. Most human infections appear to be due to type A. Types B, C and D are responsible for various diseases in animals.

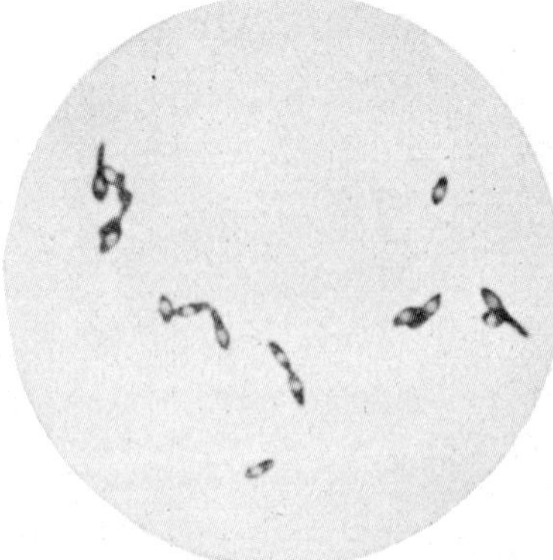

Fig. III.10. *Clostridium sporogenes* from a broth culture. ×950.

Gas gangrene

Like tetanus, gas gangrene is due to infection of severe and contaminated wounds. It occurs particularly as a result of wounds in which there has been considerable desctruction of muscle, especially of the buttock and thigh. As in the case of tetanus the source of infection is usually the soil, particularly soil which has been contaminated with animal excreta. Clostridial wound infections may also occur after surgical operations in which an adequate aseptic technique has not been maintained. This type of infection, although rare nowadays, was the most serious sequel of operations, particularly of hindquarter amputations, in pre-Listerian surgery. Gas gangrene sometimes occurs as a puerperal infection, particularly in cases of septic abortion.

Most of the systematic work on the bacteriology of gas gangrene has been concerned with the infection of war wounds. In the North African campaign of World War II MacLennan found that clostridia could be isolated from 20 to 30% of all such wounds. In the majority of cases the organisms were confined to the surface of the wound and were responsible for, at most, some slight delay in healing. The more serious clostridial infections he classified into two groups: anaerobic cellulitis in which there is considerable gas production in the tissues but the muscles remain normal and anaerobic myositis or myonecrosis in which there is considerable destruction of muscle. In MacLennan's series these conditions were comparatively infrequent, anaerobic cellulitis occurring in about 5% of wounds infected with clostridia and anaerobic myositis occurring in only 1·5%.

In addition to the primary gas gangrene organisms which have been considered above other clostridial species, notably *Cl. histolyticum, Cl. sporogenes, Cl. tertium, Cl. fallax* and *Cl. bifermentans* may be present in infected wounds. The presence of these organisms has a considerable adjuvant effect in the production of gas gangrene by creating conditions in the wound suitable for the growth of the primary pathogens. This is probably particularly true of *Cl. histolyticum* which can actively digest living tissue and of *Cl. bifermentans,* some strains of which are strongly toxigenic. (These strains are usually referred to as *Cl. sordelii.*)

Treatment and prophylaxis

The essential measures in the treatment of gas gangrene are:

1. The removal of all infected tissue.
2. Administration of antitoxin. A polyvalent antitoxin containing *Cl. perfringens, Cl. septicum* and *Cl. novyi* antitoxins is usually employed. In severe cases large doses are given every four to six hours by the intravenous route. Since it is not possible to identify the infecting organisms on clinical grounds a polyvalent antitoxin must be employed.
3. Chemotherapy. The best results have been obtained with tetracyclines, which may be combined with penicillin. Penicillin alone has given excellent results.
4. Recently, treatment with hyperbaric oxygen at a pressure of three atmospheres, in a special chamber, has been favourably reported on.

In the prophylaxis of gas gangrene the most important single measure is the early surgical toilet of the wound. Antitoxin may be used in prophylaxis but has been largely replaced by prophylactic penicillin. It should be noted that tetanus prophylaxis must be carried out in all persons with severe wounds of such a type as to favour the growth of clostridia.

III.19

The Spirochaetes

The order Spirochaetales comprises a group of spiral, actively motile, flexuous organisms. They are non-flagellated, their motility being due to gyratory and bending movements. Morphologically the spirochaetes are more complex than bacteria. They possess a cell wall and between this and the cytoplasmic membrane are a number of fibrils, the number depending on the genus. The fibrils are attached to knobs at the end of the organism. They constitute an axial filament which holds the organism in a spiral conformation. The pathogenic spirochaetes are also much thinner than bacteria. Consequently they are far more difficult to see and in wet preparations are visible only under dark ground illumination. Except for the borreliae they stain poorly with aniline dyes. They can, however, be stained with Giemsa's stain or by silver impregnation methods, e.g. Fontana's or Levaditi's.

Three genera of pathogenic spirchaetes are distinguished—*Borrelia, Treponema* and *Leptospira.* These are differentiated on morphological grounds. Of these genera only the leptospirae are cultivable in inanimate media.

THE BORRELIAE

Borreliae differ from other spirochaetes in possessing larger, more open coils and in being more readily stainable. Two groups are associated with disease in man—the borreliae of relapsing fever and *Borrelia vincentii.*

A variety of species of *Borrelia* has been described as being responsible for different geographical types of relapsing fever but it is doubtful whether each should be regarded as a separate species. Two main species are distinguishable—*Borrelia recurrentis,* the cause of European louse-borne relapsing fever, and *Borrelia duttonii,* the cause of African tick-borne relapsing fever. Although differing in the insect vectors by which they are spread under natural conditions and in their antigenic properties these two species are very similar in their laboratory characteristics. Of the two *Borr. duttonii* is the more pathogenic for laboratory animals.

Borrelia recurrentis (Borrelia obermeireri)

Borrelia recurrentis measures from 8 to 20 μm in length and 0·3 to 0·4 μm in diameter. Its spirals may number from three up to as many as ten in the long forms. At least six axial fibrils are demonstrable by electron microscopy. The organisms are highly flexible. *Borr. recurrentis* stains fairly well with aniline dyes but much better with Giemsa's or Leishman's stain. When stained by Gram's method it is Gram-negative.

Borrelia recurrentis has been cultivated in hydrocele or ascitic fluid containing a piece of fresh rabbit kidney but serial subculture has not been possible by this or any other in vitro method. It can, however, be propagated by inoculation on to the chorioallantoic membrane of the developing chick embryo and in various laboratory animals.

Relapsing fever

Louse-borne relapsing fever may occur in sporadic or epidemic form—the latter occurring under conditions which favour louse infestation, e.g. cold and overcrowding. Consequently epidemics of louse-borne relapsing fever are frequently associated with epidemics of louse-borne typhus. Lice become infected on biting but

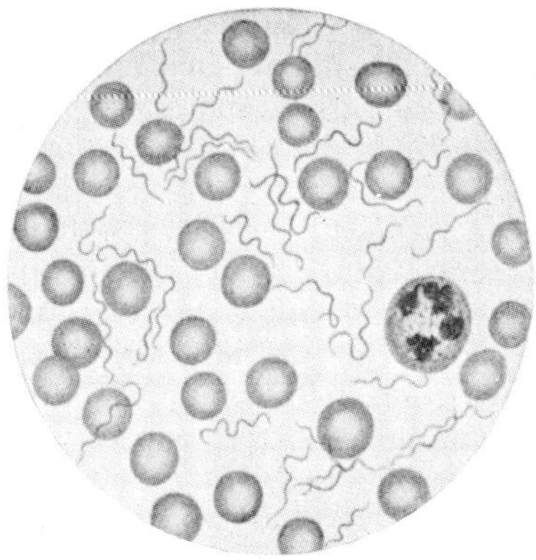

Fig. III.11. *Borrelia recurrentis* in a blood film.

for a few days after biting are non-infective. During this period the spirochaete multiplies in the body of the louse, eventually finding its way into the coelomic cavity. Infection may be caused by the feeding of the insect but more commonly by contamination of the wound with the coelomic fluid of lice which have been crushed in its vicinity. Transovarian passage of the spirochaete does not occur in lice which can therefore only become infected by biting. There is no known animal reservoir of infection.

In contrast to louse-borne relapsing fever the tick-borne variety occurs as a sporadic disease. The parasite is transmitted transovarially in the tick. Some species of ticks are able to transmit infection by biting but with others infection is due to contamination of the bite with the secretions of the coxal glands. The main natural reservoir of infection is in wild rodents but a variety of other wild animals are also naturally infected.

Clinically the tick-borne and louse-borne varieties of relapsing fever are indistinguishable. The onset of the disease is acute with the appearance of fever about three to ten days after infection. The fever usually lasts for about four days and then suddenly subsides; after about a week another bout of fever occurs and equally suddenly gives place to a normal temperature. In all, three to ten bouts of fever may occur each of less severity and of shorter duration than the preceding one. The mortality is low in sporadic cases but may be as high as 50% in epidemics.

The spirochaete appears in the blood in cases of relapsing fever shortly before the temperature begins to rise, and the number increases until the fever is at its height. They remain numerous throughout the greater part of the period of pyrexia, but shortly before the crisis they disappear and none can be found until the beginning of the relapse. Spirochaetes which survive the crisis do so because they have undergone a change in antigenic type and each relapse is associated with a further antigenic change.

The tetracyclines, especially chlortetracycline, are the drugs of choice in treatment. Penicillin is also effective.

Diagnosis

The diagnosis is most readily made by the demonstration of the borreliae in blood films stained by Giemsa's or Leishman's methods or by carbol fuchsin or by the examination of preparations under dark ground illumination. Sometimes the diagnosis of relapsing fever is first suggested during the performance of a blood count when the red cells are found to exhibit an unexpected motility as a result of being jostled by spirochaetes.

Mice and rats are susceptible to both tick- and louse-borne strains, while guinea-pigs are susceptible only to tick-borne strains. For diagnostic purposes the white rat is the most suitable animal; after inoculation of blood from a case, organisms usually appear in its circulation within 48 hours and can be demonstrated in stained films.

The Wassermann reaction may be positive in both types of relapsing fever; in the louse-borne variety the Weil–Felix reaction may also be positive against *Proteus* OXK. There are, however, no specific serological diagnostic tests.

Borrelia vincentii

Borrelia vincentii measures from 5 to 25 μm in length and has a variable number of loose irregular spirals. It stains readily with aniline dyes and in wet preparations is actively motile. It occurs as a normal mouth commensal but is present in particularly large numbers in Vincent's angina—an ulcerative condition of the gums and throat—and in acute ulcerative gingivitis. It is generally found in association with fusiform bacilli (*Fusobacterium fusiforme*) (p. 231). A combination of the two organisms is also found in

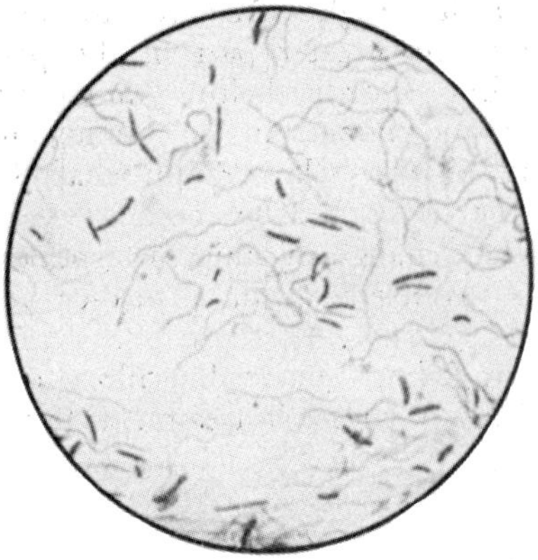

Fig. III.12. Spirochaetes and fusiform bacilli from the throat of a patient with Vincent's angina. ×950.

tropical ulcer and in gangrenous conditions of the lung. It has been suggested that they are merely secondary invaders and that the primary condition may be a virus infection or the result of a dietary deficiency.

Borrelia vincentii has been grown under anaerobic conditions in media containing ascitic fluid. The organism is very difficult to grow in pure culture. Better growth is obtained in mixed cultures with fusobacteria.

THE TREPONEMATA

Treponema pallidum

This organism, which is usually about 8 μm in length, may measure from 3 to 18 μm, and is extremely slender, rarely exceeding 0·25 μm in thickness. It is characterized by sharp, deep, regular spirals, the number of which depends on the length of the individual organism. The size of each spiral is fairly constant, a distance of about 1 μm separating the crest of one from that of the next, while the depth of the curve is approximately 1 μm. The axial filament of *Tr. pallidum* is composed of four thin fibrils, two of which are attached at each end of the organism.

So far *Tr. pallidum* has not been cultivated in inanimate laboratory media. It can, however, be grown in vivo in the testis and, less profusely, in the shaved skin of the rabbit. It can be propagated similarly but less readily in the hamster. Suspensions of *Tr. pallidum* from infected rabbit testis maintain their viability and activity for some time if kept in special media under highly anaerobic conditions. Suspensions prepared in this way are used in the *Treponema* immobilization test.

Treponema pallidum has a very low resistance to inimical agencies and dies out rapidly outside the body. It is very susceptible to the action of heat being destroyed in one hour at 42°C. Under optimal environmental conditions, however, it will survive in the cold for an appreciable time. It has been reported to be viable in refrigerated serum for about 24 hours and in tissues of the cadaver for a period of up to five days.

Syphilis

Syphilis is most commonly acquired in sexual intercourse, but non-venereal infections also occur. It is unlikely that the organism can penetrate the unbroken skin or mucous membrane, but contact of the slightest break in the surface with material containing the organism may be sufficient to cause infection. A most important type of innocent or non-venereal infection is that seen in congenital syphilis, the disease being acquired in utero from an infected mother.

The clinical evolution of syphilis is divided into three stages—primary, secondary and tertiary. The typical primary lesion is an indurated ulcer or *chancre* which appears in from nine to 90 days after infection. In venereal infections the primary lesion is usually on the genitalia. Extragenital non-venereal lesions may occur but are uncommon. Although the clinically obvious lesion of the primary stage is localized, the organisms at this stage are nevertheless distributed widely through the body, as a result of lymphatic and blood spread. Lymphatic spread to the local glands probably occurs within a few hours of infection.

The main characteristic of the secondary stage is the appearance of multiple lesions in the skin and mucous membranes. In some cases there are also clinical signs of involvement of the central nervous system. About 50% of patients progress to the tertiary stage, which may only appear after the lapse of several years. This stage is characterized by a variety of chronic lesions which may occur in any organ. In the remaining 50% of patients the disease either persists without clinical manifestation or undergoes apparent spontaneous cure.

Syphilitic infection confers some degree of immunity. This immunity is sufficiently developed to protect against further infection only if the disease has persisted at least as far as the secondary stage. Immunity to reinfection appears to be due to the antibody demonstrable in the treponema immobilization (TPI) test. Unlike reaginic antibody this antibody is capable of sensitizing the organism to the killing action of complement.

Diagnosis

In the primary stage a diagnosis of syphilis is made by demonstration of the spirochaete in the chancre by dark ground examination (p. 26). Other spirochaetes may also be present in the genitalia, but these are for the most part thicker with coarser and less regular curves than those of *Tr. pallidum.* In addition *Tr. pallidum* usually exhibits, when in motion, an angular bending of the body which is highly characteristic.

Serological tests are of paramount important in the diagnosis of syphilis. The tests employed fall into two groups: *reagin* tests and *treponemal* tests.

Reagin tests (standard tests for syphilis, STS)

These are tests which measure antibody to the *cardiolipin* antigen. This substance, which has been characterized as a diphosphatidyl glycerol, is a normal mammalian tissue component, heart muscle being a particularly potent source, but appears to occur also as a component of *Tr. pallidum.* In the standard serological tests for syphilis highly purified preparations of cardiolipin have, to a considerable extent, replaced the cruder preparations formerly employed. The reactivity of the antigen is considerably increased by the addition of lecithin and cholesterol, and these substances are therefore included in the formulations of cardiolipin currently employed. Cardiolipin antigen has been synthesized, and in preliminary studies synthetic antigen has given results similar to those of the natural product. Reagin tests are of two types—the Wassermann test which is a complement fixation test and various types of flocculation test, the oldest and possibly the best known of which is the Kahn test. Other flocculation tests in common use differ only in technical details from the Kahn test.

Reaginic antibody is not normally detectable until at least one week after the appearance of the chancre; thereafter positive serological results are obtained with increasing frequency until in the florid secondary stage positive reactions should be obtained in all cases. In late syphilis the antibody is usually, but not invariably, demonstrable. In cases showing involvement of the central nervous system reagin tests are usually positive in the cerebrospinal fluid as well as in the blood. As a rough rule, reaginic antibody persists for about twice as long as the condition has been present prior to therapy.

Treponemal tests

These are tests in which the antigen employed is of treponemal origin. A great variety of treponemal tests have been described, the most valuable of which are the treponema immobilization (TPI) test, the fluorescent treponemal antibody (FTA) test—in both of these the antigen employed is *Tr. pallidum*—and the Reiter protein complement fixation (RPCF) test.

The TPI test was first described by Nelson and Mayer in 1949. In this test dilutions of the patient's serum and complement are mixed with a suspension of *Tr. pallidum* obtained from infected rabbit testis. The mixtures are then incubated at 37 °C and examined by dark ground illumination; if the patient's serum contains antibody it will immobilize the spirochaete. The TPI test measures antibodies to species-specific *Tr. pallidum* antigens. It is technically difficult and available only in specialized reference laboratories but it is the most specific of the tests currently used. It does not become positive as early as reagin tests but persists for longer after clinical cure.

The FTA test is a fluorescent antibody test carried out by the antiglobulin or sandwich procedure. A film prepared from a suspension of *Tr. pallidum* is first exposed to the serum under test, washed and then exposed to a fluorescent antibody against human gamma globulin. The FTA test measures two distinct antibodies: a group-specific antibody probably identical with that detected in the RPCF test and an antibody specific for *Tr. pallidum.* The latter appears to be distinct from the immobilizing antibody detected in the TPI test.

In a modification of the FTA test, known as the FTA-ABS test, a group-specific antibody is absorbed out or inhibited with an extract of the Reiter strain of treponeme. A specificity as high as that of the TPI test is claimed for this modification.

A passive haemagglutination test using red cells coated with an extract of the Nichols strain of *T. pallidum* has recently been introduced; its value has still be be assessed.

The antigen used in the RPCF test is a protein extracted from the Reiter strain of non-pathogenic *Treponema.* The antibody reacting in the test appears to develop somewhat earlier than *Treponema* immobilizing antibody but later than reaginic antibody. Its duration after cure appears to be intermediate between that of reagin and *Treponema* immobilizing antibody. Like the TPI test, therefore, it is of limited value as a test for cure. Because of its technical simplicity, however, it is much more suitable for routine use.

Treatment

For many years organic arsenical compounds and preparations of bismuth were the only drugs available for the treatment of syphilis. They were highly effective but their use has been rendered obsolete by the introduction of penicillin—a drug which is actively spirochaeticidal, virtually non-toxic and effective after a much shorter period of treatment. Unfortunately, a small proportion of patients treated with penicillin relapse after apparent clinical cure. It has been suggested that these failures may be due to the persistence of spirochaetes in a quiescent state in which they are insusceptible to the action of the drug. Cephaloridine is also effective and has been recommended for the treatment of patients allergic to penicillin.

Treponemata of the endemic treponematoses

The endemic treponematoses constitute a group of non-venereal treponemal infections which are endemic in various parts of the world. They are transmitted by close contact and only occur in communities with low standards of hygiene. In such communites infection normally occurs early in life and is usually contracted from an infected parent. Congenital infection does not appear to occur. Three distinct types of endemic treponemal infection are recognized: yaws, due to *Tr. pertenue,* pinta due to *Tr. carateum* and endemic syphilis.

The treponemata responsible for these conditions are morphologically indistinguishable from *Tr. pallidum* and like the latter are non-cultivable on inanimate laboratory media. Their antigenic resemblance to *Tr. pallidum* is shown by the fact that the sera of infected patients contain both treponemal and reaginic antibodies.

Yaws

Yaws is almost exclusively a disease of tropical countries, occurring in a belt extending roughly from the Tropic of Cancer to the Tropic of Capicorn.

The primary lesions of yaws occur on exposed areas, usually on the lower legs or the feet; these appear as painless papules which progress to ulceration. In the secondary stage, which usually supervenes after an interval of two or three months from the onset of the disease, widespread lesions similar in appearance to the primary lesions appear in the skin and mucous membranes. Corresponding to the tertiary stage of syphilis, ulcerative lesions in the skin and bone appear as late manifestions. The disease is contracted by direct contact with the ulcerating lesions in which the treponemata are present in large numbers.

Like *Tr. pallidum, Tr. pertenue* can be propagated in rabbits and hamsters. According to Turner and Hollander there are slight but significant differences in the pathogenicity of the two species, *Tr. pertenue* strains producing on the whole less indurated lesions in rabbits and more spreading lesions in hamsters than *Tr. pallidum.*

Pinta

Pinta appears to be almost exclusively a disease of Central and South America. The first sign of the disease is the appearance on an exposed area of the body of a primary papule or 'pintid'. Unlike primary lesions of yaws this does not progress to ulceration. It is usually followed in from five to 18 months by the appearance of crops of flat erythematous skin lesions which ultimately become hyperkeratotic and depigmented.

Treponema carateum has not so far been serially propagated in laboratory animals but the disease has been transmitted to human volunteers by inoculation.

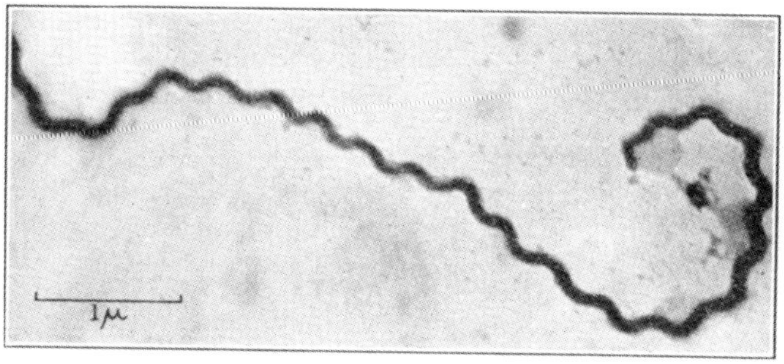

Fig. III.13. *Leptospira icterohaemorrhagiae* shown on an electron micrograph.

Endemic syphilis

In the eighteenth and nineteenth centuries endemic syphilis was widespread in Europe. It was known by a variety of names, e.g. sibbens in Scotland, button scurvy in Ireland. It now occurs mainly in certain parts of Africa and in the Middle East, particularly in the more primitive communities. In its major clinical manifestations endemic syphilis is indistingishable from venereal syphilis but differs from the latter in the rarity of a clinically obvious primary lesion.

Many workers consider that endemic syphilis is caused by *Tr. pallidum* and that it is epidemiologically the precursor of venereal syphilis, its non-venereal transmission being due to low hygienic standards. With improvement in hygienic conditions the venereal route would remain as the only method of transmission. Turner and Hollander have found, however, that in their pathogenicity for animals and in their immunological properties strains of treponemata isolated from cases of endemic syphilis are intermediate in characteristics between *Tr. pertenue* and *Tr. pallidum*. If this is correct it would seem that the transformation of syphilis from a non-venereal into an exclusively venereal infection is associated with some change in the properties of the treponema.

Non-pathogenic treponemata

Non-pathogenic treponemata are frequently present about the genitalia and in the mouth; they can be distinguished morphologically from *Tr. pallidum* but the differentiation requires considerable experience.

The Reiter strain of non-pathogenic treponeme is of importance as the source of antigen for the RPCF test (p. 300). Unlike the pathogenic treponemata this organism can be grown in media containing serum. Although originally isolated from a case of syphilis it differs sharply from *Tr. pallidum* in morphology and in being non-pathogenic for rabbits or hamsters.

THE LEPTOSPIRES

The leptospires are a group of spirochaetes with very shallow close-set spirals, whose characteristic morphological feature is that the ends are curved to form terminal hooks. They measure from 5 to 20 μm in length and about 0·15 μm in width. The spirals have an amplitude of about 0·5 μm and measure 0·5 μm from crest to crest. Leptospires possess two axial filaments, one anchored at each end of the organism.

The group comprises free-living and parasitic varieties, but these have proved very difficult to classify. Currently only one species is defined, *Leptospira interrogans*. Parasitic and free-living strains are assigned to the complexes *interrogans* and *biflexa* respectively. Various serotypes are recognized, the strains previously defined as *L. icterohaemorrhagiae* and *L. canicola* forming distinct serotypes. These two serotypes are of world-wide distribution. A number of other serotypes are found in different geographical areas, e.g. the *hebdomadis* serotype has been reported only from Japan. Parasitic species are harboured normally in the kidneys of various vertebrate hosts, frequently without causing overt infection in the host species, and are excreted in the urine.

Leptospires can be cultivated in various media, parasitic strains requiring the addition of protein.

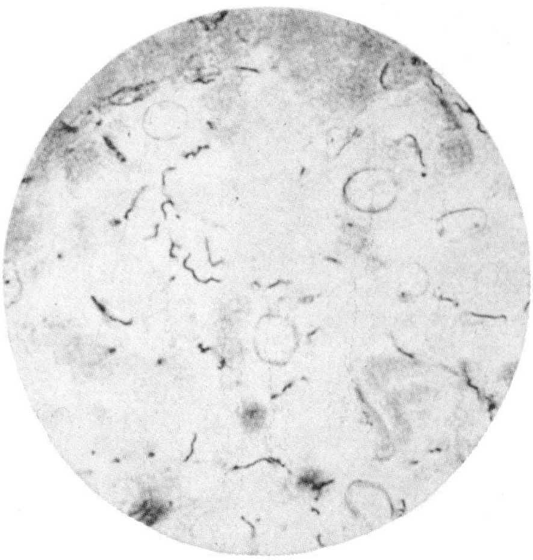

Fig. III.14. *Leptospira icterohaemorrhagiae* in a section of the kidney of an infected guinea-pig.

They are highly aerobic and have an optimum temperature of 28 to 32°C. All strains have a requirement for long-chain fatty acids, which they cannot synthesize for themselves and which must be supplied in bound form. Strains from human sources are frequently isolated in Fletcher's semi-solid medium containing diluted rabbit serum and 1% agar.

Leptospirosis

The classical manifestation of leptospiral infection—Weil's disease, sometimes known as epidemic jaundice—is most frequently caused by the *icterohaemorrhagiae* serotype. It is characterized by the occurrence of pyrexia, gastrointestinal symptoms, haemorrhages, either into the tissues or from mucous membranes, and albuminuria. Conjunctivitis is an early and characteristic feature. A high proportion of leptospiral infections are anicteric; this is especially so of infections due to the *canicola* serotype. In some cases the organisms localize in the central nervous system producing the symptoms of an aseptic meningitis which is clinically indistinguishable from aseptic meningitis caused by the enteroviruses. Leptospiral infection also occurs frequently in a mild form which can be recognized only by serological methods.

The parasites are present in the blood stream during the first week of the disease. From the second week onwards, they are demonstrable in the urine in which they may continue to be found for several weeks after convalescence.

Human infection is most frequently due to contact with water contaminated by the urine of infected animals. Infection with the *icterohaemorrhagiae* serotype is usually derived from rats, and therefore occurs mainly in persons whose occupation brings them into contact with water which has been contaminated with rat urine (sewer workers, miners and those employed in cleaning fish and preparing tripe). Infection can also occur as a result of bathing in infected water, usually that of a canal or river. Infection probably occurs mainly through cuts and abrasions of the skin but the organism can probably gain access to the tissues also through the conjunctiva or through the mucous membrane of the nose and mouth. Infections due to the *canicola* serotype are contracted mainly from dogs, for which the organism is a natural pathogen. In this case infection presumably occurs through contamination of the hands with the animal's urine. Other domestic animals may also harbour leptospires but appear to be infrequent sources of infection.

Diagnosis

In the diagnosis of leptospiral infection serological tests are of primary importance. The agglutination test is most frequently employed. It may be carried out as a microscopic procedure, the results being read by dark-ground microscopy, or as a macroscopic slide agglutination test. In Great Britain the *icterohaemorrhagiae* and *canicola* serotypes, which are the most prevalent, are most commonly used as antigens. Complement fixation tests with antigen from a number of serotypes may also be employed. In contrast to the agglutination tests, these detect antibody to group-specific antigen. Other techniques available are the indirect immunofluorescence technique and an erythrocyte lysis test.

Isolation of the organisms may be attempted by the intraperitoneal inoculation of a guinea-pig or suitable fluid medium with the patient's blood in the first week, or later with the centrifuged deposit of the patient's urine. Leptospires may be demonstrable in the peritoneal cavity within a few days of inoculation. When this occurs some of the blood of the animal is taken by cardiac puncture and inoculated into a suitable medium. Since some human strains have little pathogenicity for animals, direct culture from the same material should be attempted. Serological characterization of isolates is difficult and is only available in specialized reference laboratories. Microscopic examination of the deposit from the urine with the dark field condenser is often sufficient, but care must be exercised since other spirochaetes may be present. If the urine is acid in reaction the organisms disintegrate rapidly. Dark ground examination of the blood is not reliable since fragments of red blood cells may present an appearance very similar to that of the leptospires.

Treatment

Penicillin is highly active against the leptospires in vitro but has given disappointing results in treatment. Streptomycin and the tetracyclines are also used.

III.20

The Rickettsias, I: Chlamydiales

CHLAMYDIA

Organisms of this group are, like the viruses, obligatory intracellular parasites. Because of this they have in the past been classified as viruses. They show, however, a much closer resemblance to bacteria. Like bacteria, they possess a murein-containing cell wall. They contain both DNA and RNA, do not undergo a genuine eclipse phase and multiply by a type of binary fission. Their obligatory parasitism appears to be due mainly to a deficiency in energy-producing metabolic equipment associated with extreme permeability of the cell membrane. Their properties are consistent with an evolution, through a loss of synthetic properties, from highly parasitic bacteria.

There is still no general agreement on the nomenclature of the group. It has been widely suggested they should all be called *Bedsonia* after Sir Samuel Bedson, who did much of the fundamental work on the psittacosis agent. The generic name *Chlamydia,* which strictly applies only to the agents of trachoma and inclusion conjunctivitis, has also been extended to include the organisms formerly in the genus *Miyagawanella,* psittacosis/ornithosis, lymphogranuloma venereum and certain similar organisms infecting animals. In view of the fact that the agents of trachoma and inclusion conjunctivitis can now be cultivated in the yolk sac of the developing chick embryo and that they contain the same group antigen as the *Miyagawanella* group, the argument for having two separate genera is much weaker. It would seem better to put them all together in a single genus and to call this *Bedsonia* to avoid confusion with the earlier classification. However, this view has not prevailed and the group are now all placed in the genus *Chlamydia,* which constitutes the only genus in the family Chlamydiaceae, which is the only family in the order Chlamydiales. Certain ill-characterized agents apparently responsible for conjunctivitis in ruminants, pigs and poultry may be included later if they can be shown to have similar properties to the chlamydiae.

All are Gram-negative, non-motile more or less spherical organisms of identical morphology, which exhibit a characteristic cycle of intracellular growth and which possess a common group antigen. As observed in the electron microscope, chlamydiae show considerable heterogeneity in size and structure. At one extreme is a small particle of about 0·3 µm in diameter with a dense central body and presenting the so-called 'wrinkled pea' appearance. At the other extreme is a large particle from 0·5 to 1 µm in diameter of granular appearance and without a central body (Fig. III.15). Between these extremes there is a continuous spectrum of intermediate forms. The dense central body of the small form does not appear to be a nuclear equivalent since, on treatment of small particles with desoxycholate, the DNA is lost without any alteration in the appearance of the central body. It probably represents some, yet unexplained, condensation of cytoplasmic material within the cell wall. At the end of the growth cycle, in infected cells, small forms predominate. They appear to be forms particularly fit for survival in an extracellular environment and for the penetration of fresh cells. The large particles, on the other hand, seem specially adapted to an intracellular environment and are much less infectious.

Both types of particle are surrounded by a cell wall which is appreciably thinner than that of a bacterium. The cell walls contain murein and, like the Gram-negative bacteria, a high concentration of lipid. The particles may be stained by

Fig. III.15. The trachoma organism shown on an electron micrograph in a shadowed preparation. × 22 000. (*R. C. Valentine and L. H. Collier*)

Machiavello's, Castaneda's or Giemsa's methods. When stained by Machiavello's method, which is that most generally used, the small particles are stained red and the large particles blue.

Observations with the electron microscope have given more precision to the concept of an intracellular development cycle. The time required for a complete cycle of intracellular multiplication varies from 24 to 72 hours, depending on the type of host cell involved. Following penetration of the cell by one of the small 0·3 µm particles, probably by phagocytosis, the particle begins to enlarge to form a pale staining initial or reticulate body 0·7 to 1·0 µm in diameter during the first half of the cycle. During the second half of the cycle, the large particles multiply by binary fission and budding to yield an increasing number of smaller and smaller particles until finally the majority are of the dense staining small type (elementary bodies). The population of intracellular particles constitutes an inclusion body demonstrable by light microscopy (Figs III.16, III.17).

In some cases the particles are embedded in a homogeneous glycogen-containing matrix which can be demonstrated by staining with iodine. The glycogen appears to be produced by the organism itself, rather than by the host cell, since its production is inhibited by penicillin. Glycogen-containing inclusions are much more compact than inclusions lacking glycogen. They are produced by the lymphogranuloma venereum, and trachoma and inclusion conjunctivitis (TRIC) agents, but not by the majority of psittacosis or ornithosis strains or most other chlamydiae isolated from animals. Glycogen-producing chlamydiae are in general much more susceptible to sulphadiazine and cycloserine than non-glycogen-producers. Because of these associations

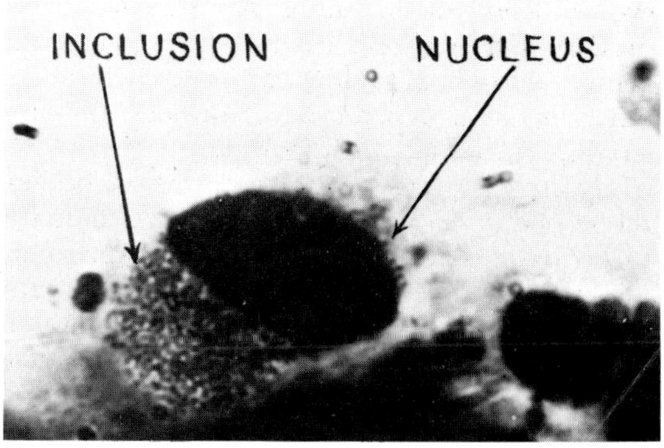

Fig. III.16. The inclusion body of *Chlamydia trachomatis* in an infected conjunctival cell. The inclusion body fills most of the cytoplasm and has compressed the nucleus. Gimensa–May–Grünwald. × 1820. (*L. H. Collier and J. Sowa*)

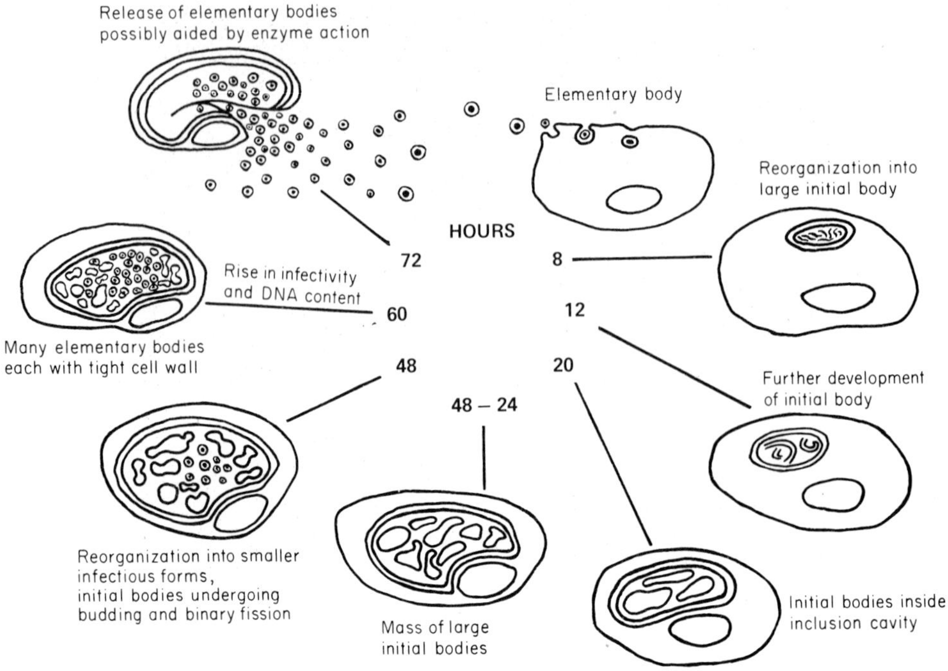

Fig. III.17. The life cycle of *Chlamydia*. (*Dr Judith Woolfit*)

it has been suggested that the capacity to produce glycogen may be of phylogenetic significance. In this case, the lymphogranuloma venereum, TRIC agents and chlamydiae isolated from rodents would be closely related and would be distinct from avian chlamydiae and chlamydiae isolated from other animals. This raises the possibility that the characteristically human chlamydiae may have evolved from rodent strains. A distinction based on these characteristics has now been made. The genus *Chlamydia* at present comprises two species only: the first, *C. trachomatis,* includes all those strains, whatever their origin, which form compact, glycogen-containing inclusion bodies; the second, *C. psittaci,* includes those which do not form glycogen and in which the membrane round the inclusion seems to rupture relatively early allowing the developmental forms to spill into the cytoplasm, so giving the appearance of a diffuse inclusion body.

Antigenic structure

The chlamydiae possess both group and specific antigens. The specific antigens are certainly components of the cell wall and the group antigens probably so. The group antigens, however, are less firmly attached and can be removed by treatment of the organisms with desoxycholate, which does not remove the specific antigens. Group antigens are heat-stable and are inactivated by periodate and lecithinase. They are therefore probably complexes of polysaccharide and lecithin. Some chlamydiae produce a haemagglutinin which appears to be due to a lecithin–protein–DNA complex. The specific antigens are heat-labile and are probably protein. The psittacosis, lymphogranuloma venereum and trachoma agents possess different specific antigens, though with some sharing (see below).

All the chlamydiae possess heat-labile toxins which are lethal for mice on intravenous inoculation. The toxins have the same serological specificity as the specific antigens. Since, however, the isolated cell walls, though capable of absorbing antitoxic antibody, are not themselves toxic, the specific antigen is probably only a component of the toxin. The specific antigens stimulate the production of protective antibody, while the group antigens do not.

Metabolism

Since chlamydiae grow only intracellularly it has not been possible to define their synthetic capabilities in detail. However, it has been clear from the differential incorporation of labelled phosphorus that they can carry out the terminal stages, at least, of DNA and RNA synthesis. Moreover, they have been found to possess ribosomes and can synthesize their own protein. They must also be able to synthesize murein and its constituent muramic acid and agents sensitive to the sulphonamides must be able to synthesize folic acid.

The principal metabolic deficiency in the chlamydiae appears to be in energy-yielding metabolism. They are deficient in cytochromes and other components of the electron transport chain and are therefore incapable of carrying out aerobic oxidations. Anaerobic glycolysis, though for many years thought not to occur, has recently been demonstrated. The capacity of the chlamydiae in this respect is only partial, however, since they do not appear to be capable of generating ATP or of reoxidizing reduced NADP. These functions must therefore be supplied by the host cell. The chlamydiae must therefore, unlike bacterial cells, be permeable to ATP and NADP.

Chlamydiae are susceptible to many of the chemotherapeutic drugs active against bacteria, e.g. penicillin and cycloserine, which inhibit cell wall synthesis, and erythromycin, chloramphenicol and the tetracyclines, all of which appear to act at ribosomal level. Some chlamydiae are also susceptible to the sulphonamides, sulphadiazine being the most effective. As therapeutic agents the tetracyclines are the drugs of choice. Chlamydiae are resistant to streptomycin and this is the antibiotic of choice in the preparation of material for their attempted isolation.

The way in which chlamydiae damage their host cells is not understood. Whether the toxin, which is lethal for mice on intravenous inoculation, has a damaging effect when produced within infected tissue cells is not known. A possible method of damage to the cell is by nutritional deprivation. Thus there is some evidence that the intracellular growth of chlamydiae can inhibit host cell DNA synthesis, an effect reminiscent of the action of many viruses on host cells. When released from infected cells in the intact animal, chlamydiae stimulate inflammatory reactions which are frequently of polymorphonuclear type.

Chlamydiae have a considerable tendency to establish latent infections. Under experimental conditions latency has been induced by starvation and exposure to a drug to which the chlamydiae are susceptible, or by rapid serial passage. It has been suggested that these various procedures act by interfering in some way with the reorganization of the small particles into the larger multiplying forms.

Outside the body the chlamydiae are highly unstable. They are susceptible to the range of disinfectants active against bacteria. To ensure maximum viability infective material should be stored below $-50°C$ or be lyophilized. None of the chlamydiae is known to have an arthropod vector. This is in contrast to the rickettsiae, all of which are or can be arthropod transmitted.

Chlamydia psittaci: the psittacosis/ornithosis group

Organisms of this group were first isolated from psittacine birds and these are probably the most important sources of human infection. Similar organisms are also present in a great variety of non-psittacine birds including pigeons, ducks and domestic fowl. Strains isolated from non-psittacine birds are often referred to as ornithosis agents. Psittacosis-like agents have also been isolated from cat, mouse, calf, sheep and opossum. These show significant differences in range of pathogenicity from strains isolated from human and avian sources and are probably of negligible importance in human infections. *C. psittaci* isolated from some cases of pneumonitis in man are not identical to psittacosis/ornithosis although very similar. They are presumed to be of avian origin although the source of infection is unknown.

Chlamydia psittaci grows readily in the chick embryo, particularly in the yolk sac. Mice, guinea-pigs and rabbits are also susceptible. The mouse is the most generally useful animal and may be infected by the intranasal, intraperitoneal or intracerebral routes. *C. psittaci* isolated from different sources shows significant differences in animal pathogenicity. In general, strains from psittacine birds and turkeys exhibit the widest range of pathogenicity and give rise to the most severe lesions. Following intranasal inoculation of mice they produce areas of pulmonary consolidation. Following intraperitoneal inoculation they usually kill the mouse in three to 10 days; at post mortem an exudate in which elementary bodies may be demonstrated is found on the surface of the peritoneum, the spleen is enlarged and the liver shows areas of necrosis. Following intracerebral inoculation they produce an encephalitis. *C. psittaci* can also be propagated in a variety of tissue cultures but the yields obtained are generally low. Tissue cultures have, however, been of particular value for experimental purposes. Recently much better results have been obtained with irradiated cells.

By the use of toxin neutralization, infectivity neutralization and complement fixation tests, together with fluorescent antibody studies of cell walls extracted with desoxycholate to remove group antigen, considerable serological heterogeneity has been demonstrated in psittacosis and ornithosis agents and in similar agents isolated from mammalian sources. Some of the specific antigens are widely distributed while others are characteristic of strains isolated from particular sources. Thus avian strains have been shown to possess specific antigens absent from mammalian strains, and vice versa, and strains from psittacine birds have been found to possess antigens not present in strains from non-psittacine birds.

A haemagglutinin acting on mouse red cells has been described in some strains. The haemagglutinin which is separable from the virus particle appears to be a complex of lecithin and nucleoprotein. It possesses group serological specificity. Dermal hypersensitivity to psittacosis or lymphogranuloma venereum antigen sometimes develops as a result of infection with chlamydiae of the psittacosis group but not as frequently as in lymphogranuloma venereum. The antigen responsible is group specific.

Psittacosis

It should be noted that the term psittacosis is applied to infection by *C. psittaci* occurring in psittacine birds and in man; the term ornithosis is used for infections occurring in non-psittacine birds.

The disease in birds usually occurs as a latent infection and appears to be transmitted from the mother to the young in the nest. In the case of parrots this latent infection can be transformed into active infection by stress dependent on capture and transportation. Consequently newly imported birds frequently develop a severe form of the disease in the course of which they excrete large amounts of the agent and are more liable to transmit infection to man.

The great majority of human infections are derived from avian sources and particularly from psittacine birds. They appear to be contracted mainly as a result of the inhalation of infected dust and droppings. Some infections have occurred in individuals who have been bitten by infected birds. Infection by ingestion of infected birds, e.g. of poultry, has not been described. The disease can also be transmitted from one human case to another via the respiratory secretions and a considerable number of infections have occurred in nurses and attendants of psittacosis patients. Many laboratory infections from the handling of infected material have been described.

In man psittacosis may give rise to inapparent infection or to a relatively mild illness clinically resembling influenza. In severe cases the presenting syndrome is that of pneumonia, usually of the lobar type, and on occasion resembling primary atypical pneumona (p. 283) but without the presence of cold agglutinins. Following recovery the agent can apparently persist in the respiratory tract in a latent form for a considerable time; the frequency with which this occurs, however, is not known.

The tetracyclines, especially chlortetracycline, are the drugs of choice in treatment. Chemotherapy, however, though controlling the clinical manifestations of the disease, is not always successful in eliminating the organisms from the body. The development of antibiotic resistance has been reported.

An effective vaccine is not available. As a control measure, because of the predominance of psittacine birds as sources of infection, the quarantine of imported psittacine birds is of great importance although this does not eliminate the infection which can still be reactivated if the bird becomes ill or stressed. Total prohibition of importation has also been used as control measure.

Diagnosis

The organism may be isolated from the blood during the first week of the disease and from the sputum in pneumonic cases. Isolation is usually attempted by yolk-sac inoculation of the chick embryo or by intraperitoneal or intranasal inoculation of mice. The organism may be demonstrated microscopically in the yolk sac of infected eggs and in smears from the liver and spleens of infected mice. Final identification is carried out by the use of complement fixation and toxin and infectivity neutralization tests. Isolation is not without risk: the organism has an unsavoury reputation for causing laboratory infections.

Antibody may be detected in the patient's serum by the complement fixation test using a heat-inactivated and phenolized yolk-sac antigen. Two specimens of serum should be examined: one taken as early as possible in the acute stage and the other at a later date. If the case is one of psittacosis, a rise in titre should be demonstrable between the two examinations. On account of the sharing of antigens, positive complement fixation reactions with psittacosis antigen are also obtained in cases of lymphogranuloma venereum and infection with TRIC agents. Specific complement fixation may be obtained by the use of a living antigen and sera which have been exhausted of group antibody by absorption with heated antigen. This procedure is not, however, suitable for ordinary diagnostic use. The complement fixation test remains positive for long periods after recovery, psittacosis differing in this way from most virus infections.

In birds infected with *C. psittaci* the conventional complement fixation test is often negative as the antibody present is not capable of fixing complement when combined with antigen. However, it is possible to demonstrate antibody by the so-called indirect complement fixation test. In this test the antigen is first allowed to react with the test serum. Complement and a serum known to contain complement-fixing antibody are then added and allowed to react. The test is completed in the usual way by adding sensitized red cells. If antibody is present in the test serum it will have reacted with the antigen in the first stage, so that no antigen is available to react with the complement-fixing antibody and the complement remains available to lyse the sensitized red cells. In contrast to the conventional test, a positive result is indicated by lysis of the cells and a negative result by their failure to lyse.

Chlamydia trachomatis

The lymphogranuloma venereum agent is more difficult to propagate than the psittacosis agent but can be isolated in the yolk sac of the chick embryo and by intracerebral inoculation of mice. It can also be grown in HeLa cell tissue cultures and more readily in irradiated or IUDR-treated McCoy cells. Three serotypes can be distinguished, although there is a good deal of cross-reaction between them (see below).

The agents responsible for trachoma and inclusion conjunctivitis have long been known, on morphological grounds, to be chlamydiae. They were not isolated, however, until 1957 when Tang showed that agents from trachoma could be isolated by inoculation of the yolk sacs of six- to eight-day chick embryos. Only a few strains appear to be capable of growth in normal tissue culture. However, the use of irradiated McCoy cells combined with centrifugation to bring particles in the inoculum rapidly into contact with the cell sheet has afforded a method for the isolation of these agents which is much more sensitive than the yolk-sac and which is now the method of choice for routine work. It has recently been shown that cells treated with IUDR (5-iodo-2′-deoxyuridine) are at least as sensitive as irradiated cells and much more convenient. Claims by some workers to have found normal untreated cells as sensitive as irradiated or IUDR-treated cells have not been generally confirmed. Lesions can be produced in the eyes of monkeys, but no other laboratory animal is susceptible. In their morphology, developmental cycle, common group antigen and toxicity they resemble the psittacosis lymphogranuloma venereum agents. A completely species-specific antigen has not been demonstrated.

Since, trachoma and inclusion conjunctivitis are epidemiologically clearly distinct it is reasonable to assume that there is some difference between the agents involved. Yet the strains from each condition produce the same type of lesion on inoculation onto the conjunctivae of human volunteers and of monkeys. At present they are therefore grouped together without distinction as TRIC agents. Nevertheless, it is now clear that TRIC agents can be subdivided into a number of groups on serological grounds. While certain serotypes occur particularly in cases of classical trachoma and other serotypes in genital infections and in inclusion conjunctivitis, the restriction of particular serotypes to particular syndromes is not absolute. There is also sharing of antigens with lymphogranuloma strains. TRIC agents resemble the lymphogranuloma venereum agents and differ from the psittacosis–ornithosis group in producing glycogen-containing inclusion bodies.

Lymphogranuloma venereum

Lymphogranuloma venereum is a venereal disease affecting only man. It is of world-wide distribution, but commonest in tropical countries. Three stages of the disease are distinguished. In the first stage a primary lesion in the form of a small papule which progresses to ulceration appears on the genital mucosa or in the anterior urethra in from three days to three weeks after infection. In the secondary stage, which appears about two weeks after the development of the primary lesion, the patient becomes febrile and there is enlargement of the local lymphatic glands. In most cases the glands involved suppurate and discharge through the skin. Apart from local lesions, some cases show signs of generalized infection with widespread lymph gland involvement; occasionally arthritis and meningitis are observed. The tertiary stage, which may be highly chronic and which is commoner in the female than in the male, is characterized by the appearance of granulomatous lesions around the genitalia and in the perineal region. Lymphatic involvement in this region may lead to the development of localized elephantiasis. The discharging granulomatous lesions of the tertiary stage are very liable to secondary bacterial infection.

Though infections are usually of venereal origin, non-venereal infections may occur; a number of such infections in which the virus has gained entry through the conjunctiva have been described.

Very satisfactory results have been obtained in the treatment of lymphogranuloma venereum with the tetracyclines and these are the drugs of choice. The sulphonamides are also reported to be of value.

Diagnosis

The organism may be demonstrable microscopically in smears of pus from infected glands or in material taken from the glands by biopsy. An attempt may also be made to isolate it by intracerebral inoculation of mice or inoculation into the chick embryo yolk sac but is frequently unsuccessful.

The complement fixation test using inactivated yolk-sac antigen may be used for the demonstration of antibody. Positive reactions with lymphogranuloma venereum antigens are, however, also given by the sera of patients suffering from psittacosis, and on occasion by sera from patients with other *C. trachomatis* infections and from apparently uninfected individuals. However, the titres are seldom as high as in lymphogranuloma being generally below 1/32.

A skin test using antigen prepared from infected yolk sac inactivated by heat is widely used as a diagnostic method. This test, which is known as the Frei test, depends upon a delayed-type allergy. In positive cases, following intradermal injection of the antigen an inflammatory swelling appears within a few days around the site of inoculation. Hypersensitivity is usually sufficiently established to give a positive skin test within two to six weeks after infection and persists for long periods, possibly for life. Positive reactions in the Frei test may also be obtained from patients who have been infected with psittacosis and ornithosis, and is negative in a proportion of patients with lymphogranuloma venereum. The antigen responsible for the skin reactions appears to be distinct from the group complement fixing antigen. It is reported that a skin test antigen specific for lymphogranuloma venereum can be obtained by extraction of crude yolk-sac antigen with 0·02 N hydrochloric acid.

Trachoma and inclusion conjunctivitis

Trachoma is a form of follicular conjunctivitis which is of world-wide distribution. It is particularly prevalent in the Middle East but is rare in the British Isles. The organism is very susceptible to drying and spread of the disease from one person to another requires close contact and occurs mainly under unhygienic conditions. Trachoma is a serious disease, since after the acute stage has passed it develops into a chronic form with the formation of scar tissue in the eyelids, conjunctiva and cornea, leading ultimately to blindness. In communities in which it is prevalent infection usually occurs within the first year of life, the disease as a rule being acquired from the mother. The condition is considerably more severe in the adult than in the child. Secondary bacterial infection, which increases the amount of damage caused by the organism, is usual.

By contrast the lesions of inclusion conjunctivitis are milder than those of trachoma and occur mainly in the lower lid; the cornea is rarely involved and when it is the lesions are transient and show no residual scarring. The disease is of world-wide distribution. Inclusion conjunctivitis appears to be but one manifestation of what is essentially a venereal disease. The epidemiology of the infection is very similar to that of gonorrhoea (see p. 225). The primary infection seems to be in the genital tract: the urethra of males and the cervix and the urethra of females. In this situation the organism produces either no symptoms or only mild ones (non-specific urethritis or cervicitis).

Inclusion conjunctivitis proper occurs in two clinical forms:

1. As a neonatal infection, the infant's eyes being infected during delivery by organisms present in the mother's genital tract. The condition is an acute one with the accumulation of pus cells in the conjunctival exudate. It must be distinguished clinically from gonococcal ophthalmia from which it differs in that it usually appears between the fifth and fifteenth day after birth.
2. As an infection of adults this variety is less acute than the neonatal infection and the conjunctival exudate usually shows the presence of lymphocytes as the predominant cell type. The infection appears to be contracted mainly in swimming baths contaminated with infected genitourinary secretions. Like the agent of inclusion conjunctivitis, *N. gonorrhoeae* can also affect the adult eye, though usually only when damaged in some way as, for example, by an old trachoma infection.

The sera of patients from both conditions may give positive complement fixation tests with chlamydial group antigen but these tests have no diagnostic value. Although isolation of the agent is possible by chick embryo inoculation, diagnosis is more readily achieved by demonstration of the typical cytoplasmic inclusion bodies in infected conjunctival cells. In trachoma these are most numerous in cells from the upper tarsal conjunctiva and in inclusion conjunctivitis from those of the lower tarsal conjunctiva.

Isolation is much more sensitive than the demonstration of inclusions for the diagnosis of established trachoma infection. Chlamydiae are not susceptible to streptomycin and this drug can therefore be used for the control of bacterial contamination in the isolation of these agents in tissue culture. The examination of smears by the fluorescent antibody technique has been favourably reported on.

Sulphonamides and tetracyclines are effective therapeutic agents in both conditions.

Much remains to be learnt about the natural history of *C. trachomatis* infections in man. Genital infections are extremely common in this country. Up to 50% of unselected female patients attending a venereal disease clinic and up to 25% of the men will yield *C. trachomatis* at their first visit. The difference between the sexes may be more apparent than real, simply reflecting greater ease in isolation from the cervix than from the urethra: indeed if isolations from the female urethra only are considered the figures for the two sexes are comparable. In what are presumably less promiscuous populations the isolation rate is much lower, seldom exceeding 1 to 2% in women. Studies have been carried out in hospital personnel, antenatal patients and those attending a family planning clinic. In spite of the prevalence of the infection it is still not clear to what extent it is responsible for symptoms. There is some not altogether conclusive evidence that it may be responsible for a proportion of non-gonococcal urethritis in men. It is commoner in patients with other demonstrable genital infections, e.g. gonorrhoea, *Trichomonas vaginalis,* candidiasis, than in patients with no evidence of another infection. This fact can be interpreted in various ways, but one possibility is that the infection may have been latent until reactivated by another infection. This idea perhaps gains support from one recent study by the author in which the incidence of infection did not seem to be related to the individual's current levels of promiscuity. As with gonorrhoea the infection has been detected in the rectum and pharynx. Many of the adults presenting with inclusion conjunctivitis will often also have genital infections, as will their sexual partners.

The introduction of the fluorescent antibody method for serotyping isolates has been valuable. At the time of writing three serotypes of lymphogranuloma venereum (types I, II and III) and 11 serotypes of trachoma/inclusion conjunctivitis (types A–K) have been described. Although classical trachoma is mainly due to types A, B and C, cases occur from which other types are isolated and the classical trachoma serotypes may be incriminated in the milder self-limiting inclusion conjunctivitis or isolated from the genital tract. It is becoming clear that the clinical presentation of a *C. trachomatis* infection does not depend solely on the serotype of the infecting organism. Other factors which have been considered important are the age at infection, the route of infection (classical trachoma is usually spread by articles such as shared towels), the frequency of reinfection, the incidence of secondary infection and environmental factors such as humidity and dust in the air. However, it is not at present possible to assess the relative importance of any of these factors. In spite of considerable sharing of antigens with oculogenital strains, especially types F, G and K, lymphogranuloma strains seem to be restricted to cases of lymphogranuloma venereum (lymphogranuloma inguinale) which is not a common condition in this country. The picture is further complicated by some sharing of antigens between *C. trachomatis* strains and strains of *C. psittaci* (this is distinct from the group antigen) and by the existence of strains intermediate in their properties between the two species.

Non-bacterial regional lymphadenitis (cat scratch disease)

This condition, which usually follows the scratch of a cat, after an interval of one to seven weeks, is characterized by a mild degree of fever and a regional lymphadenitis which may progress to pus formation. The syndrome has also resulted from other types of scratch, e.g. from thorns. A possible relationship of the agent to the chlamydiae is suggested by the observation that some patients give a low-titre positive complement fixation test with chlamydia group antigen. A skin test using pus or macerated lymph node tissue is claimed to be highly specific, but the Frei test (see p. 310) is negative. Many workers have doubted whether cat scratch fever is due to a member of this group and a recent electron microscope study of the affected lymph nodes revealed particles resembling a herpesvirus.

III.21

The Rickettsias, II: Rickettsiales

The order Rickettsiales comprises a large and very mixed collection of organisms and it is probably true to say that there is no one property which is common to all the species within the order and at the same time serves to distinguish them from all other micro-organisms. Although many are as yet poorly characterized, there seems nothing to suggest any close affinity with the true viruses, even though many are apparently obligate intracellular parasites. They are best regarded as small bacteria: they are sensitive to broad-spectrum antibiotics and, in some cases, to penicillin or sulphonamides. One feature which appears to characterize those genera which have been studied in detail is an extreme permeability of the cell membrane to medium-sized molecules such as ATP, DPN, coenzymes and Krebs cycle intermediates, and even to some macromolecules. This appears to be special adaptation to the intracellular parasitic mode of existence. The organism is relieved of the necessity of synthesizing certain molecules because they can diffuse in from the host protoplasm. However, this same property of leakiness effectively precludes extracellular multiplication or even prolonged extracellular survival in many cases. Fortunately for the student of medical microbiology, those species which produce disease in man are, with the exception of *Bartonella bacilliformis* all restricted to three of the genera: *Rickettsia, Rochalimaea* and *Coxiella.* In addition to human pathogens, the order Rickettsiales contains many important pathogens of wild and domestic animals. Many species are transmitted by arthropods which serve as alternative hosts and four of the genera are apparently restricted to arthropods. Because of this it has been suggested that this group of organisms may have originated from the commensal flora of arthropods. The species *Eperythrozoon coccoides* is of interest in that many strains of the mouse hepatitis virus (a coronavirus) only exert their full pathogenic effect if the mouse is also infected by the eperythrozoon, which otherwise seems to produce a virtually sumptomless infection.

RICKETTSIAE

Properties of the group

The rickettsiae are a group of small organisms which, because of their apparently obligatory intracellular parasitism, have in the past been regarded as intermediate between viruses and bacteria. There is now, however, no doubt that like the chlamydiae they are in fact very small bacteria. Like bacteria they possess a murein-containing cell wall; they contain both RNA and DNA, multiply by binary fission and have considerable independent metabolic activity. Under in vitro conditions they have been found to be capable of carrying out certain transamination reactions and of oxidizing some small molecular weight substrates, notably glutamate. Unlike the chlamydiae they can generate high-energy phosphate bonds. An electron transport chain has been demonstrated and they possess the enzymes of the Krebs cycle. However, unlike bacteria they cannot break down glucose. Apart from *Rochalimaea quintana,* which has recently been reported to grow on a blood agar medium, rickettsiae have not so far been cultivated in the laboratory on cell-free media. Like the chlamydiae they possess an unusually permeable cell wall. This permeability is not wholly unselective since they cannot apparently take in citrate although they can pass glutamate and ATP. Their main energy source would appear to be the oxidation of glutamate with the formation of ATP. They can synthesize protein only if they are supplied with all the amino acids.

Rickettsiae may be demonstrated in infected tissues by Machiavello's, Castaneda's or Giemsa's methods. In Giemsa-stained preparations they appear as bluish-purple structures; when stained with Machiavello's stain they appear bright red showing up in sharp contrast against a blue background. They also stain by Gram's method and are Gram-negative. Gram's stain is, however, not a satisfactory routine stain for demonstrating them.

Organisms in smears from infected tissue show considerable pleomorphism, coccal, oval and bacillary forms being observed. The average size of the rickettsiae is from 0·3 to 4 µm in width and up to 1·5 µm in length. In electron micrographs a central structure analogous to a nucleus, a cell wall and a peripheral layer resembling a bacterial capsule have all been demonstrated. This capsular layer probably contains the group-specific soluble antigens. None of the rickettsiae is motile. In smear preparations they are frequently found as pairs of flame-shaped organisms, somewhat resembling pneumonococci, and occasionally in short chains. In infected tissue cells they usually appear in large clumps.

Rickettsiae are heat-labile, being destroyed at 50°C in half an hour, and are faily susceptible to the action of chemical disinfectants; they are readily destroyed by low concentrations of phenol and formalin. They undergo inactivation rapidly at room temperature but may be preserved for long periods by freeze drying. Suspensions of rickettsiae in saline solutions at room temperature readily lose their infectivity. This can be largely reversed if ATP is added to the system together with DPN and coenzyme A.

The rickettsiae appear to be primarily parasites of arthropods, particularly of ticks and mites and to a lesser extent of fleas and lice, and infect animals on which their insect host is parasitic. Ticks and mites are capable of transmitting rickettsiae transovarially to their offspring. In these cases though the animal host of the arthropod is also infected, the arthropod should perhaps be regarded as the primary natural reservoir of infection. Epidemic typhus and trench fever are unusual in that the only mammal naturally infected appears to be man. In their arthropod vectors the rickettsiae multiply in the epithelium of the intestinal tract; they are excreted in the faeces but occasionally gain access to the salivary glands. *Rochalimaea quintana* apparently replicates extracellularly in the louse gut and it can be grown on a bacteriological medium containing a high concentration of lysed blood and of serum. There appears to be a requirement for haemin and for a detoxifying agent (charcoal can be substituted for the serum). Like the members of the genus *Rickettsia,* it appears unable to metabolize glucose and to derive its energy from the oxidation of glutamate or glutamine. However, in view of its capacity for independent replication, its metabolism must differ from that of the rickettsiae: at least one transaminase has been detected. Because of these differences, this organism, which is the cause of trench fever (see p. 319) has recently been transferred to a genus of its own within the tribe Rickettsieae; it was formerly in the genus *Rickettsia.* In their vertebrate host rickettsiae are found principally in the endothelium of the small blood vessels, particularly in those of the skin, brain and heart. In infected endothelial cells they produce proliferative and, in the more severe infections such as Rocky Mountain spotted fever, necrotic changes. These endothelial changes result in vascular thromboses and in haemorrhages.

The rickettsiae are pathogenic on experimental inoculation for a wide range of animals. For their isolation guinea-pigs and mice have been most used. They will grow in the developing chick embryo, the highest concentrations being obtained on inoculation into the yolk sac. They can also be propagated in cultures of certain rat and mouse tissues. In general, growth of the rickettsiae is maximal when the metabolism of the host cell is proceeding at a low rate as in growth in eggs or tissue cultures at suboptimal temperatures or in the presence of metabolic inhibitors. In infected cells the rickettsiae of the spotted fever group and *R. akari* are found in the nucleus where they appear to multiply. Other rickettsiae are not found in the nucleus and appear to multiply in the cytoplasm.

Concentrated rickettsial suspensions are lethal for mice on intravenous inoculation. In addition they are haemolytic for the red cells of a variety of animal species. Toxicity and haemolytic activity are both intrinsic properties of the organism and on storage are lost concomitantly with infectivity. There is some evidence that these properties, though not requiring multiplication of the rickettsiae, depend on enzymatic activity, since suspensions which have become non-haemolytic and non-toxic on storage regain these capacities, as well as their infectivity if suspended in NAD.

Three major types of rickettsial antigen have been identified:

1. Group-specific soluble antigens. These are demonstrable in the environment of the organism and appear to be derived from the surface capsular layer. The soluble antigens of the typhus fever group and of the spotted fever group are group-specific, the same, or similar, soluble antigens being present in different members of the group. The soluble antigens of the scrub typhus group, on the other hand, show considerable heterogeneity.

2. Species-specific or, in the case of the scrub typhus rickettsiae, strain-specific antigens associated with the bodies of the rickettsiae. In order to demonstrate these antigens, well-washed suspensions, from which the group antigens present in the capsular layer have been removed, must be employed.

3. Antigens shared with various strains of *Proteus*. Sharing of antigens has been shown between various rickettsiae and the O-antigens of *Proteus* strains X19, X2 and XK. These antigens appear to be polysaccharides. The rickettsial antigens do not react in agglutination tests and therefore do not occur as surface components.

In the diagnosis of rickettsial infections serological tests are of primary importance. The most generally applicable test is the Weil-Felix reaction, an agglutination test which measures the titre of antibody against the *Proteus* antigens OX19, OX2 and OXK. The Weil-Felix reaction is specially valuable in the diagnosis of epidemic and murine typhus, in both of which there is a high titre of antibody to OX19 and in cases of scrub typhus where the antigenic heterogeneity of the organism makes more specific tests almost useless. In some populations numerous false positive reactions make interpretation of the test extremely difficult. For the differentiation of the two types of typhus, however, complement fixation or agglutination tests with well-washed rickettsial suspensions are required for the identification of antibody to the specific antigens. Complement fixation tests with soluble antigens have a particular value in the diagnosis of infections with the spotted fever group. Here again, however, species differentiation requires the use of complement fixation or agglutination tests with specific antigens.

A substance capable of sensitizing red cells so that these are agglutinated specifically by rickettsial antibody has also been prepared from *R. prowazeki* and *R. typhi*. The sensitizing antigen can be extracted by heating the rickettsiae under alkaline conditions. It does not fix complement; haemagglutination tests, however, give results which parallel those obtained in complement fixation tests with the group-specific soluble antigen. A similar but serologically distinct red cell sensitizing antigen, also possessing group specificity, has been isolated from certain of the tick-borne spotted fever group. The demonstration of antibody to red cell sensitizing antigens has been used as a diagnostic procedure.

For diagnostic purposes an attempt may be made to isolate rickettsiae by the intraperitoneal inoculation of guinea-pigs or mice with blood taken during the febrile stage of infection. For this purpose defibrinated, heparinized or citrated whole blood or ground-up clot may be used. If sampled later than the first week ground-up clot is preferable. Isolation of rickettsiae is technically difficult and, because of their high infectivity, should be attempted only in specialized laboratories where adequate safety measures can be employed.

In identifying a *Rickettsia,* the clinical response of the animal is of considerable importance. Thus all rickettsiae will give rise to fever in the guinea-pig following intraperitoneal inoculation, but the rickettsiae of spotted fever and murine typhus will produce also inflammation of the testes. It may be possible to demonstrate rickettsiae microscopically in the organs of animals which have been sacrificed or which have died from the infection. When guinea-pigs are used confirmation of the presence of rickettsiae can be obtained by demonstration in their sera of complement fixing antibody. Inoculation into the yolk sacs of developing chick embryos is a feasible but not very satisfactory method for isolation since satisfactory growth can only be obtained after the organism has been adapted to the chick embryo by a number of blind passages.

Like the psittacosis–lymphogranuloma group, rickettsiae are sensitive to chloramphenicol and the tetracyclines and these agents have proved of great value in the treatment of rickettsial infections. The rickettsiae are not inhibited by penicillin or the sulphonamides. The latter in fact have a definite stimulatory effect and when given to a patient adversely affect the course of the disease. *Para*-amino benzoic acid (PABA) has been found to have an inhibitory effect and prior to the introduction of the wide-spectrum antibiotics was used with some degree of success in the treatment of infections of the typhus group.

Three groups of rickettsiae may be distinguished:

1. The typhus fever group comprising *R. prowazeki*, the cause of epidemic typhus, and *R. typhi*, the cause of murine typhus. These rickettsiae share a common soluble antigen.
2. *R. tsutsugamushi*, the cause of scrub typhus.
3. The spotted fever group.

Epidemic typhus

Epidemic typhus caused by *Rickettsia prowazeki* was at one time a common disease of world-wide distribution but has now been eliminated from Western Europe, North America and many other parts of the world where it was formerly prevalent. Endemic foci of infection, however, still exist in Eastern Europe and parts of Asia, Africa and South America. These endemic foci may be the site of origin of large-scale epidemics if conditions favour the spread of the organism.

Typhus is spread from case to case by lice and on this account epidemics are particularly associated with war, cold and dirt—conditions which favour louse infestation. It is no exaggeration to say that this disease has been the crucial factor in determining the outcome of many military campaigns in the last 500 years. During World War II a very large epidemic occurred in Naples. The disease first appeared in air-raid shelters in which many people were closely confined under highly insanitary conditions and in which they rapidly became infested with lice. Lice do not become infective until four to six days after biting an infected patient. Thereafter they remain infective until they die, which usually occurs within a further two weeks. The rickettsiae are present in considerable numbers in the intestinal contents of the louse but do not apparently gain access to its salivary glands. Infection is therefore probably due to the contamination of scratches and abrasions by louse faeces rather than to the bite of the insect. Infection by the respiratory route has been shown

Table III.9. Some features of rickettsial infections

*Organism**	*Disease*	*Vector*	*Reservoir*	*Proteus agglutinins*	*Isolation (intraperitoneal)*
R. prowazekii	Epidemic typhus	Louse	Man	OX19	Guinea-pig: fever
R. typhi	Murine typhus	Flea	Rats	OX19	Guinea-pig: scrotal swelling
R. tsutsugamushi†	Scrub typhus	Mites	Wild rodents	OXK	Mice: ascites
R. rickettsi	Rocky Mountain spotted fever	Tick	Wild rodents and dogs		Guinea-pig: scrotal necrosis
R. conori	*Fièvre boutonneuse* (and African tick typhus)	Tick			Guinea-pig: scrotal swelling
R. sibirica	North Asian tick typhus	Tick	Wild rodents Farm animals Birds	OX19 and/or OX2	Guinea-pig: scrotal swelling
R. australis	Queensland tick typhus	Tick	Wild rodents and marsupials		Guinea-pig: scrotal swelling
R. akari	Rickettsial pox	Mite	House mice	—	Mice: ascites
R. quintana	Trench fever	Louse	Man	—	—
C. burneti	Q fever	Tick	Cattle, sheep, goats, gerbilles, bandicoots	—	Guinea-pig: fever

* Brackets at left of organisms indicate species sharing group complement fixing antigens.
† Numerous serotypes with little or no cross immunity.

to occur in the laboratory and may also occur under epidemic conditions from rickettsiae in dried louse faeces.

Typhus is a severe disease marked clinically by considerable prostration; a characteristic rash which in severe cases is purpuric in character appears about the fifth day. The mortality from typhus has been found to vary considerably in different epidemics and is highest in individuals over forty.

During the course of an epidemic, as well as cases of the classical type, relatively trivial and asymptomatic latent infections also occur. Latent infection may persist for many years. The epidemiological variety of typhus known as Brill's disease is a well-documented example of such persistence. Brill's disease is a form of typhus observed on the Eastern seaboard of the United States in European immigrants who were almost certainly latently infected before they left Europe. Persistent latent infections of this type may be important as a reservoir of the rickettsiae under natural conditions. Some such reservoir is necessary since *R. prowazeki* has not so far been identified as a natural parasite of any animal host, although strains with properties intermediate between those of *R. prowazeki* and *R. typhi* have been identified in Mexico and it has been claimed that their existence indicates that under suitable conditions *R. typhi* may change into *R. prowazeki*. A species *R. canada* possesses properties intermediate between the typhus rickettsiae and the spotted fever rickettsiae. It has been suspected of causing a spotted-fever-like illness in man.

Prophylaxis

A formolized egg yolk vaccine is widely used for immunization against typhus and is of value in preventing the disease and in considerably diminishing its severity in inoculated individuals. The immunizing potency of the vaccine appears to depend on soluble antigens as well as on antigens present in the bodies of the rickettsiae. Consequently, in modern methods of vaccine preparation, the techniques used are designed to retain a considerable amount of soluble antigen. Following inoculation of rickettsial vaccine there is an increase in specific rickettsial antibodies, as detected by rickettsial agglutination and complement fixation tests; there is, however, no increase in the antibody demonstrable in the Weil–Felix reaction. Although there is a high degree of cross-immunity, under natural conditions of infection, between epidemic typhus and murine typhus, inactivated *R. typhi* vaccines do not protect against epidemic typhus and inactivated *R. prowazeki* vaccines do not protect against murine typhus. Immunization against epidemic typhus has also been carried out on a limited scale with the living attenuated E strain of *R. prowazeki*. For the control of epidemics disinfestation, particularly by the use of DDT, is of primary importance.

Murine typhus

Like epidemic typhus, murine typhus is a disease of world-wide distribution. The rickettsia responsible—*R. typhi,* also called *R. mooseri*—is a natural parasite of the rat. *R. typhi* is also parasitic on fleas and lice and these appear to be mainly responsible for transmitting the organisms from rat to rat. Transmission of the disease to man is due to the rat flea, *Xenopsylla cheopis*. There is some evidence that body lice infected from patients with murine typhus may cause epidemic spread of the disease.

Clinically, murine typhus is a much milder

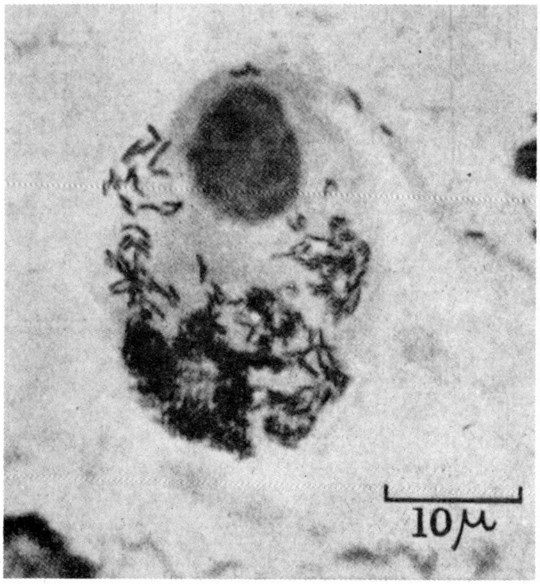

Fig. III.18. *Rickettsia typhi* shown on an electron micrograph in an endothelial cell of a guinea pig.

disease than epidemic typhus. Involvement of the central nervous system, heart and kidneys is considerably less frequent. Except in individuals over 50 the mortality is negligible.

The control of the disease depends essentially on the use of DDT and on measures designed to control the rodent population. The insecticide must be applied before the rat poison is exposed; otherwise the swarms of fleas leaving the dying rats will greatly increase the risk to human beings. Immunization, though feasible, is not considered necessary.

Scrub typhus

This disease, which is caused by *R. tsutsugamushi,* also called *R. orientalis,* occurs in the Far East and Japan, where it is known as Tsutsugamushi, in Malaya and in the Pacific area. The disease is transmitted to man by the bites of the larvae of certain trombiculid mites. The parasite is transmitted from one generation of mites to the next transovarially. Wild rodents, especially rats and, in Japan, the vole *Microtus montbelloi* on which the mites are parasitic, constitute an animal reservoir of infection. Scrub typhus occurs particularly in areas where the jungle has been cut down and replaced by low scrub—an environment in which rats and mites multiply rapidly. It was very common amongst troops in the Pacific theatre of World War II, where these conditions frequently prevailed.

In its main clinical features scrub typhus resembles epidemic typhus. A distinctive characteristic of the disease is the appearance in most cases, either just before or at the onset of symptoms, of a small, punched-out ulcer covered by a dark scab at the site of the insect bite.

Rickettsia tsutsugamushi is antigenically very heterogeneous. Because of this, in endemic areas, second attacks are common. The antigenic heterogeneity of the group has so far made it impossible to develop an effective vaccine.

The sera of patients suffering from scrub typhus may show a high titre of agglutinins for *Proteus* OXK. No rise in titre is usually demonstrable against *Proteus* OX19 or OX2. A considerable proportion of cases, however, fail to produce detectable *Proteus* agglutinins. Because of the antigenic heterogeneity of the species rickettsial agglutination and complement fixation tests are of very limited value in diagnosis.

Spotted fever group

Under this heading are included a number of rickettsial infections caused by rickettsiae possessing a common group of soluble antigen. The infections fall into three groups: Rocky Mountain spotted fever, tick-borne rickettsioses of the Eastern hemisphere and rickettsial pox.

Rocky mountain spotted fever

This condition, which is much the most severe of this group of rickettsial infections, is due to *R. rickettsi,* first identified as the cause of Rocky Mountain spotted fever by Ricketts in Montana. *R. rickettsi* is now known to be of wide distribution in North, Central and South America. It occurs as a parasite of various ticks in the Rocky Mountain region. The main vector is *Dermacentor andersoni* which is parasitic on a variety of wild animals. Ticks, e.g. *Dermacentor variabilis* and, in Mexico, *Rhipicephalus sanguineus,* which infest dogs, are probably of considerable importance in spreading the disease, because of their close association with man. There would seem to be no justification for trying to subdivide the tick-borne rickettsiae of the New World into more than one species corresponding to the different local manifestations of the disease, e.g. Tobra fever (Colombia), São Paulo fever (Brazil).

Rickettsia rickettsi produces marked necrotic changes in the endothelium of the small blood vessels and frequently also in the smooth muscle cells of the arterioles. In the skin these vascular changes result in an extensive haemorrhagic rash which in severe cases may progress to necrosis of large areas of skin. There is usually extensive involvement of the blood vessels of the brain; in recovered cases this may lead to permanent neurological sequelae. The mortality of Rocky Mountain spotted fever in untreated, unvaccinated individuals varies from about 40% in children to 80% in adults. The sera of the majority of patients show a high titre of agglutinins against one or more *Proteus* strains, the highest titres usually being found against OX19. Complement fixation tests with group antigens will differentiate Rocky Mountain spotted fever from the typhus group. Differentiation from other members of the spotted fever group, by the use of specific rickettsial antigens, is not normally required.

Inactivated vaccines of *R. rickettsi* prepared from

infected yolk sacs or from the tissues of infected ticks are available and have been found to give a considerable degree of protection. Effective vaccines have not yet been prepared from the other members of the spotted fever group.

Tick-borne rickettsial infections of the eastern hemisphere

These are much less severe than Rocky Mountain spotted fever and characteristically show a local lesion at the site of the bite. Three closely related species are involved each with a characteristic geographic distribution:

1. *R. conori* produces infection in Africa (South African tick fever, Kenya tick fever), India (Indian tick typhus) and along the Mediterranean littoral (*fièvre boutonneuse*).
2. *R. australis* is the cause of Queensland tick typhus.
3. *R. sibirica* is prevalent in Siberia, Armenia and the Central Asian Republics.

These rickettsiae can be distinguished from one another, and from *R. rickettsi* and *R. akari,* by complement fixation tests with specific antigens and by toxin neutralization tests. Agglutinins develop to *Proteus* OX19 and OX2 but to a lower titre than in Rocky Mountain spotted fever. All these rickettsial species are maintained in nature in wild rodents. Dogs and marsupials also appear to be important in the maintenance of *R. conori* and *R. australis,* respectively.

Rickettsial pox

This is a mild disease, usually lasting less than a week, which was first recognized in New York in 1946. The condition has since been reported from Africa and the USSR. In the majority of cases a local lesion appears at the site of the bite. A maculopapular rash later becoming vesicular as a characteristic feature. The disease shows a mainly urban distribution.

The causative organism, *R. akari,* is transmitted by the mite *Allodermanyssus sanguineus*; house mice appear to be the animal reservoir from which human infection occurs. *R. akari* is closely related to the tick-borne rickettsiae considered above with which it shares a common group-specific soluble antigen. It is consequently usually classified as a member of the spotted fever group in spite of the different clinical picture and the different vector.

Agglutinins for *Proteus* are not produced during the course of infection, but specific rickettsial antibodies may be detected by complement fixation tests. *R. akari* is usually isolated by inoculation of mice.

Trench fever: Wolhynian fever

Trench fever is a febrile influenza-like disease associated with a rash. The disease was first recognized in troops in World War I in whom it occurred in epidemic form. It appeared again in troops on the eastern front in World War II. Apparently it is never fatal. There is no known animal reservoir. The disease was spread by lice, and *Rickettsia*-like organisms were demonstrated, by a number of workers, in the intestinal epithelium of infected lice. To these organisms the name *Rochalimaea quintana* has been given. *R. quintana* cannot be grown in ordinary laboratory animals, in tissue cultures or in the chick embryo, but lice bred in the laboratory have been infected by intrarectal inoculation. Its isolation on a blood agar medium has been described (see p. 314).

Q fever: Coxiella burneti

Q fever (Query fever) was first recognized by Derrick of Brisbane in 1935 and the causative organism first isolated by Burnet and Freeman from meat workers in Queensland in 1937. The disease appears to be of world-wide distribution. It is usually a mild febrile illness clinically resembling influenza and with a very low mortality. In contrast to other rickettsial infections there is no rash. In about half the cases there is radiographic evidence of lung involvement. In some of these the consolidation shows the patchy distribution characteristic of primary atypical pneumonia but in the majority it has a more segmental or lobar distribution. Occasionally the infection assumes a chronic form presenting the picture of subacute bacterial endocarditis, the vegetations being particularly prominent on the aortic valves. Granulomatous involvement of the liver has also been described.

The causative organism was first classified as a *Rickettsia (R. burneti)* but since it shows a number of distinctive characteristics it is now assigned to a separate genus *Coxiella* as *C. burneti. C. burneti* is appreciably more resistant to heat, desiccation and chemical agents than the rickettsiae proper. It resists heating to 60°C for one hour and will survive for several years in dried tick faeces. It is, moreover, not toxic for mice on intravenous administration, and soluble complement fixing antigens are not produced in infected tissues.

Two antigenic phases of *C. burneti* can be distinguished on the basis of their reactivity with guinea-pig convalescent sera. On first isolation the organisms are in phase 1 in which they react in complement fixation tests only with late convalescent serum. On passage in the chick embryo they pass into phase 2 which reacts with both late and early convalescent serum. Phase 1 is regained on passage of egg-adapted phase 2 strains in guinea-pigs. Minor antigenic differences have been demonstrated in strains of *C. burneti* isolated from different geographical areas. These differences are not, however, of immunological importance.

Coxiella burneti grows readily in the yolk sac of the developing chick embryo, the infection killing the embryo in seven to ten days. Of the ordinary laboratory animals guinea-pigs are the most susceptible and are readily infected by the peritoneal route. The infection in the guinea-pig is not associated with any particular signs, although the animal may develop fever. However, the diagnosis can be established some five to six weeks later by the demonstration of antibody production in the infected animal.

Coxiella burneti is widely distributed in nature and is responsible for infection in various wild animals and in birds, cattle, sheep and goats as well as in man. Amongst wild animals it is probably maintained by ticks, many species of which have been found to be infected. Human infection appears to be derived almost exclusively from infected domestic animals—cattle, sheep and goats—to which *C. burneti* is probably transmitted from a wild animal reservoir by ticks. In animals other than man it produces latent asymptomatic infection. Infected domestic animals excrete the organism in milk, urine and faeces, but in much larger amounts in the placenta and uterine discharges and the latter therefore appear to constitute the main sources of infection for other domestic animals and for man. Considerable numbers of rickettsiae are also present in the faeces of ticks parasitic on infected domestic animals.

Human infection appears to occur mainly through the respiratory tract. Q fever occurs principally in persons whose occupation brings them into contact with infected animals or with the products of infected animals, e.g. wool, hair, hides and meat. Some human infections may occur as a result of ingestion of rickettsiae in infected milk but in general it would appear that the gastrointestinal route is of minor importance in the spread of the disease, although in a number of cases of *Coxiella* endocarditis seen by the author milk was undoubtedly the vehicle of infection. Case-to-case infection does not normally occur but some infections have been described in hospital personnel infected during the performance of a post-mortem examination. A number of laboratory infections have been described, as a result of the manipulation of infected materials. A considerable number of cases of subacute endocarditis clinically resembling subacute bacterial endocarditis but due to *C. burneti* has now been described and a complement fixation test for this organism should form part of the routine investigation for all cases of suspected subacute bacterial endocarditis. Treatment has proved difficult although some success has been obtained in cases in which the underlying abnormality is amenable to surgical treatment. This must be carried out under antibiotic cover.

The wide-spectrum antibiotics, of which the tetracyclines are preferred, though valuable in the treatment of Q fever, appear to be appreciably less effective than in the treatment of rickettsial infections. For endocarditis lincomycin appears to be the agent of choice. Trimethoprim–sulphonamide (cotrimoxazole) has also been recommended. Either should be combined with tetracycline and therapy should be prolonged.

A killed vaccine against Q fever is effective but is not generally available. It should be given to any one planning to work with the organism which has a bad reputation for causing laboratory infections.

Diagnosis

Though it may be possible to isolate *C. burneti* from the blood in the early febrile stages of the disease, or from the sputum, urine or cerebrospinal fluid, the diagnosis is usually established by serological methods using either complement fixation or agglutination tests. The complement fixation test, using an egg-adapted phase 2 strain is the more widely used. In acute human infections, only antibodies reacting with phase 2 organisms are produced. In chronic human infections such as endocarditis, however, antibody may be present both to phase 2 and phase 1 organisms. The Weil–Felix reaction has no value in diagnosis. In *Coxiella* endocarditis, the diagnosis can sometimes be established by inoculating blood into guinea-pigs which are subsequently tested for the production of antibody.

Bartonella bacilliformis

The genus *Bartonella* belongs to the family Bartonellaceae, a group of organisms which are parasites of the red blood cells of a variety of animal species. The Bartonellaceae are classified by Bergey in the order Rickettsiales but differ from the other members of this order in being able to grow in inanimate laboratory media.

Bartonella bacilliformis is a small Gram-negative, motile, non-sporing organism. It exhibits marked pleomorphism, particularly in the blood of infected individuals in which bacillary, coccal and even ring forms may appear.

The organism is aerobic and grows optimally at about 28°C. 'Sloppy' agar containing 10% rabbit serum and a low concentration of rabbit haemoglobin is the medium most frequently used for its isolation. In this medium, growth becomes apparent in about 10 days as a narrow band about 1 cm wide just beneath the surface of the agar.

Bartonella baciliformis appears to be restricted in distribution to the mountainous parts of Peru, Ecuador and Columbia, where it gives rise to latent infection in a considerable proportion of the population. In these people the organism is present in the blood, from which it may be isolated. In a small proportions of infected individuals it gives rise to overt disease, producing two symptomatically distinct clinical syndromes—Oroya fever and Verruga peruana.

Oroya fever is a febrile haemolytic anaemia; it is a serious disease and in the absence of treatment has a high fatality rate. The parasite may be demonstrated in the peripheral blood both in red cells and in the cytoplasm of mononuclear leucocytes. The organisms are also widely distributed throughout the body and are demonstrable in the endothelial cells lining lymph and blood capillaries. The infected endothelial cells are grossly distended and bulge out markedly into the lumen of the vessel. *B. bacilliformis* has not been shown to possess a haemolysin and the mechanism by which it causes haemolysis is not known.

Verruga peruana is a much less serious condition than Oroya fever. It is characterized by warty granulomatous lesions in the skin. The organisms are demonstrable in the skin lesions, particularly in the proliferating endothelial cells which are invariably present. The condition usually develops as a sequel to an attack of Oroya fever, but may appear as the only manifestation of infection. No differences have been demonstrated between the strains of *Bartonella* responsible for these two conditions. Consequently it is believed that the relatively mild character and the symptomatology of Verruga peruana are determined by immunity to the parasite. The identity of the aetiological agents of Verruga peruana and Oroya was first demonstrated in 1885 by Carrion, who submitted himself to inoculation with material from a Verruga patient and died six weeks later from Oroya fever.

The organism is transmitted by sand-flies—*Phlebotomus*—which bite only at night. Consequently prophylaxis by the use of DDT and the application of measures to prevent the attack of night-biting insects is relatively simple and effective. Diagnosis is by microscopic demonstration and isolation of the parasite. In both conditions isolation is most readily achieved by blood culture. Penicillin has been found to be of considerable value in the treatment of Oroya fever, but is of little value in the prevention or treatment of Verruga peruana.

III.22

The Pathogenic Fungi

Of the several thousand species of fungi that are known, less than 100 are pathogenic for man. Most of these belong to the Deuteromyces or fungi imperfecti, which are asexual and reproduce by the formation of various types of asexual spores. Recent work has shown, however, that by special methods of cultivation it is possible to isolate the perfect or sexual stages of the ringworm fungi or dermatophytes, the sexual stages belonging to the class Ascomycetes. It seems likely that many of the fungi imperfecti are fungi whose sexual stage has disappeared during the course of evolution.

Structurally and in their method of reproduction fungi differ markedly from bacteria. Like bacteria they possess a cell wall, but this lacks the characteristic bacterial peptidoglycan. In many, rigidity depends on the macromolecule chitin, which is a β-1-4 polymer of N-acetyl glucosamine. In yeasts it depends on glucan, which is a β-1-6 polymer of D-glucose. The walls of fungi also contain polysaccharrides and proteins; lipids are present in the walls of some yeasts.

Fungi are widespread in nature, the natural habitat for most species being the soil. In the soil fungi, like bacteria, play a vital role in the degradation of dead animal and plant tissue. The only fungi which have a primarily parasitic existence are those which infect epidermal structures. Many of these are primarily animal parasites and only a few, the anthropophilic dermatophytes, are primarily human pathogens.

Morphology

On the basis of their microscopic morphology the fungi imperfecti can be conveniently divided into four groups:

Filamentous fungi. The most important of these are the dermatophytes or ringworm fungi. The basic morphological elements of the filamentous fungi are long branching filaments or hyphae which intertwine to produce a tangled mass of filaments or *mycelium.* When such fungi are grown on solid media some of the mycelium, known as *vegetative* mycelium, penetrates deeply into the medium while some grows above the surface of the medium as an *aerial* mycelium. The filamentous fungi reproduce by a variety of asexual spores—*arthrospores, chlamydospores* and *conidia.* Arthrospores are spores formed by fragmentation of the hyphae into small more or less rectangular segments. Chlamydospores are spores formed by migration of the protoplasm from a number of hyphal segments into a single segment which becomes much enlarged and acquires a thick cell wall. They are in some ways analogous to bacterial spores, being produced under conditions unfavourable to the growth of the fungus proper and capable of remaining dormant for long periods. Conidia are spores formed at the tips or sides of the hyphae. They may be directly attached to the hyphae or may be carried on special supporting structures known as conidiophores. Conidia may be small and single (*microconidia*) or large and multicellular (*macroconidia*). The various types of asexual spores referred to above are seen only when the fungi are grown in the laboratory at room temperature. In the body or when grown at 37°C only arthrospores are produced.

Because of their mycelial character the colonies of the filamentous fungi are strongly adherent to the medium and, unlike most bacterial colonies, cannot be emulsified in water. The surface of the colony may be warty, velvety or powdery or may show a cottony aerial mycelium. Pigmentation of the colony itself and of the underlying medium is frequently present. The macroscopic appearances of the colonies are very susceptible to slight differences in cultural conditions and since their interpretation in species identification requires considerable experience will not be considered in detail.

Yeasts. These occur in the form of round or oval bodies which reproduce by the formation of buds known as blastospores. Yeast colonies resemble bacterial colonies in appearance and consistency. The only pathogenic yeast is *Cryptococcus neoformans.*

Yeast-like fungi. These are fungi which occur both in the form of budding yeast-like cells and as chains of elongated unbranched filamentous cells which present the appearance of broad septate hyphae. New buds are formed at the points of constriction of the hyphae. The latter intertwine to form a *pseudomycelium* which has some capacity to penetrate into the underlying medium. The yeast-like fungi are grouped together in the genus *Candida.*

Dimorphic fungi. These are fungi which exhibit a filamentous mycelial morphology—the saprophytic phase—when grown at room temperature, but have a typical yeast morphology —the parasitic phase—in the body and when grown at 37 °C in the laboratory.

The pathogenic fungi most frequently encountered in medical practice are the dermatophytes which cause the various conditions known colloquially as ringworm. These are entirely superficial infections, the fungus growing in the skin or its appendages and showing no tendency to invade the deeper tissues. In contrast to the dermatophytes the yeast-like and dimorphic fungi are capable of invading the tissues and of giving rise to generalized or systemic infections. Fungal infections of this type are very uncommon in the British Isles. Since fungi do not produce toxins the tissue reactions found in the systemic mycoses are characteristically of a granulomatous type. In general, fungal diseases differ from bacterial diseases in being much more chronic in character and, except in certain types of ringworm in children, in not occurring in epidemic form.

Types of lesion

The lesions produced by fungi are known as mycoses. They may be conveniently classified according to the extent to which invasion of the body occurs:

1. *Superficial* mycoses in which only the superficial layers of the skin and hair are infected.
2. *Cutaneous* mycoses or dermatomycoses in which, depending on the particular fungus, there is invasion of the skin, hair or nails.
3. *Subcutaneous* mycoses in which the infection involves primarily the subcutaneous tissues, the lesions being granulomatous in character. The fungi may be spread by lymphatic channels to the neighbouring lymphatic glands, in which granulomatous lesions are also produced. Invasion of the blood stream may occur but is very uncommon. Soil and vegetation are the usual sources of infection, which is much more common in tropical climates than in temperate zones and in rural areas than in towns and cities. Subcutaneous mycoses are usually contracted through infection of wounds and scratches in the skin or by direct implantation, as by the prick of a thorn.
4. *Systemic* mycoses. In these the primary lesion appears most often to occur in the respiratory tract as a result of the inhalation of fungal spores, the soil being the principal source of infection. The pulmonary lesions may be mild, resolving spontaneously, or they may be chronic, progressive granulomas which may be easily confused, both clinically and radiologically, with those of tuberculosis. As in tuberculosis, infection of the respiratory tract is often asymptomatic. Infection may also occur through the skin or through the mucous membrane of the mouth, pharynx or alimentary tract, the particular route depending on the fungus involved. Following the primary infection the viscera or central nervous system may be infected through dissemination of the fungus by the blood stream.

Diagnosis

An attempt should be made to demonstrate the fungus by direct microscopic examination of the suspected material. In the identification of the filamentous fungi microscopic examination is usually carried out with unstained wet preparations. Fungi which invade the tissues can usually be demonstrated in sections stained with haematoxylin and eosin or with Gram's stain, but better results are obtainable with the periodic acid Schiff technique. With this method, which stains the polysaccharride in the cell wall, most fungi are stained a bright red.

All the pathogenic fungi grow readily on the ordinary media used for the cultivation of bacteria. They are aerobic in character and grow best under humid conditions. Growth, however, is frequently slow and cultures may require incubation for one or more weeks. As a routine medium for fungal isolation Sabouraud's glucose agar is mainly used. This medium is highly acid in reaction (pH 5·5) and is thus strongly inhibitory to bacteria. For the isolation of the yeast-like and dimorphic fungi, blood agar, which may be rendered selective for fungi by the incorporation of penicillin and streptomycin or of chloramphenicol, should be used. Cycloheximide, which is bacteriostatic, is also employed as a selective agent, particularly in the isolation of dermatophytes. Certain other media are used for special purposes, e.g. corn meal agar to encourage the development of chlamydospores by *Candida albicans.* For the isolation of filamentous fungi, cultures should be incubated at room temperature or in an incubator at 28°C, at which temperature the characteristic morphology of the fungi is best seen. In order to demonstrate the yeast forms of dimorphic fungi, incubation should be carried out at 37°C.

When a fungus has been isolated on culture morphological examination is of paramount importance for its identification. For the examination of filamentous fungi and of dimorphic fungi in the saprophytic phase wet preparations are essential since the organized structure of the fungus is broken up in the preparation of fixed films of the type used for the demonstration of bacteria. For this purpose a small portion of the growth should be gently teased out on a slide in a drop of lactophenol blue and the preparation covered with a coverslip. A permanent preparation can be made, if required, by sealing the edges of the coverslip with nail varnish. The morphology of filamentous fungi is most satisfactorily studied in situ. For this, various types of slide culture, which are inoculated from the primary culture, may be used. For the demonstration of yeast forms satisfactory preparations may be obtained by simply emulsifying a small amount of the growth in a drop of water.

Except for the differentiation of *Candida* and *Cryptococcus* species biochemical tests are of no value in species identification. Pathogenicity tests usually on mice are of value in the identification of *C. albicans,* in the differentiation of *Cryptococcus neoformans* from non-pathogenic cryptococci and in the isolation and characterization of the fungi causing systemic infections. Serological tests are not of value in fungal identification.

Antibodies detectable by ordinary in vitro tests are not produced during the course of superficial and cutaneous fungal infections but are sometimes demonstrable in the sera of patients suffering from systemic mycoses. Delayed-type hypersensitivity, demonstrable by the intradermal injection of an appropriate fungal extract, is, however, frequently present in most fungal infections. Its demonstration may sometimes be of value in diagnosis.

Chemotherapy

Fungi are quite insusceptible to the chemotherapeutic agents used in the treatment of bacterial infections. The following substances, which are inactive against bacteria, are however effective in the systemic treatment of certain types of fungal infection. Their action appears to be purely fungistatic.

Iodides. These are reported to be of value in the treatment of sporotrichosis and certain forms of phycomycosis.

Griseofulvin. This antibiotic, which is produced by various species of *Penicillium,* possesses a high degree of activity against all species of dermatophytes. It is well absorbed from the alimentary tract following oral administration and attains a high concentration in newly formed keratin. Since, however, it is not taken up significantly by keratin already formed, administration of the drug must be continued until the latter has been allowed to grow out. In general the length of treatment required depends on the thickness of the epidermal structure affected; thus infections of the nails as a rule necessitate administration of the drug for several months. Griseofulvin has no action against *Candida albicans,* a fungus which can cause a variety of superficial infections resembling those produced by the dermatophytes. A precise mycological diagnosis is therefore an essential preliminary to its clinical use. In its action griseofulvin is fungistatic rather than fungicidal and although it has revolutionized the treatment of the dermatomycoses it has not eliminated the necessity for various forms of ancillary local treatment. The mechanism by which it inhibits fungal growth is unknown, but it is presumed to be by some type of specific interference with the synthesis of the fungal cell wall. It is significant that it acts only against fungi whose cell walls contain chitin as the main structural component. In high concentration griseofulvin is a potent inhibitor of mitosis but, apart from the occasional development of a transitory leucopenia which disappears when the drug is withdrawn, no serious toxic reactions following its administration have been described.

Polyenes. This group which acts by combining with sterols in the cytoplasmic membrane, includes nystatin (mycostatin), amphotericin B, candicidin, trichomycin and pimaricin. Of these nystatin and amphotericin B have been the most widely used. Polyenes are of value by surface application in the treatment of superficial *Candida* infections. Amphotericin B is the drug which has been principally used in the treatment of the systemic mycoses in which because of its poor absorption from the alimentary tract it must be given intravenously. The drug, however, is highly toxic and is only used for patients who are seriously ill. Trichomycin, which is also active against *Trichomonas* infections, is used for the treatment of vaginal candidiasis.

Newer drugs for the treatment of fungal infections are 5-fluorocytosine and clotrimazole. 5-fluorocytosine has an established place in the treatment of systemic infections. The value of clotrimazole has still to be assessed.

OPPORTUNISTIC INFECTIONS

This term is applied to infections with organisms normally occurring as commensals or present in the environment in patients whose resistance has been lowered either by the administration of drugs—immunosuppressives or broad-spectrum antibiotics—or who are suffering from certain conditions, notably leukaemia, lymphosarcoma, diabetes or terminal wasting diseases. The most important of the opportunist fungi is *Candida albicans.* Less frequent are *Aspergillus* species and the Phycomycetes *Rhizopus* and *Mucor,* but the latter are more serious invaders than *Candida* since once they have established themselves they have a marked capacity to grow into blood vessels and to cause metastatic lesions in the viscera.

Candida albicans

This organism occurs both in the form of oval yeast-like bodies and as thick septate pseudohyphae (p. 323). *C. albicans* stains well by Gram's method and is Gram-positive.

Candida albicans occurs as a normal inhabitant of the mouth and intestinal tract. It is the cause of thrush, a condition at one time common in children, in which white patches containing the fungus are found on the mucous membrane of the mouth. A similar condition may occur on the vaginal and vulval mucosa giving rise to vaginal irritation and discharge. This type of infection is particularly likely to occur in pregnancy and in diabetic subjects. *C. albicans* occasionally causes infections of the skin, especially intertriginous lesions of the hands in people whose occupation involves frequent immersion of the hands in water. The fungus may also cause an infection of the nails resembling that produced by the dermatophytes but which is less destructive and is usually associated with a considerable degree of paronychia. It is an occasional cause of athlete's foot. During the course of wide-spectrum antibiotic therapy there may be considerable overgrowth of *C. albicans* in the lungs and in the alimentary and urogenital tracts. It is doubtful whether this overgrowth is associated with any significant degree of tissue invasion. Pulmonary invasion may, however, occur in individuals with prior lung disease, e.g. tuberculosis or cancer. Allergic vesicular skin lesions—*monilids*—may occur in any type of *C. albicans* infections.

The diagnosis of mucosal lesions is readily achieved by the examination of Gram-stained films, the oval budding bodies being highly characteristic. For the diagnosis of cutaneous infections the methods described for the diagnosis of ringworm are employed.

Candida albicans grows readily on blood agar at 37°C, producing after 24 to 48 hours' incubation small greyish bacteria-like colonies. On

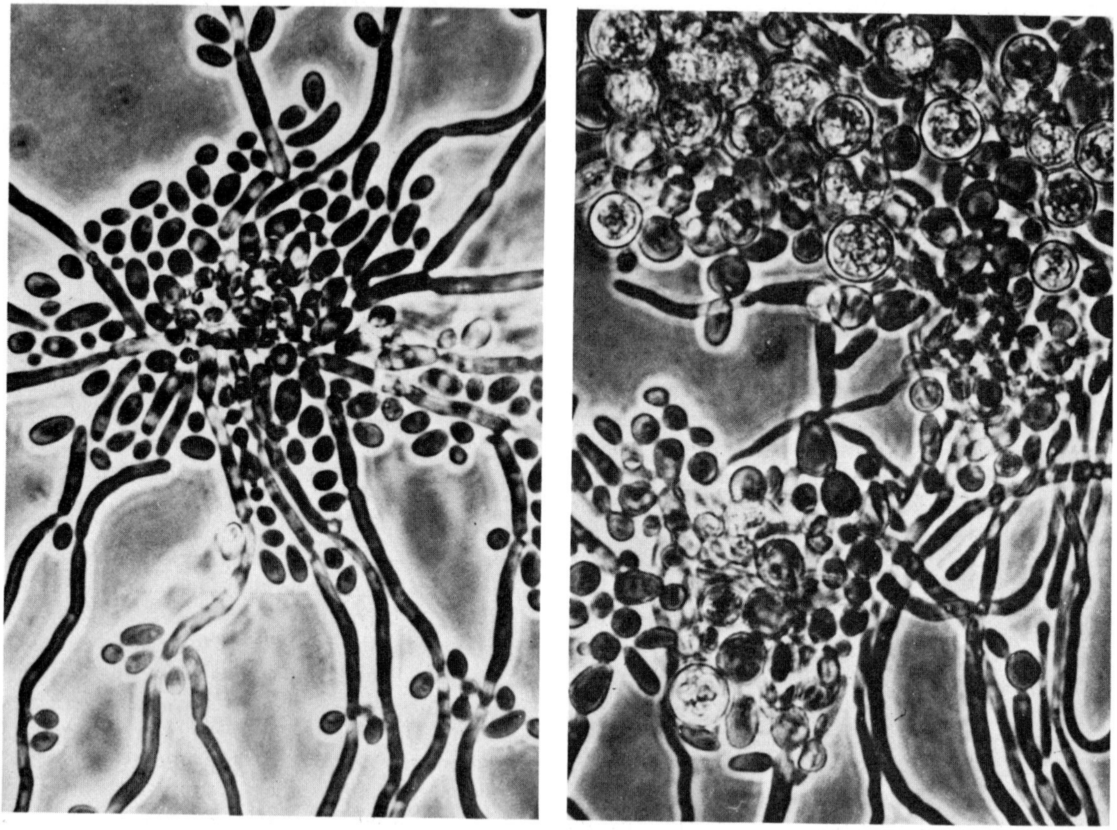

A B

Fig. III.19. Cultures of *Candida* on corn meal agar, shown by phase contrast. A, Non-pathogenic species. B, *Candida albicans*, note the chlamydospores.

Sabouraud's medium the colonies are cream-coloured and pasty and on prolonged incubation develop a honeycombed appearance as a result of glucose fermentation. A number of tests have been described for the differentiation of *C. albicans* from other *Candida* species. The most valuable of these are the demonstration of chlamydospore production on corn meal agar and sugar-fermentation reactions. The chalmydospores, which are large, round and thick-walled, are produced at the tips of the hyphae. *C. albicans* ferments glucose and maltose, producing acid but no gas; it does not ferment lactose. In cases of doubt a pathogenicity test may be carried out in rabbits or mice. Animals inoculated intravenously with a suspension of the organism die in four to five days and at post mortem show multiple small white abscesses in the cortex of the kidney.

SUPERFICIAL MYCOSES

The most widespread of the superficial mycoses is tinea versicolor, a condition characterized by the occurrence of scaly areas on the trunk. It is believed to be caused by a yeast-like fungus *Pityrosporum furfur* which has frequently been isolated from the lesions.

In white and black piedra nodules are formed in the hair of the scalp and beard due to the non-keratolytic fungi *Trichosporum beigelii* and *Piedraia hortai* respectively.

CUTANEOUS MYCOSES

The fungi of the dermatophyte group attack the keratinized surface of the body, the skin, hair and nails, producing the various conditions known

SUBCUTANEOUS MYCOSES

Sporotrichosis

This condition is due to *Sporotrichum schenckii,* a dimorphic fungus which occurs in the soil and on plants and wood. The infection occurs commonly through the scratch of a thorn. An ulcer develops at the site of inoculation, the infection tending to spread by the lymphatics in which it produces granulomatous lesions. Further dissemination is rare, the meninges being exceptionally involved.

Chromomycosis

The lesions in this condition are warty fungating ulcers, brownish in colour usually occurring on the legs and feet. Chromomycosis is not uncommon in developing countries where it is found in barefooted labourers. It is caused by dimorphic fungi of the genera *Phialophora* and *Cladosporium.*

Mycetoma

This granulomatous condition of the foot and hand occurs mainly in tropical countries. It can be produced by a number of species of fungi of the genera *Madurella, Allescheria, Phialophora* and *Aspergillus* and by actinomycetes of the genus *Nocardia* (p. 281). Pus from the discharging sinuses characteristically contains granules consisting of colonies of the causative organism. Mycetoma due to fungi is often called maduromycosis.

Subcutaneous phycomycosis

In this condition subcutaneous granulomas which do not ulcerate form in the subcutaneous tissues. They are due to phycomycetes of the genera *Basidiobolus* and *Entomophthora. Basidiobolus* infections occur mainly on the trunk in children. *Entomophthora* causes lesions on the face around the nose. Infection with these fungi occur mainly in tropical countries.

Rhinosporidiosis

In this condition friable pedunculated tumour-like masses are found in the skin and mucous membranes. The commonest site of infection is the anterior nares. The condition is endemic in India and Ceylon but also occurs in other parts of the world. The fungus responsible, *Rhinosporidium seeberi,* occurs in the tissues in the form of large sporangia measuring up to 300 µm in diameter. When fully developed the sporangia contain large numbers of small endospores measuring from 5 to 8 µm in diameter. The fungus has not been grown in laboratory culture.

SYSTEMIC MYCOSES

Cryptococcus neoformans

This fungus occurs both in the tissues and in cultures in the form of round yeast-like cells surrounded by a large gelatinous capsule. In the tissues the individual cells measure from 5 to 20 µm in diameter but in culture they are considerably smaller, usually measuring from 2 to 5 µm in diameter. No mycelium is produced either in the tissues or in culture but elongated tube-like cells are occasionally found. On Sabouraud's medium at room temperature *Cryptococcus neoformans* produces a glistening mucoid colony of brownish colour which in course of time flows down the slope to the bottom

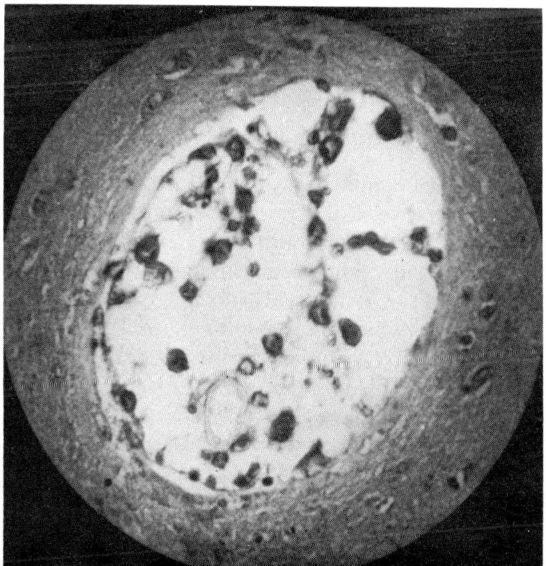

Fig. III.24. *Cryptococcus neoformans* in the brain. Cryptococci are free in a cystic cavity in the brain substance. Haematoxylin and mucicarmine.

of the tube. The fungus may be distinguished from *Candida* species and non-pathogenic yeasts by its positive urease reaction and from non-pathogenic cryptococci by its capacity to grow well at 37°C and by its lethal effect on mice on intraperitoneal inoculation.

The organisms may be isolated from the soil, particularly when it has been contaminated by pigeon droppings. Pigeons are possibly a natural reservoir of infection. Human infection occurs through the respiratory tract and low-grade inflammatory lesions are produced in the lungs. From the lungs the organism may spread throughout the body but it has a marked predilection for the central nervous system in which it produces chronic meningitis. Cryptococcosis is the only systemic mycosis of world-wide distribution.

In cases of meningitis the organism can be demonstrated in the centrifuged deposit from the cerebrospinal fluid and is readily recognized by its conspicuous capsule; the latter is best demonstrated in wet India ink preparations. The diagnosis of cryptococcal meningitis is, however, sometimes missed, the fungi being mistaken for lymphocytes. The cerebrospinal fluid may give a positive precipitin reaction with immune rabbit serum. Antibodies may be demonstrated by immunofluorescence, complement fixation and passive haemagglutination, but are not of diagnostic value.

Histoplasma capsulatum

This is a dimorphic fungus occurring in the tissues as small oval yeast-like bodies of from 3 to 5 μm in length. In the body it multiplies mainly within the cells of the reticuloendothelial system. On Sabouraud's medium, at room temperature, it produces a mould-like colony in which may be seen filaments, conidia, borne singly on short branches or sessile on the sides of the hyphae, and large chlamydospores—up to 15 μm in diameter—with characteristic radiating processes. The tissue form is best obtained by growth in sealed tubes on Francis's cystine glucose blood agar. *H. capsulatum* is often difficult to grow and colonies frequently take some weeks to develop. The fungus is highly pathogenic for mice.

Histoplasmosis is a common disease in certain parts of the United States and Central America but appears to be uncommon in other parts of the world. The infection primarily involves the lungs and in most cases is asymptomatic. In symptomatic infections the lung lesions are usually multiple and heal slowly by calcification. A generalized progressive infection occurs in a small proportion of cases. In this type of infection the liver, lymph nodes and spleen are particularly involved. In endemic areas the fungus is present in the soil, particularly soil contaminated with bird droppings.

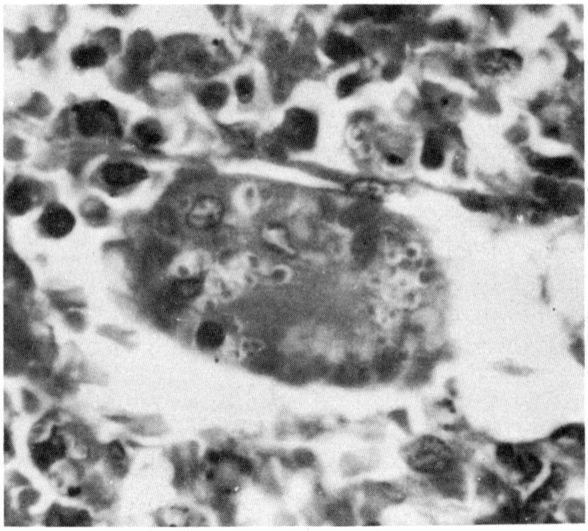

Fig. III.25. The tissue phase of *Histoplasma capsulatum* in a macrophage.

Antibodies detectable by complement fixation tests appear within a few weeks of infection. Cutaneous hypersensitivity is also present and may be detected by the intracutaneous injection of a filtrate of a fluid culture of *H. capsulatum* known as *histoplasmin*. The reaction is of delayed type and is indicative of past as well as of current infection.

A diagnosis of histoplasmosis may be made by demonstrating the intracellular fungus in Giemsa-stained smears of peripheral blood, sternal bone marrow, sputum or infected lymph nodes, the histological appearance of macrophages filled with capsulated yeasts being highly characteristic. Isolation of the fungus should also be attempted.

Blastomyces dermatitidis

In the tissues this fungus appears either free or within phagocytic cells as a round yeast-like body 8 to 15 μm in diameter with a thick highly refractile cell wall. The cells, which in the tissues reproduce by the formation of a single bud, are often seen as pairs of cells of unequal size. On Sabouraud's medium at room temperature the growth is at first yeast-like but later assumes a typical mould form showing filaments, arthrospores and conidia resembling those of *H. capsulatum.* The yeast phase may be obtained on blood agar at 37°C when, in addition to the yeast form, characteristic short undeveloped mycelial fragments may also be seen.

Blastomyces dermatitidis is the cause of a condition generally known as North American blastomycosis, since the majority of infections initially recognized occurred in Canada and the United States. However, the disease has since been found to be widely prevalent in Africa and many workers consider it should be called simply blastomycosis. The source of infection is not known with certainty but it is believed to be the soil in which the fungus can be cultivated. The initial infection is in the lungs, from which the organism may be disseminated throughout the body, the skin being a particularly important target.

Delayed-type hypersentivity develops as a result of infection and may be detected by the intradermal injection of an extract of the fungus, blastomycin. Positive reactions to blastomycin are also given by persons with histoplasmosis and less frequently by those suffering from coccidioidomycosis. Antibodies may be demonstrated by complement fixation tests but these are of slight value in diagnosis since they show the same cross-reactions as the skin tests.

Paracoccidioides brasiliensis

The tissue form of this organism is somewhat larger than that of *B. dermatitidis* being usually from 20 to 30 µm in diameter and occasionally larger. It reproduces both in the tissues and at 37°C by the simultaneous formation, around the periphery of the cell, of several buds each about 1 µm in diameter. Mycelium, conidia similar to those of *B. dermatitidis,* and arthrospores are present in cultures grown at room temperature.

The organism appears to gain access to the body mainly through the mucous membrane of the mouth or alimentary tract. It may produce local lesions on the oral mucosa and adjacent skin, lymph node involvement particularly in the neck region or a systemic infection in which the initial lesion is usually in the ileocaecal region and in most cases involves the lungs. The source of infection is not known.

Coccidioides immitis

Coccidioides immitis is a dimorphic fungus which in the body appears as a large spherule or sporangium of 15 to 70 µm in diameter with a thick wall measuring about 2 µm in cross section. Characteristically the spherules are filled with endospores measuring from 5 to 8 µm in diameter, but some of the spherules may be immature and these do not contain spores. The fungus reproduces by bursting of the spherule thereby releasing the spores which in turn germinate into fresh spherules. In the body the parasite is found mainly within multinucleate giant cells. In culture only a mycelial form which reproduces by the formation of arthrospores is found. Cultures of *C. immitis* are extremely infectious and require very careful manipulation. The fungus is pathogenic for mice and mouse inoculation, which is safer than culture, can be used for diagnosis.

Coccidioidomycosis occurs as an endemic infection in certain desert areas in the south western United States, particularly in the San Joaquin Valley in California. The fungus vegetates in the soil and human infection appears in most cases to be due to inhalation of contaminated

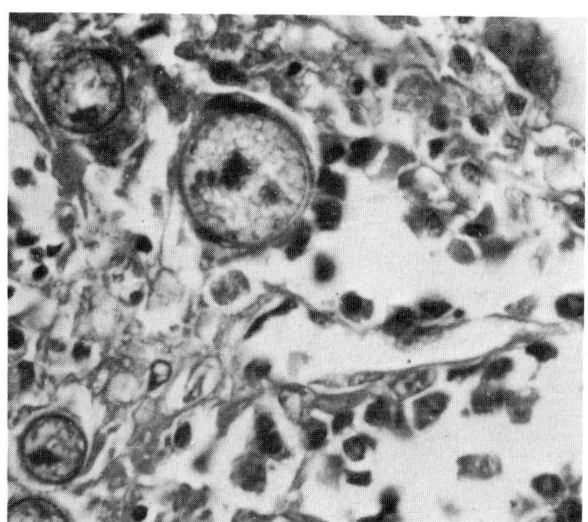

Fig. III.26. The tissue phase of *Coccidiodes immitis* in the lymph gland. Spherules containing endospores.

dust. The disease is not transmitted from case to case. In endemic areas it shows a maximal prevalence in the dry months of the year. It occurs in two forms:

1. A primary type usually involving the lungs, less frequently the skin. Primary pulmonary infection is frequently asymptomatic or may appear as a relatively mild upper respiratory tract infection.
2. A disseminated type known as coccidioidal granuloma. This type occurs in about 1% of white individuals infected but in about 10% of infected Negroes and has an overall mortality of about 60%. The main lesions, which are in the lungs, are granulomatous in character and are similar to those found in miliary tuberculosis. In addition to the lungs the subcutaneous tissues, skin and central nervous system are also frequently involved.

During the course of coccidioidomycosis a delayed-type hypersensitivity which may be detected by the intradermal injection of coccidioidin—obtained from cultures of the fungus in fluid media—develops. The active agent in coccidioidin appears to be a polysaccharide. The skin reaction is positive in individuals with asymptomatic infection. In severe primary infections and in coccidioidal granuloma positive precipitin and complement fixation tests with coccidioidal antigen are also obtained.

Part IV
VIROLOGY

IV.1

The General Properties of Viruses

Although many important infectious diseases of Man and other vertebrates are caused by viruses, we have no monopoly of these microbes. Essentially similar organisms infect invertebrates, particularly arthropods, higher and lower plants and bacteria. The bacterial viruses are dealt with in Chapter IV.13. Any extensive treatment of the viruses of invertebrates or of plants would be out of place here. Those interested may be referred to the excellent textbooks of Smith and of Bawden. Nevertheless, some reference will be made to insect, plant and bacterial viruses by way of examples and in reading the following chapters it should be remembered that in considering only viruses which infect Man we are deliberately restricting ourselves to a small, artificially contrived, group, all of whose members share properties with viruses which do not infect Man. Indeed, we actually share many of our viruses with other vertebrates and, in the case of the arboviruses, with invertebrates too. No plant viruses are known to infect Man, but some important plant viruses are very like the reoviruses and others closely resemble rabies virus in their morphology. Likewise, no bacterial virus is able directly to infect Man, although both diphtheria and scarlet fever are indirectly due to temperate bacteriophages which infect the bacteria concerned (see pp. 214, 229 and 466) and produce the phenomenon of phage (lysogenic) conversion.

As currently defined, viruses are a group of micro-organisms differing from bacteria in the following respects:

1. They are in general much smaller, although the smallest viable cells of the mycoplasmas overlap the largest viruses in size. The animal pathogenic viruses range in size from the small viruses of the parvovirus group, which measure about 20 nm in diameter, to the large viruses of the pox group which measure from 250 to 300 nm in their longest axis. Since the limit of resolution of the ordinary light microscope is about 250 nm it is only the viruses at the upper end of this range which can be demonstrated by ordinary microscopic methods.

2. They cannot be cultivated on inanimate laboratory media. Viruses are obligatory intracellular parasites and will grow only within other living cells. Their obligatory parasitism is linked with their small size. They are in fact so small that they could not accommodate the enzymatic machinery which would allow them to maintain an independent free-living existence. They can reproduce only if they can make use of the metabolic machinery of the host cell for the synthesis of virus components. They are therefore not independent organisms in the sense in which bacteria are independent organisms. Although other organisms may be obligate intracellular parasites, none, so far as is known, is as totally dependent as viruses on the metabolic machinery of the host because none is as totally bereft of any metabolic machinery of its own.

The criteria considered above have for many years been regarded as sufficient for the definition of the viruses. The definition is too wide, however, since it admits the inclusion as viruses of micro-organisms—the rickettsiae and chlamydiae—which are now more usually regarded as very small bacteria. In current usage therefore the term virus includes the following additional criteria.

1. They possess only a single type of nucleic acid, which may be either RNA or DNA.

2. They do not multiply by binary fission. Instead, when a virus infects a cell the viral particle is broken down with the liberation of the

viral nucleic acid. The latter then proceeds to replicate and its genetic information is translated into the synthesis of virus-specific protein. Because of this loss of virus integrity, there is a phase, following infection and concomitant with the breakdown of the virus particle and before the assembly of new virus, during which fully infective virus cannot be recovered from the cell. This is known as the eclipse phase. For largely technical reasons the eclipse phase is much more easily demonstrated with bacterial viruses than with animal or plant viruses, which usually show a substantial drop in infectivity shortly after infection rather than a complete loss of infectivity.

The obligatory parasitism of the viruses has aroused considerable speculation and controversy concerning their evolutionary status. One view which has been widely held is that they have arisen by a process of retrograde evolution from bacteria. As has been pointed out before (Chapter I.1), it is possible to trace a gradual loss of synthetic ability in bacteria as we pass from the less to the more highly parasitic types. It is therefore possible that at least some viruses are forms of life in which this process has been carried to a stage where the organism is completly devoid of independent synthetic power.

A second possibility is that viruses are derived ultimately from self-reproducing constituents of the cells of higher organisms which have in some way acquired an independent status. Thus it seems not impossible that certain of the DNA viruses may at some time have originated from the host cell nucleus and that the RNA viruses may have originated from host cell cytoplasmic components. As yet, however, no one has demonstrated the transformation of any host cell component either of nuclear or of cytoplasmic origin into a virus. Nevertheless the distinction between temperate bacteriophages and plasmids such as F factors, transfer factors and Col factors (see Chapters I.1 and IV.13) is a very narrow one and if comparable genetic elements are found to be common in mammalian cells this will face animal virologists with similar problems of definition.

A possible diarch origin for animal viruses is supported by recent studies which have shown a high degree of homology between host and virus nucleic acids in certain cases, especially the tumour viruses. In such cases it is possible that the viruses may have arisen from host-cell nucleic acids. In other cases, where the degree of homology is very much less, it is possible that the viruses may have arisen from some other organism quite unlike its current host, such as a bacterium.

MICROSCOPY

Although the majority of viruses cannot be seen by ordinary microscopic methods, the light microscope nevertheless has some important applications in virology.

1. The large viruses of the pox group are just above the limit of resolution with ordinary light and can be demonstrated by the use of staining procedures which, as a result of deposition of stain on the surface of the particle, increase its apparent size (Fig. IV.1).
2. The fluorescent antibody technique (p. 141) has been of considerable value for the demonstration of viral antigens in tissues. Antigenic components of the smallest viruses can be demonstrated by this technique.
3. For the demonstration of virus inclusion bodies (p. 372).
4. For the recognition of virus induced degenerative changes (cytopathogenic effect, cytopathic effect, CPE) in infected tissue cultures.
5. For the recognition of characteristic histological and cytological changes in infected animals or Man, e.g. poliomyelitis, glandular fever and the characteristic myositis produced in baby mice by the Coxsackie viruses.

The great advances of recent years in our knowledge of the structure of virus particles have been achieved through the use of the electron microscope. In this instrument a beam of electrons is used instead of light rays and, because of the very short associated wavelength of the high velocity electrons, the limit of resolution is enormously increased. The electron microscope is capable of resolving particles with diameters of less than 1 nm and its use has consequently brought even the smallest viruses into the visible range.

Although positive staining methods and shadow-casting procedures have been valuable in revealing overall virus morphology, they are incapable of revealing details of virus substructure. For this purpose the greatest single advance has been the introduction of the negative-

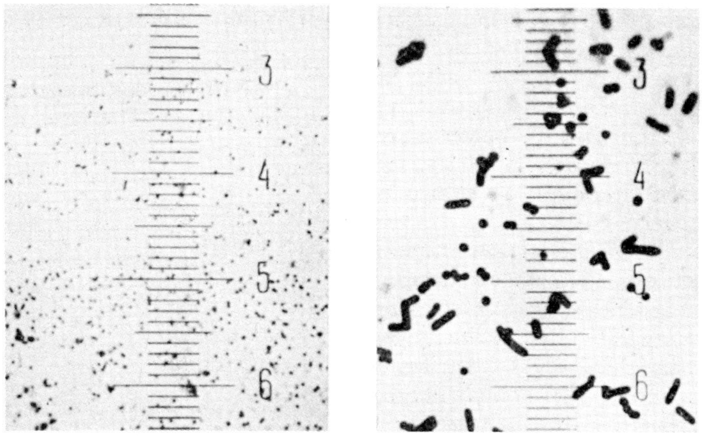

Fig. IV.1. Viruses and bacteria. A, Elementary bodies of vaccinia virus from the conjunctiva of the rabbit. B, Staphylococci and *E. coli*. Each division on the micrometer scale equals 1·1 μm. Paschen's stain. × 1200.

staining technique of Brenner and Horne. In this technique the virus-containing material is mixed with a neutral solution of phosphotungstate. The latter forms an electron dense glass around the particle, producing a reversal of contrast in the final image. The method not only allows visualization of the virus particle but also permits the demonstration of its structural sub-units (see Figs IV.2 and IV.34); it is analogous to the negative-staining techniques used in the demonstration of bacterial capsules (p. 13).

Electron microscopy has also been used for counting the number of virus particles in a suspension. For this purpose the suspension is mixed with a suspension of polystyrene latex particles of known concentration. From the relative numbers of latex particles and of virus particles seen in droplets of the mixture the concentration of virus in the original suspension can be estimated.

PURIFICATION OF VIRUSES

The starting point in virus purification is usually an infected tissue which has been broken down by a process of homogenization. Where available, infected fluids, e.g. allantoic fluid or tissue culture fluid, may also be used. They constitute as a rule a purer source of virus than tissue homogenates. Purification of virus from an infected tissue presents considerable technical problems since the concentration of virus rarely amounts to more than 10^{-8} to 10^{-9} of the wet tissue. Because of this the preparation of highly purified virus normally requires the combined use of a number of different procedures. Even then it is difficult to be completely sure that the virus is entirely free from tissue components, some of which may remain adsorbed to the virus surface throughout the purification procedure.

The more important techniques currently employed for purification are the following.

Filtration

This may be used to separate virus from larger particles and has been of particular value in the preparation of viral suspensions free from tissue fragments, cells and contaminating bacteria. Before the introduction of electron microscopy, filtration through special collodion membranes of known average pore diameter was very valuable for the estimation of virus size.

Centrifugation

The marked disparity of size between bacteria and viruses permits their separation by centrifugation. Centrifugation in relatively low centrifugal fields of 1000 to 2000 *g* will sediment bacteria but not viruses. Viruses can be sedimented by very rapid centrifugation. For this purpose elaborate and expensive centrifuges capable of producing centrifugal fields of 50 000 *g* or higher are employed. By suitable cycles of alternate slow- and high-speed centrifugation it is possible to obtain extremely pure preparations of virus which are free both from large particles—bacteria and tissue debris—and from smaller particles and soluble substances; this technique is known as *differential* centrifugation.

The introduction of *density gradient centrifugation* has resulted in an enormous improvement in the readiness with which particles of different sizes and densities may be separated by centrifugation. In this procedure the material is centrifuged through an aqueous solution of a high density solute, the concentration of which increases progressively from the top to the bottom of the centrifuge tube. The substances most used for gradient preparation are sucrose and glycerol, with which the concentration gradient is preformed by layering, and caesium chloride and rubidium chloride, with which the concentration gradient is formed during centrifugation. The rate of sedimentation of the particle through the gradient depends on its size, shape and density. If centrifugation is continued for long enough the particle eventually comes to rest in a zone determined only by its density. Density gradient centrifugation is carried out in plastic tubes from which the different zones can be withdrawn through a hole pierced in the bottom.

Precipitation

Many substances capable of precipitating proteins may be used to precipitate viruses from tissue suspensions; ammonium sulphate and methanol have been of particular value. For the removal of contaminating proteins protamine sulphate has been widely used; this precipitates the protein but leaves the virus particles in suspension. Isoelectric precipitation in which the pH is adjusted to the isoelectric point of the virus is also valuable. The virus may, however, be precipitated at the isoelectric point of protein impurities, to which it is adsorbed on precipitation. A combined immunological and enzymic method may be used by which the virus is precipitated with specific antiserum, the precipitated virus then being freed from antibody by tryptic digestion to which most viruses, though possessing protein capsids (p. 341), are relatively resistant.

Adsorption-elution procedures

A biological elution method is of considerable value for the preparation of pure suspensions of the myxoviruses. The virus is adsorbed onto red cells in the cold and allowed to elute, through the action of the virus enzyme, at 37°C (p. 424). Adsorption–elution procedures employing ion exchange resins, such as DEAE cellulose, and inorganic adsorbents, such as calcium phosphate and diatomaceous silica, are also extensively used.

Other methods

Treatment with organic solvents (ether or fluorocarbons) is of value in removing tissue lipids from non-lipid-containing viruses. Lipid-containing viruses such as the myxoviruses and arboviruses are, however, broken down by treatment with lipid solvents.

Zone electrophoresis permits the separation of particles on the basis of charge.

Separation in two-phase systems using water-soluble polymers, in which the viruses are distributed in the two phases in concentrations differing from those of the impurities.

Molecular sieves, which sort molecules according to size, may be used in virus purification. The most valuable are the polymeric carbohydrates (cross-linked dextrans) known commercially as Sephadex. These are available in a number of pore sizes and, while molecules below the pore size are retained in a Sephadex column, those which are larger pass through unhindered. Diethylaminoethyl and carboxymethyl functions can be attached to Sephadex so that particles may be fractionated on the basis of both size and charge on a single run.

THE STRUCTURE OF VIRUSES

In discussing the structure of viruses we must always remember that we are considering the intact virus particle (virion) which represents the extracellular form of the virus. As already indicated, on entering a susceptible cell, the virus particle is disassembled and can no longer be recognized. New virus particles will not be demonstrable for some time and may, indeed, only be completed as they actually leave the host cell.

Not unexpectedly, the chemical complexity of viruses depends, to some extent, on their size, the smallest viruses consisting only of nucleic acid and protein. The protein constitutes a coat for the virus nucleic acid, protecting it from inactivation by inimical substances, e.g. tissue nucleases, which may be present in the environment. Once the virus has gained access to the cell the protein coat generally seems to play no further part in viral multiplication. The protein coat, however, has an important role in the initial attachment of the virus to susceptible cells. Such a function has, for example, been postulated for the poliomyelitis protein coat. In this case the coat also imposes a limitation on infectivity since the intact virus will not infect cells lacking receptors which bind with the coat protein, although such cells can be infected by protein-free viral nucleic acid. The protein coat of a virus is generally referred to as a *capsid* and the combination of nucleic acid and capsid as the *nucleocapsid*. In some cases the nucleocapsid constitutes the entire virus. In others it is in turn surrounded by an outer layer or *envelope* which is added as the virus passes through or emerges from the host cell and much of which seems to be derived directly from host cell membranes. The envelope contains lipid as well as protein.

In the smaller viruses the capsid, as originally postulated by Watson and Crick, is built up, in the interests of genetic economy, from a number of identical subunits. Such a subunit structure reduces the amount of genetic information required for capsid synthesis, the information needed being simply that involved in the synthesis of one type of subunit. This economy is necessary, since the nucleic acid of the smaller viruses is in itself insufficiently large to code for more than a few medium-sized proteins and this range must include, in addition to structural proteins, enzyme proteins involved in the replication of the viral genome. It is now known in fact that certain viruses, e.g. the tobacco necrosis satellite virus, some strains of the Rous sarcoma virus (p. 459) and the adeno-associated viruses (p. 451) do not carry all the information necessary for their own replication. Such viruses, referred to as 'defective', can multiply only with the cooperation of a 'helper' virus which, in the course of its own replication, can carry out the missing functions. Sometimes a virus will appear to be defective in one host-cell but not in another. In the case of certain plant viruses, different fragments of the genome are carried in different virus particles, so that productive infection, with the formation of progeny virus, only occurs if the cell is more or less simultaneously infected with at least one of each sort of particle. Such viruses are known as 'split genome viruses'.

Construction from prefabricated subunits makes possible a process of self-assembly of the viral capsid in which the subunits themselves come together to form a capsid, in a manner analogous to crystallization. The form of the capsid will in this case be a function of the geometry of the subunits. In the capsid structure the subunits will be held together by intermolecular forces, without the formation of covalent bonds, and the final position they take up will be that characterized, as in crystal formation, by a minimum of free energy. Because of this requirement the resultant structure will have considerable stability. Self-assembly is obviously a highly economical process since it eliminates the necessity for a multiplicity of enzymatic steps which might otherwise be required. It is in addition a highly efficient method for the construction of a complex structure since it is self-checking, subunits which may have structural defects being automatically rejected. There is evidence that a similar process of self-assembly is involved in the synthesis of bacterial flagella from flagellin subunits (p. 9). In the completed capsid, bonding between the nucleic acid and the overlying protein subunits seems to contribute to the stability of the final structure.

The subunit composition of the capsid is also important in relation to the release of the viral nucleic acid in infected cells. The removal of the protein coat is clearly a simpler process if it involves only the disassembly of non-covalently bonded subunits than if it involves the degradation of an integral protein sheath. In the case of the icosahedral viruses the removal of only a relatively few subunits could result in the release of the contained nucleic acid.

Many viruses when present in high concentration form characteristic crystals. The crystals consist of aggregates of virus particles which are released again when the crystal is dissolved. Crystal formation is a consequence of the regular size and symmetry of the virus particle. It occurs most readily with purified preparations of virus but in some cases—tobacco mosaic virus and the adenoviruses—virus crystals may be observed in infected cells.

Capsid symmetry

Three types of capsid symmetry are described: complex, helical and icosahedral.

Complex symmetry

This describes the structure of complex viruses, e.g. the poxviruses and the bacteriophages, in which a number of major structural components may be identified. In these the geometry of the total particle cannot be treated directly as a function of the geometry of repeating subunits. Such a treatment, is, however, possible in respect of the helical and icosahedral viruses.

Helical symmetry

In this the protein subunits are packed in helical array around the viral nucleic acid. The helical virus whose structure was first determined in detail was the tobacco mosaic virus. This was achieved entirely by X-ray diffraction studies which can be accomplished only for viruses that can be crystallized. The completely assembled virus consists of rigid rods; the helically disposed structure units—except for those at the ends of the particle—are completely equivalent, i.e. each subunit has identical relations with its neighbours. There is a central, apparently empty core and the nucleic acid which runs in helical fashion along the length of the particle rests in a groove formed by indentations in the subunits. The complete equivalence of the subunits of the tobacco mosaic virus makes self-assembly a straightforward process and this has been achieved in vitro. It is possible, under appropriate conditions, to disaggregate the protein subunits of the capsid and the nucleic acid and, by reversing these conditions, to reaggregate the complete virus. The protein subunits will also spontaneously reaggregate in the absence of the nucleic acid, to form either the normal helical capsid or a stacked disc structure. The occurrence of the latter in the absence of nucleic acid, and the finding that nucleic acid-free helical capsids are more readily disaggregated than native or reconstituted virus, indicate that some degree of bonding between the nucleic acid and the protein subunits can act as a

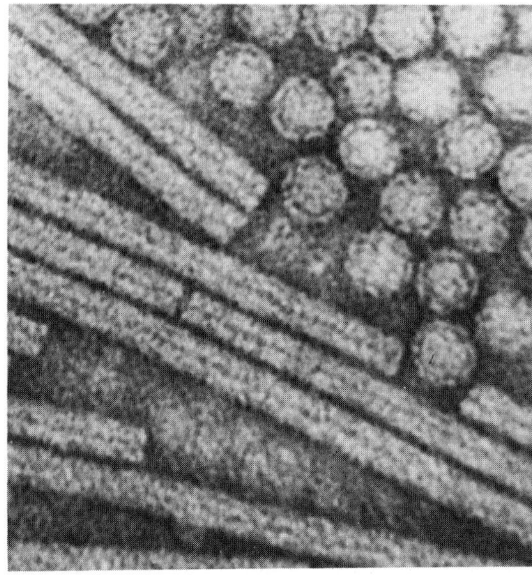

Fig. IV.2. Tobacco mosaic rods and spherical turnip yellow mosaic virus mixed in one preparation. (*Dr R. W. Horne*)

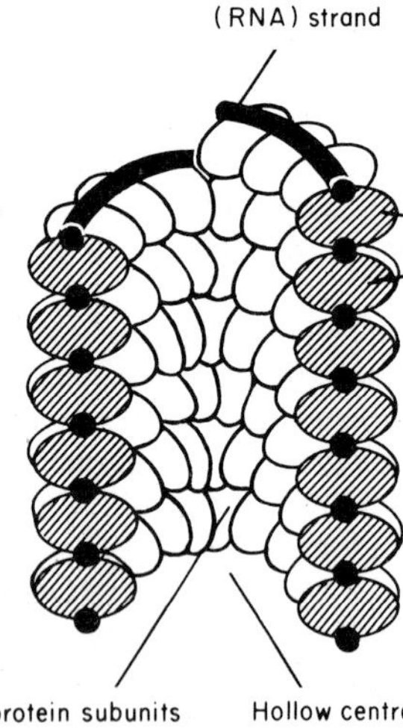

Fig. IV.3. Longitudinal section of a helical rod-shaped virus. Note that it is not strictly correct to refer to the protein subunits of a helical virus as capsomeres. The term is employed in this diagram only to emphasize the essential similarity in the structural components of helical viruses and of viruses with icosahedral symmetry: nucleic acid and protein units. (*After Dr L. B. Quesnel*)

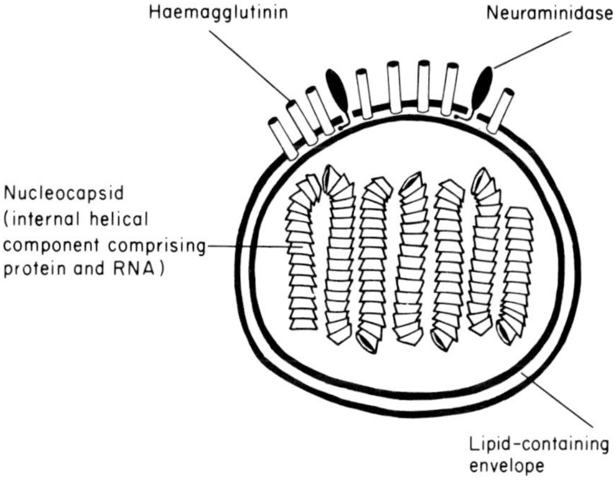

Fig. IV.4. The structure of the influenza virus. Note that the nucleocapsid is in a number of separate pieces. Among the myxoviruses this appears to be peculiar to the influenza virus, the other members of the group having an uninterrupted nucleocapsid. Compare with Figs IV.26 and IV.28.

check on the self-assembly process and contribute materially to the stability of the finished product. Although the rods of tobacco mosaic virus are rigid, those of other helical plant viruses, such as potato virus X, are flexible.

For a long time it was thought that rod-shaped helical viruses of mammals did not exist. However, it has now been shown that one group at least, the myxoviruses, consists of a flexible, helical rod similar to the flexible plant viruses rolled up like a ball of wool and enclosed in an envelope so that the complete particle is more or less spherical (see Figs IV.4 and IV.26). Recent evidence indicates that other groups of viruses also possess helical symmetry. It is significant that, so far, the viruses possessing helical symmetry have all been RNA viruses. It is possible that this type of capsid would be quite unsuitable for the more rigid DNA double helix. Filamentous, DNA-conaining bacteriophages do exist, but the nucleic acid is single stranded and their structure appears to be quite different from that of the helical RNA plant viruses (see p. 355). Considered simply as containers, helical capsids are less efficient than icosahedral capsids. On the other hand being capable of exhibiting complete equivalence of subunits they are more readily assembled. Moreover, it is possible that the larger area of contact of the capsid with the environment, and the more intimate association of the capsid subunits with the nucleic acid, may, under certain circumstances, have a specific biological value.

Icosahedral symmetry

In early electron micrographs many viruses appeared to have a spherical shape. Further studies by X-ray diffraction and later by electron microscopy using the phosphotungstate negative-staining procedure showed, however, that some apparently spherical viruses were in fact icosahedral (adenoviruses and tipula iridescent virus). An icosahedron is a polyhedron with 12 vertices, 20 faces and 30 sides, providing axes of five-fold, three-fold and two-fold symmetry respectively. The implications of such a structure for viral capsids was examined by Caspar and Klug, who demonstrated that it was in fact the preferred arrangement for the construction of capsids consisting of a large number of asymmetric subunits, the subunits being arranged in such a way that they possessed a maximum degree of equivalence in their relationships with each other—an essential requirement for self-assembly. All the apparently spherical viruses, whether of plant or animal origin, appeared to have icosahedral symmetry, so it is apparent that icosahedral symmetry does fulfil some essential biological and structural requirements.

The nucleic acid comprises a densely packed mass at the centre of the particle (the core) and the protein constitutes a shell round it (the capsid is assembled from the capsomeres). A lipid-containing envelope may or may not be present outside the capsid (Fig. IV.5).

The geometry of icosahedral capsids is complex. The simplest capsid is one in which subunits are located at the 12 vertices of the icosahedron. Such a structure, with 12 subunits, is found in the bacteriophage ϕX174. Further structures can be built up from this simplest structure, based on families of regular solids known as icosadeltahedra, in which all the faces are congruent equilateral triangles. The simplest of these families is derived by repeated subdivision of the faces of the basic icosahedron, as shown in Fig. IV.6. However, other, more complex, families of icosadeltahedra also exist. Each family is characterized by class number P which is an integer such that $P = h^2 + hk + k^2$ (h and k being integers). Within each family, each individual

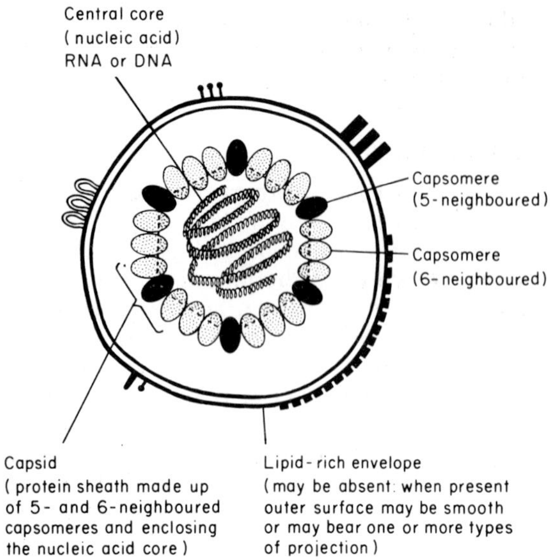

Fig. IV.5. The structure of a virus with cubic (icosahedral) symmetry. Note that the internal component is nucleic acid alone, unlike the internal component of the myxoviruses which is made up of nucleic acid and protein. See also Figs IV.4 and IV.28.

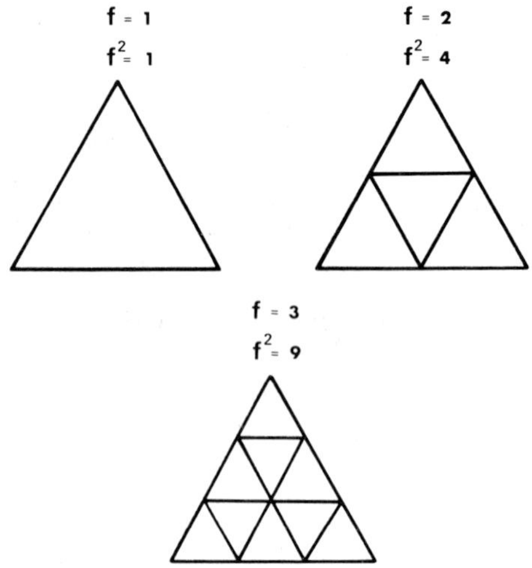

Fig. IV.6. Successive subdivisions of the faces of an icosadeltahedron.

member is characterized by a number, f, which represents the degree of subdivision of the basic equilateral triangles. For any given solid, the number of vertices is equal to $10Pf+2$. In general, the capsomeres visible under the electron microscope seem to be centred on the vertices of the icosadeltahedron on which the structure is based. In such cases, the number of capsomeres is also given by the above formula. On the other hand, it has been suggested that the capsomeres may be centred on the triangular faces of the icosadeltahedron, in which case the number of capsomeres is equal to $20Pf^2$. It has been proposed that the picornaviruses (Chapter IV.4) and the reoviruses (see p. 452) are of this type. In the case of turnip crinkle virus the capsomeres appear to be centred on the mid-points of the sides of contiguous triangular facets of the icosadeltahedron. If this is the case, the number of capsomeres will be equal

Table IV.1. The capsid structure of some viruses showing icosahedral symmetry

Name of virus	P (Class)	f	$T=Pf^2$	*No. of capsomeres* Pentons	Hexons	Total	*No. of fundamental structure units*
ϕX174	1	1	1	12	0	12	60
Parvoviruses*	3	1	3	12	20	32	180‡
Picornaviruses	3	1	3	All trimers		60	180
Papovaviruses	7	1	7	12	60	72	420
Reoviruses†	1	3	9	All trimers		180	540
Herpesviruses	1	4	16	12	150	162	960
Adenoviruses	1	5	25	12	240	252	1500§
Tipula iridescent virus	1	9	81	12	800	812	4860

* Some parvoviruses may have 12 capsomeres, thus resembling ϕX174 and in some the number is still unknown.
† The Reoviruses have a double capsid; the details given relate to the outer capsid only. The detailed structure of the inner capsid is unknown, but it appears to have cubic symmetry.
‡ This structure is effectively equivalent to a rhombic triacontahedron, a solid figure with 30 rhombic sides and 32 vertices; it has been so described.
§ The penton capsomeres are structurally and antigenically different from hexons (see p. 450).

to 30Pf². It is a feature of all icosadeltahedra that 12 of the vertices are connected by edges to five other vertices. If there are more than 12 vertices, the additional ones are connected to six others. In consequence of this, where the capsomeres are centred on the vertices, 12 of the capsomeres have five neighbours and the rest six neighbours. Where the capsomeres are centred on the triangular faces, they all have three neighbours.

X-ray diffraction studies, where feasible, have, in general, indicated that the number of subunits of viral capsides is 60 or some multiple of 60. However, as is clear from the examples given in Table IV.1, the numbers of capsomeres obey this rule only in the case of the Picornaviruses and the Reoviruses. This apparent paradox has been resolved by distinguishing between the morphological subunits, or capsomeres, which are visible under the electron microscope and the fundamental structure units which are responsible for the X-ray diffraction patterns. The latter comprise the building-blocks from which the capsomeres are constructed. A six-neighboured capsomere would be made up of six fundamental structure units and a five-neighboured capsomere of five. The two types of capsomere are consequently termed *hexons* and *pentons* respectively. By analogy, the three-neighboured capsomere is assumed to be a trimer made up of three of the fundamental structure units and the subunits of turnip crinkle virus to be dimers.

This rationalization fits the observed electron microscopic data well for the herpes virus, the vertical capsomeres of which are clearly pentagonal in outline and those on the sides and faces hexagonal in outline. If in such a case the structure units which make up the capsomeres are all identical, the six-component capsomeres should be the most stable arrangement and the five-component capsomeres should involve some distortion in the normal bonding arangements and therefore involve more free energy and possess less stability. They must nevertheless have sufficient stability to function as integral capsid components even if they cannot exist in the free state. It is possible that in the capsid they may be rendered more stable by association with some underlying component.

Table IV.1 gives details of the structure of the capsids of eight viruses or groups of viruses. Although a great many viruses have yet to be examined in detail, it already seems clear that the type of capsid structure just described is very widespread amongst viruses of animals, plants and bacteria. The current theories of capsid structure were originally put forward to explain how a closed shell could be assembled from identical subunits, all of which would, in the final structure, be as nearly as possible equivalent to one another in the geometrical sense. Nevertheless, it has now been shown that the pentons of at least one group, the adenoviruses (see Fig. IV.34), are structurally and antigenically distinct from the hexons. Whatever the significance of this finding, it does not solve the problem of explaining how it is that the capsids of any given virus are so completely uniform in size, shape and structure. The evidence all indicates that these properties are not determined by the size and shape of the nucleic acid core. In most virus preparations one can observe apparently empty particles consisting of capsids of exactly the same size and structure as those of the normal, filled particles. It is likely that the final structure of the capsid is determined by the inherent properties of the ultimate protein structure units.

Virus envelopes

As has been indicated some viruses, notably the myxoviruses and the herpes viruses, possess an outer layer or envelope surrounding the nucleocapsid. Both the herpes- and myxoviruses possess a high concentration of lipid, the lipid occurring mainly, if not exclusively, in the envelope. In the case of the influenza virus the lipid has been shown to resemble in composition that of the host cell in which the virus is grown and to be derived mainly from lipid present in the cell before infection. Influenza virus also contains polysaccharide which resembles that of the host tissue. It seems likely therefore that the envelope of the influenza virus is derived from the host cell membrane. However, in addition to host cell material the envelope also contains virus specific material, notably the neuraminidase and the haemagglutinin, which appear to be associated with the spikes seen projecting from the surface of the virus and which are identical with the surface strain-specific antigens involved in haemagglutination inhibition and infectivity neutralization tests (p. 373). Morphologic evidence on its acquisition indicates clearly that the envelope of the herpes virus is derived from the host cell nuclear membrane and this is supported by the finding that enveloped herpesviruses are agglutinated by anti-host cell serum, whereas non-enveloped viruses are not agglutinated. That the arboviruses also possess envelopes is suggested by electron microscopic data but the viruses are so small that resolution of their detailed structure has been extremely difficult (but see Chapter IV.5). The infectivity of all three groups is destroyed by treatment with ether and in the case of the herpes and myxoviruses, this has been shown to be associated with disintegration of the viral envelope—a finding clearly consistent with the prominent role of lipid in the composition of the envelope.

NOMENCLATURE AND CLASSIFICATION OF VIRUSES

Unfortunately, no entirely satisfactory or universally accepted system of classifying viruses has yet been devised. For a long time viruses were simply named after the disease which they caused and in many cases these casual names are still the most widely used. Well known examples are smallpox virus, foot-and-mouth-disease virus and tobacco mosaic virus. However, with the development of tissue culture techniques for virus isolation in the early 1950s and the subsequent developments in electron microscopy, viruses began to be discovered which had no demonstrable connection with any disease or which caused many different diseases, or which caused a syndrome which could also be produced by a lot of other viruses. This obviously raised considerable problems of nomenclature which were only partly solved by the introduction of such names as ECHO (enteric cytopathic human orphan) and Coxsackie (the name of the small town in up-state New York where the first examples were isolated).

However, the biggest problem arose not in the field of nomenclature but in attempts to arrange a classification of viruses. Originally it seemed logical to group viruses on the basis of the type of disease caused; for example neurotropic viruses, dermotropic viruses. As more and more came to be learnt about individual viruses themselves, this basis of classification became increasingly unsatisfactory.

At the Fifth International Congress for

Table IV.2. A classification of animal viruses on the basis of nucleic acid type and particle morphology

Nucleic acid	*Capsid symmetry*	*Envelope*	*Viral particle size* (nm)	*Acid lability*	*Virus group*
RNA	Cubic	Naked	18–38	Stable	Enteroviruses
				Labile	Rhinoviruses
			70–7	Stable	Reoviruses*
		Enveloped	30–75	Labile	Togaviruses
	Helical	Enveloped	90–300	Labile	Myxoviruses†
			37–218	Labile	'Myxovirus-like'‡ agents
DNA	Complex		200 × 325	Labile	Poxviruses
	Cubic	Enveloped	120–250	Labile	Herpesviruses
		Naked	70–80	Stable	Adenoviruses
			40–57	Stable	Papovaviruses
			18–26	Stable	Parvoviruses (Picodnaviruses)

Based on B. I. Wilner (1969) *Classification of Human and Animal Viruses*, 4th ed., p. 2. New York: Burgess.
* RNA is double-stranded.
† Includes Paramyxoviruses and Orthomyxoviruses.
‡ Comprises a variety of morphologically and biologically distinct groups including Coronaviruses, Rhabdoviruses, Arenaviruses, RNA tumour viruses and a number of poorly characterized etherdsensitive viruses believed to contain RNA and to possess a helical nucleocapsid surrounded by a lipid-containing envelope.

Microbiology in 1950 a serious effort was made to make the properties of the virus particles themselves the basis of classification. Host range and tissue affinity were still to be taken into account but were to be regarded as of less importance than chemical composition, particle morphology, susceptibility to physical and chemical agents and antigenic structure. Although a classification based on the properties of virus particles has proved to be an enormous advance on earlier systems and, in particular, has drawn attention to similarities between viruses of vertebrates, invertebrates, plants and bacteria, it has not proved entirely satisfactory. Table IV.2 is modified from one in a recent monograph on the classification of viruses of vertebrates. Almost all the well characterized viruses of vertebrates can be fitted into this scheme, although the group entitled 'myxovirus-like' includes a very heterogeneous collection of organisms, many of which are poorly characterized, but which appear to have a helical component enclosed in a lipid-rich envelope. It is by no means established that in all such cases the internal component is entirely comparable to that found in the myxoviruses. For various reasons this scheme has proved less suitable for plant and bacterial viruses, although there does seem general agreement that some sort of classification based on the characteristics of the particles is likely to be the most satisfactory at this stage. It has recently been suggested that, ultimately, a system of classification based on the base sequence of the viral nucleic acid may become feasible. One of the most hotly contested issues has been whether viruses should be given latinized binomial names like plants and animals. The difficulties are that present systems of classification can make no pretence to a phylogenetic basis and that the usual criteria for determining what constitutes a species or what constitutes a genus simply cannot be applied to viruses. Nevertheless, various schemes have been put forward none of which has been widely used. The Provisional Committee for the Nomenclature of Viruses has proposed a system which at least has official status. This has recently been further developed.*

A different way of classifying viruses, based on the type of nucleic acid and the mechanism of mRNA synthesis, has been proposed by Baltimore. It is dealt with in some detail in the next chapter (see p. 352).

*See F. Fenner (1976) *J. gen. Virol.*, **31**, 463.

RESISTANCE

Heat is the most reliable method of virus disinfection. The human pathogenic viruses are, with some exceptions, inactivated by exposure to 60°C for 30 minutes. As in the case of bacteria the inactivating effect of heat depends on the environmental conditions, the presence of organic material affording the virus considerable protection. Viruses, like bacteria, are much less susceptible to dry than to moist heat. Viruses are inactivated by exposure to ultra-violet irradiation and this method is sometimes used in the preparation of viral vaccines. The most active wave-lengths appear to be in the region of 260 nm. Ionizing radiations on the other hand appear to be much less active than against bacteria, a reflection of the small target-size of viruses.

As a general rule viruses are more resistant than bacteria to chemical disinfection, presumably because of their deficiency in metabolic enzymes. Of the disinfectants commonly used against bacteria, phenol, the lower alcohols and the quaternary ammonium compounds have very low virucidal activity. On the whole the most generally active agents are chlorine, the hypochlorites, iodine and formaldehyde. Other compounds with appreciable activity against one or more viruses are mercuric chloride, silver nitrate, ethylene oxide and the nitrogen mustards. Ether and sodium desoxycholate have a highly selective action. They have a marked activity against viruses with a high content of lipid, e.g. the arbo- and influenza viruses. The selective action of these compounds may be of value in virus identification.

INFECTIOUS NUCLEIC ACID

As we have seen, viruses are completely devoid of metabolic enzymes and are therefore dependent on the metabolic machinery of the host cell. They must, however, carry all the genetic information required for the synthesis of viral material and this information is contained in the nucleotide sequence of the viral nucleic acid. The nucleic acid can therefore be regarded as the essential virus, the function of the capsid and the envelope being primarily, and in some cases exclusively, that of protecting the nucleic acid. The first clear indication of the essential role of the viral nucleic acid was in the demonstration by Hershey and Chase that, on infection of *E. coli* with T2 bacteriophage, the phage protein coat remained attached to the outside of the cell, only the nucleic acid gaining access to the interior. This finding was soon followed by the extraction—by Girer, Schramm and Fränkel–Conrat—of an infectious protein-free nucleic acid from tobacco mosaic virus and, later, by the extraction of infectious nucleic acid from a number of animal viruses. RNA viruses have been particularly potent in yielding infectious nucleic acids. These have now been obtained from a considerable number of the picorna and arbo groups. Infectious DNA has been obtained from the papovaviruses.

So far, infectious nucleic acid has not been obtained from the larger DNA or RNA viruses. The failure to extract infectious nucleic acid from these larger viruses might be due to the fact that their larger nucleic acid threads may be broken down on extraction, but in many cases is due to the fact that enzymes, particularly RNA-dependent RNA polymerase, are necessary for the initiation of infection. These enzymes are present in the complete virus particle, but absent from purified nucleic acid and from host-cells (see Chapter IV.2).

The methods most used for nucleic acid extraction have involved as a first stage the rupture of the capsid by treatment with phenol or with a combination of phenol and a detergent, e.g. sodium dodecylsulphate. The nucleic acid is then obtained as the sodium salt, by addition of sodium acetate, and finally precipitated by ethanol. During extraction the nucleic acid must be protected from the action of contaminating nucleases. This can be achieved by the addition of bentonite, which adsorbs these enzymes, and of the chelating agent Versene, which removes ions necessary for their activity. The recovery of nucleic acid with these procedures is high but the infectivity of the extracted nucleic acid is low, usually less than 1% of that of the original virus, even under optimal conditions. This low infectivity seems to be due to a low intrinsic capacity of the naked nucleic acid to penetrate susceptible cells. Infectivity of such preparations must be due to nucleic acid since this is lost after treatment of the preparation with the appropriate nuclease and it is insusceptible to antiviral antibody.

Although it possesses a low infectivity, infectious nucleic acid usually exhibits a wider host range than does the intact virus. Thus poliovirus RNA will infect a variety of cells which are not infectible by intact poliovirus. This is presumably because, for the cells to be infected with intact virus, they must possess the appropriate receptor capable of binding to the viral capsid. This limitation on infectivity is removed when the capsid-free nucleic acid is used. With the latter, however, only one cycle of virus replication is possible since the product of growth is a complete virus to which the cells are, of course, not susceptible.

IV.2

Virus Replication

With certain important qualifications, one can say that replication appears to proceed along similar lines in animal, bacterial and plant viruses. Until recently direct evidence on the replication of animal viruses was scanty. However, this deficiency is now being remedied and we are no longer as dependent as formerly on the analogy of bacterial viruses when trying to explain the replication of animal viruses. Nevertheless, much remains unexplained and, as with the bacterial viruses, what we do know derives from the intensive study of only a few of the known viruses. For descriptive purposes it is convenient to divide the process of replication into discrete stages and to treat each one separately. These stages are:

1. Attachment.
2. Penetration.
3. Uncoating.
4. Redirection of cellular metabolism (retooling).
5. Synthesis of viral components.
6. Assembly.
7. Release.

ATTACHMENT

This appears to be an almost essential preliminary to infection and, perhaps surprisingly, seems to constitute the most specific step in virus replication. Study of bacterial viruses indicates that for them attachment is a two-stage process. An initial reversible stage, which takes place irrespective of the susceptibility of the bacteria to the particular virus, is probably due to electrostatic forces. If the bacteria are in fact susceptible to the virus, this stage is followed by a second irreversible attachment due to highly specific binding of chemical groups on the surface of the virus particle or its tail-fibres to complementary receptor groups in the cell wall. Removal of the cell wall gives protoplasts or sphaeroplasts which are resistant to infection by viruses to which the intact cells are fully susceptible. The receptors have been lost and the virus particles cannot form the attachment which is the essential prerequisite of infection. That different viruses make use of different receptors is clear from the ease with which mutant bacteria can be obtained which have lost their susceptibility to only one of a large number of different viruses to which the wild-type cell is susceptible. Animal cells, of course, have no cell wall, but it seems that the process of attachment is essentially similar, except that the specific stage of attachment may perhaps be more complex. The specific receptors are in the cytoplasmic membrane and a cell which lacks the receptors for a particular virus is necessarily resistant to that virus, although it may be susceptible to the nucleic acid from it.

In many cases viruses are capable of binding to the surface of the red cells of one or more species. A virus particle lying between two red cells can bind to both, so that high concentrations of virus can act as a kind of cement binding the red cells together in clumps. This is the phenomenon of haemagglutination dealt with more fully on p. 369. There is evidence that, at least in some cases, the receptors on the red cells are very similar, if not identical, to those on the surface of susceptible cells. Haemagglutination has been most intensively studied in the myxoviruses, especially influenza viruses. It now appears that this group is atypical in possessing an enzyme, neuraminidase, which specifically destroys the receptors. In most cases the virus and red cells can be disassociated by such means as alteration of the pH or ionic constitution of the medium, leaving the receptors intact. Although haemagglutinating viruses attach specifically to red cells, these cells do not support virus replication. On the other hand, all animal viruses seem to form specific attachments to susceptible cells, irrespective of whether they also haemagglutinate.

When cells are taken from the animal and propagated in tissue culture, they may gain or lose receptors for particular viruses. The mechanism is unknown, but the phenomenon may reflect the adaptation of the cell membrane to the abnormal conditions of tissue culture.

PENETRATION

The penetration of bacteria by those bacterial viruses with contractile tails is described on p. 464. These, as well as most other bacterial viruses do not enter the host cell intact. Only the nucleic acid and a very small porportion of the protein actually get inside the cell wall, so that in this group the stages of penetration and of uncoating occur as one. Most animal viruses, on the other hand, enter the cell intact and the protein coat is removed after penetration. There is some evidence that the lipid-rich envelope, when present, may initiate penetration by fusing with the chemically similar cytoplasmic membrane. The mechanism whereby the virus particle actually enters the cell seems identical with the process known as pinocytosis or 'cell-drinking' (cf. phagocytosis = 'cell-eating'). This is a process by which cells engulf droplets of the fluid bathing them which are then carried into the cell and migrate towards the nucleus to the region of the Golgi apparatus. Any particle attached to the cytoplasmic membrane at the site of formation of a pinocytic vesicle will inevitably be carried into the cell inside the vesicle. It has been suggested that virus particles attached to a cell are carried inside quite accidentally because pinocytosis happens to occur just where the virus is situated. If this were so, it is difficult to see why *specific* attachment should be quite so crucial. It seems more likely that the attachment itself must stimulate pinocytosis locally. At least it is known that proteins stimulate pinocytosis whereas nucleic acids have no such effect. Exactly how the virus particle is liberated from the vesicle is not clear.

It will be obvious from the foregoing that the energy for penetration must come from the cell. This is confirmed by the observation that animal viruses cannot penetrate dead cells. The tailed bacteriophages, on the other hand, provide their own energy from molecules of ATP in the contractile tail sheath and will even penetrate isolated bits of bacterial cell wall.

UNCOATING

The protein surrounding the nucleic acid of the virus serves to protect it from attack by nucleases but it also effectively prevents the interaction of the nucleic acid with the metabolic machinery of the host cell. Before such interaction can begin the protein has to be removed. This being so, the uncoating enzymes must be coded for by the cell and not the virus. The process of uncoating has been most carefully studied in vaccinia virus (the virus used for smallpox vaccination). It seems that this organism which is structurally more complex than smaller viruses, is uncoated in two stages. The first stage is effected by a host enzyme which exposes the DNA sufficiently for some transcription to occur. A messenger RNA is produced which interacts with the ribosomes of the host to produce a second, virus-coded enzyme which completes the uncoating process. There is also evidence that adenoviruses may begin transcription before uncoating is complete. However, most viruses seem to rely exclusively on host enzymes for uncoating. The pox viruses such as vaccinia replicate entirely in the cytoplasm, so they must also provide the transcriptase necessary for the production of messenger RNA since the host enzymes for this function are restricted to the nucleus and the mitochondria. It has recently been shown that vaccinia virions contain a DNA-dependent RNA polymerase (transcriptase) which is capable of transcribing viral DNA while it is still complexed with protein. The uncoating of other viruses has not been studied in such detail as that of vaccinia, but the known association of cytoplasmic vesicles with lysosomes which contain active proteases and other lytic enzymes suggests a ready source of uncoating enzymes. It is perhaps noteworthy that poliovirus, which is normally resistant to trypsin, is sensitized to it when the virus interacts with the cell membrane.

REDIRECTION OF CELLULAR METABOLISM

Because of its own total lack of metabolic machinery, a virus must make use of the machinery of the host to effect its replication. However, the metabolism of the cell is organized and regulated to meet the requirements of the cell and not to make virus particles. It must therefore be adapted before it can fulfil the needs of the virus. This adaptation can be likened to the 'retooling' which is necessary when a factory production-line is turned over to making a new product.

Early students of the replication of the T-series of bacterial viruses* found that shortly following infection, it was impossible to detect any infectious virus if the cells were broken open. They called this the 'eclipse' or 'dark phase' (see Fig. IV.12). It ended about half way through the replicative cycle (latent period). Then new virus particles began to accumulate within the bacterial cell prior to release. It was shown that, during the second half of the eclipse phase, viral components were produced and it was correctly assumed that there was a delay before they were assembled to form infectious virus. However, the significance of the first half of the eclipse phase long remained obscure. It was known that protein synthesis occurred, that the proteins formed were quite unlike anything found in the uninfected cell and that, with one exception, they were not incorporated into the viral particles. The exception is the so-called internal protein which accompanies the DNA in the head of the phage particle. Hitherto thought of as something of an unnecessary by-product of virus replication, the true significance of these 'early proteins' became clear when Flaks and Cohen, Kornberg and his collaborators and others showed that amongst them were newly synthesized, phage-coded enzymes needed for replication of the viral DNA. The DNA of the T-even phages (T2, T4, T6) has unusual features. The cytosine is replaced by 5-hydroxymethyl cytosine (HMC) and glucose and gentiobiose molecules are attached to a proportion of the HMC. Obviously special enzymes must be required to synthesize this atypical DNA. In fact, well over a dozen enzymes have been recognized amongst the early protein of T4 bacteriophage. Although some of the new enzymes appear to have the same specificities as enzymes which are already present in the host and which, in vitro, can participate in the replication of viral DNA; it seems that, in vivo, these viral enzymes are essential for replication. Gel electrophoresis has revealed other, so far uncharacterized, early proteins. Some of these are undoubtedly enzymes, but others appear to be regulatory or inhibitor proteins acting on bacterial or viral genes or on their products. It is obvious that, so far from being useless by-products, the production of the early proteins represents the first essential step in virus replication; the 'retooling' of the cell to enable the formation of new viral components to begin. It has recently become clear that some mechanism operates to ensure the orderly expression of viral genes. Not only does the synthesis of viral components not begin during the phase of early protein synthesis, but as soon as the production of viral components does begin early protein synthesis ceases abruptly. The switch from early to late functions seems to be somehow related to the start of DNA synthesis. Many mutations which block DNA synthesis also allow early protein synthesis to proceed well beyond the normal time. Even the early functions may not all be expressed simultaneously. With some of the smaller viruses there does not appear to be an 'early' and a 'late' phase of gene expression, all the genes are transcribed simultaneously. Nevertheless, even in the case of the small RNA bacteriophages such as MS2 whose genome comprises no more than three cistrons, there is a mechanism to ensure that the three gene products are produced in different quantities and at different rates.

The situation so far as animal viruses are concerned is far less clear. Early proteins are undoubtedly formed but are only now beginning to be identified. One of the most constant effects of virus infection in animal cells is the abrupt switching off of host macromolecular synthesis, involving RNA, DNA and protein. The evidence suggests that this switching off is mediated by protein inhibitors, but it is not clear whether these are coded for by the virus or are normal regulatory proteins of the cell, production of

* A series of seven *E. coli* bacteriophages which have probably been more intensively studied than any other bacterial viruses with the possible exception of 'lambda', another *E. coli* phage. The T-even phages, T2, T4 and T6, are closely related to one another; the T-odd phages T1, T3, T5 and T7 are a heterogeneous group. The T-phages are virulent (see p. 46).

which is stimulated by the virus. Several new enzymes coded for by the virus have been described, including the uncoating enzyme and replicase of vaccinia already mentioned. In cells there is no known mechanism for the formation of RNA from a pre-existing template of RNA. All RNA is stranscribed from DNA. Obviously RNA viruses have somehow got to make copies of their RNA and will have to generate the machinery for doing this. The most obvious mechanism would be for the single-stranded RNA to act as a template for the synthesis of a complementary strand of RNA; then for the two strands to separate and each act as a template for further RNA synthesis. In this case the process would mirror the normal mechanism of DNA synthesis. Whatever other enzymes such a mechanism called for, an RNA-dependent RNA polymerase would clearly be needed. Such an enzyme was detected and characterized by Spiegelman and his colleagues in *E. coli* infected with a small RNA phage. The enzyme, which they called a replicase, proved to be highly specific for the RNA of its own virus and with it they were able to set up an in vitro system in which a small amount of phage RNA added at the beginning of the experiment served as a template for a large amount of new RNA which proved to be infectious. Earlier, Baltimore and Franklin had found evidence of an RNA-polymerase bound to an RNA template in mouse fibroblasts infected with Mengo virus and, afterwards, in HeLa cells infected with poliovirus. The search for the expected double-stranded 'replicative form' (RF) of viral RNA has been successful in cells infected with animal viruses, plant viruses and bacteriophages. However, one puzzling aspect of the replication of certain RNA viruses has been its sensitivity to the antibiotic actinomycin D. This group includes influenza virus and certain oncogenic RNA viruses. Actinomycin D blocks transcription from DNA to RNA and should not, therefore, interfere with the formation of new viral RNA by the mechanism just described. The discovery in particles of Rous sarcoma virus and animal leukaemia viruses of an entirely new enzyme, RNA-dependent DNA polymerase, and an earlier observation that double-stranded viral RNA may be absent from cells infected with oncogenic RNA viruses would point to a second method of replication for RNA viruses involving the production of DNA from an RNA template—a complete reversal of the usual process. Once formed, the new DNA could generate viral RNA by much the same mechanism as that by which normal host cell RNA is made. This step would, of course, be sensitive to actinomycin D.

SYNTHESIS OF VIRAL COMPONENTS

The fundamental step in virus replication is the production of new molecules of nucleic acid which are exact replicas of the nucleic acid in the infecting particle. In the case of certain defective viruses which have lost the capacity to form coat protein and have to 'borrow' the coat of another virus which is replicating in the same cell, this is effectively the final step. However, complete virus particles comprise protein besides nucleic acid, and this has to be synthesized as well. The envelope, when present, appears always to be derived from some pre-existing host cell membrane. But the envelope and the membrane are not identical; for example the envelope of the influenza virus carries the haemagglutinin and the neuraminidase neither of which is present on the normal host cell. Study of influenza viruses indicates that, in this case at least, modification of the cytoplasmic membrane takes place before it enwraps the nascent virus particle. This modification involves the production and incorporation of new molecular species, virus encoded and probably, for the most part, proteins.

In considering the synthesis of viral components it is convenient to consider separately double-stranded DNA viruses, single-stranded DNA viruses, double-stranded RNA viruses and single-stranded RNA viruses. Amongst the latter we can recognize an atypical group which may differ from the generality of single-stranded RNA viruses in replicating by way of a DNA intermediate.

Baltimore has proposed a classification of viruses based on their nucleic acid and the way in which viral protein is synthesized. His scheme is summarized in Fig. IV.7. He identifies five classes as follows:

Class I Double-stranded DNA; mRNA is transcribed from this in much the same way as host mRNA is transcribed from host DNA.

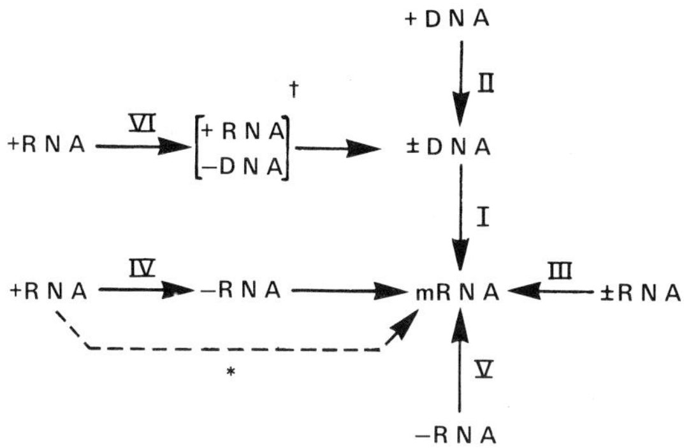

Fig. IV.7. Baltimore's classification of viruses, based on the nucleic acid type and the mode of formation of mRNA.

The first step (†) is the formation of an RNA–DNA hybrid molecule. It appears that the input RNA, although plus-stranded, cannot function as mRNA.* On first entering the cell * the input RNA acts as its own messenger. Compare with Classes V and VI.

Class II Single-stranded DNA; a double-stranded DNA is first synthesized and this then generates mRNA as in Class I.

Class III Double-stranded RNA; this cannot function as a messenger, therefore single-stranded transcripts are made and these act as mRNA. The genome is segmented.

Class IV Single-stranded RNA of the same sense (plus) as mRNA; this can act as its own messenger and initiate protein synthesis (e.g. of RNA polymerase). Replication of RNA will involve the synthesis of complementary minus strands from which in turn more plus strands can be transcribed to provide the nucleic acid for the progeny particles. These plus strands could function as messengers, but in some cases such as Semliki Forest virus (an arbovirus) smaller RNA molecules are formed corresponding to only part of the genome. These are presumably messengers.

Class V Single-stranded RNA of the opposite sense (minus) to mRNA; this cannot act as a messenger itself and it must first be transcribed to give one or more plus strands. To be infective, the virus particle must contain at least one molecule of RNA-dependent RNA-polymerase. Some viruses of this class have segmented genomes.

Class VI Single-stranded RNA usually of the same sense (plus) as mRNA, which replicate by way of a DNA intermediate; the fact that the particles contain polymerase (reverse transcriptase) suggests that mRNA is only formed after the initial transcription of the input RNA on to double-stranded DNA. Genome apparently segmented or duplicated.

Double-stranded DNA viruses: Baltimore Class I

In this group the mechanism of replication does not differ essentially from that of the host cells. The key enzyme in the formation of new DNA is a DNA-dependent DNA polymerase (duplicase) which synthesizes new strands complementary to each of the parent strands so that two molecules of DNA are generated, each comprising one of the original strands and one newly synthesized strand. These new molecules then interact with the enzyme to form further double-stranded molecules which themselves then act as templates. The process continues so long as viral DNA continues to be formed. This method of replicating the DNA is called 'semiconservative' and it will be noted that that the two parental strands are conserved and should appear in the progeny but in different particles. If a 'mistake' is made in the copying process, this constitutes a mutation. If it occurs at an early step in the replication, many of the progeny particles will show the mutation; if it occurs late only one or a few particles will carry the mutation. In a classic experiment, Luria was able to show that the mechanism of DNA replication was as described above by noting the distribution of different numbers of mutant particles amongst the progeny particles derived from single bacterial cells infected with bacteriophage T2.

As indicated above, the parental DNA strands should appear in two of the progeny particles from a given infected host cell. This is by no means always true and portions of the input strands are distributed amongst more than two of the progeny particles. This is only explicable if it be postulated that the molecules of DNA, after their formation but while they are as yet uncoated, are able to undergo some process of breakage and

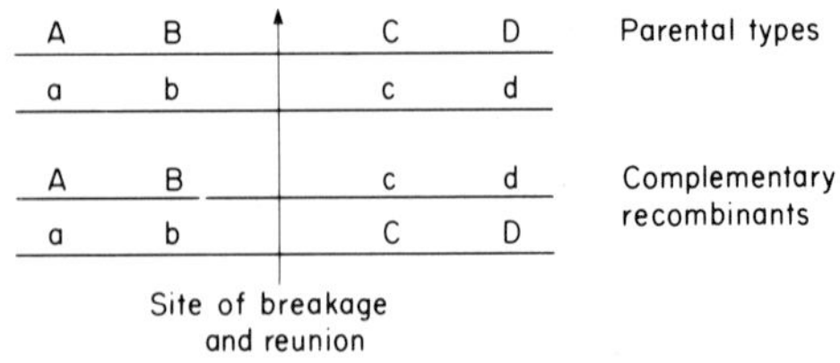

Fig. IV.8.

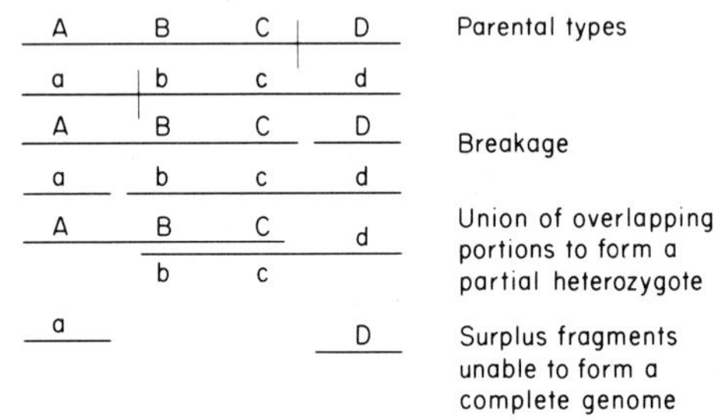

Fig. IV.9.

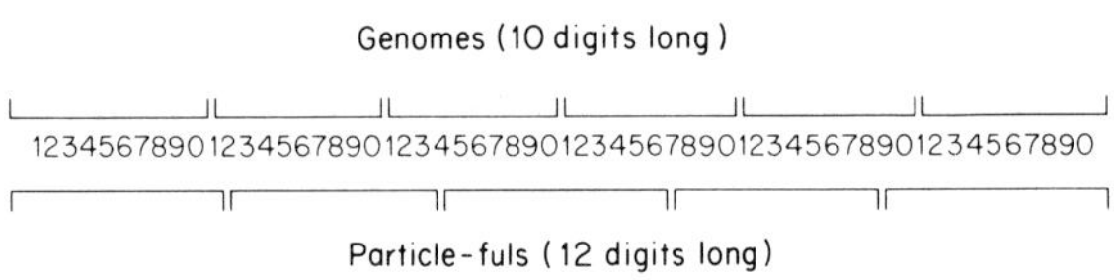

Fig. IV.10.

reunion whereby a broken fragment of one molecule can unite with part of another molecule to form a complete genome. If the two molecules involved are identical this phenomenon will have no genetic implications, but if the two molecules differed at several loci there would obviously exist the possibility of recombination. Many DNA viruses form recombinants quite readily and it seems likely that the breakage and reunion of DNA molecules represents the physical basis for this. Although recombination in viruses seems to fulfil the same genetic function as in higher organisms, the mechanism appears to be different. If the recombinants were produced by a simple process of crossing over, one would expect the formation of complementary recombinants (Fig. IV.8). In practice, these do not occur and it seems probable that the mechanism involves the formation of partial heterozygotes. These are virus genomes which are diploid at a few loci and haploid at the rest. Partial heterozygotes form only a small proportion of the DNA virus and they are unstable. The mechanism suggested for their formation is shown in Fig. IV.9. It is further suggested that, unless 'frozen' in the partially heterozygous state by being incorporated into a completed virus particle, these structures do not persist but replicate to give two normal haploid double-stranded molecules. The full details of the mechanism of recombination have yet to be elucidated, but there is general agreement that these so-called internal, heteroduplex or recombinational heterozygotes form the basis of the process. It has been established that at least two steps are involved. One requires protein synthesis but not DNA synthesis and involves hydrogen bonding of DNA strands; the other involves the formation of phosphodiester bonds and appears to utilize the repair enzyme which is responsible for filling in gaps resulting from the excision of thymine dimers.

Like those of bacteria, the genetic maps of double-stranded DNA viruses appear generally to be circular. However, this does not necessarily imply circularity of the DNA molecule itself. While in certain cases the DNA molecule is in fact circular, at least at certain stages of replication, other viruses owe the circularity of their genetic maps to so-called circular permutation. It seems that while all the particles in a particular virus population may possess the same genes, in the same order; the starting point is randomly determined. The sequences *abcd, bcda, cdab* and *dabc* all contain the same sequence of letters, it is merely the starting point which differs. These arrangements of letters show circular permutation: note that whereas the letters *a* and *d* are as far apart as they can be in the first arrangement, they are adjacent in the others. Some DNA viruses exhibit what is known as terminal redundancy, that is duplication of a few of the genes at the ends of the sequence. It is claimed that this implies that the DNA is formed as a long continuous strand comprising many genomes and that this is then cut up into lengths which may be thought of as 'particle-fuls' each slightly longer than one genome. Such a process, if it occurred, would obviously produce both terminal redundancy and circular permutation (Fig. IV.10).

Long strands of DNA known as 'concatenates' have been demonstrated in extracts of bacteria infected by certain bacteriophages, but it is by no means certain that these are not artefacts of the extraction process. The DNA molecules of at least one bacterial virus which is neither circularly permuted nor terminally redundant in the ordinary sense, have short single-stranded extremities which are complementary. By base pairing of these regions a circular molecule can be formed either in vivo or in vitro. It is also possible for molecules to link with one another to form long chains. It is not clear whether this concatenate formation represents a reversal of the mechanism whereby the individual DNA molecules were formed in the first place.

The formation of viral proteins seems to take place by essentially the same mechanism as the formation of host cell proteins. Messenger RNA is formed by transcription from the viral DNA and the messengers associate with preexisting ribosomes to form polypeptide chains. In at least one bacterial virus, lambda, it is not meaningful to speak of sense and nonsense strands of the DNA. In one part of the genome one strand acts as the sense strand and is transcribed and in the rest, it is the other strand which so functions. The switch of strands seems to occur at the point where transcription is initiated, so transcription can proceed from this point in either or both directions. Although similar to the mechanism of host protein synthesis, the mechanism of viral protein synthesis is not identical with it. Although the earliest viral functions may rely on host transcriptase for their expression, later functions probably require the production of a virus-coded transcriptase or a protein modifying the specificity of the existing enzyme. There is also a modification of the properties of the ribosomes which become less able to translate host mRNA and yet can still translate virus messengers. Many viruses induce the formation of new species of transfer RNA and the corresponding aminoacyl transferases to go with them. These new species of tRNA seem to duplicate the function of transfer RNAs already in the host cell and there is no obvious need for them. The number of viral proteins which can be made obviously depends on the amount of information which can be included in the viral genome. This in turn depends on the length of DNA in each particle which will be greater in larger viruses. This implies that the smaller a virus particle is, the more completely it must depend on unmodified host cell mechanisms for its replication.

Single-stranded DNA viruses: Baltimore Class II

This type of nucleic acid is found in at least two groups of bacterial viruses and one group of animal viruses. Most of our knowledge of their replication comes from the study of two of the bacterial viruses, the *E. coli* phages ϕX174 and S13. These are small (20 nm), more or less spherical viruses consisting of a core of single-stranded DNA enclosed in a capsid of 12 five-neighboured capsomeres, the lowest member of the icosahedral series. The animal viruses with single-stranded DNA have a similar structure and are included in the parvovirus group. The second group of bacterial viruses with single-stranded DNA are morphologically quite different. The virus particle is a long filament consisting of two strands lying close together and joined at the ends of the filament: a structure which has been likened to a stretched elastic band. The DNA appears to be enclosed in a sheath of protein and in the intact particle forms a closed loop. The best studied members of this group are able only to infect male (F+) strains of *E. coli* because infection involves attachment of the tip of the phage particle to the tip of the F-pilus (see p. 22). The viral DNA and protein both enter the cell, the latter ending up in the cell membrane. It has been suggested that the pilus with the attached phage is withdrawn into the cell. These phages are also unusual in their mode of release. There is no evidence of the formation of a lytic enzyme and the newly formed virus particles simply leak from the bacteria as the latter continue to multiply, albeit at a reduced rate and with the loss of the F+ character.

Investigation of the mechanism of DNA synthesis for these viruses reveals two interesting differences from what has been described for double-stranded DNA viruses. The DNA is only infective if in the form of a closed ring, and the parental DNA does not appear amongst the progeny. Immediately after entering the host cell the single stranded molecule is converted into a base-paired, double-stranded ring, by the synthesis of a complementary strand. The enzymes involved are of host origin and pre-exist in the cell as blockage of protein synthesis does not prevent the formation of this double-stranded ring, the so-called 'replicative form' (RF). The RF can apparently exist in the cell in two forms; with both rings covalently closed, or with one ring closed and one ring opened. These two forms appear to be interconvertible but the equilibrium is normally well over towards the doubly closed form. After its formation if the RF is to initiate further replication it apparently has to become bound to some structure within the bacterial cell, localized within the 'membrane fraction'. The number of such sites in the cell depends on its physiological state and in previously starved bacteria is often only one. Also required for replication is virus-coded protein. Once this has been formed and the RF is attached to the specific site in the bacterial cell, replication proper begins. The original input strand remains bound to the 'site' and the newly synthesized complementary strand separates from it. A new strand identical to that in the original virus is now formed to produce another RF. If an unoccupied 'site' is available this may bind to it and itself become the source of further RFs. If not, one of two things may happen. It may serve as a template for viral mRNA synthesis; or it may begin to generate single, plus strands of DNA for incorporation into the developing progeny virus particles. This process of asymmetric generation of strands begins after about 12 minutes when there are about 30 RFs per cell and it appears to depend on the production of one or more of the capsid proteins which has a second function, that of blocking the formation of minus strands while allowing the plus strands to be formed at an increased rate. Host DNA synthesis is not stopped immediately after infection but ceases about the same time as the asymmetric production of plus strands begins. Replication of the circular DNA is by the 'rolling circle' method.

Obviously, the process of replication cannot be exactly the same as has just been described, for all the single-stranded DNA viruses. The filamentous phages do not stop cell multiplication, so they cannot stop host DNA synthesis entirely. The enormous differences between a bacterium and a mammalian cell must imply differences between the mechanism of replication of ϕX174 and animal viruses with single-stranded DNA such as the 'minute virus of mice' (MVM).

Single-stranded RNA viruses: Baltimore Classes IV, V and VI

Unlike DNA, the RNA of bacteria and of higher organisms is single-stranded and does not replicate itself. The RNA molecules are assembled from information transcribed from DNA. Elucidation of the mode of replication of those viruses which contain single-stranded RNA as their genetic store, a group which includes all but a few of the known plant viruses and well over half of the known animal viruses, had to await the discovery of RNA bacteriophages. Those so far discovered resemble the filamentous *E. coli* phages described above in being restricted to sex-pilus bearing strains. In their morphology, however, they more closely resemble the picornaviruses, having cubic symmetry and a capsid comprising 12 five neighboured capsomeres and 20 six-neighboured capsomeres. Their specific attachment sites appear to be on the sides rather than the tip of the sex-pilus.

One important piece of information was available before the discovery of the RNA bacteriophages: this was that the RNA carries all the information necessary for the replication of the virus including the formation of all the viral proteins. In 1956 Frankael-Conrat, and Gierer and Schramm independently demonstrated the infectivity of the purified RNA of tobacco mosaic virus. Subsequently, infectious nucleic acid has been obtained from a considerable number of plant and animal viruses; mainly but not exclusively, those containing RNA. The discovery in bacteria infected with RNA phages, of a RNA-dependent RNA polymerase and of double-stranded RNA has already been referred to (p. 352). Many details of the mechanism of RNA replication are still far from clear. It is now generally accepted that base pairing is involved and that some kind of double-stranded intermediate is involved. Unlike the case of the single-stranded DNA viruses there appears to be no requirement for circularity of the molecule, although in some cases a small proportion of circular RNA molecules can be isolated from infected cells. Parental nucleic acid strands have never been found in progeny particles. It seems likely that replication involves three essential steps which may in fact occur more or less synchronously. Firstly, after uncoating, the viral RNA acts as a messenger and interacts with ribosomes to generate some molecules of the RNA-dependent RNA polymerase. These enzymes appear to have almost complete specificity for their own RNA to which they bind strongly. They are apparently quite unable to initiate the replication of any of the wide variety of host RNA molecules in the cell. Once formed, a polymerase molecule binds to one end of the viral RNA molecule and proceeds along it synthesizing a complementary, 'minus' strand. Soon afterwards a second polymerase molecule attaches itself to the binding site on the 'plus' strand and then moves along it, simultaneously synthesizing a second 'minus' strand and displacing the first one. In this way a structure is produced comprising a double-stranded RNA molecule with one or more single-stranded tails consisting of partially displaced 'minus' strands and complexed with protein (the polymerase molecules). Finally, the newly formed 'minus' strands bind further molecules of the polymerase and new 'plus' strands are synthesized. These are almost immediately incorporated in new virus particles and the 'minus' strands remain to act as templates for the synthesis of more 'plus' strands. At some stage the rest of the viral proteins have to be synthesized. It is not clear how this is accomplished; although it has been shown that the RNA of poliovirus Type 1 (Mol. Wt. approx. 2.6×10^6 daltons) interacts with ribosomes as a polycistronic messenger and produces a giant protein which is subsequently cleft to form the four structural proteins and three others. Of these, one is the RNA-dependent RNA polymerase, one inhibits formation of host mRNA and one blocks its translation. It is difficult to imagine that the ribosomes can accept double-stranded forms of RNA and unlikely that the newly formed 'plus' strands have time to act as messengers before they are incorporated in virus particles. However, it has been established that the RNA of a para-influenza virus when complexed with the internal protein to form the nucleocapsid can still interact with ribosomes to incorporate amino acids, and indeed does so at a rate about 1·5 times that of the uncomplexed nucleic acid. The great majority of newly formed viral RNA in infected cells is indistinguishable in molecular weight and base composition from that found in the virus particles. However, in some systems fractions are consistently found which are either larger or smaller than the particle RNA.

While the mechanism described above accounts adequately for the observed features of the replication of the majority of single-stranded RNA viruses, it does not do so for all. The para-myxoviruses and the rhabdoviruses, and possibly others, contain single-stranded RNA which cannot be read by the host ribosomes because they are 'nonsense' or 'minus' strands. Before any translation can occur, the viral strand has to be transcribed on to a complementary 'plus' strand which can act as a messenger and initiate protein synthesis. It has been observed that in cells in which Newcastle disease virus (a paramyxovirus) is replicating, several different RNA molecules smaller than the complete particle RNA are found. Most at least of these are single-stranded and complementary to some part of the viral genome. They are presumed to be messenger RNAs.

Because the input RNA is a 'nonsense' strand, it cannot directly initiate the formation of any protein. However, its complementary strand can only be formed by the action of the viral enzyme, RNA-dependent RNA polymerase (replicase). The virus escapes from its dilemma by bringing into the cell with it one or more molecules of the enzyme, possibly already complexed with the nucleic acid. This can perform the essential first step of replicating the input RNA by way of a double-stranded intermediate. It is obvious why purified RNA from such viruses is not infectious; the process of purification removes the vital enzyme from the RNA.

Unlike the paramyxoviruses, the true myxo-viruses, human and animal strains of influenza, perform most of their replication in the nucleus, only entering the cytoplasm to reach the cell surface and pass to the exterior, acquiring the envelope as they do so. It has now been established that the RNA of influenza viruses is in seven separate and distinct pieces and it is suggested that the readiness with which these viruses form recombinants, which is in sharp contrast to most other RNA viruses, is due to the fact that these pieces can be exchanged bodily, almost like pairs of chromosomes segregating at meiosis. The actual mechanism of RNA replication remains somewhat obscure as does the mechanism of viral protein synthesis. The inner helical component (nucleocapsid) appears to be formed in the nucleus along with the RNA: the envelope proteins are not detectable except at or near the cytoplasmic membrane. Of course, the site of first appearance of detectable quantities of a protein is not necessarily its site of formation: if newly formed protein were removed as fast as it was formed it would have no opportunity to accumulate. Recent evidence indicates that the protein of the nucleocapsid is made in the cytoplasm and very rapidly transferred to the nucleus. Replication of influenza viruses involves an early step which is sensitive to actinomycin D. However, once production of new viral RNA is under way, the antibiotic has no effect. The concentration of actinomycin necessary to block viral replication is somewhat larger than that needed to inhibit the production of cellular RNA, and the mechanism of its action is still unresolved. The discovery of RNA-dependent RNA polymerase and of double-stranded forms of viral RNA of a size to be expected if they were intermediates in the formation of the pieces of RNA destined for incorporation into the virus particle, indicates that, apart from the early, actinomycin-sensitive step, replication of the pieces of influenza virus RNA probably proceeds along essentially the same lines as the replication of the nucleic acid of smaller RNA viruses.

In considering RNA tumour viruses such as the Rous sarcoma virus, we seem to get right away from what we may think of as the usual mechanism for the replication of RNA viruses. The recent discovery of a RNA-dependent DNA polymerase and the consistent failure by almost all workers to demonstrate any formation of double-stranded RNA in infected cells suggest that the mechanism by which this group of viruses replicate is quite different from the mechanisms discussed above. More recently further polymerase activities have been demonstrated. These include one which acts most effectively on a double-stranded template, one strand of which is RNA and one is DNA, which is exactly the structure that would be produced by the operation of the RNA-dependent polymerase; one which will duplicate normal double-stranded DNA, one which will replicate double-stranded RNA and possibly a ligase. In addition there is an endonuclease for DNA. It is not clear how many different polypeptides are involved in these various activities. We can account for the continuing sensitivity of these viruses to actinomycin D, almost throughout their replicative cycle, if we postulate that the actual multiplication of the viral genome involves DNA transcribed from the input viral RNA, such DNA being itself subsequently transcribed to give single-stranded RNA copies for incorporation into viral particles or to function as mRNA in the production of viral proteins. As in the normal cell, such transcription would be blocked by actinomycin D.

The full details of the mechanism of the replication of the RNA tumour viruses have yet to be elucidated. However, the rate of progress since the RNA-dependent DNA polymerase was first described in the middle of 1970, has been so rapid that we may hope that a fuller understanding will not be long delayed. The discovery that a DNA step is involved in replication has at least offered a possible explanation of how the genome of a RNA virus might be incorporated in and replicate with the genetic material of a mammalian cell (see p. 459). A search for RNA-dependent DNA polymerase, now called 'reverse transcriptase' in material from human patients has revealed such an enzyme in the lymphoblasts of patients with acute lymphoblastic leukaemia but not in control lymphoblasts stimulated by phytohaemagglutinin, in some sarcomas, but in only one of seven cases of Burkitt's lymphoma (see p. 441). The enzyme is said to differ in important respects from that found in the particles of animal tumour viruses. Nevertheless, this is a finding of considerable interest.

Another remarkable feature of the RNA tumour viruses is that they appear to contain two complete copies of the genome (i.e. they are diploid). The particles also incorporate a small RNA molecule which appears to be a host transfer-RNA and which pairs with the virus genome to produce a short double-stranded region at or close to its 5′ end. This is the wrong end for it to act as a primer (nucleic acid strands are begun at the 5′ end and synthesis proceeds towards the 3′ terminus: as the two strands of double-stranded molecules are anti-parallel, the template strand is copied from the 3′ end). Evidence suggests that the input RNA, although closely resembling host mRNA, cannot initiate protein synthesis; this only begins after the formation of the DNA 'provirus' from which effective mRNA can be transcribed. The presence of the short double stranded segment at the 5′ end of the genome could perhaps account for this. Claims have been made that these viruses contain a minute fragment of DNA: if this were to be confirmed, it would raise difficulties in connection with the definition of a virus (see p. 337).

This method of replication is not confined to tumour viruses but has been confirmed in the 'foamy agent' which is an extremely common contaminant of monkey kidney tissue cultures and in a pair of very closely related slow viruses which are called *visna* and *maedi* and affect the central nervous system and lungs respectively of sheep in Iceland.

Double-stranded RNA viruses: Baltimore Class III

This type of RNA is found in the Reoviruses and in the morphologically similar viruses of plants, wound tumour virus of clover and rice dwarf virus, as well as in other viruses of mammals, of insects, of fungi (mycophages) and at least one bacteriophage (the unique, lipid containing $\phi 6$ of *Pseudomonas phaseolicola*) which differ somewhat in their morphology from the reoviruses. The whole group of double-stranded RNA viruses have been referred to as diplornaviruses: it seems likely that many more await identification. The RNA is in a number of distinct pieces, like that of the influenza viruses. The number of pieces varies from two to as many as 15, but for the reoviruses is probably 10. Because of this the reoviruses also resemble influenza viruses in showing a high rate of recombination. A number of facts have been established about the replication of reoviruses. There is no requirement for DNA synthesis: 25–40% of newly formed RNA is double-stranded; the rest is single-stranded and shows no tendency to become double-stranded: there is a requirement for protein synthesis for RNA formation, but this is a continuing requirement only for the production of double-stranded RNA; if protein synthesis is blocked after replication has begun, production of double-stranded RNA ceases whereas production of the single-stranded RNA continues. Taking this evidence into consideration, it seems likely that replication of the viral genome occurs by a process essentially similar to the replication of double-stranded DNA. Replicase activity has recently been detected in the virus particles, but it is not clear whether this initiates replication of double-stranded RNA or, as seems more probable, makes single-stranded transcripts from the parental RNA which can function as messenger RNA and initiate the production of viral proteins, including the replicase which produces the double-stranded copies of the parental genome. In either case it seems clear that it is the single-stranded RNA, newly formed in the infected cell which fulfils the messenger functions. The reovirus particles have a double capsid, and it appears that, as with vaccinia, transcription can begin before uncoating is complete.

ASSEMBLY

After production of the components of the virus particles these have to be assembled. Virologists have tended to look for self-assembly, particularly with the smaller viruses, arguing that the genomes of such viruses were not large enough to accommodate the coding for any machinery for assembly if they were to also find room to code for the structural virus proteins and for such enzymes as might be required for replication of the nucleic acid. Rod-shaped plant viruses, such as tobacco mosaic virus, can quite readily be dissociated into RNA and protein and then, by altering the physical conditions, reassembled spontaneously. It has proved much more difficult to demonstrate spontaneous self-assembly with spherical viruses, although some limited success has been achieved with small DNA and RNA bacteriophages and with a few plant viruses. However, it is worth noting that even the very small RNA phages, which can code for no more than three proteins, produce one which is neither necessary for RNA replication nor incorporated in the virus particle, but is needed for the assembly of infectious virus. Considerable study has been devoted to elucidating the assembly mechanism of the tailed bacteriophages, particularly T4 and λ. Genetic studies indicate that many more genes are involved in assembly than can be accounted for by the number of proteins present in the phage particles. Moreover, mutations at some of these excess loci give rise to abnormalities of assembly such as the production of 'polyheads', large tubular structures made up of normal head protein. The way in which these assembly genes act is not known, but they presumably code for proteins which are in some way necessary for assembly, functioning either as enzymes or as some kind of 'jig' on which assembly can take place. Once formed, heads and tails can be made to assemble spontaneously in vitro.

Although difficult to carry out in vitro, the assembly of viruses appears to be a curiously random process. For example, if two distinct phages are replicating in the same bacterial cell, hybrid particles, having the DNA of one virus and the head and tail proteins of the other may be found amongst the progeny. This phenomenon is

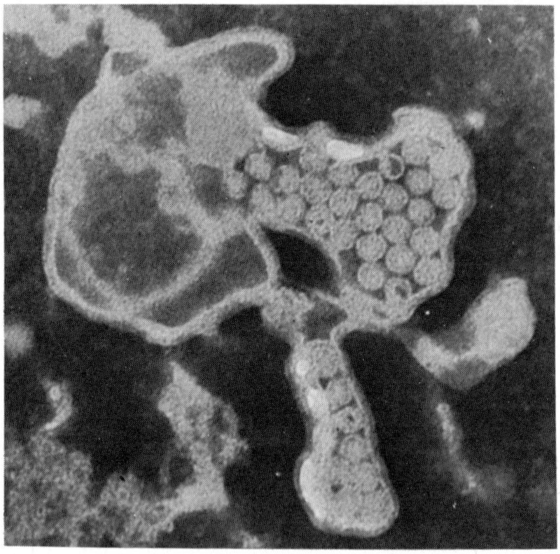

Fig. IV.11. Polioviruses in a cytoplasmic fragment. Possibly the vesicle in which virus particles are synthesized. Negative staining with phosphotungstate. (*Horne and Nagington, J. molec. Biol.*)

known as phenotypic mixing. The phenomenon of phage mediated transduction (see p. 467) is, of course, dependent on the inclusion within the phage particle of bacterial DNA. It has even proved possible, in vitro, to produce particles which consist of tobacco mosaic virus RNA coiled inside the capsid of a small spherical plant virus called cowpea chlorotic mottle virus, and such particles are infectious.

The assembly of animal viruses has not been successfully achieved in vitro. The virus which has been most extensively studied is probably influenza. Formation of the helical inner component, the nucleocapsid, takes place in the nucleus. This migrates out into the cytoplasm and comes to lie beneath the cytoplasmic membrane, which then becomes modified by the incorporation of viral proteins, particularly the neuraminidase and the haemagglutinin. The final coating of the nucleocapsid with the modified cell-membrane to form the envelope of the finished particle occurs by the formation of an evagination of the modified cytoplasmic membrane into which the nucleocapsid passes. The neck of the evagination constricts and the particle is at once completed and released from the cell. Another enveloped virus which has received detailed study is the herpes simplex virus. This is a DNA virus which develops in the nucleus and obtains its envelope from the nuclear membrane as it passes through. There is some evidence that additional material may be added to the envelope either in the cytoplasm or at the cytoplasmic membrane. It has been established that the envelopes of both influenza and herpes contain host cell antigens which differ according to the species of animal providing the cells for culture. When multiplying in cells to which they are poorly adapted, influenza viruses often develop as long filaments. These are not naked and extended nucleocapsids which would be very much thinner, but are long tubes of modified cell membrane. Once adapted to the new host, the virus produces much more regular, more or less spherical particles (see Figs IV.26 and IV.27).

RELEASE

This is the final step in viral replication. In the case of the bacteriophages, except for the filamentous phages containing single-stranded DNA, it occurs by bursting of the cell. A lysozyme-like enzyme is formed as a late viral function and this attacks the cell wall and weakens it. The final rupture is probably due to osmotic forces. No such dramatic event usually occurs in the case of animal viruses and cells. The newly assembled virus generally begins to leave the cell almost as soon as it is formed and the process of release continues over a period of several hours. Even when the cell finally dies and virus replication necessarily comes to an end, much of the infectious virus may remain within the cell or attached to cell debris. The mechanism of release of influenza virus has been described above. Much the same mechanism is probably used by most viruses which complete their assembly by acquiring an envelope at the cell membrane. This group includes not only myxoviruses and paramyxoviruses but also the bullet-shaped viruses such as rabies. However, the most common mechanism for release seems to be by a kind of reversed pinocytosis. The virus particles collect in vesicles just beneath the cell surface and these vesicles presumably discharge to the exterior. As well as non-enveloped viruses, some enveloped viruses such as rubella and arboviruses appear to use this mechanism. Studies in vitro indicate that some viruses such as varicella have no specific release mechanism. What happens in vivo is not clear; possibly release only occurs when

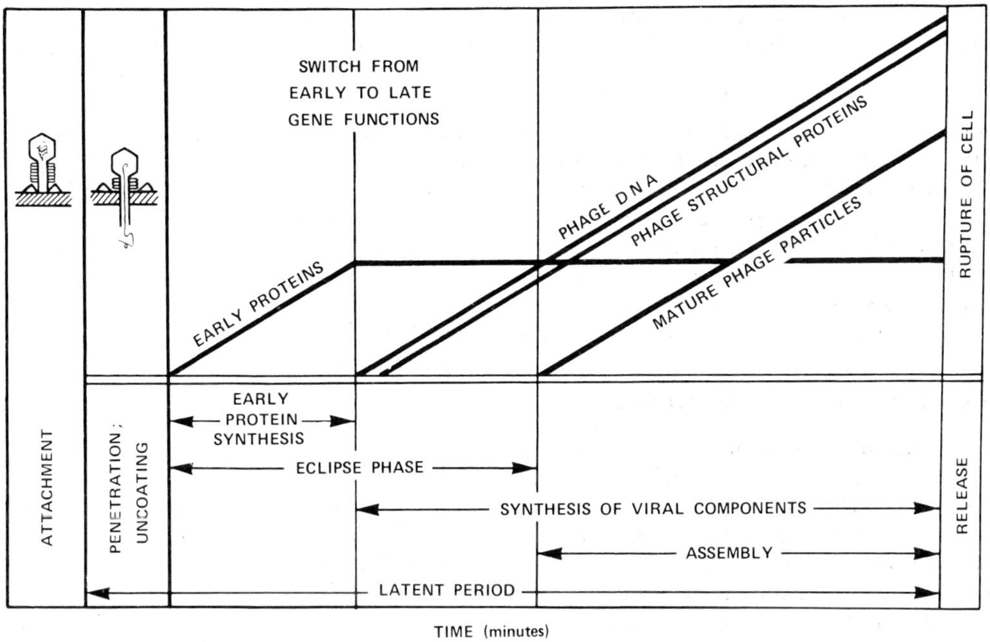

Fig. IV.12. The time relationships of various stages in the replication of a typical bacteriophage.

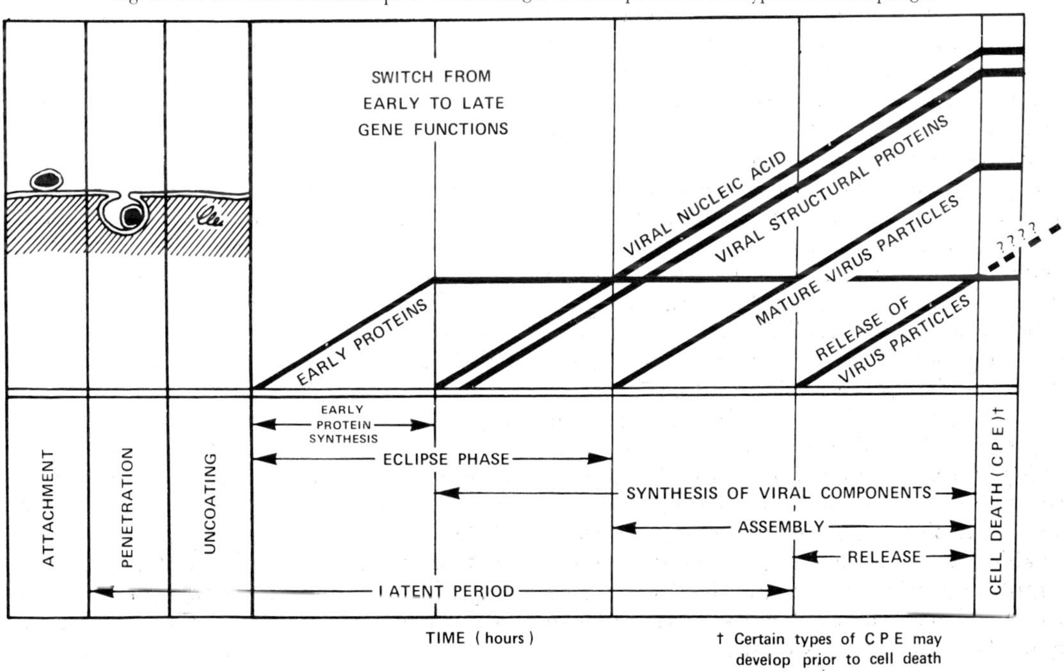

Fig. IV.13. The time relationship of various stages in the replication of a typical animal virus. Note that the time scale is in hours, whereas that in Fig. IV.12 is in minutes.

the cell dies and is broken down by the combined action of phagocytic cells and its own lysosomal enzymes. Some viruses possess the capacity to pass from cell to cell without emerging by making use of transient contacts between adjacent cells. Indeed viruses such as herpes, parainfluenza and respiratory syncitial virus actually bring about extensive cell fusion which presumably facilitates this process.

Figs IV.12 and IV.13 illustrate the time relationships of the various steps in the replication of a bacterial and an animal virus. Some of the terminology used in the animal virus diagram has been taken from bacteriophage terminology in order to emphasize the essential similarity between the two. In simplified diagrams of this kind it is impossible to incorporate all the variations which are observed with different viruses.

INHIBITORS OF VIRAL GROWTH

As already indicated the replicating virus makes extensive use of the metabolic machinery of the host cell, particularly those aspects concerned with the production of energy and the synthesis of protein. Any agent which intereferes with these common areas of metabolism will act indiscriminately against host and parasite. However, as we have seen, viral replication involves enzymes distinct from any found in the uninfected cell and in many cases (e.g. RNA viruses) take place by mechanism fundamentally different from anything observed in the normal host. It is these specialized pathways which offer the greatest hope for the development of effective viral chemotherapy. Although the compounds available for chemotherapy are still extremely limited, antiviral compounds and substances such as actinomycin D, which may selectively attack the host, have proved most valuable in elucidating the mechanisms of viral replication. There is one particular problem which will always tend to make viral chemotherapy less effective than bacterial chemotherapy. By the time the characteristic symptoms of a disease like measles have developed and a confident diagnosis is possible, virus multiplication has often virtually ceased. This is why almost all the compounds currently available are only effective if used prophylactically rather than therapeutically. Theoretically, virus infection could be halted by interfering with any of the stages of virus replication.

Absorption and penetration

These are the stages at which antibody and nonspecific serum inhibitors act. It is probably specific attachment rather than penetration which is blocked. Penetration appears to be prevented by amantadine (1-adamantanamine), a compound active against influenza A but not influenza B and against rubella virus. Controlled trials reveal a definite, if limited effect on influenza A_2 infections when used prophylactically in man. The mechanism of action of certain isoquinoline derivatives which are active against a number of viruses, including influenza A and B, has not been fully worked out, but they too probably interfere with absorption or penetration. Limited trials have protected against influenza in man.

Uncoating

It has recently been suggested that amantadine may act by preventing uncoating rather than attachment or penetration.

Inhibition of nucleic acid synthesis

Various compounds which block nucleic acid synthesis have been shown to have antiviral activity. However in most cases this is quite nonspecific and the compounds are far too toxic for clinical use. One compound which has been extensively used is 5-iodo-2′-deoxyuridine (IUDR). Originally introduced for cancer chemotherapy, this compound is 'mistaken' for thymidine and is incorporated into developing DNA in place of it. This fraudulent DNA will, of course, not function normally. Although highly toxic, the drug has been used systemically with questionable benefit in cases of herpetic encephalitis and generalized cytomegalovirus infection of the newborn. Unfortunately IUDR is too toxic to use in advance of a firm diagnosis and the delay involved in getting one clearly militates against a successful outcome to the treatment. There is experimental evidence that the multiplication of RNA viruses may actually be enhanced by IUDR. It has found its main use in the treatment of recurrent herpetic keratitis. However, a recent report indicates that, except in cases previously treated with topical steroids, IUDR has no advantage over phenol. A high concentration of the drug dissolved in dimethyl sulphoxide has given encouraging results in severe and extensive herpes simplex and herpes zoster. Its use has been advocated in the routine treatment of primary herpetic stomatitis in children, even in the absence of a definite diagnosis. The use of this drug in any condition which will clear up without special treatment is indefensible, especially in children. The compound is incorporated into the DNA of the cells exposed to it; it is a known mutagen and, as such, must be presumed to be at least potentially carcinogenic. Herpesvirus strains resistant to IUDR have been described; the mechanism of resistance has not been determined. Another somewhat toxic compound, cytosine arabinoside, has recently been used systemically in herpetic infections. Other compounds which interfere with DNA synthesis are under investigation. These include 5-bromo-2′-deoxyuridine, adenine arabinoside, 5-iodo-2′-deoxycytidine, 5-methylamino deoxyuridine and trifluorothymidine. Strains of herpes resistant to IUDR are sensitive to some of these.

Inhibition of protein synthesis

The production of infective virus can be inhibited by exposure of infected cells at a sufficiently early stage of growth to puromycin, which is an analogue of amino-acyl RNA and to *p*-fluorophenylalanine, which is an analogue of the amino acid alanine. Both of these compounds are quite non-specific in action and are therefore not suitable for chemotherapy.

Two compounds, 2α-hydroxybenzyl-benzimidazole and guanidine, which have been found to inhibit the multiplication of certain picornaviruses, appear to act by interfering with the production of viral RNA polymerase. They have not been found to be of value in the treatment or prophylaxis of viral infections in the intact animal. This appears to be primarily due to the rapid emergence of drug resistant viral mutants.

Interferon, which acts by interfering with viral protein synthesis, is more fully considered below. Actinomycin D, which acts by binding DNA and blocking transcription, has proved a valuable biochemical tool but has no therapeutic application. Rifampicin (rifamycin SV) is a semisynthetic compound derived from rifamycin, an antibiotic derived from *Streptomyces mediterranei*. In bacteria, it blocks the initiation of translation. It appears to have some action in vitro against pox—viruses and adenoviruses and an in vivo activity has been claimed against a hamster parvovirus, H1.

Inhibition of virus maturation

The maturation of the poxviruses can be inhibited by isatin β-thiosemicarbazone and its N-methyl derivative, methisazone. Even in cells which have been exposed to the drug at an early stage, the synthesis of viral DNA and of a number of the viral antigens is not affected. Because of this the cells develop degenerative changes even though they fail to yield infective virus. Spread of virus to other cells therefore does not occur. This compound appears to have a highly specific effect on the maturation of the poxviruses the only other susceptible viruses being adenoviruses which show a limited response in vitro. Rifampicin may exert its antiviral action by interfering with maturation rather than by inhibiting protein synthesis.

INTERFERENCE AND INTERFERON

It has been known for many years that infection with one virus may prevent subsequent infection with another one. This effect, known as viral interference has been demonstrated in intact animals, in chick embryos and in different kinds of tissue culture. The phenomenon is quite distinct from classical immunity and the two viruses concerned may be quite unrelated serologically. Even if the first virus has been rendered non-infectious by ultra-violet irradiation, the interference may still be demonstrable, indeed in certain circumstances it may even be enhanced. In 1957 Isaacs and his colleagues published the result of their studies on interference produced in fragments of chorio-allantoic membrane by heat-inactivated influenza virus. After treatment with the inactivated virus for three hours at 37°C, the pieces were rinsed and placed in fresh buffered salt solution and held for 24 hours at 37°C. The original pieces of membrane were then discarded and fresh pieces of membrane added. After a further 24 hours these were found to be resistant to challenge by active influenza virus. They called the soluble substance responsible for transferring the interference, interferon. It proved to be a low molecular weight protein (20 000–40 000 daltons) with remarkable properties. It showed almost total lack of specificity so far as the challenge virus was concerned, but a high degree of specificity in respect of the cells: interferon produced in the cells of one species has little or no activity in the cells of other species. For the development of interference the cells must be exposed to interferon for a minimum of 6–12 hours; incubation of the virus with interferon leaves it unaffected and, in some cases, viruses show an inverse relationship between virulence and capacity to produce interferon. In view of its possible implications for antiviral therapy, a great deal of study has been given to interferon in the past eighteen years. It now appears that the cells of a given species do not produce only a single interferon but a series of them, that a wide variety of living and non-living materials can give rise to interferon production, that production of interferon in intact animals far exceeds anything that can be produced in in vitro systems and that there probably exist at least two different mechanisms of interferon production. Viruses cause cells to produce interferon only after a lag and protein synthesis is necessary. Bacterial endotoxins, on the other hand, when administered to animals, lead to the almost immediate production of interferon which tends to be of higher molecular weight and does not require protein synthesis for its production. It is postulated that in the latter case a preformed molecule is simply activated and released. A large part of current research on interferon is devoted to the discovery of non-toxic, non-infective interferon inducers which might be used therapeutically. Following the observation that double-stranded RNA is particularly active, tests were carried out with synthetic polymers. Amongst the most active proved to be one usually referred to as 'poly-I, poly-C' this is a double-stranded molecule in which one strand comprises a chain of cytidylic acid residues and the other, which is base-paired with it, consists of inosinic acid residues (a purine nucleotide containing ribose and hypoxanthine which is found in transfer RNA and behaves as guanine in base-pairing). Unfortunately this substance is extremely toxic. Even if a safe and potent interferon producer were to be found a limitation to its use would be imposed by the fact that cells cannot be made to produce interferon indefinitely: after 20–50 hours production ceases and cannot be reinitiated for some time. It seems that cells in tissue culture must undergo at least two cell divisions before they can again make interferon.

Clearly interferon acts on the cell and not on the virus and, as one virus can stimulate different interferons in cells of different species and different viruses can produce the same interferon in a given cell type, it is cell-coded and not virus-coded. Interferon does not appear to act directly, because protein synthesis must take place in a cell exposed to exogenous interferon before interference is established. It has now been established that interferon interacts with host cell DNA to stimulate the production of a species of mRNA which gives rise to a 'translation-inhibitory protein' (TIP). This binds to ribosomes and modifies them in such a way that they become much less apt to form polysomes with viral mRNA, while retaining their capacity to translate host mRNA virtually unchanged. (It will be recalled that viruses themselves may affect ribosomes in just the opposite way, see p. 355.) It seems that interferon production tends to be maximal under conditions which limit or prevent viral nucleic acid synthesis and that when a virus enters a cell treated with exogenous interferon, it cannot replicate and instead stimulates the cell to produce interferon itself.

The probable biological role of interferon in viral infections has been much discussed. In view of the impotence of passively acquired antibody to modify the course of an established virus disease, it seemed unlikely that antibody plays any significant part in halting a viral infection. The same conclusion has been drawn from the more or less normal course of most acute virus diseases in the majority of patients with extreme hypogammaglobulinaemia: vaccinia seems to be an exception. It therefore seemed at one time that interferon might well be the agent responsible for bringing infection to an end. However, the position is now far less clear-cut than it seemed to be. It has become obvious that immune responses deemed to be primary may occur in far less than the traditional seven to ten days and immune deficiency syndromes have now been described in which viral infections do not run a normal course. The defect seems to involve the development of cell-mediated immunity rather than of humoral immunity; and as the importance of the former in both the pathogenesis of infectious disease and defence against established infection (these may be different aspects of the same thing) is only now coming to be fully recognized, it seems unwise to claim too much for interferon. The demonstration that lymphocytes are potent sources of interferon when stimulated either specifically by the appropriate antigen or non-specifically by mitogens, means that interferon production and immunity cannot be regarded as independent and quite separate responses of the host to a virus infection.

Although the discovery of interferon provided the impetus for an immense research effort, it has not explained every instance of interference. Cases are still encountered in which it is necessary to postulate some other mechanism. An interferon-like substance has been described in plants. Although interferon shows activity in man under highly artificial experimental conditions, early hopes of its value in antiviral therapy have not been realized and it is not available for routine therapeutic use.

LATENT AND INAPPARENT VIRUS INFECTIONS

It has long been known that infection and infectious disease are not synonymous. This is as true of viruses as it is of bacteria, if not more so. However, these inapparent infections are not all of the same kind. The most obvious class are individuals in the incubation period of a virus disease such as smallpox or measles. Studies using rabbitpox and mousepox as models have revealed that following initial infection and perhaps some limited multiplication at or close to the portal of entry, there is a transient viraemia. The blood is rapidly cleared by the reticuloendothelial system and multiplication continues within the cells of that system. If there are any symptoms at the time of the primary viraemia, they are trivial and transient and probably correspond to the 'illness of infection' recognized by clinicians. Infection continues in the reticuloendothelial system and virus builds up until it, as it were, overflows into the blood stream and a secondary viraemia develops. This time the blood is not cleared so rapidly and the viraemia reaches a higher level and persists longer than the primary viraemia. Symptoms appear at this stage and virus is distributed to susceptible tissues throughout the body. Multiplication of virus in these produces the characteristic lesions of the disease in question, e.g. the exanthem in smallpox or measles.

A proportion of infected individuals, particularly those infected with enteroviruses such as poliomyelitis virus, never develop an overt illness at all. In these cases it seems that either the virus never reaches a target such as the central nervous system, involvement of which is essential for the characteristic syndrome to develop, or the virus never reaches a level at which symptoms are produced (see p. 373). Studies on various strains of yellow fever virus in different lines of mice have shown that initially strains which are going to kill the animal, strains which will produce illness with recovery and strains destined to produce only symptomless infection all multiply at about the same rate following intracerebral inoculation. However, the final levels of virus attained differ, the most virulent continuing to increase until death supervenes and the others ceasing to increase as soon as they reach some level below that finally attained by the lethal strains. Moreover, strains which produce no symptoms cease to multiply at a lower level of virus than those which produce symptoms. Where a particular strain of virus produces severe disease in one line of mice and mild or inapparent infection in another, the same differences in the final level of virus are discernible. The mechanism behind the differences just described is obscure; but, as indicated in the previous section, interferon could be responsible. The facts that interferon is virtually non-toxic and that attenuated viruses tend to produce more interferon than more virulent strains provide circumstantial support for this possibility.

In none of the foregoing examples of inapparent infection is there any reason to suppose that there is anything unusual at the cellular level. The same is true of certain chronically infected cell cultures in which virus production, interference and cell multiplication seem to be so balanced that the virus is not eliminated and the cells are not overwhelmed. However, examples do occur in which the infection seems to be different at the cellular level. There is now abundant evidence that when an individual suffers from attacks of recurrent herpes simplex (cold sores) each attack does not represent a fresh infection but a recrudescence of the same infection. Nevertheless it is most unusual to be able to detect any virus in these patients between their attacks. Even if they are killed in an accident so that more extensive studies can be carried out, no virus is detectable by conventional methods. A somewhat similar situation exists with herpes zoster. Following an initial attack of chickenpox the virus becomes latent, only to reappear years later and produce herpes zoster. Many tumour viruses, especially those which contain DNA, disappear once the cell they infect is transformed into a malignant cell (see p. 456). No infective virus can be detected, yet it can be shown that the viral genome persists in whole or in part and replicates with the cells, so that each daughter cell gets at least one copy. The parallel with lysogeny (see p. 466) is an obvious one, but it is important to remember that the differences between eukaryotic and prokaryotic cells, particularly as regards their nuclear and chromosomal structure, are so considerable that the mechanism of latency must be somewhat different in bacteria and in cells of higher organisms.

IV.3

Miscellaneous Properties: Virus Diseases

CULTIVATION

Human pathogenic viruses may be propagated in the laboratory in the chick embryo, in tissue cultures or in the tissues of some laboratory animal.

The chick embryo

Inoculation may be into the allantoic or amniotic cavities, into the yolk sac or on to the chorio-allantoic membrane, the precise route used depending on the particular virus being cultivated (Fig. IV.14). The amniotic route is the method of choice for the isolation of the influenza and mumps viruses. The allantoic route, which is technically the simplest, is used mainly for the passage of influenza viruses that have already been established in the embryo. Yolk sac inoculation is of particular value for the propagation of rickettsiae, including *C. burneti,* and chlamydiae. Inoculation on to the chorio-allantoic membrane is used for the isolation and characterization of the poxviruses and herpes simplex.

Tissue cultures

Most of the human pathogenic viruses may be propagated in tissue cultures derived from a variety of animal species. Since, however, a number of common viruses, e.g. the enteroviruses and the adenoviruses, will grow only in the cells of primate tissues, the tissue cultures used in diagnostic virology are generally derived from monkey or human sources.

Tissue culture techniques may be divided broadly into three groups; fragment cultures, cell cultures and organ cultures.

Fragment cultures

The simplest form of fragment culture is the Maitland type culture which consists of fragments of tissue suspended in a fluid medium. The cells remain viable for several days—sufficiently long to permit virus growth—but do not multiply. In plasma clot cultures the tissue fragments are fixed by a plasma clot to the sides of tubes or bottles; new cells, mostly fibroblasts, then grow out from the tissue fragments. Fragment cultures have no application in routine diagnostic virology today.

Cell cultures

These are prepared from cell suspensions obtained from intact tissue or from a prior tissue culture. Dispersal of cells is usually achieved by treatment of the tissue or tissue culture with a proteolytic enzyme, usually trypsin, or with the chelating agent Versene (EDTA, ethylene-diamine-tetra-acetic acid, Sequestrene); this results in the release of single cells and small aggregates of cells capable of initiating growth. Cell cultures may be prepared as suspended cultures or as monolayer cultures.

Suspended cell cultures resemble Maitland type cultures in that the cells are simply suspended in nutrient medium. They are used for metabolic inhibition tests which may, in diagnostic work, be applied to the estimation of antiviral antibody. Uninfected cells metabolize and actively produce acid; this causes a colour change in an appropriate pH indicator incorporated in the nutrient medium. If the cells are infected by virus their metabolism is interfered with and the indicator change does not occur. If however the virus is neutralized by antibody the indicator change occurs as in a normal uninfected culture (see p. 374).

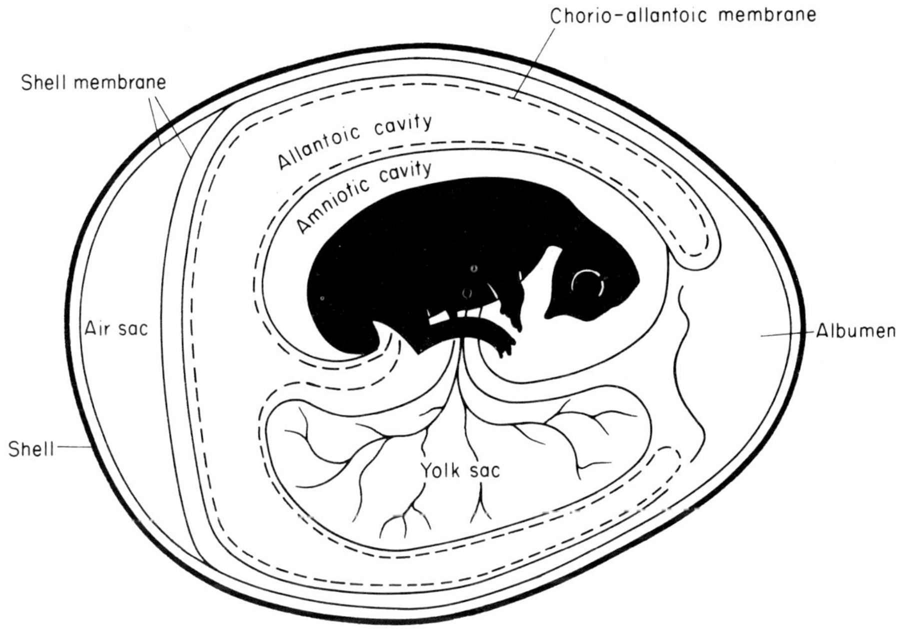

Fig. IV.14. Section of an egg containing a chick embryo, tenth to twelfth day of incubation.

In *monolayer cell cultures* the cells are allowed to settle on the sides of a tube or flat bottle and are covered with nutrient medium. Antibiotics are incorporated in the latter in order to control bacterial and fungal contamination. After two to seven days' incubation at 37°C the growing cells will have formed into a continuous sheet or monolayer, so called because it is one cell thick, adherent to the glass of the tube or bottle. The growth medium is then removed and replaced with a maintenance medium which is nutritionally less rich than the initial growth medium but adequate to maintain the viability of the tissue cells. Then the virus inoculum is introduced and the culture again incubated. During incubation tubes are maintained in a slightly sloped, and bottles in a horizontal, position; as a result the cells are continuously bathed by the nutrient medium. Alternatively they may be rolled in a special drum or mechanically shaken. Many viruses when propagated in monolayer cultures produce degenerative changes in the tissue cells readily visible under the lower power objective or, if the area involved is sufficiently large, to the naked eye; this is known as a cytopathogenic effect (cytopathic effect, CPE) (see Fig. IV.19). The type of cellular change produced differs with different viruses. It may consist simply of a loss of normal staining properties. In some cases there is the formation of giant cells and syncytia. In others there may be extensive destruction of the tissue culture sheet. When a virus producing a cytopathogenic effect has been isolated its identity is finally confirmed by neutralization tests with specific immune serum—its cytopathogenic effect being neutralized by homologous serum.

A modification of the monolayer technique, which has been introduced by Dulbecco, will permit the isolation of pure lines of virus and very accurate virus titrations. In this, a sheet of cells, which has been grown as a monolayer culture in a flat bottle or Petri dish, is seeded with virus and covered with a layer of agar incorporating the nutrient medium employed.* Many viruses produce cytopathogenic effects readily visible to the naked eye as degenerated areas in these plate cultures. If the number of virus particles inoculated into the culture is small, the degenerated areas will appear as discrete plaques analogous to the plaques of bacteriophage (Chapter IV.13). By subculture from one of these plaques a pure culture of the virus can be obtained which is derived from a single virus

* The plaques are more readily seen if a second agar overlay containing neutral red is added. This vital dye stains living but not dead cells.

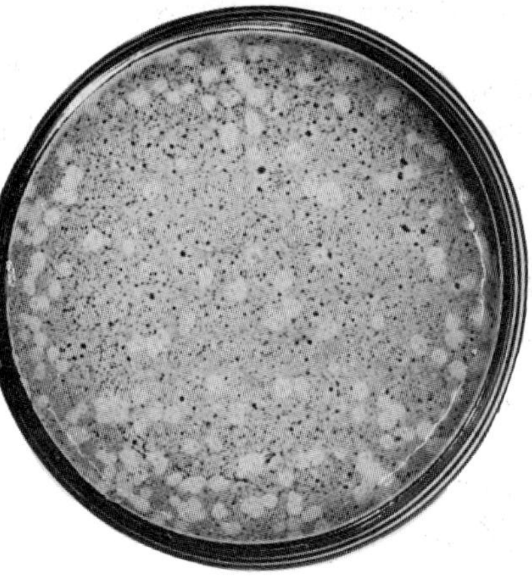

Fig. IV.15. Plaques produced by fowl plague virus on a chick embryo monolayer tissue culture. (*Dr H. G. Pereira*)

particle. The plaques produced in this type of culture frequently have a distinctive appearance which is valuable in identification (Fig. IV.15).

Cell cultures may be divided into three types according to the passage history of the cells used for their preparation.

Primary and secondary cell cultures

Primary cell cultures are prepared from cells obtained directly from the tissues. A secondary cell culture is the first subculture of a primary cell culture, the cells of which are dispersed by treatment with trypsin or Versene for the preparation of the secondary culture. This procedure is particularly convenient for the preparation of monkey kidney cell cultures since the primary cultures can be prepared in a central laboratory, which alone needs to maintain a stock of monkeys, and these cultures are then distributed to satellite laboratories for the production of secondary cultures. Secondary monkey kidney cultures are, however, more likely to show the growth of latent simian viruses (p. 378) than are primary cultures.

Continuous cell lines

As a rule when tissue cultures are prepared directly from an animal tissue the cultures cannot be serially propagated, the cells dying out after a few subcultures. A number of lines of mammalian cells are, however, available (mainly derived from fetal and malignant tissues) which can be serially propagated more or less indefinitely. These established cell lines have the advantage of obviating the necessity for procuring fresh animal tissue for each set of cultures. They have the disadvantage, on the other hand, that on prolonged subculture the tissue is liable to undergo spontaneous changes in its susceptibility to infection. Moreover, many cell lines which have been directly propagated for long periods have been found to be contaminated with mycoplasmas.

The cells of continuous cell lines differ from the cells of primary and secondary cultures in the following respects. They all tend to present, regardless of source tissue, a similar morphology, they are genetically aneuploid, i.e. they possess a variable number of chromosomes and they fail to show contact inhibition (see Chapter IV.12). Cells with these characteristics are said to be *transformed*. The established cell lines most frequently employed are the HeLa cell culture which was derived originally from a human cervical carcinoma, and the HEp-2 cell derived from a carcinoma of the larynx.

Diploid cell cultures

The cells of certain tissues, notably human embryo cells, can, as first shown by Hayflick, be serially propagated for about 50 subcultures without transformation. Genetically they differ from continuous cell lines and resemble normal cells in being diploid, not aneuploid. Diploid cell cultures are highly susceptible to infection by viruses such as the H rhinoviruses and the cytomegalovirus which may be difficult to propagate in other systems. One particular human diploid cell line, WI-38, is now beginning to be used for the production of virus vaccines, thus removing the need to use primary or secondary cultures of animal cells which are inevitably variable and liable to contamination with infectious agents, including viruses, derived from the animal. The diploid cells afford a uniform substrate for the growth of vaccine viruses which has been subjected to the most exhaustive scrutiny to exclude extraneous agents and to establish the absence of any tendency of these cells to undergo spontaneous transformation or to promote transformation in other cells. Preliminary results indicate that vaccines prepared in WI-38 cells are at least as potent as those made by conventional methods.

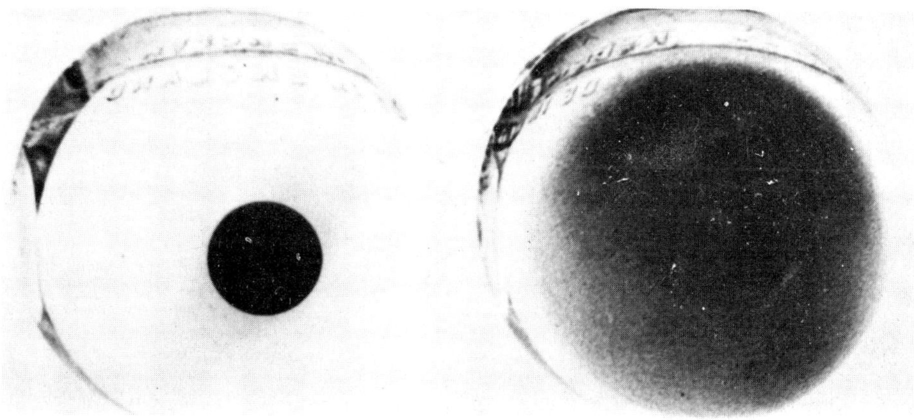

Fig. IV.16. Haemagglutination detected by the pattern method. The cell–virus mixtures are placed in the round-bottomed cups of a Perspex haemagglutination tray or in round-bottomed tubes. Where haemagglutination does not occur, the cells settle to the bottom of the cup to form a small dense button (negative, *left*). However, where haemagglutination does occur, the cells stick to one another and do not slide to the bottom of the cup but form instead a granular layer over the whole of the concavity (positive, *right*).

Organ cultures

In one procedure, recently introduced by Hoorn, and adapted by Tyrrell and Hoorn to the cultivation of viruses from the respiratory tract, small pieces of human embryonic nasal and tracheal epithelium in the intact state are grown in Petri dishes containing a thin layer of nutrient medium. Virus growth may lead to impairment of ciliary function, which may be detected by microscopy in situ, and to cellular degeneration, which may be demonstrated in stained sections. The growth of certain respiratory viruses which fail to produce these effects convincingly, may be demonstrable by the interference phenomenon. Organ culture has been recommended for the isolation of cold viruses which cannot be propagated by other procedures. The technique is being extended, using tissue from the appropriate organ for the isolation of viruses other than those affecting the respiratory tract.

Animal inoculation

Of the ordinary laboratory animals the mouse is the most generally useful and, depending on the virus, can be infected by nasal instillation or by intraperitoneal or intracerebral inoculation. The presence of virus in the inoculated material is shown by the development of appropriate symptoms, diagnostic phathological changes or specific antibody.

HAEMAGGLUTINATION AND HAEMADSORPTION

A number of viruses possess, or are associated with the production of, haemagglutinins which agglutinate the red blood cells of a variety of animal species. In certain cases notably in the myxovirus group the detection of viral haemagglutinins is of considerable value as a method of demonstrating the presence of the virus (Fig. IV.16). Haemagglutination is also used as a basis for antibody estimation since antibody specific for the viral haemagglutinin neutralizes its capacity to agglutinate red cells. Such tests are known as haemagglutination inhibition tests (Fig. IV.17). They are used both for the serological identification of an unknown virus and for estimating the antibody content of a patient's serum (see p. 379).

Viral haemagglutinins are of two types:

1. Haemagglutinins occurring as integral components of the virus particle. Haemagglutinins of this type are found in the myxoviruses, arboviruses, adenoviruses and certain of the ECHO and Coxsackie viruses.

2. Lipid haemagglutinins distinct from the

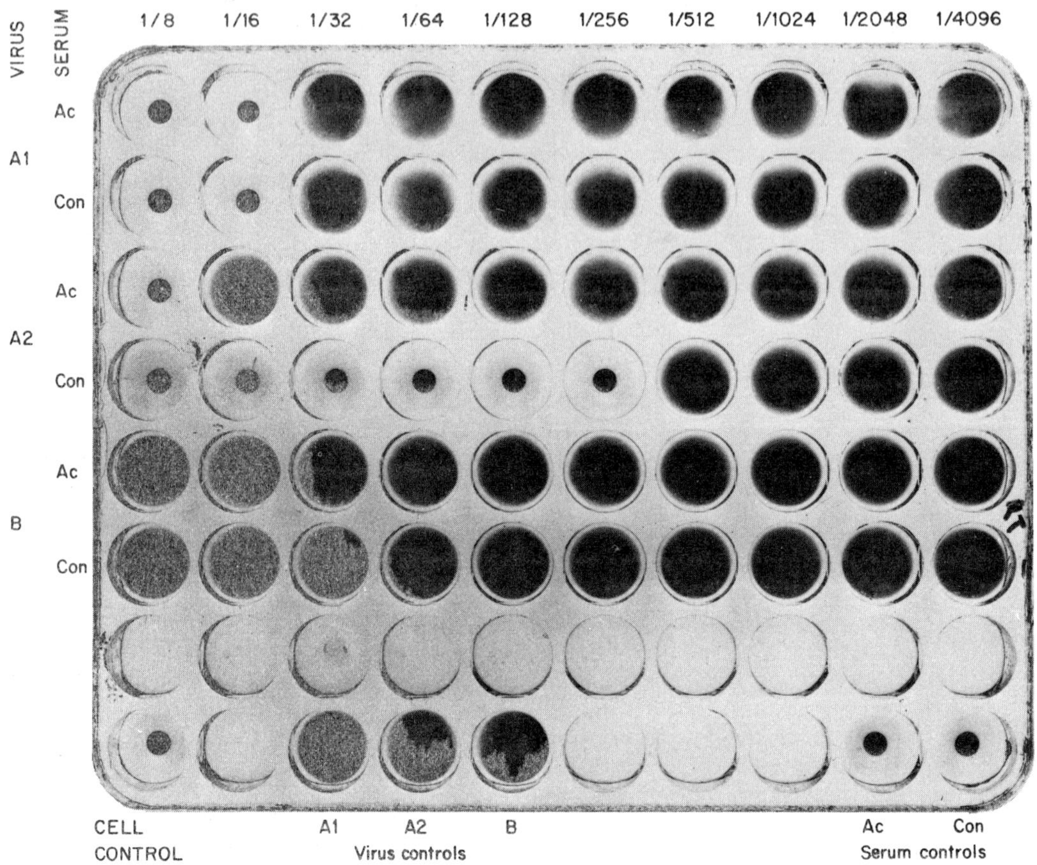

Fig. IV.17. A typical haemagglutination inhibition test using acute phase and convalescent serum from a patient with suspected influenza and influenza A1, A2 and B. Note that both sera have inhibited haemagglutination by influenza A1 to a titre of 1/16. With A2 virus a titre of 1/8 in the acute phase serum has risen to 1/256 in the convalescent serum. Neither serum contains detectable antibody to influenza virus B. Note that the three virus controls show haemagglutination and that the cell and serum controls do not. The test indicates infection with A2 or a closely related type A strain.

virus particle. This type of haemagglutinin occurs in tissues infected with the vaccinia and variola viruses. The mechanisms involved in the production of lipid haemagglutinins are at present unknown.

If a cell-sheet in which a haemagglutinating virus is multiplying is exposed to a dilute suspension of the appropriate red cells, these may adhere to the tissue culture cells. This phenomenon is known as haemadsorption. It is of considerable value in the detection of haemagglutinating viruses, especially myxoviruses, paramyxoviruses and poxviruses. Because the effect can be blocked by exposing the cells to specific antiserum before adding the red cells, haemadsorption inhibition can be used in the identification of viruses.

PATHOGENICITY

The animal viruses show considerable variability in their host range. Some are highly specific and are capable of infecting only one or a few animal species. Others will infect a wide range of species. Under the conditions of natural infection viruses exhibit a considerable degree of tissue *tropism,* i.e. an affinity for the cells of a particular tissue. In some cases inability of a virus to grow in the cells of a particular tissue may be due to the absence of specific receptors (p. 349); in other cases it is probably due to environmental factors governing the route of access of the virus to that organ, rather than to the fact that it is the only organ in which the virus can multiply.

In general it can be said that viruses damage cells by interfering, during the course of intracellular growth, with their metabolism. In some cases, for example in infection with the papovaviruses (p. 457), this interference leads to cellular hyperplasia. The more usual observable effect of infection, however, is a progressive degeneration of the cell and it is this degeneration which results in clinical disease.

The extent to which viruses possess an intrinsic toxicity distinct from the effects of virus multiplication is not clear. It has not been established that any virus produces a substance separable from the virus particle analogous to the exotoxins produced by certain bacteria. But some viruses, notably the myxoviruses, appear to be inherently toxic in that they may be rapidly lethal to experimental animals if injected in very large amounts or by a route which does not clearly lead to virus multiplication. Nevertheless it has not been possible to establish in such instances that some restricted virus multiplication has not occurred. In the intact animal overt virus infections are normally associated with pyrexia and general constitutional disturbance—symptoms which in bacterial infections are attributed to the action of bacterial endotoxins. These symptoms are, however, not necessarily due to an inherent virus toxicity. Pyrexia may be due to endogenous pyrogens liberated by leucocytes in response to the stimulus of phagocytosed virus. Other constitutional symptoms might be due to the action of breakdown products of tissue cells infected with

Table IV.3. Viral haemagglutinins

Virus	*Most susceptible species*	*Notes on haemagglutinin (HA)*
Adenovirus	Rat and/or rhesus monkey[2]	Both pentons and fibres have HA activity but not hexons or penton bases. Maximal at 37 °C[5]
Arbovirus	Chick (24 hours), goose and pigeon	Very sensitive to pH and non-specific serum inhibitors[6]
Coxsackie and echoviruses (certain serotypes)[1]	Human group O	May elute at 37 °C with or without receptor destruction. Receptors probably lipoglycoprotein and distinct from myxovirus receptors. Most strains give highest titres with cell from the newborn and at room temperature (ECHO type 19 at 4 °C)
Reoviruses	Human group O[3]	Receptors are polysaccharides insusceptible to receptor-destroying enzyme (RDE)
Vaccinia–variola	Fowl[4]	Soluble lipoproteins separable from the infectious particle
Myxoviruses (except RSV and measles)	Human Group O, fowl, guinea-pig	Component of envelope spikes. Virus elutes at 37 °C through destruction of red cell receptors by viral neuraminidase (RDE). Highly susceptible to non-specific serum inhibitors[7]
Measles virus	Rhesus monkey	Maximal at 37 °C. No elution[8]
Rubella virus	Chick (24 hours), goose and pigeon	As for arboviruses[9]

[1] Considerable strain variation in HA capacity, Coxsackie A7 only agglutinates fowl RBC.
[2] Susceptible cells vary with virus serotype.
[3] Bovine cells agglutinated by type 3. Bovine receptors susceptible to RDE.
[4] Red cells of certain fowl only susceptible. Susceptible cells agglutinable also by cardiolipin and certain other lipids.
[5] May agglutinate only with heterotypic serum. They act as red cell sensitizing antigens.
[6] Removed by treatment with lipid solvents or kaolin.
[7] Mucoproteins susceptible to RDE, KIO_4 and trypsin, and proteins (see Chapter IV.8).
[8] HA preparations highly heterogeneous in size, varying from intact virus to membrane fragments. HA therefore not a satisfactory measure of virus content.
[9] Serum inhibitors removed by treatment with kaolin, with manganous chloride and heparin or with dextran sulphate and calcium chloride.

virus. It has been suggested that severe shock-like systemic reactions, such as are found in smallpox, may be due to some multiplication of virus in vascular endothelium.

In contrast to pyogenic bacteria, viruses do not generally provoke a very marked polymorphonuclear response either locally or systemically, although frank pus accumulates in the skin lesions of smallpox, vaccinia, varicella, herpes simplex and herpes zoster—particularly, but not exclusively, when secondary bacterial infection supervenes. Polymorphonuclear leucocytes are the predominant cell in the cerebrospinal fluid in the earliest stages of paralytic poliomyelitis and are prominent in the parenchymal lesions in fatal cases. When polymorphonuclear leucocytes are present in viral lesions they frequently appear very degenerate and often disintegrate to form haematoxyphil debris. It is not clear to what extent neutrophil polymorphs are not attracted to viral lesions and to what extent they are rapidly destroyed once they get there. In circulating blood a significant increase in granulocytes is not usual. There is usually a neutropenia with an absolute or relative lymphocytosis. The lymphocytes may include some which are abnormal, 'virucytes'. This abnormality of the lymphocytes finds its most florid expression in glandular fever, where the white cells may be so numerous and so abnormal that the condition has been misdiagnosed as acute lymphatic leukaemia.

It is becoming increasingly clear that the tissue damage and the symptoms associated with virus infections are not due solely to the destructive action of the virus on the cells in which it is multiplying. The role of immune mechanisms, especially cell-mediated immune responses, in both resisting the parasite and in causing tissue damage is only now coming to be recognized. These two effects are, of course, likely to be but different aspects of the same process. It is noteworthy that diseases such as kuru and scrapie, due to agents which apparently do not stimulate any kind of immunological response, pursue a relentless downhill course and present none of the clinical or pathological features which have come to be associated with an infection. Indeed, until their transmissability to experimental animals was established beyond any doubt there were many who could not bring themselves to accept that they were the result of an infection. Many, perhaps all, true viruses alter the surface antigens of the cell which they parasitize. This creates antigenic novelty, the affected cells are treated as 'foreign' and provoke a host-versus-graft reaction which will have the effect of killing cells in which virus is multiplying. This will restrict virus multiplication but will also cause tissue damage and provoke an inflammatory response with its associated signs and symptoms. The fact that antigens can stimulate lymphocytes to release interferon (see p. 363) will also be likely to influence the outcome of viral infections. Any agency, e.g. immunosuppressive drugs, which depresses the immune mechanisms can activate latent infections such as herpes simplex, herpes zoster and cytomegalovirus, and may influence the course of new infections. Conversely, some virus infections are themselves immunosuppressive, e.g. measles (see also p. 432).

In many virus diseases, structures known as *inclusion bodies* are present in the cells of infected tissues. Inclusion bodies are homogeneous masses of varying size, showing characteristic staining properties. They may be present either in the cytoplasm or the nucleus and frequently have a very characteristic appearance. Some inclusions have been removed from infected cells by dissection and have been shown to be infective. Many consist of large numbers of virus particles, apparently embedded in a gelatinous matrix (Fig.

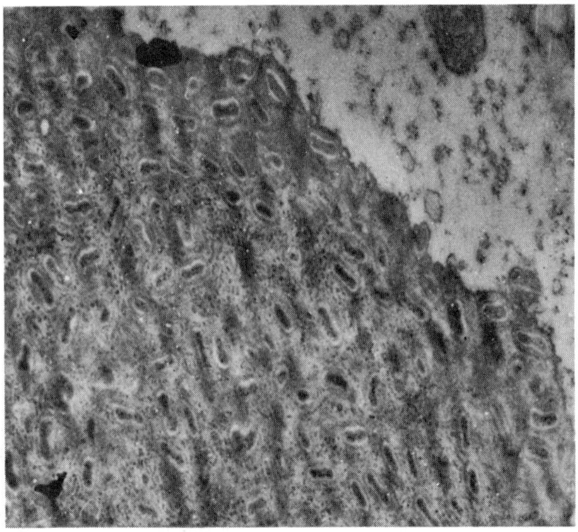

Fig. IV.18. Fowlpox inclusion body in chorio-allantoic membrane. The virus particles appear to be embedded in a homogeneous matrix material. Electron micrograph of an ultra-thin section. × 17 500. (*Dr T. H. Flewitt*)

IV.18). These therefore appear to be intracellular colonies of virus. Others, however, e.g. the late intranuclear inclusions of herpes simplex, are found which contain no demonstrable active virus particles. Yet others, e.g. some adenovirus intranuclear inclusions and the inclusions in the cells of human warts, seem to consist of ordered crystalline arrays of mature virus particles.

Types of infection

In some diseases—influenza, the common cold, adenoviruses—the virus appears to remain localized during the entire course of the infection to its initial site of entry. Although the recovery of influenza virus from the urine indicates that even in this infection spread can occur from the site of initial infection. The lesions characteristic of these diseases are the result of local growth of the viruses in the area or organ initially infected. Spread by the blood stream from the initial site of infection with secondary multiplication in some organ remote from the primary site is, however, a feature of many virus diseases, the major symptoms of infection being the result of this secondary multiplication (see p. 381). In this latter group the frequency with which viraemic spread and secondary multiplication supervene on initial infection differs with different viruses. In smallpox, measles, chickenpox and mumps, viraemic spread with secondary multiplication is almost invariable. With other viruses which exhibit this pattern of infection, however, the frequency of secondary multiplication is very low, the virus in most cases remaining localized and causing only mild or inapparent infection. Thus it has been estimated that involvement of the central nervous system occurs in only a small proportion of individuals infected with the poliomyelitis virus, while in the remainder the virus remains localized to the intestinal or pharyngeal mucosa. In certain of the arthropod-borne encephalitides it has been estimated that the frequency of secondary multiplication is even less than in poliomyelitis. Rabies virus spreads by way of the peripheral nerves and poliomyelitis may also adopt this route of spread, at least under experimental conditions. This does not necessarily imply spread along the axons: spread could occur by way of the neural lymphatics or by progressive involvement of the neurilemmal cells. Within the central nervous system, poliomyelitis shows a remarkable tendency to spread by way of anatomically defined neuronal pathways.

Like bacteria, viruses are antigenic and each virus particle possesses, as a rule, a number of serologically distinct antigens. Some viral antigens are characteristic of a particular group or 'species' of virus and these are of considerable importance in viral identification. In many cases, e.g. amongst the enteroviruses and myxoviruses, it may be possible to distinguish a number of different serological types within a particular virus species. In certain instances there may be sharing of antigens by viruses of distinct pathogenic properties. Thus the measles virus shows a serological relationship to the rinderpest and canine distemper viruses.

In so far as they have been chemically characterized viral antigens appear to be proteins, including glycoproteins and lipoproteins. Some viruses of which the myxoviruses and adenoviruses are of particular importance produce or are associated with so-called 'soluble' antigens. These are antigens distinct from the virus particle which may be demonstrable in infected tissue cultures, eggs or animals. In some cases soluble antigens appear to be serologically identical with components of the viral nucleocapsid. These may represent excess material which has not been incorporated into mature virus particles or may be products of virus breakdown. Those which are not components of the virion (virus particle) probably correspond to the 'early proteins' of bacterial viruses (see p. 351). Tumour viruses give rise to new antigens, both in the nucleus ('T' antigens) and on the surface. The latter function as transplantation antigens. Both appear to be virus coded.

Antibody to viral antigens may be demonstrated by the conventional in vitro techniques of complement fixation and precipitation (Chapter II.5). Because of the small size of virus particles, however, precipitation techniques are relatively insensitive. Since extremely large numbers of particles are required to give a visible agglutinate because of their small size, the ordinary agglutination test finds no application in virology. However, agglutination revealed by electron microscopy has been used to establish the identity of objects believed to be virus particles. Complement fixation tests are more useful since they give positive results with much smaller amounts of antigen. In respect of viruses that possess haemagglutinins as intrinsic components of the virus particle or which produce soluble haemagglutinins, haemagglutination inhibition tests may be used for the demonstration of antiviral antibody. In these the serum under examination or an appropriate dilution thereof is mixed with a standard amount of virus or of soluble haemagglutinin and, after a suitable period to permit combination of antibody with haemagglutinin, susceptible red cells are added. If the serum contains antibody to the viral haemagglutinin haemagglutination will be inhibited (see p. 370).

Neutralization tests are undoubtedly the most sensitive of the methods available for the detection of antiviral antibody. They determine or measure the capacity of antibody to prevent the infection of susceptible cells. The possible mechanisms involved are discussed below. Neutralization tests may be carried out in experimental animals, in chick embryos if the virus is capable of growing in the chick embryo, but most conveniently in tissue cultures. They can be carried out with quantities of virus which though sufficient to initiate an infection are insufficient to produce any demonstrable in vitro reaction. They will only detect antibody to virus particle antigens. Antibodies to soluble antigens are not demonstrable. Tissue culture neutralization tests may be performed either by demonstrating the capacity of the antibody to prevent the production of a cytopathogenic effect (p. 367) or by *metabolic inhibition tests* (see p. 366). The latter are based on the fact that the metabolism of cells infected by virus is inhibited, whereas if cell infection is prevented by the presence of antibody the cells metabolize normally. The occurrence of cell metabolism is detected by the incorporation of a pH indicator in the nutrient medium. Actively metabolizing uninfected cells produce acid which causes a colour change in the indicator. Non-metabolizing infected cells do not produce acid and therefore fail to produce the colour change. The most discriminating type of neutralization test is the quantitative plaque reduction test in which a known amount of virus is mixed with antiserum and the mixture sampled at intervals. The residual virus in the samples is determined by the plaque count produced (see pp. 368 and 461) and from the results the inactivation constant can be calculated. This test when carried out on two viruses may reveal minor antigenic differences which are quite inapparent on conventional testing, e.g. different strains of herpes virus or of the same type of poliomyelitis virus.

IMMUNITY

Acquired immunity

Humoral antibodies are produced as a result of virus infection and, since their administration both to animals and man will protect against infection, they are presumed to play an important part in specific actively acquired immunity. The precise role of antibody in antiviral immunity is not clear. Neutralizing antibodies can abolish the infectivity of viruses on their own without the intervention of other cellular or humoral factors, and they may act principally in this way. In some systems, but by no means all, the neutralizing activity of a serum is enhanced by complement. In a number of conditions virus has undoubtedly been demonstrated in leucocytes but has been present in these in a fully infective form. Recent work has tended, however, to attribute an antiviral role to the macrophages of the reticuloendothelial system. The cells which appear to be of greatest importance in this connection are the fixed macrophages or Kupffer cells of the liver. The capacity of macrophages to remove virus from the blood stream appears to be considerably enhanced in the presence of antibody, and in the case of the poxviruses there is evidence that the opsonized virus is digested by the macrophage.

Under laboratory conditions antiviral antibody can be shown to have a marked virus-neutralizing effect. It will prevent the infection of susceptible cells by virus, an effect which, in neutralization tests, is the basis of an important method of measuring antibody concentration. Work reported by Mandel, on poliovirus, indicates that two phenomena are involved in virus neutralization:

1. Certain antibodies prevent the adsorption of the virus to susceptible cells and *ipso facto* prevent infection. The prevention of adsorption appears to be particularly a function of 19S antibodies.
2. 7S antibodies, on the other hand, permit adsorption of the virus, but interfere with the normal release of the viral nucleic acid. They appear in fact to trigger a reaction in the cell by which the latter is degraded instead of replicated.

That there may also be a cellular component in acquired antiviral immunity appears to be indicated by the finding that persons with hypogammaglobulinaemia not only recover normally from virus infection but appear in most cases to possess a normal resistance to reinfection. Similar evidence has been obtained from the finding that the vaccination of guinea-pigs whose capacity to produce humoral antibody has been destroyed by X-irradiation produced a specific immunity of considerable duration. Such an immunity could clearly not have been due to interferon production since the protection afforded by the latter is of short duration and is non-specific. On the other hand, hypogammaglobulinaemics do produce a small amount of gammaglobulin and this, if it were highly avid, might serve to protect against virus infection. In this case it would not, of course be necessary to invoke the participation of delayed hypersensitivity or any other type of cellular immunity. The possibility that delayed hypersensitivity may contribute substantially to antiviral immunity and may be the type of immunity demonstrable in cases of hypogammaglobulinaemia has received considerable support from the observation of Kempe, that a child who had failed to produce antibody after vaccination, and who had, as a result, developed progressive vaccinia (p. 416), was cured by the injection of leucocytes from a previously vaccinated subject. The possibility is also borne out by the demonstration of parallelism in hypogammaglobulinaemic children between immunity to live vaccinia virus and hypersensitivity to dead vaccinia virus. In the di George syndrome (thymic aplasia) the structures developed from the third and fourth branchial arches fail to develop. The gammaglobulin levels may be approximately normal and there is at least some capacity to form circulating antibody; but there is a gross deficiency in the capacity of the affected individuals to develop delayed hypersensitivity and they frequently succumb to viral or fungal infections. Patients with extreme hypogammaglobulinaemia, on the other hand, generally respond normally to virus infections, although prone to recurrent bacterial infections.

The precise way in which delayed hypersensitivity might be responsible for antiviral immunity is entirely a matter for speculation. It is possible that the resultant inflammatory reaction—by lowering pH, increasing local temperature and diminishing oxygen tension—may have a significant antiviral effect, possibly by increasing interferon production. Alternatively, the reaction might be mediated either through sensitized lymphocytes, as has been suggested for homograft rejection (p. 85) or through sensitized macrophages, as is possible in antituberculous immunity (p. 272). A finding of considerable interest is that sensitized lymphocytes and macrophages may respond to virus infection by interferon production.

Many virus diseases are followed by a long-lasting immunity, one attack of the disease apparently being sufficient to protect the patient for life. In communities in which a particular virus is endemic and in which, after the first attack, the patient will frequently encounter the virus, such encounters will serve to boost the immunity which he has derived from the clinical attack. There is, however, evidence that antiviral immunity can persist without further contact with the virus, and therefore without further antigenic stimulation. Thus epidemiological evidence has been obtained for the persistence of immunity against measles on the Faroe Islands for at least 65 years, and demonstrable antibodies have been shown to persist for 70 years after an attack of yellow fever in a patient who could not possibly have been exposed to the virus after the initial attack. Antibody to polioviruses have been found to persist amongst Eskimos without any identifiable contact with the virus for 40 years. This long persistence of immunity in the absence of external antigenic stimulation raises the possibility that the virus may continue to be present though in a latent form, in the tissues.

The virus infections in which there is a persisting solid immunity are characteristically systemic infections with long incubation periods, and in which the virus has spread through the body via the blood. Under these conditions the invading virus would have considerable opportunity to evoke a secondary immune response in a person possessing some degree of potential immunity. In the case of surface infections with short incubation periods, e.g. influenza, the common cold and adenovirus infections, the virus can establish itself and produce overt infection before it succeeds in stimulating a secondary response.

A further factor operating in the case of surface infections is that the virus is much less accessible to antibody than is virus which must first be disseminated by the blood stream. Surface infections of this type are almost entirely due to viruses which exhibit considerable antigenic heterogeneity, e.g. influenza, the common cold and adenoviruses. Because of this, an individual may have a number of attacks of the same disease, each attack being produced by a different antigenic type of virus. For such conditions, therefore, immunity to the homologous antigenic type is much more long-lived than immunity to the disease. Recent findings have drawn attention to the crucial role of secretory IgA (see p. 107) in protecting mucous membranes against superficial virus infections such as influenza.

Immunization

Passive immunity can be produced against certain virus diseases by the injection of the gammaglobulin fraction prepared from pooled adult plasma. This is the preparation issued commercially as gammaglobulin. Its prophylactic value is due to the fact that since adults have been exposed to a number of virus infections their plasma contains a variety of antiviral antibodies. The main use of commercial gammaglobulin is for the prevention of measles in very young and debilitated children. It is also of limited value for passive immunization against poliomyelitis and infective hepatitis. For the prophylaxis of mumps and smallpox, for which commercial gammaglobulin is of little or no value, gammaglobulin prepared from convalescent serum or, in the case of smallpox, from the serum of persons who have been recently vaccinated, is sometimes employed.

Active immunity to a number of virus diseases may be produced by the inoculation of virus vaccines. The vaccines currently available are described in the sections dealing with the corresponding viruses. Smallpox and yellow fever vaccines are living vaccines. From neither of these viruses is it possible to produce a satisfactory non-living immunizing antigen. This is possibly due to the high lability of the viral antigens responsible for the stimulation of protective antibody to the various inactivation procedures used in vaccine preparation. Alternative possibilities are that the live virus, as a result of multiplication in the body, provides a greater antigenic stimulus than killed virus or that it acts by producing a delayed-type hypersensitivity. Both living and dead polio-vaccines are available, and both are highly effective in producing immunity. Inactivated measles vaccine has caused individuals given repeated doses to develop an abnormal and very severe reaction when subsequently exposed to the virus. A satisfactory live vaccine, adequately attenuated for routine use, at least in healthy children, is now available instead.

Viral vaccines are variously obtained by growth of the virus in the chick embryo, in tissue cultures or in infected animals. It should be noted that when vaccines prepared in the chick embryo are employed they must be administered with caution to individuals showing hypersensitivity to eggs.

SPREAD

The majority of the human pathogenic viruses are essentially human parasites, but some are primarily animal pathogens. Thus rabies occurs as a result of the bite of a rabid animal, usually a dog. The arthropod-borne encephalitis viruses appear to be normally parasitic on a variety of bird and animal hosts. The lymphocytic choriomeningitis virus causes a natural infection in mice. The main reservoir of yellow fever infection is in monkeys, although in epidemics the virus is spread by case to case transfer through the intermediary of a mosquito.

In most of the common virus diseases infection occurs through the respiratory tract. The common virus diseases spread by the respiratory route—influenza, the common cold, measles, rubella, mumps—are highly infectious; this is probably because of the large numbers of virus particles in the respiratory secretions. Faeces appear to be a less common source of infection than the respiratory secretions, the enteroviruses and the virus of infective hepatitis being amongst the main viruses transmitted by this route. As with bacterial infections, the extent to which infection is transmitted by the faecal–oral route depends on the standards of personal and community hygiene in the population concerned, being greater the lower the standard of hygiene. The common wart is the result of direct inoculation of virus into the skin. Rabies occurs as a result of wound infection and yellow fever, dengue, sandfly fever and the arthropod-borne encephalitides are transmitted by arthropod vectors. In most cases of arthropod-borne disease, transmission is biological, i.e. the virus undergoes a stage of multiplication in the body of the vector. Rabbit myxomatosis provides a clearly established case of mechanical transmission, the virus being mechanically transferred from rabbit to rabbit, without multiplication in the arthropod—mosquitoes in Australia, rabbit fleas in the British Isles. Some viruses, the most important of which is the virus of rubella, can be transmitted across the placenta from the maternal to the fetal circulation. The homologous serum jaundice virus is transmitted artificially, mainly by injection or transfusion.

Most virus diseases occur in epidemic form, the epidemics usually exhibiting a seasonal incidence and with an interval of one or more years between each major episode. The tendency of respiratory diseases such as influenza to occur during the winter months can probably, in part, be attributed to the fact that it is during these months that people are in closest contact with each other, thereby providing optimum opportunity for spread of virus from one person to another. In general, however, the factors determining the seasonal incidence of different virus diseases are not fully understood. We do not know, for example, why poliomyelitis shows a maximum incidence in the late summer and autumn. It is possible that seasonal alterations in basic host resistance may be involved. The periodic recurrence of many virus diseases—the classic example is measles, which in urban communities recurs at two- or three-year intervals—can be explained on the basis of alterations in susceptibility of the individuals comprising the population. Immediately after an epidemic the immunity of the whole population is high; as time passes the immune population is diluted by recruitment of children born since the previous epidemic and old enough (6 months) to have lost passively acquired maternal antibody. Once these become sufficiently numerous the scene is set for a new epidemic.

A great deal has still to be learnt as to what happens to the virus in inter-epidemic periods. In some, e.g. influenza, it appears to migrate from one susceptible population to another. In others it persists in the community and causes inapparent infection or, possibly, remains in a completely latent form. In the case of infections which are exotic for a particular community, e.g. smallpox in the British Isles and United States, the outbreak of an epidemic depends on the adventitious introduction of the disease from outside.

DIAGNOSIS

Although the identification of the aetiological agent of a virus infection is of limited value in indicating the use of a particular chemotherapeutic agent, viral diagnosis is of practical importance for a number of reasons:

1. The identification of a particular infection as a viral rather than a bacterial infection eliminates the necessity for chemotherapy.
2. In certain cases, notably smallpox, rabies and rubella, the information obtained may be of prophylactic value in respect of individuals exposed to the infection.
3. When used on a large scale, diagnostic procedures provide a valuable epidemiological record of the infection experience of the community—a record which is of considerable value for the planning of preventive measures on a community scale.
4. In diseases such as smallpox confirmation of the diagnosis is essential before appropriate administrative action to contain the infection can be put in hand.

Microscopy

Microscopy is of limited but definite value in the diagnosis of viral infections. In smallpox, virus may be demonstrable in cutaneous lesions by special staining procedures as well as by electron microscopy. Electron microscopy has also been used to establish the diagnosis in herpesvirus infections and adenovirus infections. Fluorescent antibody techniques are of considerable potential value and one of their most successful applications has been in the post-mortem diagnosis of rabies (p. 404) in infected animals. In the live subject they have been used for the diagnosis of herpes simplex, especially herpetic encephalitis and smallpox. In general, microscopy is unlikely to be of value in systemic infections in which there are no localized accessible virus-containing lesions. Cytological examination for the presence of characteristic giant cells and/or inclusion bodies is of value in the diagnosis of herpes simplex, varicella–zoster and cytomegalovirus infections and the post-mortem diagnosis of rabies.

Detection of viral antigen in lesions by serological methods

In smallpox the virus may be demonstrated in material derived from the skin lesions by serological methods. Virus-containing material may give positive precipitation tests by gel diffusion tests or positive complement fixation tests with immune antivaccinial serum.

Isolation

An attempt may be made to isolate the virus by inoculation of suitable material, derived from the patient, into the chick embryo, into an experimental animal or into an appropriate tissue culture. Since viruses are very susceptible to drying, swabs should be transmitted to the laboratory in an appropriate transport medium. If the material cannot be examined immediately it should be refrigerated, preferably at −70°C. Most of the human pathogenic viruses can now be isolated by tissue culture. The prospect of isolation is considerably enhanced if a number of different types of tissue culture are inoculated, the particular tissues chosen depending on the identity of the suspected virus. At a minimum the material should be inoculated into primary cultures of monkey or human kidney and into a human cell line such as HeLa cells. When primary cultures are employed the possibility of activating a latent virus present in the tissue is a considerable diagnostic hazard. The risk appears to be maximal with monkey kidney cultures, from which over 50 latent viruses have so far been isolated. Except for the isolation and identification of the poxviruses, chick embryo inoculation has been largely superseded by the use of tissue culture methods.

Following inoculation the presence of virus may be inferred from the development of cytopathic effects in tissue culture, lesions on the chorio-allantoic membrane or typical symptoms or death in an infected animal, or the detection of viral haemagglutinins or complement fixing antigens in infected egg or tissue culture fluids or the demonstration of haemadsorption in tissue culture. As a final stage in the identification of an animal virus its effects on an experimental animal, tissue culture, or chick embryo or its haemagglutinating capacity should be shown to be neutralized by a specific antiserum.

It should be noted that the isolation of a virus from a patient does not necessarily mean that the patient's symptoms are being caused by that virus. This difficulty arises particularly in the case of patients from whom viruses of the ECHO, Coxsackie (Chapter IV.2) and adenovirus groups (Chapter IV.8) are isolated. Proof of infection may require a demonstration that the patient has developed antibody against the virus, but even this

Table IV.4. Isolation of viruses

Tissue culture or animal	*Virus*
HeLa cells or HEp-2 cells (continuous cell line)	Polio Cocksackie B Adenovirus Respiratory syncytial virus* Herpes simplex
Monkey kidney (primary or secondary culture)	Polio Echovirus Cocksackie B Myxovirus† Reovirus M rhinovirus§
Human amnion (primary or secondary culture)	Polio Cocksackie B Echovirus Varicella–zoster Measles Rubella‖
Human diploid cells (lung)	Zoster–varicella H rhinovirus Cytomegalovirus
Chick embryo: amnion	Influenza Mumps
Chick embryo: chorio-allantoic membrane	Herpes simplex Variola Vaccinia
Chick embryo: yolk sac	Rickettisae Coxiellae Chlamydiae
Mouse	Arboviruses Cocksackie A‡ Rabies Herpes simplex Encephalomyocarditis Lymphocytid choriomeningitis

* Will grow only in certain sensitive cell lines (Bristol HeLa).
† Parainfluenza and mumps grow best in rolled cultures at 33°C.
‡ Only newborn mice are susceptible.
§ Usually require low pH and low temperature (33°C).
‖ The CPE produced by rubella appears very late and is irregular; other cell lines, e.g. RK 13 are better for isolating this virus.

does not prove that the patient's symptoms are due to the infection which has been demonstrated.

Serological methods

Where possible two specimens of serum should be examined, one taken during the acute phase of the infection and the second about two weeks later, usually during convalescence. As a rule a diagnosis can be made only if there is at least a four-fold rise in antibody titre between the two examinations.

Complement fixation is on the whole the most useful technique for the estimation of antiviral antibody. Since, however, complement fixing antibodies are directed against soluble antigens as well as against constituent antigens of the virus, complement fixation tests are usually less type-specific than neutralization tests. Complement fixing antibodies tend to persist for a shorter time after infection than neutralizing antibodies; as a rule, therefore, they are of more value as an index of current or recent infection than neutralization tests. In some cases, notably in poliomyelitis, the maximum titre may have been reached by the time the acute phase specimen of blood has been taken. Under these circumstances a serological diagnosis may have to be made on a single serum examination.

The ability to measure separately the different classes of immunoglobulin specific for a particular virus, chiefly by the indirect immunofluorescence technique, has, by enabling us to determine the level of IgM antibody relative to IgG, provided an additional criterion for judging whether antibody is due to a recent or a remote infection. This can be extremely helpful if no early specimen of serum is available, where the two samples both show high levels of antibody but there has not been a significant rise and in the case of pregnant women who have been exposed to rubella.

It cannot be too strongly emphasized, however, that isolation of a virus or the demonstration of an immune response only establishes the existence of an infection: whether the infection can be accepted as the cause of all or any of the patient's symptoms is a matter of clinical judgement.

TREATMENT

The treatment of viral infections presents quite different problems from those involved in the treatment of bacterial infections. The agents available and the limitations on their use and efficacy have been discussed on p. 362.

In most systemic virus diseases recognizable symptoms only appear after the peak of virus multiplication has passed. In such conditions, therefore, chemotherapy, even if an agent suitable for systemic administration were available, would not appear to be practicable. Conditions which might respond to a specific antiviral agent are those in which there is a prodromal period of non-specific symptoms before viral localization, e.g. smallpox, poliomyelitis and measles, or those showing a prolonged slow development, e.g. herpetic keratitis, progressive vaccinia and virus-induced tumours.

Although viruses are not susceptible to anti-bacterial chemotherapeutic agents, such agents can nevertheless be of considerable value in controlling the secondary bacterial infection which is liable to complicate certain viral diseases, particularly those, such as influenza, and measles which involve the respiratory tract.

IV.4

The Picornaviridae

Picornaviridae are small icosahedral non-enveloped RNA ether-resistant viruses, with diameters in the range of 25 to 30 nm. Such viruses are extremely widespread in man and animals. They have been divided into two main groups: the *Enteroviruses* and the *Rhinoviruses*. Some would recognize a third group, the *Cardioviruses,* comprising the EMC (*encephalomyocarditis*) virus of mice and related viruses. Similar in many ways to the other groups and included in the family *Picornaviridae* are the *Caliciviruses*. The particles of these viruses are larger, 35–40 nm, and they are not stabilized by divalent cations: they affect pigs and cats, but not man, so far as is known.

ENTEROVIRUSES

As their name implies enteroviruses are viruses which are to be found in the alimentary tract, in some cells of which they are able to multiply. Three groups are distinguished: polioviruses, Coxsackieviruses and Echoviruses.

Enteroviruses are occasionally disseminated through the body to specific target areas by the blood stream and in some cases perhaps also by neural pathways. The most important of these target areas is the central nervous system, which can be invaded by viruses of all three groups but particularly by the polioviruses. Children are more frequently infected than adults and infection, particularly in younger children, is often asymptomatic. Except for the Coxsackie A group, enteroviruses grow readily in tissue cultures of monkey and human cells. In infected cells they produce a characteristic cytopathic effect, the cells becoming round in shape, smaller and with highly pyknotic nuclei. In agar overlay monolayer cultures they produce distinctive plaques. They are amongst the most stable of the human pathogenic viruses outside the animal body. Since they are resistant to many of the commonly used disinfectants, material which has been contaminated with the more highly pathogenic enteroviruses should be sterilized by heat. Susceptibility to heat inactivation is considerably reduced in the presence of high concentrations of certain cations, particularly Mg^{++}. Advantage is taken of this property in the preparation of live attenuated poliomyelitis vaccines which are stabilized by the addition of magnesium chloride.

Polioviruses

Properties of the viruses

Three antigenically distinct types of poliovirus—types 1, 2 and 3—are known. Although immunologically distinct, the types appear to show some minor antigenic overlap, which is most readily demonstrable in complement fixation tests with virus which has been inactivated by heat or by ultra-violet irradiation. In general the type I virus is that most frequently responsible for clinical infection and is the cause of most epidemics. By the use of strain-specific antisera it is possible to demonstrate minor antigenic differences between different strains of the same type. These differences are of value as marker characteristics. They do not appear to affect the response to immunization since vaccine prepared from any one strain will immunize against all other strains of the same type.

In addition to infective virus, infected tissue culture fluids contain a considerable amount of non-infectious, nucleic acid-free, 'incomplete' virus, which can be separated from infectious virus by centrifugation in a sucrose density gradient. This non-infectious or 'incomplete' virus possesses mainly type-specific antigens which are antigenically somewhat different from the type antigens of infectious virus. This non-infectious antigen, known as the C antigen, can be obtained from the D antigen and the antigen of infectious virus by heating. Unlike the C antigen, the D antigen stimulates the production of protective antibody. Antigen obtained by heating shows appeciable species (group) specificity in complement fixation tests and a corresponding loss of type specificity. There is some limited cross-reactivity in neutralization tests between types 1 and 2 and types 2 and 3 but not between 1 and 3. Previous infection with type 2 confers some protection against paralysis due to type 1 virus.

Freshly isolated polioviruses will produce infection only in primates; these may be infected by the intracerebral and intraspinal routes. Chimpanzees and cynomolgus monkeys may be infected also by the oral route. Certain strains of all three types, but particularly of type 2 have been adapted to mice; these animals are, however, of no value for isolation. Apart from some specially adapted strains the poliomyelitis viruses will not grow in the chick embryo.

The discovery in 1949, by Enders, Weller and Robbins, that polioviruses could be cultivated in tissue cultures of a variety of primate extraneural cells initiated a revolution in the study and control of poliomyelitis. For purposes of isolation, monkey kidney cells and HeLa cells have been most used. In the cells of infected tissue cultures all the types produce a well marked cytopathogenic effect.

By passage in rodents and in tissue cultures, strains of attenuated virulence have been derived, some of which, notably those developed by Sabin, are suitable for vaccine production. These strains possess identifiable marker characteristics by which they can be differentiated from virulent strains:

1. They possess neurovirulence for monkeys only when inoculated intraspinally—the most sensitive route of inoculation.

2. They grow poorly at 40°C as compared with 34 to 36°C. Virulent strains grow well at 40°C. This is the basis of the RC_{t40} or temperature marker.

3. Their capacity to produce plaques in agar overlay monolayer cultures is depressed by a reduction in bicarbonate concentration of the nutrient medium.

4. They show minor antigenic differences which may be demonstrated by intratypic neutralization tests.

5. They are stable and show only a limited tendency to revert towards the wild type even on repeated passage in man.

Poliomyelitis

It was at one time thought that poliomyelitis was exclusively an infection of the central nervous system. It is now realized, however, that this is not so and that paralytic symptoms are in fact of infrequent occurrence. According to one widely accepted estimate they appear on average in only one out of every hundred persons infected with the virus. At least 90% of infections are completely asymptomatic. In others, the only evidence of infection is the occurrence of a mild illness clinically resembling influenza. These symptoms usually disappear after a few days and the patient makes an uneventful recovery. This form is usually known as *abortive* poliomyelitis. It is due to multiplication of the virus in the alimentary tract and possibly also in other extraneural foci. In only a few cases is this first phase of infection followed by signs of involvement of the central nervous system. These usually appear some six to seven days after the onset of symptoms, and are sometimes preceded by an interval during which the patient feels quite well. In most cases they are simply meningitic in character and the condition progresses no further. In this type of infection, which is known as *non-paralytic poliomyelitis,* the cerebrospinal fluid shows an increase in protein and in leucocytes. In the early stages the cellular reaction is predominantly polymorphonuclear in type, the cell count usually being in the range of 50 to 150 cells/μl, but after a few days becomes predominantly lymphocytic. Non-paralytic poliomyelitis is clinically indistinguishable from aseptic meningitis caused by a number of other viruses and by leptospirae. In a minority of cases (about 1 to 2%) meningitic symptoms are followed by the development of the typical flaccid paralyses which are the classical signs of the disease. In some cases, especially young children, the earlier stages may be unnoticed and paralysis may be the first sign to be observed.

The characteristic lesions of paralytic poliomyelitis are degenerative change occurring in the anterior horn cells of the spinal cord and in the motor nuclei of the cranial nerves. In fatal cases there is usually extensive involvement of the central nervous system with widespread lesions in the cord, medulla, cerebellum and motor areas of the cerebral cortex. By the time the first paralytic symptoms appear, extensive degenerative changes have already occurred in the central nervous system. Since the degenerative process is usually arrested within a few days of the appearance of symptoms, paralysis reaches its maximum extent early in the disease, being complete usually within three days of onset of the paralytic phase. Pain and muscular spasm are often prominent just before the onset of the paralysis.

The most widely accepted view on the pathogenesis of poliomyelitis is that proposed initially by Bodian, on the basis of experimental oral infection in chimpanzees. This suggests that the virus multiplies first in the tonsils and/or the Peyer's patches of the small intestine. Coincident with this multiplication there may be, particularly in young children, a short pyrexial phase of a few days' duration, beginning about three days after initial infection. During this pyrexial phase the virus gains access to the lymphatics by which it is transported to the local lymph nodes, cervical and/or mesenteric. Further multiplication of virus in lymph nodes is followed by a viraemia by which the virus is carried to the central nervous system. Although difficult to demonstrate, the evidence for viraemia in man—occurring about the time of the initial general symptoms of infection—is considerable. Some workers consider, however, that it only represents an overflow of virus from extraneural sites of multiplication and is not an essential preliminary to invasion of the central nervous system. On this view the virus is believed to gain access to the nervous system by neuronal routes.

The distribution of lesions in the central nervous system is suggestive of neuronal spread but the fact that gammaglobulin and active immunization protect against paralytic infection is more readily explained by the viraemic theory. As yet, however, there is no decisive evidence in favour of either the viraemic or neuronal theories.

In poliomyelitis, virus is demonstrable in the pharynx and in the faeces. It is present in both these situations immediately prior to and during the acute stages of infection and in the faeces for a variable period, on occasion for some weeks after the appearance of paralytic symptoms. It is therefore presumed that the disease can be spread both by droplet infection and by infected faecal material. The relative importance of these two routes, however, has still to be determined. Since epidemics of poliomyelitis occur mainly in the summer and autumn—times when gastro-intestinal infections are most frequent—it has generally been assumed that of the two the gastro-intestinal route is the more important. In countries with relatively low standards of hygiene, particularly tropical countries, it is probable that considerable dissemination of the virus occurs, as in the enteric fevers, through large-scale excretal contamination of the environment. In contrast to diseases like typhoid and infective hepatitis, large scale water- or food-borne epidemics of poliomyelitis do not seem at all common, if indeed they occur at all. It seems likely that in poor and overcrowded communities with low standards of personal and community hygiene, enteroviruses, including poliomyelitis viruses, spread freely by the faecal–oral route. In such communities, especially in tropical areas, children under five years of age show very high carriage rates for enteroviruses and poliomyelitis, when it occurs, affects this age group and really is 'infantile paralysis'. Serological surveys indicate that individuals over the age of five or six almost all have antibody to all three types of poliomyelitis virus; the disease is virtually never seen in adult members of the indigenous population and large epidemics do not occur. As standards begin to improve, more and more young children escape infection in the first few years and the proportion of cases at older ages begins to increase. This, in turn, allows the accumulation of a large enough number of susceptible individuals for an epidemic to be possible. When this occurs seems to depend on the introduction of a strain of virus with an exceptional capacity to spread, a property which appears generally to go hand in hand with a high level of virulence: in epidemics the paralysis rate amongst those infected may greatly exceed the 1% given above which is an average figure based on all infections, both in epidemics and inter-epidemic periods. Epidemics of paralytic poliomyelitis are generally, but not exclusively, due to type 1 virus. By the time standards of personal and, particularly, of community hygiene reach those general in the countries of Europe and North America, faecal–oral spread seems to be largely restricted to spread within households, particularly those containing young children. Interfamilial spread seems to be more or less confined to droplet infection. The capacity to colonize the pharynx as well as the intestine varies from strain to strain, being virtually absent from the strains used as live vaccines. If, as seems not improbable, one of the properties which characterize epidemic strains of the virus is a capacity to multiply to high levels in the nasopharynx and to spread by droplets to individuals in close contact, such strains are the only ones likely to spread at all widely in communities with a high standard of hygiene. Strains of low or moderate virulence will cease to circulate, the majority of children will escape infection in their early years and from time to time epidemics will occur. By the time that poliomyelitis vaccination became generally available, we had reached the point in the United Kingdom at which roughly one-third of the cases still occurred in children under five, another third in children aged five to 15 years and the rest in individuals over 15 years of age. A disquieting feature in the current world-wide picture is the rising incidence of the disease in the countries of Africa, tropical Asia and South America which have traditionally shown the low incidence and absence of epidemics characteristic of the least favoured communities.

There is no doubt that the severity of the disease and its incidence are both greater in older children and adults, not already immune, than in young children. When epidemics occur in hitherto isolated communities in which no one has any previous experience of the virus, the highest incidence is generally in the 5 to 20 year age range. The case fatality rate, whilst high amongst the small number of cases which occur under one year, falls to a low level throughout the rest of childhood, begins to climb again at about the age of 12, and then rises progressively with increasing age. It is difficult to give precise case fatality rates as much will depend on the extent to which the abortive, the non-paralytic and the mildest paralytic cases are recognized, the virulence of the organism and the age of the patients. It is the higher incidence of the more dangerous bulbar form of the disease which largely determines the increased mortality amongst older patients and much will depend on the proportion of such cases and the facilities available for their treatment. One can say that, in general, over 80% of cases should survive. Of the survivors, 2–5% will remain totally disabled, 10–15% will be left with significant disability requiring appliances or surgery, a similar proportion will have residual muscle weakness which does not seriously inconvenience them and the remainder, about half of which will have been non-paralytic, will recover completely.

There is now clear evidence that various extrinsic factors—in particular, fatigue, trauma, pregnancy, tonsillectomy and the intramuscular administration of prophylactics containing alum (Chapter II.8)—may favour the development of paralytic poliomyelitis in individuals exposed to infection. These factors are presumed to act by transforming asymptomatic and abortive infections into paralytic ones, rather than by increasing the overall susceptibility to infection, but precisely how they do so has not been satisfactorily explained.

Individuals with inapparent infections so greatly outnumber diagnosed cases that they are the principal source of infection; indeed the very fact that they are not ill means that they continue about their daily business and have a wider range of contacts than those ill enough to take to their beds. The rarity of case to case transmission long delayed universal acceptance of the view that poliomyelitis is an infectious disease.

Diagnosis

Usually, a diagnosis of paralytic poliomyelitis can be readily made on clinical grounds. It should be remembered, however, particularly with the declining incidence of poliomyelitis, due to vaccination, that a paralytic syndrome can be produced by certain ECHO and Coxsackie serotypes, particularly Coxsackie A7, and by mumps. Tick-borne encephalitis may also mimic poliomyelitis and not uncommonly leaves residual paralysis (see p. 394). The Guillain–Barré syndrome, acute infective polyneuritis, can usually be distinguished once the characteristic changes have appeared in the CSF. The correct diagnosis can only be made by using laboratory procedures, which are particularly important in the diagnosis of non-paralytic poliomyelitis. This condition, presenting as an aseptic meningitis, may be caused not only by other enteroviruses and by LCM virus, but also by a number of non-neurotropic viruses, particularly the mumps virus, and also by *Leptospira canicola*. Sporadic cases of aseptic meningitis, occurring during the winter and spring are usually due to agents other than the enteroviruses.

Where possible an attempt should be made to isolate the virus by culture. It is essential to culture the faeces as virus may be recovered from this source when none is demonstrable in the CSF or a throat swab. Rectal swabs give a significantly lower isolation rate than faeces. Antibodies to the poliovirus may be demonstrated by neutralization tests in tissue culture and by complement fixation tests. Neutralizing antibody appears early during the course of infection and may be maximal at the time of onset of paralysis. It is therefore frequently impossible to demonstrate a significant, i.e. four-fold, rise in antibody titre between the acute and convalescent phase sera. Neutralizing antibody persists for a long time, sometimes apparently for life; this reduces the diagnostic significance of such antibody detected by a single examination. Complement fixing antibody appears somewhat later in infection than neutralizing antibody and in the majority of cases it is possible to demonstrate a rise in antibody titre. For its demonstration live virus is preferable as an antigen, since the C antigen of heated virus may react strongly with acute phase serum and shows less type specificity. Since complement fixing antibody is of short duration, examination of a single specimen may yield a diagnostic titre. Overall, the complement fixation test permits a serological diagnosis in up to 80% of cases. However, as a means of diagnosing infection by poliomyelitis viruses, virus isolation is much preferable to serological tests. It must be emphasized that isolation of the virus or a serological response establish the existence of the infection. Since infection by enteroviruses is so often without symptoms, it is always a matter of clinical judgement to decide whether the infection which has been demonstrated can account adequately for the patient's clinical condition.

Prophylaxis

In the control of poliomyelitis at the community level, isolation of cases and quarantine of contacts are no longer believed to be of any value. It is possible that some restriction on large gatherings such as football matches may have some beneficial effect although there is little evidence for this. It is, perhaps, more logical to close public swimming baths combining as they do, strenuous exercise with particularly favourable opportunities for faecal–oral transmission of infection.

Gammaglobulin prepared from pooled adult plasma has some protective value if given early in the incubation period; the protection obtained is, however, of short duration, lasting at most for some four to five weeks. Its occasional use has, however, been recommended for susceptible persons introduced into an environment in which the disease is prevalent, but it is only effective if given very soon after exposure, or just before.

Two types of vaccine are available for active immunization: Salk killed vaccine and live vaccine prepared from attenuated strains of virus.

Salk vaccine. This consists of virus of all three antigenic types grown in monkey kidney tissue culture and inactivated by exposure to a low concentration of formalin (1/10 000 HCHO at 37°C for 10–12 days). Three injections of the vaccine are usually given with an interval of three to six weeks between the first and second injections and an interval of six to seven months between the second and the third. This long interval is required in order to ensure a satisfactory secondary antibody response.

In short-term follow-up a full course of three injections with the original vaccines gave 80–90% protection against paralytic infection. Subsequently more potent inactivated vaccines have been developed giving even higher levels of protection particularly against type 1 virus. Although potent inactivated vaccines, which are given by injection, do not produce the same degree of resistance to gut infection as living attenuated vaccines, experience in Sweden where high potency inactivated vaccines are in use and live vaccines have never been used do indicate that there is some effect. Not only has clinical poliomyelitis declined to negligible proportions, but circulation of poliomyelitis viruses has also virtually ceased. So clearly some degree of resistance to infection has been engendered by the vaccine. Unlike live vaccine, inactivated vaccines are unaffected by interference. Inactivated vaccines are, however, more expensive to produce and to administer and, as they must be given by injection, more highly trained personnel are needed for their administration. They are of no value in the face of an epidemic. Killed vaccine may be combined with diphtheria, tetanus and pertussis antigens without any diminution in antigenic potency.

Live vaccine. The feasibility of using live attenuated polioviruses for human immunization was first established by Koprowski who showed that children could be immunized apparently without risk, by the oral administration of virus attenuated by passage in rodents. Subsequently strains suitable for oral administration were developed by Cox and by Sabin. Sabin's strains derived by plaque selection in tissue culture are the basis of most of the live vaccines currently in use. Live vaccine is now accepted as the antigen of choice for immunization against poliomyelitis in most countries. It is easier to administer, cheaper to produce, stimulates an immunity which, resembling that produced by natural infection, develops more rapidly and is at least as effective in protecting against paralytic infection. Where it has been used on a community-wide scale wild-type polioviruses have been virtually eliminated. In addition to protecting against paralytic infection live vaccine produces a local immunity of the intestinal tract. At first this local immunity is due to viral interference (p. 363) but later local antibody production or a local cellular immunity is involved. Whatever its basis it develops sufficiently rapidly to permit the use of live vaccine as a prophylactic during the course of an epidemic and as administration of a vaccine prepared from virus one type has been shown to check an epidemic due to another type (e.g. type 2 vaccine in a type 1 epidemic in Singapore) the effect is likely to be due mainly to interference rather than to specific immunity.

There are, however, certain potential disadvantages inherent in the use of live vaccine:

1. Live attenuated virus is not completely avirulent. It multiplies in the intestinal tract. This multiplication is in fact essential to provide an antigenic mass sufficient for the production of a solid immunity. The type 2 component, at least, can produce a viraemia. In addition some residual neurovirulence is demonstrable by intraspinal inoculation of monkeys. It is therefore possible that, on occasion, the vaccine strains might invade the central nervous system and produce paralytic disease, though their diminished neurovirulence would be expected to make this an extremely rare event. From careful scrutiny of data in vaccine associated cases of poliomyelitis occurring in the United States shortly after vaccine administration, it seemed probable that the type 3 component at least was in fact responsible for some of the cases. These infections occurred only in adults, and with an estimated incidence of 1 in 1000 000 vaccinations. Because of this it may be considered preferable not to give live type 3 vaccine to adults not previously immunized. Under present conditions, however, the risk is not such as to modify the use of live attenuated vaccines for childhood immunization.

2. The possibility that the vaccine strains, which are to a limited extent transmissible from the individuals to whom they are administered to others, might, as a result of serial passage through the human intestinal tract, regain some of their human pathogenicity. The reality of this danger has been shown particularly by Dick and his associates, who found that intestinal passage led to an increase in the neurovirulence of the vaccine strains for monkeys. Studies by Russian workers, however, suggest that the degree of reversion which occurs on even repeated human passage is limited. From the evidence available it seems unlikely that the slight increases demonstrated in this way are associated with significantly increased risk. Any danger to the community of an increase in the virulence of the vaccine strains would presumably be largely eliminated if the vaccine were given to all or most of the susceptible members of the community simultaneously. This procedure has not been adopted in the British Isles, although the vaccine is included in routine immunization schedules; there is no evidence that failure to do so has had any untoward effect.

3. The possibility that other enteroviruses prevalent in the community might interfere with the establishment of the vaccine strains in the intestinal tracts of a proportion of the population. This risk would be minimized if the vaccine were given only at times of low enterovirus prevalence.

4. The danger of disseminating other viruses, pathogenic to man, which might be present in the tissue cultures used for vaccine preparation. This danger could probably only be totally eliminated by the propagation of the vaccine in established cell lines proved to be free from contaminating viruses. So far the only virus known to have been

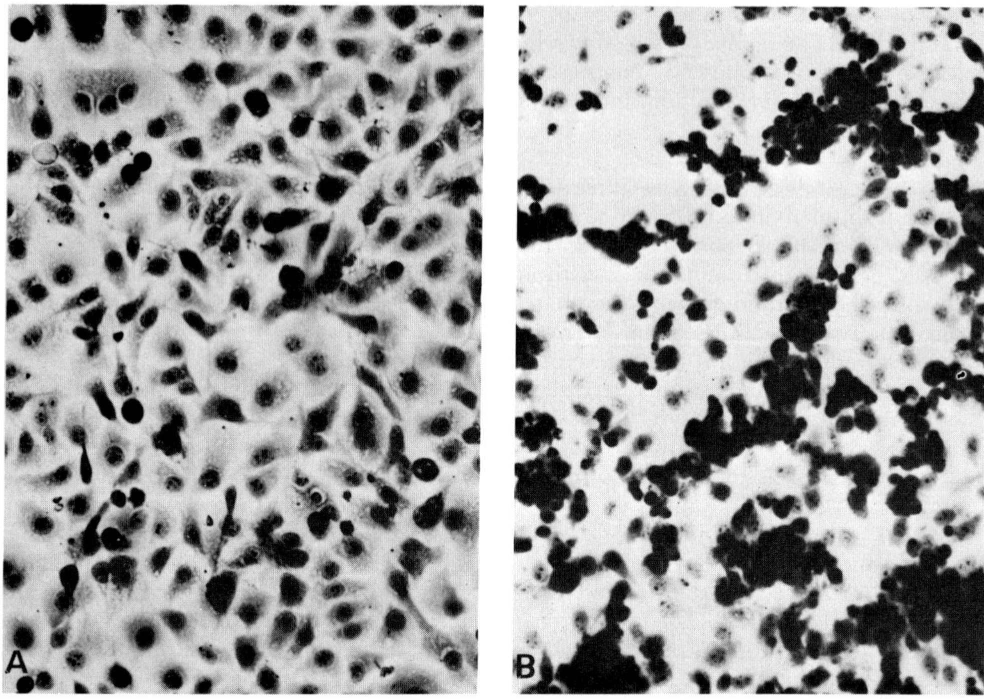

Fig. IV.19. Cytopathogenic effect in HeLa tissue culture infected with a Coxsackie B3 virus. A, Uninfected culture. B, Infected culture.

transmitted in this way, the so-called vacuolating agent SV_{40}, is, as far as is known, non-pathogenic for man. The recent reports of the isolation of a virus closely similar to if not identical with SV_{40} from two cases of progressive multifocal leucoencephalopathy may mean that this assertion can no longer be sustained. Vaccines prepared on the human diploid cell line WI-38 are now coming into use.

The vaccine may be administered as a trivalent vaccine containing all three antigenic types of virus—this is the vaccine normally used in the British Isles—or as a monovalent vaccine containing only one antigenic type. The trivalent vaccine is given in three doses in syrup or on a lump of sugar with an interval of not less than one month between each feeding. The administration of three doses is essential because following each feeding only one type usually establishes itself, even in individuals susceptible to all three. Doses are of the order of 10^5 to $10^{5 \cdot 5}$ $TCID_{50}$ of each type contained in a volume of 0·15 ml (3 drops). If the doses are equal type 2 tends to be the first to establish itself, but because of the greater importance epidemiologically of type 1, it has been customary to make the dose of this type larger to try to ensure that it is the first. However, in view of the slight cross-protection apparently given by type 2 against the other two types, a case can be made out for allowing type 2 to get in first. To avoid the possibility of interference, it is necessary to allow enough time to elapse for the virus established at the previous feeding to be eliminated before giving another dose.

Coxsackie viruses

The name of this group of viruses is derived from the township of Coxsackie in New York State from which the first member of the group was isolated by Dalldorf and Sickles in 1948 from the faeces of children believed to be suffering from poliomyelitis. Over 30 serological types have been distinguished.

On the basis of their pathogenicity for newborn mice they are divided into two groups—A and B.

Of the serological types indicated above at least 25 belong to group A and six to group B. Cross-reactivity between the different types, due to shared group antigens, is demonstrable with human sera. Different strains within each type may show antigenic differences.

On inoculation into mice group A viruses produce generalized inflammation and degeneration of the voluntary musculature, the animals eventually dying of respiratory paralysis, and group B viruses produce localized areas of myositis, degenerative changes in the brown intercapsular fat pads and necrotic lesions in various organs, particularly in the pancreas and, especially following intracerebral inoculation, in the brain.

The only laboratory animals uniformly susceptible are suckling mice and hamsters; these may be infected by the intracerebral, intraperitoneal or subcutaneous routes. Some strains have been adapted to the chick embryo but will not grow in the embryo on primary isolation. All group B strains grow readily in monkey kidney tissue cultures with the production of cytopathogenic effects, and also, but as a rule less readily, in HeLa cells. Only a few group A strains can be propagated in tissue cultures. A9 will grow in monkey kidney and A11, 13, 15, 18, 20 and 21 will grow in HeLa cells but not in monkey kidney.

The Coxsackie A7 virus produces a haemagglutinin for fowl red cells which possess properties similar to those of the vaccinia haemagglutinin. Strains of certain other serotypes are capable of agglutinating human group O red cells, haemagglutination being inhibited by type-specific antiserum.

So far all the Coxsackie group B viruses and about half the group A viruses have been shown to be capable of producing clinical disease. Viruses of both groups can cause aseptic meningitis and mild febrile illnesses. Herpangina, a condition occurring in children and characterized by fever and the presence of small ulcers on the tonsils, uvula and soft palate, is caused almost exclusively by group A viruses. On occasion group A strains have been found to be associated with a paralytic syndrome indistinguishable from paralytic poliomyelitis. This is particularly true of Coxsackie A7, because of which this serotype was at one time classified by Russian workers as a poliovirus. A syndrome characterized by the occurrence of vesicles in the mouth and a rash, at first papular and later becoming vesicular, on the hands and feet (hand, foot and mouth disease) is caused by a number of Coxsackie A serotypes, particularly A16. Coxsackie A21 is identical to the so-called Coe virus. It causes a syndrome similar to the common cold and is more often isolated from the pharynx than from the faeces. Bornholm disease or epidemic myalgia—a febrile condition which is associated with severe attacks of pain in various voluntary muscles, particularly those of the thorax and abdomen—is caused by group B viruses. Group B viruses can also cause acute myocarditis in infants occurring mainly in those of less than ten days of age. It is possible that in certain of these cases the fetus may be infected in utero. The disease has a high fatality rate and at post mortem there is usually evidence of involvement of the liver and brain. Cases of pericarditis and, more recently, of myocarditis due to Coxsackie B viruses have been reported in older children and adults. Claims that Coxsackie viruses can cause cardiac lesions closely resembling those of rheumatic fever cannot be accepted without further evidence. The virus of swine vesicular disease possesses close antigenic and morphological similarity to Coxsackie B5, although the two can be distinguished. There is as yet no evidence that pigs acquire their infection from man, but some evidence to suggest that workers handling the swine virus may have been infected by it. Pancreatic lesions are common in baby mice infected with Coxsackie B viruses and there is some evidence to suggest that they may play a role in certain cases of diabetes in man.

Diagnosis

An attempt may be made to isolate virus from the throat or from the faeces or, in cases of aseptic meningitis, from the cerebrospinal fluid. The material is usually inoculated first into monkey kidney and HeLa tissue cultures; specimens which are negative in these are then inoculated into suckling mice. The sensitivity of mice to Coxsackie B viruses decreases markedly or disappears after 24 hours of age. Susceptibility to Coxsackie A viruses declines more gradually between about 48 hours and 15–21 days. For isolating Coxsackie B viruses, tissue cultures are generally preferred to mice. For Coxsackie A viruses mice between 24 and 48 hours old should be used as younger mice tend to show a high non-specific mortality.

Neutralizing and complement fixing antibodies

appear early during the course of infection. The serological heterogeneity of the group is, however, a considerable obstacle to the use of serologic tests as routine diagnostic methods. When a Coxsackie virus has been isolated serological tests may be used to demonstrate an antibody rise in the patient's serum against it, thereby helping to confirm its aetiological role.

Echoviruses (enteric cytopathogenic human orphan viruses)

These were originally defined as a group of viruses which could be isolated in monkey kidney tissue cultures in which they produced an enterovirus type of cytopathogenic effect; they differed from the Coxsackie viruses in not affecting mice and from the polioviruses in not being pathogenic for monkeys. Some viruses, however, which differ in some respects from this original pattern, are now included in the group. Thus many strains of type 9 can produce lesions resembling those produced by Coxsackie type A in suckling mice and it has been proposed to rename this virus Coxsackie A23. Coxsackie A9 has certain properties, including sensitivity to HBB, in common with Echo viruses and shows an antigenic relationship to Echo 9—Coxsackie A23. Following intracerebral inoculation of monkeys, some types have been found to produce neuronal lesions when large inocula are used. Monkey kidney cells are the cells of choice for isolation but echoviruses also grow well in primary cultures of a variety of human cells. They grow poorly in HeLa cells and in most continuous cell lines.

At least 30 serotypes have been differentiated by neutralization and complement fixation tests. As has also been demonstrated for the Coxsackie viruses, some strains—the so-called prime strains—possess an inherently low susceptibility to antibody in neutralization tests. These strains, however, readily stimulate the production of antibody, with a high neutralization titre against normal strains of the same serological type. There is evidence that prime strains consist of a mixture of poorly neutralizable and normally neutralizable particles. In some cases it has been shown that such strains may be transformed, on passage in tissue culture, into normally reacting strains. Because of close antigenic similarity between viruses of type 1 and type 8, the two have been combined as type 1, and type 8 no longer exists. The former type 10 was shown to differ in many ways from other echoviruses and came to form the first member of the Reovirus group. Echovirus type 28 has been reclassified as a rhinovirus.

Many viruses are capable of agglutinating human red cells. The red cell receptor is susceptible to trypsin and to reagents reacting with sulphydryl groups. It must therefore be at least partly protein, the sulphydryl groups being important too, probably in maintaining the configuration of the polypeptide. Virus elutes from the cell with increase in temperature and elution is said to be associated with breakdown of the receptor in some cases. The receptor is quite distinct from that for the myxoviruses.

Most types have now been found to be associated with clinical syndromes, both sporadic and epidemic. As is the case with all enteroviruses, the majority of infections by the echoviruses are inapparent. However, they can cause severe and even fatal disease. Probably the most characteristic syndrome is aseptic meningitis, which may be associated with a rash, usually maculopapular, though it may be vesicular, purpuric or mimic the rashes of exanthemata such as scarlet fever, measles or rubella. An enanthem indistinguishable from the Koplik's spots of measles has also been described. Similar rashes are seen without evidence of meningeal involvement, but associated with fever. Generalized lymphadenopathy, like that seen in rubella, is also not uncommon. Respiratory infections are associated with certain types, usually mild, febrile upper respiratory illnesses, but occasionally with involvement of the lung producing interstitial pneumonitis. A severe outbreak of infection due to Echovirus type 30 in Germany was associated with herpangina, follicular tonsillitis, lymphadenopathy, bronchitis, interstitial pneumonia, a morbilliform rash and Koplik's spots. Some patients showed signs of meningitis or encephalitis. Cases with orchitis or parotitis were also seen. There were six deaths amongst 590 cases. Although the human intestine is their natural habitat, it has proved curiously difficult to assess their role in the causation of diarrhoeal disease. There are a few records of outbreaks in closed communities in which the association seems convincing. But in general, all one can say is that virological studies on infants or older persons with gastroenteritis are usually negative and that when enteroviruses are isolated from cases of diarrhoea, the same viruses can generally be isolated from contacts without it. Claims have also been made for the ability of Echoviruses to cause hepatitis, pericarditis, conjunctivitis and the Guillain-Barré syndrome. Type 16 causes the so-called Boston exanthem. Very exceptionally they may cause a paralytic syndrome resembling poliomyelitis.

The distinction between the different groups of enteroviruses is not always clear-cut, and it has been proposed that, in future, human enteroviruses shall be numbered serially from 68 without any attempt to allocate them to the four subgroups.

RHINOVIRUSES

It has been known for many years that the common cold can be transmitted to human volunteers by nasal instillation of infected nasal secretions. Colds are also transmissible to chimpanzees, but no other laboratory animal is susceptible, nor has it been possible to isolate the causative agent in the chick embryo. The first major step in the isolation of a common cold virus was taken by Andrewes and his associates of the Medical Research Council Unit at Salisbury in 1953 when they succeeded in isolating a cold virus in human embryo lung tissue culture. The virus did not have a cytopathogenic effect and its presence could be recognized only by its capacity to produce colds in volunteers. The next advance was the discovery in 1960 by the same group of workers, that viruses capable of producing colds in volunteers could be readily isolated in human embryo tissue cultures maintained at 33°C rather than at the customary temperature of 36°C and with a low bicarbonate concentration and therefore a low pH. Under these conditions a recognizable cytopathogenic effect was produced. For this group of viruses the name *rhinovirus* has been proposed.

The rhinoviruses so far isolated have been subdivided into two groups: the H rhinoviruses, which will grow only in cultures of human tissues, and the M rhinoviruses, which are also capable of growing in monkey kidney tissue cultures. About 100 serologically distinct types have so far been recognized, mostly H types, and undoubtedly more remain to be discovered. These are specifically human pathogens but similar viruses have been isolated from cattle and horses. The foot-and-mouth disease viruses are extremely important animal rhinoviruses. Over 40 serological types are now recognized, falling into seven groups. They may rarely infect man, although his role in the dissemination of the virus is mainly mechanical; viruses may be carried on his boots or clothing. Unlike the enteroviruses which they closely resemble, the rhinoviruses are rapidly inactivated at pH 3 and are not stabilized to heat inactivation by high concentrations of divalent cations.

Although the rhinoviruses are almost certainly the main cause of the common cold syndrome, the H rhinoviruses being more frequently involved than the M viruses, colds may also be caused by viruses of other groups. There is clear evidence that parainfluenza viruses, influenza viruses, adenoviruses, respiratory syncytial virus, Coxsackieviruses, particularly A21 and B5, and some Echoviruses are occasionally involved. In addition, some cases of the common cold are due to viruses which can be transmitted to volunteers but which cannot be propagated in the conventional type of tissue culture. From some such cases, Tyrrell and Bynoe have isolated viruses in organ cultures of human embryonic respiratory epithelium in which the presence of virus appears to be most readily detected by electron microscopy. Two such strains, examined by Almeida and Tyrrell, appeared morphologically identical with the viruses of infectious avian bronchitis, transmissable gastroenteritis of pigs and mouse hepatitis which they resembled in size, measuring from 80 to 120 nm in diameter, and in being sensitive to ether. Viruses with this type of morphology have been designated *Coronaviruses*. A third strain showed a myxovirus morphology. Subsequently, additional viruses with the coronavirus morphology have been isolated from cases of mild upper respiratory infection. Coronaviruses are enveloped, RNA viruses with a particle diameter of 80–160 nm and characterized by the possession of large (20 nm long) surface projections, often described as petal-shaped, which give the particle a crown-like appearance under the electron microscope—hence the name (*corona*, L. = a crown).

The immunity produced by an attack of the common cold appears to be of very short duration and an individual may experience several attacks in a single year. In the experimentally transmitted disease immunity is also of short duration, immunity to the homologous virus appearing generally to last only a few weeks. It is nevertheless possible that under natural conditions the duration of immunity may be somewhat longer since in isolated communities the disease dies out rapidly. Repeated attacks may therefore be due to antigenically distinct viruses.

The agents responsible for the common cold are spread by infected respiratory secretions. They appear to be viable for a considerable time on infected clothing and this, therefore, as well as droplets, could be a source of infection. Attention has recently been drawn to the probable role of the hands in spreading respiratory infections. They are always heavily contaminated with organisms from the respiratory tract in individuals with respiratory symptoms. Children are much more susceptible to infection than adults and are of particular importance in spreading the disease. Rhinoviruses appear to differ from enteroviruses inasmuch as symptomless infections are not such a characteristic feature of the former, although they do occur. In addition to the common cold, rhinoviruses may also cause bronchitis and possibly pneumonitis. Their role in producing exacerbations of chronic bronchitis has yet to be fully assessed.

THE ENCEPHALOMYOCARDITIS (EMC) GROUP

This comprises the encephalomyocarditis virus, the Mengovirus, the Columbia-SK virus and the MM virus. They are small RNA viruses measuring about 30 nm in diameter. They are antigenically indistinguishable and, though isolated in various parts of the world and from different hosts, are probably all strains of a single species. The name of the group derives from the fact that inoculation into mice produces encephalitis and an acute interstitial myocarditis. Only few human infections have been reported; these have varied from a mild febrile disorder to severe encephalitis. EMC viruses are probably maintained in nature as parasites of rodents.

IV.5

Arboviruses: Togaviruses

The *Arboviruses,* formerly the *Arborviruses,* are a group of some 200 small RNA viruses transmitted or believed to be transmitted to man and other vertebrates by biting arthropod vectors, in which the virus also multiplies and which constitute an alternate host. Some are very restricted in their geographical distribution, others are more widespread, but the number likely to be encountered in any one area is usually quite limited. Only about a quarter have been positively established as causing overt human disease, although some of the others appear on serological grounds to be capable of producing symptomless infections. Not all the viruses have been shown to possess arthropod vectors but they are included because they are closely related serologically to other members of the group which are known to be arthropod-borne. There are also some viruses isolated from arthropods for which the natural vertebrate host has not yet been identified. Although they are of virtually world-wide distribution, they have been found particularly in equatorial Africa and in Central and South America. Two well-tried methods of detecting arboviruses are the trapping of arthropods such as mosquitoes likely to be carrying them and the exposure of caged animals such as mice where they can be bitten (sentinel animals). Twenty-eight serological groups have been defined, each comprising at least two members, and about 60 agents remain ungrouped.

It has for some time been recognized that the *Arboviruses* as defined above constitute a very heterogeneous collection of viruses. Already members of other, well-characterized genera have been identified. These include arboviruses with double-stranded RNA and double capsids obviously related to the *Reoviruses,* which have themselves been isolated from mosquitoes: these are classified as the *Orbiviruses* (see p. 452). Viruses morphologically and serologically related to lymphocytic choriomengitis virus have been placed with that virus in the *Arenavirus* genus (see p. 406): and those arboviruses which share the striking bullet-shaped form with rabies, are classed with it in the *Rhabdoviruses* (see p. 402).

Two of the largest and best characterized groups of arboviruses are those hitherto referred to as Group A (over 20 viruses) and Group B (over 40 viruses). The two groups are serologically quite distinct, but within each group all the members are related serologically to at least one other member of the same group and usually to many of them. The name *Alphavirus* has been accepted for Group A and *Flavovirus* (or *Flavivirus*) for Group B: yellow fever is a member of Group B. Several members of both groups have now been studied in detail and have been shown to be enveloped, icosahedral viruses with linear single-stranded RNA. The fact that in several cases purified viral RNA has been shown to be infective indicates that they belong to Baltimore's Class IV. These characteristics distinguish them from other defined virus groups and led to the establishment of a new genus, *Togavirus* to contain them. However, if the names *Alphavirus* and *Flavovirus* are to be accorded generic status, a higher taxon is needed to include them both and the family *Togaviridae* has been proposed. The picture has become complicated by the fact that a number of viruses meet the criteria for inclusion in the *Togavirus* genus which are not serologically related to either the alphaviruses or the flaviviruses and have no known arthropod vector. This category now includes rubella virus, as well as viruses of mice, horses, cattle and pigs. At present the taxonomic status of these viruses is uncertain.

Alphaviruses are more or less spherical,

40–80 nm in diameter; flaviviruses are 20–50 nm in diameter. Rubella, 50–70 nm, lies between the two so far as size is concerned. The envelopes of alphaviruses, flaviviruses and rubella are covered with a 'fuzz' of fine projections, possibly in the form of loops; the envelopes of the other 'non-arbo' togaviruses appear smooth.

The other large group of arboviruses is the so-called *Bunyamwera supergroup* comprising eight smaller groups (over 60 viruses altogether). These viruses are larger, 60–130 nm, and probably contain a helical nucleocapsid and possess single-stranded RNA. During their multiplication they appear to be closely associated with the Golgi apparatus. The remaining arboviruses have not been much studied, although all appear to be RNA viruses and all are to some extent ether sensitive. Some may well turn out to be togaviruses; some will resemble the Bunyamwera supergroup; others resemble orbiviruses, but have not yet been shown to possess double-stranded RNA; others may be shown to belong to non-arbovirus genera, and some may belong to new, as yet undesignated groups of viruses.

The replication of arboviruses has only been studied in detail in the case of a few viruses, mainly togaviruses. Replication takes place in the cytoplasm and the particles develop in relation to cytoplasmic membranes. They are completed as they pass into vesicles within the cytoplasm or pass to the exterior through the cytoplasmic membrane. The particles in the vesicles are released when the latter move to cell surface and discharge their contents. Although the viruses appear to belong to Baltimore's Class IV, the presence in infected cells of species of RNA smaller than the whole genome may indicate that mRNAs are formed which are distinct from the viral RNA. As with the myxoviruses, much of the lipid in the virus particles is of cellular origin, although as the lipid composition varies between viruses, the process of incorporation must be to some extent selective.

Most possess haemagglutinins which are best demonstrated against the cells of day-old chicks, ganders and pigeons. The haemagglutinin appears to be a component of the virus particle. The optimum conditions for the demonstration of haemagglutination differ considerably with different strains; pH is of particular importance, affecting both the demonstration and stability of the haemagglutinin.

Arboviruses are classified serologically by the use of haemagglutination inhibition, complement fixation and neutralization tests. Amongst the viruses of groups A and B, group relationships are more apparent in haemagglutination inhibition than in complement fixation or neutralization tests. Individual members of the groups are most clearly differentiated by neutralization tests.

The specific geographical location of many arboviruses is probably due to the fact that the presence of established endemic or enzootic infection with a particular virus will tend, by stimulating group antibody, to exclude other viruses of the same serological group as well as to geographical restriction of suitable arthropod vectors.

They are pathogenic for mice, in which they produce an encephalitis following intracerebral or intraperitoneal inoculation. As a general rule the susceptibility of mice to infection decreases with increasing age of the animal. Many will grow in the chick embryo—on the chorio-allantoic membrane and in the yolk sac—and in a variety of tissue cultures in which they produce recognizable cytopathogenic effects.

Arboviruses are maintained in nature in arthropod–vertebrate cycles and except in rare instances direct transmission from one animal to another without the intervention of a vector does not occur. With the exception of sandfly fever for which the only known vertebrate host is man the natural vertebrate hosts of the arboviruses are various wild animals and birds. Man is usually only incidentally infected and except for dengue and yellow fever is epidemiologically a 'dead end'. The main arthropod vectors are mosquitoes, ticks and sandflies. Transmission of the virus normally occurs only after its multiplication in the vector, the final stages of multiplication being in the salivary glands. Prior to this the vector is not infective; this period is known as the extrinsic incubation period and its length varies inversely with the ambient temperature. Arboviruses do not usually produce disease either in the insect vectors or in their natural animal reservoir. It is possible that arbovirus infections which, in temperate climates, are maintained during the summer in mosquito–bird cycles, may be maintained through the winter in a cycle involving small rodents and their parasitic arthropods or even in cold-blooded vertebrates such as snakes.

In infected animals and man the viruses are

present in the blood during the early stages of infection; viraemia is in fact a necessary condition for insect transfer. Following the bite of an infected arthropod it is probable that the virus multiplies first in the lymphatic glands draining the inoculation site. From the glands the virus gains access to the blood stream. As a general rule viraemia is maximal from one to five days after infection and it is only during this period that the patient is infective.

Asymptomatic (subclinical) infection is as a rule more frequent than overt symptomatic infection. It has been estimated, for example, that only one in every thousand persons infected with the Japanese B encephalitis virus shows clinical signs of infection. The type of overt infection which the arboviruses produce is varied. Some give rise to systemic illnesses without localizing signs. These illnesses vary from the relatively mild, as in almost all cases of dengue and in sandfly fever, to severe systemic infections with shock, renal failure and haemorrhagic manifestations (see p. 398). The majority produce overt infections with marked localizing signs. In most cases localization is in the brain with the production of an encephalitis or in less severe cases an aseptic meningitis but in yellow fever the virus localizes in the liver. In conditions showing localization of virus the illness is frequently diphasic with a saddle-back temperature chart, the first rise in temperature coinciding with the initial multiplication of the virus and viraemia; the second rise being associated with its localization.

Isolation of virus from human cases is frequently difficult. In some cases this can be achieved with blood taken early in the infection. In encephalitis virus can only rarely be isolated from the cerebrospinal fluid, in general most successes being obtained by inoculation of brain tissue obtained at post-mortem examination.

A serological diagnosis is achieved most readily by the use of haemagglutination inhibition and complement fixation tests. Complement fixing antibody usually persists for only a short time after clinical infection. Haemagglutination inhibiting antibody is more persistent and neutralizing antibody may be detectable for very long periods. In individuals infected for the first time with an arbovirus the reactions in complement fixation and haemagglutination inhibition tests are usually highly specific and may permit a precise serological diagnosis. In those who have suffered from a prior arbovirus infection due to a virus of the same serological group the antibody response shows a much broader pattern, reactions occurring with other viruses of the same group. For groups A and B this overlap is most marked with the haemagglutination inhibition test and least with the neutralization test.

GROUP A TOGAVIRUSES (ALPHAVIRUSES)

Equine encephalomyelitis viruses

These are often encountered and were originally recognized as the causes of encephalomyelitis in horses and mules, but they may give rise to both sporadic and epidemic human infections. The western and eastern viruses, WEE and EEE, are prevalent in the western and eastern parts of North and South America. The Venezuelan viruses are confined to Central and South America including Trinidad. The eastern equine virus is the most invasive of the three, giving rise to much fewer latent infections and, not infrequently, producing a severe disease with a mortality of up to 80 to 90%. The Venezuelan virus is the least invasive; it frequently causes latent infections and rarely produces a frank encephalitis.

The epidemiology of western equine encephalomyelitis has been most studied. The main vector of this virus appears to be the mosquito *Culex tarsalis* and birds, from which the virus has frequently been isolated, probably constitute the main reservoir of infection. Birds are probably also the main reservoir of the eastern virus and mosquitoes may be the main vectors as well. Little precise information is available on the natural history of the Venezuelan virus but it seems likely that wild animals are the main reservoir of infection. Both the eastern and Venezuelan viruses appear to be capable of being directly transmitted from one animal to another without the intervention of an insect vector.

Though belonging to the same serological group, the equine encephalomyelitis viruses are relatively easily differentiated by neutralization tests. Formalin inactivated vaccines of all these viruses have been used for the immunization of horses. Vaccines are not available for human immunization.

Chikungunya and onyongnyong viruses

The Chikungunya virus was first recognized as the cause of an outbreak of a dengue-like disease in Tanganyika in 1952. A serologically similar virus has also been identified as the cause of dengue-like disease and of severe haemorrhagic fever in Thailand. The virus is probably maintained in a primate–mosquito cycle. The Onyongnyong virus was first isolated as the cause of a febrile disease characterized by joint pains, rash and lymphadenitis which appeared in Uganda in 1959. Over the next few years infection with this virus became widespread in East Africa. The Onyongnyong virus is closely related serologically to the Chikungunya virus and may well be a mutant of the latter. The virus is thought to be spread by *Anopheles* mosquitoes.

GROUP B TOGAVIRUSES (FLAVIVIRUSES)

A number of subgroups are distinguished, the individual members of which have a closer serological relationship to one another than to other viruses of the group.

St Louis subgroup

This group includes the St Louis, Japanese B, Murray Valley fever and West Nile viruses. These viruses are closely related to each other serologically and fall into a serological subgroup of group B. Culicine mosquitoes appear to be the main vectors and wild birds the main reservoirs. The St Louis, Japanese B and Murray Valley viruses are encephalitogenic in man. The West Nile virus produces a mild systemic infection characterized by a general lymphadenopathy and rash. It rarely produces encephalitis, and then only in older people.

Infection with the St Louis virus is widespread throughout the United States and in the Caribbean. Infection with the Japanese B virus is found in China, Japan and Far eastern Russia. The West Nile virus is found from South Africa through the Middle East to southern India, but is only a significant cause of illness in the Middle East. The Murray Valley virus has been found as a cause of an epidemic infection only in south eastern Australia. The main reservoir of infection, however, is believed to be in northern Australia and New Guinea, where it is maintained in a bird–mosquito cycle. From these areas it is introduced into south eastern Australia by migratory birds, from which it is believed to be transmitted to domestic fowl and waterfowl. In these areas epidemics have been noted particularly at times of flooding, when waterfowl which carry the virus are brought into close contact with the human population.

Tick-borne subgroup

This group comprises a number of tick-borne viruses which have been isolated in various parts of the world, viz. the louping ill virus, the far eastern or Russian spring–summer and Czechoslovakian or Central European encephalitis viruses and the viruses of Kyasanur Forest disease (India) and of Omsk haemorrhagic fever (Siberia). The louping ill, Russian spring–summer and Central European viruses are very similar serologically, but can be distinguished by neutralization tests. The Kyasanur Forest and Omsk viruses closely resemble one another antigenically.

These viruses are apparently capable of being transmitted transovarially and transtadially in ticks but the extent to which this transmission is responsible for their persistence in nature is speculative. In the case of the Central European virus it has been suggested that the primary reservoir is in a cycle involving wild rodents and tick larvae. It is thought that this cycle may generate a secondary cycle involving adult ticks and larger vertebrates. On occasion, infection with tick-borne viruses may occur in the absence of a tick bite. Thus the Central European and far eastern encephalitis viruses can be transmitted by improperly pasteurized goat's milk and there is reason to believe that the development of louping ill in butchers may be the result of respiratory or alimentary infection.

The pathogenicity of the Kyasanur Forest disease virus is unusual. Like other arboviruses it may produce meningoencephalitis. In some cases, however, there are severe haemorrhagic manifestations. The disease appears to occur only in the State of Mysore in India but the range may be extending. There is believed to be a reservoir in wild birds (jungle fowl) and possibly rodents from which cattle, monkeys and man may be infected.

Transmission to monkeys and man is believed to have occurred on a significant scale only during the last decade.

The louping ill virus is the only arthropod-borne encephalitis virus which is endemic in the British Isles. It is enzootic in sheep in Scotland, northern England, Wales and Ireland. In sheep it gives rise to a condition showing marked cerebellar ataxia, from the symptoms of which the disease derives its name. There have been a few isolated human cases of infection with this virus—in a farmer and a veterinary surgeon working with infected sheep, in some abattoir workers in Glasgow and in laboratory workers. Serological evidence of infection with the louping ill virus or with an antigenically closely related virus has been obtained by Likar and Dane in Northern Ireland.

Yellow fever

Yellow fever is a disease of equatorial Africa and America, occurring in a belt extending from roughly 10° north to 10° south latitude. Its aetiology and epidemiology were first elucidated by Walter Reed and his associates in Cuba in 1911, when Reed succeeded in substantiating the theory of Carlos Finlay that the disease was spread by the mosquito *Aedes aegypti*. The virus was not isolated, however, until 1927 when Stokes, Bauer and Hudson showed that rhesus monkeys were susceptible to infection.

Properties of the virus

The yellow fever virus is a small virus measuring about 38 nm in diameter (electron microscopy). It has a dense centre and a distinct envelope. On primary isolation the virus can be propagated in monkeys by subcutaneous or intraperitoneal inoculation. In Indian monkeys it produces severe liver necrosis. In African monkeys the liver necrosis is minimal and the cells show a remarkable proliferation of mitochondria. As first shown by Theiler in 1930, the virus will also infect adult white mice to produce an encephalitis, following intracerebral inoculation. The combination of neurotropism for mice with viscerotropism for monkeys is referred to as pantropism. In serial passage intracerebrally in mice its neurotropism is increased while its viscerotropism diminishes. Chick embryos are not susceptible on primary isolation but can be infected by virus which has been passaged intracerebrally in mice.

Although all yellow fever viruses are of essentially the same immunogenic type, slight differences between American and African strains have been shown by haemagglutination inhibition tests. These differences are not, however, sufficient to interfere with the efficacy of current vaccines against both types of infection.

Pathogenesis and epidemiology

Yellow fever occurs in two epidemiological varieties: urban yellow fever and jungle yellow fever. Urban yellow fever, which is the classical form, is an epidemic disease. It assumed major proportions in Central America at the end of the nineteenth and the beginning of the twentieth centuries and even occurred as far north as Philadelphia and Baltimore. Recent epidemics have occurred mainly in West Africa. The disease is transmitted by the mosquito *Aedes aegypti*. Infection is only possible if the insect has fed from a patient in the first three days of the disease, but the insect does not become infective to another person for at least four days, and not for a much longer time if the air temperature is low. Thereafter it remains infective for life. Transovarian passage of the virus in the mosquito has not been demonstrated. Infection may also be conveyed directly from man to man by contact with the blood or serum of a patient.

When, following the work of Reed, it was found that epidemics of yellow fever could be prevented by measures directed against *Aedes aegypti* it was thought that the problem of the control of the disease had been finally solved. This was, however, found to be far from the case. The source of this continuing infection remained a problem until it was shown by Stokes that the virus is maintained in a reservoir of monkeys and possibly other jungle animals, amongst which it is spread by various mosquitoes. Sporadic human infections occur by indirect transmission from this animal reservoir; this epidemiological variety of the disease is known as jungle yellow fever. In Africa the disease is transmitted within the animal reservoir by the mosquito *Aedes africanus*. This mosquito lives in the tree tops and does not normally bite man, the disease being transmitted from monkeys to man by the link vector *Aedes simpsoni*. This mosquito, like *Aedes aegypti* has a preference for human communities. It usually becomes infected from monkeys raiding villages. Like *Aedes aegypti* it is capable of maintaining a man-to-man cycle of infection but clearly has not the same devastating epidemic potential. This variety of jungle yellow fever may affect all ages and sexes. In the forests of South America, jungle yellow fever presents a somewhat different picture. It is usually contracted by adult males who work in forests inhabited by an infected monkey reservoir. The main vector appears to be *Haemagogus spegazzinii*.

There is evidence of some degree of cross-immunity between yellow fever and the other group B arbovirus infections, to such an extent as to diminish the spread of yellow fever in areas in which such infections are enzootic or endemic, and in which an animal reservoir of yellow fever and a suitable insect vector are both present. Thus in Trinidad and Ethiopia, in recent years, there has been reason to believe that the spread of yellow fever was restrained by the presence of endemic infection with the dengue virus.

The outstanding features of a typical case of severe yellow fever are high temperature, great prostration, vomiting, haemorrhages and jaundice, the disease in most cases being of a biphasic type with a remission of a few days' duration between the major episodes. Sometimes the infection is fulminating. The disease may also occur as a mild 'flu-like infection or in a latent asymptomatic form which can be detected only by serological means. In Brazil and in Colombia serological evidence indicates the occurrence of latent asymptomatic infection in as much as 30 to 40% of the population of enzootic areas. Such areas in which the only evidence of the disease is serological but in which the infection is nevertheless widespread are known as 'silent areas'.

The virus is present in the blood of infected persons and multiplies in the vascular endothelium and in the liver cells. It is believed that following introduction into the tissues the initial multiplication of the virus occurs in the lymph nodes and that it is from these that the blood is invaded. In fatal cases the liver cells show marked necrosis. The necrotic cells, which have a characteristic hyaline appearance, are known as Councilman bodies. Occasionally infected cells show the presence of irregular acidophilic intranuclear inclusions. These are commoner in monkeys than in man; the inclusions do not contain nucleic acid and are apparently not colonies of virus.

Diagnosis

The virus may be isolated by intracerebral inoculation of mice with blood taken in the first few days of the disease. Since a considerable amount of antibody appears at an early stage of infection it is customary to inoculate a number of dilutions of blood so as to dilute out the antibody present.

A serological diagnosis may be made by the use of neutralization, complement fixation or haemagglutination inhibition tests. These tests give highly specific results in individuals who have not previously suffered from a group B arbovirus infection. In people who have had a prior group B infection the antibodies have a broad specificity reacting to a greater or lesser extent with other viruses of the B group. In this type of case antibodies are usually present in much higher titre than in those who have not had a prior group B infection. For the detection of antibody in epidemiological surveys mouse neutralization tests have been mainly used.

A diagnosis of yellow fever can often be made from the histological examination of liver tissue obtained by punch (viscerotome) through the abdominal wall over the liver in fatal cases.

Prophylaxis

As in other mosquito-borne diseases the use of residual insecticides and other anti-mosquito measures are of the greatest value in controlling epidemic yellow fever. In America *Aedes aegypti* is mainly domestic in its habits, breeding primarily in household collections of water. These breeding sites are comparatively easy to eliminate. As compared with the American varieties of *Aedes aegypti,* however, the African varieties are less domestic in habit and are therefore more difficult to control. Control of *Aedes aegypti* is in itself sufficient to prevent the development of urban epidemics of yellow fever but will not of course affect the jungle type of the disease.

Active immunity can be induced by the use of a living attenuated vaccine. The vaccine in most use is that prepared by growth of the attenuated 17D variant in the chick embryo. This variant was derived by Lloyd, Theiler and Ricci by passage from the initially fully virulent Asibi strain, which had been isolated from a patient on the Gold Coast. The virus was first propagated in mouse embryo tissue culture, then in chick embryo tissue culture and finally in cultures prepared from chick embryos from which the head and spinal cord had been removed. After a total of 176 passages the 17D variant, which was devoid of viscerotropic properties and of diminished neurotropism, appeared. That the virus retains some pathogenic potential is shown by the fact that it can regain full neurovirulence following intracerebral passage in mice and that it has, though very rarely, produced encephalitis in vaccinated infants. Nevertheless it is probably as safe as any live vaccine can be. For production the virus is propagated in embryonated eggs. It is freeze-dried.

The vaccine is inoculated subcutaneously, one injection being adequate for immunization. It does not normally produce any untoward reactions, although it may do so very rarely in the young infant. Following vaccination, immunity develops rapidly and is sufficient to give complete protection after about 10 days. The immunity produced is of long duration and can be fully relied on to protect against infection for at least 10 years. Immigration authorities of countries in which the insect vector is present, and in which the disease could therefore be easily spread once introduced, require that all immigrants from or who have recently visited yellow fever endemic areas should be vaccinated at least 10 days prior to entry. Yellow fever vaccination should not be carried out within 21 days after a primary smallpox vaccination. If smallpox vaccination is required, yellow fever immunization should be carried out first; smallpox vaccination can then be given after an interval of four days, except in the case of children of less than nine months of age, when the interval should be 21 days. It is generally considered unnecessary, however, to immunize children in this age group. Care is necessary in administering the vaccine to those sensitive to egg. It must be kept on ice when reconstituted from the freeze-dried state and even then is stable for only one hour. The dried vaccine is stable for about three months at +4°C. Long term storage should be at temperatures below −20°C.

A neurotropic variant obtained by intracerebral passage in mice has been used extensively by the French in West Africa. It is inoculated by a scratch technique which has considerable advantages for mass immunization. The immunity produced appears to be at least as good as that produced by the 17D variant but in the past it has shown a much greater tendency than the 17D virus to produce encephalitis. There is some evidence that this complication is much less frequent with the vaccine currently used. It should probably not be given to children under 14 years of age. It is more stable than 17D vaccine.

Dengue fever

The dengue viruses are probably the most widely distributed of all the arboviruses. They are, however, extremely difficult to isolate, success having been achieved only by intracerebral inoculation of suckling mice. Four serotypes are recognized: types 1 and 2 are world wide in distribution, but types 3 and 4 have so far been recognized only in the Philippines, Thailand and (type 4 only) India. The viruses produce two distinct clinical syndromes:

1. A febrile disease of several days' duration characterized by severe pains in the back and limbs, because of which it is known as breakbone fever. A similar syndrome is produced by infection with certain other arboviruses, notably West Nile fever, Chikungunya and Onyongnyong.
2. A much more severe disease with marked haemorrhagic manifestations. This is an important form of the disease in the Philippines and Thailand.

The chief vector is *Aedes aegypti,* which is the sole vector in the western hemisphere. In the east, however, other *Stegomya* mosquitoes, notably *Aedes albopictus,* are of importance and, being less domestic in habit than *Aedes aegypti,* render the disease more difficult to control.

Immunization with attenuated vaccine is still in the experimental stage. The epidemiology of dengue closely resembles that of yellow fever. Dengue is, however, much more readily spread than yellow fever and may be maintained on an epidemic scale by a much smaller population of *Aedes aegypti.*

OTHER GROUPS OF TOGAVIRUSES

Amongst the other arboviruses causing disease in

man are several members of the Bunyamwera supergroup which produce a febrile illness sometimes with encephalitis, especially in children, Rift Valley fever, primarily a disease of sheep, goats and cattle, and sandfly or *Phlebotomus* fever.

Sandfly fever

Sandfly or *Phlebotomus* fever is a mild febrile condition usually of from two to four days' duration. The virus is spread by the sandfly *Phlebotomus papatasi.* Only the female of this species bites, usually during the night or in the early hours of the morning. Insects do not become infective for some seven to 10 days after biting.

Sandfly fever occurs in Europe, Africa and Asia in a belt, corresponding to the distribution of the insect vector, from 20 to 45° north latitude. It has not been recognized in America. The disease does not cause death and the pathological changes produced are therefore unknown.

The viruses so far isolated fall into two antigenic groups—the Neopolitan and Sicilian—the prototypes of which were first isolated by Sabin. A number of viruses isolated in South America seem to be related to the Neapolitan group. The viruses can be isolated only by intracerebral inoculation of suckling mice but, when established, strains have been propagated in a variety of tissue cultures. Isolation is usually only possible from blood taken during the first or second day of the disease.

Haemorrhagic fevers

A considerable number of different arthropod-borne viruses may give rise to a severe haemorrhagic febrile illness with a purpuric rash, widespread internal haemorrhages, shock and renal failure. This syndrome has been recorded especially in infections due to the following agents: Dengue types 1, 2, 3 and 4, Chikungunya, yellow fever, Kyasanur Forest, and Omsk and Crimean haemorrhagic fever. The same clinical picture can be due to Arenaviruses (p. 406) and the Marburg agent (p. 406). A form of this disease referred to as epidemic haemorrhagic fever has been encountered in Manchuria, eastern Siberia and Korea. It is thought to be transmitted by a mite, but this has not been proved and it may not even be due to an arbovirus. It has not proved possible to transmit the disease to experimental animals or produce a cytopathic effect in tissue culture. However, an agent has been derived from man and mites which will interfere with the growth of poliovirus in human embryo tissue culture.

Rubella

The virus of rubella was first isolated by Weller and Neva, in 1962, by the inoculation of blood and urine of rubella patients into tissue cultures of primary human amnion. On passage of the virus, though not on primary isolation, a cytopathogenic effect characterized by the development of focal areas of degeneration, and involving only a limited number of cells in the culture, was observed after 17 to 21 days' incubation. Affected cells showed nuclear disruption and acidophilic cytoplasmic inclusions. A similar cytopathogenic effect, but without the development of inclusion bodies was found by McCarthy et al. in rabbit kidney and human thyroid tissue cultues. The virus will grow in grivet, patas and rhesus monkey kidney but without the production of a demonstrable cytopathogenic effect. Under these conditions its presence may, however, be demonstrated by showing that the culture is protected from infection with another potentially cytopathogenic virus. The rubella virus cannot be isolated in the chick embryo, but a strain has been adapted to growth in chick embryos. Rubella virus has a great capacity to produce chronic infection in tissue culture. The cells continue to multiply and small amounts of rubella virus are formed continuously. Such cultures show a marked degree of interference when challenged with other viruses but interferon is not necessarily detectable. The virus is 50–60 nm in diameter and consists of a rather ragged outer coat surrounding a dense core about 40 nm in diameter. The symmetry of this core appears to be icosahedral, so that its structure resembles that of a togavirus rather than that of a myxovirus. This together with the properties of its haemagglutinin and its ability to multiply in suckling mouse brain have led some workers to suggest it is an arbovirus which has lost its dependence on its arthropod intermediate host. Its sensitivity to amantadine (see p. 363) would suggest a relationship to influenza.

However, it is now classified as a togavirus (see p. 391). It is ether-sensitive and as its growth in tissue culture is not inhibited by bromo- or fluoro-deoxyuridine, it is presumed to be an RNA virus. Because of the long-lasting and solid immunity which follows infection there is probably only serological type. It was originally thought that rubella virus did not haemagglutinate; but it has since been shown that, provided a non-specific inhibitor, present in the serum of the tissue culture medium used in the growth of the virus, is first removed, the virus carries a haemagglutinin active against the cells of day-old chicks, pigeons and geese. The haemagglutinin differs from that of most arboviruses in being destroyed by lipid solvents and having an acid pH optimum (pH 6·2). Haemagglutination inhibition tests are satisfactory provided that the non-specific inhibitor is removed (see Table IV.3, p. 371).

In contrast to measles, rubella is characteristically a disease of the older child and young adult. When it occurs in postnatal life the disease is usually a mild one, although in adults distressing and disabling arthralgia may be a feature and recently cases of subacute sclerosing panencephalitis (p. 432) apparently due to rubella virus and not to measles virus have been described. If, however, a mother is infected during the first four months of pregnancy the fetus may develop a severe systemic infection resulting in death or, if it survives, in congenital defect. The most important defects observed are deaf mutism, congenital heart disease, cataract, mental retardation and microcephaly. The association between maternal rubella and congenital defect in the fetus was first observed by Gregg in Australia in 1941. Gregg's survey was a retrospective one and estimated the hazard for the birth of a child with a congenital defect to a mother contracting rubella in the first four months of pregnancy at 70%. Prospective surveys in which the diagnosis can be much more accurately determined, however, have given somewhat lower figures. In a survey carried out by the British Ministry of Health, covering the period 1950–52, of 279 pregnancies in women who had had contracted rubella in the first 16 weeks of pregnancy, 11 aborted, 11 gave birth to stillborn children and 16 of the children born alive died before the age of two years. These figures represent a total fetal loss of 7·3%. Of the remainder, 37 showed major congenital anomaly and 37 a minor anomaly, giving an overall figure for death or some degree of congenital defect of 40%.

In general, the more severe congenital defects have been found in women infected in the first six weeks of pregnancy. A raised incidence of congenital defect was found by Coffey and Jessop in women who had been in contact with cases of rubella but who had not actually developed the disease, indicating that the fetus may be exposed as a result of asymptomatic infection of the mother, although this observation has not been confirmed in other studies. In the severe epidemic of rubella that affected the United States in 1964 many cases were described in which congenital rubella infection resulted in a severe systemic disease which showed, in addition to previously recognized defects, splenomegaly, pneumonitis, thrombocytopenic purpura, anaemia, hepatitis and bone involvement.

Virus can be isolated from throat swabs, nasopharyngeal washings or urine of the great majority of children showing the congenital rubella syndrome. In some cases of congenital infection, excretion of the virus has been found to continue up to the age of 12 months. Such children are therefore an important community source of infection. Excretion of virus has also been found in normal children born of mothers who had contracted rubella in early pregnancy. In the development of fetal rubella there is evidence that infection of the fetus is associated with infection of the placenta, which may therefore be the primary site of multiplication of the virus in the fetus. In fatal cases of the rubella syndrome, virus can be isolated from virtually every organ.

Immunological tolerance does not develop as a result of fetal rubella infection. There is in fact considerable evidence that infected children can produce antibodies against the virus in utero. This is shown by the facts that (*a*) they frequently have a serum antibody titre which is higher than that of the mother and (*b*) a considerable amount of the antibody is of the IgM type, which is not transmitted across the placenta from the maternal to the fetal circulation. Why there should be a persistent excretion of virus in the presence of a high antibody titre is not known, nor why some infected children escape with minimal damage while in others the damage is severe, nor why fetal damage is restricted to infection in the first four months of pregnancy.

Rubella infection in postnatal life is believed to occur via the respiratory secretions. The virus can be isolated from nasopharyngeal washings and throat swabs from about a week before to a week after the appearance of the rash. The virus can also be demonstrated in the urine and faeces; these may be important sources of infection in infants with the congenital rubella syndrome. The primary site of multiplication is presumed to be in the respiratory tract. Dissemination is clearly by the blood stream, viraemia being demonstrable before the development of the rash. The incubation period under the conditions of natural infection is from 12 to 23 days. Epidemiological, virological and immunological studies suggest that inapparent infection occurs with about the same frequency as overt infection.

Laboratory diagnosis is not normally required for the individual case of postnatal rubella. It cannot be too strongly emphasized that the clinical history is not a reliable guide to immunity to infection. Many cases diagnosed as rubella are apparently due to other agents and many rubella infections are inapparent. Only a serological test can determine the immunity of a given individual. The identification of an outbreak of rubella is desirable so that adequate prophylactic measures to protect pregnant women exposed to infection can be undertaken. For the isolation of the virus one can use cultures of grivet monkey kidney using the production of interference as evidence of the presence of virus. However, it is more convenient to use on of the cultures RK_{13},* SIRC,† Vero† or BHK21‡ in which a cytopathic effect is produced by rubella. Antibody can be demonstrated by neutralization tests or haemagglutination inhibition. Complement fixation tests, utilizing a soluble complement fixing antigen which is demonstrable in infected tissue cultures, are also employed. The precise relationship of the complement fixing antigen, which appears to be a protein, to the virus particle has not been determined. Immunofluorescence using infected tissue culture cells as antigen has also been used for antibody detection. This technique, using specific anti-human IgM fluorescein-labelled serum or haemagglutination inhibition on serum fractionated on sucrose density gradients may be valuable in confirming or excluding recent infection in pregnant women with stationary antibody titres and in the diagnosis of congenital rubella. The haemagglutination inhibiting antibody usually has a much higher titre than the neutralizing antibody. Both appear to persist for years after infection.

In view of the association of congenital defect with rubella in pregnancy it is generally considered advisable to ensure that all girls should have been exposed to rubella before they reach the childbearing age. To achieve this some workers have deliberately infected girls by instillation of infective nasopharyngeal washings. Gammaglobulin is of no value in preventing the development of rubella in contacts. Whether a suppressive effect on the development of clinical rubella in pregnant women who have been in contact with the disease is sufficient to prevent transmission to the fetus is not yet known. In the United Kingdom something like 15% of pregnant women will be found not to have antibody to rubella. This proportion will drop, it is hoped, as vaccination begins to take effect. Several attenuated strains have become available for use as live vaccines. The most most widely used have been the HPV77 strain attenuated by passage in green monkey kidney and subsequently in dog kidney cells and duck embryos and the 'Cendehill strain' attenuated by passage in primary rabbit kidney cultures. The RA27/33 grown in the WI-38 line of human embryonic diploid cells has been licensed in this country and was the first vaccine to receive approval in which the virus has been propagated in human cells. The vaccines are administered by subcutaneous injection although, in a sufficiently large dose the RA27 strain is infective by intranasal inoculation. After vaccination by either route, a small amount of virus is recoverable from the nasopharynx but the vaccine viruses show no tendency to spread. However, as their total lack of teratogenicity has not been established they should not be given to those in contact with pregnant women and they should not, of course, be given to pregnant women themselves.

A proportion of those vaccinated, especially adults, may show mild symptoms, arthralgia, slight fever, a transient rash or some lymph node enlargement a few days after vaccination. With HPV77, such symptoms may be unacceptably severe in adults and this vaccine has therefore been restricted to individuals under 14 years of

* Continuous rabbit line.

† Continuous green monkey kidney lines.

‡ Baby hamster kidney cell line.

age. With the others such symptoms are not usually severe enough to interfere with normal activities. The duration of immunity is not known but it appears likely to be several years, even though neutralizing antibody titres are considerably lower after vaccination than after infection.

Two vaccination policies are possible. That adopted in the United Kingdom is to try to protect pregnant women by vaccinating girls at about the age of puberty and to offer vaccination to women who have just been delivered and are without antibody. Such women must take stringent precautions to avoid pregnancy for at least eight weeks after vaccination. In the United States the policy is to try to reduce the circulation of the virus in the community to a negligible level by vaccinating boys and girls at about the age of two years. If a woman does contract rubella in the first trimester of pregnancy it will be necessary to give careful consideration to the possibility of termination.

IV.6

Rhabdoviruses, Arenaviruses

RHABDOVIRUSES

The rhabdoviruses comprise a group drawn from such diverse hosts as mammals, insects, fish and plants and resembling one another in possessing a characteristic bullet-shaped morphology. Rabies virus is 75–80 nm wide and up to 180 nm long with one rounded and one flat end. The surface is covered with projections and there is evidence of an inner component said to be a helical nucleocapsid with a diameter of 18–20 nm. There appears to be only one serological type. Goose erythrocytes are agglutinated at 0°C at pH 6·4. In tissue cultures of several types some multiplication occurs. Large cytoplasmic eosinophilic inclusion bodies are formed but relatively little complete virus is released. Most of the cells in the cultures tend to survive with the production of carrier cultures. In addition to the virus particle a soluble complement fixing antigen of about 12 nm in diameter is demonstrable in infected tissues. The nucleic acid appears to be RNA.

Rabies

Properties of the virus

All the ordinary laboratory animals are susceptible but mice, particularly, have been used for isolation. They are normally infected by intracerebral inoculation. Strains of rabies virus from different sources show appreciable differences in mouse virulence, the most virulent being those derived from canines. The virus does not infect the chick embryo on primary isolation but rabbit- or mouse-adapted strains will do so. It is a typical rhabdovirus.

Pathogenesis and epidemiology

Rabies is primarily a disease of carnivorous animals. Except for certain island countries, e.g. the British Isles, Australia and New Zealand, it is widely prevalent in wild carnivores throughout the world. Notably it occurs in wolves, which were at one time the main source of human infection in Europe and are now a common source of infection in Iran, foxes, jackals in India and Israel, the mongoose in India and the Caribbean, in North America various species of insectivorous and frugivorous bats, raccoons, skunks and foxes and in Trinidad and Brazil the vampire bat. It has been suggested that the main reservoirs of infection are the mustelidae* and viverridae† in which the disease is maintained as a latent infection, with only sporadic overt cases, which protects them from their natural canine enemies. From this reservoir, the disease is transmitted first to wild canines which in turn infect domestic animals—dogs, cats and cattle, and occasionally man. In Trinidad and Brazil the vampire bat, which suffers an asymptomatic infection, can directly infect cattle and man.

In canines rabies is characteristically encephalic. The virus invades the central nervous system and the main symptoms produced are those of encephalitis. In infected cells characteristic inclusion bodies known as Negri bodies may be found. These are particularly numerous in the pyramidal cells of Ammon's horn of the hippocampus. In addition to the central nervous system the mucus-secreting glands may contain the virus. Of these the salivary glands, from which the virus gains access to the saliva, are the most

* Weasels, stoats, ferrets, martens, badgers, wolverines, skunk, otters, mink, etc.

† Mongooses, civets.

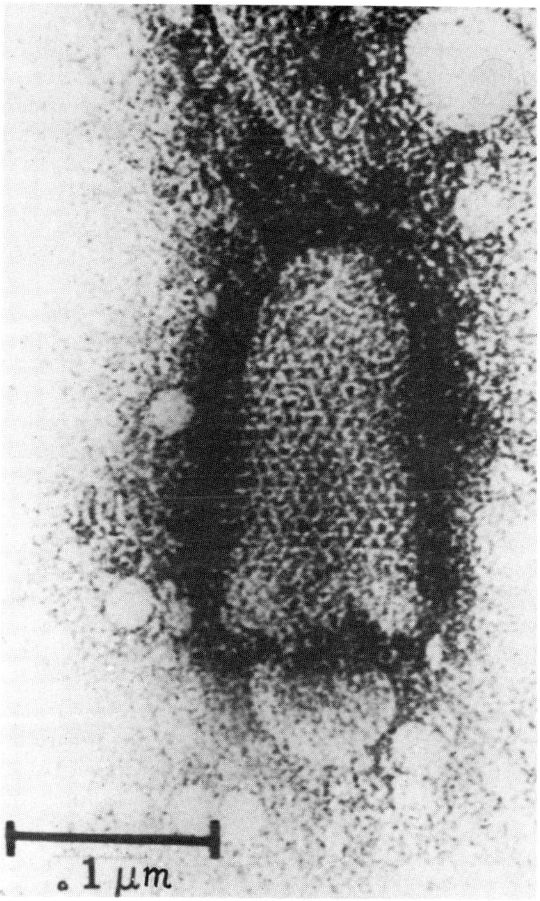

Fig. IV.20. A rabies virus particle. Negatively staining. × 225 000. (*Dr H. Koprowski*)

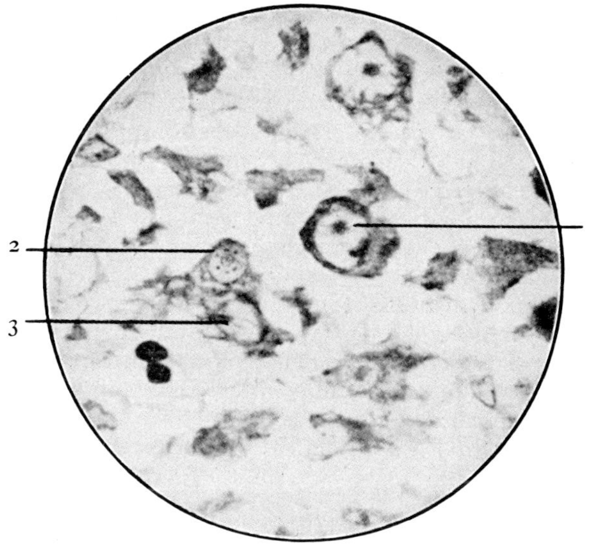

Fig. IV.21. A Negri body in a nerve cell. 1, Nucleus of nerve cell. 2, Negri body showing a rosette-like arrangement of its granules round a central structure. 3, Degenerated nucleus of the nerve cell. × 850.

important for the transmission of the disease. In mustelidae the central nervous system is only infrequently involved. Involvement of the salivary glands, however, ensures that the infection may be spread, via the saliva, from one animal to another.

Since about 1950 rabies has been extending widely in Europe, Russia and North America. In Europe the main reservoir appears to be foxes and cases of fox rabies have been reported as far west as Belgium, Holland and France. Rabies has been effectively kept out of the United Kingdom by six months quarantine for cats and dogs. However, within the past few years cases of rabies have occurred in dogs released from quarantine and a considerable tightening of the regulations for these species which now receive vaccination as well as six months quarantine will, it is hoped, counter the mounting threat from the increasing incidence of rabies in other countries, particularly continental Europe. Quarantine regulations have also been introduced for exotic animals liable to be infected.

Human infection usually occurs as a result of the bite of a rabid or pre-rabid animal—the virus present in the animal's saliva coming into contact with exposed nerve endings. Not all persons who are bitten by rabid animals, however, develop rabies. The risk is greatest in cases with severe and multiple wounds and particularly with wounds involving the head, neck and face. Infection may also occur from a lick of a rabid or pre-rabid animal, the virus present in the animal's saliva gaining access to nerve endings through scratches and abrasions of the skin. The incubation period of rabies in man is highly variable; it may be anything from 10 days up to two years but is usually less than two months. The infection has been transmitted to animals and man by the infected dust and aerosols produced by infected bats in large caves.

The exact route by which the virus gains access to the central nervous system is not known with certainty but as viraemia has not been demonstrated, either in experimental animals or in man, it seems certain that it travels by the peripheral nerves. Positive evidence in favour of this view is the finding that, in mice, section of the nerves prior to inoculation prevents infection with virus injected into the foot pad. Similarly, division of efferent nerves interferes with spread of virus centrifugally from the central nervous system to the salivary glands. Thus the centrifugal, as well as the centripetal, movement of the virus probably occurs by neuronal routes. Exactly how the virus travels along the nerves remains in dispute. The fact that the Negri bodies are found in neurons and the demonstration within them of viral antigens indicate that these cells do become infected and it is natural to suggest the axon itself as transmitting the virus. However, other evidence, including the presence of inflammatory changes in the peripheral nerves, points to transmission by the cells of the neurilemma (Schwann cells). In carnivorous animals infection may also result from eating the flesh of an infected animal. In this case the virus presumably gains entry through either the nasopharynx or the intestinal tract. It is not known whether any local multiplication of the virus occurs at the site of inoculation. In experimental animals there is clear evidence that stress and the administration of ACTH can transform a latent into an overt infection.

In man rabies is virtually always fatal; only two cases of recovery have ever been recorded. The earliest symptoms are fever, headache and general malaise with, in the majority of cases, pain at the site of the bite. Difficulty in swallowing is the most characteristic symptom of the established disease. When water reaches the fauces it is violently expelled as a result of forcible contraction of the pharyngeal muscles. Because of this the patient develops the fear of water which gives the condition its common name—*hydrophobia*. This symptom is associated with a progressively increasing excitability of the musculature which gives rise to generalized convulsions in response to minimal stimuli, together with signs of increased stimulation of the sympathetic nervous system, e.g. salivation, perspiration and dilatation of the pupils. Death usually occurs during this phase of the disease but occasionally follows a terminal phase of paralysis and coma. Although life can be prolonged by modern techniques of treatment the mortality remains total for practical purposes.

Diagnosis

The diagnosis of rabies in dogs is of considerable importance as a means of deciding whether a human being, bitten by a suspected animal, should receive prophylactic treatment. As in man, the incubation period in the dog may be very long, extending on occasion up to about nine months. The dog may show one or other of two contrasting clinical syndromes: the furious, characterized by symptoms of excitation, and the dumb, characterized by the early appearance of paralytic symptoms. If the dog is dead, an attempt should be made to demonstrate the presence of Negri bodies in the hippocampal cells. In addition, the brain should be examined for rabies antigen by the fluorescent antibody technique. If microscopy is negative it may be possible to isolate the virus by intracerebral inoculation of mice. If the dog is alive and has not developed paralysis, it should be kept alive for 10 days, by which time, if it is infected, paralysis should have developed. If it develops rabies the diagnosis should be confirmed as indicated above.

The diagnosis of rabies in man is mainly a clinical problem and usually gives little difficulty. If the case comes to post-mortem examination, the procedure is the same as in the diagnosis of rabies in the dog. In a considerable number of human cases, however, Negri bodies are not demonstrable. During life an attempt may be made to isolate the virus from the saliva, throat swabs or urine by intracerebral inoculation of mice.

Prophylaxis

As a general measure the most important step is to abolish rabies in dogs. This can be done by the killing of stray dogs, enforced muzzling, annual vaccination and canine registration. When, as in Britain, the disease has been eliminated, it has hitherto been possible to keep it out by strict quarantine of dogs and cats. In countries where there is much wild-life rabies attempts should be made by trapping and other means to reduce the wild animal reservoir. It is doubtful, however, whether it will ever be possible by these measures to eliminate the disease completely. A disturbing feature of the present position in the United Kingdom is the extent to which the fox has become an urban, or at least a suburban, species.

Often living on the abandoned railway lines along which they originally made their way into the town to reach their new habitat, they live a scavenger existence and must have frequent opportunities for contact with the large number of feral or inadequately controlled dogs which are also a feature of the present-day urban scene. It is to be hoped that the recently increased penalties for attempting to smuggle animals into the country and so avoid quarantine will have a real deterrent effect.

The feasibility of preventing rabies in persons who have been bitten by rabid animals, by active immunization during the incubation period, was first shown by Pasteur. The original Pasteur vaccine, known as *virus fixé,* was derived by serial intracerebral passage in rabbits, the preparations used being obtained from infected spinal cords which had been dried for various periods over potassium hydroxide. Pasteur believed that drying reduced the virulence of the virus. In fact it appears to cause the death of the virus, very little, if any, remaining alive after 10 days' drying.

Pasteur's method of vaccine preparation is now seldom employed. Most of the vaccines now in use are prepared by Semple's method and consist of a 4% suspension of sheep or rabbit brain inactivated by exposure to 0·5% phenol saline. The virus strains used in these vaccines, however, are all derived from Pasteur's *virus fixé*. The immunization course consists of from 14 to 21 injections each of 2 ml of vaccine given by the subcutaneous route at daily or twice daily intervals. In the current standard prophylactic procedure rabies hyperimmune horse serum* is given immediately after exposure, followed 24 hours later by the first injection of vaccine. There is some evidence that such combined active and passive immunization is more effective than active immunization alone. However, it has been shown that the immunological response to rabies vaccines is severely depressed by the administration of hyperimmune serum. As a result, it is recommended that extra doses of vaccine (booster doses) be given to patients who have received hyperimmune serum. Prophylaxis is, however, undertaken only if there is clear evidence, or a strong suspicion, that the person has been bitten by a rabid animal.

Anti-rabies vaccination is not without risk: a small proportion of individuals inoculated develop a condition known as a neuroparalytic incident with symptoms referable to damage of the brain, spinal cord or peripheral nerves. This complication usually appears about two weeks after the initiation of treatment. Its reported frequency varies from 1 in 800 to 1 in 8000 inoculations. In general it has been less frequent with phenolized vaccine than with the Pasteur type of rabbit cord vaccine. The pathogenesis of the condition is discussed on p. 454.

Because of the risk of neuroparalytic incidents after immunization with nervous tissue vaccines considerable work has been carried out in recent years on the feasibility of using living attenuated vaccines propagated in the chick embryo. The most successful of these have been vaccines prepared from the attenuated Flury strain. Two types of vaccine have been prepared from this strain:

1. A low egg passage (LEP) vaccine consisting of virus that has received less than 50 passages in the embryo. LEP vaccine is non-pathogenic for dogs by the intramuscular route but is still pathogenic for mice. It has been used extensively for the immunization of dogs, in which it stimulates a better antibody response than nervous tissue vaccines. Because of this it is now generally accepted as the vaccine of choice for dog immunization.

2. A high egg passage (HEP) vaccine consisting of virus that has been maintained for from 117 to 210 passages in the chick embryo. The vaccine is non-pathogenic for both dogs and mice by the intramuscular route. The HEP vaccine has been used on a limited scale for human immunization. It does not stimulate as good an antibody response in man as nervous tissue vaccines but has the enormous advantage for human immunization that it does not apparently produce neuroparalytic incidents. It is considered by many to be the vaccine of choice for the selective immunization of individuals whose work brings them into contact with rabid animals and for whom frequent inoculation with nervous tissue vaccines would carry considerable risk of allergic sensitization. The vaccine, however, cannot replace nervous tissue vaccine for individuals who have been bitten by rapid animals. A vaccine prepared from virus grown in duck embryos and inactivated with B-propiolactone is also being used for prophylaxis in individuals deemed to be

* Human antirabies immunoglobulin is now available and is preferable.

at special risk and is now beginning to be used for post-exposure treatment. A vaccine prepared in human diploid cells has also been introduced and is showing promise.

In the prophylaxis of rabies energetic local treatment of the wound is of the greatest importance. The wound should be thoroughly cleaned with soap or a detergent solution. It should, be noted that the application of ordinary antiseptics has no effect on the rabies virus. There is evidence that injection of hyperimmune serum around the wound is of considerable value. In a recent series of 581 persons bitten by animals proved to be rabid and receiving a full course of vaccine, 49 or 8·4% died. Of 153 individuals who did not receive vaccine 77 or 50% developed rabies and died.

Other rhabdoviruses: Marburg agent

Of the mammalian rhabdoviruses the best studied is vesicular stomatitis. There are at least two serological types of this virus which share a common core antigen. The disease, which resembles foot-and-mouth disease but is milder, can be distinguished from that condition by the fact that horses are susceptible. Man may be affected; the disease is mild and resembles influenza. The virus was thought to be transmitted by arthropods, but this has been questioned; it was originally grouped with the arboviruses. The disease is essentially restricted to the New World; it has been introduced into Europe and South Africa, but does not appear to have become established. A number of rhabdoviruses isolated from wild animals in West Africa have been shown to be related to but distinguishable from rabies. They have not been associated with disease in man. Other rhabdoviruses, not serologically related to rabies, have been isolated from mosquitoes, bats and fish, and from cattle suffering from bovine epizootic fever.

A somewhat questionable member of the group is the Marburg agent which was transmitted by African Green (Vervet) monkeys to laboratory workers and some of their contacts in Germany and Yugoslavia. The primary cases occurred in those handling monkey tissues. The batch of monkeys concerned in the German outbreak showed no characteristic symptoms and no excess mortality; though some excess mortality was observed in Yugoslavia. Of the 31 cases, seven were fatal. From the patients an agent was isolated in guinea-pigs, with a bizarre morphology. Essentially it is a rod 75–80 nm in diameter and varying in length from 130 nm to 4000 nm, often curved into remarkable shapes such as hooks and figures-of-eight. However, closer study has revealed features characteristic of rhabdoviruses—one end is rounded and the other flattened, often with an apparently empty bladder-like structure attached. Prominent cross-striations and a helical inner component add to its resemblance to rhabdoviruses. It appears to contain RNA.

The disease is of sudden onset with high, often biphasic, fever. Gastrointestinal upset is associated with prostration and the second febrile phase is accompanied by renal failure, central nervous system signs, rashes and haemorrhages. Serological evidence indicates that the agent is a fairly common infection of non-human primates in Africa. More recently a second outbreak has been reported from South Africa. A young man who had been hitch-hiking in Rhodesia passed on his infection to his travelling companion and to a nursing sister who looked after him in hospital. He remembered nothing unusual during his trip through Rhodesia except for a particularly painful bite by an unidentified insect while he was sitting by the roadside. This may indicate an arthropod vector in Africa, although there was nothing to indicate that the European cases were infected in this way. The young man died from massive haemorrhage associated with disseminated intravascular coagulation. Both the secondary cases recovered, possibly as the result of prophylactic heparin therapy. However one developed severe pancreatitis and the other, two months after the original illness and apparent complete recovery, developed an acute uveitis; Marburg agent was cultured from fluid aspirated from the anterior chamber of the eye. The symptoms were otherwise essentially similar to those in the European outbreak and serve to emphasize the ease with which this infection, like the clinically similar Lassa fever, can infect those looking after patients. One secondary case in the German outbreak appeared to have been due to persistence of the agent in the semen after apparent recovery.

ARENAVIRUSES

A new taxonomic group, the *Arenaviruses,* has

recently been proposed. The group includes lymphocytic choriomeningitis virus, the Lassa virus isolated from patients with a severe illness occurring in Nigeria during the civil war and a group of agents formerly classified as Arboviruses and known as the Tacaribe complex. They take their name (*arenarius,* L. = sandy) from the electron-dense intraviral granules seen in thin sections under the electron microscope. The members of this group are ether-sensitive, contain RNA and in thin sections appear as round, oval or irregular particles containing a varying number (usually two to five) of granules 20–30 nm in diameter and resembling in appearance host ribosomes. The particles, which vary in diameter from 50 to over 300 nm, are bounded by a unit membrane with closely spaced projections and an unstructured interior. There is no evidence of an internal, helical nucleocapsid. All show some sharing of the antigens. Their natural hosts appear to be rodents and they all probably produce persistent tolerant infections in their natural hosts (see p. 408).

Lymphocytic choriomeningitis

The lymphocytic choriomeningitis virus is a natural parasite of mice in which it produces a latent infection and from which man is infected. Its natural history in mice is discussed below.

In man the virus can produce aseptic meningitis, a severe influenza-like illness, bronchitis, pneumonia or be inapparent, but infections only occur in individuals in contact with wild mouse populations which are infected. Human infection probably occurs by the respiratory route and also possibly by ingestion of contaminated food. Virus is present in nasal secretions, urine, faeces and semen of infected mice. It is most readily diagnosed by the demonstration of complement fixing antibody in the patient's serum appearance of which may be considerably delayed (e.g. three weeks after onset), or by isolation of the virus by the inoculation of CSF into mice from a stock free of LCM, or by inoculation of guinea-pigs with subsequent passage to mice. Alternatively the presence of viral antigen in the spleens of infected guinea-pigs can be detected by the complement fixation test using a standard antiserum. The virus measures about 50 nm in diameter. It can be propagated in mice and in guinea-pigs and will grow in a variety of tissue cultures, in which it rarely produces a cytopathogenic effect.

Lassa fever

An apparently new disease appeared in 1969 and took its name from the place whence the first case came, Lassa in north-eastern Nigeria. The disease characteristically exhibits high fever, malaise, body pains, headache, cough, sore throat, vomiting and a bleeding tendency. Physical signs include lymphadenopathy, injection and ulceration of the mouth and pharynx and epigastric tenderness. In some cases there may be pneumonia with pleural effusion, cardiac failure, renal failure or encephalopathy with coma and convulsions. Deafness, which is also a feature, may persist. No jaundice has been observed, but at post mortem the liver has shown changes like those seen in yellow fever. The mode of transmission is not known, but the epidemiological evidence does not indicate a requirement for an arthropod vector, even if one exists. The reservoir host is believed to be the multimammate rat, *Mastomys natalensis,* which in many areas competes with the ordinary domestic rats for the domestic and peridomestic environment. The disease carries a high mortality and has acquired a particularly sinister reputation from the readiness with which infection is transferred to those in attendance on the patients and those handling the agent in the laboratory. In one case convalescent human serum seemed to be of benefit; there is no other specific therapy and the treatment is that of any acute haemorrhagic fever. The World Health Organization has set out a recommended code of practice to be followed in handling of Lassa fever patients and their contacts. As the longest recorded incubation period is 17 days, contacts should be kept under surveillance for at least three weeks. Virus has been demonstrated in urine and saliva as well as in the blood, so body fluids as well as aerosols are hazardous. One patient continued to excrete virus in the urine for 32 days. The precautions recommended would be appropriate to any infection liable to pass to the attendants, for example Marburg agent disease. Diagnosis is by isolation of the virus from blood, secretions or organs at post mortem; it is extremely hazardous and should only be attempted in specially equipped laboratories. As in lymphocytic choriomeningitis the complement fixation test may not become positive for more than 14 days and it has been suggested that human infection with Lassa fever generally occurs by the contamination of food by rodent urine.

Prolonged tolerant infections

The infant mouse is infected by its mother before or at about the time of birth. There may be a slight excess neonatal mortality at this stage but thereafter the affected mice appear perfectly healthy. However, examination of the blood reveals that there is a continuing viraemia amounting to about 10^4 to 10^5 infective units per ml. The concentration of virus in the spleen is about 100-fold higher. No antibody is detectable. The mice, however, do ultimately succumb, not to any neurological effect of the virus, but to chronic renal failure associated with the accumulation of deposits containing gammaglobulin on the basement membranes of the glomeruli. It is believed that small quantities of antibody are formed but that under conditions of antigen excess, this combines with some of the circulating virus and the complex is deposited in the kidney. The course of events in a mouse which escapes infection until it is older is quite different. The virus can produce a severe illness and the mouse may die with meningoencephalitis. If it recovers, antibody appears in the blood, although as LCM, like other arenaviruses, is a very poor antigen, its appearance may be considerably delayed. It is suggested that the chronic infection in the animals infected at birth is an example of immune tolerance (see p. 175) and the infection has been given the name 'prolonged tolerant infection' (PTI). However, this does not seem to be the whole explanation. Tolerant mice can be caused to produce antibody by adoptive immunization with lymphocytes derived from another mouse of the same in-bred line which has become immunized as the result of a postnatal infection with LCM. Antibody appears in the recipient mouse and the level of virus in the blood drops sharply, but it does not disappear. One slow virus, Visna, which affects sheep in Iceland causing a neurological illness, has been shown to be associated with polymerase activities similar to those in the RNA tumour viruses; so it is possible that immune tolerance is not the whole explanation and that the slow viruses, or some of them, can establish a chronic, non-cytocidal infection similar to that of the RNA tumour viruses except that no malignant transformation is produced.

IV.7

The Poxviruses

This is a group of viruses with a marked affinity for the skin, in which the majority produce the characteristic lesions or pocks which gives the group its name. They are the largest and, both structurally and chemically, by far the most complex of the animal pathogenic viruses. Nevertheless, they are true viruses, containing only one type of nucleic acid (DNA) and exhibiting an eclipse phase after infection. In electron micrographs they appear as oval or brick-shaped particles measuring from 200 to 300 nm in length.

They may be divided into subgroups on the basis of both morphology and antigenic constitution. The members of a given subgroup usually show close antigenic similarity and considerable cross-protection and cross-neutralization. All poxviruses so far tested, whatever their subgroup, have contained a common nucleoprotein antigen which is not concerned in neutralization or protection. Several members are associated with a haemagglutinin which, unlike that of the myxoviruses, is separate from the virus particle. The haemagglutinin is associated with a particle about 15 nm in diameter which contains about 75% of lipid. Cultures infected with haemagglutinating poxviruses demonstrate the phenomenon of haemadsorption (see p. 370). There are six subgroups, the sixth containing those viruses not yet allocated to one of the other subgroups. The subgroups and their members are as shown in Table IV.5.

Vaccinia virus

Vaccinia is the name given to the strains of virus used for vaccination against smallpox, all of which have had a long history of passage in calves and/or sheep. The origin of many of these strains is obscure but it is believed that most have been derived originally from naturally occurring cowpox virus or naturally occurring smallpox virus. Since it has not, however, so far been possible to transform a wild-type variola virus or a wild-type cowpox virus into a vaccinia virus, this proposed lineage cannot be regarded as fully established. The strain in current use in Britain seems likely to have been derived from a virus isolated from a German soldier during the Franco-Prussian war. Another possibility that has been recently suggested is that vaccinia viruses may have arisen by recombination between variola and cowpox viruses. Recombination between different strains of vaccinia and between cowpox and vaccinia has been demonstrated in the laboratory and the recombinants have been found to possess considerable genetic stability. It is, therefore, distinctly possible that at least some vaccinia viruses have arisen in this way. Such an origin, followed possibly by the formation of further hybrids would provide a neat explanation for the differences encountered between different strains of vaccinia. The wide host range of vaccinia is more like that of cowpox than of variola.

The main component of the vaccinia virus is protein but it has a lipid content of approximately 5%. Highly purified preparations of the virus in addition contain copper, flavinadenine dinucleotide and biotin. There is still considerable doubt as to whether these should or should not be regarded as integral components of the virus particle. On the whole the evidence suggests that they should, but their function is, in this case, unknown. They may have a role in relation to some, so far unidentified, viral enzymes or they may simply be the vestiges of an evolutionary past. Lipid can be extracted from poxviruses by treatment with ether. Infectivity is generally

Table IV.5. The poxviruses of vertebrates

Subgroup I	*Subgroup II*	*Subgroup III*	*Subgroup IV*	*Subgroup V†*	*Subgroup VI*
Vaccinia*	Orf*	Goatpox	Fowlpox	Myxomatosis	Buffalopox
Variola*	Milkers' nodes (paravaccinia)*	Sheeppox	Canarypox	California myxoma	Horsepox
Alastrim*		Lumpy skin disease of cattle	Pigeonpox	Hare fibroma	Rhinoceros-pox
Cowpox*	Bovine papular stomatitis*		Turkeypox	Rabbit fibroma	Swinepox
Monkeypox*			Sparrowpox		Molluscum contagiosum*
Rabbitpox	Chamois contagious echthyma		Starlingpox	Squirrel fibroma	Tanapox*
			Juncopox		
Ectromelia (Mousepox)					Yaba tumour virus of monkeys*
Camelpox*					

* Capable of infecting man.
† Members of this group cause fibromatous or myxomatous tumours.

reduced rather gradually, in contrast to viruses such as influenza and some members of the group are ether resistant although they are susceptible to other lipid solvents. Vaccinia is relatively resistant to ether.

Structurally the vaccinia virus is highly complex but the details of the structure have still not been fully resolved. The poxviruses do not conform to either of the main symmetry types and are therefore sometimes referred to as complex. A rather irregular outer layer containing lipoprotein tubules or threads lies on a second layer which in turn encloses a layer of less electron-dense mate .i which surrounds a central, biconcave nucleoid and two so-called elliptical bodies which lie in its concavities. The nucleoid is covered by a dense membrane which has a palisade-like appearance. The nucleic acid is contained within the nucleoid and consists of double-stranded DNA with a molecular weight of approximately 150 million and a G + C ratio of only 36%. At least 20 viral proteins have been identified, at least two of which are enzymes, a DNA-dependent RNA polymerase (transcriptase) and a nuclease. The lipid appears to be virus-specific, thus being unlike the lipid of enveloped viruses such as herpes. Uncoating takes place in two stages. The outer lipoprotein layer is removed almost immediately after penetration. Transcription by the viral transcriptase then begins and some viral mRNA is formed. This when translated appears to give an enzyme which completes the uncoating and liberates the DNA. Replication takes place exclusively in the cytoplasm in 'virus factories' which appear in conventional histological sections as juxtanuclear eosinophilic inclusion bodies which often appear to be lying in a small vacuole and in epithelial cells of infected animals generally lie on the basal side of the nucleus. One of the structural viral antigens—the LS antigen of the literature—is a complex protein consisting of heat-labile and heat-stable components. The partial purification of an antigen responsible for stimulating neutralizing antibody has been reported. In addition, vaccinia possesses a

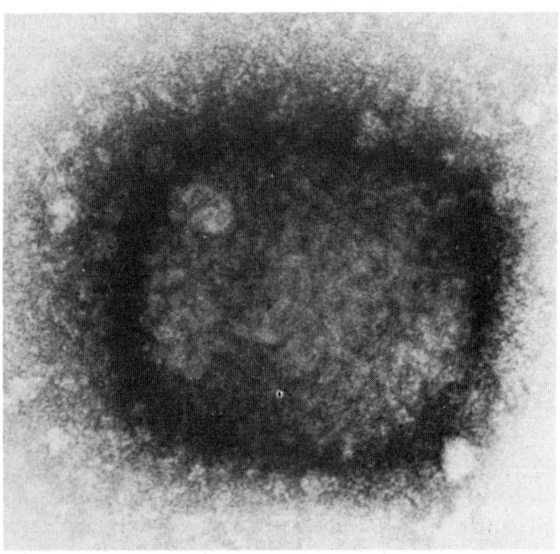

Fig. IV.22. Vaccinia virus from vaccine lymph, showing the mulberry appearance of the surface. Electron micrograph. Negative stain. (*Dr T. H. Flewett*)

number of other antigens which it shares with other members of its subgroup, but not with members of other subgroups.

In infected tissues the virus produces a haemagglutinin which is distinct from the virus particle. The haemagglutinin is active against fowl red cells but the cells of only about 50% of fowl are susceptible.

The vaccinia virus grows in a variety of tissue cultures in which it produces a well-marked cytopathogenic effect. It also grows readily on the chorio-allantoic membrane of the chick embryo, producing large, irregular opacities with necrotic centres known as pocks. It will infect a variety of animals, particularly calves and rabbits, on intradermal inoculation. By intracerebral passage in rabbits strains of increased neurovirulence, known as neurovaccinia, may be produced. The virulence of the virus for rabbits and calves is diminished by serial passage on the chorioallantoic membrane.

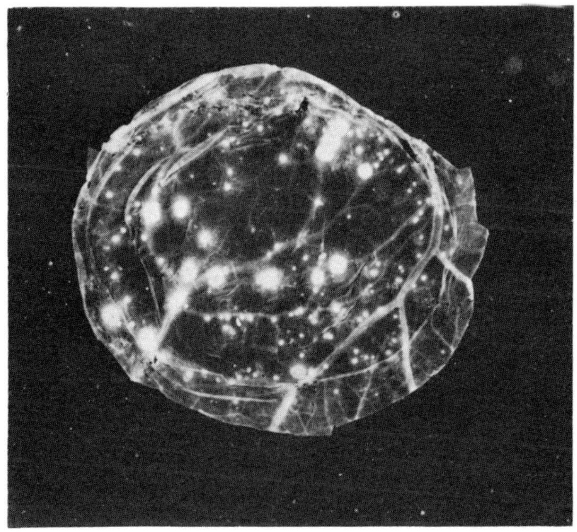

Fig. IV.23. Pocks produced by variola and vaccinia viruses on the chorio-allantoic membrane. The larger pocks are those of vaccinia and the smaller of variola. (*A. W. Downie*)

Variola virus

The variola virus has been examined in much less detail than the vaccinia virus but it is very similar in its properties. Like the vaccinia virus it will grow in a variety of tissue cultures, though the cytopathogenic effect is slower to appear and less extensive. Inclusion bodies can also be demonstrated in infected cells. Haemadsorption of fowl red cells to the tissue culture can be demonstrated apparently before the appearance of a significant cytopathogenic effect. Haemadsorption is presumably due to viral haemagglutinin.

The variola virus has a much narrower host range than the vaccinia virus. It will infect monkeys, producing symptoms resembling those of smallpox in man, and most strains will infect rabbits when inoculated onto the cornea. Occasional strains will cause skin lesions in rabbits on intradermal inoculation but serial passage of these strains has not been found possible. Dumbell and Bedson, however, have reported the serial intradermal passage in rabbits of virus which had been previously passaged in rabbit kidney tissue culture. Serial passage is also possible in suckling mice by the intracerebral route. Variola virus produces pocks on the chorio-allantoic membrane of the chick embryo but these are smaller than those produced by vaccinia, have a smoother surface, are less necrotic and have a uniformly greyish white appearance.

Smallpox virus from cases of variola minor or alastrim are in general less pathogenic for the chick embryo and the pocks are not haemorrhagic, but they more readily infect rabbits than strains from variola major. The two can be more precisely differentiated, on the basis of growth temperature, on the chorio-allantoic membrane, variola major strains being capable of producing pocks at both 35 and 38·5°C while alstrim strains will produce pocks at 35°C, but not at 38·5°C. This is consistent with a general principle that virulent viruses will grow at higher temperatures than viruses of low virulence.

Smallpox

Smallpox occurs in two clinical forms: variola major which is a serious disease with a fatality rate of 20–25% and variola minor or alastrim, which is a relatively mild disease with a negligible fatality rate. Each type breeds true and there must therefore be some intrinsic differences in the viruses responsible. This is consistent with the differences in pathogenicity that have been demonstrated in the laboratory.

In both types of smallpox, infection usually occurs through the respiratory route by the inhalation of infective material. This material may be droplets from the respiratory tract of a patient, dust containing dried droplets or material derived from the skin lesions. The danger of infection from such material is considerable since the virus has been shown to be capable of surviving in dried crusts for longer than a year. Epidemiological evidence indicates that cases of smallpox are not infective until just before the appearance of the skin lesions. These are numerous and each contains a large amount of virus. On the other hand, the virus present in them does not readily gain access to the exterior and it is more likely that the main sources of dissemination are lesions of the mouth and upper respiratory tract.

The disease appears after an incubation period of from 12 to 14 days (limits 9 to 16 days). It has an acute onset with the development of a high temperature; the characteristic rash appears on the third or fourth day and when it does the temperature falls. The rash goes through papular, vesicular and pustular stages, its evolution to the postular stage taking about five to seven days. When the pustular stage develops, the patient's temperature again rises; this stage is usually associated with secondary pyogenic infection of the skin lesions. Finally the pustules dry up, giving rise to scabs which fall off in two to four weeks. Since the lesions involve the skin to a considerable depth they leave permanent scarring which causes the characteristic disfigurement.

The severity of smallpox varies considerably. The majority of cases show a discrete rash and a mortality of under 10% but milder and more severe cases also occur. In cases with a confluent rash, the mortality rises to about 40%. Fulminant infections also occur, with early appearance of a haemorrhagic rash. Almost all such cases are fatal and as death may occur before the development of the characteristic rash, such cases may be misdiagnosed as purpura until secondary cases of smallpox occur amongst contacts. Cases of alastrim do not differ from the mildest cases of variola major: the difference between the two conditions is that the more severe types of illness with the high mortality only occur in variola major. Patients vaccinated years previously often develop a very mild infection with no rash at all, *variola sine eruptione* or more commonly with a very sparse rash on the limbs which is overlooked or misdiagnosed as, for example, chickenpox or septic insect bites. They often have very little constitutional upset and are quite fit to travel. Such cases arriving in this country by air constitute the usual way by which smallpox is introduced.

The site of initial viral multiplication is not known with certainty but it is presumably some part of the respiratory tract. Since the virus ultimately reaches the skin from the respiratory tract a viraemic stage must occur. Viraemia is readily demonstrable in very severe cases but in others has been found only in the early stages of the disease. From analogy with results obtained by Fenner with mouse ectromelia it is believed that there are in fact two viraemic phases in smallpox. In the first, occurring shortly after infection, reticuloendothelial cells throughout the body are invaded, multiplication of the virus in these leading to a second and more intense viraemic stage associated with the development of clinical symptoms.

In fatal cases of smallpox widespread focal lesions are found in the respiratory tract, kidney, liver, spleen and intestines. Infected skin cells show marked degenerative changes. The vesicles have a multilocular character due to the fact that they are traversed by strands of degenerating cells. In the pustular stage large numbers of polymorphonuclear cells are present; these probably accumulate initially as a result of the stimulus of dead or dying epithelial cells, but their presence is also associated with secondary pyogenic infection. Characteristic acidophilic cytoplasmic inclusion bodies—Guarnieri bodies—are present in infected epithelial cells. In the early stages of the disease the inclusion bodies are usually oval homogeneous masses, lying close to the nucleus and surrounded by a clear unstained area. In older lesions the inclusions are usually granular with an irregular outline, and may occupy most of the cytoplasm. The inclusions are identical with the 'virus factories' demonstrable by electron microscopy (p. 371 and Fig. IV.8).

Though it had been endemic in parts of Asia and Africa for centuries, smallpox was only introduced into Europe in the Middle Ages, probably by the Saracens. From the sixteenth to the eighteenth centuries it was a common disease in the British Isles. From the early nineteenth century onwards, however, following the adoption of vaccination as a general prophylactic measure, it showed a gradual decrease in incidence until, by the later part of the nineteenth century, it ceased to exist as an endemic infection. Nowadays, cases occur only when the disease has been introduced from other countries, as in the extensive 1962 epidemic of variola major, when it was introduced by immigrants from Pakistan. Due to the efficacy of control measures, however, this remained largely an institutional outbreak. As smallpox is solely a human infection and a potent vaccine is available there is no reason why it should not be eliminated altogether from every country. The eradication campaign organized and managed by the World Health Organization is at the time of writing apparently on the verge of a successful conclusion. The disease is now believed to be endemic in only two countries, Bangladesh and Ethiopa, and it is anticipated that by the end of 1976 these countries will be clear of the disease as well. With the disappearance of smallpox the practice of vaccination will presumably cease. It will then be interesting to see whether other related poxviruses, such as monkeypox and camelpox, which are capable of producing in man an illness resembling mild smallpox, will move in to fill the ecological vacuum. Much will depend on the capacity of these other viruses to spread directly from man to man; this appears to be limited.

Diagnosis

The clinical diagnosis of smallpox in a typical case is relatively easy. Most difficulty is encountered in cases of mild smallpox in vaccinated persons and in cases of the severe haemorrhagic types which do not show the typical rash. Mild cases may be confused clinically with chickenpox, which is therefore usually made notifiable when a smallpox outbreak occurs. Important features differentiating mild smallpox from chickenpox are that in smallpox the lesions have a predominant impact on the extremities, are all at the same stage of development and involve the skin more deeply than do chickenpox lesions.

In the initial febrile stage, prior to the development of the rash, it may be possible to demonstrate the virus in the blood by inoculation onto the chorio-allantoic membrane of the chick embryo. Persistence of the virus in the blood after the second day of the disease is found only in severe haemorrhagic cases. The complement fixation test has been used to demonstrate the presence of virus antigen in the blood but is only likely to give a positive result in very severe cases.

When the rash appears the following methods of diagnosis are available:

1. Virus particles may be demonstrable microscopically in scrapings from papular or vesicular skin lesions. This is best achieved by electron microscopy, using the negative-staining procedure in which the characteristic morphology of the virus is demonstable. Alternatively the material may be examined by ordinary light microscopy, using preparations stained, for example, by Gutstein's method. Stained preparations have little value during the pustular stage, when it is virtually impossible to distinguish the virus from the particles of debris present in pustular lesions. The fluorescent antibody technique has also been used.

2. The virus may be isolated by inoculation of scrapings from skin lesions, taken at any stage, onto the chorio-allantoic membrane or into an appropriate tissue culture. A positive result in tissue culture may be confirmed by neutralization or haemadsorption inhibition.

3. Viral antigen may be demonstrable in scrapings from vesicular and pustular skin lesions and in crusts by the complement fixation test. This method is highly reliable and will give a more rapid result than egg inoculation. Virus antigen may also be demonstrated by the Ouchterlony technique for which the microtest procedure of Dumbell and Nizamuddin has been recommended. This procedure is highly sensitive for the detection of antigen in vesicle fluid and pus and permits a specific identification in from four to five hours.

4. Antibody may be demonstrated in the blood of smallpox patients by the use of haemagglutination inhibition, complement fixation or neutralization tests. The antigens used in these tests may be prepared from vaccinial skin lesions in the rabbit or from chorioallantoic membrane infected with vaccinia. Haemagglutination inhibition and neutralization tests are more sensitive than complement fixation tests and may detect antibody as early as the fifth or sixth day in patients who have not been vaccinated; the complement fixation test is not usually positive before the tenth day. In those who have not been vaccinated a diagnosis may be made on the basis of a single serum examination. For patients who have been previously vaccinated the demonstration of a rising titre of antibody in paired sera is essential. Serological examination is of limited value in the diagnosis of smallpox because of the delay in establishing the diagnosis. Its main value

is in the diagnosis of mild atypical cases particularly those presenting without a rash. None of the tests described above, except inoculation of the chorio-allantoic membrane, will distinguish between the different members of Subgroup I; however, they will exclude chickenpox and other infections not due to Group I poxviruses.

Prophylaxis

As has been mentioned smallpox is not now an endemic disease in the British Isles, the cases occurring being the result of the introduction of the disease from abroad. This state of affairs can be attributed in the first place to the adoption of universal vaccination. Vaccination is, however, not now compulsory and immunity from epidemics of the disease is almost entirely dependent on the vigilance of Port Health Authorities, whose duty it is to prevent its reintroduction. When a case of infection is detected it is immediately isolated in an infectious disease hospital. All contacts are vaccinated and are kept under surveillance for a period of 16 days.

Vaccination consists of the intradermal inoculation of vaccinia virus. This results in a local vaccinial lesion which protects against smallpox because of the very close antigenic relationship between the smallpox and vaccinia viruses. Originally vaccination was carried out by arm-to-arm transfer. Nowadays the virus is used in the form of lymph; this is obtained from the lesions produced by inoculation of the virus onto the skin of calves or sheep. The seed virus is maintained by alternate passage in calves and rabbits. For the preparation of lymph the vesicular lesions are curetted, ground up to form a pulpy mass and suspended in 1% phenol. The material is stored for 48 hours at 22°C during which time there is a considerable fall in its bacterial content. Glycerol is then added to give a final concentration of 40%. The glycerinated lymph is stored at −10°C for a further period when there is a further fall in its bacterial content. Before issue it is tested for potency by inoculation into the skin of rabbits or onto the chorio-allantoic membrane, and for the absence of pathogenic organisms. If stored at −10 to −20°C, lymph retains its potency for months but if kept at room temperature must not be used after one week from the date of issue. Dried lymph is now available and is being used on a wide scale. It has the advantage of having a much longer 'life'—up to two years at 4°C—than ordinary vaccine lymph.

Vaccines prepared from virus grown on the chorio-allantoic membrane of the developing chick embryo have been in use in certain areas of the United States for some years. They appear, however, to be less efficient antigens than calf lymph vaccine and have not therefore been adopted for general use in Europe. Vaccines prepared in tissue cultures of bovine skin are increasingly being used and have given good results. Chick embryo and tissue culture vaccines have the advantage that they can be prepared under sterile conditions. It has not yet been shown, however, whether they are less likely to give rise to post-vaccinal encephalitis than the calf lymph vaccine.

It was thought at one time that the hazard of post-infection encephalitis increased progressively with increasing age. It is now generally accepted, however, that the risk is less at two years of age than in early infancy. Two years is therefore the age recommended for primary vaccination of children in the British Isles. Children who are debilitated or who have been exposed to an infectious disease or who are suffering from infantile eczema should not be vaccinated. Elective vaccination should not be carried out during pregnancy or within 21 days before or less than four days after yellow fever inoculation.

A substantial degree of immunity is present by about the eighth or ninth day after vaccination. Since the average incubation period of smallpox is 12 days, individuals who have been in contact with a case are therefore usually protected if vaccination is carried out within the first two or three days of the incubation period. It can usually be assumed that vaccinated individuals are completely immune to infection for at least three years, and some degree of immunity persists for life. In countries where smallpox is prevalent it is, however, advisable to carry out revaccination at least every two years. Immunity to natural smallpox infection in vaccinated individuals is much more persistant than immunity to revaccination.

Vaccination is usually carried out by the multiple-pressure method. In this procedure a Hagedorn needle is repeatedly pressed through a drop of lymph placed on the skin, the pressure being applied with the needle parallel to the skin surface. The virus gains entry to the skin through the minute pricks produced by the needle in the epidermis. For primary, elective vaccination use no more than five pressures. This should be increased to 15–25 for revaccination or for individuals who have already been exposed to infection and in whom there would not be time to revaccinate if the first vaccination were to fail to take. This method has largely replaced the scratch method formerly in use. Some authorities consider, however, that the latter may give more effective immunity and advise its use in the face of an epidemic. Evidence has been presented that the multiple pressure technique may be more successful than the scratch method for the revaccination of male subjects. More recently a specially designed multipointed needle has been widely employed. Its use is more easily taught to unskilled or semi-skilled personnel than is the multiple pressure technique. After vaccination the vaccination site should be covered with a light but not airtight dressing to prevent dissemination of the virus.

Following a successful primary vaccination a papule appears within three to four days at the site of inoculation. The lesion then goes through the stages of vesiculation, pustulation and scabbing, taking approximately three days for each stage. The peak of the reaction, which in some subjects may be associated with considerable swelling of the arm and high fever, occurs at about the ninth or tenth day. The scab falls off finally after about 20 days. This type of reaction to vaccination, which is that occurring in subjects with no previous immunity, is known as a 'primary take'. In those who have previously been vaccinated the reaction appears sooner, evolves more rapidly and is milder in character than the typical primary reaction. Some show an immediate reaction to vaccination, a small papule appearing within 24 hours; in such cases there is usually no vesicle formation and the reaction fades within about three days. This type of reaction has often been referred to in the past as the reaction of immunity. This designation is, however, incorrect; the reaction may be due to immunity but may occur following vaccination with an inactive lymph either as a result of trauma or as an allergic response to the virus or to some other component of the lymph. In some cases no reaction follows vaccination. It should be emphasized that unless there is evidence of a successful recent vaccination the complete absence of a reaction or the development of an immediate-type reaction should be regarded as due to a faulty technique or to the use of inactive vaccine lymph and the vaccination should be repeated.

Vaccination not only protects against infection but also lowers the fatality rate tremendously in patients who, in spite of vaccination, contract the disease. Thus in cases of smallpox occurring in Britain between 1942 and 1953 the mortality in unvaccinated patients was 48·4% while in vaccinated ones the mortality was only 13·5%. All deaths in the latter group occurred in patients who had been vaccinated at least 13 years previously. The routine vaccination of children is no longer policy in the United Kingdom; the risks of vaccination manifestly outweigh the risks of smallpox and if the World Health Organization's programme is successful even the small risk which exists today will soon be removed. However, until the disease has been completely eradicated, it will be important to maintain in each area a group of medical, nursing and ancillary staff, including ambulance personnel and laboratory workers, whose vaccinations are kept up to date and who can deal with suspected cases as they arise. Those who handle Subgroup I viruses in the laboratory should be vaccinated before starting work and at intervals thereafter. Unvaccinated personnel should not be allowed into the working area under any circumstances.

Gammaglobulin prepared from the sera of recently vaccinated individuals has definite protective value and is recommended for household smallpox contacts and for persons who cannot be vaccinated within three days of contact. There is some evidence that its administration may be valuable in preventing the development of encephalitis in adults who are being vaccinated for the first time, and for the protection of eczematous children who have been in contact with recently vaccinated persons. It has also been recommended for the treatment of complications of vaccination, e.g. corneal vaccinia, but its value for this purpose has still to be determined.

Complications of vaccination

Post-vaccinal encephalitis

This is a very rare but grave complication with an overall mortality of about 10%. The frequency with which it occurs has shown considerable variability in different parts of the world. In Scotland in 1942 it occurred once in 8000 vaccinations, whereas in Durban in 1944 no cases were observed in 450 000 vaccinations and in Wales in 1962 there were 17 cases amongst 900 000 estimated to have been vaccinated. Most cases have occurred after primary vaccination of adults, and the hazard is clearly related to age.

There is some evidence that the risk of post-vaccinal encephalitis following late primary vaccinations may be considerably reduced if gammaglobulin, prepared from the sera of recently vaccinated persons, is given at the same time. Thus in Holland, where all army recruits are vaccinated, the incidence of encephalitis was found to be reduced by 75% by concomitant administration of immune globulin. It has also been suggested that the risk may be reduced by the administration of ultra-violet or formalin-inactivated vaccine and of attenuated live vaccine, prior to immunization with calf lymph. These procedures are still under examination. There is, however, some evidence from work in experimental animals that the administration of inactivated vaccine may induce a state of hypersensitivity which might result in an unduly severe reaction to the live vaccine.

Generalized vaccinia

In this condition, which usually appears from 9 to 14 days after vaccination, there is a widespread dissemination of the vaccinia virus through the body by the blood with the production of a generalized vaccinial eruption. The condition is rare, occurring about once in 50 000 vaccinations. Recovery is usually rapid and there is little scarring. The condition usually responds promptly to antivaccinial gammaglobulin.

Eczema vaccinatum
(Kaposi's varicelliform eruption)

This is the development of secondary vaccinial lesions on eczematous or burnt skin areas. Because of the risk of such infections children with eczematous lesions or burns should not be vaccinated unless they have been in contact with a case of smallpox. Most cases appear to be the result of contact infection from an individual who has been recently vaccinated. A clinically indistinguishable condition may result from the infection of eczematous individuals by herpes simplex virus (*Herpesvirus hominis*). Eczema vaccinatum carries a significant mortality (30–40% in untreated cases) which has been reduced as low as 7% by treatment with antivaccinial gammaglobulin.

Progressive vaccinia

In very exceptional cases the vaccinial lesion fails to heal. In such cases there is a progressive necrosis of the skin at the vaccination site and the development of metastatic lesions in other areas of the skin and in the viscera. Some of the individuals in whom this condition has been described were agammaglobulinaemic and the absence of much cellular response in the majority of cases suggests a defect in cell-mediated immunity in addition to the failure of the humoral response. Fortunately the condition is rare and seems to be confined to individuals with an immunological defect. In all cases examined there has been a failure to form any neutralizing antibody and a persistent viraemia has also been noted. When persons with any immunological defect have to be vaccinated, it must be done under cover of antivaccinial gammaglobulin. Reports suggest that methisazone combined with antivaccinial gammaglobulin may be of value in treatment.

Chemoprophylaxis

As an alternative to vaccination for the protection of smallpox contacts, the administration of methisazone (p. 363) has been favourably reported on; the degree of protection obtained is of the same order as that following vaccination. When methisazone was combined with vaccination the degree of protection, as shown in data reported by Bauer, was much greater than that following vaccination alone. Thus of 1987 treated vaccinated contacts only two developed smallpox whereas of 2380 contacts who were not given methisazone 62 developed smallpox. The protective effect of methisazone is of short duration and administration of this drug cannot therefore be substituted for vaccination as a long-term prophylactic procedure. More recent results with this drug have not all been as encouraging as the earliest trials, but there is little doubt that it does form a

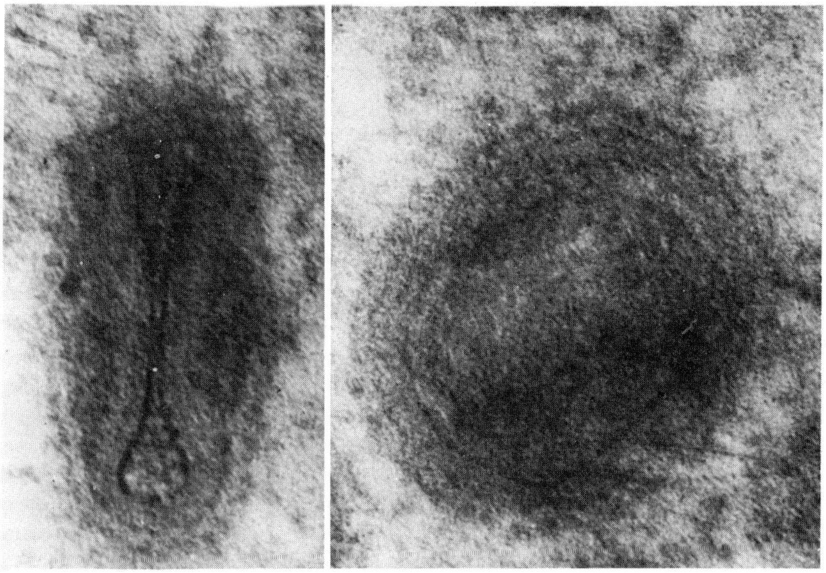

A B

Fig. IV.24. Molluscum contagiosum. Sections across (A) and along the length (B) of the particle. Note the dumb-bell shaped nucleoid in A. Electron micrograph. Negative stain. (*Dr T. H. Flewett*)

valuable addition to vaccination in the control of smallpox.

Cowpox

The virus of cowpox closely resembles those of variola and vaccinia but exhibits minor antigenic difference. Man is occasionally infected from cattle, with the production of localized skin lesions. The pocks on the chorio-allantoic membrane are strikingly haemorrhagic.

Molluscum contagiosum

This condition (see also p. 458) is characterized by the occurrence of painless pearly white nodules in the skin of various parts of the body. Only man appears to be susceptible.

In infected epithelial cells typical inclusion bodies are seen. These are large acidophilic structures frequently almost completely filling the cell. In electron micrographs the virus appears as a brick-shaped particle, morphologically similar to the other poxviruses. The viruses possess a soluble complement fixing antigen serologically distinct from those of variola and vaccinia. In some cases of molluscum contagiosum antibody to this antigen is demonstrable in the serum but not sufficiently frequently to make its demonstration of diagnostic value. The virus has not been propagated in the laboratory.

Contagious pustular dermatitis (orf)

This is a disease of sheep and goats of worldwide distribution. Human infection is not uncommon in those whose work brings them into contact with sheep or goats. In man it usually takes the form of a papule on the finger, hand, forearm or, rarely, face which develops slowly into a rather thick-walled pustule which can rupture to give an ulcerated lesion. The lesion is usually single, but multiple lesions can occur.

The virus (Fig. IV.25) has a characteristic morphology. It is somewhat smaller than the vaccinia and variola viruses, and in electron micrographs, in contrast to the mulberry appearance of the latter, shows a criss-cross pattern formed by two sets of parallel threads. The virus shows no immunological relationship to the viruses of the variola group and because of this and its different morphology has been grouped

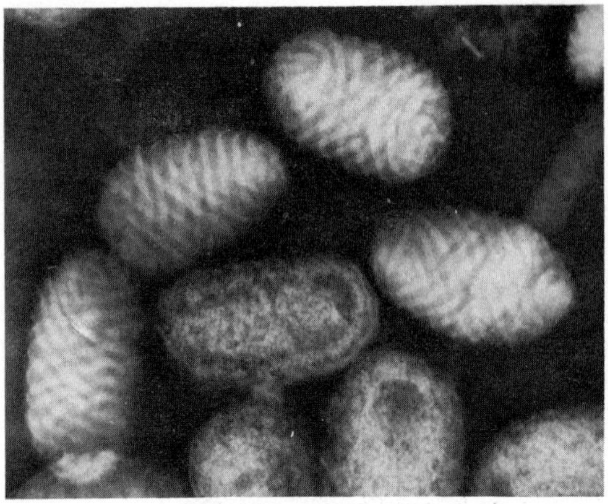

Fig. IV.25. Of virus. Note the criss-cross structures on the surface. The internal structure of some particles may be seen. Electron micrograph. Negative stain. (*Horne et al., Virology*)

with other similar viruses, including the virus of milker's nodes as the paravaccinia group (Subgroup II in Table IV.4). It has not been propagated in the chick embryo or in any of the ordinary laboratory animals but lesions can be produced in the skin of sheep by the inoculation of material obtained from a human lesion. It has been grown in fibroblast monolayer cultures prepared from sheep skin.

Milkers' nodes (paravaccinia)

These are red, papular lesions which occasionally develop on the hands of individuals milking infected cows. The lesions differ from those of vaccinia in not having a vesicular stage and in not being associated with systemic symptoms. A generalized rash is occasionally present. The disease occurs in many countries but appears to be particularly common in Norway. In infected cows the lesions occur on the teats and udder.

Morphologically the virus resembles the Orf virus. It will not grow in the chick embryo or in ordinary laboratory animals but will infect cows on inoculation into the skin. It has recently been propagated in tissue cultures of bovine kidney. It shows no antigenic relationship to the vaccinia or cowpox viruses.

Tanapox

Recently a new human disease caused by a poxvirus has been described amongst communities living along the Tana river in Eastern Kenya. Several hundred people were affected in two separate outbreaks. Victims suffered a short febrile illness associated with headache and prostration. There was a single pock-like skin lesion on the upper part of the body from which it was possible to isolate a poxvirus unrelated to the vaccinia–variola group. The disease has been called tanapox and the agent tanapox-virus. It appears to be transmitted to man from monkeys which constitute its natural host. It resembles closely a virus isolated from monkeys in a number of North American primate colonies in 1966. It is unrelated to monkeypox virus, but shows some cross-reaction in complement fixation tests with Yaba virus, another poxvirus of monkeys which causes progressive tumours. Both outbreaks of tanapox occurred during periods of flooding when men and domestic and wild animals were crowded together on the few remaining islands of dry land. Mosquitoes were extremely active, biting rates in excess of 700 per hour on man were recorded during one outbreak. It seems likely that the agent is transmitted from the natural host, monkeys, to man by biting mosquitoes in much the same way that myxoma virus is spread amongst Australian rabbits. The virus is restricted in its host range to man and primates and will only produce cytopathic effect in human or primate tissue cultures. Histological examination of tanapox lesions reveals the eosinophil intracytoplasmic inclusion bodies characteristic of the poxviruses, in hypertrophied epithelial cells. There is pronounced hyperplasia of the epidermis but little of the destruction seen with vaccinia and variola and relatively little inflammatory infiltration.

IV.8

The Myxoviruses

The myxoviruses were originally defined as a group of viruses capable of agglutinating the red blood cells of a variety of animal species through combination with mucoprotein receptors present on the red cell surface, and possessing at the same time an enzyme capable of breaking down the receptors. Subsequently it was found that all viruses possessing these properties were essentially similar in structure, and that the same structure was shown by certain viruses without demonstrable haemagglutinating properties. As a result the group is now defined by structural criteria. As defined in this way, myxoviruses are single-stranded RNA viruses possessing an internal component or nucleocapsid of helical symmetry. The nucleocapsid is surrounded by an envelope with radially disposed spike-like projections (see Fig. IV.4). These projections carry the viral haemagglutinin. The myxoviruses have a high content of lipid which constitutes about 20% of the virus particle. The lipid appears to be confined mainly, or exclusively, to the envelope which, as a result, is disrupted when the viruses are treated with ether. This causes liberation of the internal component or nucleocapsid and a concomitant loss of infectivity. The particles also contain about 5% of carbohydrate. The internal component released by ether treatment shows an appearance in negatively stained preparations consistent with a helical arrangement of protein subunits surrounding a coiled strand of RNA, an arrangement essentially similar to that found in the tobacco mosaic virus. As they fall into Baltimore's Class V, infectious nucleic acid cannot be extracted from the virus particles.

Three subgroups of myxoviruses have been distinguished.

Subgroup I: The influenza virus group: orthomyxoviruses

This group includes the influenza viruses of man and animals, including the virus of fowl plague. These measure from 80 to 120 nm in diameter, the envelope spikes measuring 90 by 3·5 to 4·5 nm. The internal component is a helix 90 Å wide by 6000 to 1000 nm in total length. The RNA content is about 1%. Filamentous forms, which may be up to 1 μm in length, are frequently found, particularly in recently isolated strains. The viral RNA of influenza seems to exist as seven distinct pieces which replicate separately and are loosely held together by hydrogen bonds, possibly involving the associated protein rather than the nucleic acid itself. When influenza viruses are passed at high concentration in chick embryos, a high proportion of defective particles lacking one or more of these pieces of RNA is found amongst the progeny (the von Magnus phenomenon). As such defective or 'incomplete' virus particles haemagglutinate normally, the haemagglutinin: infectivity ratio of such preparations is higher than normal.

Subgroup II: The paramyxoviruses

This comprises the parainfluenza viruses, mumps and Newcastle disease of fowls (fowl pest). The viruses of this subgroup are distinctly larger than those of subgroup I, measuring from 100 to 300 nm in diameter; the spikes are longer, measuring 12–15 by 2–3·5 nm and the internal component is larger—18 nm wide and 3500 to 5000 nm in length. These viruses also have a concentration of RNA of around 1%. The genome

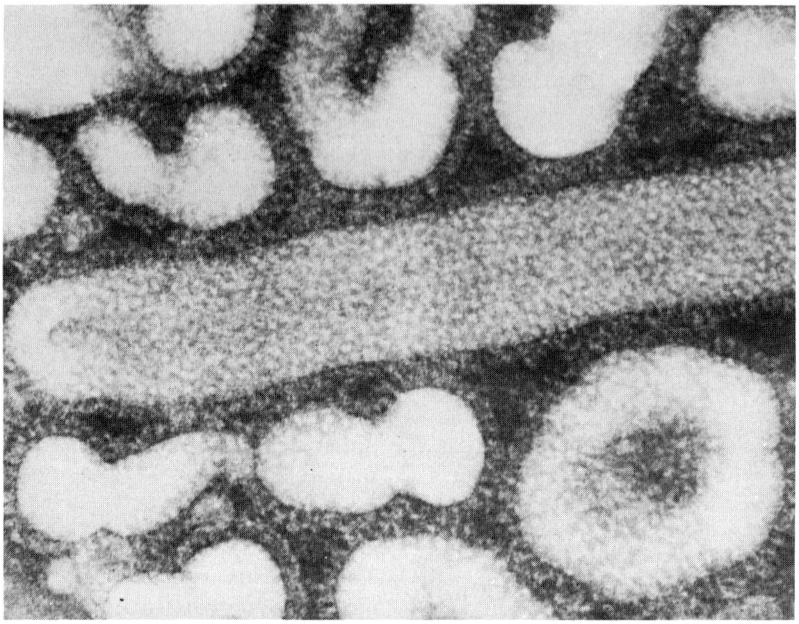

Fig. IV.26. Influenza virus, showing filaments and particles. Electron micrograph. Negative stain. (*Dr T. H. Flewett*)

appears to be in a single piece and not segmented as in the orthomyxoviruses. The membranes of Subgroup II viruses undergo spontaneous rupture, releasing the internal component, much more frequently than the membranes of subgroup I viruses.

Subgroup III: Pseudomyxoviruses

This group includes measles with the related viruses of canine distemper and rinderpest (cattle plague) and the respiratory syncytial virus. Although differing in a number of respects from Subgroup II viruses, measles has a haemagglutinin and haemolytic activity but lacks neuraminidase while respiratory syncytial virus apparently lacks all three, Subgroup III viruses are still classified officially as paramyxoviruses and the term pseudomyxoviruses is not official.

SUBGROUP I: THE ORTHOMYXOVIRUSES

The influenza viruses

Two distinct types of influenza virus antigen may be distinguished:

1. *Soluble antigens (S antigen).* These can be readily extracted from infected tissues and are present also in infected fluids, e.g. allantoic fluid. Although designated as soluble they occur as small particles of 10 to 15 nm in diameter. Chemically they are ribonucleoproteins, owing their antigenic specificity to the protein component. Antigens of identical chemical and serological specificity can be obtained by rupture of intact virus particles by ether treatment or by ultrasonic disintegration. There is little doubt that the soluble antigens found in infected tissues are in fact fragments of the internal component or nucleocapsid of the virus, their presence in free form in infected tissues and in tissue fluids representing either excess material which has not been incorporated into the completed virus particle or material which has been released by spontaneous rupture of the virus envelope. They are non-infective and non-haemagglutinating but are demonstrable by complement fixation tests. On the basis of the soluble antigens they possess influenza viruses are subdivided into three completely distinct antigenic types A, B and C.

2. *Strain-specific surface antigens (V antigens).*

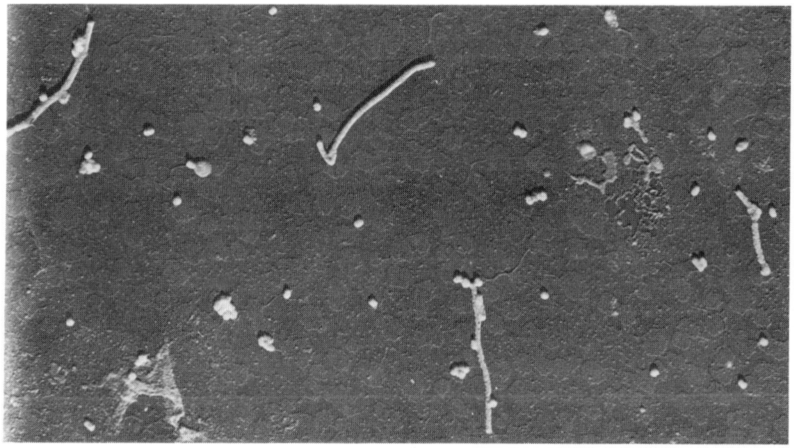

Fig. IV.27. Filaments and spheres of influenza virus type A, Melbourne strain. Platinum shadowed. × 10 000. (*Dr A. Isaacs and Dr R. C. Valentine*)

These have been identified with the spikes on the surface of the virus envelope, which are also responsible for haemagglutinating activity and for neuraminidase activity (see p. 345). Because of this, antibody to V antigens will inhibit haemagglutination and, since the virus makes contact with the susceptible cell through the spikes, will neutralize infectivity. Hence this is the antibody responsible for immunity against influenza. It has now been conclusively demonstrated that the neuraminidase and the haemagglutinin are quite distinct molecules. Recombinants have been produced with the neuraminidase of one parent and the haemagglutinin of the other. The two are distinct antigenically and morphologically. The haemagglutinin consists of rod-shaped particles, triangular in cross-section, with a hydrophobic end which is embedded in the lipid layer and a hydrophilic end which projects outwards and carries the haemagglutinating activity. The neuraminidase consists of a pear-shaped head and a tail fibre tipped by a small knob. As the total length exceeds the sum of the observed length of the spikes on the viral surface and the thickness of the lipid layer, it has been suggested that the fibres run more or less hosizontally in the surface layer of the virus. The lipid of the virus particle is derived from the host, but the original host proteins seem to have been entirely displaced by the new viral proteins, although there is a host-derived carbohydrate antigen in the envelope. It has been calculated that a typical influenza virus particle carries about 120 neuraminidase units and about 600 haemagglutinin units, but the total number of units and the ratio between the two types seems to vary considerably with the strain. The reaction between V antigens and antibody can also be demonstrated by complement fixation tests. Specific reactions, however, are only obtained if purified preparations of virus, freed of soluble antigens, are used. Such preparations can be readily obtained by adsorbing the virus onto red cells and allowing it to elute at 37°C. The supernatants of fully absorbed preparations will contain the soluble antigen.

When influenza viruses of types A and B are examined by haemagglutination inhibition and neutralization tests they exhibit considerable antigenic heterogeneity. This heterogeneity is due to differences in strain-specific antigen composition and is most marked in type A strains. Its precise basis is not known. Francis and Davenport, however, have brought forward evidence for the presence in various type A strains of at least 18 distinct antigenic specificities and have suggested that the antigenic differences between the various strains are due to an arrangement of the antigens on the virus surface in which antigens inaccessible in some strains are prominently sited on the surface of other strains and *vice versa*. Evidence in favour of this view is provided by Davenport and his associates in their finding that the viral haemagglutinins obtained by disruption of the virus by ether show a broader antigenic reactivity and a capacity to stimulate a broader immune response than does the intact virus. They suggest that in the intact virus the dominant strain-specific antigens are those located at the tip of the spike. This is consistent with the results obtained from electron microscopy of virus particles exposed to ferritin-labelled antibody. Davenport has proposed that, in addition to these antigens, other specificities are located along the sides and at the bases of the spikes; these are normally inaccessible to antibody but become accessible when the virus particle is inactivated by ether treatment. This is also consistent with the evidence obtained from electron microscopy, which in disrupted preparations shows fragments of membrane with attached spikes occurring in the form of rolled-up hedgehog-like balls or rosettes in which the spikes are considerably more splayed out and separated than in the intact virus and with the observation that repeated immunization of an animal with one strain of influenza type A virus may ultimately result in the production of antibody which will react with antigenically quite different type A strains. A second possibility is that the antigenic differences depend on progressive replacement of existing antigenic specificities by entirely new ones. Whatever the nature of the variation, all the antigenic specificities, except that of the host-specific carbohydrate, appear to be associated with either the neuraminidase or the haemagglutinin.

Whatever the basis of these differences may be, more or less continuous antigenic variation has been observed in human strains of influenza A virus since the first isolation by Smith, Andrewes and Laidlaw in 1933. One can recognize a progressive and relatively gradual change which is going on all the time, even during a single epidemic, whereby different strains, although all reacting to ferret antiserum against any one of them, in haemagglutination inhibition tests, nevertheless show differences inasmuch as any given antiserum will give a higher titre with the strain used for immunization than with other strains. This type of change is referred to as 'antigenic drift'. The smallest changes may only be recognizable if the most refined techniques are used, involving purified haemagglutinin (H) and neuraminidase (N) antigens and the corresponding antisera in gel diffusion and neuraminidase inhibition, as well as haemagglutination inhibition tests.

Of quite a different order are more fundamental antigenic changes which occur periodically. What is essentially a new virus so far as the surface antigens are concerned, more or less suddenly appears and, finding the world's population relatively susceptible because they have had no previous experience of the new strain, produces a global epidemic (pandemic). Changes of this magnitude are referred to as 'antigenic shift'. The development of techniques for separating the haemagglutinin and the neuraminidase from one another and from the rest of the virus components has made it possible to understand the nature of the changes which have occurred. The haemagglutinin and the neuraminidase antigens are referred to as H and N respectively and Table IV.6 shows how these have altered over the years since the 1930s. It will be noted that major changes have occurred more often in the haemagglutinin than in the neuraminidase. There is now persuasive circumstantial evidence to suggest that the fundamentally new dominant surface antigens which appear when antigenic shift takes place may be derived from strains of influenza A infecting other mammals or even birds.

Table IV.6. The surface antigens of human strains of influenza type A

Years of prevalence	*Old designation*	*Haemagglutinin*	*Neuraminidase*	*Antigenic formula*
1933–47	A_0 or A	H_0	N_1	H_0N_1
1947–57	A_1	H_1	N_1	H_1N_1
1957–68	A_2	H_2	N_2	H_2N_2
1968–	A_2*	H_3	N_2	H_3N_2

The suffixes, H_3N_1 etc. represent the different haemagglutinin (H) and neuraminidase (N) antigens.

* This is the Hong Kong variant: it is still designated A_2 because, although the haemagglutinin has changed, the neuraminidase has not (but cf. A_0 and A_1).

Antigenic shift will always render vaccines prepared from earlier strains ineffective; antigenic drift will vary in its effect, depending on the extent of the change. For example, vaccines prepared from early A_2 (Asian) isolates remained fairly effective up to 1964; thereafter the extent of the drift became such that only vaccines prepared from later isolates were satisfactory. The Hong Kong, H_3N_2, variant, first isolated in 1968, has been changing so rapidly that new vaccines have been needed almost every year.

Influenza type B has so far shown only relatively minor changes, although in recent years the rate of change has seemed to accelerate somewhat. Relatively few strains of influenza type C have been studied, but there is now evidence to suggest that, contrary to the previously accepted view, some variation may have occurred.

In contrast to the surface antigens, the nucleocapsid antigens characteristic of the three types A, B and C appears to be stable, and the same type A antigen is present in all type A isolates, whether from man, other mammals or birds. Influenza types B and C seem so far to be restricted in nature exclusively to man and will only infect laboratory animals with difficulty, if at all, rarely causing any symptoms.

It is not sufficient for a strain of influenza to possess antigenic novelty for it to establish itself as a pandemic strain: in addition it must obviously possess the other properties of an epidemic strain, such as high infectivity.

Evidence has been presented by Francis and Davenport that the dominant antibodies in an individual's serum are specific for the antigenic components of the particular virus to which he or she was first exposed. Thus people who were first exposed to influenza during the period of prevalence of the A_1 types, i.e. from 1947 to 1957, possess antibodies reacting particularly with these varieties. Older people who were first exposed to the pre-1947 or A_0 virus possess antibodies reacting particularly with A_0 components. Since the sera of individuals born before 1918 were distinguished by the presence of antibodies reacting particularly with the swine influenza virus these workers believe that the virus of the 1918 epidemic was antigenically related to the swine virus Hsw1 N_1 isolated during the 1930s. Similarly a high titre of antibodies to the 'Asian' virus A_2 was found by Mulder in the sera of persons old enough to have experienced the influenza pandemic of 1890. This suggests that there may be some antigenic relationship between the Asian virus and the virus responsible for the 1890 pandemic and supports Francis's theory of a fixed set of antigenic components which vary quantitatively in different strains. In addition to differing antigenically, strains of influenza virus of precisely the same antigenic composition may differ in the avidity with which they combine with antibody. This type of heterogeneity was first observed by Mulder in A_1 strains. Mulder distinguished three phases—P, Q and R—differing in their behaviour in haemagglutination inhibition tests. P phase strains are normal strains which are inhibited to high titre by homologous antiserum. Q phase strains are poorly inhibited by homologous serum. R phase strains are inhibited by homologous serum and by certain heterologous sera. It is suggested that Q phase strains may be of importance in permitting the virus to persist in immune populations, since they are relatively insusceptible to specific antibody. P→Q variation can be induced under laboratory conditions by growing the virus in the presence of immune serum. Similar differences have been demonstrated for A_2 strains, which have been found by Choppin and Tamm to consist of varying proportions of two types of particle: avid particles with a high combining power for antibody and non-avid particles with a low combining power. The relative proportions of the types of particle present determines the overall behaviour of the strain. Avid particles also appear to have a higher affinity than non-avid particles for red cell receptors and mucoprotein inhibitors.

Growth of the virus

The influenza virus grows readily in the chick embryo. On primary isolation most strains grow only in the amniotic cavity. When established by amniotic inoculation they may then be propagated by the allantoic route (except type C). Allantoic inoculation is usually carried out by using 10- or 11-day-old embryos, the infected allantoic fluid being harvested two days later. The concentration of virus in the allantoic fluid of infected eggs is high the infectivity being of the order of 10^{10} EID_{50}/ml of fluid. The virus will also grow in chick embryo tissue culture, in fragments of chorio-allantoic membrane suspended in nutrient medium and in cultures of human amnion and monkey kidney cells. In tissue cultures the cytopathogenic action of the virus is normally slight though it may be appreciable with some strains particularly A_2. Infection of tissue cultures with influenza virus is most readily detected by the *haemadsorption* technique in which a suspension of red cells is added to the culture; if virus is present the red cells adhere to the infected tissue cells. This technique is more sensitive for the detection of virus than the demonstration of haemagglutinin in the culture fluid.

If allantoic cells are simultaneously infected with two strains of influenza virus which differ in properties that can be readily distinguished, the progeny of infection may contain hybrid or recombinant forms. The formation of these hybrids is probably not analogous to the recombination phenomenon in bacteria and in bacteriophages but is due to reassortment of the pieces of genome. It is not known whether or to what extent recombination can occur under natural conditions but it is potentially a mechanism which might result in the emergence of variants of enhanced virulence or of altered antigenic composition (see p. 422).

A defect in maturation is observed when virus propagation is attempted in mouse brain (except with adapted strains), in certain types of tissue culture, notably HeLa cells, or in the chick chorion as distinct from the allantois. In all these cases virus nucleic acid and protein are formed but are not assembled into mature infective virus.

Haemagglutination

The influenza virus has the property of agglutinating the red cells of a variety of animal species. The cells which have been most used in the study of virus haemagglutination are those of fowl, man and guinea-pig. Haemagglutination is a function of the envelope spikes which adsorb to the appropriate receptor on the red cell surface, the virus particle then forming a bridge joining one cell to another.

When they are first isolated type A_1 viruses are frequently in the phase called by Burnet the O or original phase; in this phase the virus has a low haemagglutinin titre for fowl red cells relative to its titre for human and guinea-pig cells. On further propagation by the allantoic route it assumes the D or derived phase which has a high haemagglutinin titre against fowl cells. The O phase is presumed to be the form of the virus as it exists in nature and to be the only form which is naturally infective for man.

If red cells which have adsorbed virus are allowed to stand for some time, the virus is released or eluted from the cell. Elution takes place more rapidly at 37°C than at lower temperatures. It is due to the action of a virus enzyme which breaks down the receptor substances on the red cell with which the virus combines. Red cells whose receptors have been broken down to the extent that they are no longer agglutinable by the homologous virus may nevertheless be agglutinated by another myxovirus. It is in fact possible to arrange viruses in a series or 'receptor gradient', such that each virus is capable of removing receptors for itself and for the viruses below it in the series, but not for the viruses above it. The precise basis of the phenomenon is not understood but, presumably, depends on differences in the way in which different strains of virus can bind to the receptors and the completeness with which they can destroy them.

In addition to destroying the red cell virus receptors the virus enzyme effects a number of other changes in the cell:

1. Virus-treated cells are agglutinable by the so-called 'incomplete' Rh antibodies; this change is also produced by treatment of the cells with trypsin and various proteolytic enzymes. It has not been shown, however, that influenza virus possesses any protease activity.
2. Treated cells differ from normal cells in electrophoretic mobility.
3. Treated cells possess an antigenic specificity different from that possessed by untreated cells. It is not known whether this change is due to chemical modification of the receptor or to the unmasking, by removal of the receptor, of some antigen normally present in the cell.

Because of the capacity of the virus enzyme to break down the virus receptors on the red cell surface it is known as the receptor-destroying enzyme (RDE). The enzyme has been shown by Gottschalk to have the properties of a neuraminidase or sialidase: splitting off N-acetyl neuraminic acid from various mucopolysaccharides. Enzymes with properties similar to those of the virus RDE have been detected in culture filtrates of a number of bacterial species, particularly potent preparations being obtained from the cholera vibrio. The role, if any, of the enzyme in infection has not been determined. The initial stage of infection is undoubtedly the adsorption of the virus on to the receptor substances of susceptible cells; these receptors have been shown to have very similar properties to those of red cells. It would therefore seem likely that breakdown of the receptors by the virus enzyme permits penetration of the virus into the cell. There is evidence, however, that the virus can infect cells whose receptors have been modified in such a way that though they can combine with virus they are not broken down by it. It is more likely that the virus RDE is more important in securing the release of virus from infected cells than in facilitating its entry.

The receptor for the virus appears to be a mucoprotein. Mucoproteins from many sources—saliva, various types of mucus, serum and urine—are capable of inhibiting haemagglutination; this they do by combining with the virus and thereby preventing its adsorption to the cells. Inhibitory mucoprotein has a much more marked effect on virus which has been heated sufficiently to inactivate the receptor destroying enzyme than on unheated active virus. This is presumably due to the fact that inactivated virus can combine with the mucoprotein but is unable to elute from it.

The inhibitory mucoprotein present in normal mammalian sera is usually known as α inhibitor. It is readily broken down by RDE, by trypsin and by periodate. A second inhibitor known as β inhibitor and distinguished from α inhibitor by its insusceptibility to RDE and periodate and by its heat lability is also present in normal human serum. β inhibitor is an α-globulin and is readily broken down by trypsin and by proteases present in crude *Vibrio cholerae* filtrates. A third inhibitor, γ inhibitor, is present in horse serum and acts specifically against A_2 viruses, preventing experimental infection in mice when administered intranasally as well as inhibiting haemagglutination. It is an α_2-macroglobulin. Serum inhibitors are of considerable practical importance since they cause difficulty in the interpretation of haemagglutination inhibition tests (p. 369).

Influenza

Influenza is a disease of acute onset and short duration, usually lasting, in the absence of complications, for about three days. The essential lesions are degenerative changes in the ciliated epithelium of the bronchi and bronchioles, the lesions found being very similar to those encountered in experimental infections of the ferret. There is now clear evidence that influenza virus can, in severe cases, cause a specific type of pneumonia associated with degenerative changes in the alveolar wall. It is not clear, however, that this syndrome is due to invasion of the alveolar cells by virus. It may be due to the toxic effect of a large number of virus particles on the alveolar epithelium, an effect which has been observed in experimental infections of mice. The most serious effect of influenza is, however, the fact that it predisposes, especially in older people, to secondary pulmonary infection by pyogenic organisms, particularly by pneumococci and *Staph. aureus*: a high proportion of staphylococcal pneumonias occur, in fact, as complications of influenza. Staphylococcal pneumonia complicating influenza is particularly serious and has a high fatality rate. It usually develops during the acute phase of the disease. Pneumonia due to the pneumococcus, on the other hand, usually occurs after the acute phase of the infection has subsided. Influenza is a particular hazard to patients suffering from pre-existing pulmonary, cardiovascular or renal disease and epidemics are always associated with an excess mortality in patients with these diseases and in the elderly.

Influenza is a disease that affects all age groups but those aged 5–14 years and 25–34 years show maximum susceptibility. Following the introduction of a new antigenic variant, as occurred with the spread of the A_2 virus in 1957, the main brunt of the disease is borne by the younger age groups.

Influenza is mainly an epidemic disease but occurs also as a sporadic infection. Serological evidence indicates that during the course of an epidemic, in addition to overt infections, a considerable number of associated inapparent infections also occur. Of the three antigenic types, type A is epidemiologically the most important having a particularly marked capacity for epidemic spread. Epidemics of type A infection tend to occur with a periodicity of two to three years. Epidemics of type A infection are explosive in onset and tend to spread rapidly by contiguity from the initial focus of infection. Occasionally the disease assumes a pandemic form, spreading rapidly from one country to another until the entire world is involved, as in the great pandemics of 1918 and 1957. The type B virus is more frequently responsible for sporadic cases and small localized outbreaks, large epidemics occurring much less frequently than with type A. Type B epidemics tend to occur with a periodicity of three to six years. The type C virus has not so far been shown to be of any great epidemiological importance and is not known to have given rise to large-scale epidemics. Antibody to virus is, however, present in the sera of a high proportion of normal persons. It is consequently believed that inapparent infection is frequent and is the normal result of infection with this type. Experimentally, in volunteers, type C influenza virus has produced the symptoms of the common cold rather than classical influenza.

There has been considerable speculation as to the way in which the influenza virus persists in inter-epidemic periods when no virus is recognizable. It is possible that it assumes a dormant or masked form which, though capable of causing sporadic cases, would be incapable of large-scale epidemic spread. It has been suggested that this dormant form might be the Q form referred to above, since this form seems particularly adapted to survival in immune populations. The observation that some large type A epidemics appearing in the autumn were preceded by a summer 'flurry' of infection involving a relatively small number of cases is in support of the masked virus theory. This sequence of events was observed in Sicily in 1948 and in Scandinavia in 1950. Following the minor summer outbreak the virus apparently went 'underground' and became dormant, only to be lighted into activity in the following autumn. An interesting sequence of events has been revealed by the work of Shope on the epidemiology of swine influenza. The swine virus can apparently persist for long periods in lung-worms. The larvae of the lung-worms are ingested by earthworms which are in turn swallowed by pigs. The lung-worms with their contained virus become parasitic in the pig's lung in which the virus remains latent; if, however, the pig develops a respiratory infection with *Haemophilus suis,* the latent virus is activated and causes an attack of influenza. There is, however, no evidence that *H. influenzae* plays the role of a precipitating factor in the pathogenesis of influenza in man, although it may give rise to secondary pulmonary infection during the course of the disease.

The isolation from various animal sources of viruses possessing the type A soluble antigen has suggested the possibility of an animal reservoir for human influenza viruses. However, as the world-wide surveillance of influenza is extended, it is becoming clear that human cases are almost always occurring somewhere; so it is perhaps less essential to consider the possibility of latency. Nevertheless, there is much about the epidemiology of influenza which remains obscure; for example, the delay which may occur between the introduction of a new strain to a population and the development of an epidemic and the factors responsible for the marked seasonal incidence usually seen in temperate climates.

The capacity of the virus to undergo antigenic variation plays an important part in the epidemiology of influenza. As has been mentioned above the type A virus appears to be undergoing a progressive antigenic variation, the virus of each epidemic differing detectably from that of the previous epidemic and more markedly from the viruses of still earlier epidemics. Consequently the virus responsible for each major epidemic is to a greater or a lesser extent insusceptible to the immunity persisting from earlier epidemics. Because of this the influenza type A virus is able to give rise to epidemics which recur at relatively short intervals.

The great severity of the influenza pandemic of 1918–9 is well known. The mortality in this pandemic was very high; it was in fact responsible for more deaths than World War I, killing in all from 10 to 20 million persons. A particularly serious feature of the pandemic was the great severity of the disease in young adults. It has been suggested that the virus responsible differed sharply in antigenic properties from the viruses present before 1918, against which the population possessed appreciable immunity. There is, however, some indirect serological evidence that the 1918 virus antigenically resembled the swine influenza virus and that it was prevalent for some time prior to 1918. If this was the case, the severity of the epidemic could not have been solely due to antigenic variation. An alternative explanation is that the virus, although showing some normal antigenic variation from preceding viruses, was of very much greater virulence. Andrewes suggests that increase in virulence might have resulted from a recombination between two distinct strains of virus producing a vicious new hybrid. In support of this possibility he points out that the virulent form of the disease first appeared around Boston and in Brest—ports of embarkation and disembarkation for American troops going to the European war theatre. At these ports there might well have been a fusion between an American and a French strain of virus. Both these explanations are, however, entirely hypothetical and the great virulence of the 1918 pandemic is likely to remain for ever an unsolved problem.

Several studies have purported to show an increase in various congenital abnormalities and in malignancies of the lymphoid and haemopoietic systems in children born to mothers having influenza during pregnancy. Taken as a whole the evidence is not conclusive.

Diagnosis

The virus may be isolated from the throat during the acute stage of infection. For this purpose chick embryos inoculated by the amniotic route have been used, the presence of virus in the amniotic fluid being detected by haemagglutination tests. Rhesus monkey kidney cultures are also satisfactory, infection being recognized by haemadsorption, and more convenient in a routine diagnostic laboratory.

For clinical purposes serological tests are more valuable than isolation of the virus but have the disadvantage that they only permit a retrospective diagnosis. Two samples of serum are taken—the first, if possible, within the first three days of the disease and the second during convalescence about a fortnight later. If the case is one of influenza there should be a rise in antibody titre of at least four-fold between the two examinations. Antibody may be detected either by haemagglutination inhibition or by complement fixation tests. The latter are best carried out with preparations of soluble antigen freed of virus particles by absorption with red cells. Soluble antigens of both type A and type B should be used but need only be obtained from a single strain of each type. Haemagglutination inhibition tests are carried out using allantoic fluid as a source of virus. In this case it is important that the virus strain employed should be of the prevalent antigenic type. Best results are obtained in haemagglutination inhibition tests if the patient's serum is first treated with a preparation of the receptor-destroying enzyme from *V. cholerae* or with trypsin and with heat so as to break down the non-specific inhibitors present. Serial dilutions of the serum are made in the cups of a perspex WHO haemagglutination tray and mixed with a standard amount of virus, usually four agglutinating doses. Mixtures are allowed to stand for some time and red cells are then added. The cups are examined for agglutination either by a densitometric or a pattern method after a further period. Complement fixing antibody declines more rapidly after infection than haemagglutination inhiting antibody and is consequently more valuable as an index of recent infection. In addition, it is not produced, as is the latter, by vaccination.

Prophylaxis

A great deal of experimental work has been carried out on the value of killed vaccines in active immunization against influenza. In general the results have been satisfactory; when the virus used for vaccine preparation has been closely related antigenically to the virus responsible for the next epidemic, significant, though by no means complete, protection has been obtained. Thus in the Medical Research Council trials of A_2 vaccine in England in 1957, protection rates of from 52 to 75% were achieved. A significant degree of protection was demonstrable 8 days after the administration of asingle dose of vaccine.

The antibody response to a given amount of antigen can be greatly increased by incorporating the antigen in an adjuvant consisting of mineral oil and an emulsifying agent (p. 98). Limited trials with such mineral oil vaccines have been highly satisfactory, high levels of antibody being found in a considerable proportion of vaccinees for up to two years after vaccination. Adjuvant vaccines have the additional advantage of permitting the employment of a much smaller amount of antigen than with the conventional saline vaccines. They have, however, shown a tendency to give unacceptable reactions and they are no longer used in the United Kingdom. More recently an adjuvant based on peanut oil has been tested and, although less potent than mineral oil emulsions, seems less prone to produce untoward reactions.

The most recent vaccines are made from partially purified and disrupted virus particles. Such vaccines may still give some reactions, but these are less frequent and less severe than those given by conventional killed vaccines. It has only recently been recognized that the level of antibody in the secretions of the respiratory tract is important in immunity to influenza. In an attempt to stimulate local antibody formation, a vaccine has been introduced which is administered by nasal spray. Whether such a vaccine, which is essentially similar to those given by injection, can confer worthwhile immunity remains to be seen, especially as it has recently been shown that the level of circulating antibody is also important in immunity to influenza.

A more satisfactory immunity should be produced by live, attenuated vaccines which mimic the natural infection. These have been quite widely used, especially in Russia, and appear to be safe both to the individual and to the community. They give reasonable immunity to infection with strains antigenically close to the vaccine strain. They are given as nasal drops in two doses a fortnight apart. Up to now the vaccines have comprised a single strain of type A and so cannot be expected to give much immunity to unrelated type A strains or any immunity to type B. The time necessarily taken to produce and test a suitably attenuated strain of virus and then undertake production is such that the current virus is likely to have changed significantly by the time that the vaccine is generally available. Three methods have been used to derive attenuated strains. These are growth of the virus in the presence of horse serum to give inhibitor-resistant strains, growth at reduced temperatures to give temperature-sensitive (ts) mutants and genetic crossing with a 'laboratory strain' to give hybrids with the surface antigens of the 'wild' strain and the growth properties and lack of virulence of the 'laboratory strain'. The latter technique is also of value in killed vaccine production. When first isolated most strains grow poorly in eggs and it takes time to adapt them to grow well in this host. However, by crossing the 'wild' with an egg-adapted strain, a hybrid can be produced which is suitable almost at once for making a vaccine. Live vaccines, on the other hand must still undergo an exhaustive programme of testing for potency and safety, even when an apparently suitable virus has been obtained. Whether the programme can safely be curtailed it is not at present possible to say.

Though the organization of vaccine programmes on a community scale is not at the moment practicable it is advised that vaccination should be offered annually to persons suffering from certain chronic diseases—chronic pulmonary disease, chronic heart disease, chronic renal disease, diabetes—in which an attack of influenza might aggravate disability or prove fatal.

SUBGROUP II: THE PARAMYXOVIRUSES

The morphological features distinguishing these from Subgroup I have already been considered. Paramyxoviruses are more difficult to propagate than those of Subgroup I. Only two members of the group—the mumps virus and the Newcastle disease virus—can be isolated in chick embryos. The viruses appear to be formed exclusively in the cytoplasm and incomplete virus is rarely produced. All carry a haemagglutinin and a neuraminidase. They show a capacity to lyse red cells.

The parainfluenza viruses

Four serotypes are distinguished on the basis of soluble antigen composition. The types show some sharing of envelope antigens with each other and with mumps virus and Newcastle disease virus. No single envelope antigen is, however, common to all, although each has antigens common to at least one other member of the paramyxovirus group. The haemagglutinins are best demonstrated with chick cells at +4°C or with guinea-pig cells at +25°C. For primary isolation, monkey kidney tissue cultures have given best results. Only type 2 produces a well-marked cytopathogenic effect in monkey kidney but a parainfluenza virus is readily recognized by the use of the haemadsorption technique.

The parainfluenza viruses have been isolated mainly from mild upper respiratory tract infections in children. Not infrequently, however, they invade the lower respiratory tract where they produce laryngotracheobronchitis (croup) and very occasionally bronchopneumonia. Types 1 and 3 have been found to produce cold-like illnesses in human volunteers but whether, and to what extent, they contribute to the total incidence of colds has still to be determined. Similar but antigenically distinct viruses have been isolated from animal sources. The Sendai virus which antigenically resembles type 1 was isolated in Japan from pigs and mice and viruses resembling type 3 have been isolated from a respiratory infection—shipping fever—in cattle assembled for transport as well as from sheep and horses. Viruses similar to type 2 have been isolated from monkeys.

The mumps virus

The mumps virus is serologically homogeneous but shares minor surface antigens with some of the parainfluenza viruses. It is completely distinct from the influenza viruses in both V and S antigen content. Like the parainfluenza viruses, it possesses a haemolysin which appears to act on the same cell receptors as the receptor-destroying enzyme from which, however, the haemolysin appears to be distinct. The enzymic activity of the mumps virus is appreciably less than that of the influenza virus. After elution at 37°C there still remains a considerable amount of irreversibly bound virus, which is insufficient to cause significant agglutination of the red cells but whose presence can be detected by the occurrence of haemagglutination when the cells are exposed to antiviral serum.

Mumps

Parotitis or inflammation of the parotid gland is the most frequent manifestation of infection with the mumps virus; in addition to the parotid, the sublingual and submaxillary glands are also occasionally involved. In about 20% of infections in males of over 13 years there is associated orchitis developing from about four to seven days after the appearance of the parotitis. This may result in a pressure atrophy of the affected testis but leads to sterility only if, as only occasionally happens, the condition is bilateral. In the female oophoritis may develop but this is a very much less frequent complication than is orchitis in the male. Other organs which may more rarely be involved include the pancreas, the mammary glands in either sex and the myocardium. It is thought that these conditions may be the result of an autoimmune process triggered by the infection. Aseptic meningitis or meningoencephalitis may also occur as complications of mumps. The other manifestations of mumps may develop at the same time as the parotitis, before it, after it or, rarely, in its absence. In the United States and Britain the mumps virus appears to rank next to the enteroviruses as the most important single cause of aseptic meningitis. Serological evidence shows that the mumps virus frequently gives rise to latent infection. It has been estimated from such data that during the course of an epidemic about 40% of infected persons are without obvious symptoms.

Mumps occurs mainly as an infection of the 5–15-year age group. The disease is followed by a high degree of immunity and second attacks are uncommon.

In mumps the virus is present in the saliva, from which it may be isolated just before and for a few days after the onset of parotitis. The virus has also been demonstrated in the saliva of persons who are latently infected and who are therefore of considerable importance in spreading the disease. The virus has been detected in the blood during the first few days of infection, and it is generally believed that involvement of the parotid glands as well as of the central nervous system, testis and ovary are the result of viraemic spread. This would be consistent with the long incubation period of 18 to 21 days.

Diagnosis

The laboratory is not usually called on for assistance in the diagnosis of mumps since the condition is usually quite obvious clinically. When laboratory assistance is required the virus may be isolated from the saliva by amniotic inoculation or in tissue culture. A convenient method of collecting saliva is by the insertion of cotton wool pledgets into the mouth. In cases of aseptic meningitis it may be possible to isolate the virus from the cerebrospinal fluid. Antibodies may be demonstrated by complement fixation or haemagglutination inhibition tests. Of these the complement fixation test is on the whole the more satisfactory. The antibody titre to both V and S antigens should be estimated by complement fixation. Antibody to the V antigen does not usually appear until some eight or nine days after infection but persists for a long time after recovery. Antibody to the S antigen is usually demonstrable in the first two or three days of the disease and declines rapidly after recovery. Where possible an attempt should be made to demonstrate a rising antibody titre, but the demonstration in the early stages of infection of a high S titre in the presence of a low V titre is virtually diagnostic. Serum amylase regularly increases during the course of mumps and its estimation may be of value in differential diagnosis.

In the course of mumps a delayed type of skin hypersensitivity develops which may be detected by the intracutaneous injection of an antigen prepared from infected allantoic fluid. The skin test has no value as a diagnostic test, as it usually does not become positive until some three to four weeks after the onset of the disease. It may, however, be used for the detection of immunity. The antigen involved has not been identified.

Prophylaxis

The administration to contacts of gammaglobulin prepared from the serum of convalescents gives some degree of protection; it is claimed that if given early in the disease gammaglobulin considerably diminishes the risk of the development of orchitis.

Immunization with an inactivated chick embryo vaccine appears to give considerable protection but has not been generally adopted as a prophylactic procedure. Some degree of immunity can also apparently be produced by the use of an attenuated vaccine sprayed intra-orally or, as practised by Russian workers, injected intradermally.

Newcastle disease virus

This virus, which shares antigens with some of the parainfluenza viruses, causes a severe and economically important respiratory infection in poultry. It is quite distinct from fowl plague virus which is an avian influenza virus of type A. Newcastle disease virus may attack man with the production of conjunctivitis, often unilateral, and sometimes more severe general symptoms.

SUBGROUP III: THE PSEUDOMYXOVIRUSES

These viruses have the same morphology as the paramyxoviruses but as they lack neuraminidase they can be regarded as a distinct group.

Respiratory syncytial virus

Although it lacks haemagglutinating properties the respiratory syncytial virus (RSV) shows, on electron microscopy, the general morphology of the paramyxovirus group. Three distinct antigens are demonstrable in infected tissue culture fluids:

1. An antigen which stimulates the production of neutralizing antibody and which appears to be analogous to the envelope antigens of the influenza virus.
2. A soluble complement fixing antigen which does not stimulate the production of protective antibody, and which is presumed to be an internal component or nucleocapsid.
3. An antigen which appears to be lipoprotein in nature but which has not so far been identified as any particular component of the virus particle.

RSV shows no antigenic relationship to other myxoviruses. A number of different serotypes has now been recognized, although there is considerable antigenic overlap between them.

The virus does not grow in the chick embryo but can be propagated in cultures of various human tissues and in monkey and bovine kidney. The cultures most used have been those of HeLa cells, particularly the Bristol line, and HEp-2 cells. In infected tissue cultures a characteristic cytopathogenic effect with marked syncytium formation is produced but the yield of virus is very poor. The virus is very unstable outside the body and clinical material should be cultured as soon as possible. Infectivity is rapidly lost even at −20°C.

RSV is particularly pathogenic for children and has emerged as the single most important cause of serious respiratory infection in children of under one year of age. In Newcastle it was isolated by Gardiner et al. from approximately 60% of serious respiratory infections of infants and in a Bristol study 18 of 21 viruses isolated from children of less than five months of age with respiratory infection were RSV. These findings confirm and extend the original observations of Chanock in the United States on the seriousness of RSV as a respiratory pathogen of childhood. The most characteristic clinical syndrome in children is acute bronchiolitis, but pneumonia, bronchitis with pharyngitis and upper respiratory tract infections also occur. In older children and adults infection takes the form of a mild upper respiratory infection or is asymptomatic, although there is evidence that exacerbations of chronic bronchitis may be due to this virus.

Prophylaxis by vaccination is still in an experimental stage. A vaccine producing appreciable antibody response has been prepared by formalin inactivation but its value has still to be assessed. Vaccination would be required most for very young children in whom the immunological response is poor due to the presence of maternal antibody. This may be expected to interfere seriously with the efficacy of vaccination at such an early age. However, there is no doubt that infection can occur in individuals with preexisting antibody and it has been suggested recently that the severe manifestations of respiratory syncytial virus infection only occur in subjects with partial immunity. Caution is therefore necessary in proceeding with the killed vaccine.

The measles virus

The measles virus is an RNA virus with the typical structure of the paramyxoviruses. It has haemagglutinating properties but only for monkey red cells, maximal at 37°C. Unlike influenza virus, the virus does not elute from the agglutinated cells nor are the red cell receptors destroyed. Like most of the paramyxoviruses it has haemolytic properties. It is uncertain whether the haemolysin is also the factor responsible for cell fusion. It appears to be distinct from the haemagglutinin. Only one serological type of the virus is known. There is, however, sharing of antigens with the rinderpest virus of cattle and with the virus of canine distemper.

The virus will grow, though rather slowly, in a variety of tissue cultures but isolation is usually difficult and has been most successful in primary cultures of human, monkey and dog kidney. In infected cultures a cytopathogenic effect characterized by multinucleated giant cells and by syncytium formation is produced; numerous acidophilic, cytoplasmic and nuclear inclusion bodies are formed.

Measles

Measles is primarily an epidemic disease, the period of maximal incidence being from November to March. Major epidemics show a marked periodicity recurring regularly in urban communities at two to three yearly intervals, although in Nigeria, where the disease shows a high prevalence, annual epidemics are the rule. Epidemic periodicity appears to be due to the accumulation of susceptibles in the population since the last epidemic. This in turn is due mainly to the addition of susceptibles by births in the intervening period. When the proportion of susceptibles reaches a critical level conditions are suitable for epidemic spread. The virus is present in the respiratory secretions and in the urine prior to the development of the rash. It is during this period that infectivity is at a maximum. The main route of infection is almost certainly via the respiratory tract but there is some suggestion that the virus may also gain entry through the conjunctiva. Inapparent infections are rare.

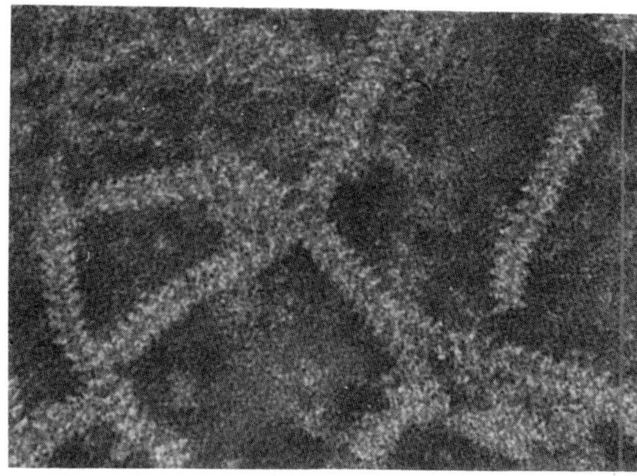

Fig. IV.28. Internal component of the measles virus showing the helical structure. Electron micrograph. Negative stain. (*Dr June D. Almeida*)

The disease has a long incubation period of 12–14 days. During this period the virus probably multiplies in some part of the respiratory tract from which it gains access to the blood stream. During this phase lymphoid tissue throughout the body is invaded. Multiplication of virus could then be responsible for a secondary viraemia during the prodromal phase by which the virus eventually reaches the skin. The disease presents with the onset of fever, malaise, coryza, cough and conjunctivitis. The characteristic rash appears on the fourth day. The so-called Koplik's spots, an enanthem comprising a collection of bluish-white spots on the inside of the cheeks, usually appear a day or so before the rash. Abdominal symptoms are quite common in the first few days of the illness and have led to operations for appendicitis. In such cases the appendix contains Warthin-Finkeldy giant cells (see below). The pathogenesis of the skin lesions is not fully understood. They are probably not due to multiplication of the virus but are the result of damage inflicted by an antigen–antibody reaction or by a cell-mediated hypersensitivity reaction.

Children between the ages of one and five years are mainly affected. In the more severe cases there is evidence of extensive involvement of the respiratory tract with symptoms of laryngo-tracheitis, bronchitis or interstitial pneumonia. On rare occasions the disease assumes a severe haemorrhagic form with a high mortality rate.

In more highly developed countries mortality from measles has declined dramatically over the last 50 years. Hence, in Glasgow the mortality in children aged one to five years was approximately 5% in 1908, but fell to about 0·2% in 1960. In England and Wales in general mortality has been maximal in children of under one year. In West Africa, however, the disease showed, in the period 1958–61, a case mortality for all age groups of over 20%, the highest mortality being in children of over one year. There is considerable evidence that high mortality rates are associated with low nutritional status. The most important complications of measles are secondary bacterial infection, especially pneumonia and otitis media, and encephalitis. The incidence of the former seems to be little influenced by the prophylactic use of antibiotics, which should probably, therefore, not be administered unless a bacterial infection develops. Measles can be complicated by at least two distinct types of encephalitis. Firstly there is the acute demyelinating leucoencephalitis, post-infection encephalitis (see p. 454), which may follow any virus infection or rabies vaccination. No virus can be isolated from the brain and the lesions are similar to those seen in experimental allergic encephalitis of animals, produced by injecting brain extracts. Clinically obvious encephalitis occurs in one or two cases per thousand. The onset is usually about four to five days after the appearance of the rash. Mortality is over 10% and a high proportion of the survivors show permanent sequelae such as epilepsy, mental changes and paralysis. However, about 50% of cases of apparently uncomplicated measles show electroencephalographic abnormalities. The significance of this is not clear, the changes usually disappear on recovery. The condition known as subacute sclerosing panencephalitis (SSPE) is now known to be a late manifestation of measles infection. Measles antigen is detectable in the affected brain tissue and recently the virus itself has been demonstrated by cocultivation of brain cells with cells more susceptible to measles virus. The condition seems to develop years after an attack of measles and has the characteristics of a slow virus infection. Many of the brain cells contain eosinophilic intranuclear inclusions and electron microscopy shows characteristic nucleocapsid (see Fig. IV.28). A recent study revealed evidence of another type of virus, possibly a papovavirus, as well as measles in a case of SSPE. The disease lasts six to 12 months with progressive signs of loss of cerebral function and spastic paralysis. It has recently been claimed that cells chronically infected with measles, RS and certain arboviruses contain DNA which will hybridize with the viral RNA: uninfected cells do not contain such DNA sequences. The origin of the reversed transcription mechanism involved is not clear. One intriguing possibility is that it is supplied by a latent RNA tumour virus present in the same cells another that it is a host function normally involved in the process of 'gene amplification'. Subsequently, two groups have claimed that the DNA from the chronically infected cells was infectious in other, suitable cells and gave rise to normal RNA-containing virus. If confirmed these observations would have obvious relevance to the problem of subacute sclerosing panencephalitis as well as other conditions, including multiple (disseminated) sclerosis. Koprowski has recently reported the detection of a latent infection with a paramyxovirus-like agent in the latter condition, closely related to but not identical with parainfluenza type 1 (Sendai).

There is now convincing evidence that most, if not all cases of a rare, but invariably fatal disease of children—giant cell pneumonia—are caused by the measles virus. This condition has been found only in children who have been unable to develop antibodies to the virus although they may apparently react normally to other antigenic stimuli; virus proliferation may continue for weeks until the death of the patient. Most reported cases have been in children with leukaemia. The absence of a rash in these cases supports the hypothesis that the measles rash is due to an immune mechanism. Giant cells are also found in the lungs of patients dying during the acute stage of measles. Meales in pregnant women produces abortion in up to 20% of cases but the virus does not seem to have the teratogenic potential of rubella virus. The tuberculin test may become negative during an attack of measles and any tuberculous lesion may become more active.

One attack of measles produces lifelong immunity. Consequently, in isolated communities, the disease tends to die out. When reintroduced into such populations is has been found to assume a very severe form with a high incidence and a high mortality in the adult population. The disastrous effect of measles on a virgin population was clearly shown in the 1875 epidemic in the Fiji Islands and more recently in a somewhat similar epidemic in Greenland.

The simplest diagnostic test for measles is the demonstration of multinucleated giant cells, known as Warthin-Finkeldey cells, in sputum and nasal secretions taken during the prodromal period. These cells, which may measure 100 μm in diameter and contain up to 100 nuclei, are very similar to the multinucleated giant cells seen in infected tissue cultures; they appear to be derived from lymphoid tissues. Their demonstration may be of some diagnostic value. Isolation of the virus in tissue culture is not a routine diagnostic procedure in view of the ease with which a clinical diagnosis is possible and the rarity of symptomless infections.

Complement fixing, neutralizing and haemagglutination-inhibiting antibodies develop within two to three days of the rash, reaching a maximum in about three weeks. The former declines quite soon after infection whereas the two latter persist indefinitely, even in the absence of exposure to infection. The finding of high levels of measles antibody in the cerebrospinal fluid is of diagnostic value in cases of suspected subacute sclerosing panencephalitis.

Prophylaxis

Measles can be partly or completely prevented by the administration of gammaglobulin prepared from pooled adult plasma. Children who have been fully protected will again become susceptible when their passive immunity wears off. On the other hand, children who are partly protected and who develop a mild attack acquire an active immunity and will therefore be insusceptible to future infection. Complete protection is considered desirable only for the very young (under one year) or for debilitated children. It can be achieved only if gammaglobulin is given within five days of contact, the amount required depending on the weight of the child. After the fifth day only attenuation is possible and little if any protection can be achieved by passive immunization after the ninth day.

The first attempts at active immunization against measles were with the attenuated Edmonston strain of virus developed, by Enders and his colleagues, by passage in tissue culture. Although this strain proved to be a highly efficient immunizing agent it was by itself unacceptable for general immunization because of the severity of the reactions it produced. Since then further attenuated variants have been developed but in general the greater the degree of attenuation the less effective is the virus as an immunizing agent. Some further attenuated variants have been derived, however, which appear to produce an acceptable degree of immunity. Thus the Schwarz variant of the Enders-Edmonston strain was found in a Medical Research Council trial to give a protection rate of approximately 85% over a follow-up period of six months. Children who developed measles in spite of vaccination experienced only a mild form of the disease. The duration of the immunity produced by this vaccine has, however, still to be determined. If this were only for a short period immunization might simply defer an attack to an age at which it could be less readily withstood by the patient. Immunization with live attenuated vaccine is nevertheless now recommended by the British Ministry of Health for all children not previously protected and ultimately when the majority of children have been protected it is envisaged that routine immunization should be carried out in the second year of life. A small number of cases of encephalitis have followed the administration of live measles vaccine. It is difficult to know whether this risk can ever be entirely eliminated while retaining the potency of the vaccine. However, one strain formerly used in this country was withdrawn on the ground that more cases of encephalitis followed its use than followed the use of other strains. It must be remembered that even if every case of encephalitis following measles vaccination with currently available attenuated vaccines be ascribed to the vaccines, the mortality and morbidity is only a fraction of the mortality and morbidity which natural measles would have produced in the same populations had they not been vaccinated.

Killed vaccines have also been tried. They give rise to negligible immediate reactions but have proved to be poor immunizing agents, the antibody levels produced being low and transient. It has been suggested that killed vaccine might be inoculated as a preliminary to live vaccine in order to reduce the number and severity of the reactions produced by the latter, thereby permitting the use of a less attenuated, more highly immunogenic live strain. It has been reported from the USA, however, that children who had been given repeated injections of killed vaccine developed severe local and systemic reactions on subsequent exposure to measles or following the injection of live vaccine. Under these circumstances combined immunization with killed and live vaccines cannot be recommended.

IV.9

The Herpesviruses

This is a group of moderately ether-sensitive DNA viruses of apparently identical morphology. The important human pathogens are the herpes simplex virus, varicella–zoster virus, the cytomegalovirus and the Epstein Barr virus. Of these only the first can be propagated in animals, chick embryos or animal tissue cultures. A wide variety of animal pathogenic herpes viruses is known. These include herpesviruses of Old World monkeys (virus B), New World monkeys (marmoset virus), dogs and rabbits (virus III); pseudorabies virus of pigs; equine abortion virus; infectious laryngotracheitis virus of fowls (ILT); cytomegaloviruses of several species and at least two tumour viruses, the Lucké leopard frog agent and the agent of Marek's disease (visceral lymphomatosis) of fowls. Of these only B virus is known to cause disease in man. Cells infected with all members of the group develop characteristic intranuclear inclusions of the so-called Type A of Cowdry (see below). Some members also give rise to the formation of cytoplasmic inclusion bodies.

Herpes simplex

In infected tissues and fluids the herpes simplex virus is found in two morphological forms, an enveloped form and a naked non-enveloped form, both of which appear to be infective. Enveloped forms are roughly spherical, measuring from 130 to 180 nm in diameter. The envelope has a high content of lipid and has been shown to possess host cell antigens. Like the myxovirus envelope it is probably derived from the host cell but from the nuclear membrane and not the cytoplasmic membrane, although it has been claimed that the latter does contribute an additional layer to the envelope when the virus particle passes through it. The non-enveloped forms represent the naked viral nucleocapsid. The nucleocapsid possesses icosahedral symmetry and contains 162 capsomeres. The 12 apical capsomeres are pentagonal in outline and the remainder, on the edges and faces of the icosahedron, are hexagonal. Many virus particles, both of enveloped and naked types, appear in electron micrographs to be empty. They are presumably incomplete or non-infective particles lacking DNA. The virion probably contains about 12 polypeptides and may contain a layer of protein beneath the capsid.

Although all strains of herpes simplex virus closely resemble one another antigenically and may therefore be regarded as belonging to a single species, minor antigenic differences are demonstrable between strains. Strains have been divided on the basis of antigenic and other properties into two types. Of these type II is generally isolated from lesions of the genitalia or neighbouring areas, whereas type I is isolated from lesions in any situation but particularly from lesions above the waist. More recent work casts some doubt on the validity of this simple division into two types, as intermediate strains undoubtedly occur, and further studies are needed to clarify the situation. However, most virus laboratories are now attempting to type isolates by neutralization tests and many are finding that most strains can be allocated to one or other type on this basis, although many of the isolates do not possess all the other attributes usually regarded as characteristic of their particular type. In addition to antigens associated with the virus particle, a number of soluble antigens have been demonstrated in infected fluids by complement fixation and agar gel diffusion.

The herpesvirus is highly unstable and is rapidly

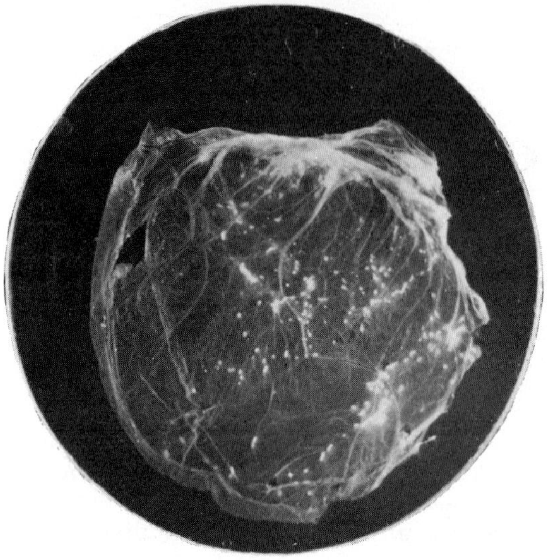

Fig. IV.29. Pocks of herpes simplex virus on the chorio-allantoic membrane.

inactivated on standing. It is susceptible to ether. This would be readily understandable if only the enveloped forms were infective. Since, however, the naked forms are also apparently infective their ether sensitivity suggests that they contain some lipid as an integral component of the nucleo-capsid.

The virus can be readily propagated on the chorio-allantoic membrane of the chick embryo on which it produces discrete pocks visible to the naked eye (Fig. IV.29). Type I strains characteristically produce small, superficial lesions which do not exceed 0·75 mm in diameter, even after seven days incubation: type II strains, on the other hand, produce pocks which continue to enlarge until they exceed 2 mm in diameter, which involve the whole thickness of the chorio-allantoic membrane and undergo necrosis. The rabbit is highly susceptible to the herpesvirus and can be infected by intradermal or by corneal inoculation. Suckling mice may be infected by both intracerebral and intraperitoneal inoculation. The virus grows in a variety of tissue cultures with the production of a cytopathic effect, one distinctive feature of which is multinucleated giant cells which may form a syncytium particularly with type II strains.

Infected cells show characteristic intranuclear inclusions, the development of which can be followed in preparations stained with haematoxylin and eosin. The inclusions are at first basophilic, rich in virus and have a high concentration of DNA; later they become acidophilic and at this stage are devoid both of DNA and of virus. The latter, which are known as Cowdry type A inclusions, fill most of the nucleus and appear to compress the nuclear chromatin into a peripheral ring. A clear area or halo is usually demonstrable between the inclusion body and the compressed chromatin; this appears to be a fixation artefact. Infected epidermal cells show in addition a characteristic ballooning degeneration.

By the time they reach adult life the majority of people have been infected with the herpesvirus. Antibody surveys show that infection occurs earlier in the lower social groups amongst whom the infection is almost universal by the time adult life is reached. Amongst the more well to do, infection tends to occur later and a proportion of adults may fail to show evidence of infection. Primary infection usually occurs before the age of five years, almost certainly through contact, e.g. in kissing, with virus present in the saliva of an infected person. It is probable that in most cases children are infected in this way by their mothers. When primary infection occcurs in the adult it appears to be quite often through a wound, e.g. of the fingers, with the formation of a herpetic whitlow or through the genital tract, with the development of vesicles on the external genitalia or uterine cervix. There is ample evidence that genital herpes is usually venereally transmitted and some entirely circumstantial evidence to suggest that it may predispose, in the female, to the subsequent development of carcinoma of the cervix. This evidence is difficult to summarize briefly. That there is an association between genital herpes and abnormalities of the uterine cervix, both malignant and premalignant, is not in dispute: patients from whose genital tracts herpesvirus is isolated show a relatively high incidence of cytological abnormalities and herpesvirus antigens have been demonstrated in biopsy specimens from women with abnormal cervical smears. However, as genital herpes and carcinoma of the cervix are both positively correlated with poverty and promiscuity, it could be that both are simply independent indices of the socio-sexual characteristics of the women concerned. Attempts to establish relationship between the two by serological surveys depend critically on the possibility of distinguishing between antibodies due to the two types of virus. Not all authorities are agreed that is possible; but even those workers who claim that the distinction can be made, have not always found any difference between the carcinoma subjects and controls, especially if the latter are very carefully matched for such things as age, socio-economic factors, parity, age at first pregnancy and marital status.

Evidence of an entirely different kind is afforded by the undoubted oncogenic potential of other members of the *Herpesvirus* genus and by recent confirmation of the ability of herpes simplex virus itself to transform hamster and rat cells in tissue culture under certain conditions.

The majority of primary infections are asymptomatic but in some cases primary infection may be quite severe. The commonest type of symptomatic primary infection is acute gingivostomatitis, a condition in which there is an extensive eruption of vesicles in the buccal mucosa; the regional lymph glands are usually enlarged and fever is generally present. Keratoconjunctivitis, vulvovaginitis and, perhaps, aseptic meningitis may also occur as primary herpetic infections. In cases of herpetic encephalitis it is possible that the virus may gain access to the brain from the blood stream since viraemia has been demonstrated in primary herpes: however, in view of the fact that latent infections probably persist in peripheral ganglia, especially the trigeminal, neural pathways are probably of at least equal importance. Herpetic encephalitis frequently shows a degree of localization, most often to one temporal lobe. Except when it occurs as part of a generalized infection in a neonate infected during passage through the birth canal, herpetic encephalitis seems generally to be due to type I strains. Some cases of Kaposi's varicelliform eruption (see p. 416) are also caused by the herpes simplex virus. This condition occurs most frequently as a primary herpetic infection in children with eczema but involving the normal, as well as the eczematous, skin. In premature infants and in neonates born of mothers without circulating antibody the virus may, though rarely, produce a generalized infection with lesions in the viscera, particularly in the liver, brain and adrenals. The fact that in such cases the virus is often of type II suggests that the infant is infected from the mother during birth, although it is not possible to discount the possibility that an infant born to a non-immune mother may be infected by other adults, especially if premature.

Following the primary infection the virus in many cases, possibly in all, assumes a latent form in which it persists in the tissues without the production of obvious lesions. In this state it is liable to be aroused into activity by a variety of debilitating factors. Persons latently infected are therefore liable to recurrent herpetic lesions. The commonest manifestation of recurrent herpes is the appearance of small vesicles at various mucocutaneous junctions particularly around the lips and nostrils. Vesicles may also appear on the genital mucosa, conjunctiva and cornea. One of the most effective precipitating factors in stimulating an attack of recurrent herpes is fever, hence the name 'fever blisters' formerly applied to herpetic vesicles occurring in the course of such bacterial infections as lobar pneumonia. Attacks are also commonly precipitated by exposure to sunlight.

The nature of the immunity in herpes is of some interest. Individuals who are subject to recurrent attacks nevertheless possess a considerable amount of antibody in their blood; they can also be experimentally infected by their own strain of virus. Under these circumstances the development of overt lesions is probably due to the fact that the virus can pass directly from one cell to another (p. 362) and thereby escape contact with antibody in the extracellular fluid. Herpes and varicella–zoster provide the best examples in man of persistent latent viral infections. The precise relationship between the persisting virus and the infected cells is not understood. It is possible that the virus genome becomes integrated into the host cell genome in a manner analogous to prophage in lysogenic bacteria (p. 466). If it were so, environmental factors might interfere with the control exerted by the cell over virus multiplication, in a manner analogous to prophage induction.

The multiplication of the herpesvirus in tissue cultures can be inhibited by exposure of the culture to 5-iodo-2'deoxyuridine (IUDR, p. 362). This compound has been found to have a considerable ameliorative effect on superficial herpetic infections, e.g. of the cornea, in experimental animals. In general, however, withdrawal of the drug results in renewed proliferation of the virus. The drug has been used in the treatment of human herpetic infections but initial reports of its value have not been entirely confirmed. Resistant strains of virus may emerge during treatment or may be present *ab initio*. A high concentration of the drug (30 to 40%) in dimethyl sulphoxide has given encouraging results when applied topically in severe cases of cutaneous herpes simplex and herpes zoster. Systemic treatment is not without risk and claims for its value in the treatment of herpetic encephalitis are difficult to evaluate. IUDR should not be used in the treatment of trivial, self-limiting infections, especially in children (see also p. 363). Cytosine arabinoside and adenine arabinoside have given evidence of being superior to IUDR in the treatment of herpetic encephalitis. Reports of the effectiveness of an ultra-violet irradiated vaccine in cases of recurrent herpetic infection still require evaluation.

Diagnosis

The diagnosis of recurrent herpes is mainly a clinical problem. Laboratory procedures used include:

1. Isolation of the virus from the lesions or from the throat, saliva or cervical discharge.
2. The demonstration of giant cells and inclusion bodies in scrapings obtained from the base of the vesicles.
3. The direct demonstration of virus by electron microscopy.
4. The demonstration of viral antigen by the fluorescent antibody technique. Antibodies are best demonstrated by the neutralization test although a complement fixation test is also available. Because of the wide prevalence of herpetic infection the demonstration of a rising antibody titre is of particular importance.

The diagnosis of herpetic encephalitis can only be confirmed by brain biopsy: if there is to be any hope of effective treatment this must not be delayed. The fluorescent antibody test affords a rapid (one to two hours) and, in experienced hands, reliable method of demonstrating the virus in the specimen, although the diagnosis should always be confirmed by inoculation of tissue cultures.

'B' virus

In morphology, growth characteristics and antigenic structure, this virus closely resembles the herpes simplex virus. The most important distinguishing features are the failure of B virus to infect adult mice and the fact that, though it stimulates antibodies against herpes simplex virus, it is poorly neutralized by herpes simplex antiserum. The virus is a natural parasite of Old World monkeys, in which it is the analogue of the herpes simplex virus. Human infection usually occurs as a result of a monkey bite but it has also been recorded after contamination of a wound from an infected monkey kidney tissue culture.

Zoster and varicella

Although they present quite distinct clinical syndromes there is now conclusive evidence that zoster and varicella are caused by the same virus, the differences in the clinical and epidemiological features of the two conditions depending on host factors. The virus, which is known as the varicella–zoster virus, has been found to be morphologically identical with the virus of herpes simplex from which, however, it differs considerably in growth characteristics and in antigenic structure. The virus will grow on primary isolation in a variety of human tissue cultures. Its cytopathic effect consists of focal lesions much less marked than those of herpes simplex; these are characterized by cell rounding, formation of multinucleate giant cells and the occurrence in infected cells of intranuclear inclusions resembling those produced by the herpes simplex virus. In tissue cultures the virus appears to spread directly from cell to cell and infective virus is not found free in the tissue culture fluid, except in cultures of human thyroid. The fluid, however, contains large numbers of non-infective particles with a ragged appearance on electron microscopy. By contrast, large numbers of infective non-ragged particles are present in the vesicle fluid of human lesions.

Serological examination using human immune sera has failed to reveal any difference between strains of virus from cases of zoster and those from cases of varicella. Although a number of antigens has been demonstrated in agar gel diffusion tests, all strains so far isolated have belonged to a single antigenic type.

Varicella or chickenpox is a mild febrile disease of childhood characterized by the presence of a generalized rash. Adults are occasionally infected, the disease being more severe than in childhood. The rash is at first papular, later vesicular, the lesions finally drying up with the formation of crusts. The lesions, which are most marked on the trunk and face, tend to appear in successive crops; these features are of value in distinguishing varicella from mild smallpox in which there is usually but a single crop of vesicles and maximum density of lesions on the extremities. The histology of the skin lesions is very similar to that of herpes simplex, infected cells showing a ballooning degeneration and intranuclear inclusions which are at first basophilic, later becoming acidophilic with accumulation at the base of the lesion of multinucleate giant cells. The mortality of the condition is extremely low except in the case of neonatal infections contracted from the mother in which it may be as high as 20%, and in patients under treatment for leukaemia. Varicella has a well marked seasonal incidence, epidemics occurring mainly in the winter and spring. It is a disease of very high infectivity. Infection is via the respiratory tract, where initial multiplication of the virus probably occurs and from which the virus is disseminated by the blood stream. Lung involvement seems to be more common than was realized and may result in permanent fibrosis. Up to 50% of cases in adults may show radiological evidence of varicella pneumonia, although only 15–20% show obvious clinical signs. Numerous other complications of varicella have been described: all are rare. These include encephalitis and a haemorrhagic rash. The disease is apt to be especially severe in patients receiving steroid therapy and a fatal generalized infection with widespread lesions in the viscera has been described in such cases. Secondary bacterial infection with subsequent scarring or even septicaemia may also occur in varicella or zoster.

In contrast to varicella, zoster is a disease mainly of adults, in which after a few days of fever a vesicular eruption appears in the area of distribution of one or more sensory nerves. Rarely the rash subsequently becomes generalized. The thoracic nerves and the ophthalmic division of the fifth cranial nerve are those most frequently involved. The condition is usually unilateral. In zoster of the seventh cranial nerve there are pain and sometimes vesicles in the external auditory meatus and facial paralysis (Ramsay Hunt's syndrome). The virus appears to multiply mainly in the posterior root ganglia from which it spreads centrally to the cord and peripherally along the nerve trunks to the skin. Occasionally the anterior horn cells of the cord are involved with the development of paralysis in the corresponding muscles. The disease shows no seasonal incidence and its frequency is not increased during varicella epidemics. No evidence has been obtained of direct transmission of zoster from one person to another.

The epidemiological evidence supporting the identity of the varicella and zoster viruses is considerable. Children experimentally infected with zoster fluid have been found to develop varicella; contacts of these children have also developed varicella. Contacts of zoster cases frequently develop varicella. The incidence of varicella in contacts of zoster cases is, however, lower than it is in the contacts of varicella cases, presumably because of the absence of virus from the respiratory and salivary secretions in zoster.

The differences in the clinical and epidemiological manifestations of varicella and zoster are believed to be due to differences in the immune status of the host. In non-immune persons the virus produces a generalized disease—varicella—because, in the absence of humoral antibody, it is able to spread widely throughout the body. In the immune, on the other hand, the virus is not able to spread widely and because of this produces zoster—a localized disease. Zoster resembles recurrent herpes simplex in being due to the activation of a latent virus. This is apparently maintained after primary childhood infection in the posterior root ganglia, although, unlike the situation in herpes simplex, second attacks of zoster are so rare that the diagnosis should be called in question. There is no absolutely decisive evidence in support of this hypothesis, but the fact that a history of having had varicella in childhood can be elicited from most adults with zoster, the finding that in zoster antibody generally rises more rapidly and attains a higher level than in varicella—suggesting a secondary antibody response—and the absence in most cases of zoster of any history of recent contact with a case of zoster or of chickenpox make it plausible. Post-herpetic neuralgia is a distressing sequela of zoster which afflicts subjects over about 60 with arteriosclerosis.

Microscopic examination of smears from the skin lesions in both zoster and varicella is of considerable diagnostic value. These show the presence of characteristic giant cells—Tzanck cells—with intranuclear inclusions. Similar cells are found in cases of herpes simplex but not in material from cases of smallpox or vaccinia. The identification of viral antigen in vesicle fluid by immunodiffusion, fluorescent antibody and complement fixation tests, isolation of the virus by tissue culture and the demonstration of complement fixing antibodies in the patient's serum have also been used as diagnostic procedures.

Cytomegaloviruses

Cytomegaloviruses are a group of viruses which, both in the intact animal and in tissue culture, give rise to infections, the characteristic feature of which is the presence of greatly enlarged cells with large nuclei containing prominent acidophilic inclusions. The inclusion fills most of the nucleus but is separated from the nuclear membrane by a clear well-defined halo. Acidophilic juxtanuclear cytoplasmic inclusions may also be present. Cytomegaloviruses have been isolated from a variety of animals, in all of which they have a particular predilection for the salivary glands. The strains isolated from different animal species possess a high degree of species specificity and grow, in culture, only in cells of the species which they naturally infect.

Cytomegaloviruses are classified in the herpes group on the basis of particle morphology, and the fact that the nucleic acid has been identified as DNA. Electron micrographs show the nucleocapsid of the human virus to measure about 100 nm in diameter and to contain 162 hollow capsomeres, many of the particles surrounded by a loose irregular envelope similar to that of the herpesvirus (Fig. IV.30).

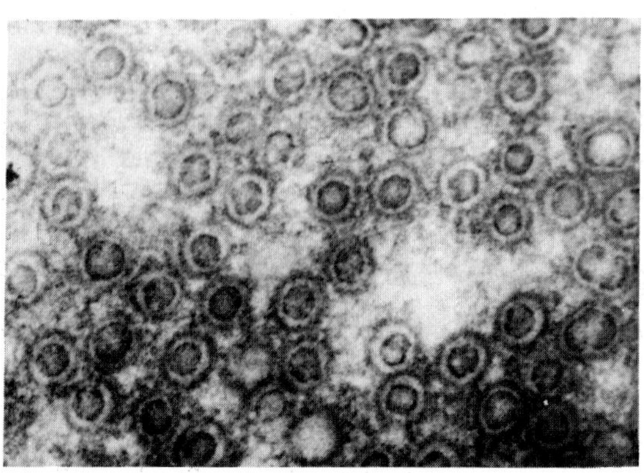

Fig. IV.30. Intranuclear form of cytomegalovirus. (*Dr T. H. Flewett*)

Cytomegaloviruses of human origin can be cultivated in tissue cultures of human fibroblasts and human muscle cells; they grow poorly in epithelial cell cultures. On primary isolation growth is slow, with the production of focal cytopathic lesions in which the virus, like that of zoster, is transmitted directly from cell to cell. Also as in zoster, the extracellular fluid contains mainly non-infective virus with a ragged appearance on electron microscopy. So far, antigenic analysis has been possible only with human sera, the results obtained indicating considerable sharing of antigens, with at the same time appreciable strain differences.

There is a fair amount of evidence indicating that in man infection with cytomegaloviruses is widespread:

1. The frequent recognition of enlarged cells with inclusions, resembling those known to be produced by cytomegaloviruses, at routine post-mortem examination, in the salivary glands of children who have died from a variety of causes.
2. The presence of complement fixing antibodies in a high proportion of adult human sera. In a study in the London area the proportion of positive sera ranged from 4% in the six-month to five-year age group, to 54% in the 35–75-year age group. The precise nature of the latent infections responsible for these findings is not known but many are presumably infections of salivary glands.
3. Isolation studies indicate that from 1 to 2% of children actively excrete the virus in the saliva and/or urine.

The results of infection depend on age. In newborn infants it gives rise to a severe and often fatal generalized disease with symptoms referable particularly to involvement of the haemopoietic tissues. This type of infection clinically resembles erythroblastosis fetalis and, like the latter, presents at post mortem evidence of extensive extramedullary erythropoiesis. Since in these cases virus has been isolated from the placenta and from the urine of the mother it seems likely that the infection is contracted in utero, although cases have also occurred following exchange transfusion in the newborn. Children who recover show microcephaly or hydrocephalus, cerebral calcification, mental defect and chorioretinitis: all features also produced by congenital infection with *Toxoplasma gondii.* In older children and in adults clinical evidence of infection is infrequent and appears to occur only in the presence of some debilitating factor. Under these conditions the symptoms depend on the organ primarily involved, usually the intestine, liver, kidney, lungs or brain. Cytomegaloviruses may be significant causes of hepatomegaly, with or without jaundice in older children, and of infectious mononucleosis not associated with a positive Paul-Bunnell reaction in adults.

A diagnosis of cytomegalic inclusion body disease is most simply made by the demonstration, in urine and saliva, of the characteristic enlarged cells with intranuclear inclusion bodies. Where possible, the virus should, however, be isolated by tissue culture. The development of antibody may be demonstrated by the complement fixation test. As judged by the results of complement fixation tests, most infections are inapparent and may occur at any age. Evidence of infection may be found in patients on renal dialysis or after renal transplants or open heart surgery. Such infections seem usually to be inapparent, even in patients on immunosuppressive therapy. However, a mild glandular-fever-like illness following blood transfusion has been shown to be associated with a cytomegalovirus infection. Sometimes a more severe condition, pneumonitis or a generalized infection, may develop in patients on immunosuppressive therapy or with such diseases as leukaemia and other reticuloses. It is not clear to what extent these adult conditions represent new infections and to what extent they are due to reactivation of a latent infection. Some at least seem to be primary, particularly, those after blood transfusion.

The Epstein–Barr virus: glandular fever (infectious mononucleosis)

In 1958 Burkitt described a form of malignant lymphoma which afflicted children in East Africa. It subsequently transpired that the disease was widespread in Africa and some other tropical regions and study of hospital records showed that it was not a new disease. As it seemed to be restricted to the hotter and wetter regions in

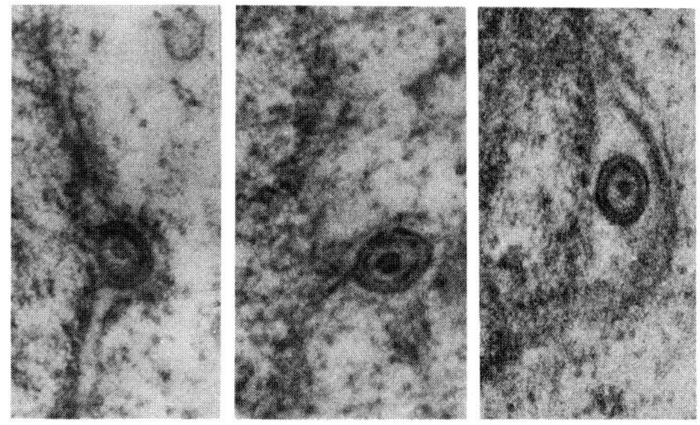

Fig. IV.31. Herpes-like virus particles in cultured Burkitt tumour cells. The photographs show the passage of the virus particle through the nuclear membrane, acquiring an outer envelope as it goes. (*Epstein et al. (1967) Br. med. J.*)

Africa, it was suggested that the condition was due to a transmissable agent carried by a biting arthropod. An intensive search for the aetiological agent began and a number of candidate viruses were discovered. The two most promising appeared to be reovirus type 3 and a new herpesvirus seen by electron microscopy in cultures prepared from lymphoma cells. The latter virus cannot be cultivated except in lymphoma cells or in cells derived from the human haemopoietic system which, on contact with the virus, become transformed (see p. 456), can then be cultivated indefinitely in vitro and will continue to produce small quantities of the virus. As the evidence accumulated, the case for the new herpesvirus, named the Epstein-Barr virus (EB) after its discoverers, became much stronger than the case for the reovirus. Virtually every case of Burkitt's lymphoma was shown to possess antibody to the virus by complement fixation or fluorescent antibody test, using infected tissue culture cells as antigen; whereas the incidence of antibody was substantially less than 100% in control children. Unfortunately, serological studies showed that the infection is world-wide, while the lymphoma is very restricted in its distribution. While the EB virus has now been shown to most people's satisfaction to be the long-sought aetiological agent of glandular fever (infectious mononucleosis) its role in lymphoma is still far from clear. It has been suggested that the virus does cause lymphoma, but only when acting in concert with some other factor which damages or stimulates the reticuloendothelial system. It is further postulated that this second factor is usually chronic malignant tertian malaria, which is, of course, mosquito-borne, but that other factors may sometimes operate, since very rare cases of lymphoma are observed in areas such as the United Kingdom where malaria is virtually unknown.

Morphologically, chemically and in its mode of replication, EB virus appears to be a typical member of the herpes group, although serologically it seems to be unrelated to any other member. As second attacks of glandular fever do not seem to occur it is presumed that there is but a single antigenic type. The virus, or its antigens, can be demonstrated in cultured lymphoblasts from cases of Paul-Bunnel-positive glandular fever, a proportion of apparently normal individuals, some cases of leukaemia and almost every case of Burkitt's lymphoma.

Although glandular fever is one of the most protean of all diseases, the great majority of cases conform to a fairly narrow pattern of sore throat, lymphadenopathy, fever and malaise. Laboratory tests reveal the characteristic blood picture with an absolute and relative lymphocytosis; the presence of characteristic abnormal mononuclear cells; the presence of heterophile antibody to sheep red cells; and often some evidence of impaired liver function. The test for heterophile antibody, the Paul-Bunnel test, has to be interpreted with care. A proportion of normal individuals has sheep-cell agglutinins and they are common following injections of horse serum. The normal antibody is absorbed by guinea-pig kidney, the antibody from horse serum injections is removed by guinea-pig kidney and ox red cells and the glandular fever antibody is removed only by ox cells.

The same clinical and haematological picture can be produced by a number of agents including *Toxoplasma gondii,* cytomegalovirus and infective hepatitis virus, but in these infections the Paul-Bunnel test is negative. In true glandular fever the generally accepted diagnostic titre of 1/80* may not be obtained when the patient is first seen and repeat testing may be necessary. Antibodies to the EB virus also appear, but their level does not necessarily reflect that of the Paul-Bunnel antibody and, unlike the latter which disappears in one to two months, they seem to persist indefinitely.

Undoubtedly, the majority of infections with EB virus are symptomless, or at any rate unrecognized as glandular fever. Unusual manifestations of the disease include jaundice, rupture of the spleen, haemolytic anaemia, thrombocytopenic purpura, pneumonitis, abnormal ECG findings, meningitis, peripheral neuritis, myelitis, encephalitis, abdominal pain and a typhoidal state. Any of these may or may not be associated with the triad of sore throat, malaise and lymphadenopathy. A variety of skin rashes and an enanthem at the junction of the hard and soft palate are not uncommonly seen. The incubation period is about 10–14 days in children, but may be appreciably longer in adults. A feature of the disease is the prolonged convalescence, during which the patients complain of such symptoms as asthenia, depression, inability to concentrate and being easily fatigued. There is no specific treatment for the disease itself, but antibacterial therapy† may be required for secondary infection in the throat, frequently due to Vincent's organisms, and appropriate measures to sustain life may be required in cases with neurological involvement. The mode of transmission is not known but it appears to require fairly close direct contact. The disease particularly affects adolescents and young adults: it seems that infection in children is much less likely to be manifest as glandular fever.

* If the antibody behaves as glandular fever antibody in absorption tests, a lower titre may be acceptable.

† Ampicillin should never be exhibited in suspected cases of glandular fever: fully 50% of cases of glandular fever so treated develop a severe skin eruption, a complication only rarely seen in untreated patients.

IV.10

Hepatitis

Two main types of transmissible hepatitis are distinguished; originally this was done on the basis of the length of the incubation period. These are infective hepatitis (viral hepatitis type A) with an incubation period of 20 to 40 days and serum hepatitis (viral hepatitis type B) with an incubation period of 60 to 180 days. Not all cases of acute hepatitis can be put into these two categories. Hepatitis may also be caused by a variety of well categorized agents including yellow fever virus, cytomegalovirus, glandular fever (EB) virus, adenoviruses, herpes simplex virus, *Coxiella burneti* and *Leptospira icterohaemorrhagiae*. The clinical pictures in infective and serum hepatitis are so similar that differentiation on clinical grounds alone is not possible. Nevertheless, there are statistical differences: for example, the former tends to have a more abrupt onset and to be more often febrile (temperature over 38°C). In recent years enormous progress has been made in understanding the epidemiology of both types of hepatitis, especially hepatitis type B, and the accounts current even five years ago are now seen to have been totally inadequate.

VIRAL HEPATITIS TYPE A (INFECTIVE HEPATITIS; CARARRHAL JAUNDICE)

This is probably the commoner of the two types of hepatitis and it can occur in large epidemics or sporadically. Epidemiologically it has been likened to poliomyelitis; its primary impact is on children, inapparent infections are common and the faecal–oral route of transmission is important. Large-scale epidemics initiated by contaminated water and food, especially shellfish taken from sewage-contaminated waters, have been described, e.g. the water-borne epidemic in Delhi in 1955–6. In this context the relative resistance of the virus to chlorine disinfection is important. It cannot be too strongly emphasized that the virus is present in the blood of pre-icteric, icteric and anicteric patients and that this infection is, in theory, as readily transmitted as hepatitis type B by such procedures as tattooing, dental treatment, injections of blood or blood products and the use of shared and improperly sterilized syringes. The incidence of hepatitis type A following blood transfusion naturally mirrors the incidence amongst the population from which the blood donors are drawn. Now that the disease is notifiable throughout the country, it is easier to monitor the incidence more accurately. However, inasmuch as the infection particularly affects those too young to act as blood donors and many of the adults likely to receive blood will already be immune, the incidence of hepatitis type A following transfusion is, in practice, generally lower than it might otherwise be.

In cases contracted by normal contact the disease is usually mild with complete recovery and the mortality is of the order of one to four per thousand cases. Anicteric infections probably exceed jaundiced cases. There is some evidence that the disease tends to become more severe with increasing age and in patients infected parenterally. There is no convincing evidence for any natural host other than man. There are several reports of small outbreaks in people in close contact with non-human primates, particularly chimpanzees, but there is good reason to suppose that these animals themselves contracted the infection from a human source. Chronic carriers would appear on presently available evidence to be rare and it seems likely that in the great majority of cases infection is acquired from another individual undergoing an acute infection, icteric or non-icteric.

Up to now research on the disease has been hampered by the absence of any means of propagating the virus in the laboratory and of any reliable serological test for infection. Chimpanzees occasionally develop a mild illness following inoculation of material from human cases but this response is irregular and it seems that once in contact with man the animals rapidly become immune to infection. As long ago as 1967, Deinhardt and his collaborators showed that a mild hepatitis could be produced in various species of marmoset by the intravenous injection of blood and faecal extracts from cases of infective hepatitis. However, other workers failed to confirm these results and concluded that the response was due to the activation of a latent marmoset virus. Now, however, there is general agreement that the marmoset really is susceptible. Attempts to grow the agent in tissue culture have thrown up a number of candidate viruses, none of which can be accepted. The electron microscope has also revealed particles claimed to be the virus of hepatitis type A. Until recently none of these could be accepted either. However, 27 nm particles observed in stool extracts since about 1970 would now appear to have a reasonable claim. The particles are aggregated by convalescent serum. Similar particles have since been seen in the livers of infected marmosets and human beings. The marmoset particles react with guinea-pig sera raised against the human faecal particles. The particles have been detected in the stools of volunteers from five days before the onset of symptoms and disappear as the transaminase levels reach their peak, when antibody appears in the serum. The agent has recently been alleged to contain RNA and would seem to be present in preparations of Krugman's MS1 virus (see below). At present the tests for particles and for antibody are not suitable for routine use and depend heavily on material from marmosets which are in short supply. One question which probably can be answered with presently available techniques is whether hepatitis A viruses are antigenically homogeneous and, if not, how many serotypes there are. It will also be possible to establish more accurately the ratio of inapparent to overt infections and to determine the role, if any, of hepatitis virus A in chronic liver disease and in the fulminant and rapidly fatal condition known as acute yellow atrophy of the liver.

Treatment consists of bed-rest with general nursing care and an appropriate diet. This is adequate for all but the most severe cases which may need more energetic supportive treatment. Resistance to the disease has been regularly observed for up to three months following the injection of pooled gammaglobulin. In some cases the protection lasts a great deal longer, possibly because it has allowed an inapparent but immunizing infection.

There still remains much to be learnt about hepatitis type A, but if the laboratory techniques currently available can be adapted to give methods suitable for routine use progress should be rapid. In temperate climates the disease is characteristically seasonal, with the greatest number of cases in mid-winter, which is surprising if the main method of transmission is by the faecal–oral route. In countries where adequate records are available the disease has generally shown a tendency to fluctuate in incidence year by year with a periodicity, sometimes strikingly regular, of from three to ten years.

VIRAL HEPATITIS TYPE B (SERUM HEPATITIS: SYRINGE JAUNDICE)

Serum hepatitis was first distinguished from infective hepatitis in 1926 when it was recognized that cases of hepatitis occurring in patients attending a diabetic clinic were due to an infectious agent transmitted from one patient to another by improperly sterilized syringes and that the incubation period was much longer than that of infective hepatitis. It gradually came to be recognized that this type of hepatitis was by no means uncommon and could occur following any medical procedure in which there was the possibility of the transfer of blood or tissue fluid from one individual to another. Probably the most extensive outbreaks were those associated with the administration of batches of yellow fever vaccine stabilized with human serum. Following the great expansion of blood transfusion and the use of freeze-dried plasma during the Second World War, it became clear that a certain proportion of apparently healthy individuals are chronic carriers of the agent responsible for serum hepatitis. Experiments with volunteers enabled some progress to be made with regard to the characterization of the agent. It was shown to pass bacteria-retaining filters, to be relatively resistant to physical and chemical agents compared to most viruses and to be present in the blood of carriers at very high titre: as little as 0·0001 ml of serum can transmit infection. Certain non-medical procedures, notably tattooing, were also shown to carry a risk of hepatitis. For years the disease was something of a mystery—a disease apparently without any natural mode of transmission. Nevertheless, it remained a matter of concern and energetic steps were taken to reduce its incidence. The proper sterilization of syringes and later their replacement by disposable items; the proper sterilization of all surgical instruments, even those used for minor procedures; the restriction of blood transfusion to essential cases; the careful selection of blood donors to exclude those with a history of jaundice; the keeping of adequate records so that donors whose blood had apparently caused jaundice could be identified and eliminated; and the preparation of blood products such as plasma from small pools of not more than ten donors—all helped to reduce the risk to some extent. But cases continued to occur and latterly to increase again, especially in countries where paid blood donors are used rather than volunteers. This rise seemed to run parallel to the illicit use of parenterally administered drugs.

The picture would probably have remained much as it was, were it not for two important advances. The first of these was the discovery by Blumberg and his colleagues in 1965 of a novel antigen in the serum of an Australian aborigine, which reacted with antibodies present in the serum of a number of much transfused haemophiliacs. At first this antigen was not connected with hepatitis at all, but when it was shown that it could be acquired and was not, therefore, a typical serum protein isoantigen, and when electron microscopy of positive sera revealed virus-like particles, attention was given to the possibility that it was an infectious agent. The prevalence of the antigen, named Australia antigen, proved to vary considerably in different populations, being usually low in Europe and North America, but much higher in many tropical areas. It was also found in a significant proportion of cases of lepromatous leprosy, Down's syndrome and certain types of leukaemia. However, it was only found consistently in cases of serum hepatitis. About the same time Prince, working independently, described an antigen in the blood of cases of serum hepatitis which he called the SH (serum hepatitis) antigen. This proved to be identical to the Australian antigen. The antigen (see Fig. IV.32) consists of a mixture of more or less spherical particles about 20 nm in diameter and elongated particles of about the same diameter but varying in length. The particles appear to be built up of subunits resembling capsomeres and they contain lipid but apparently no nucleic acid. They have been likened to the aggregates of capsomeres which may be formed when attempts are made to reassemble certain isometric small plant viruses. There seems little doubt that the particles just described cannot be infectious. However, most (? all) preparations of Australia antigen contain a varying proportion of larger particles, Dane particles, which are about 42 nm in diameter. The outer layer of these carries the same antigenic determinants as the smaller particles, but the 28 nm core has a distinct antigen not found in the smaller particles. It has also been shown recently to contain DNA polymerase and a circular molecule of DNA of about $1{\cdot}6 \times 10^6$ daltons. It therefore has a substantial claim to be the elusive hepatitis B virus. The infection can be transmitted to chimpanzees which have provided an experimental model. It has been found that cores can be demonstrated in the nuclei of liver cells and that they acquire additional layers in the cytoplasm to give the complete Dane particle. Cores have also been seen in human liver cells. Recently, the terminology relating to Australia antigen has been revised. The antigenic determinants on the surface of the Dane particle and on the smaller particles are called HB_SAg, the core antigens are HB_CAg. Antibody to the first of these is anti-HB_S and to the second anti-HB_C. The surface antigen has been shown to be complex. All samples examined carry a determinant **a**, but there are a number of other determinants as well. Most important and best studied are two pairs of 'allelic' specificities **d–y** and **w–r**. Additional specificities and 'subtypes' of **a** and **w** have still to be worked out. On the basis of the principal determinants it is possible to distinguish four types: **adw, ayw, adr** and **ayr**. It has been shown that the specificities are carried on every particle

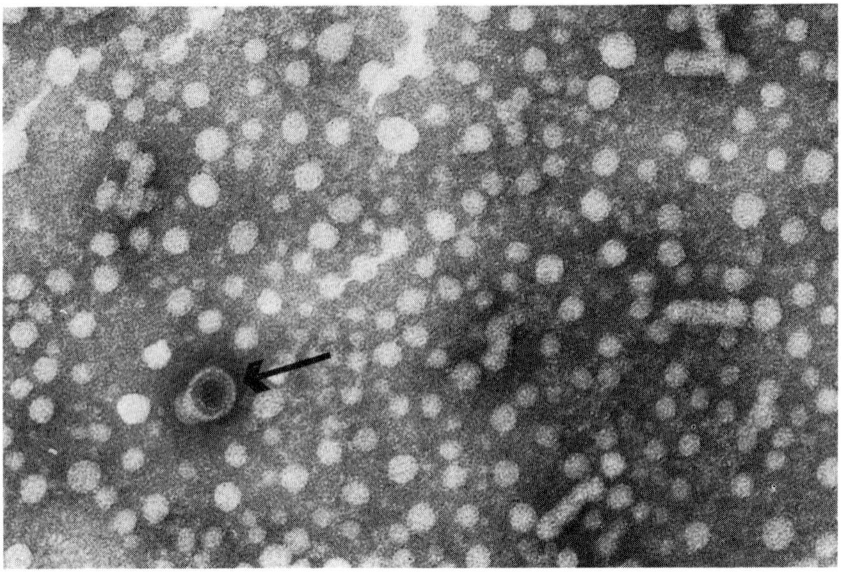

Fig. IV.32. Australia (hepatitis B) antigen from human serum. Note the numerous, more or less spherical, particles, somewhat variable in size, rod-shaped particles and a single larger Dane particle (arrowed). Negative staining. × 150 000. (*Dr G. M. Cleator*)

and that the type 'breeds true'. This being so, it is reasonable to conclude that the specificities are virus coded. Typing has been of value in epidemiological studies and it transpires that the geographical distribution of the subtypes is by no means uniform.

Shortly after the recognition of Australia antigen, reports began to appear of the work on hepatitis being carried on by Krugman and his colleagues at the Willowbrook Schools for mentally handicapped children. Hepatitis was endemic and Krugman deemed it ethically justifiable to infect newly-admitted children deliberately, on the ground that they would inevitably acquire the infection anyway. Working with this population, he distinguished two types of hepatitis which were circulating naturally in the institution. One of these, designated MS1, caused hepatitis with a short incubation period; the other, MS2, also produced hepatitis when fed to children, but less readily than MS1 and with an incubation period of over 60 days (30–45 days if given parenterally). Preparations of MS2 but not MS1 contained the Australia antigen which also appeared in the blood of infected children. Clearly MS2 was serum hepatitis, but yet it was circulating naturally and was not dependent on parenteral inoculation.

Armed with the tests for HB_sAg which were being developed and adapted for routine use and with the knowledge that hepatitis B did have a 'natural' mode of transmission, workers all over the world have undertaken studies which, in the past five years, have transformed our understanding of the disease. It is clear that transmission is both lateral and vertical. Individuals in close contact, especially sexual contact, transfer the infection quite readily and hepatitis B is an important venereal disease, especially amongst homosexuals. Cases of apparent transmission of infection to the spouses of individuals with serum hepatitis have hitherto been dismissed as the coincidental contracting of infective hepatitis by the person apparently infected secondarily.

In certain parts of the world, especially tropical Africa, the prevalent endemic form of hepatitis is hepatitis B rather than hepatitis A. The antigen has been detected in mosquitoes. Experimentally it can be detected immediately after feeding and it persists until the blood meal has been completely digested. There is no evidence that the agent multiplies in the insects, but they could clearly transfer infection if interrupted while feeding on a carrier and subsequently completing their meal on a susceptible individual.

Although claims to have demonstrated HB_sAg

in faeces have been made, it seems that the faecal–oral route of infection may be of only limited importance; the same is true of urine. However, the antigen has been consistently demonstrated in saliva and in 'sneeze' specimens from chronic carriers. It has also been demonstrated in semen and menstrual blood.

Some 40–70% of babies born to mothers with antigenaemia in the last trimester of pregnancy or the two months following delivery will themselves develop antigenaemia and a substantial proportion will become chronic carriers.

Hepatitis B leads to chronic liver disease far more frequently than hepatitis A and even apparently symptomless carriers frequently show biochemical and histological evidence of liver damage. Infection in a susceptible adult may result in an acute illness with complete recovery and disappearance of the antigen from the blood. The antigen appears several weeks before the onset of symptoms and persists for some time afterwards. Its disappearance coincides with the appearance of anti-HB_S. On the other hand, this antibody does not appear in cases which go on to become chronic carriers. The other antibody, anti-HB_C, appears in both groups at about the time symptoms develop. It is not known what determines the outcome of a hepatitis B infection in apparently healthy individuals: however, it is known that renal failure predisposes to both a relatively silent infection and the development of chronic carriage. This would seem to explain the very sinister reputation of this infection in renal dialysis units and the fact that the most severe illnesses and the deaths occur in the staff rather than the patients. It is now clear that immunological mechanisms are involved in the symptomatology of the acute illness and in the development of chronic liver disease subsequently. This being so, it is not surprising that chronic infection is common in all patients with natural or therapeutically induced immune deficiency; such patients are increasingly being recognized as posing the same risks to staff as renal patients. Recent studies have indicated that there may be an inverse relationship between the severity of the initial illness and the probability of developing the condition known as chronic active hepatitis, which almost inevitably progresses to cirrhosis. This is greatest following anicteric infections and least following the most severe primary illness consistent with recovery (submassive necrosis). It has also been established that prolonged carriage is not necessarily lifelong. Patients have been observed to lose their antigen and develop anti-HB_S after as long as 73 months. The overall importance of hepatitis B virus in chronic liver disease seems to depend on the geographical area; being greatest in those areas in which chronic carriage is commonest. In view of the importance of anicteric infections and the likelihood that they remain undiagnosed, it is not easy to assess the long-term outcome of infection. In a recently published series of 492 cases ill enough to be hospitalized 90% recovered completely and 10% became carriers: of the latter, 3% developed chronic active hepatitis and the rest showed persistent hepatitis, a benign condition which ultimately heals although the course may be very protracted.

Diagnosis depends on the demonstration of HB_SAg in the serum and biochemical tests of liver function. The most sensitive test for the antigen is radio-immunoassay, but this is not suitable for large-scale routine use and simpler, if less sensitive, methods such as countercurrent immunoelectrophoresis and passive haemagglutination have to be used. Fortunately, the level of antigen is high in most chronic carriers so that the majority will be readily detected by these methods. However, in chronic active hepatitis levels are often so low that they are detectable only by radio-immunoassay.

Treatment is as for hepatitis A, but the disease tends to be more severe, especially in older people, and many of the sufferers have other illnesses. Consequently, more cases will require supportive treatment.

The prognosis is significantly worse than for hepatitis A; a mortality of 10–12% is quoted but this relates to patients infected parenterally in hospital and the overall mortality is much lower than this. Patients who survive the initial illness may die, possibly years later, from cirrhosis of the liver.

Prevention of the disease altogether is not possible, but the measures outlined on p. 445 can do much to reduce the incidence in hospital patients in whom the prognosis is especially poor. It is essential that all those who work in hospitals should appreciate the risks of infection and take the necessary steps to protect themselves. Recommended procedures for dialysis and transplantation units and for clinical laboratories have been published by the World Health Organization. Regular screening of staff and patients deemed to be at special risk can both identify possible sources of infection and act as a check on the precautions in force. Careful selection of blood donors and the collection and retention of adequate records are vital. The incidence of chronic carriers amongst the general population in the United Kingdom is probably about 0·1% whereas in certain tropical areas the level may rise as high as 10%. However, while 50–70% of United Kingdom patients transfused with antigen-positive blood are likely to develop icteric or anicteric hepatitis, few, if any, of those from areas with high carriage rates will do so, presumably because they are already immune. Within a given country considerable differences may exist between different sections of the population, the rate being particularly high in drug addicts and urban drop-outs of various kinds. Apart from the obvious risks associated with the parenteral administration of drugs, it is likely that the way of life of these people is conducive to the non-parenteral spread of infection. Where such people are allowed to sell their blood this will be reflected in a relatively high incidence of post-transfusion hepatitis, especially if rigorous screening is not carried out.

Passive immunization has not been conspicuously successful: most of the studies were done some years ago using pooled gammaglobulin. However, at least in the United States of America the titre of anti-HB_S has been rising from 1 in 8 in 1967 to 1 in 256 in 1973, so the problem would seem to merit reinvestigation. Work is in hand to try and develop a vaccine from HB_SAg, so far without any great success. But this is an exciting development and an effective vaccine would be of great value, especially in hospital personnel. It has been claimed that anti-**a** is protective in chimpanzees: if this were confirmed it would seem to indicate that a single vaccine could protect against every type of hepatitis B virus. Recently a new antigen, *e*, has been detected in patients with HB_SAg. It is suggested that the presence of this antigen raises the probability that the patient will develop chronic liver disease following his initial infection.

Evidence is beginning to accumulate that cases of post-transfusion jaundice occur which are not due to hepatitis viruses A or B, nor to any of the other agents listed on p. 443. It may well be, therefore, that additional hepatitis viruses await recognition, but present evidence suggests that, if they exist, they play a much smaller part than hepatitis A and B in causing clinical illness.

IV.11

Miscellaneous Viruses

THE ADENOVIRUSES

Adenoviruses are a group of ether-resistant, non-enveloped viruses possessing classical icosahedral symmetry. They were first recognized by Rowe et al. in 1953 in tissue cultures of human adenoids and tonsils undergoing spontaneous degeneration and in which they had been present as a latent infection. They were subsequently independently isolated by Hilleman and Werner from cases of upper respiratory tract infection. Similar viruses have since been isolated from monkeys, cattle, dogs, pigs, mice, opossums and birds. Over 30 serotypes distinguishable by neutralization tests have been recognized amongst strains derived from human sources. All the human types possess a common complement fixing antigen. The group antigen is also present in adenoviruses derived from simian, bovine and canine sources.

The adenovirus particle is a regular icosahedron, 70–80 nm in diameter. There is no envelope. The capsid surrounds an inner core consisting of double-stranded DNA and a number of polypeptides rich in arginine; it comprises 252 capsomeres. The 12 apical capsomeres, called pentons, differ from the remaining 240 which are called hexons. Attached to each of the pentons is a fibre ending in a small knob. The fibre and knob constitute the haemagglutinin, whereas the penton base is responsible for an immediate toxic cytopathic effect which can be produced by high concentrations of adenoviruses in less than an hour and it is identical with the 'cell detaching factor' (Figs IV.34 and IV.35). The antigenic composition is complex: all adenoviruses except avian types cross-react in complement fixation tests. The principal common antigen is associated with the hexons, although a different shared antigenic specificity may reside in the penton bases. The fibre antigens appear to be mainly type specific, although minor cross-reactions in haemagglutination inhibition tests have been reported between viruses in the same HA group (see below). Neutralizing activity seems to depend on antibody reacting with an antigenic determinant on the hexons. This is quite distinct from the group antigen and is largely type-specific, although minor cross-reactions with viruses in the same HA group are observed. The human adenoviruses have been divided into groups on the basis of their ability to agglutinate the red cells of the rat and the rhesus monkey. Four groups can be distinguished. As stated above, the members of each group show minor cross-reactions in neutralization and haemagglutination inhibition tests. In many preparations, a non-infectious haemagglutinin can be demonstrated; this is smaller than the virus particle and appears to consist of a cluster of 12 pentons with their fibres. This structure, in which the penton bases are centred on the five-fold axes of symmetry of an icosahedron, has been called a 'dodecon'. Adenoviruses do not elute from the red cells, the human types react optimally at 37°C and haemagglutination inhibition tests can be used to type those adenoviruses which haemagglutinate.

Adenoviruses will grow in a variety of tissue cultures, but growth is best in primary, e.g. kidney, or continuous, e.g. HeLa cells of human epithelial origin. Although a few types can be propagated serially in non-human cells, in most cases infection is abortive, although the cells may be transformed (see Chapter IV.2). They multiply in the nuclei of infected cells in which basophilic inclusion bodies can be demonstrated by conventional staining procedures. With certain serotypes—3, 4, 7 and 8—the intranuclear virus

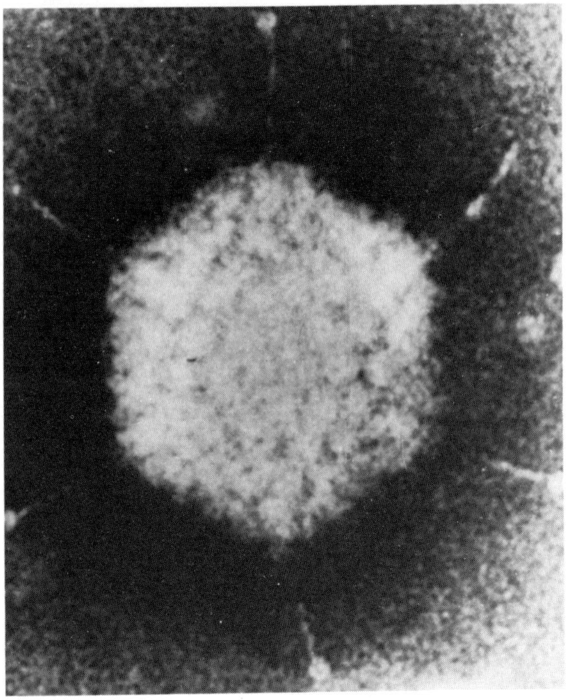

Fig. IV.33. An adenovirus particle showing the knobs and fibres attached to the pentons. (*Valentine and Pereira (1965) J. molec. Biol.*)

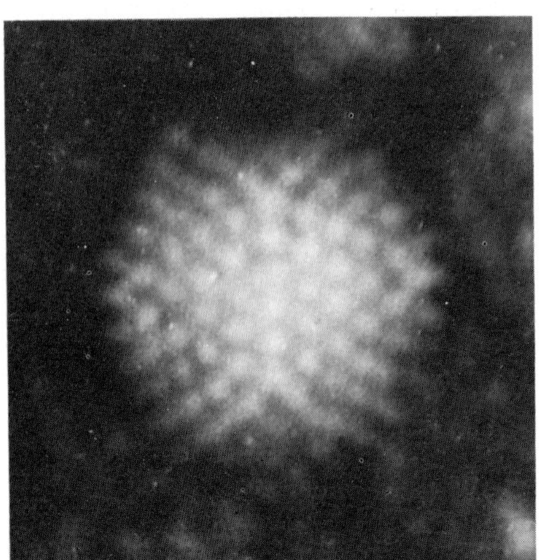

Fig. IV.34. Adenovirus, showing the arrangement of the capsomeres in equilateral triangles with six capsomeres along each side. Negative staining with phosphotungstate. (*Horne et al., J. molec. Biol.*)

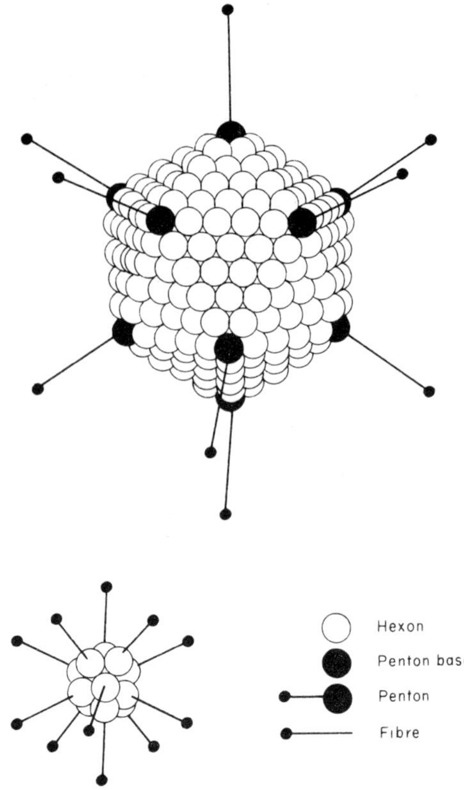

Fig. IV.35. The complete adenovirus particle, components of the capsid and the structure of the 'soluble' haemagglutinin which is made up of 12 pentons. (*From B. I. Wilner (1969) A Classification of the Major Groups of Human and Other Viruses. Burgess*)

aggregates in the form of a crystal (Fig. IV.36). The evolution of the nuclear changes in cells infected with these serotypes has been found to be recognizably different from those occurring in cells infected with types 1, 2, 5 and 6, which do not produce crystals and which, in HeLa cells, exhibit a longer eclipse phase before the appearance of virus. In tissue culture, attachment and penetration of the virus takes four to five hours. Newly infectious virus appears 10 to 15 hours later and increases for five to six hours. However, only about 1% of this newly formed virus is released, even from cells showing advanced CPE. Except for the oncogenic serotypes, adenoviruses of human origin are not pathogenic for laboratory animals, though under certain conditions they may give rise to latent infection. In suckling rats and

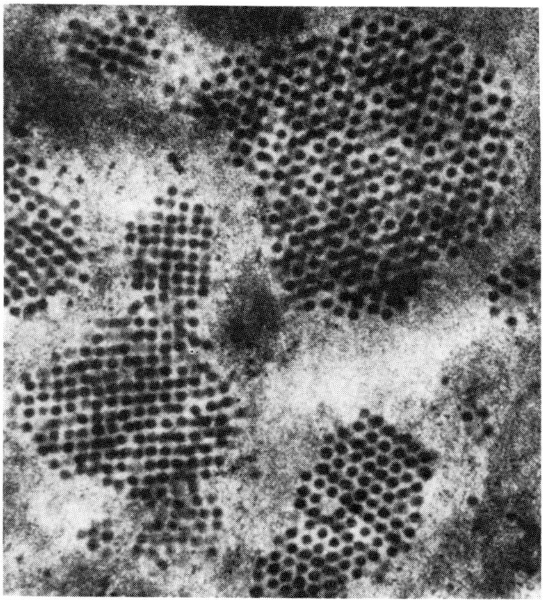

Fig. IV.36. Intranuclear crystals of adenovirus type 8. (*Dr T. H. Flewett*)

hamsters a number of serotypes are oncogenic and can induce the production of sarcomas (p. 458). On the basis of their capacity to induce tumours, four groups of adenoviruses have been recognized: Group A (types 12, 18, 31) are highly oncogenic; Group B (types 3, 7, 14, 16, 21) are oncogenic but less potent than Group A; while Groups C and D (types 1, 2, 5, 6 and types 9, 10, 13, 15, 17, 19, 26 respectively) are non-oncogenic but can be distinguished by the specificity of the T (nuclear) antigen with which each group is associated. All four groups can transform hamster or rat cells in vitro.

There have been numerous reports of the occurrence in stocks of adenoviruses of virus-like DNA-containing particles, of about 20 nm in diameter. The particles are defective in that they cannot replicate by themselves, but multiplication occurs when the cells are also infected with a 'helper' adenovirus. In this respect they resemble the small tobacco necrosis virus and the Rous sarcoma virus (p. 459). So far no role in pathogenicity has been identified for these particles to which the name adenosatellite virus has been applied, although they have been isolated together with adenoviruses from the nasopharynx. They are classified as parvoviruses and at least four distinct serotypes exist. There is no relationship between the serotype of the adenosatellite virus and the adenovirus with which it is associated. The presence of the adenosatellite viruses depresses the multiplication of the adenovirus.

Human adenoviruses will not replicate in vervet monkey cells and behave as though they were defective. However, if the monkey cells are concurrently infected with SV_{40} virus (see p. 457) the adenoviruses replicate normally. Occasionally such mixed infections yield hybrid particles containing an SV_{40} genome together with a more or less defective adenovirus genome in an adenovirus capsid. Such particles are necessarily defective themselves, but can replicate if a normal adenovirus is present as well. Because of the presence of the SV_{40} genome in the hybrid the normal virus can replicate too, even in monkey cells.

Antibodies to many adenovirus types are widely prevalent in adult human sera. Adenovirus infection must therefore be of frequent occurrence. Some of the types are capable of causing latent infections, particularly of children; others tend to be associated with specific disease syndromes, of which the most frequent is some type of mild upper respiratory tract infection. This may be a simple pharyngitis or a febrile influenza-like illness usually lasting three to four days and often with well-marked constitutional symptoms. Respiratory infection may be associated with conjunctivitis, as in pharyngoconjunctival fever, or conjunctivitis may on occasion be the only symptom of infection. Adenoviruses may produce a condition indistinguishable clinically from streptococcal tonsillitis. In rare cases, the virus produces a pneumonia resembling primary atypical pneumonia, but wihout cold agglutins. The serotypes particularly involved in these infections are 3, 4, 7 and 14. Latent infection of tonsils and adenoids in children are mainly caused by types 1, 2 and 5. Type 8 is uniquely involved in the production of epidemic keratoconjunctivitis. There is evidence suggesting that adenoviruses of a variety of serotypes can cause mesenteric lymphadenitis, and that this condition may in turn be the precipitating cause of an intussusception. Claims have also been made that adenoviruses can cause diarrhoea and, more recently, evidence has been adduced which indicates that they may cause a syndrome indistinguishable from whooping cough. The pathogenicity for man of a number of adenovirus types, notably types 11, 12, 13, 15, 16, 17, 18 is still in doubt.

Inactivated vaccines have been prepared from some of the antigenic types—3, 4 and 7—but are still in the experimental stage. However, they do appear to be of some value in military recruits who, especially in the United States, show a high incidence of adenovirus respiratory infection when they first join the colours.

Diagnosis may be achieved by isolation of the virus from throat swabs, conjunctival swabs or faeces or by the demonstration of a rising titre of antibody in the complement fixation test against the group antigen. Adenoviruses are not infrequently isolated from patients, especially children, with a very wide variety of clinical syndromes. Even when the evidence afforded by isolation is supplemented by the demonstration of a significant rise in neutralizing antibody to the virus type isolated, it is usually impossible to establish the aetiological role of the adenovirus with any confidence: adenovirus infections are common, especially in children, and are often asymptomatic or accompanied by no more than mild upper respiratory symptoms. In spite of their oncogenicity in experimental animals, no substantial case has so far been made for their having any aetiological role in malignant disease in man.

REOVIRUSES

Reoviruses were first classified as ECHO virus type 10 but have now been assigned to a distinct group. They are considerably larger than picornaviruses—with a diameter of 60 to 75 nm. They are non-enveloped RNA viruses possessing icosahedral symmetry; the capsid appears to be double. The outer capsid probably comprises 180 capsomeres (trimers) (see p. 344). The structure of the inner capsid has not been determined. The nucleic acid of the viruses is unusual in being a double-stranded RNA.

Three serotypes are demonstrable by neutralization tests, the types sharing a common complement fixing antigen. Reoviruses have the property of agglutinating human red cells, the red cell receptors being destroyed by periodate but not by RDE. In addition, type 3 strains are capable of agglutinating bovine cells.

Reoviruses will grow in tissue cultures of a variety of human and animal cells. They produce a cytopathogenic effect which is recognizably distinct from that of the enteroviruses. Suckling mice and a variety of laboratory animals can be infected, but infection is often asymptomatic. Reoviruses have a wide host range and have been isolated not only from man but also from a variety of animal hosts. Most human isolations have been from the faeces but the viruses have also been recovered from the throat. In man they have been found in association with both respiratory disease and gastrointestinal disease, hence their name. They can, however, be isolated from apparently normal individuals. Their precise pathogenic role is still far from clear. A very limited tumorigenic activity in mice can be demonstrated only under special circumstances, but reoviruses of type 3 have been isolated from a number of cultures of cells from cases of Burkitt's lymphoma. Their role in this condition is far from clear and the case for regarding them as the causative agent is much weaker than the case for the EB virus.

OTHER DOUBLE-STRANDED RNA VIRUSES

The reoviruses are by no means the only double-stranded RNA viruses. The host range of the group extends over vertebrates, intervertebrates, higher and lower plants, and bacteria.

Orbiviruses

Borden et al. have distinguished a group of arthropod-borne viruses of vertebrates with double-stranded RNA which differ in their physicochemical and morphological properties from the reoviruses. This group, for which they have coined the name *Orbiviruses,* comprises the bluetongue virus of sheep (biting midges and gnats: multiple serotypes), African horse sickness virus (biting midges and gnats: multiple serotypes), Colorado tick fever virus (ticks: affects man) and a variety of hitherto unclassified arboviruses distinguished by their relative resistance to lipid solvents and deoxycholate compared with the typical togaviruses. Included are two small groups, Kemerovo (ticks) and Changuinola (sandflies); as well as a number of ungrouped arboviruses. Like the reoviruses, the orbiviruses have a segmented genome and a double capsid. However, the outer layer of the latter appears diffuse and seems to be much more easily removed than that of the reoviruses.

Viruses of acute gastroenteritis

In a substantial proportion of cases of acute, apparently infective gastroenteritis of infants, no bacterial or protozoal cause can be found. This inevitably raises the possibility of a viral aetiology; support for this was afforded by the demonstration that, in some cases, the disease could be transmitted by bacteria-free faecal filtrates. However, repeated attempts to demonstrate the viruses responsible proved unsuccessful. The recognition of the enteroviruses in the early 1950s raised hopes that members of this group might prove to be the missing viruses. Curiously, this group seems to play, at most, a very small part in the production of acute gastroenteritis, although there is some evidence that echoviruses, especially types 6, 11, 14, 18, 20 and 22, may be capable of causing gastroenteritis, sometimes associated with respiratory symptoms.

In 1972 Kapikian and his colleagues reported that they had found virus-like particles in the stools of cases of acute non-bacterial gastroenteritis during an outbreak of 'winter vomiting disease' at Norwalk, Ohio, in 1968. The particles, seen by electron microscopy, resembled picornaviruses or parvoviruses but could not be cultured. Subsequent studies led them to the conclusion that they were in fact parvoviruses. Filtrates of stools containing the agent reproduced the symptoms in volunteers and the convalescent but not the acute phase sera of naturally occurring and experimental cases agglutinated the particles. While examining jejunal biopsy specimens from cases of infantile gastroenteritis for virus particles and for disaccharidase activity which is known to be depressed in cases of infantile gastroenteritis, Bishop and her colleagues in Australia identified particles of quite a different kind in cells of the duodenal epithelium. Later the same year similar particles were identified in stools by Flewett et al. in Birmingham. The association of these particles with infantile gastroenteritis is now being confirmed all over the world. The particles, are morphologically identical with the agents of infectious calf diarrhoea and of epizootic

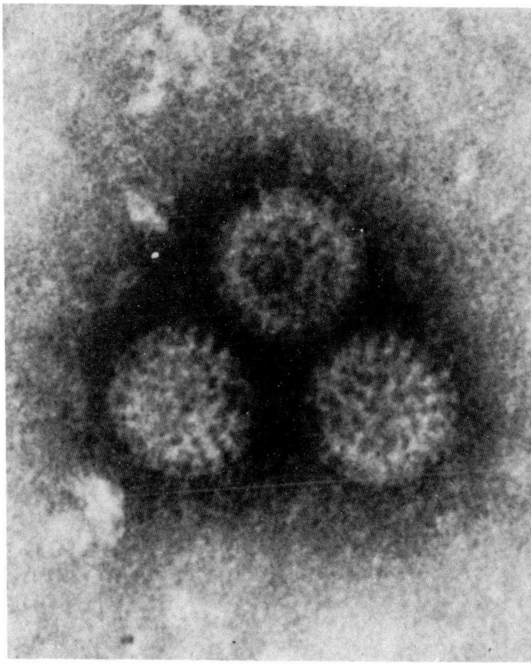

Fig. IV.37. Rotavirus (duovirus) particles from the stool of a baby with gastroenteritis. Negative staining. ×330 000. (*Professor J. Banatvala*)

diarrhoea of an infant mice: they are also related serologically to these two agents, especially the former. The particles have a striking appearance: they consist of a central core 36–38 nm in diameter, bounded by a membrane from which project short, cylindrical capsomeres supporting a sharply defined outer layer of capsomeres: giving the whole particle the appearance of a continuous outer rim supported on short radial spokes projecting from a central hub (Fig. IV.37). The whole particle is about 60–65 nm in diameter. The particles are agglutinated both by convalescent human sera and by convalescent sera from calves infected with the calf agent and the latter, which has been adapted to grow in tissue culture has been used as antigen in a complement fixation test for human sera. The two agents also cross-react in immunofluorescence tests. Cross-neutralization tests with human sera and the calf virus have given somewhat equivocal results, indicating that the two agents are not identical, a factor confirmed by the apparent inability of the human agent to infect calves. Final characterization of the particles seen in human stools must await successful laboratory cultivation. However, the close association with infantile gastroenteritis, the agent being practically never isolated from children without this syndrome; its world-wide distribution; and its obviously close relationshp with two other viruses which unequivocally cause diarrhoea in young animals—all go to make up a very strong case for accepting that this is one of the causes of non-bacterial gastroenteritis, especially of cases occurring in the winter. Unfortunately many different names have been given to this group of viruses. While morphological considerations would group them with the double-capsid, double-stranded RNA viruses, they would seem to be distinct from reoviruses and orbiviruses, although they have been classed with both these groups. The Birmingham workers have proposed the name *Rotavirus,* to emphasize the wheel-like appearance; the Australian group have objected to this and suggested the name *Duovirus,* emphasizing both the *dual* capsid and the *duo*denal habitat. Neither designation has yet received official recognition. The name infantile gastroenteritis virus has also been proposed, but would seem to be objectionable as it could be applied equally well to the Norwalk agent and to any other viral agent found to cause infantile gastroenteritis.

The disease caused by rotaviruses has a short incubation period of about 48 hours and an acute onset with diarrhoea, fever and usually, vomiting. Damage to the jejunal mucosa is associated with reduced disaccharidase activity, which probably accounts for the recurrence of diarrhoea which may occur when an apparently convalescent baby is put back on to milk feeds. The disease can be fatal. While rotaviruses have been detected in children from a few days old to over five years, they are commonest in children between six and 12 months old. Excretion of virus usually reaches a maximum about the third or fourth day and it has generally ceased to be detectable by the eighth day after the onset of symptoms. IgM antibody may be detectable as early as two days after onset. Conclusions about the absence of particles from a particular specimen should be drawn in the knowledge that the electron microscope is an extremely insensitive method of detecting viruses. In temperate climates at least, detection of rotaviruses is consistently more frequent in the winter than in the summer. Treatment consists of the control of dehydration and the avoidance of feeds likely to exacerbate the symptoms. It has been suggested that the calf virus could be used as a vaccine.

Adenoviruses are not infrequently isolated from children with diarrhoea, but in several studies the isolation rate from normal children has been about the same and there is no real evidence that they cause gastroenteritis. Coronaviruses (p. 389) have been detected in some outbreaks of gastroenteritis in adults.

Encephalitis lethargica

This condition occurred extensively in epidemic form in Europe in the early 1920s, but since 1928 no epidemics have appeared. The condition is believed to have been almost certainly of virus origin. The virus of herpes simplex was isolated on a number of occasions from rabbits inoculated with material from fatal cases, but is not believed to have been the causative agent.

Post-infection encephalitis

This type of encephalitis may follow infection with a number of viruses but occurs particularly after measles and after vaccination against smallpox (p. 415). A similar condition may also occur following anti-rabies treatment, rarely after other prophylactic inoculations and after bacterial infections. The pathological changes in post-infection encephalitis consist in a demyelination in the white matter of the central nervous system. These changes are quite different from those occurring in true viral encephalitis, in which the fundamental lesion is a degeneration of nerve cells. This difference renders it unlikely that post-infection encephalitis is the direct consequence of viral infection. More likely alternatives are that it is due to an allergic reaction or to autoimmunization. A possible autoimmune aetiology is suggested by the finding that a similar pathological picture can be produced in animals by inoculation of brain tissue combined with Freund's adjuvant. This type of experimental encephalomyelitis would appear to have its strictest analogue in the encephalitis following rabies vaccination with nervous tissue vaccine (p. 405). It nevertheless raises the possibility that when encephalitis follows any viral infection it develops in response to antigens liberated from the brain tissue by the virus infection.

IV.12

Viruses and Tumours; Slow Viruses

A number of viruses are capable of producing tumours in experimental animals and, in many cases, of inducing a malignant transformation of primary or secondary tissue culture cells in vitro, some also cause tumours naturally. The RNA tumour-producing (oncogenic) viruses include the avian leucosis group (to which the Rous sarcoma virus belongs), the murine leucosis group and the mouse mammary cancer (Bittner) agent. The DNA viruses include the polyoma, SV_{40} and papilloma viruses (including the rabbit papilloma and human wart viruses) which are known collectively as the papovaviruses, various oncogenic adenovirus serotypes, members of the herpesvirus group and certain of the poxviruses.

Virus-induced tumours carry new antigens not present in the corresponding normal tissue cells. Two types of such antigen are described:

1. Intracellular antigens—in the case of the papovaviruses and adenoviruses these antigens are found in the nuclei—demonstrable by complement fixation tests. These antigens are virus-specific in that they are present in all cells transformed by the same virus and the antigens produced in response to any one oncogenic virus are distinct from the antigens of cells transformed by other such viruses. The new antigens, which are known as T antigens, are distinct from any antigens which have been identified as components of the virus particle and they usually persist in transformed cells which no longer produce infectious virus.

2. Antigen components of the surfaces of transformed cells. These were demonstrated in the first place by transplantation tests as histocompatibility antigens, but in some cases have also been visualized by the fluorescent antibody technique.

The genetic and biochemical basis of the transformation of cells by oncogenic viruses is still far from clear. In tissue culture the transformed cells show a number of characteristics which distinguish them from normal, untransformed cells. They fail to show the phenomenon of contact inhibition whereby normal cells are inhibited from multiplication when in contact with one another. Because of this, transformed cells go on dividing even when the monolayer is complete, cells piling up on one another in a disorderly mass, to give a far higher final density of cells than can be obtained with untransformed cells. In addition, transformed cells are capable of indefinite subcultivation, will grow in suspension culture and in suspension in soft agar or methocel, are agglutinated by plant lectins such as concanavalin A, require less serum for growth and show increased uptake of nutrients. The shape of the cells is usually different and the nucleus : cytoplasm ratio is likely to alter. It is not usually easy to demonstrate the malignancy of the cells because the novel surface antigens induced by the virus cause any animal into which the cells are inoculated to reject them as foreign, even though the animal is of the same inbred line as that from which the cells were originally derived. This problem can be overcome by the use of immunosuppressed animals or by implantation in immunologically 'privileged' areas such as the hamster cheek pouch. Not all transformed cells will demonstrate all the properties listed above. Because of the small size of the genomes of many transforming viruses and the close similarity between cells transformed by different viruses, it has been suggested that much of the information needed for transformation is already present in the cell: this can be derepressed either by a transforming virus or by a chemical carcinogen. Assuming that it exists, this genetic information could be part of the essential genetic make-up of the cell or constitute part of the genome of a latent virus present in the cell already.

PAPOVAVIRUSES

This name is applied to a group of small, non-enveloped double-stranded DNA viruses,

40–57 nm in diameter, possessing icosahedral symmetry. The capsids of the more intensively studied members of the group have been shown to contain 72 capsomeres. Capsids with this structure show skew symmetry and can exist in either a right- or left-handed form—the two forms are called enantiomers. The human wart virus is said to be the right-handed enantiomer and rabbit papilloma the left-handed. They multiply in the nuclei of infected cells and crystal formation is frequent. The group includes the polyoma virus, the papilloma viruses of rabbits, man and a variety of other animals, as well as simian virus $(SV)_{40}$. These viruses differ antigenically and in growth properties but share the capacity of inducing tumours in susceptible animals. They are exceptional amongst the animal viruses in possessing a circular chromosome, resembling in this respect bacteria and bacteriophages.

The polyoma virus

This virus, which, of the group, has received most attention, was isolated from the tissues of leukaemic mice. It is apparently a natural parasite of wild mice, which excrete it in the saliva and urine without showing any evidence of clinical disease. The disease is imported into laboratory stocks of mice by contaminated grain. The virus can be propagated in mouse embryo tissue culture in which it produces a cytopathic effect. It possesses a haemagglutinin as an integral component of the virus particle. The haemagglutinin appears to combine with red cell receptors similar to those for the myxovirus haemagglutinins, since, like the latter, they are destroyed by RDE and myxovirus neuraminidase. Unlike the myxoviruses, the polyoma virus does not possess a receptor destroying enzyme itself.

Tumours are most readily produced by inoculation of animals when very young. In mice the commonest tumours are cancer of the parotids and of the breast. In rats, hamsters and ferrets the virus produces sarcomas. Malignant transformation can also be demonstrated in cells that have avoided the cytocidal effect of the virus in tissue culture. The transformed cells appear triangular in form, assume a random arrangement in irregular masses—instead of forming cell sheets—as a result of loss of normal contact relationships and show a high rate of glycolysis. Infective virus cannot be isolated from transformed cells and, since the cells can be reinfected with fresh virus, the virus initially present would appear to have been lost. In tissue culture only about 3% of cells are transformable by polyoma virus even with virus doses of many hundreds of virus particles per cell. Cells can be transformed not only by intact virus but also by viral DNA. The transformation, clearly, is not due to the selection of spontaneous mutants since these have not been demonstrated by cloning experiments. The polyoma virus has been of great interest by providing a possible model for the development of malignancy in man. The outcome of infection with polyoma depends on the cell type. In so-called *permissive* cells, e.g. mouse, the virus undergoes the normal replicative cycle with production of infectious virus and death of the cell: on the other hand, in *non-permissive* cells the virus undergoes an abortive cycle in which it enters the cell and can cause the synthesis of certain early gene products without multiplying. A small porportion of abortively infected cells will undergo transformation: the process is very inefficient in that it usually takes an inoculum of 10^4 to 10^5 infectious units per cell transformed. For the establishment of stable transformation the cell needs to undergo one cell division. Studies with temperature-sensitive mutants have indicated that continued production of an early gene product is necessary for the maintenance of the transformed state: there is evidence to suggest that this may not be either the T antigen or a virus-specific surface antigen.

Evidence has been obtained that some portion of the viral genome does persist in transformed cells from which no infectious virus can be isolated. Thus Benjamin has found that some of the RNA of polyoma transformed cells will hybridize with viral DNA and is therefore presumably virus specific and transcribed from persisting viral genome in the transformed cell. The persistence of some of the viral genome would explain the continued production of viral antigen in transformed but non-virus producing cells. It is possible in fact that the new antigens are enzymes coded for by the viral genome and required by the virus in the early stages of virus synthesis, since they are also produced in cytocidal infections. The latest evidence indicates that each transformed cell contains one to two copies of the viral genome, presumably incomplete because cell hybridization experiments such as those successful with SV_{40} have not been successful in inducing the formation of infective virus.

Simian virus (SV_{40})

Simian virus 40 was first isolated as a contaminant of rhesus and cynomolgus monkey kidney tissue cultures in which it produces a rather trivial cytopathic effect which is difficult to detect unless the cells are fixed and stained. In vervet monkey kidney tissue cultures it causes a destructive lesion with marked vacuolization of the cytoplasm, to which it owes its name of simian vacuolating agent. On inoculation into suckling hamsters it produces sarcomata. With one important difference the in vitro and in vivo behaviour of this virus is very like that of polyoma. While the whole genome is unnecessary for transformation and damaged or defective virions may even be more active as transforming agents than intact complete virus, infective virus can be recovered from many clones of cells transformed by complete virus and which do not themselves yield virus, by hybridization of the transformed cells with others which support a cytocidal infection such as green monkey kidney cells. Cell fusion is brought about by adding sendai virus irradiated with ultra-violet light. As with polyoma, one or two molecules of the viral DNA seem to be present in every transformed cell. The T and surface antigens of cells transformed by SV_{40} are serologically distinct from those of cells transformed by polyoma. It is now clear that in a stably transformed cell the viral genome, or part of it, is inserted into the host cell genome in a manner reminiscent of the insertion of prophage into the bacterial genome and, in the case of SV_{40}, there is experimental evidence that the insertion site is on a particular chromosome, No. 7, which apparently also carries any other genetic information which may be needed for transformation. The experiments involved the formation of hybrid cells from transformed human cells and mouse cells. Even when the hybrid contained a complete diploid set of mouse chromosomes and no human chromosome apart from 7, it exhibited the transformed phenotype.

SV_{40} has considerable practical importance because of its presence in monkey kidney tissue cultures used for the preparation of poliomyelitis vaccine. It has been found as a contaminant of live polio vaccine and, since it may resist formaldehyde inactivation, has been found also in inactivated polio vaccine. It appears to be capable of multiplying in the human subject and of stimulating the production of antibodies. So far, however, there is no evidence that it can produce neoplastic proliferation in man, but it is conceivable that it might do so after a prolonged incubation period. Because of this hazard it is essential that tissue cultures used for the preparation of poliovirus vaccine should be completely free of SV_{40} (see also p. 385).

Papilloma viruses: human warts

The papilloma viruses (53–57 nm) are somewhat larger than polyoma and SV_{40} (40–47 nm) and cause benign wart-like epithelial tumours. The rabbit papilloma virus has been extensively studied, particularly by Shope, who was responsible for its isolation in 1933. The virus is readily recoverable from the papillomas of wild rabbits but only infrequently from the papillomas of domestic rabbits. Infectious nucleic acid, however, can be recovered from the latter and the infected cells continue to synthesize a viral antigen. The papillomas readily undergo malignant change. This change may, as in the case of polyoma-induced tumours, be associated with the disappearance of infective virus. Bovine papilloma virus produces fibromas and fibrosarcomas in newborn hamsters.

That the common wart—verruca vulgaris—is due to a virus is rendered very probable by the occurrence of intranuclear inclusions and virus crystals in infected cells and by the fact that warts can be transmitted experimentally to human volunteers. The virus is provisionally classified with the papilloma group of the papovaviruses. It has a high degree of host specificity and the reported isolation of a common wart virus in tissue cultures has still to be confirmed. The condition appears to have a very long incubation period and susceptibility to infection appears to depend to some extent on psychological factors. All the different varieties of wart are thought to be caused by the same agent. The different types depending on the location, the reaction of the host and probably the level of immunity. Experimentally transmitted warts have an incubation period of one month to two years.

Other papovaviruses

Recently there have been reports of the isolation of papovaviruses from human sources, particularly cases of progressive multifocal leucoencephalopathy and patients on immunosuppressive therapy. Some of these seem to resemble SV_{40} closely, although the individuals concerned gave no history of poliomyelitis vaccination. Other isolates are serologically unrelated to SV_{40} or show only minor sharing of antigens. Although some of the patients concerned had malignant diseases, there is no evidence to relate these viruses to human cancer.

ADENOVIRUSES

Several human adenovirus types have now been shown to have oncogenic potential in varying degree. This can be demonstrated by the inoculation of newborn hamsters or other rodents. The tumours are sarcomas and the cells do not yield infective virus either spontaneously or on hybridization. This is hardly surprising, as it has been shown that only part of the viral genome is incorporated in that of the host; in some cases as little as 14%.

Human adenoviruses will not multiply in monkey cells unless the latter are infected concurrently with SV_{40}. Occasionally hybrid particles are formed which contain the SV_{40} genome (see p. 451). Such hybrids induce the formation of SV_{40} T-antigen, whereas cells transformed by complete adenovirus virions, develop T and surface antigens which are specific. Exhaustive searches for adenovirus T-antigens in human tumours have proved negative.

HERPESVIRUSES

In view of the propensity of the two best known members of the group—herpes simplex virus and the varicella–zoster virus—to produce latent infections, it is not surprising that they have been regarded as potential tumour viruses. However, there is still no absolutely convincing evidence of their capacity in this respect. The most acceptable cases are the agent causing Lucké's renal adenocarcinoma of the leopard frog and the agent detected in poultry with Marek's disease, visceral lymphomatosis. The role of the EB virus in Burkitt's lymphoma (see p. 441) has still to be clarified and the evidence that herpes simplex virus plays a part in cervical cancer is wholly circumstantial. Experimentally *Herpesvirus saimiri* from squirrel monkeys produces malignant lymphomas in virtually 100% of marmosets and owl monkeys. Transformation in vitro has been obtained with ultra-violet irradiated *Herpesvirus hominis* (herpes simplex virus), and more recently, with cytomegalovirus (see also Chapter IV.9).

POXVIRUSES

Molluscum contagiosum is believed on the evidence of histology and electron microscopy to be due to a poxvirus. The agent has never been cultured. The disease is contagious and is characterized by small pearly white tumours in the skin, each with a central pore from which can be squeezed a small amount of cheese-like material loaded with virus particles. The disease usually clears up after six to 12 months but may persist longer. The disease frequently extends by autoinoculation. The exact mode of transmission is unknown but it can undoubtedly be spread by close contact such as occurs in wrestling, by kissing or venereally. There is no evidence of any serological relationship to other poxviruses (see also p. 417).

The rabbit fibroma virus is serologically closely related to the myxoma virus. These agents are transmitted by biting arthropods, but unlike the arboviruses, they do not replicate in the vector which simply acts as a 'dirty syringe', transmitting the infection on its contaminated mouth parts. There is therefore no extrinsic incubation period and the infectivity of the vector is soon lost.

Other oncogenic poxviruses are the fibroma viruses of hares and of squirrels and the Yaba virus of monkeys. The myxoma virus behaves as a fibroma virus in its natural host, cotton-tail rabbits, although causing the fatal acute disease, myxomatosis in European rabbits (see also Chapter IV.7).

RNA TUMOUR VIRUSES

A number of RNA-containing viruses are known which have the capacity of inducing tumours in a variety of experimental animals. They include the Rous sarcoma virus and the viruses of avian and mouse leukaemias. The Rous sarcoma virus has been particularly studied. In contrast to the papovaviruses infectious virus continues to be produced by most transformed cells. The Bryan strain and some other strains of Rous sarcoma virus are unusual in that, although capable of initiating malignant transformation, they cannot by themselves produce infective progeny. This they can only do in association with a helper virus known as Rous-associated virus (RAV). The RAV is some member of the fowl leucosis group which provides the complete outer layer of the virus so that a phenotypically mixed particle (see p. 359) is produced, having the host range and serological specificity of the RAV and the genome of the Rous sarcoma virus.

The RNA tumour viruses, *oncornaviruses,* can be divided into two groups: those with so-called B-type particles, which cause mammary carcinomas and possibly other carcinomas, and those with C-type particles, which cause leukaemias and sarcomas. The two types of particle are essentially similar in construction, comprising an outer double membrane and a dense nucleoid. The difference lies in the size of the nucleoid: in type-B particles the nucleoid is eccentric and has a diameter about half that of the space within the outer membranes, but in type-C particles the nucleoid is central and more or less fills the space. Both particles average about 100 nm in diameter and the nucleoids of types B and C average 35 and 68 nm respectively. (These figures relate to mouse viruses specifically, but the viruses of the fowl leukosis–sarcoma complex are essentially the same as mouse type C particles and particles resembling mouse mammary tumour viruses have been observed in other mammals.)

The classical mouse mammary tumour virus is the so-called *Bittner agent.* This is transmitted to the young in the milk and determines a high incidence of mammary carcinoma in later life in genetically susceptible mice. There is also evidence of increased incidence in the offspring of infected males and uninfected females. Although it is possible to prepare tissue cultures from tumours which continue to form virus, it has not so far proved possible to infect cultures in vitro. The Bittner agent is probably only one of a number of similar agents in mice. Recently similar particles have been found in a mammary tumour in a monkey and claims have been made that they can be demonstrated in human milk, especially in women with a family history of mammary cancer.

The leukaemia–sarcoma viruses have been divided into leukaemia viruses and sarcoma viruses. However, they do not always 'breed true' when inoculated into animals and they are now distinguished on the basis of whether they do (sarcoma) or do not (leukaemia) produce transformation of tissue culture cells. The leukaemia viruses multiply in the untransformed cells without destroying them. The transforming viruses are much more efficient than the DNA viruses; up to 90% of the cells in a culture can be transformed within 24 hours. The majority, if not all, of the mouse sarcoma viruses and some of the avian sarcoma viruses appear to be defective, requiring a leukaemia virus as helper before they can produce infective virus. Both the leukaemia and the sarcoma viruses affecting a given species all share a common core antigen. The great majority of the work on C type viruses has been done on the viruses of fowls and of mice. However, C type particles have been reported in a wide variety of species, including man, but there is as yet no absolutely convincing evidence of their aetiological role in any natural disease in these other species. The mode of replication of this interesting group is discussed on page 358.

In summary, we can say that while there are a number of well authenticated cases of viruses causing malignant conditions in lower animals, both experimentally and in nature; this is not the case in man. However it would be strange if man was an exception in this respect, and, as recorded in this and earlier chapters, circumstantial evidence is accumulating in respect of a number of viruses, particularly EB virus. Tumour viruses are, by their nature, well able to establish a more or less permanent relationship with the cells of their host and to be transmitted vertically. Such 'resident viruses' can alter the properties of the cells they inhabit and can interact with other viruses coming into the cell. Unless the 'resident viruses' can be induced to form free infective particles, it is difficult to see how viral genes can be distinguished from cellular genes; indeed, when a virus has persisted in a given race of animals for many generations such a distinction may really be meaningless. It is becoming very clear that the properties of bacteria are determined by the prophages which they carry to an extent not dreamed of even a few years ago: it is perhaps not too far fetched to speculate that the properties of animal cells may likewise be determined by viruses capable of entering the host genome and becoming latent, such as the RNA and DNA tumour viruses.

SLOW VIRUSES

This is a most unusual and interesting group of infective agents affecting man and animals and deriving their name from the fact that the diseases with which they are associated have a very long incubation period, measured in months or years rather than days. There is some discussion as to which agents should actually be included: some would include such viruses as LCM and Riley's lactic-dehydrogenase virus which produce prolonged tolerant infection (PTI). Even rabies has been proposed as a member of the group, as has hepatitis type B. However, it is probably best to restrict the term to the agents associated with the so-called spongiform encephalopathies.

The best studied of these is the scrapie agent which causes the disease of the same name in sheep. This disease is characterized by severe itching (from which it takes its name), ataxia and finally paralysis. The incubation period is of the order of many months to two years and the infection can be transmitted to sheep, goats and rodents but not primates by intracerebral inoculation. The mouse is a particularly valuable tool. The agent has remarkable properties, including resistance to boiling and ultra-violet light. It is small, under 30 nm by filtration, and has never been visualized under the electron microscope. However, infectivity is associated with a membrane fraction in cell homogenates and recent evidence suggests that it is associated with the cytoplasmic membrane rather than the endoplasmic reticulum. It appears to stimulate no immunological reaction of any kind and this is in keeping with the total absence of any of the conventional signs of an infective process in the central nervous system—the histological changes were regarded as purely degenerative in nature.

A human infection of this type, now mainly of academic interest, is kuru, a variety of fatal cerebellar degeneration restricted to a small group of people in New Guinea. The disease, which at one time caused about 50% of the deaths in the tribe, is apparently due to a transmissabl agent present in the brains of the dead. It was formerly the custom of these people to show their respect for their deceased relatives by eating them almost raw. The brain was regarded as an especial delicacy and was usually given to the children. The cannibalism has now ceased, at least officially, and the incidence of the disease seems to be declining. The disease has been transmitted by intracerebral inoculation to chimpanzees and to certain species of monkey.

Another human infection transmissible to chimpanzees and certain other primates is one form at least of Creutzfeld–Jakob disease, a form of presenile encephalopathy. It has been suggested that these agents possess a genome which is not nucleic acid at all. This proposition is perhaps less daunting now that the RNA tumour viruses have demonstrated that what has been inaccurately called the 'central dogma' of molecular biology, that information flow is unidirectional from DNA to RNA to protein to other molecules, is not as universally valid as had been supposed.

One of the most remarkable recent developments in virology has been the identification of the 'viroids'. These are short lengths of RNA of a molecular weight of no more than 75 000 to 100 000 daltons. Although mainly single-stranded, they may possess short double-stranded regions, presumably due to folding. Although capable of coding for no more than 70–80 amino acids, when introduced into a suitable host they replicate and may induce symptoms of disease. Those identified so far are parasites of plants such as the agent of potato spindle tuber disease, but it has been suggested that the agents of scrapie and kuru, which are likewise very small, have not been identified under the electron microscope, are non-antigenic and very resistant to ultra-violet light, may be similar entities. How such small molecules could possibly programme their own replication and take over the metabolism of a host cell for the purpose remains totally unknown.

The findings in respect of scrapie, kuru and other slow virus infections have naturally led to attention to other chronic neurological conditions in man. The most promising seem to be motor neuron disease, in respect of which claims have been made by Russian workers, and disseminated (multiple) sclerosis. There is a substantial body of evidence relating the latter to environmental factors. The disease is extremely rare in South Africa, except among European immigrants who have left their country of origin after the age of about 15 years. These individuals show a much higher incidence of the disease and it seems that the incidence reflects that in their country of origin: in Europe the incidence falls from north to south. One group claimed to have isolated a transmissible agent from cases of disseminated sclerosis, but it subsequently turned out to be indistinguishable from scrapie, with which they were also working. Particles resembling those of papovaviruses have been described in the oligodendroglia in progressive multifocal leucoencephalopathy.

The balance of evidence seems to favour an autoimmune basis for the condition known as acute post-infective demyelinating encephalomyelitis which occurs as a late complication of many quite trivial virus infections, particularly vaccinia and measles, after rabies vaccination and more rarely after bacterial infections.

IV.13

Bacteriophages and Bacteriocines

BACTERIOPHAGES

Bacteriophages or phages are viruses that are parasitic on bacteria. As a rule they exhibit a marked specificity of action, each phage being capable of attacking only groups of closely related bacteria. In general, phage specificity follows taxonomic boundaries closely—a particular phage attacking only organisms belonging to a single species. In certain cases phages are not only restricted in action to a particular species but will attack only certain strains of that species. In such cases susceptibility to a particular bacteriophage forms a very convenient basis for species subdivision or typing and one which has been found to be of considerable epidemiological value.

With a few exceptions, bacteriophages either multiply freely in the host cell with its ultimate lysis and the liberation of a considerable number of newly formed infectious particles (lytic cycle) or they exist in a latent form in the host cell being transferred to each daughter cell at cell division and only very rarely reverting to the lytic cycle and causing lysis and death of the host cell (lysogenic cycle). Some bacteriophages, referred to as *virulent,* can only manifest the lytic cycle; others, called *temperate* can exhibit either, and a third group, the *defective lysogens,* can exist only in the lysogenic state or, at most, undergo an abortive lytic cycle. Bacteriophage undergoing replication as part of the lytic cycle is called *vegetative phage* and when in the lysogenic condition it is referred to as *prophage*. Bacterial cultures carrying prophage are said to be *lysogenic cultures*. Some bacteriophages do not fit into the above classification. The single-stranded, filamentous DNA phages (see below), for example, multiply freely without lysing or even killing the host cells: indeed the latter may continue to multiply.

Propagation

Bacteriophages may be demonstrated in two ways:

1. By inoculation of material containing the phage into a young broth culture of a sensitive strain. In a short time, usually from half an hour to one hour after infection, the culture will undergo lysis. When infection is due to a lytic phage the culture will become completely clear. If it is due to temperate phage, a partial clearing will occur and the culture will then become turbid through growth of lysogenic organisms.

2. Plate method. A bacterial lawn is spread on the surface of an agar plate either over the entire surface or in small patches. The lawn is allowed to dry and a loopful of a preparation containing phage is deposited on its surface. After incubation a patch of clearing will be seen in the part of the culture attacked by the phage. If the phage is a lytic one and its concentration has been high, there will be complete clearing or confluent lysis over the inoculated area. If the concentration of phage in the preparation is low, there will instead be a number of very small areas of lysis; each of these areas is the result of growth of a single phage particle and is known as a plaque. The plaques produced by lytic phages are completely clear; areas of confluent lysis are clear but may occasionally show a few colonies of phage-resistant mutants. The plaques produced by temperate phages, on the other hand, show an area of lysis containing scattered microcolonies of bacteria, especially towards the centre (turbid plaques); most of the organisms in the plaque are lysogenic (Fig. IV.38).

In order to isolate a phage in pure culture a

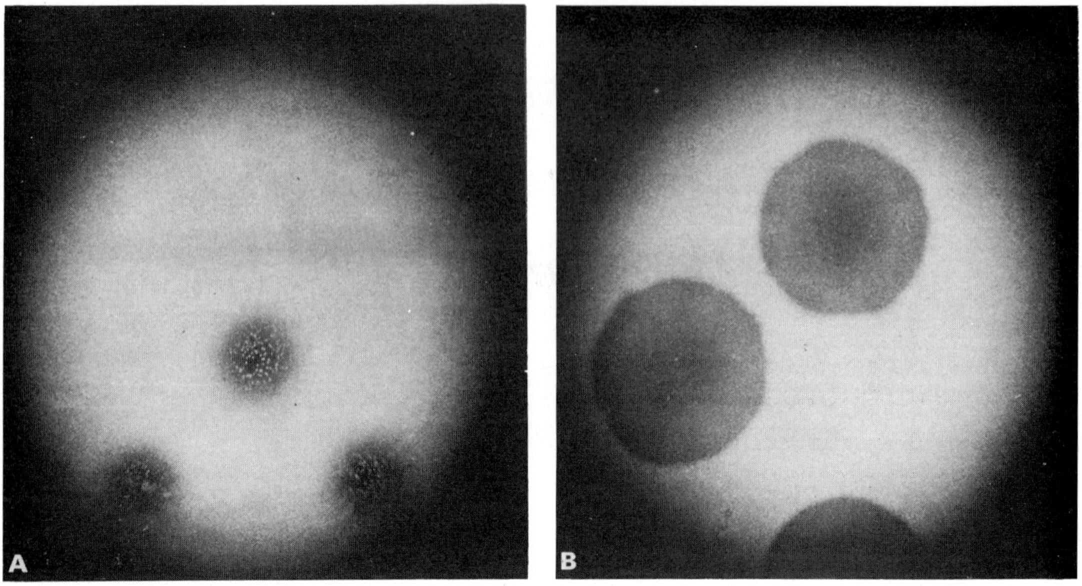

Fig. IV.38. Single plaques of bacteriophages. A, A temperate phage active against *E. coli*. Note the growth of lysogenic organisms in the centre of each plaque. B, A Virulent phage active against *E. coli*.

small piece of agar is cut from a single plaque; the agar should include some of the area of lysis and some of the growth of the indicator strain at the periphery of this area. The agar is inoculated into broth and after incubation should contain a considerable amount of phage, of the order of 10^8 to 10^9 phage particles per ml. Single plaque isolation is essential if a pure strain of phage is to be obtained; its role in work with phage is precisely comparable to single colony isolation in bacteriology. Most phages are less sensitive to heat than their host organisms. In such cases phage preparations free from residual unlysed bacteria can be obtained by heating. Where heat cannot be used bacteria-free preparations of phage may be obtained by filtration, or treatment with chloroform.

Morphology

The majority of phages are tadpole-shaped structures, which possess a head, functioning as a container for the phage nucleic acid, and a tail. The heads are of variable shape but all exhibit some sort of cuboidal symmetry; they may be bipyramidal, hexagonal prisms as in the T-even coliphages, or octahedra or icosahedra. The heads appear to be built up from protein subunits, approximately 1000 of such units comprising the heads of the T-even phages (see p. 351). The phage tails, which are of variable length, are normally surrounded by a sheath; in some cases this is contractile, in others non-contractile. In the T-even coliphages, which possess contractile tail sheaths, there is a collar between the base of the head and the top of the sheath and, in addition, a base plate with projecting spikes at the lower end of the tail. In the normal untriggered state of the phage a network of fibres extends from the base plate to the collar. Like the head, the tail sheath has a subunit structure. The sheaths show a lateral striation due to the arrangement of the component subunits in parallel rings along the length of the sheath. In the sheaths of T2 and T4, in the uncontracted state, each ring appears to consist of six subunits; on contraction of the sheath the number of subunits per ring is increased to 12, thereby allowing a halving of the total length of the sheath. The core of the tail is hollow with a central channel of about 2 nm in diameter. It is through this channel that the phage injects its nucleic acid into the bacterium. Each of the protein subunits of the contractile sheaths of the T-even phages is associated with one molecule

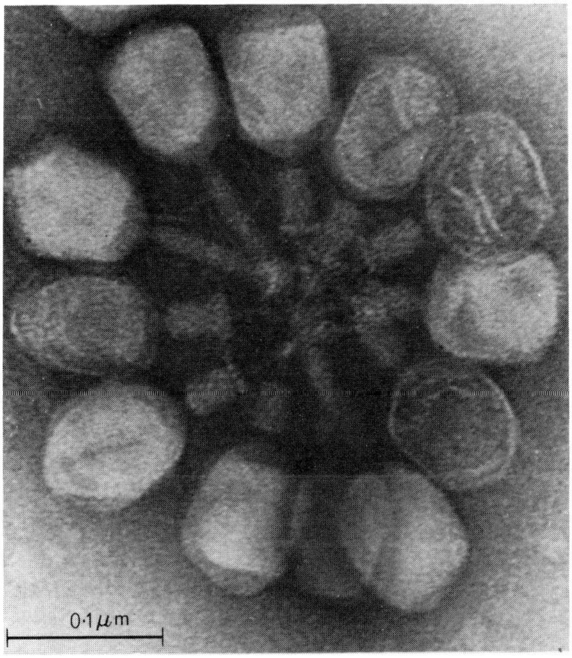

Fig. IV.39. T2 coliphage particles showing heads and sheaths. Electron micrograph. Negative stain. (*Brenner and Horne, J. molec. Biol.*)

of the nucleoside triphosphate, ATP, the breakdown of which, by the appropriate phosphatase also present in the sheath, provides the energy for its contraction. The tail structure of the T-even phages also contain a lysozyme-like enzyme which, by causing a local breakdown of the mucopeptide layer, facilitates the penetration of the cell wall in infection. Not all phages, however, possess obvious tails. Some, e.g. T3, have very short tails, while others, e.g. the coliphage ϕX174, appear to possess no tail at all. ϕX174 is a phage with icosahedral symmetry, possessing specialized apical subunits which appear to fulfil the function of a tail. An unusual filamentous bacteriophage, which is specific for F^+ bacterial cells (p. 22), and morphologically quite distinct from those considered above, has been described (see p. 355).

In the majority of phages that have been investigated, including the T phages, the nucleic acid is a double-stranded DNA. In the case of T2 this has a molecular weight of circa 10^8, contains about 2 by 10^5 base pairs and, when fully extended, is about 50 μm in length. Hence its length is between 1 and 2% of that of the bacterial chromosome. The T-even phages contain an unusual pyrimidine–5-hydroxymethyl cytosine —the demonstration of which has been of considerable value for the detection and estimation of intracellular phage-specific DNA. The pyrimidine is present in a glycosylated form, the extent of glucosylation varying with different phages (see p. 470).

A lipid-containing phage with double-stranded RNA has been described: it is the ϕ6 (phi 6) phage of *Pseudomonas phaseolicola,* the cause of halo blight of kidney beans and related plants. The phage genome is in three fragments and is contained in a polyhedral head 60 nm in diameter, surrounded by a flexible membranous structure essential for infectivity. This envelope attaches to the bacterial pili, being drawn out into a tail-like structure in the process. The particles contain 25% lipid, 13% RNA and 62% protein. They are, of course, sensitive to detergents and lipid solvents.

Growth

As it occurs within the phage head, T2 phage DNA is in a highly condensed form. It is associated with a small amount of protein—the *internal protein*—whose function appears to be to bind to, and maintain, the DNA in the condensed state. In addition to the internal protein, the head contains a polypeptide and a small amount of the polyamines spermidine and putrescine. The latter appear to be of importance in neutralizing the high negative charge on the phage DNA. Some very small phages, e.g. ϕX174 contain a single-stranded DNA, and in a few cases the nucleic acid is RNA. The most interesting of the latter are the icosahedral phages specific for F^+ cells which adsorb to the sides of the specialized F pili.

The intracellular growth of bacteriophages has been most studied in connection with the T coliphages. The first stage in infection is a random collision between phage and bacteria, as a result of which the phage adsorbs to specific receptors on the surface of the bacterial cell wall. The receptors for each of the T phages are different. This is shown by the fact that bacteria which have mutated to resistance to a particular phage may still retain the power to absorb other T phages. Adsorption, in the case of the T-even phages, appears to occur in the first place through the fibres attached to the tail plate. Some strains of T-even phages adsorb only in the presence of amino acid co-factors, the most effective of which is tryptophane. Adsorption co-factors appear to act by allowing the tail fibres to free themselves from the sheath and collar of the phage. Tailless phages, such as ϕX174, appear to adsorb through the specialized subunits located at the apices of the icosahedron.

Not all bacteriophages have their attachment sites in the bacterial cell wall. The sex pili afford attachment sites for the filamentous single-stranded DNA phages which attach to the tip and for the isometric RNA phages which attach to the sides. These viruses are specific for particular types of sex pilus and the isometric phages are of particular value to bacterial genetecists and electron microscopists in distinguishing the different types of sex pilus (see p. 476). Some phages bind, at least initially, to the bacterial capsule and others attach themselves to flagella by special coiled tail fibres and then seem to slide along the flagellum to reach the cell body.

Following adsorption the phage proceeds to inject the nucleic acid contained in its head membrane into the bacterium. The whole structure of the phage—head membrane and tail—can in fact be regarded as a micro-syringe developed precisely for this purpose. Neither the protein head nor tail enters the cell; they remain attached to the surface and play no further part in infection.

In the case of phages with contractile sheaths injection is associated with a contraction of the sheath, the energy for which is supplied by the breakdown of the nucleoside triphosphate associated with the tail subunits. Presumably the contraction of the sheath assists in forcing the tip of the tail through the bacterial surface. In some cases the phage tail carries a lysozyme-like enzyme which causes a local breakdown of the cell wall mucopeptide at its point of contact with the tail. Following this local break in the wall the phage DNA is believed to be forced into the bacterium by mechanical pressure resulting from its condensation in the phage head. Active participation of the cell is not required since injection occurs following adsorption of phage to killed bacteria. It does not now appear that the phage tail physically breaches the cytoplasmic membrane. However, it does seem that it alters its permeability to allow entry of the phage DNA.

If a large number of phages adsorb to a single cell the damage to the wall produced by the phage lysozyme may be sufficient to result in the death of the cell. This is known as *lysis from without.* Under normal conditions of infection with one or a few phage particles, repair of the damaged wall is initiated immediately after infection, and usually results in a considerable increase in the strength of the wall. As a result further phages, which may subsequently adsorb, are unable to inject their DNA into the cell but instead release it into the surrounding medium. This is referred to as *superinfection breakdown.*

Though the tail assembly is necessary for injection of the phage DNA, infection has, in certain cases, been produced by extracted protein-free phage DNA. Thus T2 and ϕX174 DNA will infect protoplasts and lambda (p. 466) phage DNA will infect cells exposed at the same time to intact 'helper' phages. *B. subtilis* is reported to be susceptible to infection with phage DNA when it is in a state of competence in which it is transformable by transforming DNA (p. 23).

The period elapsing between the initial contact of the bacterium with phage and the lysis of the infected cell, with the release of the new phage progeny, is known as the *latent period.* For the T-even phages it lasts about 25 minutes at 37°C. The number of phage particles liberated at the end of the latent period varies with the bacteriophage and the host strain but is usually of the order of 100 phage particles per cell. This is known as the *burst size.* Neither the duration of the latent period nor the burst size is affected by the number of particles initially infecting the cell.

The sequence of events occurring during the latent period can be studied in cells which have been artificially disrupted, e.g. by lysis with another phage, by ultrasonic irradiation or by exposure to chloroform. During approximately the first half of the latent period, no infective intracellular phage is detectable. This period is known as the *eclipse phase.* Around the middle of the latent period the first infective phage particles start to appear and thereafter increase in number until lysis. During the eclipse phase there is, however, an active synthesis of phage components and, to an extent depending on the particular phage, of the enzymes necessary for the synthesis of these components. The T-even phages are outstanding in their independence of host cell synthetic enzymes, many of the enzymes they require being coded for by the phage DNA. In such cases the first obvious effect of infection is disruption of the nucleus. With many other phages, e.g. T1, T3, T7, the nucleus is not disrupted and the phage utilizes enzymes such as DNA polymerase coded for by the host cell DNA. Because of the disruptive effect of T-even phages on the nucleus these phages block all cellular synthesis. Though highly independent of host synthetic mechanisms they nevertheless depend on the host cell for their energy.

In T-even phage infection the first phage-specific component to appear is an RNA whose phage specificity is shown by the fact that it is capable of hybridizing with phage DNA. The newly synthesized RNA is a messenger RNA which participates in the synthesis of the 'early' enzymes. These are enzymes necessary for the synthesis and replication of the phage DNA. At least 12 such enzymes have been described, including deoxy-cytidylate hydroxymethylase required for the synthesis of hydroxymethyl cytosine, a DNA polymerase specific for the synthesis of phage DNA and a glucosyl transferase responsible for the transfer of glucose from UDP-glucose to hydroxymethyl cytosine. In T2 infected cells the next stage is the synthesis and replication of the phage DNA which is distinguishable from the host cell DNA because of its content of hydroxymethyl cytosine. Phage DNA begins to appear in about the sixth or seventh minute after infection and continues to increase in amount until the end of the latent period. Since mature phage starts to appear at about the middle of the latent period there is, when the cell eventually undergoes lysis, an excess of DNA over that which is incorporated into mature phage.

In the early stages of DNA synthesis the DNA can be seen in electron micrographs as delicate fibrils in the bacterial cytoplasm. Shortly afterwards the fibrils condense. Condensation appears to be a consequence of binding of the DNA strands with the internal phage protein which is synthesized along with the early enzymes. At about the time of appearance of the condensates the phage head protein, which can be detected serologically, begins to make its appearance. It then proceeds to coat the condensates (p. 359). The heads next acquire tails and fibres. It now seems that heads, tails and tail fibres are put together independently on separate production lines. Heads and tails then appear to unite spontaneously, but attachment of the tail fibres requires the presence of a particular gene product, presumably a protein. Although some 50 genes have been shown to be involved in the assembly of T4 particles, no more than about 20 polypeptides have been found in the actual particle itself. The morphogenetic function of the excess genes has not been elucidated. The final stage of the maturation process appears to be the setting of some cementing substance which confers stability on the completed structure. Rupture of the cell before this occurs leads to disintegration of the nascent phage. This final stage in maturation is inhibited by proflavine.

The final release of the phage from the cell appears to be achieved as a result of the action of the lysozyme produced in the later stages of growth.

Phage genetics

The study of the genetics of bacteriophages (see also Chapter IV.2) has been of enormous value in elucidating fundamental genetic mechanisms. That this should be so is not surprising since phages are essentially genes which can be transferred from one organism to another, and whose effect can be precisely evaluated under defined conditions. Bacteriophages differ from one another in a great variety of genetically controlled properties. Those which have been of most value in the study of genetic mechanisms have been:

1. Host range, that is the range of organisms against which the phage is active.
2. The type of plaque produced.

Both of these properties can be readily determined and are therefore particularly suitable for experimental work. Thus by the use of host range mutations in T2 (r II mutants) Benzer and his colleagues have been able to establish the unit of mutation and of recombination, in mixed infections, as being a single nucleotide pair.

If bacteria are infected with two recognizably different phages or with two different mutants of the same phage, recombination with the formation of hybrids occurs not only between the parent phage DNAs but also between the daughter phage DNAs. The genetic evidence indicates that, prior to withdrawal of the DNA for naturation, multiple rounds of mating occur in the replicating DNA pool. There is evidence that mating occurs by breakage and reunion of paired DNA strands (but see Chapter IV.2). When bacteria have undergone mixed infection with two phages the phenomenon of *phenotypic mixing* may also occur. In this process the DNA of one phage is accidentally coated with the head protein of another. The issuing phage has the host specificity of the second phage but on replication in a new host regains its original specificity.

Lysogeny

The elucidation of the precise relationship of a temperate phage with its lysogenic host bacterium first became possible with the discovery that *E. coli* K12, which is capable of undergoing conjugation (p. 21), carried a temperate phage called lambda (λ). Strains of K12 carrying λ are known as K12 (λ). In crosses between a non-lysogenic F^+ parent and a lysogenic F^- parent, non-lysogenicity for λ was found to be linked to the *gal* locus. Similarly in crosses between an F^+ or Hfr parent—both lysogenic for λ phages carrying different markers—with a non-lysogenic F^- parent, the same linkage with the *gal* locus was found. These findings clearly established the location of the λ prophage on the bacterial chromosome.

The model currently accepted for the association of the prophage with the chromosome is that proposed by Campbell. This envisages recombination between areas of homology on a circular prophage and a circular bacterial chromosome, as a result of which the prophage becomes inserted linearly into the continuity of the bacterial chromosome. When non-lysogenic susceptible bacteria are infected with a temperate phage capable of lysogenizing them, the phage undergoes a lytic cycle of development in the majority of the infected cells but, in a small number of cells, becomes incorporated into the chromosome. The evidence indicates, however, that it is usually not the infecting phage DNA which is incorporated, but a replica of the infecting DNA. For some prophages more than one chromosomal locus at which the phage can be inserted has been identified. For others, e.g. the P1 coliphage, the chromosomal site is not known. This is believed to be due to the fact that in conjugation experiments—the only means of determining the chromosomal locus—the phage is released on conjugation from its chromosomal location.

When non-lysogenic F^- cells are mated with lysogenic F^+ or Hfr cells lysogenic recombinants are never found. Instead the prophage embarks on a lytic cycle of multiplication in the infected F^- cells. The investigation of this phenomenon, known as *zygotic induction,* established that the maintenance of the prophage state depends on the production, under the direction of the prophage genome, of a cytoplasmic repressor which prevents its transcription. When the prophage gains access, on conjugation, to a non-lysogenic cell which does not possess the cytoplasmic repressor, transcription is derepressed and the lytic cycle of multiplication can occur. The phage repressor not only represses the genes of the resident prophage, but those of any phage sensitive to it which superinfects the host cell. Since the repressors are highly specific, only a phage identical or closely similar to the prophage will be repressed. This is the basis of the 'immunity to superinfection' exhibited by lysogenic bacteria. The DNA of the superinfecting phage penetrates but cannot replicate unless the host cell is simultaneously induced, in which case the prophage will enter the vegetative state and, unless too defective, will replicate alongside the superinfecting phage so that both viruses, and possibly recombinants as well, will appear in the progeny. Derepression or *induction* of prophage can be achieved in a variety of ways, e.g. by exposure to various chemicals, but particularly by ultra-violet irradiation. Ultra-violet radiation does not seem to act directly on the repressor. Instead it leads to the accumulation of an adenine derivative, possibly as a result of the inhibition of DNA synthesis. This compound, termed 'active inducer', binds to the repressor and alters its conformation so that it will not bind to the phage operons which it normally represses. In the small proportion of cells in a lysogenic culture which undergo lysis and liberate phage, induction of the prophage is a spontaneous event occurring, possibly, as a result of the action of some inducing substance liberated during bacterial growth. The establishment of the lysogenic state in a small proportion of the infected cells, when non-lysogenic bacteria are exposed to phage, is best interpreted as a race between the production of repressor with its effective control of phage genome and the translation of the genome into the synthesis of phage enzymes and phage components. In most cases the latter is dominant and the phage undergoes a lytic cycle of infection, but in some cells the repressor wins and the phage is incorporated into the bacterial chromosome.

A temperate phage may give rise to mutants which have lost their capacity to lysogenize. This may occur as a result of mutations in different parts of the phage genome:

1. The phage may lose its capacity to synthesize repressor as a result of mutation in the regulator gene.
2. The phage operator may become insensitive to the repressor.
3. A mutation may occur affecting the area of the phage genome which is homologous with, and pairs with, the bacterial genome; recombination between the bacterial and phage genomes cannot then occur. Mutations in the so-called *xis* (excision) gene give rise to phages which can lysogenize normally but, once integrated with the host chromosome, cannot be liberated again, so these phages are not inducible. This is one class of defective lysogen: others are those with mutations affecting functions essential for completion of the lytic cycle. The degree of defect varies; at one extreme are cases where, following induction, lysis occurs normally with the production of non-infectious phage particles which may show greater or lesser degrees of morphological abnormality; at the other extreme are cases where apparently nothing happens following induction, although the phage genome may be excised from the host genome and remain in the cytoplasm as a non-replicating entity which will pass to one daughter at cell division. The term *abortive lysogeny* is used to describe this situation which can also arise when a phage genome enters a cell and both produces and is sensitive to the repressor but does not or cannot integrate with the host genome.

Transduction

This is a process in which genetic material is transferred by the agency of bacteriophage from one bacterium to another. Transduction may be effected by both virulent and temperate phages. They may acquire host cell genetic material in one of two ways:

1. Through an exchange between homologous areas on the phage and host cell chromosomes; as a result some of the host cell chromosome is incorporated into the phage chromosome.
2. By a process of phenotypic mixing in which the host cell genetic material is incorporated in the phage head.

Transduction was first described by Ledeberg and Zinder who found that certain auxotrophic mutants of *Salm. typhimurium* were converted into prototrophs by growing them in mixed culture with other auxotrophs having different growth-factor requirements. That this was not due to conjugation was clear from a classical experiment in which the two strains were grown separately in the two halves of a U-tube divided by a bacteria-tight filter. Intermittent positive and negative pressure caused fluid, but not cells, to pass from one side to the other. Wild-type prototrophs appeared in one half of the tube. The effect was shown not to be due to transformation (see p. 23) when it was found that the agent responsible for the phenomenon was not sensitive to deoxyribonuclease. It subsequently transpired that one of the two strains of *Salm. typhimurium* used was lysogenic for a phage P22 to which the other strain was susceptible. As is virtually always the case, cultures of the lysogenic strain contained small numbers of mature, infectious phage particles. These passed through the filter and infected the second strain. Most of the newly infected bacteria lysed and liberated phage particles which could pass back through the filter to the side containing the lysogenic bacteria. These, of course, were immune to superinfection but the DNA is able to penetrate in such cases although it cannot replicate because of the repressor already present in the cell. However, in a small proportion of cases the DNA introduced by the phage particle was not the phage genome but an equal length of host genome. This was not, of course, subject to the repressor and could recombine with the host genome to produce a recombinant. In this way genes could be transferred from the non-lysogenic to the lysogenic strain including the wild-type alleles of those which were defective in the latter. Transduction has also been demonstrated in *E. coli, Shigella, Pseudomonas, Staph. aureus* and *B. subtilis*.

Two types of transduction may be distinguished: generalized or unrestricted and localized or restricted.

Generalized transduction

Transduction in *Salm. typhimurium* with phage P22

is an example of generalized transduction. In this, any gene of the donor cell can be transduced. In respect of any particular marker only 10^{-5} to 10^{-7} P22 particles are capable of transducing. The fragment of host cell genome transferred appears, in the case of P22, to be incorporated into the phage head as a result of phenotypic mixing. On infection of the recipient strain the chromosomal fragment is integrated into the recipient's chromosome but the recipient does not become lysogenic for the transducing phage, nor can the latter replicate vegetatively because it probably contains no phage DNA at all. In some cases, the chromosomal fragment is not integrated but is maintained in the cytoplasm and is transmitted to one daughter cell only at each subsequent division. In the cytoplasm it is genetically active and determines the phenotypic expression of the transduced property; this can of course only occur in the single cell in which the fragment is present. This phenomenon, known as *abortive transduction,* was first observed by Stocker, who found flares of growth at various distances from the site of inoculation of a strain of *Salm. typhimurium* which he had attempted to transduce for motility, the flares being due to the movement over the surface of the medium of a single abortively transduced and, consequently, motile cell leaving behind it a trail of non-transduced, non-motile cells, which multiplied to form small colonies visible to the naked eye.

Localized or restricted transduction

This type is effected by temperate prophages which have a permanent locus on the bacterial chromosome and which, on induction, are capable of transducing only those genes which are adjacent to this locus. Thus lambda phage is capable of transducing only the genes for galactose fermentation. When lambda prophage is induced some of the prophage particles incorporate a chromosomal segment carrying the *gal* locus, at the same time leaving behind a segment of their own genome in the bacterial chromosome. On infection of the recipient cell the prophage carrying the donor chromosomal fragment is integrated into the recipient chromosome. Since, however, the prophage is defective, lacking some of its own chromosome, it is incapable of yielding infective phage on induction. On induction it may, however, if the defect in the genome is relatively slight, determine the synthesis of phage components and cause lysis of the cell. If the defect in the phage genome is more severe, lysis does not occur on induction and phage components are not synthesized. In this case the presence of the prophage on the bacterial chromosome may be demonstrated by the immunity of the cell to superinfection with fully infective phage. It is possible, however, by simultaneously infecting a susceptible cell with transducing phage and normal phage, to induce a state of double lysogeny: the two phage genomes are inserted in tandem in the host genome. When such a double lysogen is induced, about 50% of the particles are transducing phage and 50% are normal infective phage. Provided that the defects of the transducing phage are not too gross, the complete phage genome can act as a *helper* and complement the defects of the transducing phage. In contrast to the LFT (low frequency of transduction) lysates produced by inducing singly lysogenic cultures, the HFT (high frequency of transduction) lysates produced by the induction of these double lysogens contain transducing particles which are virtually all identical in respect of the exact piece of host DNA which they contain and the segment of virus DNA they lack. This is because they are all copies of the transducing particle which originally participated in the double lysogenization.

Lambda transductants are diploid for the *gal* locus and segregate out *gal*-positive and *gal*-negative cells. Partial diploids of this type are known as *heterogenotes,* and appear to represent an intermediate phase in which the phages associate with, but are not stably integrated into, the chromosome.

Transducing phages can transfer episomes or plasmids as well as chromosomal genes. Col factors (p. 474), F factors (p. 20) and R factors (p. 20) have all been transferred in this way, as have penicillinase plasmids in staphylococci (p. 209). Transduction is apparently the only way in which these latter plasmids can be transferred.

Phage conversion

This phenomenon consists in the genetic alteration of the cell, either by a temperate phage which can establish the lysogenic state or by lytic variants of temperate phages which can only undergo a lytic cycle of infection. The best known example of the former is the control of toxin production in *C. diphtheriae* by β phage (p. 230). In this case, toxin is coded for by the *tox* gene of the bacteriophage, although expression of the gene, that is toxin production, is critically dependent on the physiological state of the host, particularly the iron concentration of the medium. In *Salmonella* spp. phage conversion is reflected in the antigenic composition of the cell wall. Indeed several pairs of named serotypes differ only in that one member of the pair carries a prophage which modifies the antigenic composition of the cell wall. One example is the pair *Salm. gallinarum* and *Salm. pullorum*: the former possesses a phage-specific antigen, O1, absent from the latter. It is interesting to speculate whether the differences in pathogenicity and epidemiology between these two are also phage specified. Sometimes the phage specifies a substitution rather than an addition as in the case of *Salm. anatum* (O3, 10) and *Salm. newington* (O3, 15), where the substitution of antigen 15 for antigen 10 is phage determined. When phage conversion is due to virus in the lytic cycle, the change of phenotype is necessarily transient, lasting only until the cell lyses.

One of the most dramatic recent examples of phage conversion is the discovery that toxigenic *Clostridium botulinum* type C can be rendered non-toxigenic and cured of its lysogeny by heating the spores at 70°C. Relysogenization of the culture restores the toxigenicity and the serotype. However, lysogenization with another phage converts the non-toxigenic *C. botulinum* into a toxigenic type D, and, more remarkably, a third temperate phage can convert it into a totally different organism, *Clostridium novyi* type A. It is becoming clear that many bacterial characters are phage determined, including some which have been used to separate different species or even different genera. This has obvious and far-reaching consequences for bacterial taxonomy: however, in cases where the prophage is so defective that it cannot be excised on attempted induction, there is no valid way of distinguishing viral genes from bacterial genes.

Sexduction

A process of genetic transfer basically similar to transduction (p. 467) has been shown to be mediated in *E. coli* K12 by substituted or intermediate—F′—sex factors. These are sex factors that have undergone recombination with an area contiguous to the F locus on the bacterial chromosome. Substituted sex factors appear to migrate rapidly to and fro, between the chromosome and the cytoplasm. They resemble F in being transmissible to F^- cells and they resemble Hfr (p. 22) in determining high frequency chromosomal transfer.

Phenotypic or host-controlled modification

Bacteriophages may, in addition to genotypic modification or mutation, also undergo a non-genetic form of variation known as host-controlled modification. This is a variation of a temporary nature in the host range of the phage, occurring as the result of a single cycle of growth in a new bacterial host. It appears to be due to a modification produced in the phage DNA by the new host and manifesting itself in an increased capacity of the phage to infect and multiply in the host and/or a decreased capacity to infect and multiply in an alternative host. If one host is designated as A and the second as B the process may be illustrated as follows. When propagated on host A the phage has a high capacity to grow on host A but a low capacity to grow on host B. This is most conveniently represented by the term efficiency of plating (EOP). Taking the EOP on host A to be 1 and assuming that only 1 in 100 000 particles growing on host A propagates on host B then the EOP on host B is 10^{-5}. If propagated on host B, however, the EOP of the issuing phage in host B is equal to that on host A, i.e. 1. From these observations alone it would clearly be impossible to distinguish between an extension of host range produced by phenotypic modification and an extension produced by mutation. The distinction may, however, be made by examination of the fate of the modified phage when propagated again on host A. Phenotypically modified phage when so propagated regains its original specificity, i.e. it again has a low EOP on strain B. Mutated particles, on the other hand, retain their mutant host range with a high EOP on strain B.

The phenomenon is dependent on a pair of closely related enzymes, which in some cases even apparently share some polypeptide subunits. One, the *modification enzyme,* modifies the host cell DNA immediately it is formed by the methylation of certain bases. This highly specific pattern of methylation prevents the DNA being attacked and broken by the second enzyme, the *restriction endonuclease*. The modification enzyme will also label and so protect any other DNA being formed in the cell, e.g. virus DNA, but preformed DNA introduced into the cell from outside is vulnerable to the endonuclease. This enzyme breaks both DNA strands, but not usually at the same point, so that the DNA each side of the break has short single-stranded tails which are complementary in respect of base sequences. It is the extreme specificity of the restriction endonucleases for particular base sequences and the fact that they leave these complementary, single-stranded tails which have made them such valuable tools in so-called genetic engineering in which fragments of DNA from different sources can be joined end-to-end to give what are entirely novel combinations of genetic material. The repair of the two single-stranded 'nicks' which remain after the complementary tails have paired are repaired by a ligase. The possibilities of this technique for good and ill are so great that scientists have voluntarily agreed to restrict such work and only carry it out under the most stringent safeguards.

In nature, the modification–restriction system seems to function as a self-recognition system for DNA, enabling a bacterial cell to destroy foreign DNA, while leaving its own DNA unscathed. What seems to happen in host-controlled modification is that a very occasional molecule of phage DNA is modified before it can be broken down by the restriction enzyme. All its progeny DNA will, of course, be modified as it is formed. The glucosylation of the DNA of the T-even phages of *E. coli* serves a similar purpose; but in that case it protects the virus DNA and not the host DNA which is actively broken down by virus-coded nucleases.

The most extensively studied example of phenotypic modification is that which occurs in phage λ when propagated in a strain of *E. coli* K12 lysogenized by the symbiotic phage P1, the lysogenic strain being known as K12(P1). If the phage is first grown on non-lysogenic K12 its EOP on *E. coli* K12(P1) is approximately 2×10^{-5}. There is considerable evidence that the majority of the particles fail to grow on K12(P1) because the DNA of the infecting particles is, shortly after infection, degraded to small molecular weight components, as a result of the action of a nuclease synthesized under the influence of the P1 genome. The few particles which succeed in infecting K12(P1) are not mutant particles and differ in no way from the rest of the λ population derived from K12. They multiply in K12(P1) because they have infected a small number of cells in the K12(P1) population which have, for some reason, a diminished phage-inactivating capacity. This capacity is markedly influenced by environmental factors and, as recently shown by Schell and Glover, is heat-labile. It therefore seems likely that the diminished capacity of a minority of K12(P1) cells to inactivate the phage is due to the influence of physiological conditions within the bacteria cells. The particles issuing from K12(P1), in contrast to those coming from K12, are unrestricted and plate efficiently both on K12(P1) and K12. They have been modified as a result of growth on K12(P1) so that they are insusceptible to inactivation by K12(P1) because their DNA has been methylated by the K12(P1) modification enzyme. Whatever the exact nature of the modification, non-lysogenic K12, by reversing it, can be regarded as producing a modification of its own which renders the phage normally unacceptable by K12(P1). In this particular case, as in many others that have been studied, restriction is related to lysogenization which is not, however, necessary for restriction since λ, when propagated on another strain of *E. coli,* is found to be restricted for non-lysogenic K12. From the above it will be seen that the modifying host has two actions: it can break down unmodified phage and it can modify the progeny of the survivors so that they are resistant to this breakdown.

A particularly important example of host-controlled modification is that observed in the adaptation of *Salmonella typhi* Vi II phage which has been grown on *Salm. typhi,* type A, and which as a result is active only on type A, by growth on certain other non-lysogenic *Salm. typhi* phage types (p. 242). Thus, if phage grown on type A and with a low EOP against, for example, non-lysogenic type C1 is grown on C1, the issuing particles after a single cycle of growth plate efficiently on C1. If, however, they are again propagated on type A they revert in a single cycle of growth to the wild type Vi II phage. Type adaptation due to phenotypic modification can in practice be distinguished from type adaptation due to mutation by the fact that the phenotypically modified phage when propagated on type A reverts to type A; a host range mutant on the other hand, when propagated on type A, retains its mutant specificity.

Bacteriophage typing

Bacteriophages are of considerable practical value to the bacteriologist as epidemiological tools since certain bacterial species can be subdivided on the basis of their susceptibility to various phages into a number of stable phage types. The value of phage typing in epidemiological studies has already been discussed (Chapter I.2).

Salmonella typhi

The phages used in the typing of *Salm. typhi* are active only on Vi-containing strains and are consequently known as Vi phages. All are derived from a parent phage—Vi phage II—isolated by Craigie and Yen. These workers found that by propagating this phage on different strains of *Salm. typhi* it could be rendered specific for the strain on which it was propagated; the adapted phages produced in this way are the specific typing phages. The phages and their corresponding bacterial types and subtypes, of which a total of more than 72 have been described, are designated by letters, A, B, C, D, E etc. and numbers 25 onwards.

In order to determine the phage type of a typhoid bacillus the phages are first propagated on a broth culture of the homologous type so as to yield a high titred preparation of the phage. The culture is then heated to 58°C for half an hour in order to kill any unlysed organisms present. The highest dilution of the resulting preparation which causes confluent lysis of the homologous type—the routine test dilution or RTD—is used in the test. In most cases confluent lysis is produced by a few hundred phage particles. Some of these may, however, be able to produce isolated plaques on heterologous types. In routine tests such plaques are more likely to be found with adapted phages which produce very small plaques since with these a much greater number of particles are required to produce confluent lysis on the homologous type. The strain to be typed is spread either as a lawn, or in a number of areas each about 1 cm in diameter on the surface of an agar plate. The inoculum is dried and a loopful of each of the test phages is placed on it. After incubation the type of the organism is shown by the presence of an area of confluent lysis with the homologous phage or, in the case of subtypes, by a pattern of lysis with a number of related phages.

The reactions of a representative series of typing phages and of the corresponding phage types are shown in the table below. It will be seen that type A differs from other types in being lysed by all the typing phages. On the other hand the type A phage, which is in fact the wild-type Vi II phage, is the most specific of the typing phages. In contrast to type A many types are lysed by only a single typing phage while others occupy an intermediate position in being lysed by a number of related phages. Such related phages define subgroups, namely C, D, E, F, J, K, L, M, the first member of which shows the same relation to the other members of the group as does type A to all other phage types in that it is lysed by all the phages defining the subgroup.

Each of the Vi phage types of *Salm. typhi* is capable of absorbing the full range of typing phages. Type specificity cannot therefore be due to differences in the nature of the phage receptors on the bacterial surface. In fact it would appear that all the typing phages are adsorbed to the cell through the same receptor, i.e. the Vi antigen. In some cases the occurrence of transduction in the absence of multiplication has established not only that the phage is adsorbed but that its DNA gains

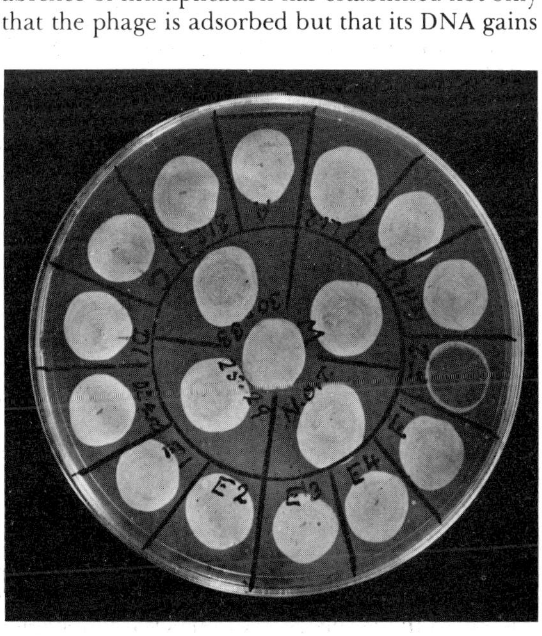

Fig. IV.40. Bacteriophage typing of *Salmonella typhi*. The strain tested was of phage type F2. (*Dr E. S. Anderson*)

Table IV.7. Vi phage typing of *Salmonella typhi*

Salm. typhi	*Reactions of representative typing phages**												*Structural*	*Change in*
types	A	B1	B3	C1	C2	C3	D1	D2	D6	F1	F2	29	*formula†*	*phage‡*
A	CL	CL	CL	CL	CL	CL	CL	CL	CL	CL	CL	CL	A	
B1	+	CL	++	−	−	+	+	−	−	+	+	+	B1	G
B3	−	−	CL	−	−	−	+	+	−	−	−	−	A(b3)	G
C1	+	++	−	CL	CL	CL	+	−	+	+	+	+	C1	P
C2	−	−	−	−	CL	−	−	−	+	−	−	−	C1(d6)	P/G
C3	−	−	−	+	CL	CL	−	−	+	−	−	+	C1(f2)	P/G
D1	−	+	−	+	CL	−	CL	CL	CL	−	+	−	A(d1)	G
D2	−	−	−	−	−	−	−	CL	−	−	−	−	D2	G/P
D6	−	−	−	−	SCL	++	+	−	CL	−	++	+	A(d6)	G
F1	−	−	−	−	−	−	−	−	−	CL	CL	−	F1	P
F2	−	−	−	−	−	−	−	−	−	+	CL	−	F1(f2)	P/G
29	−	+	++	+	CL	CL	+	−	CL	+	CL	CL	A(f2)	G

Data from A. Bernstein and E. M. J. Wilson (1963) *J. gen. Microbiol.*, **32,** 349.
* CL=confluent lysis, SCL=semi-confluent lysis, + and ++=increasing numbers of isolated plaques.
† Structural formula is indicated as non-lysogenic precursor, in capitals, followed by designation of type determining phage, in parentheses. The latter are identified according to phage type of strain from which they were isolated.
‡ Change in phage indicates the change involved in adaptation of the wild-type Vi II phage to yield the specific typing phage. P=phenotypic or host controlled modification, G=genotypic modification, i.e. mutation.

entry to the cell. Specificity must therefore depend on differences in the capacity of the different phage types to permit the growth of the typing phages. The mechanisms involved have been extensively investigated by Felix and Anderson. These workers have shown that the specificity of several of the types can be related to the presence in them of symbiotic type-determining phages. At least 16 different type-determining phages have so far been isolated. With these it is possible to alter deliberately the phage type of any strain. The type-determining phages would appear to act by preventing the growth of heterologous typing phages. A number of the phage types, however, are not demonstrably lysogenic or not lysogenic with a phage capable of type determination. The non-lysogenic types would appear to be the precursors from which other types are derived by lysogenization with the appropriate type-determining phage. The different non-lysogenic types differ, however, in the way in which their type specificity is modified by infection with a particular type-determining phage. Thus infection of the non-lysogenic type A with the type-determining phage (d6) from type D6 yields type D6 but infection of the non-lysogenic type C with the same phage yields type 33. Consequently a limited number of non-lysogenic precursor types and a limited number of type-determining phages can give rise to a large number of Vi phage types.

Of the various non-lysogenic types, type A appears to be in a special position. It is the non-lysogenic precursor of many lysogenic phage types, many of which appear to be capable of spontaneously reverting to it. The fact that it is lysed by all typing phages suggests that it is the most primitive of all the types and that other non-lysogenic precursor types have also been derived from it. This might conceivably occur as a result of a mutation affecting the capacity of the organism to be infected by, or to support the multiplication of, the wild-type Vi II phage. In some cases it is possible that the apparently non-lysogenic precursor type is in fact lysogenic for a defective prophage (p. 461) or for a phage with a high tendency to lysogenization and which cannot as a result be detected by ordinary lytic indicator tests. This may be the case in respect of types such as B1 (see Table IV.7), phage adaptation to which resembles adaptation to types whose specificity is determined by type-determining phages in being purely genotypic in nature.

The nature of the adaptation of wild-type Vi II phage to yield the specific typing phages has been examined extensively by Anderson and his co-workers. The changes involved may be purely genotypic, i.e. they may be host range mutations or purely phenotypic, in other words they may be host-controlled modifications achieved by one cycle of growth on the new type, the adapted phage reverting to the wild type when propagated on type A, or they may represent a combination of genotypic and phenotypic modification. When adaptation occurs to types derived from type A by lysogenization with type-determining phages the changes involved are purely genotypic. In most cases of adaptation to non-lysogenic types other than type A, the change is phenotypic while adaptation to types derived by lysogenization of non-lysogenic precursors other than type A involves a combination of genotypic and phenotypic modification.

Other salmonellae

Phage typing is also applied to the typing, for epidemiological purposes, of *Salm. paratyphi A, Salm. paratyphi B, Salm. typhimurium, Salm. enteritidis* and *Salm. dublin,* amongst others.

With these organisms, however, the types are defined by the pattern of reactions obtained with a number of phages rather than by specific reactions with type-specific phages. Many of the phages used are symbiotic. Phage typing has been used most for *Salm. paratyphi B,* in which, in the scheme elaborated by Felix and Callow, a total of 30 types, subtypes and variations have been described and for *Salm. typhimurium* in which 190 types have been described. There is some evidence that as in the case of *Salm. typhi* certain of these types are determined by the presence of type-determining symbiotic phages.

A different technique has been described by Boyd for the typing of *Salm. typhimurium*; in this the types are defined not by the susceptibility of the organisms to various typing phages but according to the nature of the symbiotic bacteriophages which they carry. This procedure is technically more difficult and slower than the conventional technique and has had as a result only limited epidemiological application. It has the additional defects that it can be applied only to lysogenic strains and that it will not differentiate strains carrying the same symbiotic phage but which may be derived from different non-lysogenic precursors.

Staphylococcus aureus

The phage typing of *Staph. aureus* is technically considerably more difficult than the typing of the salmonellae. The phages employed are symbiotic phages derived from lysogenic strains of *Staph. aureus.* Twenty-one of these phages constitute the basic set of typing phages in the scheme described by Williams and Rippon. They fall into four groups of antigenically related phages. An individual strain of *Staph. aureus* is usually lysed only by phages of one serological group. Virtually all strains derived from human sources belong to groups I, II and III. In this scheme strains are distinguished by the pattern of results they give with the typing phages and are designated by formulae indicating the individual phages to which they are susceptible; they cannot, as can *Salm. typhi,* be assigned to types on the basis of reactions with type-specific phages. The precise basis for this pattern specificity has not been determined but since many strains of *Staph. aureus* are lysogenic it is probable that in such cases it is determined by symbiotic type-determining phages.

Other species

Bacteriophage typing has also been found to be of value for *Ps. aeruginosa, V. cholerae, Shigella* spp., *Mycobacterium* spp. and the *Yersinia–Pasteurella* group of organisms.

BACTERIOCINES

This name is applied to a group of antibiotic proteins produced by strains of a number of bacterial species and which show a highly specific action against other strains of the same or closely related species. Bacteriocine-producing strains are resistant to their own bacteriocine, though they may be sensitive to bacteriocines produced by other strains of the same species. So far bacteriocine production has been demonstrated in the following genera: *Escherichia, Klebsiella, Arizona, Serratia, Pseudomonas, Pasteurella, Bacillus, Listeria* and in *Strep. faecalis* and *Staph. aureus.* In *E. coli* bacteriocinogeny is determined by the presence of cytoplasmic genetic determinants or episomes. Whether this is true of all bacteriocines remains to be seen.

Bacteriocines have not been used, so far, as chemotherapeutic agents and it is on the whole unlikely that they have any significant therapeutic application:

1. Their molecules are too big for free diffusion in the body and for traversing membrane barriers.
2. They would almost inevitably stimulate antibody production.

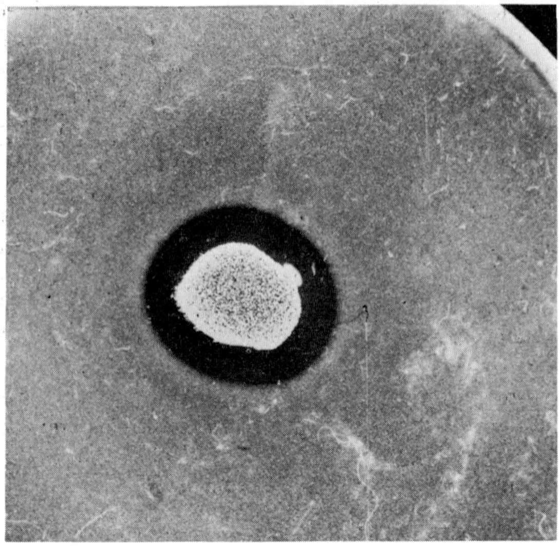

Fig. IV.41. Action of a bacteriocine produced by one strain of *Streptococcus faecalis* on a second indicator strain of *Str. faecalis*. The producing strain (centre) is grown first as a macrocolony on the surface of an agar plate. After incubation a second layer of agar incorporating the indicator strain is poured over the first.

3. Their spectrum of action is extremely restricted.
4. Their susceptibility to proteolytic enzymes would prevent their being administered orally.

Bacteriocine susceptibility is, however, proving of considerable value as an epidemiological marker, e.g. in the typing of strains of *Sh. sonnei* and of *Ps. aeruginosa*.

Colicines

The production by strains of *E. coli* of bactericidal substances active against other strains of *E. coli* was first demonstrated by Gratia in 1925 and much of our present knowledge has derived from the work of Gratia and of Fredericq. Approximately 30–40% of randomly selected strains of *E. coli* are colicinogenic, and 60–70% of *E. coli* strains are sensitive to one or more colicines.

Colicine production is most readily demonstrated by the macrocolony technique. In this, the producing strain is inoculated over an area of about 1 cm in diameter on the surface of an agar plate. After overnight incubation the organisms present in the macrocolony are killed by exposure to chloroform vapour. This is most readily achieved by pouring some chloroform into the lid of the plate and holding the plate in an inverted position over the lid for half an hour. The plate is then seeded by flooding with a broth culture of the appropriate indicator strain, the excess fluid being pipetted off. After further incubation, colicine activity produces a clear zone of growth inhibition around the macrocolony, the organisms originally seeded in this zone having been killed by the action of colicine. As might be expected from diffusion of colicine through the agar, colicines are also demonstrable in cell-free filtrates.

Colicines appear to be liberated by only a minority of the cells of a colicine-producing culture. In these, colicine liberation is associated with the death of the cell. Because of this, colicine production has been described as a lethal biosynthesis, but the mechanisms involved are not understood. In many cases colicine liberation can be induced in all of the cells of a colicinogenic culture by exposure of the culture to ultra-violet light. This phenomenon shows a marked resemblance to the induction of symbiotic bacteriophage and, like the latter, is associated with lysis of the producer.

Colicines adsorb to specific receptors on the surface of susceptible cells. An organism may acquire resistance to a particular colicine by mutation. This resistance has been shown to be due to loss of the appropriate receptor. On the basis of their spectrum of action and their cross-resistance patterns, Fredericq has distinguished a total of 17 colicines. The differentiation of types of colicines and of different colicine receptors therefore makes possible the use of colicinogeny and colicine sensitivity as strain marker techniques in epidemiological procedures.

In some cases the same receptor is involved in the adsorption of a colicine and of a bacteriophage. Thus bacteriophage T6 has the same receptor as colicine K. In this case there is a complete cross-resistance between the bacteriophage and the colicine. There is also evidence that some colicines may adsorb to receptors involved in conjugation.

The genetic determinants of a number of colicines have been shown to be extra-chromosomal and therefore to come within the general category of plasmids. They are known as Col factors. In most cases there is no evidence that Col factors can become integrated into the bacterial chromosome, although this is not true in every case as some Col factors can form stable Hfr clones.

In certain cases Col factors can transfer spontaneously from organisms possessing them to organisms lacking them, the latter becoming colicinogenic as a result. Spontaneous transfer requires cell contact and is analogous to F transfer in conjugation. Some Col factors are, however, incapable of mediating their own transfer but may be transferred, as in the case of Col E1 by the F factor or by other Col factors. Self-transfer of Col factors is most readily achieved with newly acquired Col factors, a phenomenon known as high-frequency colicinogeny transfer (HFCT).

Col factors create immunity to the corresponding colicine and, as a general rule, to other colicines of the same type. The lack of cross-immunity between colicines of similar specificity has provided the basis for a further subdivision of colicines. Thus, type E can be divided into E1 and E2, both combining with the same receptors but differing in that the E1 Col factor confers resistance only against E1, and the E2 Col factor only against E2. The resistance produced by Col factors shows considerable resemblance to the resistance produced by prophage to superinfection with the same or with a closely related phage. It presumably involves a block, probably at an early stage, in the colicine biosynthetic pathway.

In cases in which their composition has been investigated, Col factors have been found to be composed of DNA, that of colicine E2 having been estimated to contain 3×10^4 nucleotide pairs. This would be sufficient DNA to code for about 20 proteins of molecular weight 40 000. It seems clear, therefore, that Col factors possess genetic information capable of determining properties other than colcinigenicity. In the case of those Col factors capable of initiating the formation of sex pili, the factors clearly possess the genetic information necessary for this function.

One can only speculate as to the role of bacteriocines in nature. Presumably they are the weapons in an internecine struggle between closely related organisms for the same ecological niche and it may be that they play a part in that resistance to newcomers so often exhibited by resident bacterial flora.

Colicines appear to have a primarily bactericidal action. In the case of colicines K and E2, the colicines whose action has been examined in most detail, adsorption of the colicine is followed very rapidly by a cessation of all macromolecular syntheses. After adsorption there appears to be an appreciable period during which the lethal effect of the colicine can be prevented by treatment of the cells with trypsin, which destroys the adsorbed colicine. It seems likely, therefore, that the biochemical lesion which results in the death of the cell is produced by the colicine while adsorbed. This lesion has not been identified with any demonstrable alteration in the permeability of the cytoplasmic membrane, though the possibility of the occurrence of a highly specific membrane lesion cannot be excluded. In some cases it would appear that the cell can be killed by the action of a single colicine particle, a finding which suggests an analogy between colicine action and the action of complement. In the case of those colicines which act on the cell membrane this implies some kind of domino effect whereby a local disturbance, presumably allosteric, is produced and somehow spreads to affect the function of a wider area of membrane or even the whole membrane. Not all colicines act at the cell membrane; some undoubtedly penetrate the cell to act on such structures as ribosomes.

Colicinogeny determinants clearly show considerable resemblance, on the one hand, to symbiotic bacteriophages and, on the other, to sex factors. They can occur as cytoplasmic determinants, they can replicate in phase with the nucleus and they are transmissible. Their resemblance to bacteriophage genomes and the occasional sharing of receptors by bacteriophages and colicines have prompted Fredericq to speculate that they are defective bacteriophages synthesized by a genome which is incapable of synthesizing the entire bacteriophage. Evidence in support of this view has been obtained by Bradley, who has found, on electron microscopy, bacteriophage-like particles sedimenting with certain colicines at 15 000 *g*. Other colicines were not sedimented under these conditions, however, and the group would appear to be heterogeneous. The non-sedimentable type might possibly be colicines, like colicine K, which have been found by Goebel to be associated with the somatic antigens of the producing strains into which they appear to be incorporated as additional components. Bradley also found structures resembling bacteriophage tails in preparations from bacteriocinogenic strains of *Ps. aeruginosa,* the central channel of the tail containing what may have been a DNA strand. He has suggested that death of the cell could be due to injection of this DNA which, however, unlike phage DNA, is incapable of replication. Structures resembling bacteriophage tails have also been identified in a bacteriocinogenic strain of *V. cholerae.*

Recently a number of Col factors has been found to determine the production of specialized pili (sex pili) morphologically distinguishable from the common pili (fimbriae) frequently found in Gram-negative bacteria. The specialized pili so far described fall into three categories:

1. *F pili,* which adsorb male-specific RNA and DNA bacteriophages and are coded for by the sex factor F and by fi+R factors.
2. *I pili* adsorb male-specific filamentous DNA phages and are coded for by some Col factors, including Col I, and fi−R factors.
3. *Miscellaneous.* It is becoming increasingly clear that there exist a number of sex pili which do not fall into either of the above categories. Such pili have not been studied in the same detail as F and I and as new types are still being discussed the number of types must still be regarded as indeterminate.

F and I pili are antigenically distinct from one another and from category 3 pili but all F and all I pili so far studied appear to be antigenically homogenous.

It is clear that a rational classification of colicines requires not only the identification of those whose production is associated with the presence of bacteriophage components but also the recognition of Col factors determining the production of specialized pili as well as the identification of the type of pilus produced.

Index

Index